Introduction to Mineral Processing

Introduction to Mineral Processing

ERROL G. KELLY
University of Auckland

DAVID J. SPOTTISWOOD
Colorado School of Mines

A Wiley-Interscience Publication

JOHN WILEY & SONS, New York   Chichester   Brisbane   Toronto   Singapore

Copyright © 1982 by John Wiley & Sons, Inc.

All rights reserved. Published simultaneously in Canada.

Reproduction or translation of any part of this work beyond that permitted by Section 107 or 108 of the 1976 United States Copyright Act without the permission of the copyright owner is unlawful. Requests for permission or further information should be addressed to the Permissions Department, John Wiley & Sons, Inc.

Library of Congress Cataloging in Publication Data

Kelly, Errol G.
   Introduction to mineral processing.

   Includes index.
   1. Ore-dressing.   I. Spottiswood, David J.
II. Title.

TN500.K44   622'.7   82-2807
ISBN 0-471-03379-0   AACR2

Printed in the United States of America

10 9 8 7 6 5 4

Preface

This book is intended primarily as a university text for a general undergraduate course in mineral processing. For completeness, however, a considerable amount of material has been included that normally is covered in more advanced undergraduate courses. Therefore extensive references are provided to allow the contents to be easily expanded. For this reason, the book should also be of use to practicing engineers who feel that they need a more up-to-date understanding of the subject.

Unlike most of the (few) published texts available on mineral processing, we give little attention to the "art" of mineral processing. Readers requiring information on how equipment is operated, or how specific ores are processed, are referred to handbooks and to the older text books. Such information has its usefulness, but it is of secondary importance to a thorough understanding of the basic *principles* of mineral processing and it is these principles that we concentrate on. Even descriptions of equipment have been kept comparatively brief; we have relied heavily on a restricted range of self-explanatory illustrations of the basic types of equipment. Virtually all these illustrations have been taken from manufacturers' catalogues, which today constitute such a valuable source of information that we consider their examination by students to be as important in their education as plant visits.

It is now more than 40 years since Gaudin's book made its valuable contribution to the science of mineral processing. Even today some of his enlightened work remains neither substantiated nor disproved. Considerable advances have been made in our knowledge in those 40 years, and we have tried to put these into perspective. Even so, several equally plausible analyses exist for some areas, and in these instances, rather than select one approach, we have presented a survey of the field, believing that this will better serve the reader.

It must be admitted that in many instances it is still not yet possible to make quantitative analyses in mineral processing. However, it is often desirable and frequently possible to make semiquantitative calculations. To this end we have included a number of worked examples, to show how order-of-magnitude solutions can be obtained.

The newcomer to mineral processing must learn to appreciate the interdependence of the various unit operations. To aid in this, a considerable number of the worked examples are integrated. To eliminate extensive iterations and to keep the calculations within reasonable bounds, it has sometimes been necessary to impose starting constraints that are not entirely realistic. However, this does not alter the prime value of these examples.

In keeping with current practice throughout the world, SI units have been used. A brief set of conversion factors is given in Appendix I.

We received help from many sources in preparing this book. While the manufacturers who provided us with illustrations are acknowledged elsewhere, we also thank all those manufacturers who supplied the range of catalogues from which we made our final comparatively narrow selection.

As with any work that reviews previously published material, we have had to adapt diagrams and data from many sources. Where diagrams have been adapted, this has been done with the permission of the publisher named in the reference given in the legend. Where tabular material has been used with permission, the reference is given in the text or in the Acknowledgments. Chapter quotations and appended material are credited in the Acknowledgments.

The help we received through travel assistance and leaves from Colorado School of Mines, Auckland University, and Michigan Technological University is gratefully acknowledged.

Discussions with friends and colleagues in industry, research institutions, and universities throughout the world have assisted in numerous ways. D. F. Kelsall and W. Schaap provided helpful criticisms of Chapters 10 and 25, respectively. While their input was invaluable, any inaccuracies in these chapters are our responsibility. E.G.K.'s personal thanks are given to E. Volin, whose longtime friendship and support led to the initiation of this project.

We also thank the typists who survived the preparation of the manuscript, especially Pauline Robinson and Nina Eads.

The constant understanding and tolerance of our wives, Jan Kelly and Irene Spottiswood, during the period of writing can never be adequately acknowledged.

Last, but far from least, it is no understatement to say that without Jan's help this book would not exist. She not only prepared all our diagrams but helped with all aspects of manuscript presentation. We offer our sincere thanks.

ERROL G. KELLY
DAVID J. SPOTTISWOOD

Auckland, New Zealand
Golden, Colorado, U.S.A.

January 1982

Acknowledgments

Quotations, Chapters 1–25, excerpted from the book *The Official Rules,* by Paul Dickson. Copyright © 1978 by Paul Dickson. Reprinted by permission of Delacorte Press.

Quotation, Chapter 13 (second quotation), excerpted from the book *Time Enough for Love,* by R. A. Heinlein. Reprinted by permission of Putnam and Co., Inc.

Table 20.1. Diagrams courtesy of *Chemical Engineering* Magazine.

Appendix A. Courtesy Joy Manufacturing Co.

Appendix C. Courtesy W. S. Tyler, Inc.

Appendix D. Courtesy Henry Krumb School of Mines, Columbia University, New York, NY 10027.

Appendix E. Courtesy Stearns Magnetics Inc.

Appendix F. Courtesy Joy Manufacturing Co.

Appendix G. Courtesy Barber-Greene Co.

Appendix F. Courtesy Joy Manufacturing Co.

Appendix G. Courtesy Barber-Greene Co.

Appendix H. Courtesy Warman International, Inc.

Frontispiece 1. Courtesy Rexnord Inc.

Frontispiece 2. Courtesy Joy Manufacturing Co.

Frontispiece 3. Courtesy Dorr-Oliver Inc.

Frontispiece 4. Courtesy Dorr-Oliver Inc.

Frontispiece 5. Courtesy Sala International AB.

Frontispiece Part I. Courtesy K. D. Hitching.

Frontispiece Part II. Courtesy of Skega AB.

Frontispiece Part III. Courtesy Fredrik Morgensen AB.

Frontispiece Part IV. Courtesy Outokumpu Oy.

Frontispiece Part V. Courtesy Larox Oy.

Frontispiece Part VI. Courtesy Dravo Corp.

Frontispiece Part VII. Courtesy Stearns-Roger Engineering Corp.

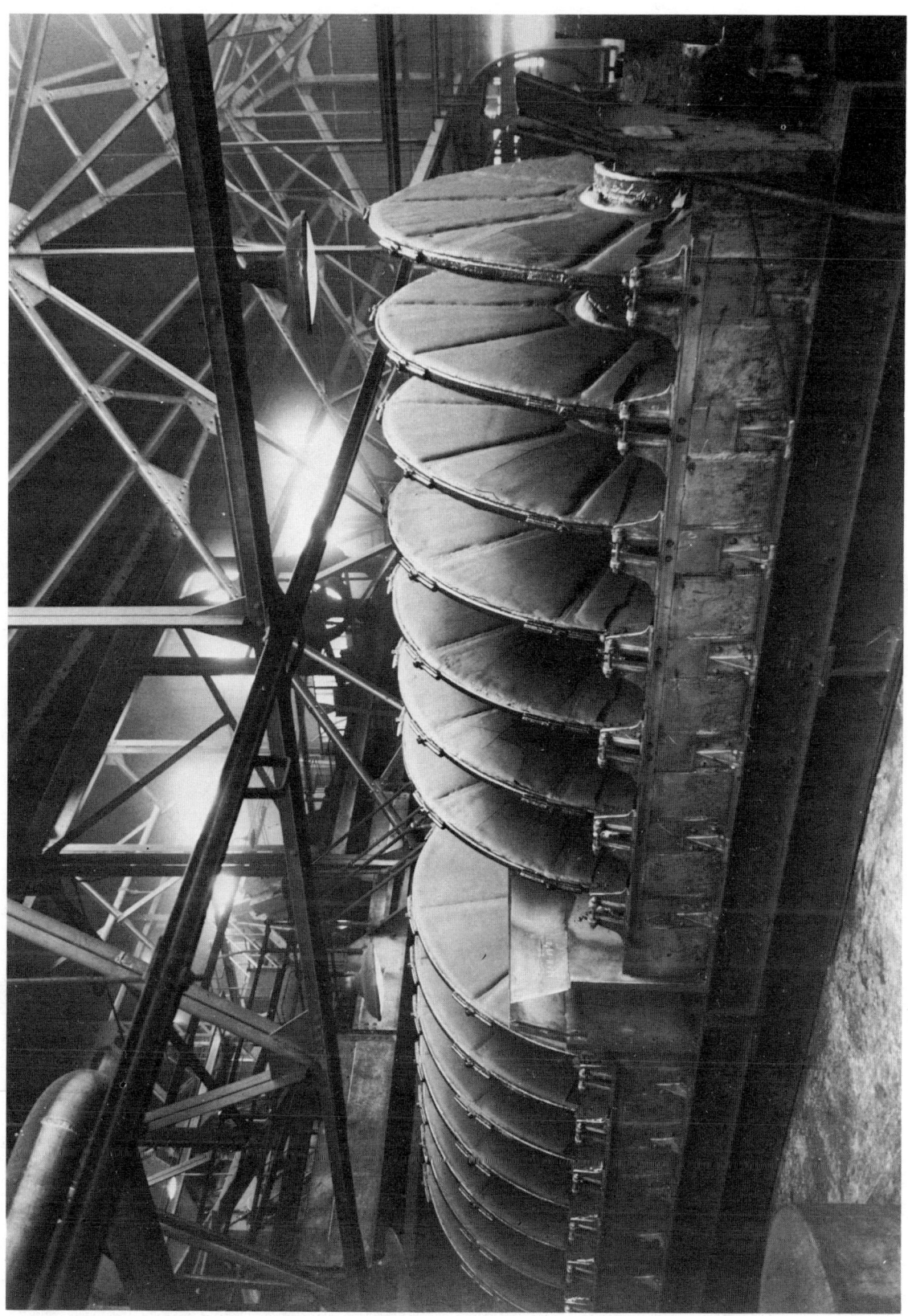

Contents

Nomenclature, xvii

PART I. FUNDAMENTALS

1. **Ores, Mills, and Concentrates, 5**
    1.1. Economic Criteria, 5
    1.2. Flow Sheets, 8
    1.3. Materials Balances, 9
    1.4. Ores, 13
    1.5. Mineralogy, 17

2. **Characterization of Particles, 21**
    2.1. Particle Size, 21
    2.2. Presentation of Size Distribution Data, 22
    2.3. Average Diameters, 27
    2.4. Particle Shape, 29
    2.5. Liberation, 32
    2.6. Determination of Particle Characteristics, 36
    2.7. Complete Particle Analysis, 43

3. **Analysis of Separation Processes, 46**
    3.1. Equilibrium Separation, 47
    3.2. Kinetics: The Rate of Separation, 50
    3.3. Mixing Patterns, 52
    3.4. Determination of Separability Curves, 55
    3.5. Separation Efficiency, 56

4. **Fluid Dynamics, 62**
    4.1. Characterization of Flow by Reynolds Number, 62
    4.2. Fluid Rheology, 64
    4.3. Friction Factors, 65
    4.4. Flow in Pipes, 66
    4.5. Flow of Bodies Through Fluids, 68
    4.6. Flow Through a Packed Bed, 75
    4.7. Fluidization and Sedimentation, 76
    4.8. Agitation, Mixing, and Particle Suspension, 82
    4.9. Flow of Liquid in Layers, 84

5. **Mechanisms and Processes of Particulate Separation, 87**
    5.1. Hindered Settling, 87
    5.2. Differential Acceleration, 88
    5.3. Trickling, 88
    5.4. Differential Velocity in a Flowing Film, 88
    5.5. Shear, 90
    5.6. Segregation of Dry Particles, 91
    5.7. Center of Gravity, 92
    5.8. Separation Limits and Mixing, 93
    5.9. General Comments, 93

6. **Surfaces and Interfaces, 95**
    6.1. Thermodynamics of Interfaces, 95
    6.2. Mechanisms of Interfacial Attachment, 107

7. **Mechanisms of Fracture, 113**
    7.1. Single Particle Fracture, 113
    7.2. The Fracture Environment, 122

PART II. SIZE REDUCTION

8. **Crushing and Grinding, 127**
    8.1. Equipment, 127
    8.2. Characteristics of Tumbling Mills, 135
    8.3. Circuits, 143
    8.4. Theory, 144
    8.5. Equipment Capacities, 157

PART III. SIZING SEPARATION

9. **Screening and Sieving, 169**
    9.1. Laboratory or Test Sieving, 169
    9.2. Screening, 171

9.3. Screen Surfaces, 173
9.4. Auxiliary Features and Equipment, 179
9.5. Measurement of Screen Performance, 180
9.6. Factors Affecting Screening, 181
9.7. Theory of Conventional Screening and Sieving, 185
9.8. Theory of Probability Screening, 190
9.9. Theory of Sieve Bends, 192
9.10. Screen Sizing, 193

10. Classification, 199
    10.1. Equipment, 199
    10.2. Classifier Performance, 202
    10.3. Sedimentation Classifiers, 207
    10.4. Hydrocyclone Classifiers, 213
    10.5. Fluidized Bed (Hydraulic) Classifiers, 223
    10.6. Capacities, 225
    10.7. Closed Circuit Grinding, 226
    10.8. Classifier Operation, 229

PART IV. CONCENTRATION SEPARATION

11. Ore Sorting, 237
    11.1. Sorting Machines and Applications, 273
    11.2. Mechanics of Sorting, 238
    11.3. Ore Sortability, 241

12. Dense Medium Separation, 243
    12.1. Equipment and Applications, 243
    12.2. Medium Control and Recovery, 246
    12.3. Performance Curves, 248

13. Gravity Concentration, 250
    13.1. Equipment and Applications, 250
    13.2. Theories of Gravity Concentrating Devices, 261
    13.3. Performance Curves, 269

14. Magnetic Separation, 274
    14.1. Equipment and Applications, 274
    14.2. Principles and Mechanisms of Magnetic Separation, 280
    14.3. Magnetic Separator Performance, 286

15. Electrostatic Separation, 291
    15.1. Equipment and Applications, 291
    15.2. Operation Environment, 295
    15.3. Mechanisms of Electrostatic Separation, 296

16. Flotation and Other Surface Separations, 301
    16.1. Flotation Equipment, 301
    16.2. Chemistry of Flotation, 307
    16.3. Mechanics of Flotation, 314
    16.4. Flotation Operations, 317
    16.5. Other Surface Separations, 319

PART V. DEWATERING

17. Sedimentation, 327
    17.1. Sedimentation Equipment, 327
    17.2. Sedimentation Behavior, 330
    17.3. Steady State Operation of a Continuous Thickener, 335

18. Filtration, 343
    18.1. Filtration Equipment, 343
    18.2. Filtration Calculations, 348

19. Dewatering: Systems and Miscellaneous Methods, 356
    19.1. Centrifugal Separations, 356
    19.2. Mechanical Dewatering Systems, 358
    19.3. Thermal Dewatering (Drying), 359

PART VI. MATERIALS HANDLING

20. Dry Solids Handling, 367
    20.1. Stockpiles, 367
    20.2. Bins and Hoppers, 368
    20.3. Feeders, 374
    20.4. Transport of Bulk Solids, 375
    20.5. Sampling, 378

21. Slurry Handling, 380
    21.1. Pumps and Pumping, 380
    21.2. Agitated Tanks, 383
    21.3. Slurry Transportation, 387

22. Tailings Disposal, 390
    22.1. Tailings Ponds and Dams, 390
    22.2. Tailings Water, 392

PART VII. PLANT PRACTICE

23. Process Integration and Flowsheet Analysis, 397
    23.1. Flowsheet Case Study, 397
    23.2. Selection Between Alternative Types of Separation Equipment, 399
    23.3. Roughing, Cleaning, and Scavenging, 400
    23.4. Plant Design, 405

24. Plant Control, 408
    24.1. Control Objectives, 408
    24.2. Principles of Automatic Process Control, 409

Contents

    24.3. Instrumentation, 414
    24.4. Control Applications, 416

25. **Economics, 421**
    25.1. Capital Costs, 421
    25.2. Working Capital, 425
    25.3. Operating and Production Costs, 426
    25.4. Investment Worth, 427
    25.5. Optimum Operating Capacity, 430
    25.6. Concluding Remarks, 431

Appendix A. Minerals and Their Characteristics, 433

Appendix B. Selected Plant Flow Sheets, 445

Appendix C. Standard Sieve Sizes, 460

Appendix D. Rosin-Rammler (Weibull) Graph Paper, 462

Appendix E. Magnetic Attractibility of Minerals, 464

Appendix F. Separation Characteristics of Minerals, 466

Appendix G. Conveyor Belt Design Data, 468

Appendix H. Pump Selection Chart, 472

Appendix I. SI Units and Conversion Factors, 473

Index, 477

Nomenclature

Osborn's Law: Variables won't, constants aren't.

⇨ Feed
→ Positive Response
∎∎∎▶ Negative Response

A = area
A_c = area of filter cake
A_e = effective area of (lamella) thickener
A_I = area of hydrocyclone cross-section inlet
A_P = projected area of an irregular particle
A_T = (surface) area of tank or pool
$\mathscr{A}_i$ = activity in solution of species i
a = acceleration
a_x = external acceleration
B = breadth (width)
$\mathscr{B}$ = magnetic induction
b_i = breakage parameter
$[b]$ = mass fraction appearing in size i as a result of breakage in size j
C = mass concentration
C_A, C_K = mass concentration in layers A and K, respectively
C_c = mass of solids deposited as filter cake per unit volume of filtrate collected
C_{crit} = critical concentration that gives rise to critical flux in thickener
C_f, C_z = limiting mass concentrations for free settling and zone settling, respectively (in sedimentation)
C_I = initial concentration (for slurry, mass/unit volume)
C_i = concentration in solution of species i
C_j = lower conjugate concentration in thickener
$C_{j,\mathrm{crit}}$ = conjugate concentration corresponding to critical concentration
C_L = linear concentration (Eq. 5.14a)
C_N = concentration in discharge of Nth device at time t
$C_V, C_{V,I}, C_{V,O}$ = volumetric concentration, initial or input volumetric concentration, and output volumetric concentration, respectively
$C_{V,c}$ = volumetric concentration of solids in wet filter cake
C_2 = see Fig. 17.12
$C_{(+)\mathrm{crit}}$ = underflow (positive response outlet) concentration when critical concentration occurs in thickener
$C_{(+)}, C_{(-)}$ = concentration in positive and negative response outlets, respectively (underflow and overflow respectively in thickeners)
Cy_{50} = cyclone characteristic number
$\mathscr{C}, \mathscr{C}_e, \mathscr{C}_n, \mathscr{C}^*$ = cost, cost of existing plant, cost of new plant, reference cost, respectively
cg, cg_I, cg_1, cg_h = center of gravity, initial center of gravity, center of gravity of light and heavy fractions, respectively
CI_e, CI_n = cost index at date of construction of existing and new items, respectively
$\Delta C_{V,I}, \Delta C_{V,O}$ = deviations from steady state of $C_{V,I}$ and $C_{V,O}$, respectively
$\mathfrak{C}$ = specific conductivity

c = constant
D = diameter
D_A = diameter of screen aperture (length of side, if square)
D_a = diameter of agitator
D_c = diameter of cyclone
D_e = effective aperture of screen
D_I = inside diameter of hydrocyclone inlet ($\sqrt{4A_I/\pi}$ for noncircular inlets)
D_l, D_{sm} = inside diameter of large and small mills, respectively
D_M = mill diameter inside lining
D_m = diameter of grinding media
D_p = piping diameter or diagonal
D_T = diameter of tank
D_W = diameter of screen wire
$D_{(+)}, D_{(-)}$ = inside diameter of positive and negative response outlets, respectively (apex and vortex finders, respectively of hydrocyclone)
$\mathscr{D}$ = diffusion coefficient
$\mathfrak{D}$ = drying rate, mass of liquid evaporated per unit time, per unit area of solid surface
$\mathfrak{D}_c$ = drying rate during constant rate period
d = differential operator
d = diameter of particle
d_A = sieve (nominal) diameter
d_a = projected area (nominal) diameter
d_d = drag (nominal) diameter
d_F, d_M = Feret's and Martin's statistical diameters, respectively
d_f = free-falling (nominal) diameter
d_G = mineral grain size
$d_{\max}, d_{\min}$ = maximum and minimum size of particles, respectively
d_{m3I} = mass median diameter of feed
d_n = nominal diameter
$\bar{d}_{na}, \bar{d}_{ng}, \bar{d}_{nh}$ = arithmetic, geometric, and harmonic mean nominal diameters, respectively
d_{nmu}, d_{mu} = nominal median diameter measured on basis u (Table 2.4)
$\bar{d}_{nqp}$ = statistical average nominal diameter (Eq. 2.8 and Table 2.4)

d_r, d_0 = diameters of reference sphere and sphere, respectively
d_s = surface nominal diameter
d_t = diameter of particle that settles in time t (Eq. 2.25)
d_{St} = Stokes (nominal) diameter
d_v = volume (nominal) diameter
d_{vs} = specific surface (nominal) diameter
d_{50} = separation size; the size of classified particle that is split equally between the positive and negative response outlets
$d_{50,a}$ = the d_{50} size on an uncorrected performance curve
$d_{80,I}, d_{80,O}$ = diameter of 80% passing sieve size in the feed and product, respectively (actually $d_{A,3,80,I}$ and $d_{A,3,80,O}$, the 80% mass passing aperture)
d^* = size modulus in distribution equation
E = energy
E_I = initial potential energy
E_p = energy acting on particle
E_S = total surface energy
E_s = potential energy after segregation (Eq. 5.28)
E_0 = specific energy necessary to cause fracture (normally assumed to be the surface energy of the new surface)
ΔE = potential energy available for segregation
$\mathscr{E}$ = electric field strength or intensity
$\mathscr{E}_a$ = streaming potential
$\mathfrak{E}$ = surface entropy (per unit area)
$e^+, e^-, e^\pm$ = electric charge
F = force
F_a, F_c, F_g, F_x = accelerating, centrifugal, gravitational, and external forces, respectively
F_b = buoyancy force
F_d, F_f = drag force and frictional force, respectively
F_e, F_i, F_m = electrical, image, and magnetic forces, respectively

Nomenclature

F_0 = applied load at fracture
$\mathscr{F}$ = Faraday's constant
$\mathfrak{F}_{b1}$ = fraction of material broken by mechanism 1
$\mathfrak{F}_{cy}$ = fraction of cycle time available for cake formation
$\mathfrak{F}_V, \mathfrak{F}_{V,sl}$ = volumetric fraction and volumetric fraction of solids in slurry, respectively
$\mathfrak{F}_{(+/-)}$ = volumetric flow split in hydrocyclone
f_c = unconfined yield strength for bulk solids
f_f = flow factor
$\mathbf{f}_p, \mathbf{f}_{sl}$ = (pipe) fraction factor, and (pipe) friction factor for slurry, respectively
$\mathbf{f}_b, \mathbf{f}_s$ = modified friction factors for packed bed, and sedimentation fluidization, respectively
$\mathbf{f}_d$ = drag coefficient (friction factor for a particle)
Fl = flow or pumping number
Fl_g = air flow number
$\mathbf{Fr}, \mathbf{Fr}_{sl}$ = Froude number and modified Froude number for slurry, respectively
fn = function $(\cdots)$
G = gape setting
G_b, G_c, G_r = grindability for ball mill, crusher, and rod mill, respectively
g = acceleration in gravitation field
H = height
H_a = height of agitator from bottom
H_c = inside distance from top of mill to top of stationary charge
H_h, H_1 = see Fig. 5.4b
H_I, H_O = inlet height, outlet height
H_V = height of free vortex in hydrocyclone (i.e., height from bottom of vortex finder to apex)
ΔH = distance center of gravity lowers
ΔH_{max} = maximum lowering of center of gravity
$H_{(+)}$ = height of positive response cut (Fig. 10.13)
$\mathscr{H}$ = applied magnetic field

$\mathfrak{H}_g, \mathfrak{H}_i$ = humidity of drying gas and humidity of gas at interface, respectively
I, I_V = input mass and volumetric flow rates, respectively
$[I]$ = input size distribution matrix
I_{crit} = critical flow rate
I_e, I_v = capacity of existing and new plants, respectively
I_g = mass flow rate of gas
I_L = mass flow across screen at distance L from feed end
I_u = unit capacity of screen
I_{Va} = volumetric flow rate through impeller or agitator
I_{Vg} = air (volumetric) flow rate to flotation cell
IEP = isoelectric point
$\vec{J}$ = vector representing state variables
$\vec{J}_M$ = vector representing measured values of state variables
$\vec{J}_{M,SP}$ = vector representing set point values
$K_A, K_B, \ldots$ = factors, constants, or parameters
K_A, K_{A11}, K_{A12} = Hamaker constants
K_B = coefficient (Eq. 5.14)
K_{Bo} = Boltzmann's constant (1.38×10^{-23} J/K)
$K_{B,i}$ = see Eq. E8.3.2
K_b = breakage coefficient
K_C = Cunningham correction factor (Eq. 4.54)
K_{Cp} = capacitance of particle (Eq. 15.3)
K_c = hydrocyclone parameter in Eqs. 10.16 and 10.27
K_{cs} = crushing strength (Eq. 8.30)
K_g = geometric form constant (Eq. 2.17)
K_L = loading factor (Eq. 8.8)
K_{La} = correction factor for lamella thickeners (Eq. 17.11)
K_{Lo} = location factor (Eq. 25.2)
K_M, K_H = mass and heat transfer coefficients, respectively
K_{Mt} = mill type factor (Eq. 8.8)
K_p = piping factor
K_R = roughness parameter of pipe

K_r = electrical resistance of particle (Eq. 15.3)
K_s = correction factor (Eq. 2.27)
K_{Sp} = speed factor for tumbling mills (Eq. 8.8)
K_{sp} = solubility product
K_w = wall coefficient (Section 4.5.7)
$K_\mathscr{D}$ = derivative or rate control constant
$K_\mathscr{G}$ = steady state gain of a process
$K_\mathscr{P}$ = proportional control constant
$K_\mathscr{R}$ = integral or reset control constant
K_ε = dielectric constant (Eq. 15.1)
$\mathscr{K}_\nu$ = constant dependent on Poisson's ratio
$K_1, K_2, \ldots$ = constants in equations
K_Σ = product of a number of factors (eg., screening Eq. 9.1)
$[K]_c$ = matrix of controller constants or operators
$[K]_{DC}$ = matrix of decoupling elements
$[K]_s, [K]_{s,DC}$ = state variable process matrix and apparent process matrix resulting from decoupling, respectively
$[K]_u$ = manipulated variable process matrix
k = rate constant
k' = rate constant independent of $\mathscr{P}^* - \mathscr{P}$
k_b, k_{ub} = rate constants for screen blinding and unblinding, respectively
k_c, k_s = rate constants for crowded and separate screening, respectively
$k_{c,i}, k_{s,i}$ = rate constants of size class i for crowded and separate screening, respectively
$k_{c,1/2}$ = rate constant for the size i corresponding to $d/D_A = 0.5$
$k_i, k_{i,\max}$ = rate constants for size i and maximum value of k_i, respectively
$k_{i,l}, k_{i,sm}$ = breakage rate of size i in large and small mill, respectively
k_p = rate constant for screening (passage)
L = length
L_c, L_v = length of hydrocyclone and vortex finder, respectively
L_{cr} = crack length

L_{mf} = length of bed at minimum fluidization
$\mathscr{L}$ = degree (fraction) of liberation
$\mathfrak{L}$ = liquid loading (Eq. 4.95)
ln = natural logarithm
l_{vap} = latent heat of vaporization of water
M = mass
M_b = mass of ball (or grinding medium)
M_c, M_w = mass of dry filter cake and mass of wet filter cake (including pore liquid), respectively
M_h = mass of hold-up in mill
M_I = initial mass
$M_{n,N}$ = mass of particles passing N layers of screens with n attempts
M_s = mass of dry solid
M_t, M_∞ = mass at time t and infinite time, respectively
$M_{(+)}, M_{(-)}$ = mass in positive and negative response streams, respectively
$\mathscr{M}$ = intensity of magnetization
m = mass fraction
m_d = mass fraction with size greater than d_t (Eq. 2.25)
m_i = mass fraction of particles in size class i
m_{iL} = mass fraction of particles in size class i in bed at distance L from feed point
m_t = mass fraction which settles out in time t
$m_{\rho I}$ = mass fraction of feed with density between ρ and $\rho + \Delta\rho$
$(mw)_g$ = molecular weight of drying gas
N = number
N_g = multiples of gravity
N_L = number of attempts per unit length of screen
$\mathscr{N}$ = speed of rotation
$\mathscr{N}_c$ = critical speed
$\mathscr{N}_{\text{op}}$ = operating speed
n = equation exponent (may in some cases become the number of attempts at passage in screening)
n_c = exponent (Eq. 9.27)

Nomenclature

O = output flow rate (mass or volumetric rate; units consistent with I)
O_V = output volumetric flow rate
$O_{(+)}, O_{(-)}$ = output flow rate in positive and negative response streams, respectively (mass or volumetric rate; units consistent with I)
$(\Sigma O)_j$ = see Eq. E8.3.3
$[O]$ = output size distribution matrix
P = pressure
P_d = disjoining pressure
P_I = inlet pressure
$P_{y,i}, P_{y,v}$ = inertial and viscous repulsive pressures, respectively
P_O = outlet pressure
PZC = point of zero charge
ΔP = pressure drop
$\Delta P_c, \Delta P_m$ = pressure drop across filter cake and filter medium, respectively
ΔP_f = filtration pressure
ΔP_{mf} = pressure drop at point of minimum fluidization
ΣP = friction losses in pipe
$\mathscr{P}$ = power
$\mathfrak{P}$ = property value
$\mathfrak{P}^*$ = separator (machine) setting
$[p]$ = probability matrix
p = probability
p_i = probability of passage of particles in the ith size fraction
Po = power number
Q = quantity with property $\mathfrak{P}$
Q_t = quantity with property $\mathfrak{P}$ recovered in time t
Q_u = quantity of material measured by property u (Table 2.4)
R = radius
$R_b, R_l, R_{s/l}$ = inside radius (in a centrifuge) of bowl, liquid surface, and solid/liquid interface, respectively
R_H = hydraulic radius
$\mathscr{R}$ = corrected recovery (unless otherwise written, this implies $\mathscr{R}_{(+)}$ the recovery in the positive response stream); normally expressed as a percentage; in the case of performance curves, it is generally expressed as a fraction
$\mathscr{R}_a$ = actual or uncorrected $\mathscr{R}$
$\mathscr{R}_{(+)}, \mathscr{R}_{(-)}$ = recovery in positive and negative response streams, respectively (see $\mathscr{R}$)
$\mathscr{R}_{(+)a}$ = actual or uncorrected $\mathscr{R}_{(+)}$
$\mathscr{R}_{(+)i}$ = mass fraction of size i reporting to undersize stream
$\mathscr{R}_{(+)l}, \mathscr{R}_{(+)lv}$ = mass fraction and volumetric fraction (respectively) of feed liquid leaving through positive response outlet (underflow in hydrocyclone)
$\mathscr{R}_B$ = dimensionless shear strain (Eq. 5.20)
$\mathscr{R}_b$ = breakage size ratio (Eq. 7.20)
$\mathscr{R}_E$ = elongation ratio (Eq. 2.14)
$\mathscr{R}_F$ = flatness ratio (Eq. 2.15)
$\mathscr{R}_L$ = mineral grain size to particle size ratio in liberation (d_G/d)
$\mathscr{R}_r$ = reduction ratio
$\mathscr{R}_s$ = ratio of specific surfaces
r = radial distance
$\mathbf{R}$ = gas constant
$\mathbf{Re}$ = Reynolds number, Reynolds number of a pipe
$\mathbf{Re}_p$ = Reynolds number of particles
$\mathbf{Re}_s$ = Reynolds number of slot (Eqs. 9.31 and 9.32)
$\mathbf{Re}_\infty$ = Reynolds number of particle, based on terminal settling velocity
S = surface area
S_p = surface area of particle
S_0 = specific surface
S_w = total wetted surface per volume of bed
$\mathscr{S}$ = shear stress
$\mathscr{S}_B$ = Bagnold shear stress
$\mathscr{S}_{B,v}$ = Bagnold shear stress (viscous conditions)
S_b, S_c, S_s = total stress, stress caused by contents, and self stress, in centrifuge, respectively
$\mathscr{S}_w$ = wall shear stress
$\mathscr{S}_y$ = yield shear stress
$\mathfrak{S}$ = stress, compressive or tensile

$\mathfrak{S}_1$ = consolidating pressure (major principal compressive stress)
$\mathfrak{S}_2$ = minor principal compressive stress
$\mathfrak{S}_c$ = critical stress
$\mathfrak{S}_f$ = stress at fracture or yield stress; strength
$\mathfrak{S}_{fr}$ = strength of reference particle of size d_r
$\mathfrak{S}_G$ = Griffith stress
$\mathfrak{S}_y$ = unconfined yield strength
s = exponent on the various forms of the Rosin-Rammler (Weibull) equations
s^* = variable in Laplace transform
SG = specific gravity
SG_{50} = separation specific gravity; particles of the SG split equally between positive and negative response outlets
$SG_{50,d}$ = separation specific gravity of particle of size d
$SG_{50,\Sigma d}$ = separation specific gravity of all sizes of particles
T = temperature
T_g, T_i = temperatures of drying gas and liquid-gas interface, respectively
t = time
t_c = time to dry to critical point
t_{cy}, t_d = cycle time and drying time, respectively
t_H = time for particle to settle height H
t_I = induction time
t_z = time to achieve zone settling
t_∞ = infinite time
U = manipulated variable (controller output
$\vec{U}$ = vector representing manipulated variables
$\mathcal{U}_i$ = chemical potential in solution of species i
u = basis for expressing size distributions (see Table 2.4)
V = volume
V_e = equivalent volume of filtrate (Eq. 18.10)
V_f = volume of filtrate collected
V_p = volume of particle
V_T = volume of tank or pool
V_w = volume of wash liquid
$\mathcal{V}$ = applied electrical potential
v = velocity
$\bar{v}$ = average velocity
v_d = deposition velocity (Eq. 4.92)
v_h = fall velocity of suspension with respect to a fixed horizontal plane (sedimentation), or superficial fluid velocity (fluidization)
v_I = initial velocity or (hydrocyclone) inlet velocity
$v_L, v_{L,A}, v_{L,K}, v_{L,Z}$ = upward velocity of layer, layers A, K, and Z, respectively
$v_{L,I}$ = settling velocity of upper interface with initial concentration C_I
v_m = tangential velocity of ball
$v_{\max}$ = maximum velocity
v_{mf} = velocity at point of minimum fluidization
v_r = velocity at radial distance r
v_s = superficial velocity (average velocity that would occur if bed were not present)
$v_{(r)}, v_{(t)}, v_{(t)p}, v_{(v)}$ = velocity components in hydrocyclone (radial, tangential, tangential at periphery, and vertical, respectively)
v_t = velocity at time t
v_δ = velocity at depth δ_H
v_ε = upward interstitial velocity (Eq. 4.77)
v_τ = horizontal velocity component in pool, resulting from residence time τ
v_∞ = terminal velocity
$v_{\infty l}$ = settling rate of limiting size particle (Eq. 10.2)
W = work
W_i = work index
W_p = work done by pump
X = bulk moisture content
X_c = bulk moisture content at critical point

Nomenclature

X_I, X_t = initial and time t bulk moisture contents, respectively
$[X]$ = see Eq. 8.17
x = horizontal distance
Y^- = cumulative fraction less than size d_n
Y_N^- = cumulative fraction passing N layers of screens (Eq. 9.30)
Y_u^+, Y_u^- = cumulative fraction, measured on basis u greater than, and less than, respectively, size d_n
$\mathcal{Y}$ = Young's modulus
y = vertical distance
Z = wetted perimeter
z = base for expressing size distribution
α = specific cake resistance
α_0 = compressibility coefficient (Eq. 18.13)
β = slope of drag friction factor curve ($\Psi = 1.0$) (Fig. 4.8)
Γ_i = absorption density or surface excess of species i
Γ_{ma}, Γ_{mw} = adsorption density at mineral-air and mineral-water interfaces, respectively
γ = surface free energy (per unit area)
$\gamma_{ma}, \gamma_{mw}, \gamma_{wa}$ = surface free energy at mineral-air, mineral-water, and water-air interfaces, respectively
$\Delta\gamma$ = change in surface free energy
δ = thickness
δ_c = thickness of filter cake
δ_D = Debye thickness
δ_h = depth from top surface of layer
δ_l = thickness of feed layer
δ_r = thickness of film at rupture
ε = porosity, fraction of total volume not occupied by solid
ε_{mf} = porosity at point of minimum fluidization
ϵ, ϵ_i = error signal
ϵ_{DB} = deadband of on-off controller
ζ = zeta potential
η = efficiency
η_A = areal efficiency (of a pool)
$\eta_{o/s}$ = screening efficiency (Eq. 9.2)
$\eta_{u/s}$ = screening efficiency (Eq. 9.1)
θ = angle
θ_B = "collision" angle (Fig. 5.3)
θ_c = contact angle
θ_{ch} = flow channel (half-) angle
θ_{crit} = critical slope
θ_{fe} = effective angle of friction
θ_{fi} = angle of internal friction
θ_{fw} = wall angle of friction
θ_h = hopper angle
θ_p = angle in Eq. 9.4
κ = magnetic susceptibility
Λ = mean free path
λ = shape factor
λ_{sn} = surface shape factor (Eq. 2.10)
λ_{vae} = equidimensional shape factor (Eq. 2.16)
λ_{vn} = volumetric shape factor (Eq. 2.11)
μ = viscosity
μ_a = apparent viscosity (Eq. 4.11)
μ_B = coefficient of rigidity (Eq. 4.9)
μ_P = consistency (Eq. 4.10)
μ_{sl} = viscosity of slurry
μ^* = coefficient of viscous friction (Eq. 3.11)
ν = Poisson's ratio
ξ = surface potential
ξ_S = potential at Stern plane
ρ = density
ρ_b = bulk density
ρ_{bh}, ρ_{bl} = bulk density of dense and light minerals, respectively
$\rho_f, \rho_l, \rho_s, \rho_{sl}$ = density of fluid, liquid, solid, and slurry, respectively
ρ_m = density of metal or medium
ρ_{50} = separation density; particles of this density split equally between positive and negative response outlets
σ = standard geometric derivation
σ_s = surface charge
τ = residence time, or nominal residence time
τ_d = time constant (Eq. 3.12)

τ_m = perfect mixing component of residence time
τ_p = plug flow component of residence time
τ_Σ = total residence time
Φ = coefficient of friction
ϕ = function in Eq. 9.5
$\chi_{i,1/2}$ = reduced rate for crowded screening constant (Eq. 9.26)
Ψ = sphericity
ψ = mass solids flux
ψ_A = flux through layer A
ψ_{crit} = critical solids flux in a thickener
ψ_d, ψ_u = net downward and upward flux in thickener, respectively
ψ_I = initial or feed flux
$\psi_{O/L}$ = overload feed flux
ψ_s = solids flux due to sedimentation
ψ_V = volumetric solids flux
$\psi_{(+)}, \psi_{(-)}$ = mass solids flux in positive and negative response outlets, respectively
$\psi_{(+),A}$ = withdrawal flux through layers
Ω_m = medium resistance
ω = radial or angular velocity
ω_e = entrapment angular velocity

Introduction to Mineral Processing

Part I

Fundamentals

Too often mineral processing is presented as an art instead of as a science. The main purpose of Part I is to emphasize that there is a vast amount of information relevant to mineral processing available from a variety of engineering and science fields. Unfortunately there are still considerable gaps in this knowledge, and perhaps more significantly, there is an appreciable difference between having fundamental knowledge and being able to apply it. This is most obvious in the so-called modeling of mineral operations, which became fashionable in the 1960s and 1970s. Although such studies have provided enormous advances in our ability to analyze mineral processing operations, most of this work relies on empirical coefficients. Gradually this situation must, of course, change, and eventually the empirical coefficients will be correlated with, or replaced by, parameters with a theoretical base. However, until that occurs, much of the fundamental material can still be useful in providing valuable qualitative information.

The first two chapters are concerned primarily with the characterization of minerals, and of particles especially. Since a large proportion of mineral processing is concerned with the separating of particles, Chapter 3 emphasizes the *commonality of separation principles,* and the measurement of separation efficiency. Rather than using single efficiency factors, as is too often the practice, the emphasis is on assessment and analysis.

Because the processing of minerals is carried out in fluids, an understanding of the interaction between particles and fluids is important, and this is treated in Chapter 4. Even for systems that are too complex to be analyzed rigorously, this information is valuable for assessing relative particle behavior.

Gravity separations must be the Cinderella of mineral processing. With the drift toward use of lower grade ores, the development of flotation, and the more obvious benefits to be gained from a better understanding of the much more expensive size reduction processes, there has been comparatively little incentive for research on gravity concentration. (The situation is also not helped by the complexity of this process.) Thus, gravity separations are still not well understood. Chapter 5 reviews the mechanisms that appear to have some relevance. Extensive and enlightened research is still needed to determine which of these mechanisms are significant.

On the other hand, a considerable amount of research has been carried out on surfaces and interfaces, and this has contributed significantly to our understanding of flotation and flocculation. It has also expanded the application of flotation to a wider range of ores and particle sizes, and led to the development of new separation processes.

The final chapter in Part I considers the fundamentals of fracture. Advances have been made in relating the breakage of minerals to energy requirements, and some success has been achieved in relating the concepts of fracture mechanics of brittle materials to complex heterogeneous ores. Unfortunately, much of the work is not receiving the attention it deserves because it is not widely available in the English-language literature.

Chapter One

Ores, Mills, and Concentrates

Farmer's Law on Junk: "What goes in, comes out."
Corollary 1: "He who sees what comes out, and why, gains wisdom."
Corollary 2: "He who sees only half the problem will be buried in the other half."

All the inorganic materials used to support our civilization, are in some way derived from the earth's crust, the thin shell of silaceous material that coats our planet to a depth of about 13 km. The various elements that make up this crust are not evenly distributed; rather they exist as a mixture of minerals, each of which has only a few major elements in its structure. (Appendix A gives some details of the more important minerals.) Further irregularities of distribution are generated by geological processes and by weathering, which result in some minerals being formed into larger concentrations in particular areas. This concentration is described as an ore body when it is large enough for the mineral to be *economically exploited*.[1]

Today, virtually no mineral as mined is suitable for conversion to a final product; rather it requires preparation, and the preparation of ores by physical methods is described as mineral processing. A wider concept of mineral processing includes chemical methods of treating minerals and therefore extends across the field of extractive metallurgy to the production of commercially pure metals. Only the physical processing of ores is considered in this book.

1.1. ECONOMIC CRITERIA

Although a certain minimum product quality is always a restriction on the production of a material, economic considerations provide the ultimate constraint (though in some situations political factors may appear to be more important). If an ore or mineral is not marketable without mineral processing, the processing operation will not be justified unless the product has a selling price greater than all the costs involved in producing it. These costs include mining, processing, transportation, and selling, as well as the costs incurred in meeting environmental regulations. Even for an ore that is marketable as mined, however, processing may be justified if its costs can be exceeded by an enhanced selling price.

On these bases, it is possible to recognize a number of situations in which mineral processing can be justified.

To control particle size. Size control may be carried out simply to make the material more convenient to handle (e.g., the crushing of an ore in an underground mine), or to make a product suitable for sale (as in the case of sized aggregate).

To produce a product of regular size and composition. Any metallurgical operation, but smelting in particular, can be performed far more efficiently if the material fed in is of regular composition and size. A classic example is the achievement of unprecedented production rates in iron blast furnaces by the use of pelletized feed.

To expose or liberate constituents for subsequent processing. Exposure and liberation (freeing) are achieved by size reduction. For subsequent leaching processing exposure of the valuable is sufficient, but when the minerals are to be separated by physical processes, an adequate degree of liberation of the different minerals from each other is a prerequisite.

To control composition. It is necessary to eliminate (or partially eliminate) constituents that would make the ore difficult to process chemically or would result in an inadequate final product. Respective examples are lowering the iron content in lead and zinc ores to avoid problems during smelting, and the removal of iron from silica sand used for glass manufacture, since even very small amounts color the glass.

The more typical situation is the elimination of the bulk of the waste minerals from an ore, thus forming a *concentrate,* a significantly smaller volume of material containing the bulk of the valuable mineral.

Generally, the production of a concentrate is the most complex and difficult form of mineral processing, in that it can involve liberation, size control, and composition control. Thus this step is the focus of our discussion, since generally this choice permits the separate situations above to be covered, as well.

1.1.1. Product Valuation

Because mineral products inherently lack uniformity, their valuation is a complex issue, involving payment credits for desirable features with penalties for undesirable features. To introduce the subject, we consider representative custom smelter schedules. A custom smelter is one that purchases concentrates and smelts (reduces) them to salable metal. The price paid by the smelter depends primarily on the market price of the metal.[2,3] Deductions are made based on all costs involved in smelting (including smelting losses), as well as penalties for constituents in the ore or concentrate that are detrimental to the smelting process.

Some simplified typical smelter schedules are given in Table 1.1. Actual figures vary from smelter to smelter, depending on the smelting method employed and its efficiency. The smelting method determines what metals can be recovered and thus what credits can be paid, while the method and its efficiency determine what fraction of the metal can be recovered from the concentrate and thus the extent of the deductions to be made from the payments. The treatment charge is seldom fixed; in-

TABLE 1.1: SIMPLIFIED SAMPLE SMELTER SCHEDULES*

A. COPPER SMELTER

Payments
Copper: Deduct 1.5 units from wet Cu assay, and pay for 97.5% of remainder at published price less a deduction of 26.4¢/kg of Cu. (No payment for less than 3.0% Cu).

Gold: Deduct 0.7 gm/tonne and pay for 92.5% of remainder at published price less a deduction of 3.2¢/gm.

Silver: Deduct 34.0 gm/tonne and pay for 95% of remainder at published price less a deduction of 0.32¢/gm.

Deductions
Treatment
Charge: $66.00/tonne

Labor
Adjustment: Smelter charge is based on an average hourly cost of labor of $10.00. Increase or decrease charge by 13.75¢ for each 1¢/hr that average hourly cost changes.

Arsenic: If 0.5 units or more, charge for all at $1.00/unit.

Antimony: Allow 0.2 units free; charge for excess at $1.50/unit.

Bismuth: Allow 0.05 units free; charge for excess at $0.50/unit.

Nickel: Allow 0.3 units free; charge for excess at $5.00/unit.

Zinc: Allow 5.0 units free; charge for excess at $0.30/unit.

B. LEAD SMELTER

Payments
Lead: Deduct 1.5 units from wet Pb assay, and pay for 90% of remainder at published price less a deduction of 9¢/kg.

Silver: Deduct 30 gms, and pay for 95% of remainder at published price less a deduction of 0.15¢/gm.

Deductions
Treatment
Charge: $55.00/tonne. Add 44¢/tonne for each unit under 25% Pb.

Labor
Adjustment: Smelter charge is based on an average hourly cost of labor of $10.00. Increase or decrease charge by 13.75¢ for each 1¢/hr that average hourly cost changes.

Arsenic: Allow 1.0 units free; charge for excess at $1.00/unit.

Bismuth: Allow 0.05 units free; charge for excess at $1.00/unit.

C. ZINC SMELTER

Payments
Zinc: Pay for 85% of Zn content at published price less a deduction of 2.2¢/kg.

Cadmium: Pay for 40% of Cd content at published price, less $3.30/kg.

Lead: Deduct 1.5 units, and pay for 65% of remainder at published price less 11¢/kg. No payment for less than 3% Pb.

Silver: Deduct 170 gms/tonne, and pay for 60% of remainder at published price less 0.18¢/gm.

Deductions
Treatment
Charge: $165.00/tonne. Increase by $2.20/tonne for each 1.0¢ increase in the Zn price above 29¢.

Labor
Adjustment: Smelter charge is based on an average hourly cost of labor of $10.00. Increase or decrease charge by 26¢ for each 1¢/hr that the average hourly cost changes.

Iron: Deduct 8 units; charge for excess at $1.50/unit.

Notes: 1. Payments are based on the average price in *Metals Week*[2] for the month following receipt of concentrate at the smelter.

2. Tonnes is a metric ton of 1000 kg.

3. Unit is 1% of a tonne, i.e., 10 kg.

4. Seller pays freight to smelter. Adjustments are also made for changes in freight rate of bullion from smelter.

Economic Criteria

stead it is tied to the metal price and/or to labor and fuel costs. Penalty elements are normally determined after a sample of the concentrate has been analyzed.

Example 1.1. A lead concentrate is produced from an ore assaying 10% PbS. In addition the PbS contains 2 g of Ag per kilogram of PbS. The concentrate assays 80% PbS.

What is the return per tonne of concentrate?
(Metal prices: Pb = $1.10/kg; Ag = $1.23/g; labor rate = $10.50/hr; PbS = 86.6% Pb.)

Solution

Basis. 1000 kg concentrate

Payments

$$\text{lead} = \text{mass lead} \times \text{return}$$
$$= 1000 \text{ kg} \left(\frac{80\%}{100} \times \frac{86.6\%}{100} - \frac{1.5\%}{100} \right) \frac{90\%}{100} \frac{\$1.10 - \$0.09}{\text{kg}}$$
$$= \$616.12$$

$$\text{silver} = \text{mass silver} \times \text{return}$$
$$= \left[\left(\frac{1000 \text{ kg}}{1} \left| \frac{80\%}{100} \right| \frac{2 \text{ g}}{\text{kg}} \right) - 30 \right] \frac{95\%}{100} \frac{\$1.23 - \$0.0015}{\text{g}}$$
$$= \$1832.31$$

Deductions

treatment charge
$$= \$55.00 + \frac{\$10.50 - \$10.00}{\text{hr}} \left(\frac{13.75 \text{ ¢}}{1} \left| \frac{\text{hr}}{\text{¢}} \right. \right)$$
$$= \$61.88$$
$$\therefore \text{return/tonne} = \$616.12 + \$1832.31 - \$61.88$$
$$= \$2386.55$$

Although the smelter schedule applies to the sale of a concentrate to another company, the concept is still quite valid for use on a concentrate being smelted by the same company, in that the concentrator in this situation is effectively selling concentrate to its own smelter.

The basic concept of credits and penalties applies to the sale of most mineral products, be they base metal concentrate, coal, or aggregates. More thorough discussions on the subject are available elsewhere.[4-8]

1.1.2. Production Costs

A more detailed discussion of production costs is given in Chapter 25; however, it is worthwhile to consider briefly here some of the important factors.

Nature of the Ore. The texture, association, and degree of dissemination of constituent minerals determine both the amount of size reduction necessary before concentration can be carried out and the extent to which it can be achieved. Size reduction is one of the most expensive items in mineral processing, so that coarse grained or soft ores are cheaper to process. In practice, it is never possible to completely liberate a valuable mineral, or to extract all of it.

Decreasing ore quality, or grade, has a twofold effect on the size of the operation. Leaner ore bodies must be operated on a larger scale simply to produce a given amount of product; but because leaner ore necessitates the treatment of a large amount of ore, the operation must be carried out on an even *larger* scale to reduce the cost per tonne, since this reduces with increasing tonnage handled. In turn this means that the investment must be further increased, with the result that lower grade ore bodies must be disproportionately larger. Today, ore bodies tend to be in isolated areas and this generally necessitates additional investment in associated infrastructures, such as towns, transportation, and service facilities.

The uniformity of the ore body itself affects processing costs, since slight changes in mineralogy either necessitate changes in processing or lead to losses of valuable material in the *tailings* or waste stream.

Mill Location. The location of the mill or plant affects the costs of fuel, power, water, transportation, and labor, as well as the cost of the mill itself.

Metal Prices. Continually changing prices necessitate a continual reappraisal of the dividing line between ore and potential ore. (See Example 3.3.) Variations in price can be large, and are affected not only by supply and demand, but also by local or world political events.

Legislation. Legislation may restrict or benefit the cost structure in a variety of ways. Government assistance in the form of subsidies (including government stockpiling) or low taxes may be used to encourage the initiation of a project, but such assistance may fade once a project is operating. Tightening environmental and safety regulations can be particularly expensive if they are not anticipated.

1.2. FLOW SHEETS

The way a plant is arranged to produce its product is best portrayed by a flow sheet, such as those shown in Appendix B. At first glance there seems to be little similarity between the examples. Closer examination will show that this is not the case, but rather that any process can be considered to be made up from three basic operations: *size reduction* (or comminution), *separations,* and *materials handling.* In turn, each of these can be further subdivided. The major subdivisions in size reduction (Part II) depend on whether the particles broken are coarse or fine, the processes being described as *crushing* and *grinding,* respectively. Separations can be either solids from solids or liquids from solids. In the former, the separation is based essentially on the composition of the solids (concentration processes, Part IV) or the size of the solids (screening and classification, Part III). The separation of liquids from solids is in most cases one of dewatering (Part V). Materials handling (Part VI) includes the auxiliary operations associated with the transport and storage of materials, in, to, and from the mill.

Products are sold to a specification, expressed primarily in terms of composition and/or size. When the material is derived from a hard rock mine, size reduction is necessary even if the product has only a size specification (e.g., a quarry producing aggregate) to meet. When a composition specification has been given, extensive size reduction is generally necessary to liberate the minerals from each other, so that separation or concentration can be carried out. However, an examination of any flow sheet (e.g., Appendix B) shows that irrespective of whether there is a size or composition specification, size separations are carried out at intermediate stages of the processing. The main reason for this is that any size reduction or concentrating device operates best on a very narrow range of particle sizes. In the case of size reduction, the product always has a wide range of sizes, even if the feed is of a "single" size. By sizing such a product, material can be removed as soon as it is formed, and can also if necessary bypass the next stage if already fine enough. Sizing before concentration is also beneficial, but because the issue is more complex, discussion of this aspect is left to the separate chapters in Part IV. It is sufficient to note at this stage that overgrinding causes losses in two ways: particles too fine to be separated are lost to the tailings (waste stream), and energy is wasted in overgrinding.

Any solid-solid separation can be represented as follows:

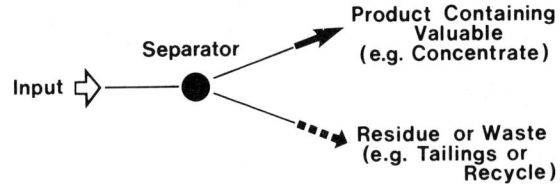

One vital characteristic of mineral processing separations is that they are *never* perfect; some of the valuable product is always in the waste stream, and some of the waste (or *gangue* minerals) is always in the valuable stream. To adequately describe the extent of the separation, two parameters are commonly considered: recovery and grade. *Recovery* measures how *effectively* the separator has extracted the valuable contained in the input stream. A suitable definition is:

$$\text{Recovery (\%)} = \frac{\text{mass of valuable in product stream}}{\text{mass of valuable in input stream}} \times 100 \quad (1.1)$$

Losses can similarly be defined by the quantity of valuable in the residue or tailings. (Alternatively, it may be convenient to consider the loss as a "recovery" of valuable in the tailings, since this consistently describes the distribution of valuable between the output streams.)

Grade is a measure of the *quality* of any stream: ideally the valuable product stream should be of high quality, and the tailings of low quality. Thus, the grade of any stream is defined by:

$$\text{grade (\%)} = \frac{\text{mass of valuable in stream}}{\text{mass of valuable and waste in stream}} \times 100 \quad (1.2)$$

Again, this definition is universal and can apply also to the input stream, or to an ore. Care should always be taken in interpreting the grade when dealing with a metallic mineral. Grades (or assays) are normally quoted as a percentage metal, whereas the metal actually exists as a mineral, with the result that the grade in principle cannot rise above that given by the stoichiometric composition of the mineral.

The point will be repeatedly made that grade and recovery represent a tradeoff; for a given sample of material, one can be obtained only at the expense of the other. Thus concentration (where the flow is large enough) is carried out in stages, described as *roughing, scavenging,* and *cleaning. Roughing* is the first stage, designed to remove the easily recoverable liberated valuable as a *rougher concentrate.* The rougher tailings then pass to *scavenging,* where the emphasis is on recovery: the aim is to extract all the remaining valuable that it is economically justifiable to recover. (The scavenger tailings are then said to be worthless with respect to that valuable mineral; they may, however, contain a second valuable mineral worth recovering in another separator.) The scavenger concentrate is in principle recycled to recover the valuable. Essentially it consists of *middlings* or *unliberated* particles; and unless further size reduction is applied to this concentrate, it is forced eventually to leave with the final tailings or the final concentrate. In the former instance, overall recovery is low; in the latter, overall grade is low.

Typically, the rougher concentrate is diluted with waste minerals, primarily in the form of *misplaced* particles, that is, liberated waste particles that have passed to the wrong stream (strictly, there are also some middling particles in the rougher concentrate, just as there are misplaced particles in the scavenger concentrate). Rougher concentrates are therefore often retreated in *cleaners* (and if necessary recleaners), where the emphasis is on grade. Since this increased grade is achieved at the expense of recovery, the cleaner tailings must be recycled. Figure 1.1 illustrates these characteristic operations. They are considered again in Chapter 23, after the various concentration processes have been described in more detail.

1.3. MATERIALS BALANCES

An important aspect of any mineral processing study is an analysis of how material is distributed whenever streams split or combine. This knowledge is necessary when a flow sheet is being designed and is also essential when making studies of operating plants. Such calculations are known as *materials balances* and are based on the principle of conservation of matter. In general,

$$\text{input} - \text{output} = \text{accumulation}$$

In a continuous system at steady state, there is no accumulation; thus the relationship reduces to:

$$\text{input} = \text{output}$$

Consider, for example, a mill represented by one of the flow sheets in Appendix B. From a single input of ore, two products are produced: a concentrate containing most of the valuable, and a tailings stream containing most of the gangue. The overall process involved is in terms of the total quantity of material entering and leaving, and the process can be represented thus:

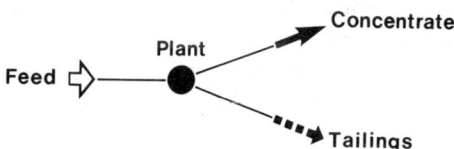

That is, the mill can be considered to be a separation point, known as a *node,* and at this stage, how the separation is achieved is irrelevant. Thus

input = output

tonnes of feed
$$= \text{tonnes of concentrate} + \text{tonnes of tailings}$$
$$M_I = M_{(+)} + M_{(-)} \qquad (1.3)$$

(M_I = mass of input; $M_{(+)}$ = mass of concentrate; $M_{(-)}$ = mass of tailings). Obviously, if any two of

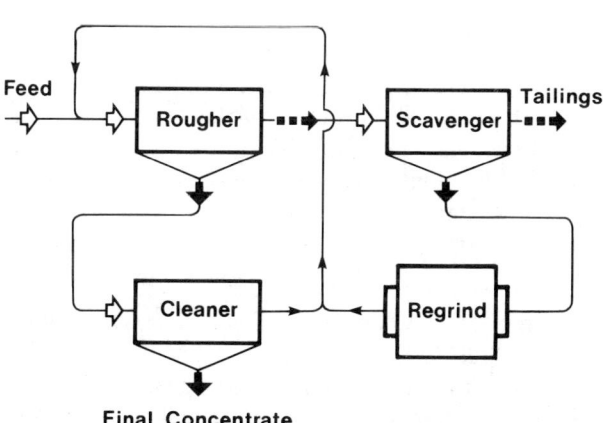

Figure 1.1. The basic concentrating circuit of roughing, scavenging, and cleaning.

the masses are known, the third can be calculated. However, in making any such analysis, it is first of all necessary to set up a reference known as the *basis*: in effect, one stream is arbitrarily fixed and all other data are *relative* to this basis. Such a reference may be a volume, mass, period of time, or flow rate. Selection of a suitable basis is certainly aided by experience, but to help the novice, the following should be considered when making the selection:

What information is available?
What information is one trying to obtain?
What is the most convenient basis?

Further clarification of the problem is always obtained by *making clear, neat sketches of the flow sheet, showing all known data*. In many instances it is convenient to take the input rate as the basis; at other times a unit base of 1 or 100 in suitable units (e.g., tonnes) is useful, since this makes subsequent quantities fractions or percentages.

Returning to the mill considered above, if from a feed of 500 t/hr it produces 10 t/hr of concentrate, a convenient basis is 500 tonnes of feed (or 500 t/hr). Thus

$$\text{input} = \text{output}$$
$$500 \text{ t} = 10 \text{ t concentrate} + M_{(-)}$$
$$\therefore \quad M_{(-)} = 490 \text{ t (or 490 t/hr)}$$

As well as completing a materials balance for the total mass in each stream, it is also possible to have a materials balance for each component of interest. Once again, convenience is the best criterion for selecting a component: for example, if the mill considered above is fed an ore that is 10% PbS and 90% SiO_2, balances can be set up for PbS and SiO_2; thus:

	input	= output	
PbS balance:	$M_{I,PbS}$	$= M_{(+)PbS}$	$+ M_{(-)PbS}$
SiO_2 balance:	M_{I,SiO_2}	$= M_{(+)SiO_2}$	$+ M_{(-)SiO_2}$
Total:	M_I	$= M_{(+)}$	$+ M_{(-)}$

Note that although there appear to be three equations here, implying that the equations can be solved for three unknowns, such is not the case. One of these equations is always redundant because of the existence of the vertical equations, such as

$$M_{I,PbS} + M_{I,SiO_2} = M_I$$

Alternatively, it is possible to take a balance of Pb and the total mass, since the mass of PbS in any sample can be calculated by the stoichiometry of PbS; then the mass of SiO_2 can be obtained from the difference between PbS and the total mass.

More thorough methods are available for estimating the number of equations required,[9] but they are beyond the scope of this book, and at this stage it is assumed that sufficient data always are available. It is worth remembering that "missing" equations can often be generated by using equations involving recovery.

So far, only the input and output of the total mill have been considered. The principles of materials balancing can in fact be applied to any node in the plant, such as a distributor, a separator, or the point at which streams combine or split. Again, it is not necessary to know how or why the material separates at a node: it is treated simply as a point. For example, the rougher flotation cells in a flow sheet

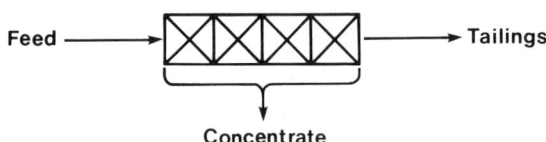

can be represented as

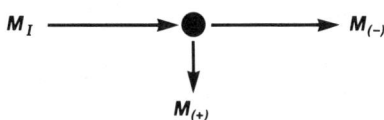

The following examples are used to illustrate the principles outlined. It is worth emphasizing at this point that with all problems of this type (and especially with more complex ones), the development of good habits is vital. Experience is of course invaluable, but the extensive use of *clear* diagrams and tables is particularly helpful.

Example 1.2. A concentrator is fed 1000 t/hr of ore assaying 10% PbS. It produces a concentrate assaying 80% PbS and a tailings assaying 0.19% PbS.

What are the flow rates of the tailings and concentrate streams?

Materials Balances

Solution

Problem:

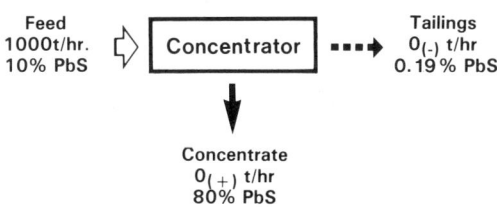

Basis. 1000 t/hr of feed

Mass Balances

$$\text{input} = \text{output}$$
$$\text{feed} = \text{concentrate} + \text{tailings}$$

Total Flow

$$1000 \text{ t/hr} = O_{(+)} \text{ t/hr} + O_{(-)} \text{ t/hr} \quad (E1.2.1)$$

PbS Flow

$$1000 \frac{t}{hr}\left(\frac{10\%}{100}\right) = O_{(+)} \frac{t}{hr}\left(\frac{80\%}{100}\right)$$
$$+ O_{(-)} \frac{t}{hr}\left(\frac{0.19\%}{100}\right) \quad (E1.2.2)$$

From (E1.2.1):

$$O_{(-)} = 1000 - O_{(+)}$$

Substituting in (E1.2.2):

$$1000 \times \frac{10}{100} = O_{(+)} \frac{80}{100} + (1000 - O_{(+)}) \frac{0.19}{100}$$

∴ $\quad 10{,}000 = 80\, O_{(+)} + 190 - 0.19\, O_{(+)}$

∴ $\quad O_{(+)} = 9810/79.81$

$\quad\quad\quad = 122.9$ t/hr

Substituting in (E1.2.1):

$$O_{(-)} = 1000 - 122.9$$
$$= 877.1 \text{ t/hr}$$

Example 1.3. A PbS concentrate is produced by a rougher-cleaner flotation circuit. The cleaner tailings assay 20% PbS and are recycled to the rougher cells, and the circulating load (recycle/fresh feed) is 0.25. The fresh feed assays 10% PbS and is delivered at the rate of 1000 t/hr. The recovery and grade in the concentrate are 98.2% and 90%, respectively. What are the flow rates and assays of the other streams?

Solution

Problem:

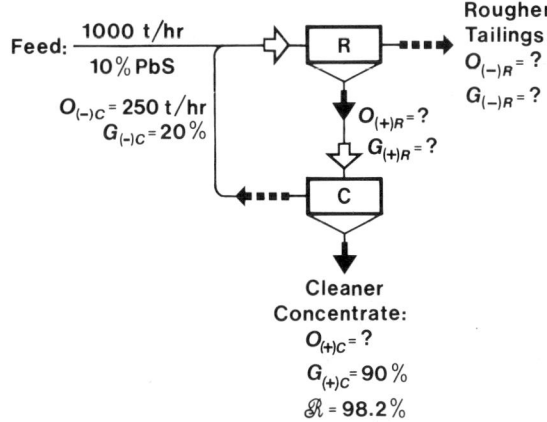

Basis. 1000 t/hr fresh feed.

Although total and PbS mass balances can be set up for the whole circuit, the rougher, and the cleaner, two of these will be redundant because the rougher + cleaner balances sum to the whole circuit balances. Thus, to allow the fifth unknown to be determined, the recovery must be used to set up a fifth equation.

$$\text{input} = \text{output}$$

Rougher total mass balance:

$$1000 + 250 = O_{(-)R} + O_{(+)R} \quad (E1.3.1)$$

Cleaner total mass balance:

$$O_{(+)R} = O_{(+)C} + 250 \quad (E1.3.2)$$

Rougher PbS mass balance:

$$1000 \times 10\% + 250 \times 20\%$$
$$= O_{(-)R} \times G_{(-)R} + O_{(+)R} \times G_{(+)R} \quad (E1.3.3)$$

Cleaner PbS mass balance:

$$O_{(+)R} \times G_{(+)R}$$
$$= O_{(+)C} \times 90\% + 250 \times 20\% \quad (E1.3.4)$$

$$\mathscr{R} = \frac{\text{mass PbS in cleaner concentrate}}{\text{mass PbS in fresh feed}} \times 100$$

that is,

$$98.2\% = \frac{O_{(+)C} \times 90\%}{1000 \times 10\%} \times 100 \quad (E1.3.5)$$

Solving Eq. E1.3.5:

$$O_{(+)C} = 109.1 \text{ t/hr}$$

Substituting in Eq. E1.3.2:

$$O_{(+)R} = 109.1 + 250 = 359.1 \text{ t/hr}$$

Substituting in Eq. E1.3.1:

$$1250 = O_{(-)R} + 359.1$$

$$\therefore \quad O_{(-)R} = 890.9 \text{ t/hr}$$

Substituting in Eq. E1.3.4:

$$359.1 \times G_{(+)R} = 109.1 \times 90 + 250 \times 20$$

$$\therefore \quad G_{(+)R} = 41.27\%$$

Substituting in Eq. E1.3.3:

$$1000 \times 10 + 250 \times 20 = 890.9 \times G_{(-)R} + 359.1 \times 41.27$$

$$\therefore \quad G_{(-)R} = \frac{10{,}000 + 5000 - 14{,}831}{890.9}$$

$$= 0.20\%$$

Example 1.4. What is the circulating load ratio in Example 10.1?

Solution

Basis. 800 t/hr, hydrocyclone feed

For the hydrocyclone:

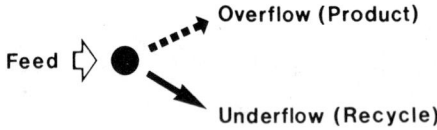

Mass Balance:

$$\text{input} = \text{output}$$
$$\therefore \quad \text{feed} = \text{overflow} + \text{underflow}$$

From Example 10.1:

$$800 \text{ t/hr} = 800 \times 0.295 \text{ t/hr} + 800 \times 0.705 \text{ t/hr}$$

For the circuit:

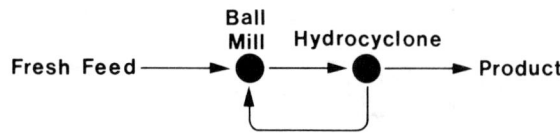

$$\therefore \quad \text{Fresh feed} = 800 \times 0.295 = 236 \text{ t/hr}$$

$$\therefore \quad \text{Circulating load ratio} = \text{recycle/fresh feed}$$

$$= 564/236$$
$$= 2.39$$

Example 1.5. If the feed stream in Example 1.4 is 30% solids by volume and 30% of the water is recycled, what is the slurry concentration of the hydrocyclone products? ($\rho_s = 3.145$ t/m³)

Solution

Basis. 800 tonnes feed solids

Volumetric Solids Balance

$$\text{input} = \text{output}$$
$$\text{feed} = \text{underflow} + \text{overflow}$$

$$\frac{800 \text{ t}}{1} \left| \frac{\text{m}^3}{3.145 \text{ t}} \right. = \frac{564}{3.145} + \frac{236}{3.145}$$

That is, $254.4 \text{ m}^3 = 179.4 \text{ m}^3 + 75.0 \text{ m}^3$

$\therefore$ volume of water in feed

$$= 254.4 \times 0.7/0.3$$
$$= 593.6 \text{ m}^3$$

$\therefore$ volume of water in underflow

$$= 0.3 \times 593.6 \text{ m}^3$$
$$= 178.1 \text{ m}^3$$

$\therefore$ solids concentration in underflow

$$= 179.4/(179.4 + 178.1)$$
$$= 50.2\% \text{ (vol)}$$

volume of water in overflow

$$= 593.6 \text{ m}^3 - 178.1 \text{ m}^3$$
$$= 415.5 \text{ m}^3$$

$\therefore$ solids concentration in overflow

$$= 75.0/(75.0 + 415.5)$$
$$= 15.3\% \text{ (vol)}$$

Comment. More likely, the feed and underflow concentrations will be set and the water recycle unknown. In this situation, the solution of Examples 1.2, 1.3, 1.4, 8.3, 10.1, and 10.4 require successive iteration.

In solving these problems, each node has been handled individually, even though the mathematical formulation is basically the same in each case. In fact, these equations are frequently presented and described as the *two-product formulas,* or in the case of a stream split into three, the three-product formulas.[10] Though appreciating that these formulas are indispensable in handling routine data or large complex flow sheets on a computer, we consider it desirable to develop a full understanding of the input/output concept, and to avoid the unthinking use of equations that can lead to a lack of appreciation of the process metallurgy.

In each example above, there was just sufficient data to solve the materials balance. Seldom is this ideal situation found in practice (and if it is, one should take care!). Sometimes the problem is insufficient data; either because data have not been obtained or because no sample point is accessible in the stream. However, a more likely problem, in operating plants in particular, is that an excess of data exists, which generally leads to conflicting results. Too often the solution is simply to collect data only sufficient to complete a materials balance, or to guess which are the more accurate data and reject the remainder. Such methods may be satisfactory if the worth of the data can be assessed, but the approach is fraught with danger. This is because major sources of error exist.

Assaying errors are the most obvious, and although their magnitude can generally be assessed, their effect on other data is easily overlooked. Sampling errors can also occur, and although they can be estimated, they are too often just neglected. However, the most significant source of error is the assumption that a circuit is at steady state. In reality, perturbations in flow and composition exist in any circuit, so that samples taken in different locations effectively represent different material.

The solution to this dilemma is to take advantage of the excess data.[11-15] The situation can be likened to having a graph show three noncollinear points that represent a straight line: to find the best estimate of the true straight line, a least-squares statistical analysis is carried out. In this case as *more* data points are included, a more reliable estimate of the true line is obtained. The application of least-squares analysis to mineral processing material balances has been considered in a number of studies, and although computation times can become long because a considerable number of streams must be treated, the increased reliability justifies the effort. Computer programs for these analyses have been made available in the literature.[12,14-16]

The concept of conservation may also be extended to energy balances. Analysis may be difficult because of heat losses, but nevertheless, such balances may be necessary in operations such as pumping and drying.

1.4. ORES

The rocks that form the earth's crust are classified as igneous, metamorphic, or sedimentary; but since the mantle is probably the starting point for all rocks, the average composition of the crust can be assumed to be similar to that of an average igneous rock. Such an estimate is shown in Table 1.2. A notable feature of these data is that the common metals are present in the crust in very small amounts. Also shown in the table are the prices of some elements, and it can be seen that there is little correlation between abundance and price. What therefore determines the price of a metal? A clue to this complex question is shown in Fig. 1.2,[17] where it can be seen that there is good correlation between selling price and annual consumption. Despite the many factors that affect the selling price and the consumption, the general supply and demand situation means that production costs are still a significant factor in determining the selling price of a material.[18-21] In turn, production costs depend on the costs of the three main processing stages: mining, mineral processing, and extractive metallurgy. (In the case of nonmetallic ore treatment the extractive metallurgy step does not occur, although there may instead be some other form of chemical processing.)

It is possible to distinguish four types of mining[1,22]: open-pit (Fig. 1.3), underground (Fig. 1.4), alluvial (Fig. 1.5), and solution mining.[22] The first two are forms of hard rock mining. Open-pit mining, which is the more economical and practical method of handling enormous tonnages of material, is becoming more common as ore grades decrease. Such a method does require the ore body to be relatively close to the surface. Underground mining is

TABLE 1.2: AVERAGE COMPOSITION OF IGNEOUS ROCK

Element	%	Cost* $/kg
Oxygen	46.6	
Silicon	27.7	1.10
Aluminium	8.1	1.20
Iron	5.0	0.22
Calcium	3.6	~4.40
Sodium	2.8	~0.80
Potassium	2.6	~6.60
Magnesium	2.1	2.22
Titanium	0.6	7.20
Manganese	0.1	1.28
Sulfur	0.06	0.05
Zirconium	0.03	37.74
Nickel	0.02	4.58
Vanadium	0.017	
Copper	0.010	1.46
Zinc	0.004	0.68
Lead	0.002	0.74
Cobalt	0.001	24.40
Tin	0.001	13.87
Ce,Ga,Li Nb,Th,Yt	10 - 100 ppm	
As,B,Ge,Hf Mo,Sb,U,W	1 - 10 ppm	
Bi,Cd,In	0.1 - 1 ppm	
Ag,Pd,Se	0.01 - 0.1 ppm	
Au,Ir,Os,Pt	<0.01 ppm	

* 1978

the most expensive method, but there is not much choice (unless solution mining can be used) when the ore is a considerable distance underground. Sometimes both open pit and underground mining are used on the same ore body.

Alluvial mining uses dredges or hydraulic water to extract ore from placer deposits (unconsolidated mineral deposits such as river beds or sand dunes) and by comparison with hard rock mining, is inexpensive.

Unlike subsequent processing, the mining method has comparatively little dependence on the mineralogical nature of the valuable minerals in the ore, although the nature of the gangue minerals may affect the actual method of hard rock mining used. Basically, mining can be considered to be an engineering problem in materials handling, dependent on the ore body rather than on the minerals.

In broad terms the mineral processing method employed depends on the mineral being processed, which in turn depends on whether that mineral is amenable to reduction to a metal. Three factors determine whether and how a metallic mineral can be reduced to a metal: the chemical stability of the metal compound, the nature of the anion(s) combined with the metal, and the ability to engineer a practical process.[23]

In general the more stable the compound the more difficult and expensive it will be to reduce it to a metal. Carbon (as CO) is the cheapest and most convenient reductant available, and so it is used wherever possible. Unfortunately, metals such as copper, lead, zinc, and nickel occur in suitable concentrations in the earth's crust only as sulfides, and because the sulfur anion does not form a stable compound with carbon, carbon cannot be used directly as a reductant. To overcome this problem, sulfides first have to be converted to oxides by a process known as roasting:

$$(\text{metal}) S + O_2 \rightarrow (\text{metal}) O + SO_2$$

Engineering limitations can be illustrated by reference to the reduction of aluminum oxide. Although this comparatively stable oxide can be reduced by carbon, the high temperatures necessary impose severe engineering problems. Consequently, in practice, lower temperatures are used, employing electrical energy to bring about the reduction. In addition, because most of the impurities are reduced at the same time, they must be removed by a chemical refining process before reduction. The expense of this chemical refining means that ores with only one major impurity are acceptable, even though there are extensive deposits of less suitable minerals.

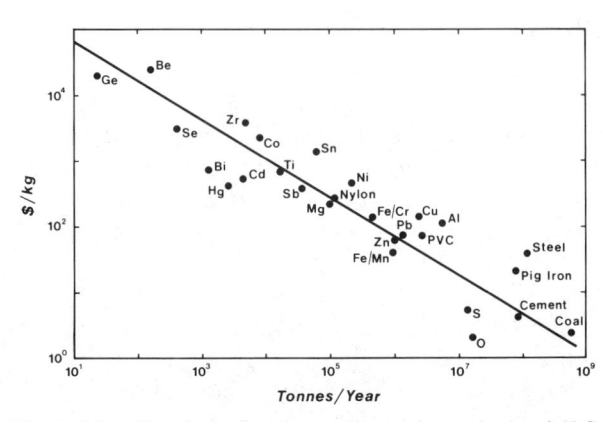

Figure 1.2. Correlation between selling price and annual U.S. consumption for a variety of materials in 1978. (Data from U.S. Bureau of Mines.[17])

Figure 1.3. Open-pit copper mine. (Courtesy Bucyrus-Erie.)

Figure 1.4. Drilling rig in underground mine. (Courtesy *Engineering and Mining Journal*.)

Figure 1.5. Alluvial mining of titaniferous beach sands with cutter suction dredge (left) and floating concentrator (center). (Courtesy Mineral Deposits Ltd.)

In summary, whether a mineral area becomes an ore body depends on the technology and economics of the combined operations of mining, mineral processing, and extractive metallurgy. If the ore body is of a high grade, more expensive operations can be tolerated in one or more stages. However, as ore grade decreases, only highly efficient operations can be tolerated. The classic example is the gradual decrease in the minimum grade of copper ores since the turn of the century.[24] This decline has been attained in face of virtually constant copper prices (in real terms) and has been achieved partly by improvements in technology, but largely by economies of scale.

Some of the minerals commonly used as ores are detailed in Appendix A. The most common nonferrous (i.e., noniron) ores are the sulfides, and typically they are low grade. The term "oxide" or "oxidized ore" is applied to minerals containing oxygen, such as the oxides, hydrated oxides, sulfates, carbonates, and silicates. These vary from relatively high grade ores (aluminum ores are given virtually no physical beneficiation, while a small proportion of iron ores are still used without concentration), down to very low grade materials containing less than 1% valuable, as in the case of some placer deposits of tin ores.

Nonmetallic ores are those that are not used for the production of a metal: they may however contain a metal.[8] Examples are ores containing MgO or Al_2O_3 used to make refractories.

Because ore bodies cannot be seen, it is necessary to qualify ore quantities or *reserves* in terms of the reliability of their existence.[1,25] *Proven* or *measured* reserves are those that have been well outlined in three dimensions by extensive drilling programs. *Indicated* reserves are those implied by a limited amount of drilling, and *inferred* reserves represent essentially educated guess work, based on the results from one or two drill holes or other exploration techniques. It is also possible to define *potential* reserves as those estimated to exist on the basis of statistical or geological data, *or* as the more thoroughly explored mineral bodies that cannot be worked economically at the present time.

1.5. MINERALOGY

Any ore can be considered to be made up of two components: the valuable minerals, and the waste or gangue components. Complex ores are those that contain more than one valuable mineral. (Even so, one should appreciate that in many situations the valuable minerals are extracted one at a time, so that with respect to a given separator, any valuable that is removed later is considered to be tailings until it is recovered.)

It is essential at this stage to grasp the fact that *each ore is unique,* and as a consequence there are no standard mineral processing procedures, even though some may appear to be very similar. A thorough knowledge of the mineralogy of an ore is therefore essential not only for the design of a plant, but also for the research program that collects data to be used for plant design.[26-30] (It is to emphasize the vitality of mineralogy that the subject is discussed here rather than in the next chapter.) Furthermore, once a mill is in operation, regular appraisal of the mineralogy is just as essential for fine tuning and for maintaining efficiency. This arises because ore bodies are not homogeneous; thus variations in feed mineralogy are normal and may occur to such an extent that major circuit modifications are required.

Certain basic mineralogical knowledge is important:

The grade of the ore in terms of the valuable minerals.

The grain size of the minerals.

The combinations of minerals present.

The relative form and association of minerals.

The existence of trace elements in the lattice of the valuable mineral.

The occurrence of minor amounts of potentially valuable minerals.

Implicit here is that knowledge of *both valuable and gangue* minerals is essential. In the case of the valuable element, chemical analysis is not sufficient. Part of the valuable may exist in a mineral form that prevents its recovery, or at least necessitates a second separation method that exploits some other property. Alternatively, it is possible for the chemical analysis of a complex ore to remain essentially constant, while the mineralogical composition changes markedly.

Details of the interrelations between valuable and gangue are also necessary. When liberation is incomplete, either grade or recovery must be sacrificed unless further grinding is practical. It is quite possible for two particular minerals to be so interlaced that no amount of grinding will adequately liberate them. Without mineralogical examination, it may be impossible to determine whether a poor separation is due to inadequate grinding or to mineral characteristics.

Identification of all constituents in the ore is essential and can lead to highly profitable by-products recovery: many a plant has been able to recover a trace mineral from a tailings stream; its recovery becomes economical because mining and grinding costs have already been met by the primary minerals. Even major mineralogical constituents have been overlooked, with the result that some of the world's major ore-bodies have remained hidden in the open for many years—witness, for example, some of the vast Australian bauxite deposits.[27]

Mineralogical Examination. The optical microscope is the major tool used in mineralogical examination,[28,31] and although it has been superseded technologically by the electron microprobe,[30] the particularly high cost and complexity of the newer instrument means that the microscope still has its place.

Two main methods of optical microscopic examination can be distinguished: reflected light (Fig. 1.6a) and transmitted light (Fig. 1.6b and c). Reflected light examination is in general simpler, requires less experience, and gives adequate information for more routine studies, such as plant control. Specimens to be examined are prepared by flattening a surface on them and polishing this surface with successively finer abrasives. Final polishing is carried out with 4–0.25 μm diamond powder. If the mineral sample is not a convenient size to handle (e.g., chips of mill products), it must be mounted in a plastic before polishing.

A number of properties can be used to identify minerals under reflected light. Color is the most obvious, but it must be used with care, first because color is affected by trace elements and second because under the microscope, color is determined largely by the mineral's refractive index. Polarized light can give further information, since many minerals have unique reactions, some showing different colors (pleochroism) as the specimen is rotated, whereas those with anisotropic crystal structures show colors that lighten and darken as the specimen rotates. Chemical tests can be used to color specific minerals, indicate texture, cleavage twinning, or the existence of trace elements.

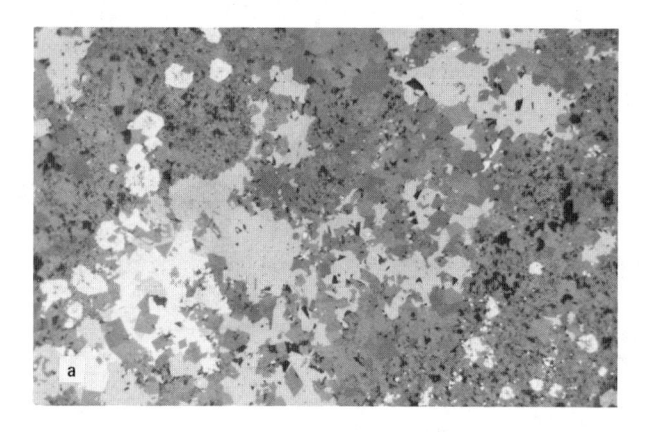

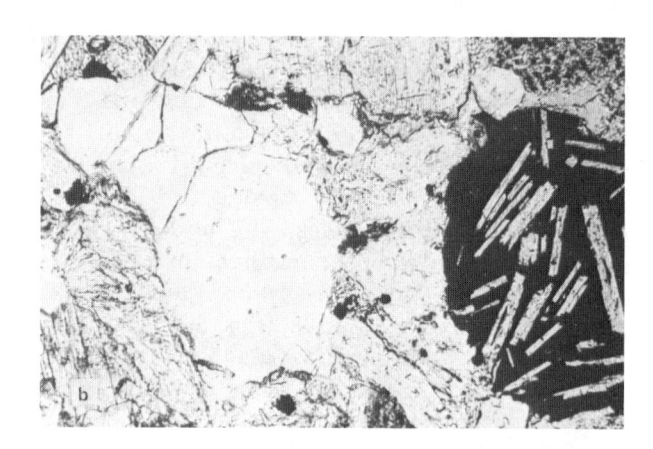

Figure 1.6. (a) Reflected light microstructure of a lead-zinc ore: white = galena; light grey = sphalerite; darker grey = dolomite. (b) Transmitted light microstructure of sandstone containing volcanic fragments of quartz, feldspar, and mica. (c) As (b) but with polarized light, and crossed Nicol prisms, illustrating how transmitted light microscopy can be used to identify the components in more complex minerals. These two-dimensional microstructures should be compared with the Part I frontispiece, which illustrates the more complex three-dimensional nature of the different phases. (Courtesy H. W. Kobe.)

Hardness can also be used as an aid to identification.[32] The preferable method is to use a microhardness tester, but when this is not available, hardness can often be estimated from the relief generated by polishing. This arises because the softer grains are polished below the harder grains. The softer grain can be distinguished by the white line test; that is, when the microscope is racked up (after being focused on the grain boundary), the line of the boundary passes from the hard to the soft mineral.

Darkfield illumination, where available, will show up translucent minerals under reflected light.

Transmitted light microscopic examination has more potential than reflected light microscopy, but the specimens are more difficult to prepare, and their analysis requires considerably more experience. Specimens in this case are sections of mineral mounted on glass; they are polished thin enough to permit light to pass through the different minerals. The appearance of the minerals differ under polarized light, and crystallographic features such as twinning are also significant in aiding identification.

The electron microprobe uses elemental analysis to identify each mineral.[30] Since mineral compositions vary in many cases and also because the instrument is incapable of detecting the lighter elements (below oxygen), identification must be supplemented by the use of standards, as well as by other techniques such as transmitted light microscopy and X-ray diffraction. These limitations are not serious, and they are far outweighed by the large amount of information that can be obtained about the distribution of the elements in the different minerals. The scanning electron microscope (SEM) with an energy dispersive analysis of X-rays (EDAX) facility is also capable of similar analysis. In general it is easier to use, but it gives more qualitative information rather than the quantitative information that can be obtained with the microprobe.

X-ray diffraction is the most reliable method of positively identifying the mineral occurring in a given ore.[33,34] A typical powder diffraction pattern is illustrated in Fig. 1.7. The *relative* distance between the different peaks is a unique measure of the crystal structure of the minerals, and the *overall* spread is a measure of the distance between the atoms. Furthermore, the heights of the different peaks are determined by the atomic number of the atoms at different points in the crystal structure. As a consequence, the diffraction pattern is unique for a given crystalline compound (mineral), and identification can be made by reference to standard sets of data.[34] The lower detection limit for a mineral is about 1–10% (depending on the mineral), but by using various separations on ground samples, concentrated samples can be obtained so that minor minerals can be identified below these limits.

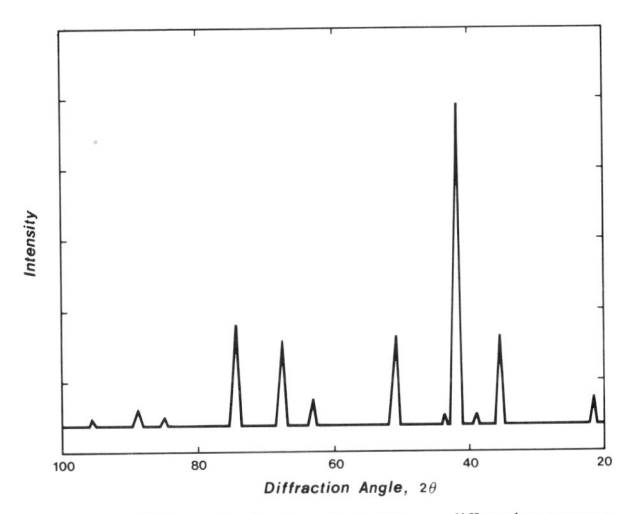

Figure 1.7. Schematic chart readout of X-ray diffraction pattern of titanomagnetite.

REFERENCES

1. W. C. Peters, *Exploration and Mining Geology*, Wiley (1978).
2. "Markets," *Eng. Min. J.* (every monthly issue).
3. *Metals Week* (published weekly).
4. P. J. Lewis and C. G. Streets, "An Analysis of Base-Metal Smelter Terms," *11th Commonwealth Min. and Metall. Congr., Hong Kong, 1978*, pp. 753–767, IMM (1979).
5. H. B. Blaber, "Smelting, Refining and Marketing for Small Scale Mining," UN. Institute for Training and Research, Int. Conf. on the Future of Small Scale Mining. Summarized in *Min. Eng.*, **31**, 149–150 (February 1979).
6. D. W. Gentry and M. J. Hrebar, "Procedure for Determining Economics of Small Underground Mines," *Min. Ind. Bull.*, **19**, No. 1 (January 1976).
7. C. F. Page, "Factors Involved in Mineral Marketing," in *Modern Mineral Processing Flowsheets*, 2nd ed. pp. 274–279, Denver Equipment Co. (1965).
8. S. J. Lefond (Ed.), *Industrial Minerals and Rocks*, 4th ed., AIME (1975).
9. M. Kwauk, "A System for Counting Variables in Separation Processes," *AIChE J.*, **2**, 240–248 (June 1956).
10. A. F. Taggart, *Handbook of Mineral Dressing*, Section 19, Wiley (1945).

11. G. J. Townsend, "Material Balancing of Pilot-Plant Data," *AMDEL Bull.*, No. 18, pp. 51–58 (October 1974).
12. R. L. Wiegel, "Advances in Mineral Processing Materials Balances," *Can. Metall. Q.*, **11**, 413–419 (1972).
13. R. L. Wiegel, "The Practical Benefits of Improved Metallurgical Balance Techniques," AIME/SME Preprint No. 79–92. Presented 1979 AIME Annual Meeting, New Orleans, 1979.
14. J. W. White and R. L. Winslow, "Flowsheet Analysis for Mass Balance Calculations in Overdefined Metallurgical Systems with Recycle," AIME/SME Preprint 79–80. Presented 1979 AIME Annual Meeting, New Orleans, 1979.
15. W. A. Hockings and R. W. Callen, "Computer Program for Calculating Mass Flow Balances of Continuous Flow Streams," AIME/SME Preprint 77-B-372. Presented at AIME/SME Fall Meeting, 1977.
16. R. L. Wiegel, personal communication, 1972.
17. "Mineral Commodity Summaries 1980," U.S. Bureau of Mines (1980).
18. W. A. Vogely (Ed.), *Economics of the Mineral Industry*, AIME (1976).
19. U. Petersen and R. S. Maxwell, "Historical Mineral Production and Price Trends," *Min. Eng.*, **31**, 25–34 (January 1979).
20. W. G. B. Phillips and D. P. Edwards, "Metal Prices as a Function of Ore Grade," *Resour. Policy*, **2**, 167–178 (September 1976).
21. G. W. Bain, "Production Growth Rate Stages as an Index for Amount of Total Resources," *11th Commonwealth Min. and Metall. Congr., Hong Kong, 1978*, pp. 13–23, IMM (1979).
22. W. C. Larson, "Uranium In Situ Leach Mining in the United States," U.S. Bureau of Mines Information Circular, IC 8777 (1978).
23. J. D. Gilchrist, *Extraction Metallurgy*, 2nd ed., Pergamon (1980).
24. J. D. Lowell, "Copper Resources in 1970," *Min. Eng.*, **22**, 67–73 (April 1970).
25. "Principles of the Mineral Resource Classification System of the U.S. Bureau of Mines and the U.S. Geological Survey," U.S. Geological Survey Bulletin 1450-A (1976).
26. R. A. Elliott, "Examination of Mineral Fragments as a Guide to Mineral Dressing Procedures," *Can. Min. J.*, **67**, 92–94 (September 1966).
27. W. R. Liebenberg, "Mineralogy and the Metallurgist," *Miner. Sci. Eng.*, **2**, 3–23 (October 1970).
28. R. D. Hagni, "Ore Microscopy Applied to Beneficiation," *Min. Eng.*, (*Trans. AIME/SME*), **30**, 1437–1447 (October 1978).
29. L. D. Muller, "Some Laboratory Techniques Developed for Ore Dressing Mineralogy," Paper 52, Int. Miner. Process. Congr., 1960, pp. 1047–1057, IMM (1960).
30. J. D. Stephens, "Microprobe Applications in Mineral Exploration and Development Programs," *Miner. Sci. Eng.*, **3**, 26–37 (January 1971).
31. E. N. Cameron, *Ore Microscopy*, Wiley (1961).
32. M. Tarkian, "A Key-Diagram for the Optical Determination of Common Ore Minerals," *Miner. Sci. Eng.*, **6**, 101–105 (April 1974).
33. B. D. Cullity, *Elements of X-ray Diffraction*, 2nd ed., Addison-Wesley (1978).
34. L. G. Berry and R. M. Thompson, "X-ray Powder Data for Ore Minerals: The Peacock Atlas," Memoir 85, Geological Society of America (1962).

Chapter Two

Characterization of Particles

Tylk's Law: "Assumption is the mother of all foul-ups."

The evaluation of particle characteristics is a vital aspect of mineral processing. In some cases, it may be simply because a product has to meet a size specification. Far more important is the use of particle size as a control measure for size reduction (comminution) processes. Sometimes material may be reduced in size to increase the surface area and so speed up a chemical process such as leaching. More commonly, size reduction is carried out to *liberate* (or free) the different minerals in an ore from each other, so that they can be separated (concentrated). Because of the comparatively high cost of size reduction, and because of the difficulties associated with separating minerals when over- or under-liberation occurs, it is essential that the correct amount of size reduction be achieved. Particle size is commonly used to measure the extent of liberation because of its relative ease of measurement, and this chapter considers some of the methods used. However, before particle size can be measured, it is essential to appreciate what is actually meant by the loosely used term "particle size"; this matter is addressed in the early parts of the chapter.

In reality, size is not an adequate measure of liberation. Typically particles from any size reduction operation have a range of characteristics, and a completely rigorous description of such a product is impossible, since it is necessary to consider:

The "size" of each particle.
The "average" size of all the particles.
The "shape" of the particles.
The range of particle sizes.
The minerals occurring in the particles.
The association of minerals in the particles.

This chapter also considers these aspects, together with some of the methods used for their determination.

2.1. PARTICLE SIZE

Only regular geometric figures can have their size conveniently described. For example, the size of a sphere is clearly defined by its diameter, and a cube by the length of its edge. A broken mineral, on the other hand, even if originally regular, consists of discrete particles of irregular shape that cannot be accurately defined. Since it is frequently desirable to use a single number to describe a particle, it is necessary to adopt an approximate description as though the particle were a definite shape. This number is known as the *nominal diameter* d_n but generally, and here, it is abbreviated "diameter."

A number of diameters can be invented. They are defined either in terms of some real property of the particle such as its volume or surface area, or in terms of the behavior of the particle in some specific circumstances, such as settling in water under defined conditions.[1] Some of the more useful nominal diameters are defined in Table 2.1. Clearly, the diameter obtained for an irregular particle will depend on the measuring technique used, and this should be relevant to the end use of the data. For example, a Stokes diameter d_{St} which is determined under laminar flow conditions, would not be applicable under turbulent flow conditions, because in the latter situation the particle orientates itself to give maximum drag, whereas in the former situation it has a random orientation. *It is therefore vital whenever mentioning particle size to define the nominal diameter used.*

TABLE 2.1: NOMINAL DIAMETERS

Symbol	Suitable Name		Basis
d_o			Diameter of a sphere
d_A	Sieve diameter		The width of the minimum square aperture through which the particle will pass
d_s	Surface diameter	$(S/\pi)^{1/2}$	The diameter of a sphere having the same surface area as the particle ($\sim 1.28\, d_A$)*
d_v	Volume diameter	$(6V/\pi)^{1/3}$	The diameter of a sphere having the same volume as the particle ($\sim 1.10\, d_A$)*
d_a	Projected area diameter	$(4A_p/\pi)^{1/2}$	The diameter of a sphere having the same projected area as the particle, when viewed in a direction perpendicular to a plane of stability ($\sim 1.41\, d_A$)*
d_d	Drag diameter		The diameter of a sphere having the same resistance to motion as the particle, in a fluid of the same viscosity and at the same velocity ($\sim d_s$ when Re_p is small).
d_f	Free-falling diameter		The diameter of a sphere having the same density and the same free-falling speed as the particle in a fluid of the same density and viscosity.
d_{St}	Stokes diameter	$= \dfrac{18\, v_\infty}{(\rho_s - \rho_f)g}$ $= (d^3_v/d_d)^{1/2}$	The free-falling diameter in the laminar flow region ($Re_p < 0.2$) ($\sim 0.97\, d_A$)*
d_{vs}	Specific surface diameter	d^3_v/d_s^2	The diameter of a sphere having the same ratio of surface area to volume as the particle.
d_F	Feret's diameter		The mean value of the distance between pairs of parallel tangents to the projected outline of the particle.
d_M	Martin's diameter		The mean chord length of the projected outline of the particle.

* Typical values only: actual value depends on particle shape.

In general, the ratio of any pair of nominal diameters (Table 2.1) is constant over a reasonably wide range of sizes for any particular material.[2] Although the magnitude of the ratios will depend on the shape of the particle, it does make it possible to correlate coarse and fine size analyses obtained by different methods (see Example 2.1).

2.2. PRESENTATION OF SIZE DISTRIBUTION DATA

Because it is normally impractical to measure each particle individually, size analysis is carried out by dividing the particles into a number of suitably narrow size ranges. Data are commonly presented in a tabular form (Table 2.2), but the pictorial or graphic forms, as plots of quantity versus particle size, give a better representation (Fig. 2.1).

Mathematical equations also may be used to describe size distributions. Although attempts have been made to relate these to actual fracture mechanics, most are empirical relationships that are simply a convenient description of data. They may be useful when data must be manipulated; but frequently this necessitates use of a computer, and under these conditions a matrix representation of the actual data is just as convenient and more reliable.

2.2.1. Size Fractions

In collecting data, the choice of size ranges is important, and a geometric progression of intervals is far more realistic than an arithmetic series. Consider the case of a sample of particles subdivided into size intervals given by the following series:

Interval	Arithmetic series (cm)	Geometric series (cm)
1	0–10	1–2
2	10–20	2–4
3	20–30	4–8
4	30–40	8–16
5	40–50	16–32
6	50–60	32–64

TABLE 2.2: REPRESENTATIVE SIEVE ANALYSIS

Aperture* mm , μm		Average Aperture**		Percent Retained	Cumulative Percent Passing
(mm)	+9.50	11.55	mm	0.03	
-9.50	+6.80	8.15		0.43	99.97
-6.80	+4.75	5.78		2.03	99.54
-4.75	+3.40	4.08		4.17	97.51
-3.40	+2.36	2.88		6.97	93.34
-2.36	+1.70	2.03		9.59	86.37
-1.70	+1.18	1.44		10.94	76.78
-1.18	+850	1.02		10.80	65.84
-850	+600	725	μm	9.91	55.04
-600	+425	512		8.42	45.13
-425	+300	362		7.12	36.71
-300	+212	257		6.10	29.59
-212	+150	181		5.23	23.49
-150	+106	128		4.38	18.26
-106	+75	90		3.85	13.88
-75	+53	64		2.91	10.03
-53	+38	45		2.14	7.12
-38 μm		19		4.98	4.98

* Diameter scale: Figs 2.1c, d, e, f.
** Diameter scale: Figs 2.1a, b.

Presentation of Size Distribution Data

Provided the size interval is small in comparison to the total size distribution, it can be assumed that the particles in any interval have an average size given by the arithmetic mean of the two size interval limits. In the case of the geometric series therefore, the average particle size in each interval always bears a constant ratio to that of the adjacent interval (2 : 1 in this example). For the arithmetic series, the averages in the first two intervals are in a ratio of 3 : 1 (15 : 5), but the ratio approaches 1 : 1 (55 : 45) for the last two intervals. This notable disadvantage of the arithmetic series becomes even more pronounced when considering the size distributions encountered in mineral processing, where the range of particle sizes of interest may involve a factor of 10^3 or greater.

The property, or *basis*, used to determine the quantity of particles in each size interval is also significant. Properties commonly used are mass (or volume), surface area, length, or number. Because of ease of measurement, the first is more practical for small particles, and the last may be suitable for very large particles. Surface area and length, being experimentally more difficult to measure, are restricted to special situations.

2.2.2. Graphic Representations

Graphically, data are conventionally presented by plotting particle size horizontally (x-axis) and the measured quantity of property vertically (y-axis). Two approaches are used to present the quantity; the first plots the amount in each size fraction (as absolute amount, fraction, or percentage), and the second plots the cumulative amount (fraction or percentage) above or below a certain size.

To illustrate some of the suitable graphing methods available, and their relative merits, Fig. 2.1 has been drawn using the size analysis in Table 2.2. The simplest method is to plot a histogram of the property frequency, in this case the percent of material in the size interval, against the size intervals, as shown in Fig. 2.1a. Provided the size intervals are small enough, the histogram can be presented as a continuous curve. Note that if this is done, the size of the particle in any given interval is taken as being the arithmetic mean of the size interval limits. Whether the property scale is plotted as frequency, fraction, or percentage is somewhat immaterial, although the latter two scales convey a more immediate quantitative picture of the relative distribution of the material over the entire size range. It is for this reason also that these scales should be linear rather than logarithmic.

A major disadvantage of Fig. 2.1a is the compression of the size scale at the fine end. Figure 2.1b uses a logarithmic scale that spreads the data points evenly along the x-direction because the size intervals are in a geometric series. Unless there is a very good reason for doing otherwise, particle size should *always* be plotted on a logarithmic scale.

In practice, plots of frequency versus size such as Fig. 2.1b are not widely used in mineral processing. They do have one particular advantage in that fracture at a characteristic grain size may show up as a secondary maximum on such a plot (Fig. 2.2). A similar feature may occur with heterogeneous materials, where each mineral can tend to fracture at its own characteristic size. For this feature to show up however, sufficiently narrow size intervals must be used, otherwise the maxima merge together.

In many cases, a size analysis may be carried out to determine the amount of material greater than or smaller than a specified size. For this reason data are more commonly plotted as the cumulative amount of material versus log size. Figure 2.3 compares some of the scales commonly used. A (linear scale) cumulative percent passing size d_n versus log size plot of the sample data is presented in Fig. 2.1c. By comparison with the preceding plots, it must be noted that the value of size used is the *actual size limit* rather than the average size. Although Fig. 2.1c appears to be a satisfactory presentation of the data, it is not widely used for two reasons. First, the "S" shape of the curve is difficult to express mathematically, and second, the two ends of the vertical scale have insufficient spread.

Most widely used is the relatively convenient plot of log size d_n versus log cumulative percent (or fraction) below (or above) size d_n; frequently referred to as the Gaudin-Schuhmann or Gates-Gaudin-Schuhmann plot (Fig. 2.1d). Although it gives a satisfactory spread of the data in the fine size region (where historically considerable attention has centered), the compression of the coarse sizes is unsuitable for many situations. In some cases size reduction data may be linear over much of the size range, particularly the fine sizes. However, this quality of fitting can be deceptive because of the compressed upper portion of the scale. The widespread use of this method, in spite of its limitations, is mainly due to its ease of use, and the ready availability of suitable graph papers.

The Rosin-Rammler or Weibull plot, illustrated in

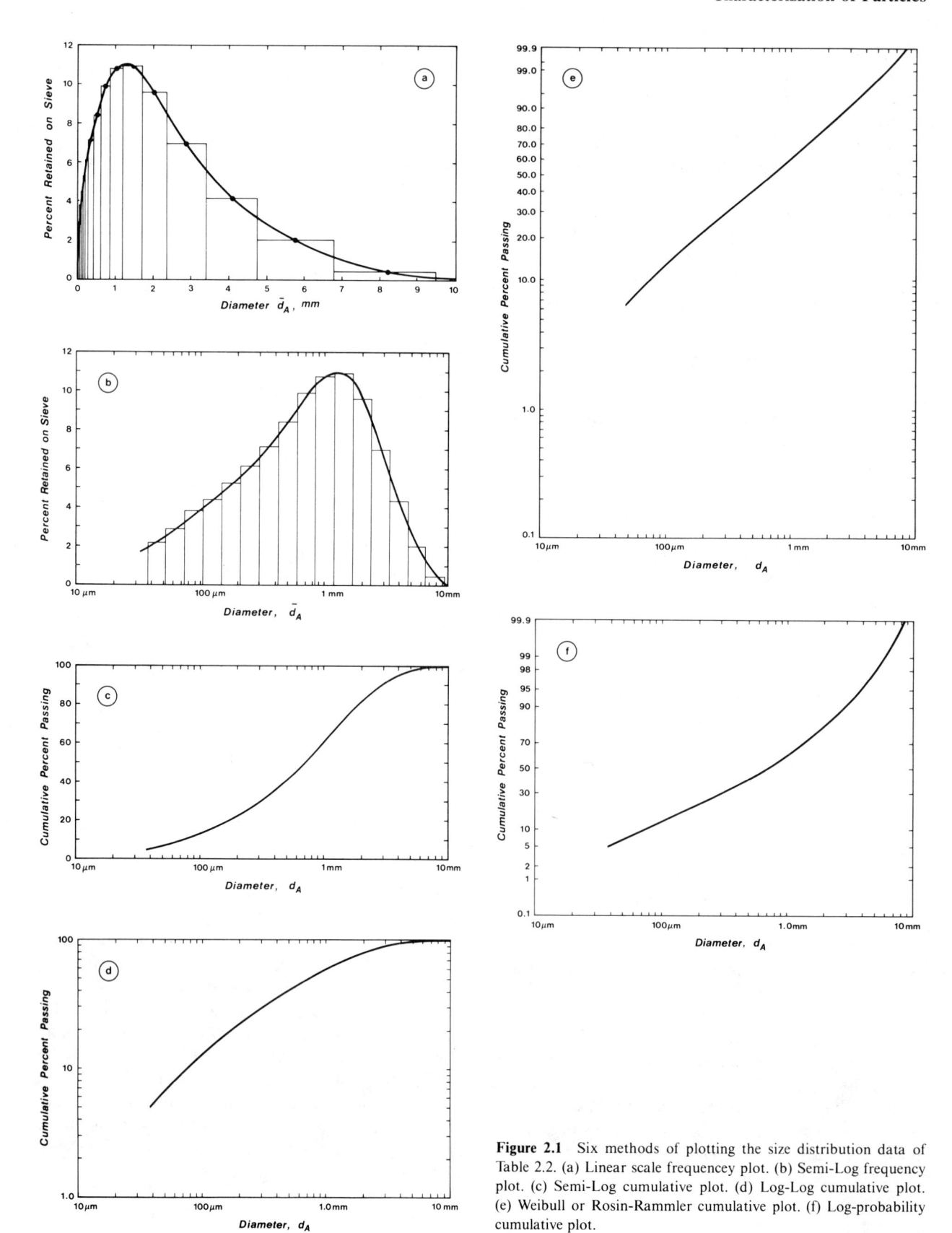

Figure 2.1 Six methods of plotting the size distribution data of Table 2.2. (a) Linear scale frequencey plot. (b) Semi-Log frequency plot. (c) Semi-Log cumulative plot. (d) Log-Log cumulative plot. (e) Weibull or Rosin-Rammler cumulative plot. (f) Log-probability cumulative plot.

Presentation of Size Distribution Data

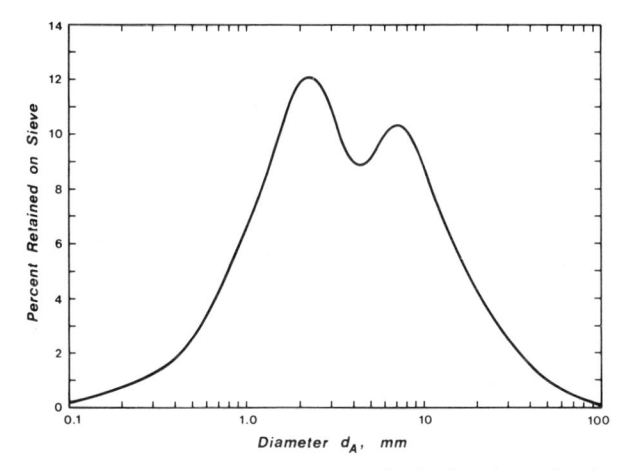

Figure 2.2. Frequency plot of size distribution data, showing secondary peak indicative of grain boundary fracture.

Fig. 2.1e, plots log ln reciprocal $(1 - Y^-)$ versus log d_n, where Y^- is the cumulative fraction passing size d_n. This method expands both ends of the Y^--scale, although the fine sizes are not expanded as much as they are on the log-log plot. A benefit of this plot is that some size reduction data are relatively linear. Moreover, the frequent criticism that the Weibull plot involves taking logarithms twice is in reality unjustified because the double logarithm is applied to a reciprocal, quite a different matter. Harris[3] has claimed that this is the preferred method for presenting size reduction data, but its widespread use appears to have been restricted by the scarcity of suitable graph paper (see Appendix D).

Figure 2.1f plots the data on a log-probability scale. Although this plot is commonly used in other fields of engineering, it is seldom used in mineral processing. It actually gives a y-scale expansion similar to the Rosin-Rammler plot (Fig. 2.3), although size reduction data generally are not so linear. Where linearity is important, it may be possible to obtain it by,[1,4] for example, plotting $d \cdot d_{max}/(d_{max} - d)$ instead of d, since size reduction data are generally asymptotic to a size d_{max}. However with less attention now being given to equation fitting, this linearity is no longer significant and is more than counteracted by the scarcity of graph paper. The method has the potential advantage that if the data are linear (as a fine size fraction of material may well be), they are amenable to convenient mathematical manipulation[2,4,5] (Section 2.3).

Example 2.1. A cyclosizer analysis is carried out on the -53 μm sieve fraction of the material shown in Table 2.2. The cyclone size fractions were then photographed with an SEM and d_M determined from the photographs. The results were:

Cyclone	d_M (μm)	(%)
1	49.2	0.04
2	42.2	1.64
3	34.1	1.50
4	23.6	1.26
5	16.1	0.97
Drain	—	1.71
		7.12%

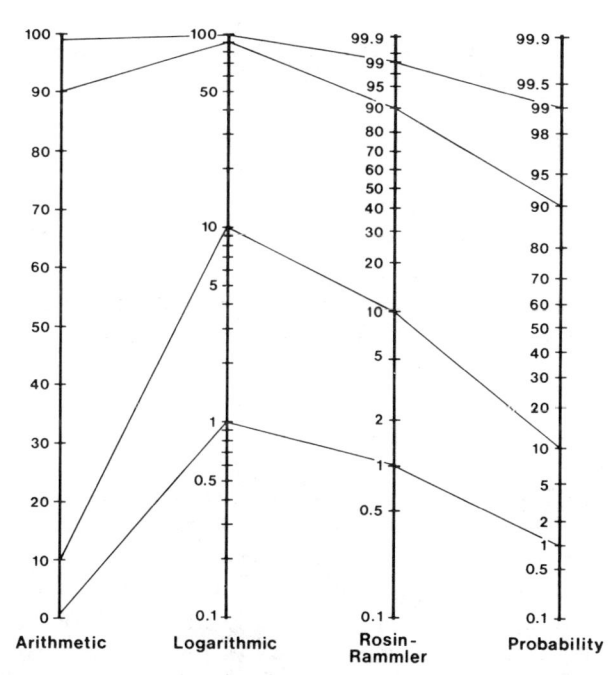

Figure 2.3. Comparison of four scales used for cumulative percentage passing size d_A.

What is the cumulative size distribution and the ratio of d_M to d_A for this material?

Solution. To prepare the cumulative size distribution it is necessary to determine the retaining cut size of each cyclone. Because of the small mass retained in cyclone 1, this size is taken as the approximate cut size. The cut sizes between the other cyclones are taken as the arithmetic mean of the material in the two cyclones. The cut size of cyclone 5 is taken from the ratio between cyclones 4 and 5. Thus:

Cyclone	Retaining Cut Size (μm)		Cumulative Amount Passing (%)
1	49		7.08
2	$\frac{42.2 + 34.1}{2}$	= 38	5.44
3	$\frac{34.1 + 23.6}{2}$	= 29	3.94
4	$\frac{23.6 + 16.1}{2}$	= 20	2.68
5	$16.1 \times \frac{16.1}{20}$	= 13	1.71

The data are plotted in Fig. E2.1.1. Reading the values of d_n at $Y^- = 5\%$, $d_A/d_M \sim 38/35.5 = 1.07$.

In selecting a method for plotting size analyses, consideration should primarily focus on *purpose*, which might be one or more of the following:

- To display data with optimum clarity.
- To establish the fit of an equation and determine the parameters.
- To compare two sets of data.

Only when such criteria have been satisfied should consideration be given to convenience.

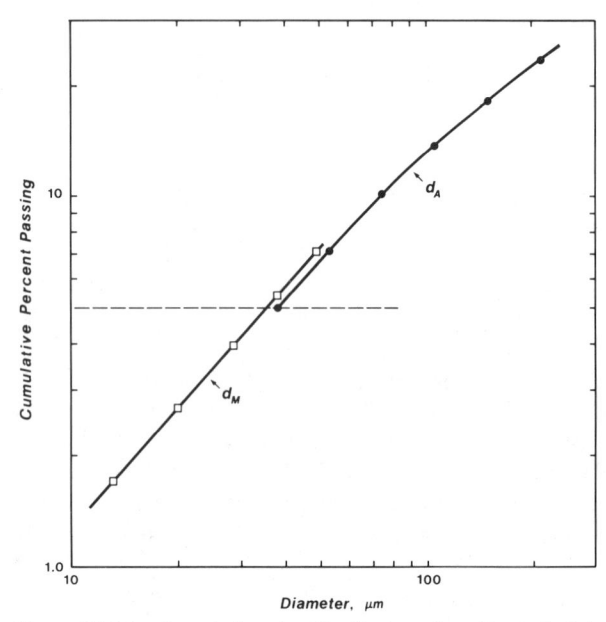

Figure E2.1.1 Cumulative size distributions from Example 2.1.

TABLE 2.3: SIZE DISTRIBUTION FORMULAE

Commonly Used Names	Ref.	Cum. Mass Fraction Passing d, Y^-	Eq. No.	Significance of d^*
Log.-Probability	4	$\mathrm{erf}\left[\frac{\ln(d/d^*)}{\sigma}\right]$	2.1a	Median particle size
Rosin-Rammler or Weibull	6,7	$1 - \exp\left[-(d/d^*)^s\right]$	2.1b	Size at which $Y^- = 0.632$
Schuhmann or Gaudin-Schuhmann or Gates-Gaudin-Schuhmann	8–10	$\left[d/d^*\right]^n$	2.1c	Maximum particle size
Broadbent-Callcott	11	$\frac{1 - \exp\left[-(d/d^*)\right]}{1 - \exp(-1)}$	2.1d	Maximum particle size
Gaudin-Meloy	12	$1 - \left[1 - (d/d^*)\right]^n$	2.1e	Maximum particle size
Harris 3-parameter equation	13	$1 - \left[1 - (d/d^*)^s\right]^n$	2.1f	Maximum particle size

2.2.3. Mathematical Representations

A number of equations have been presented to describe the size distribution of a comminution product.[6–13] Despite claims that some of them are based on fracture fundamentals, they are in reality all empirical relationships, which to a greater or lesser extent have been found capable of describing comminution size distribution. These equations are all of the general form

$$Y_u^- = \mathrm{fn}\left(\frac{d_n}{d^*}\right) \qquad (2.1)$$

(Y_u^- = the cumulative fraction [or percent], measured on basis [or property] u, smaller than size d_n; d^* = a reference size). Since d^* is an *indication* of the average particle size, it is frequently referred to as the *size modulus*. When the equation has an upper size limit, d^* is in fact the maximum particle size in the distribution. If the equation has no upper limit (e.g., Rosin-Rammler), d^* is somewhat closer to a true mean.

The equations also include a second parameter, which can be called the *distribution modulus*, since it is a measure of the spread of particle sizes. It generally appears as an exponent in the equation. Some equations have been introduced with a third parameter to account for skewness of the size distribution,[13] but this introduces additional complexity that is seldom justified.

Average Diameters

Some of the common equations are summarized in Table 2.3 and discussed in more detail in Section 7.1. With the advent of computer techniques, these equations are not as useful as they once were. They are worthy of note however in relation to the plotting methods already described, in that some represent straight lines on the appropriate graphs in Fig. 2.1.

2.3. AVERAGE DIAMETERS

Having defined a suitable diameter and measured it by a suitable method, it may be desirable to obtain a measure of the average size of a distribution of particles. Again, there are many methods available. One of the simplest is the mode, the most commonly occurring value, represented by the peak of the curve in Fig. 2.1a. It is of little use in mineral processing by comparison with the remaining average diameters, which can be considered in three groups: median diameters, geometric mean diameters, and statistical diameters.

2.3.1. Median Diameters

The general size level can be given by the median size, which can be expressed as:

$$\int_0^{d_{nmu}} dY_u^- = \int_{d_{nmu}}^1 dY_u^- = \frac{1}{2} \quad (2.2)$$

(Y_u^- = cumulative fraction of all particles smaller than size d_n; d_{nmu} = median diameter based on measured property: Table 2.4 gives a standard nomenclature for different properties u that can be measured). For example, if the amount of material of any size is determined by mass, the mass median diameter d_{nm3} means that half the mass of the material consists of particles coarser than the mass median diameter, and half of finer particles. The median diameter is therefore the 50% point on any cumulative distribution curve (Fig. 2.1) obtained in terms of the property desired. If the material obeys a log-probability relationship (Fig. 2.1f), the median diameters are related by[5]

$$d_{nmu} = d_{nmz} \exp[(u - z) \ln^2 \sigma] \quad (2.3)$$

(d_{nmu} = median diameter, basis u [Table 2.4]; d_{nmz} = median diameter basis z; σ = standard geometric deviation). Extreme care should be taken in interpreting answers when u and z differ by more than unity, since errors increase by the power of the difference between u and z.

TABLE 2.4: BASES FOR EXPRESSING SIZE DISTRIBUTION

Basis	Value of u or z
Number or count	0
Linear (or length)	1
Surface (area)	2
Volume (or mass)	3
Moment	4

2.3.2. Geometric Mean Diameters

The geometric mean diameters have no physical significance. They are based on the assumption of an even graduation in size from maximum to minimum and assume an equal number of particles in each size range. The arithmetic mean is given by

$$\bar{d}_{na} = \frac{d_{n,\max} + d_{n,\min}}{2} \quad (2.4)$$

where the bar designates a mean diameter. This mean has little use in mineral processing except as mentioned above to give the average size between two size intervals, or in the unlikely event of the material having a normal distribution.

The geometric mean is given by

$$\bar{d}_{ng} = (d_1 d_2 d_3 \ldots d_N)^{1/N} \quad (2.5)$$

$$\therefore \quad \log \bar{d}_{ng} = \frac{\Sigma(N \log d_n)}{\Sigma N} \quad (2.6)$$

(N = number of particles). It has some value with log-normal distributions.

The harmonic mean is the number of particles divided by the sum of the reciprocals of the diameters of the individual particles, that is,

$$\bar{d}_{nh} = \frac{\Sigma N}{\Sigma(1/d_n)} \quad (2.7)$$

Being related to the specific surface, it has application where the surface area of the sample is concerned.

2.3.3. The Statistical Diameters

There are a number of statistical diameters, which can be obtained from the general equation

$$\bar{d}_{nqp}^{(q-p)} = \frac{\Sigma[Q_u d_n^{(q-u)}]}{\Sigma[Q_u d_n^{(p-u)}]} \quad (2.8)$$

(Q_u = is quantity of material measured by the property u [e.g., when $u = 3$, Q_u is the mass of the size fraction represented by size d_n]; p,q = the characteristic subscripts and exponents used to define mean diameter. The values of p and q are assigned in the same way as the subscript u in Table 2.4).

These mean diameters have significance in that a distribution of irregular particles can be characterized by its total number, length, area, and volume (or mass) and can in turn be represented by a group of uniform particles having two and only two characteristics of the irregular particles (Fig. 2.4). The various statistical diameters derived from Eq. 2.8 therefore represent the possible coupling of the characteristics. For example, d_{n32} represents the mean nominal diameter, where the total volume and total surface are common to the groups of irregular and uniform particles.

Unfortunately, many of the statistical diameters have been given different names in different publications.[1,5,14,15] For this reason, the subscripting designation using the values of p and q to represent the order of the measured properties is strongly recommended. Table 2.5 shows clearly the use of this system, together with some of the commonly used symbols, names, and fields of application.

If the data fit a log-probability function, the average diameter may be calculated from[5]:

$$\bar{d}_{nqp} = d_{nmu} \exp\left(\frac{q + p - 2u}{2} \ln^2 \sigma\right) \quad (2.9)$$

TABLE 2.5: STATISTICAL MEAN DIAMETERS

Preferred Symbol	Subscripts q	Subscripts p	Order	Used by Other Authors: Names	Used by Other Authors: Symbols	Possible Fields of Application
d_{n00}	0	0	0	Geometric[5]		
d_{n10}	1	0	1	Arithmetic[15] Linear[5,15] Number[5] Length[14] Number-length[1]	d_L	Comparisons of droplet evaporation
d_{n20}	2	0	2	Surface[5,14,15] Surface-to-number[5] Number-surface[1]	d_s d_A	Absorption, crushing, light diffusion
d_{n30}	3	0	3	Volume[5,14,15] Volume-to-number[5] Number-volume[1]	d_v d_m d_V	Comparison, atomising
d_{n21}	2	1	3	Surface-to-diameter[5,14,15] Length-surface[1]	d_{sd}	Adsorption
d_{n31}	3	1	4	Volume-to-diameter[5,15]	d_{vd}	Evaporation, molecular diffusion
d_{n32}	3	2	5	Volume-to-surface[5,14,15] Sauter[5] Surface-volume[1]	d_{vs} d_{VA} d_{sv}	Mass transfer, catalytic reactions, fluid flow, efficiency studies
d_{n43}	4	3	7	Mass-surface[15] Volume-moment[1] Weight-moment De Brouckere[5]	d_{ms}	Combustion, grinding

Example 2.2. What are the $\bar{d}_{43}$ and $\bar{d}_{10}$ (strictly $\bar{d}_{A43}$ and $\bar{d}_{A10}$) mean diameters of the sample in Table 2.2?

Solution. In principle, mean diameters can be calculated by Eq. 2.8 or 2.9. Because the quantities

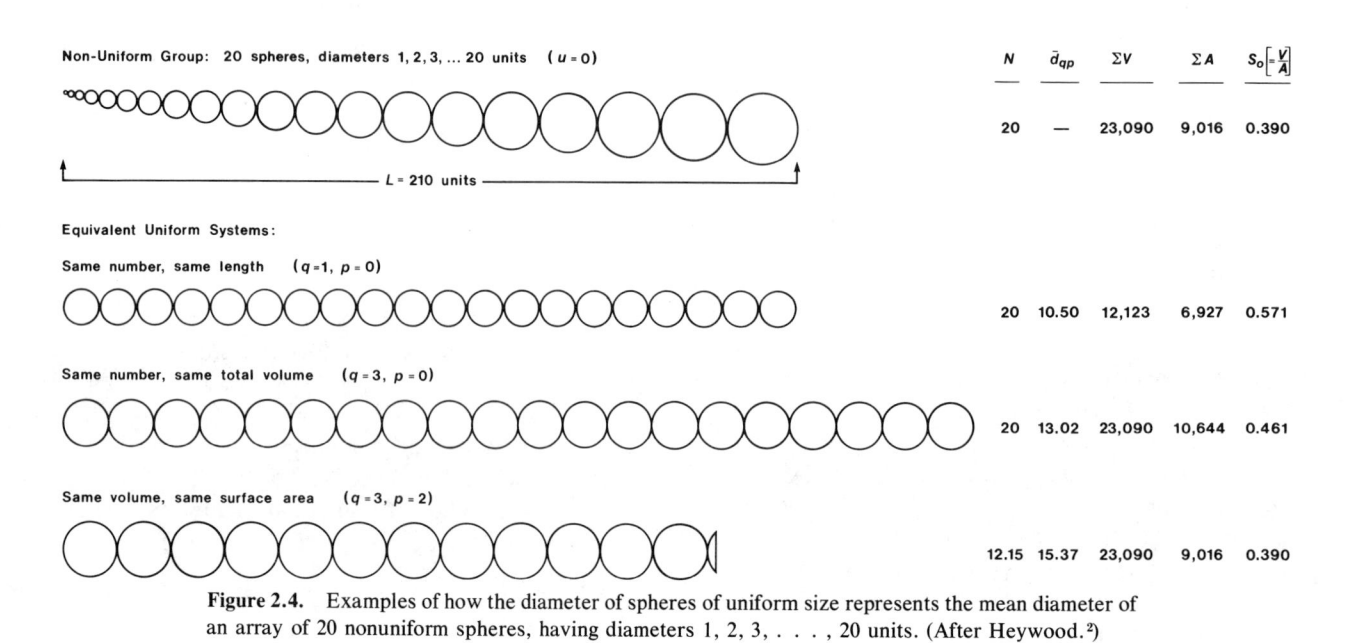

Non-Uniform Group: 20 spheres, diameters 1, 2, 3, ... 20 units ($u = 0$)

L = 210 units

N	$\bar{d}_{qp}$	ΣV	ΣA	$S_0\left[=\frac{V}{A}\right]$
20	—	23,090	9,016	0.390

Equivalent Uniform Systems:

Same number, same length ($q=1$, $p=0$)

| 20 | 10.50 | 12,123 | 6,927 | 0.571 |

Same number, same total volume ($q=3$, $p=0$)

| 20 | 13.02 | 23,090 | 10,644 | 0.461 |

Same volume, same surface area ($q=3$, $p=2$)

| 12.15 | 15.37 | 23,090 | 9,016 | 0.390 |

Figure 2.4. Examples of how the diameter of spheres of uniform size represents the mean diameter of an array of 20 nonuniform spheres, having diameters 1, 2, 3, . . . , 20 units. (After Heywood.[2])

in the size fractions in Table 2.2 are expressed by mass, $u = 3.0$ and Q_u in Eq. 2.8 becomes m_i. Values of σ and d_{Am3} for use in Eq. 2.9 can be obtained from the log-probability plot Fig. 2.1f. (In fact the relevant information can be obtained from any cumulative plot.)

1. $\bar{d}_{43}$:

using Eq. 2.8

$$\bar{d}_{43}^{(4-3)} = \frac{\Sigma m d^{(4-3)}}{\Sigma m d^{(3-3)}}$$

$$= \frac{\Sigma m d}{\Sigma m}$$

that is,

$$\bar{d}_{43} = \frac{0.03 \times 11.55 + 0.43 \times 8.15 + 2.03 \times 5.78 \cdots 4.98 \times 0.019}{100}$$

$$= 1.17 \text{ mm}$$

Using Eq. 2.9: from Fig. 2.1f, $d_{Am3} = 700$ μm and

$$\sigma = d_{84.13}/d_{50} = \frac{2200}{700} = 3.14$$

or

$$\sigma = d_{50}/d_{15.87} = \frac{700}{125} = 5.6$$

Because $\bar{d}_{43}$ is a high order mean, mass is the most significant particle parameter, and since the larger particles contain most of the mass, the value of σ determined from $d_{84.13}/d_{50}$ is a better "average" value. Thus

$$\bar{d}_{43} = 700 \exp\left(\frac{4 + 3 - 2 \times 3}{2} \ln^2 3.14\right)$$

$$= 1.35 \text{ mm}$$

which is a satisfactory approximation to the previous answer.

2. $\bar{d}_{10}$:

using Eq. 2.8

$$d_{10}^{(1-0)} = \frac{\Sigma m d^{(1-3)}}{\Sigma m d^{(0-3)}}$$

$$= \frac{\Sigma m/d^2}{\Sigma m/d^3}$$

that is,

$$\bar{d}_{10} = \frac{0.03/11550^2 + 0.43/8150^2 + 2.03/5780^2 \cdots 4.98/19^2}{0.03/11550^3 + 0.43/8150^3 + 2.03/5780^3 \cdots 4.98/19^3}$$

$$= 21.7 \text{ μm}$$

Again, because of the poor log-probability approximation, a realistic value of σ must be selected to apply in Eq. 2.9. Because $\bar{d}_{10}$ is a low order mean, weighted to the *number* of particles, the value of σ should be selected with this in mind. Since the number of particles will be high in the small size fractions, and insignificant in the large size fractions, we take $\sigma = 5.6$. Thus

$$\bar{d}_{10} = 700 \exp\left(\frac{1 + 0 - 2 \times 3}{2} \ln^2 5.6\right)$$

$$= 0.4 \text{ μm}$$

At first sight, this answer may appear to be incorrect when compared with that calculated from Eq. 2.8. In reality, it is Eq. 2.8 that is unsuitable in this instance. Examination of the calculation shows that it is the bottom size fraction that virtually determines the magnitude of $\bar{d}_{10}$, that is,

$$\frac{4.98/19^2}{4.98/19^3} = 19 \text{ μm}$$

Thus, the lower part of the size analysis needs to be determined before Eq. 2.8 can be used. Although Fig. 2.1f can be extrapolated to obtain the lower size fractions, such an approach is equivalent to using Eq. 2.9. Realistically, the $\bar{d}_{10}$ mean becomes meaningless when the sample has a relatively wide size distribution. However, these examples show that even when the sample does not fit a log-probability size distribution, judicious use of Eq. 2.9 can be just as reliable, and more convenient, than Eq. 2.8.

2.4. PARTICLE SHAPE

The shape of a particle may be important in a number of situations, in that it can affect properties such as surface area, bulk density, and flow. Such quantitative descriptions as angular, flaky, and modular are often used, but by themselves these are inadequate when mathematical analysis is required.

In general, for irregular particles, the ratios of volume to the cube of the diameter and surface area to the square of the diameter are constant and effectively define shape factors.[1] The actual shape factor will consequently depend on the nominal diameter used to measure the particle, but the value can be related to a suitable reference sphere as follows:

$$S_p = \lambda_{sn}d_n^2 = \pi d_r^2 = \pi d_s^2 \quad (2.10)$$

$$V_p = \lambda_{vn}d_n^3 = \frac{\pi}{6}d_r^3 = \frac{\pi}{6}d_v^3 \quad (2.11)$$

(S_p = surface area of the irregular particle and the reference sphere; V_p = volume of the irregular particle and the reference sphere; d_n = nominal diameter of irregular particle; d_r = diameter of reference sphere; λ_{sn} = surface shape factor [on the basis of nominal diameter d_n]; λ_{vn} = volume shape factor [on the basis of nominal diameter d_n]).

If the sample contains a large number of particles of varying size, the mean diameter is then used:

$$\overline{S}_p = \lambda_{snqp}\overline{d}_{nqp}^2 = \pi(\overline{d}_r)^2 = \pi(\overline{d}_s)^2 \quad (2.12)$$

$$\overline{V}_p = \lambda_{vnqp}\overline{d}_{nqp}^3 = \frac{\pi}{6}(\overline{d}_r)^3 = \frac{\pi}{6}(\overline{d}_v)^3 \quad (2.13)$$

where the bar superscripts and qp subscripts indicate a mean value. In practice, average values should be treated with care, since the shape factor may change with particle size.

Volume shape factors can be determined experimentally down to about 150 μm by measuring the number, mean size, mass, and density of a narrow size fraction of particles.[2] Surface shape factors are harder to evaluate, because of the difficulty in measuring the surface area of small particles.[16] Furthermore, although the surface area can be measured (Sections 2.6.5–2.6.7), care must be taken because the result obtained depends on the method used. Estimates of the surface shape factors may be made by geometric analogy from measurements of larger particles.

A more rigorous analysis of shape factors has been given by Heywood,[2,16] who considered the shape of a particle to have two distinct characteristics: the relative proportions of length, breadth, and thickness, and the geometric form. The relative proportions are evaluated by

elongation ratio, $\Re_E$ = length/breadth (2.14)

flatness ratio, $\Re_F$ = breadth/thickness (2.15)

The geometric form characteristic is a volumetric shape factor representing the degree to which a particle would, if it were equidimensional (length = breadth = thickness), approximate an ideal geometric form, such as a cube, tetrahedron, or sphere. Values based on a projected area diameter d_a have been presented for this factor λ_{vae} and are given in Table 2.6. It can be shown that this equidimensional factor is related, for this nominal diameter, to the actual shape factor by

$$\lambda_{vae} = \lambda_{va}\Re_F\sqrt{\Re_E} \quad (2.16)$$

Heywood found that the equivalent surface area shape factor λ_{sa} could be evaluated by an empirical relationship

$$\lambda_{sa} = 1.57 + K_g\left(\frac{\lambda_{vae}}{\Re_F}\right)^{4/3} \cdot \frac{\Re_E + 1}{\Re_E} \quad (2.17)$$

where K_g is a constant that depends on the geometric form. Typical values are presented in Table 2.6.[2]

Two further shape factors are commonly used. One is sphericity Ψ:

$$\Psi = \frac{\text{surface area of sphere of same volume as the particle}}{\text{surface area of the particle}} \quad (2.18)$$

From the definitions in Table 2.1, it follows that

$$\Psi = \frac{\pi d_v^2}{\pi d_s^2} = \left(\frac{d_v}{d_s}\right)^2 \quad (2.19)$$

and by substituting the appropriate forms of Eqs. 2.10 and 2.11, it can be shown that

$$\Psi = \frac{4.84(\lambda_{va})^{2/3}}{\lambda_{sa}} \quad (2.20)$$

The ratio of specific surfaces $\Re_s$ is the other shape factor, given by the ratio of the surface of the particles to the surface of a sphere of the same diameter (generally the screen diameter):

$$\Re_s = \frac{S_0}{6/\rho d_A} \quad (2.21)$$

(S_0 = specific surface, i.e., area/unit mass).

Some typical values of shape factors are given in Table 2.7.

TABLE 2.6: DATA FOR EQ. 2.17

Shape Group	λ_{vae}	K_g
Geometric forms:		
Tetrahedral	0.328	4.36
Cubical	0.696	2.55
Spherical	0.524	1.86
Approximate forms:		
Angular: Tetrahedral	0.38	3.3
Prismoidal	0.47	3.0
Sub-angular	0.51	2.6
Rounded	0.54	2.1

Particle Shape

TABLE 2.7: TYPICAL VALUES OF SHAPE FACTORS

Type of Material	λ_{va}	λ_{sa}	ψ
Rounded particles: water worn sands, fused flue-dust, atomised metals.	0.32 - 0.41	2.7 - 3.4	0.817
Angular particles of pulverised minerals: coal, limestone, sand.	0.2 - 0.28	2.5 - 3.2	0.655
Flaky particles: plumbago, talc, gypsum.	0.12 - 0.16	2.0 - 2.8	0.543
Very thin flakes: mica, graphite, aluminium.	0.01 - 0.03	1.6 - 1.7	0.216

Example 2.3. A phosphate ore is to be upgraded by flotation. Tests have shown that even though the fine phosphate particles adsorb collector, these particles do not separate from the gangue minerals during flotation. To minimize collector wastage, the flotation feed is to be deslimed before flotation, using a hydrocyclone. The hydrocyclone feed and product size distributions can be approximated by log-probability relationships:

Feed

$$d_{Am3} = 90 \ \mu m$$
$$\sigma = 1.8$$

Oversize

$$d_{Am3} = 125 \ \mu m$$
$$\sigma = 1.4$$
$$mass = 40\% \text{ of feed}$$

Assuming that the reagent consumption is proportional to particle surface area and that the grade is uniform over the entire size range, what reagent savings will the desliming operation produce?

Solution. To estimate the reagent adsorption, the size distributions can be represented by uniform sized particles (i.e., mean sized) having identical relevant properties. In this instance, the relevant properties are mass (as a reference quantity) and surface area, that is, $\bar{d}_{32}$. Thus, a given mass of mean-sized spheres will have the same total surface area as the same mass of irregular particles. Now

$$\frac{\text{surface area}}{\text{mass}} = \frac{\lambda_{sA} d_A^2}{\rho \lambda_{vA} d_A^3}$$

which for a given material is proportional to $\frac{1}{\bar{d}_{A32}}$

From Eq. 2.9, for the feed

$$\bar{d}_{32} = 90 \exp\left(\frac{3 + 2 - 3 \times 2}{2} \ln^2 1.8\right)$$
$$= 75.7 \ \mu m$$

and for the oversize

$$\bar{d}_{32} = 125 \exp\left(\frac{3 + 2 - 3 \times 2}{2} \ln^2 1.4\right)$$
$$= 118.1 \ \mu m$$

Basis. 100 g of feed to hydrocyclone

$$\therefore \quad \text{surface area of feed}/100 \text{ g of feed} \propto \frac{100}{75.7}$$

and surface area of oversize/100 g of feed

$$\propto \frac{40}{118.1}$$

$$\therefore \quad \text{reagent savings} = \frac{100/75.7 - 40/118.1}{100/75.7}$$
$$= 74\%$$

Because most of the shape factors described so far are difficult to determine, a number of simpler empirical or statistical representations of shape have been proposed.[17-20] One of the simplest involves repeated sievings using different aperture shapes.[17] Most of the other methods rely on automated equipment that scans a large number of particles, nominally of the same size. Chord length is most conveniently obtained by such equipment and its distribution is a measure of particle shape.[19] By collecting even more information, a number of different nominal diameters can be evaluated and the ratios of these are again a measure of shape.[20]

The following points are worthy of emphasis. First, a number of the common shape factors (notably Eqs. 2.10–2.17) do not tend to unity as the particle shape approaches spherical form. Second, to adequately define a shape factor it may be necessary to specify the property measured, the experimental technique used to measure this property, the nominal diameter used, and the mean diameter. Third, although there may be a clear benefit in considering the effect of shape,[18] the suitability of any shape factor to a given situation must be taken into account. In particular, some of the empirical and statistical shape factors suffer from the limitation that their physical significance may not be known. Thus, instead of measuring particle size and shape, it may be better to measure particle behavior under the envisaged conditions; such an approach can be simpler and less prone to error.

2.5. LIBERATION

An essential prerequisite for the separation of an ore into valuable and waste fractions is the liberation of the valuable mineral grains from the waste mineral grains. The degree of liberation is the percentage of a given mineral that exists as free particles, that is, particles containing only that mineral. Particles that contain both valuable and waste minerals are known as *locked* or *middling particles,* and a large proportion of the difficulties experienced in mineral separations are associated with the treatment of these particles. In this discussion, we consider middling particles containing only two minerals. It must be appreciated that the concepts can be extended to include any number of minerals in that they can always be grouped into valuable and waste minerals.

Crushing and grinding operations are employed to fracture mineral aggregates, and thus induce or increase liberation. As a result of fracture two types of liberation can be distinguished. In the first type the interface between the grains is weak, so that fracture is *intergranular*. More common is *transgranular* fracture, where fracture occurs across the grains. By applying sufficient size reduction, a high degree of liberation can be achieved by this mechanism, although complete liberation is impossible.

2.5.1. Intergranular Liberation

The concept of intergranular liberation is fracture at the grain boundaries rather than across the grains. Liberation therefore occurs at about the mineral grain size; but because any mineral has a range of grain sizes, size reduction still must reduce the particles to below the average grain size. The occurrence of intergranular liberation can be deduced from a frequency sieve analysis of a comminuted product, in that there will be a double peak in the graph, with one peak representing a preferred fracture at the grain boundaries (Fig. 2.2). Pure intergranular fracture is comparatively rare, although it has been suggested that certain types of size reduction equipment enhance its presence.[21]

2.5.2. Transgranular Liberation

Transgranular liberation is best illustrated by the concept in Fig. 2.5, which shows a mineral aggrega-

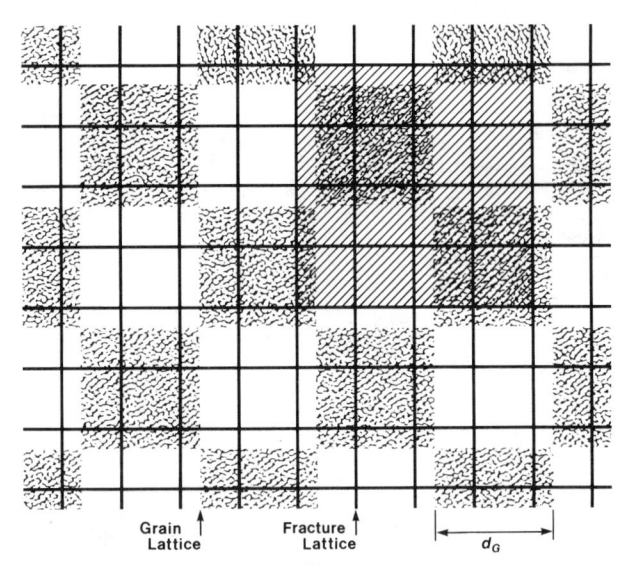

Figure 2.5. Superimposition of a fracture lattice on a grain lattice to illustrate liberation. The large shaded square shows that when the particle size is half the grain size, the degree of liberation is 2 in 16, that is, 12.5%. (Adapted from A. M. Gaudin, *Principles of Mineral Dressing.* Copyright 1939, McGraw-Hill. Used with the permission of McGraw-Hill Book Company.)

tion consisting of two uniformly sized minerals in equal quantities. It can be seen that if the aggregate is broken into square particles half the grain size, the degree of liberation for each mineral is still only 3 in 24 or 12.5%; at first sight a surprisingly low figure. Both Gaudin[22] and Wiegel[21] have extended this representation of liberation to three dimensions using the following assumptions:

1. The aggregate consists of two minerals A and B.
2. Both minerals have the same uniform size of cubic grains d_G.
3. The grains are aligned in the mineral aggregate so that the grain surfaces form continuous planes.
4. Grains of the two species are randomly located throughout the aggregate.
5. The aggregate is broken into uniformly sized particles d by a cubic fracture lattice randomly superimposed on the aggregate parallel to the grain surfaces.

On this basis the degree of liberation $\mathscr{L}$ can be expressed in terms of the grain size to particle size ratio $\mathfrak{R}_L \ (= d_G/d)$ and the volumetric fraction $\mathfrak{F}_V$ of

Liberation

each component in the original aggregate. For each component,[21] when $\Re_L > 1.0$

$$\mathscr{L} = \frac{[(\Re_L - 1)^3 \mathfrak{F}_V + 3(\Re_L - 1)^2 \mathfrak{F}_V^2 + 3(\Re_L - 1)\mathfrak{F}_V^4 + \mathfrak{F}_V^8]}{\Re_L^3} \quad (2.22)$$

while for $\Re_L < 1.0$

$$\log \mathscr{L} = \left(\frac{1}{\Re_L} + 1\right)^3 \log \mathfrak{F}_V \quad (2.23)$$

The fraction of particles that remain as locked particles of minerals A and B is obtained by difference

$$\mathscr{L}_{AB} = 1 - \mathscr{L}_A - \mathscr{L}_B \quad (2.24)$$

Figures 2.6 and 2.7 illustrate the characteristics of Eqs. 2.22 and 2.23. A number of features are notable:

1. The degree of liberation of the less abundant mineral is essentially independent of ore grade.
2. The more abundant mineral is always freer than the less abundant mineral.
3. None of the less abundant mineral is liberated until the particle size is less than the grain size.
4. For the less abundant mineral to be appreciably liberated, the particle size must be significantly smaller than the grain size.
5. If there is a very small proportion of valuable mineral (as is typical of many ores), the other (gangue) mineral has an appreciable fraction liberated when the particle size is larger than the grain size.

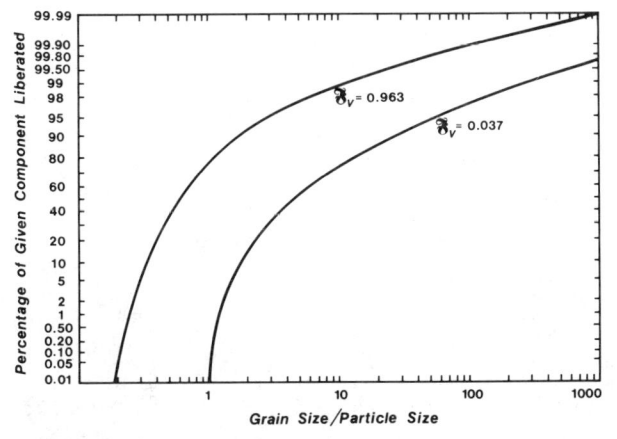

Figure 2.6. Percentage liberation predicted by Eqs. 2.22 and 2.23 when the major mineral forms 96.3% of the sample and the minor mineral forms 3.7%.

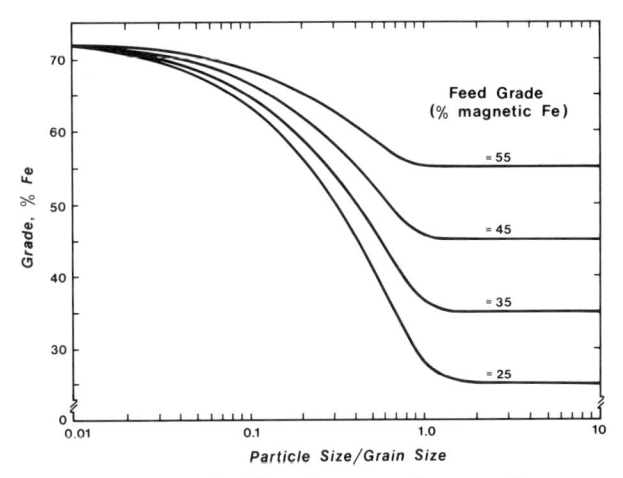

Figure 2.7. Theoretical liberation curves for magnetite-gangue system (density ratio 0.58). (After Wiegel.[21])

Of fundamental importance is the fact that incomplete liberation limits either grade or recovery. The effect on grade is illustrated in Fig. 2.7, derived from Eq. 2.22. This figure shows how the grade could be expected to change when a magnetic separation is made between magnetite (Fe_3O_4, $\rho = 5090$ kg/m³) and gangue ($\rho = 2950$ kg/m³) with 100% magnetite recovery. It can be seen that although the particles are larger than the grain size, no separation occurs (the "concentrate" retains the feed grade). As the particle decreases below the grain size, the grade rises toward the pure mineral grade of 72.4% Fe (i.e., 100% liberation).

Wiegel carried out such separations on a variety of magnetite ores and showed that the fit between experiment and theory was very good (Fig. 2.8). It should be pointed out that the theoretical curve was shifted longitudinally to give the best visual match with the data. This simply implies that the grain size taken is an *effective grain size;* but it is significant that this is sufficient to represent the range of grain sizes that must occur in a given ore. A comparison of the effective grain size with that visually estimated by microscopic examination of microstructures showed that the former was about five times larger than the latter.

The true cause for this size discrepancy is not known: it may be that the microscopic measurement does not determine the true grain size, or that locally higher magnetite concentrations affect the metallurgical result. Of the 13 ores tested in this work, three were not fitted by the theory. One had a very fine grain structure with a tendency to aggrega-

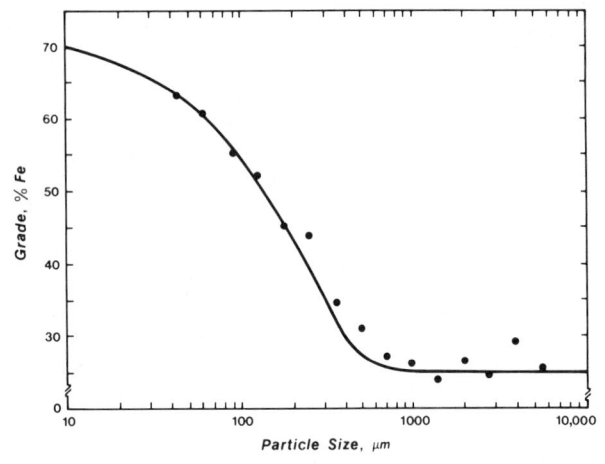

Figure 2.8. Actual Davis tube liberation data for magnetite ore: head grade 25% magnetic iron; effective grain size 460 μm. (After Wiegel.[21])

tion, and this possibly explains why this ore appeared to have several liberation stages. A second fitted the general theory but showed a considerable scatter in data, and the third had a tendency to liberate over a very narrow range of particle sizes, indicating intergranular liberation. It is possible that autogenous grinding contributed to the behavior of the latter two samples.

A corollary of this liberation analysis is that it can be combined with a size distribution to obtain the liberation of a ground product.

Example 2.4. What is the extent of liberation of the hydrocyclone overflow product in Example 10.1, if the grain size is 500 μm? The ore is 10% PbS (ρ_s = 7600 kg/m³) and 90% gangue (ρ_s = 2650 kg/m³).

Solution

Basis. 100 g of ore

The liberation can be calculated by Eqs. 2.22 and 2.23.

For the PbS:

$$\tilde{\mathfrak{F}}_{V,\text{PbS}} = \frac{10/7600}{90/2650 + 10/7600} = 0.037$$

For the size fraction $-425 +300$ μm, $\bar{d}_A = 362.5$:

$$\therefore \quad \mathfrak{R}_L = \frac{500}{362.5} = 1.38$$

From Eq. 2.22:

$$\mathscr{L}_{\text{PbS}} = \frac{(1.38 - 1)^3 \times 0.037 + 3(1.38 - 1)^2(0.037)^2 + 3(1.38 - 1)(0.037)^4 + (0.037)^8}{(1.38)^3}$$

$$= 0.001$$

$$\therefore \quad \text{fraction liberation} = \frac{\mathscr{L}}{\tilde{\mathfrak{F}}_{V,\text{PbS}}} = \frac{0.001}{0.037} = 0.03$$

and so on.

For the gangue:

$$\tilde{\mathfrak{F}}_{V,G} = \frac{90/2650}{90/2650 + 10/7600} = 0.963$$

For the size fraction $-425 +300$ μm, $\bar{d} = 362.5$:

$$\mathfrak{R}_L = \frac{500}{362.5} = 1.38$$

TABLE E2.4.1: LIBERATION EXAMPLE

					PbS				Gangue			
m_i*	Size (μm)	$\bar{d}_A$	$d_G/\bar{d}_A$	$\mathscr{L}/\tilde{\mathfrak{F}}_V$	$m_{i,\text{PbS}}$	Liberated	Locked	$\mathscr{L}/\tilde{\mathfrak{F}}_V$	m_{iG}	Liberated	Locked	Grade
(1)	(2)	(3)	(4)	(5)	(6)	(7)=(5)x(6)	(8)=(6)-(7)	(9)	(10)	(11)=(9)x(10)	(12)=(10)-(11)	Locked
0.03	-425 +300	362.5	1.38	0.027	0.003	0.000	0.003	0.859	0.027	0.023	0.004	42.9
2.58	-300 +212	256	1.95	0.129	0.258	0.033	0.225	0.914	2.322	2.122	0.200	52.9
11.05	-212 +150	181	2.76	0.276	1.105	0.305	0.800	0.946	9.945	9.408	0.537	59.8
15.50	-150 +106	128	3.91	0.428	1.550	0.663	0.887	0.965	13.950	13.462	0.488	64.5
14.85	-106 +75	90.5	5.52	0.563	1.485	0.836	0.649	0.977	13.365	13.058	0.307	67.9
13.02	-75 +53	64	7.81	0.674	1.302	0.878	0.424	0.984	11.718	11.531	0.187	69.4
10.55	-53 +38	45.5	10.99	0.759	1.055	0.801	0.254	0.989	9.495	9.391	0.104	70.9
32.42	-38	19	26.32	0.894	3.242	2.898	0.344	0.9956	29.178	29.050	0.128	72.9
100.00					10.000	6.414	3.586		90.000	88.045	1.955	64.7

* From Table E10.1.1

Liberation

From Eq. 2.22:

$$\mathscr{L}_G = \frac{(1.38-1)^3 \times 0.963 + 3(1.38-1)^2(0.963)^2 + 3(1.38-1)(0.963)^4 + (0.963)^8}{(1.38)^3}$$

$$= 0.827$$

$$\therefore \quad \text{fraction liberation} = \frac{0.827}{0.963} = 0.86$$

Columns 5 and 9 of Table E2.4.1 show the full set of values of $\mathscr{L}_i/\mathscr{F}_V$. The liberation for each size fraction is then found from $m_i \mathscr{L}_i/\mathscr{F}_V$ (columns 7 and 11); the summation of the columns represents the total mass liberated of each component. Thus, for the PbS:

$$\% \text{ liberation} = \frac{6.4}{10} \times 100 = 64.0\%$$

and for the gangue:

$$\% \text{ liberation} = \frac{88.024}{90} \times 100 = 97.8\%$$

Where recoveries are likely to be incomplete, as in a commercial separator, an efficiency factor (which will itself be a function of particle size) can be applied to allow for this. For example, Fig. 2.9 compares roughing and cleaning data with a perfect laboratory separation.

An alternative analysis of liberation has been described by King.[23] This method involves linear scanning of a polished section of ore, a process that can be automated. The interceptions of a given mineral can be statistically converted to liberation data without empirical factors such as shape factors. According to Schaap,[24] the data can also be obtained from standard laboratory tests in the form of an apparent mean grain size. Because such a size parameter allows the effect of different amounts of size reduction to be investigated, he emphasized that it is as significant a property of an ore as are grade and hardness.[24,25]

2.5.3. Characterization of Middling Particles

How the different minerals contact each other in a middling particle can be an important factor in determining the particles' behavior in a separator. In this regard, it is possible to distinguish four basic types of middling[22] (Fig. 2.10). Type I particles have two minerals making contact along rectilinear or gently curving boundaries, indicating that the grain size is considerably larger than the particle size. Increased liberation can be readily obtained by further size reduction. Veins of one mineral in another typify type II particles, which are consequently difficult to liberate. Type III particles consist of a shell of one mineral around another. Further size reduction results in the shell becoming incomplete, and although some further liberation occurs, there will always be a significant proportion of difficult-to-treat particles remaining. Occlusions of one mineral in another constitute type IV particles. Clearly the division into four types is not absolute, but rather depends to a significant extent on the relative size of the particle and the mineral grain, at the particle size where the separation is to be attempted. For example, before size reduction, type III particles would have the same appearance as type IV.

2.5.4. Behavior of Middling Particles

At this stage, it is worthwhile to examine briefly the behavior of middling particles in some common

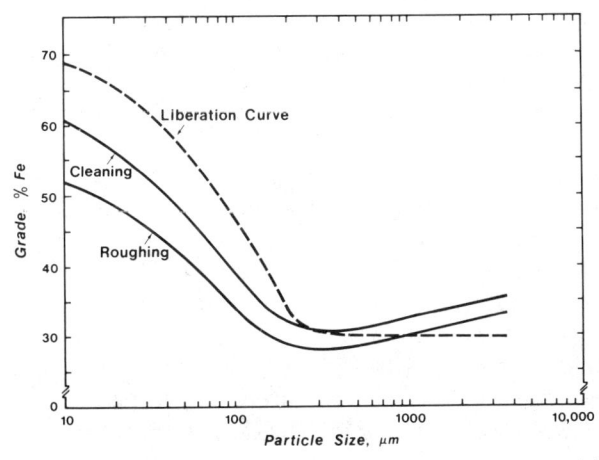

Figure 2.9. Curves showing the inefficiency of a commercial magnetic separator. (After Wiegel.[21])

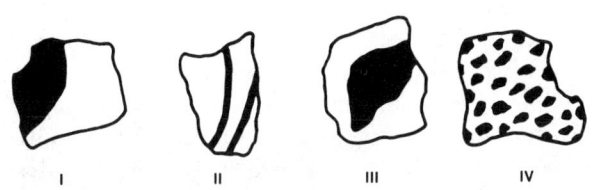

Figure 2.10. The four basic types of middling or locked particle: I, rectilinear boundaries; II, veins; III, shell; IV, occlusions.

types of separator. Any separator exploits some property of the particle, and the point of interest is how the property of the middling particle is "seen" by the separator. For example, flotation depends on the surface exposed: for a particle to be floated, a certain portion of the surface must be made up of the floatable mineral. Depending on the mineral system, this may be as low as 20% of the surface, provided it is near a corner; but even then, flotation may be slow because that particular area of the particle has to make contact with an air bubble. In the case of magnetic separations, virtually any magnetically susceptible mineral in the particle (even if it is completely surrounded) will result in recovery of the particle. Finally, in gravity separations, the particle will have an apparent density directly proportional to its volumetric composition, thus leading to a comparatively broad spectrum of behavior.

2.6. DETERMINATION OF PARTICLE CHARACTERISTICS

In this section we consider some of the methods used to measure particle characteristics.[1,5,26,27] Since particle "size" is most widely used, we emphasize these methods, but also briefly introduce other methods.

It is essential to appreciate that because an indicator property is normally determined, careful interpretation of data is vital. In turn, this means that the measuring method should be selected with care, ensuring that the data derived are relevant to the situation in question. This is particularly true with respect to measurements of particle diameter, because different methods measure different nominal diameters.

There are a number of techniques that can be carried out by automated equipment, the aims of such equipment being to provide more rapid analysis and to ensure freedom from operator bias. The former aim is generally achieved, but the second can be deceptive, and it should always be remembered that there is a place for experience in interpretation.

Five broad types of particle characterization can be recognized and these are listed in Table 2.8, together with a summary of the techniques used. (More detailed listings are available elsewhere.[1,5,26,27]) Sieving is clearly the most widely used (and often abused) method of particle sizing,

TABLE 2.8: CHARACTERISATION OF METHODS OF ANALYSING PARTICLES

Group 1	Direct Measurement of Particles
	Microscope methods
	Sedimentation methods (gravitational or centrifugal)
	Streaming methods
Group 2	Fractionation of a Sample of Particles
	Sieving
	Elutriation (gravitational or centrifugal)
Group 3	Surface Area Determination by Adsorption
	Static methods
	Gas flow methods
Group 4	Permeability Studies
	Dynamic permeameters
	Static permeameters
Group 5	Compositional Analysis
	Wet chemical analysis
	X-ray diffraction
	X-ray analysis (EDAX, fluorescence)

basically because it can produce very good data in a relatively short time. Most of the other "sizing" methods described here are referred to as *subsieve sizing*, which indicates their relative significance. Because sieving is closely related to screening, the two subjects are discussed together, in Chapter 9.

An essential prerequisite for meaningful results is sampling, and it is vital that control over sampling be extended into the plant. Sample preparation too is an easily overlooked component of the analysis, particularly when dealing with fine particles. In many situations interparticle forces can become significant and cause agglomeration of particles, which may lead to erroneous results and conclusions.

2.6.1. Microscopy

Microscopy is the most direct method of measuring particle size and has the additional advantage that qualitative information about particle shape can also be obtained. The major limitations of the method are the time taken to carry out sufficient measurements to ensure reliability and the elimination of operator bias. Less significant errors readily overlooked arise from inaccurate measurement of the magnification, edge softness near resolution limits, and parallax errors.

Three broad approaches are used.[1] Both the direct measurement method and the "count" or ultramicroscope method are applied to free particles, and in both cases, adequate dispersion of the sample is an important part of preparation, particularly where resolution is low. The third method employs polished sections.

Determination of Particle Characteristics

Direct Measurement. Measurements can be taken directly from an image of the sample[28] by actual visual examination in the microscope. To save operator fatigue, however, a preferable approach is to use some form of projected image such as a photograph or TV screen. Measurements in microscopes are taken with standard graticule scales in the eyepiece, but simpler rulers may suffice with photographs. The photographic method has the additional advantages of keeping a permanent record, and the possibility of higher resolution. Opposing these is the additional step of producing a photograph and the fact that the focus is fixed when the film is exposed.

Because it is impractical to measure the size of every particle on a microscope slide, representative areas must be chosen, and to prevent the double measurement of any particle it is best to select areas in a predetermined pattern. The number of particles that must be measured to give a statistically reliable diameter depends on the spread of sizes. With a narrow size range, relatively few particles need be measured, but with a typical size distribution from ground material, more than 600 particles may have to be measured.

A variety of microscopes are now commonly used. A conventional optical microscope is often quite adequate, even the best quality instrument is restricted by resolution limitations to particles greater than 0.5 μm. The use of ultraviolet light or phase contrast[29] attachments may extend this limit down to 0.1 μm. Transmission electron microscopes[30] (TEM) are useful for particles 0.001–5 μm. The SEM[30] is suitable for particles in the range 0.01–1000 μm and has two particular advantages. The first is that it has an exceptionally large depth of field that contributes to the particles appearing to be three-dimensional. Second, most machines are fitted with EDAX facilities that allow qualitative elemental analysis of the particles, or spots thereon, a technique that is particularly useful for heterogeneous materials.

Particle Size. One of the most difficult problems of microscopic particle sizing is to decide what to measure as the diameter of an irregular particle. Some of the diameters used are illustrated in Fig. 2.11.[31] Three of these (Fig. 2.11a–c) are *physical* diameters, since they involve what are essentially reproducible physical measurements of the particles. Consequently, they can be considered to be

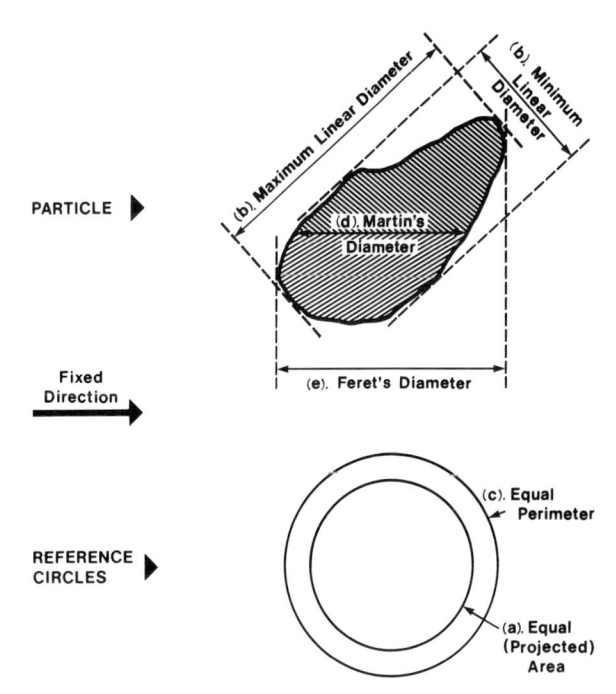

Figure 2.11. Diameters used for microscopic particle sizing. (After Kaye.[31])

more reliable; but because of the extensive amount of laborious measurements necessary they are very time-consuming to produce. The remaining two methods (Fig. 2.11d, e) are *statistical* diameters, relying on random particle shape and orientation.

The projected area diameter (Fig. 2.11a) represents the particle's diameter as that of a sphere having the same projected area. Its determination involves measuring the image area of each particle (or a sufficient number of particles). This can be achieved using approaches such as planimeters, counting squares on cross-hatching superimposed over the image, or preferably by cutting around reproduced outlines of the particle and weighing the paper.[28] Alternatively, particle areas may be estimated by comparison with standard areas, a method particularly susceptible to operator bias. The Zeiss-Endter particle size analyzer[1,32] has been developed to overcome this problem by using an adjustable spot of light for area comparisons. If the particles are opaque, a photoelectric cell can be used to determine the light blocked off by the particles, thus giving a measure of total area; this method can also be automated.[1]

The "average" diameter (Fig. 2.11b) requires averaging each particle's maximum and minimum dimensions, a method that is particularly cumbersome without a projected image. The perimeter diameter does not appear to be widely used (Fig. 2.11c).

Martin's statistical diameter d_M (Fig. 2.11d) is the length of the line that bisects the image (area) of the particle, the direction being the same for all particles. Feret's diameter d_F (Fig. 2.11e) is the distance between two opposite sides of the particle parallel to some fixed direction. Semiautomatic devices using an image splitting principle have been developed[33] for measuring this diameter to within 0.025 μm. These two statistical diameters of course rely on irregular particles having a random orientation. Although Feret's diameter may appear to be the simpler of the two to determine, it gives higher values of d, and higher deviations, than Martin's diameter, even when the bisection is estimated. However, the difference in magnitude[34] may not be a significant problem in that the ratios of different nominal diameters tend to remain constant for a given material and in fact to give a reasonable expression of shape. In particular, microscopic diameters can essentially be considered to be forms of the projected area diameter given by Kd_a (K = constant). Consequently the diameter measured can be considered as a calibrated value rather than an exact value, with the determination justifiably being carried out by the most convenient method. (The conversion between different nominal diameters has already been covered: See Fig. E2.1.1.)

Count Method or Ultramicroscopy. In this method the number of particles in a known mass of sample is counted.[28] Division of the mass by the number allows an average mass to be calculated, and thus d_v determined. Since detection rather than dimension is all that is required in this case, darkfield illumination is preferred, allowing detection down to about 0.01 μm. Again, automation of the process is possible because of the relatively simple measurement employed.

Polished Sections. Although polished section techniques are not often used to determine particle size, the method is often used to determine grain size and to estimate liberation characteristics. The notable difficulty with the technique is that the size measured is not a true size because the polished face is a random cut that intersects the grain somewhere between the middle and the edge. Theoretical and empirical methods have been used to correlate apparent sizes to true sizes.[35,36]

2.6.2. Sedimentation

There are two basic methods of determining size distribution by sedimentation: incremental and cumulative.[1] The former involves the periodic determination of the concentration of solids in the suspension at a given level; the latter is based on the mass fraction sedimented at a definite height as a function of time. In either approach the solids may start out as a homogeneous suspension or as a thin layer of slurry on top of the liquid, although the slurry form is not as widely used. It is characteristic of all sedimentation methods that they use the terminal settling velocity to determine the Stokesian diameter d_{St}, and consequently, again, produce different results from other methods.

Incremental Methods. The principle of these methods can be described as follows. In a layer of suspension at a given depth, solid is entering and leaving. A stage is soon reached at which the largest particles no longer occur in the layer because all the particles of that size above the layer have sunk through it by virtue of their higher settling rate. Knowing the maximum distance settled and the time allowed, it is possible, by assuming Stokes's law, to calculate the maximum size of particle that can exist in the layer at this time. By determining the concentration at this time, it is then possible to plot C/C_I versus d_{St}, which is a plot of the cumulative fraction undersize by mass for the solids.

The various methods used differ by the way in which the concentration is measured. One of the most widely used is the Andreasen pipette.[1] With this apparatus samples are withdrawn from a given depth by the pipette and the concentrations determined by evaporating the sample to dryness. Advantages of the method are simplicity, low cost, and with care, reproducible results. The major sources of error are associated with sample withdrawal, which causes disturbances to an unknown extent in the suspension, and leaves some of the sample in the pipette stem for the next sample. Low concentrations ($\leqslant$1%) are necessary to ensure the validity of Stokes' law, and this requires a relatively large sample, which necessitates a correction to the settling height as the test progresses.

An alternate method of determining concentration is to measure the specific gravity of the suspension with a hydrometer or diver. This method introduces new errors because it is difficult to know exactly where the specific gravity is being measured. Even though absolute results are relatively difficult to obtain, the simplicity and reproducibility of this approach makes it quite satisfactory for routine control work.

Manometers have also been used for density determination, but their accuracy is questionable. Empirical results can be obtained using the decrease in intensity of a light beam as it traverses a suspension at a given level.

Cumulative Methods. In considering these methods, it is necessary to realize that the mass fraction m_t of a suspension that settles out in time t consists of two parts: all particles of size d_t with a settling rate sufficient to fall the test height in time t, and the particles with lower settling rates that reach the measuring depth because they started below the top. On this basis it can be shown that

$$m_d = m_t - \frac{dm_t}{d(\ln t)} \quad (2.25)$$

(m_d = mass fraction with size greater than d_t). In principle the data collected are m_t and t, so that m_d can be calculated from Eq. 2.25, although integration provides some difficulties that are best overcome by graphical methods. The advantage of most cumulative methods over incremental methods is that suspensions can generally be of lower concentration, thus avoiding hindered settling effects.

One method of collecting data is to use a sedimentation balance, a number of which are available commercially.[37] All consist of a pan suspended near the base of a sedimentation chamber, with the pan connected to some device for recording mass gain. Readings from all such instruments are prone to errors because the pan is immersed in the suspension. A disadvantage of these particular devices is the need for relatively high suspension concentrations (1–5%). The Micromerigraph is a two-layer sedimentation balance that uses gas as the sedimentation fluid. Although it gives reproducible results, it is believed that a preferred loss of fines occurs because of powder adhering to the walls.

Sedimentation columns are much simpler; these glass columns are designed so that sediment can be drawn off at the bottom at appropriate intervals.[1] They are very cheap, can be operated by semi-skilled assistants, use lower suspension concentrations, and are not susceptible to disturbances during sample withdrawal. The most significant errors can arise from the taper in the lower end of the tube (causing low initial results) and possible hang-up of sediment.

Photosedimentation is a method for recording a particle's fall photographically, the length of the trace being used to calculate the size.

Decantation is a rather tedious method that has the advantage of requiring no special equipment other than beakers.[1] As such it is generally more suitable for preparing narrow size ranges of material than as a analytical method.

Centrifugal Sedimentation. Gravitational sedimentation is restricted to particle sizes greater than about 5 μm because at these prolonged settling times factors such as convection currents, Brownian movement, and ionic effects[38,39] can become significant. By using centrifugal instead of gravitational forces, the size range can be extended down to 0.02 μm, and even 0.002 μm with ultracentrifugation. The basic techniques of conventional sedimentation are still used.[1] However, because the centrifugal force varies from the center of rotation, the analysis of particle behavior is far more difficult.

Two basic types of equipment are available. The older types are based on sedimentation in a glass tube with a small diameter measuring section at the end, and use a conventional type of laboratory centrifuge. These methods are mainly limited by the tube wall effects. Shallow bowl or disc centrifuges are a relatively recent development that overcome these problems. In these, a clear liquid is centrifuged into the disc chamber and the suspension under test is injected through an entry port into the disc, where it forms a layer on the free surface. Sedimentation is followed either by withdrawing samples after given times, or by light absorption, the latter method having the advantages of being less disturbing and allowing a succession of samples to be fed through the device without stopping it. However, light scattering normally limits this method to particles larger than 0.3 μm, although X-rays can extend the limit to 0.1 μm.[40]

Photosedimentometry. By actually analyzing the light scattering phenomena (when a light beam is used to measure the rate of sedimentation), it is possible to determine size distributions to smaller sizes and as well, the specific surface and λ_{sa} can be determined.[41,42]

2.6.3. Elutriation

Elutriation is a process of sorting particles by means of an upward stream of fluid, usually air or water.[1] The grading is carried out in a series of either tubular or conical vessels of increasing size, so that the flow rate is successively decreased. Again Stokes' law is generally assumed to apply, but this in fact requires laminar flow with respect to both the particle and the vessel, conditions not always found in practice. Even so, separations can be carried out when conditions are turbulent or transitional, but it must be appreciated that a different nominal diameter is being measured (or even various diameters).

An apparent difficulty with elutriation is the velocity profile that exists across the fluid stream because of wall drag (Fig. 4.3). This suggests that particles are subjected to varying velocities, with the result that particles of varying settling rates are cut by each vessel. In reality, particles are dragged into the fastest flowing region by pressure differences on their surface and thus are eventually cut by the *maximum* fluid velocity.[1] Whether very small particles are given sufficient time for this to occur is of course another matter. It is in fact likely that another significant factor contributing to lack of sharpness in some elutriators is the occurrence of circulation currents in the column. Elutriators can be purchased as ready-made units, or simply constructed out of glass or clear plastic tubes.

Two commercially available centrifugal elutriators are worthy of note. The Bahco microparticle classifier has the powder sample introduced into a spiral shaped air current created by a hollow disc rotating at 3500 rpm. Air and powder are radially drawn inward through the cavity against centrifugal forces. By altering the air velocity, different size fractions between 5 and 100 μm can be made.[1] The Cyclosizer is a series of hydrocyclones fitted with apex chambers (Fig. 2.12) such that the particles are subjected to multiple sorting in each hydrocyclone; thus the separation becomes sharp rather than dispersed as in a conventional hydrocyclone.[43] The significant advantage of this device is the comparatively short time required for the analysis—about 10–20 min plus sample drying time. The lower size limit of about 10 μm depends on particle density, but supplementary methods have been suggested for extending this.[44,45]

2.6.4. Streaming Methods

Streaming methods can be used to determine size distributions by measuring particles individually while in a flowing stream of fluid passing a suitable detector. The essential parameter is that the stream concentration be so low that only one particle is detected at a time; otherwise coincidence errors result in small particles being oversized and large particles undersized. Consequently, at the other extreme, an average size can be measured by collecting data on a large number of particles at once.

One of the more successful applications of the technique of individual particle measurement is the Coulter counter.[1] This device has particles suspended in an electrolyte, and their presence produces a change in resistance as they pass between two electrodes. The consequent pulse signal is amplified and counted. By setting suitable detection limits, only particles greater than a certain size can be detected, and by successive adjustment of the limits, a cumulative size distribution can be built up.

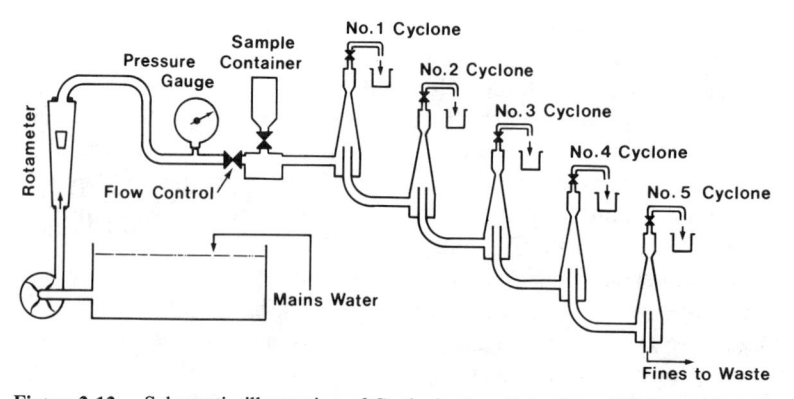

Figure 2.12. Schematic illustration of Cyclosizer particle sizer. (With permission, Warman International Ltd.)

Alternatively, limits can be set to detect a size range. Since the signal depends on the particle volume, this method determines d_v.

The initial cost of the apparatus may be comparatively high, but it has a number of advantages. Analysis by semiskilled operators is rapid, and because errors tend to cancel each other out results are very reproducible. It must be appreciated that every unit gives unique results, so that calibration is essential, preferably at more than one size. Since the machine was introduced, various modifications have been made or described to extend its usefulness.[46]

Another detection technique is called light scattering; it is possible because small particles diffract light. As the particle becomes smaller, so the angular spread of the pattern increases while the point intensity decreases. The total intensity of scattered light is proportional to Nd_a^2 (N = number of particles), and at the center of the pattern the intensity is proportional to d_a^4. One company has developed an instrument (which can be used on line) that passes a laser beam through a slurry stream, with the diffracted beam in turn passing through a suitable filter so that an Nd_a^3 term can be measured as well as the Nd_a^4 and Nd_a^2 terms.[47] By choosing suitable ratios of these three terms, d_{42} and d_{32} mean diameters are obtained. Thus, although the device is determining only an average particle size, the fact that two different averages are determined also gives some measure of the size distribution.

2.6.5. Permeability

Some information on the surface area characteristics of a powder can be obtained from the resistance to the flow of a fluid through a packed bed of the powder. In principle the specific surface of the material can be derived from such permeability data with the Carman-Kozeny equation (Eq. 4.63). However when air is passed through a bed of particles finer than about 5 μm, the equation is no longer valid and requires modification. Furthermore, it is likely that the value obtained may be in error with very wide range particle distributions, because air flow may predominate in large channels and swamp the effect of small channels.

Two commercially available instruments are fairly widely used.[1] The cheapest, illustrated in Fig. 2.13, is called the Blaine permeameter in the United States. In essence, this device draws a constant volume of air through a known mass of sample at

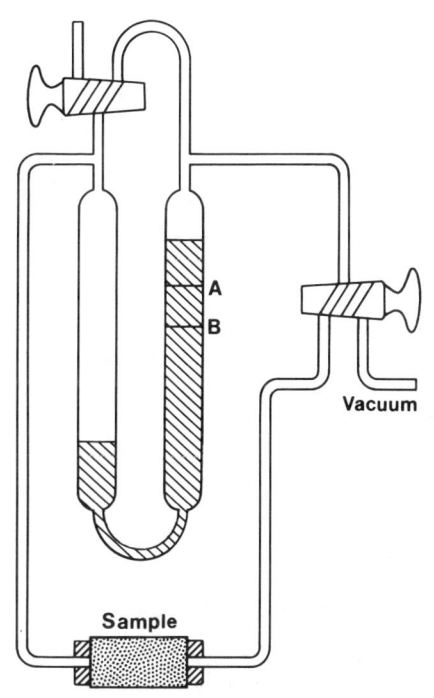

Figure 2.13. Schematic illustration of a permeameter for the determination of specific surface. The time for the liquid to fall between the lines A and B correlates with the specific surface of the sample.

known porosity. Equation 4.63 was modified by Blaine to give

$$S_0 = \frac{K\varepsilon^{3/2}t^{1/2}}{1-\varepsilon} \quad (2.26)$$

(t = time for liquid to travel from A to B on tube shown in Fig. 2.13; K = constant, which depends on the material, the instrument, and includes the constant in Eq. 4.63). In practice the specific surface is not independent of sample mass as Eq. 2.26 implies, but it has been shown that this can be compensated for by using[48]:

$$S_0 = \frac{K\varepsilon^{3/2}t^{1/2}}{1-K_s\varepsilon} \quad (2.27)$$

(K_s = a correction factor determined experimentally).

A slightly more complex device is the Fisher subsieve sizer. It draws air through a bed of powder and a capillary tube, the pressure drop across each being measured with manometers. The specific surface is then read off directly from a self-calculating chart incorporated on the unit. Although designed for a specific size of sample, corrections can be made to overcome this limitation.[49]

Although permeability gives only what is effectively an average particle size (d_{vs}), the method is very rapid and therefore useful for routine control purposes.

2.6.6. Gas Adsorption

The principle of surface area determination by gas adsorption is that the cross-sectional area of a gas molecule can be used to calculate the surface area of a powder if one measures the quantity of gas necessary to form a monolayer on the powder surface. Nitrogen is the most commonly used gas, but three experimental techniques are possible: the volumetric, gravimetric, and continuous flow methods.[1] The volumetric method is older and more cumbersome to use, and although the gravimetric methods overcome some of its difficulties, others are introduced. The continuous flow method is more recent and is a modification of gas adsorption chromatography, which eliminates the need for a vacuum system. Briefly, the process is to pass a known mixture of nitrogen and helium gases through the sample and thence through a thermal conductivity cell connected to a recording potentiometer. When the sample is cooled to liquid nitrogen temperatures, N_2 is adsorbed on the surface and the loss in the gas stream is shown on the recorder. Removal of the coolant gives a desorption peak on the recorder of equal area, but opposite direction. The apparatus must be calibrated first, and then a number of runs at different pressures are necessary because N_2 adsorbs in multilayers rather than in monolayers.

One point must be emphasized regarding surface areas determined by gas adsorption and permeability: the two methods give different results because they measure two different areas. Those determined by gas adsorption are significantly higher because this method measures area included in cracks, fissures, cavities, and regions of particle contact—areas not measured by permeability. Consequently surface area should not be measured, nor data applied, without giving consideration to its method of determination and applicability to the given situation.

2.6.7. Other Methods of Surface Area Determination

There are some other methods that can be used to determine surface area.[1] When the shape factors λ_s and λ_v are known for the material of interest, the surface area of a size distribution can be calculated by[2]:

$$S_0 = \frac{1}{\rho_s} \frac{\lambda_s}{\lambda_v} \frac{1}{d_{a32}} \qquad (2.28)$$

An approach similar to gas adsorption is fatty acid (e.g., oleic) adsorption. It is likely that this method gives low results because of coadsorption of the solvent that must be used. Similarly, dye adsorption is claimed to give low results, apparently because the surface is not fully coated with adsorbant. This is attributed to the inability of large dye molecules to enter pores (pores being the difference in surface area between the surface envelope and the total surface determined by gas adsorption). However, if this is true, then this type of measurement may be a more reliable indication of the area available for the adsorption of other large molecules such as some flotation collectors.

Surface area has also been estimated from the heat of adsorption as a gas-solid interface changes to a liquid-solid interface (i.e., the heat of immersion of a solid in a liquid).

2.6.8. Real-Time Particle Size Analysis

Real-time size analyzers are capable of carrying out a determination rapidly enough for active control purposes. This does not mean that the analyzer must be continuous in operation; determination on batch samples withdrawn from a process stream, however, must be rapid. Many techniques have been suggested for carrying out sufficiently rapid analyses, and these are well reviewed in the literature.[47,50,51] In reality, few are capable of giving the desired accuracy (generally considered to be equivalent to laboratory sieving or within 2%) with the wide range of particle sizes encountered in mineral processing flow streams. Most of those that have been successful are still somewhat limited because they determine only one point on the size distribution curve. However, in many cases (e.g., closed circuit grinding, where size distributions generally remain parallel), this may not be a significant fault. As mentioned above, one analyzer has partly eliminated this problem by measuring two different average diameters (Section 2.6.4).

2.6.9. On-Line Chemical Analysis

Chemical analysis of concentrate and tailings streams has always been an essential part of concentrator plant control. Up until the mid-1950s this was quantitatively achieved by laboratory chemical analysis, or qualitatively by visual methods such as vanning or passing the tailings stream across a shaking table. However, in the first case only massive perturbations can be removed from a circuit, and the second method is not sufficiently accurate.

Following development during the 1960s, the X-ray fluorescence analysis of process streams has become a reality, initially using extensive equipment that included delivery of part of the stream to central equipment, but more recently using in-stream probes.[52]

2.7. COMPLETE PARTICLE ANALYSIS

It must be emphasized that even "accurate" size and chemical analysis are only *indicative* representations of particle characteristics and very seldom give the exact information that one wants about the particles. For example, in a concentrator using closed circuit grinding, the grinding may be controlled using the output of an on-stream particle size analyzer, with the concentrating circuit controlled using chemical analysis of the output streams. In reality, the important questions are whether the desired degree of *liberation* has been achieved in the grinding circuit and whether the *separation* achieved in the concentration circuit has been the maximum *possible*, given the limitations imposed by the ore fed to the plant (assuming that the machine operation is kept at an optimum).

In practice, reliable determination of the degree of liberation can be quite difficult. Chemical analysis of separator products in most instances does not give reliable information because it is impossible to say whether the dilution is due to middling particles or to inadequate separator performance. There are a few separators that minimize the effect of separator performance, and the Davis tube used by Wiegel is one such example.[21]

Microscopic examination is one of the most widely recommended methods, but it has two significant limitations: data collection is tedious, and since results are obtained from a polished section, they do not truly represent three dimensions. In particular, even though a free particle cannot be sectioned so as to appear combined, it is possible to section a middling particle so that it appears liberated. Thus, the quantity of middling particles is higher than planar data indicates. Gaudin recommends correction factors.[22]

Increasing attention is now being given to the attainment of more extensive particle characterization as a means of improving process efficiency.[53] The apparent size parameter described by Schaap[24] (Section 2.5.4) is potentially useful, provided it can be conveniently evaluated, but at present only laboratory tests appear to be practical. Although a single parameter, it does in fact represent a number of characteristics relating to mineral associations in the particle. In principle analyses could be carried out for each individual characteristic; unfortunately, however, not only do successive analyses become progressively more difficult, but each one yields less than its predecessor. One approach that has possibilities is to use the SEM.[54,55] This instrument can provide information on size, texture, liberation, mineralization, shape, and agglomeration: all factors that can contribute to finer tuning of the process.

REFERENCES

1. T. Allen, *Particle Size Measurement*, 2nd ed., Chapman and Hall (1974).
2. H. Heywood, "Size, Shape and Size Distribution of Particulate Materials," Lecture Notes; Particle Characteristics Conference, University of Technology, Loughborough, Leicestershire, England, 1968.
3. C. C. Harris, "Graphical Presentation of Size Distribution Data: An Assessment of Current Practice," *Trans. IMM (C)*, **80**, C133–C139 (1971).
4. R. R. Irani, and C. F. Callis, *Particle Size: Measurement, Interpretation and Application*, Wiley (1963).
5. C. E. Lapple, "Particle-Size Analysis and Analyzers," *Chem. Eng.*, **75**, 149–156 (May 20, 1968).
6. P. Rosin and E. Rammler, "The Laws Governing the Fineness of Powdered Coal," *J. Inst. Fuel*, **7**, 29–36, and discussion 109–112 (1933–1934).
7. W. Weibull, "A Statistical Distribution Function of Wide Applicability," *J. Appl. Mech.*, **18**, 293–297 (1951).
8. R. Schuhmann, "Principles of Comminution. I. Size Distribution and Surface Calculations," AIME Technical Paper No. TP 1189, Mining Technology (July 1940).
9. A. M. Gaudin, "An Investigation of Crushing Phenomena," *Trans. AIME*, **73**, 253–316 (1926).
10. A. O. Gates, "Kick vs. Rittenger. An Experimental Investigation in Rock Crushing, Performed at Purdue University," *Trans. AIME*, **52**, 875–909 (1916).
11. (a) S. R. Broadbent and T. G. Callcott, "A Matrix Analysis of Processes Involving Particle Assemblies," *Phil. Trans. R. Soc.*, **249**, 99–123 (April 1956).
    (b) S. R. Broadbent and T. G. Callcott, "Coal Breakage Processes. I. A New Analysis of Coal Breakage Processes," *J. Inst. Fuel*, **29**, 524–528 (1956).

12. A. M. Gaudin and T. P. Meloy, "Model and a Comminution Distribution Equation for Single Fracture," *Trans. AIME/SME,* **223,** 40–43 (March 1962).
13. C. C. Harris, "The Application of Size Distribution Equations to Multi-Event Comminution Processes," *Trans. AIME/SME,* **241,** 343–358 (1968).
14. A. S. Foust et al., *Principles of Unit Operations,* 2nd ed., Wiley (1980).
15. C. Orr, *Particulate Technology,* Macmillan (1966).
16. H. Heywood, "Numerical Definition of Particle Size and Shape," *Chem. Ind.,* 149–154 (Feb. 13, 1933).
17. Y. Nakajima, W. J. Whiten, and M. E. White, "Method for Measurement of Particle-Shape Distribution by Sieves," *Trans. IMM (C),* **87,** C194–C203 (1978).
18. T. P. Meloy, "A New Approach to Characterizing Particle Shape and Size," AIME/SME Preprint 77-B-52, AIME (1977).
19. M. P. Jones and G. Barbery, "The Size Distributions and Shapes of Minerals in Multiphase Materials: Practical Determination and Use in Mineral Process Design and Control," *Proc. 11th Int. Miner. Process. Congr., Cagliari, 1975,* pp. 977–997, Università di Cagliari, (1975).
20. J. Tsubaki and G. Jimbo, "A Proposed New Characterisation of Particle Shape and its Application," *Powder Technol.,* **22,** 161–169 (1979).
21. R. L. Wiegel, "Liberation in Magnetite Iron Formations," *Trans. AIME/SME,* **258,** 247–256 (1975).
22. A. M. Gaudin, *Principles of Mineral Dressing,* McGraw-Hill (1939).
23. (a) R. P. King, "A Quantitative Model for Mineral Liberation," *J. S. Afr. IMM,* **76,** 170–172 (1975).
    (b) R. P. King, "A Model for the Quantitative Estimation of Mineral Liberation by Grinding," Research Report, Department of Metallurgy, University of the Witwatersrand, Johannesburg (1975).
24. W. Schaap, "Illustrated Liberation-Recovery Model for a Disseminated Mineral in Low-Grade Ore," *Trans. IMM (C),* **88,** C220–C228 (1979).
25. W. Schaap, "Problems in Model Formulation for Optimal Mineral Extraction," *Proc. 1979 Conf. Australian Society for Operations Research,* pp. 286–301, ASOR (1979).
26. E. A. Collins, J. A. Davidson, and C. A. Daniels, "Review of Common Methods of Particle Size Measurement," *J. Paint Technol.,* **47,** 36–56 (May 1975).
27. M. J. Groves and J. L. Wyatt-Sargent (Eds.), *Particle Size Analysis 1970,* Soc. Analytical Chemistry (1972).
28. R. P. Loveland, "Methods of Particle-Size Analysis," ASTM Spec. Tech. Publ. No 234, pp. 57–86 (1959).
29. F. Stein, "Particle Size Measurements with Phase Contrast Photography," *Powder Technol.,* **2,** 327–334 (1968–1969).
30. C. E. Hall, *Introduction to Electron Microscopy,* 2nd ed., McGraw-Hill (1966).
31. B. H. Kaye, "Determination of the Characteristics of Particles," *Chem. Eng.,* **73,** 239–246 (Nov. 7, 1966).
32. F. Lenz, "Die Bestimmung der Grossenverteilung . . . ," *Optik,* **11,** 524–527 (1954).
33. V. Timbrell, "A Method of Measuring and Grading Microscope Spherical Particles," *Nature,* **170,** 318–319 (1952).
34. H. Heywood, "A Comparison of Methods of Measuring Microscopical Particles," *Trans. IMM,* **55,** 391–404 (1945–1946).
35. W. Petruk, "Correlation Between Grain Sizes in Polished Section with Sieving Data and Investigation of Mineral Liberation Measurements from Polished Sections," *Trans. IMM (C),* **87,** C272–C277 (1978).
36. R. P. King, "Determination of Particle Size Distribution from Measurements on Sections," *Powder Technol.,* **21,** 147–150 (1978).
37. K. Leschonski and W. Alex, "Theoretical and Experimental Study of Sedimentation Balances," in *Particle Size Analysis 1970,* Soc. Analytical Chemistry, pp. 236–266 (1972).
38. E. B. Sansone and T. M. Civic, "Liquid Sedimentation Analysis: Media Conductivity and Particle Size Effects," *Powder Technol.,* **12,** 11–18 (1975).
39. B. H. Kaye and R. P. Boardman, "Cluster Formation in Dilute Suspensions," *Symposium on Interaction Between Fluids and Particles,* pp. 17–21, Institution of Chemical Engineers (1962).
40. T. Allen and L. Svarovsky, "The Ladal X-ray Centrifugal Sedimentometer," *Powder Technol.,* **10,** 23–28 (1974).
41. T. Provder and R. M. Holsworth, "Particle Size Distribution Analysis by Disc Centrifuge Photosedimentometry," 172nd American Chemical Society Meeting, ACS Organic Coatings and Plastics Chemistry Preprints, **36,** 2, 150 (1976).
42. T. Allen, "Determination of Specific Surface, Surface-Volume Shape Coefficient and Particle Size Distribution by Using the Wide-Angle Photosedimentometer," in *Particle Size Analysis 1970,* Soc. Analytical Chemistry, pp. 167–177 (1972).
43. D. F. Kelsall and J. C. H. McAdam, "Design and Operating Characteristics of a Hydraulic Cyclone Elutriator," *Trans. Inst. Chem. Eng.,* **41,** 84–95 (1963).
44. D. F. Kelsall, C. J. Restarick, and P. S. B. Stewart, "Technical Note on an Improved Cyclosizing Technique," *Proc. Australasian IMM,* No. 251, 9–10 (September 1974).
45. C. B. Daellenbach, W. M. Mahan, and F. E. Armstrong, "Rapid Particle Size Analysis by Hydrosizing and Nuclear Sensing," U.S. Bureau of Mines Report of Investigations, RI 7879 (1974).
46. (a) R. Karuhn et al., "Studies on the Coulter Counter. Part I," *Powder Technol.,* **11,** 157–171 (1975).
    (b) R. Davies, R. Karuhn, and J. Graf, "Studies on the Coulter Counter. Part II," *Powder Technol.,* **12,** 157–166 (1975).
    (c) R. Davies et al., "Studies on the Coulter Counter. Part III," *Powder Technol.,* **13,** 193–201 (1976).
47. E. L. Weiss and H. N. Frock, "Rapid Analysis of Particle Size Distributions by Laser Light Scattering," *Powder Technol.,* **14,** 287–293 (1976).
48. S. S. Ober and K. J. Frederick, "A Study of the Blaine Fineness Tester and a Determination of the Surface Area from Air Permeability Data," ASTM Spec. Publ. No. 234, pp. 279–285 (1959).
49. D. Duzevic, "On the Use of the Fisher Sub-Sieve-Sizer in

References

the Sub-micron Particle-Size Range," *Sci. Sintering* (Engl. ed.), **8**, 37–51 (1976).

50. A. L. Hinde, "A Review of Real-Time Particle Size Analysers," *J. S. Afr. IMM*, **73**, 258–268 (1973).

51. A. L. Hinde and P. J. D. Lloyd, "Real-Time Particle Size Analysis in Wet Closed-Circuit Milling," *Powder Technol.*, **12**, 37–50 (1975).

52. H. R. Cooper, "On-Stream X-ray Analysis," Ch. 30 in *Flotation: A. M. Gaudin Memorial Volume*, Vol. 2, M. C. Fuerstenau, Ed., pp. 865–894, AIME/SME (1976).

53. J. E. Johnston and L. J. Rosen, "Particle Characterization Using the Photoscan," *Powder Technol.*, **14**, 195–201 (1976).

54. J. S. Hall, "Analysis of Composite Mineral Particles Using a S.E.M.," Julius Kruttschnitt Mineral Research Centre, Technical Report, July 1975–June 1976, pp. 53–66 (1976).

55. G. C. Thorne et. al., "Modeling of Industrial Sulphide Flotation Circuits," Ch. 26 in *Flotation: A. M. Gaudin Memorial Volume*, Vol. 2, M. C. Fuerstenau, Ed., pp. 725–752, AIME/SME (1976).

Chapter Three

Analysis of Separation Processes

Flap's Law on the Perversity of Inanimate Objects: "Any inanimate object, regardless of its composition or configuration, may be expected to perform at any time in a totally unexpected manner for reasons that are either totally obscure or completely mysterious."

Mineral processing is concerned primarily with the separation of mineral particles based on variations in size or composition. Separation is brought about by suspending the particles in a medium and passing the suspension through an appropriate piece of equipment termed a separator. There a suitable force is applied to the particles; and because the materials have different properties depending on their composition or size, they are affected to varying degrees by the applied force (Fig. 3.1). *Those that are moved by the force have a positive response, those unaffected a negative response.* Obviously liberation is a prerequisite for perfect separation, although in most cases the ideal of complete liberation may not be practical.

For convenience, a separation can be considered to depend on three factors: the properties of the minerals, separator characteristics, and production requirements of grade and recovery. Some mineral properties that may be exploited are size, shape, density, magnetic susceptibility, electrical conductivity, or surface properties. These and the specific processes utilizing them are discussed in detail in other chapters; here they are discussed in general terms, and in relation to the complex issue of separation efficiency.

A separation process ideally is one that exploits a single mineral property, although in practice this situation is not achieved, thereby placing a limit on the separation. Sometimes the desired constituent is an element that occurs in more than one miner- alogical form in the ore, and it may be necessary to use a different separator to recover each mineral. For example, an iron ore may contain magnetite (Fe_3O_4), hematite (Fe_2O_3), and gangue minerals. Although both iron minerals are denser than the gangue minerals, thus providing a potential separating property, better separation is generally achieved by exploiting the magnetic properties of magnetite for its recovery.

A particle may contain more than one phase, such as several minerals, cracks, cavities, and absorbed water. As an extension of this concept, a *set* of particles consists of the particles that can be considered together because they have some feature in common relevant to the separation. For example, particles could be grouped because they contain varying amounts of the valuable to be extracted (in the case of the iron ore above, this grouping could be by content of Fe_3O_4, Fe_2O_3, or Fe, depending on the requirements). Conversely, a grouping could be those particles with varying amounts of the waste material, in which case some particles could be common to two sets if liberation is incomplete (Fig. 3.2). On some occasions, also, the feed sample can be considered to be a set (e.g., a size distribution of particles). Since particles are normally produced by a size reduction process, any set of particles will have a range of particle sizes. To give a quantitative measure of the property that is being exploited for the purpose of separation, the value of the property of an individual particle is termed ℔. The critical

Equilibrium Separation

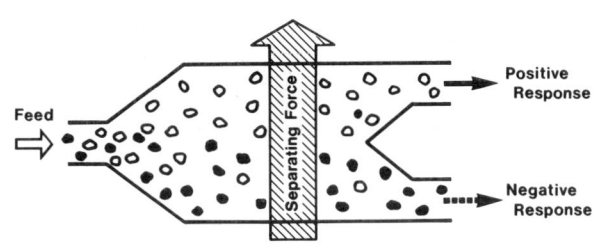

Figure 3.1. The basic principle of a separator.

value of the property that determines the response (or split) in the equipment is called the *separator setting* $\mathfrak{P}^*$.

3.1. EQUILIBRIUM SEPARATION

3.1.1. Variations in the Exploiting Property: Separability Curves

In a set of particles that show a positive response to an exploiting force, there normally is a range in the value of the property being exploited. This could be due to variations in particle size, the degree of liberation, or subtle changes in chemical and physical properties. As illustrated in Fig. 3.3, the range of property values of a given set may be presented as frequency or cumulative curves. These curves have been applied mostly to specific gravity separations, and because they were originally used to study coal washing, they are often referred to as washability curves.[1] However, since they represent the mineral property characteristics and can be applied to most separations, they are better termed *separability curves*. Size is the simplest exploited property that can be considered: Fig. 2.1 is basically Fig. 3.3 with the size d_A as the property $\mathfrak{P}$. In this case the set has been taken as the total sample and Fig. 2.1 simply

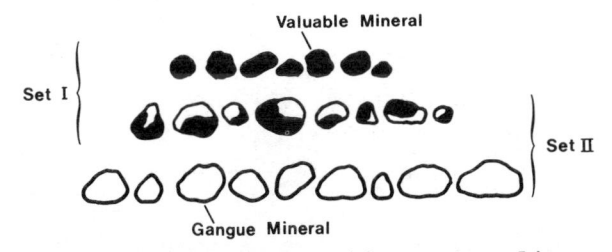

Figure 3.2. Representation of particles as sets: set I is comprised of all particles that contain some valuable mineral; set II is comprised of all particles that contain some gangue mineral. On this basis the middling particles are common to both sets.

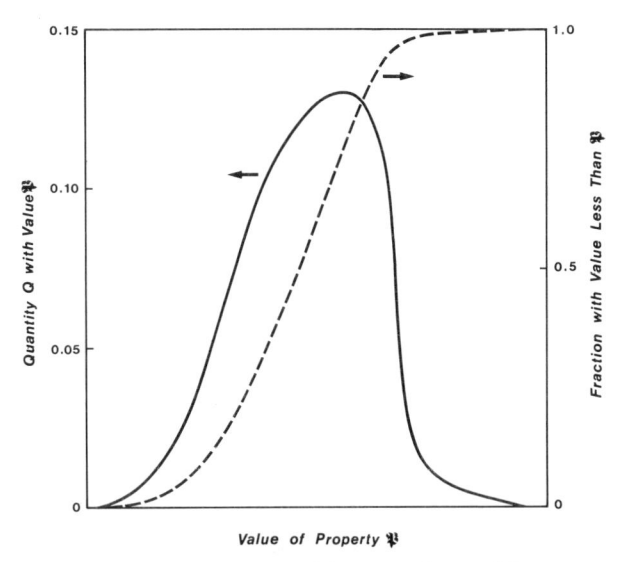

Figure 3.3. The cumulative and frequency forms of the separability curve of a set of particles. (These should be compared with the curves in Fig. 2.1, which is the specific case in which size is the property.)

states that as larger particles are included, so a greater fraction of the sample (set) is obtained.

When properties other than size, are being considered, one of the most common causes of variation in properties is incomplete liberation, although its significance depends on the exploited property. The most even gradation in response occurs when specific gravity is the property exploited, in that the specific gravity is directly proportional to the volumetric mineralogical proportions. At the other extreme, only small amounts of a magnetically susceptible mineral may be necessary to make a composite particle susceptible to a magnetic force. It follows therefore that a change in the degree of liberation changes separability curves. However, even fully liberated minerals may show variations in properties. These could be due to apparently minor variations in composition; for example, slight surface oxidation or trace elements altering the adsorption of flotation reagents.

The effect of particle-to-particle variability on separation can be illustrated using the curves shown in Fig. 3.4.[2] Since all the particles in Set A have values of $\mathfrak{P}$ less than those of set D, a separation of these two sets is possible using a separator setting between $\mathfrak{P}_3^*$ and $\mathfrak{P}_4^*$. Unfortunately, sets of particles are seldom this widely separated. Sets B and D should also be amenable to separation using separator setting $\mathfrak{P}_4^*$, but this case will be more dif-

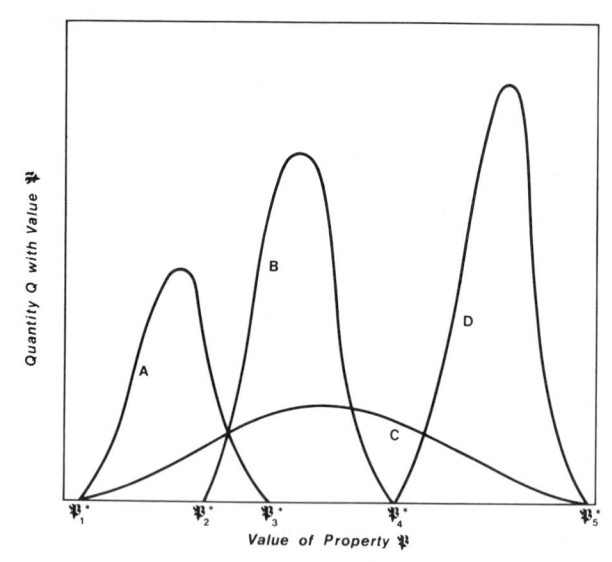

Figure 3.4. Frequency separability curves for four sets of particles. Under appropriate circumstances, sets A and B could, for example, represent sets I and II in Fig. 3.2. $\mathfrak{P}^*_1, \mathfrak{P}^*_2$, and so on are separator settings for various separations. (After Prosser.[2])

ficult because of the near overlap. A practical example of this situation would be a perfect screening of the size distribution in Fig. 2.1 into two size products (sets)—one smaller than the screen size of 1.0 mm (the separator setting) and the other larger. The two separability curves of the products are shown in Fig. 3.5. Note that suitable combination of the two product sets gives the cumulative property (size) distribution of the feed.

The separation of pairs of sets from A, B, and C in Fig. 3.4 is typical of mineral concentration operations. A good, but incomplete separation of A and B is possible using a separator setting between $\mathfrak{P}^*_2$ and $\mathfrak{P}^*_3$, the actual setting depending on product requirements. For example, a setting of $\mathfrak{P}^*_3$ would be needed for high grade B or a high recovery of A. Separations such as this may be due to incomplete liberation, where the overlap results from middling particles being common to both sets, as illustrated in Fig. 3.2.

The potential separation of A and C is less satisfactory in that no pure A can be produced, while the fraction of pure C that can be obtained depends on the relative spread of the curves. Figure 3.6 shows some flotation data in a cumulative form that represents this situation.[3]

From these examples it can be seen that using a single value of $\mathfrak{P}$ (e.g., the average) to represent the set can give misleading information; in any of the separations AB, AD, BD, or CD the average would suggest that a complete separation could be made. Conversely, B and C could have the same average value of $\mathfrak{P}$, suggesting that no separation is possible; in reality a two-stage process (at $\mathfrak{P}^*_2$ and $\mathfrak{P}^*_4$) would be capable of giving some separation. Although probably difficult, this limited separation may well be worthwhile.

It must always be remembered that the separation predicted from separability curves in this manner represents *the maximum separation that can be achieved on the material as it exists by exploiting that particular property.* If the separation is unsatisfactory, use of an alternative device exploiting the same property will generally not improve the separation because the maximum separation is limited by the properties of the particles alone. In practice, engineering limitations mean that a separator never achieves the maximum separation indicated by the

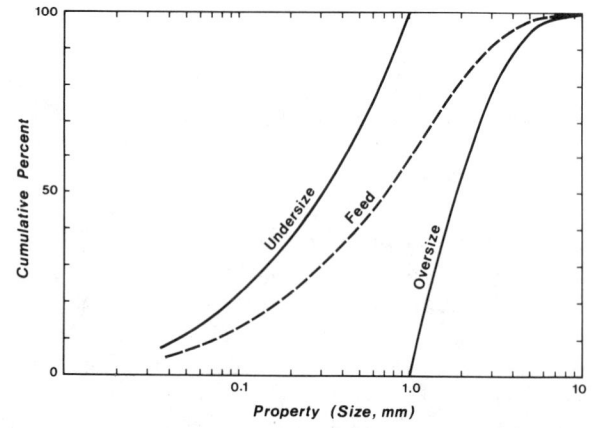

Figure 3.5. Separability curves for the two sets: oversize and undersize (cf. curves B and D in Fig. 3.4).

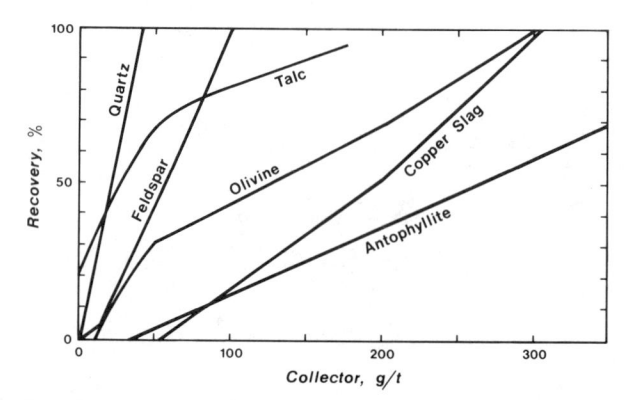

Figure 3.6. Separability curves of some silicate minerals, using collector addition as an indication of the property "floatability." Flotation separation between pairs of such minerals represents the separation A–C in Fig. 3.4. (After Lidström.[3])

Equilibrium Separation

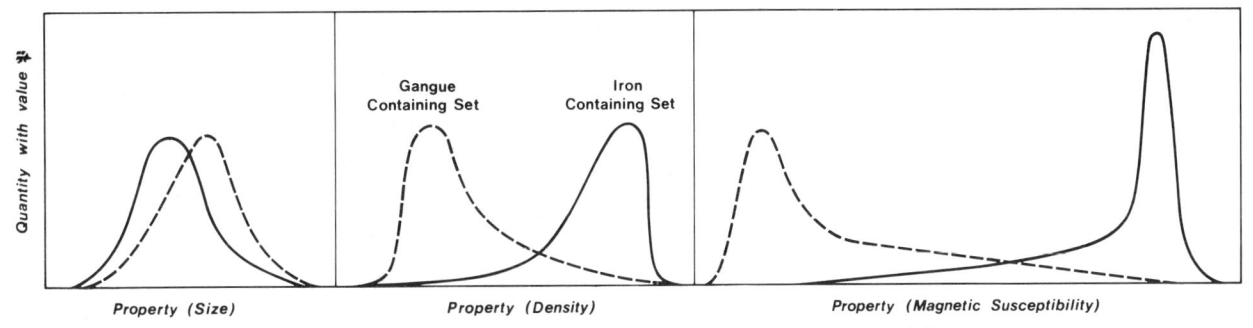

Figure 3.7. Separability curves of an iron-containing beach sand, which essentially consists of two sets: light gangue minerals and dense iron-containing minerals. Depending on the property exploited, the separation can be difficult or easy.

separability curves, but since there may be differences in the degree to which different separators approach the maximum separation, one may give a better *performance* than another.

When a satisfactory displacement does not exist between separability curves, two alternatives exist; to change the distribution of values of $\mathfrak{P}$ by some pretreatment, or to exploit a different property. Again, liberation may be used to illustrate the first point: further size reduction of the material illustrated in Fig. 3.2 would result in a higher degree of liberation and would reduce the overlap between curves such as A and B in Fig. 3.4. Exploitation of an alternate property is illustrated in Fig. 3.7, which is based on data from an iron-containing beach sand. This sample can be considered essentially as two sets: gangue, and heavy minerals that contain iron. Some separation is possible by screening (Fig. 3.7, left) but a better separation is obtainable with a relatively difficult gravity separation (Fig. 3.7, center). However, an easy separation of the magnetic fraction provides a very satisfactory recovery of the iron, although it is incomplete because some of the iron is in a heavy, lower grade, nonmagnetic mineral (Fig. 3.7, right).

Example 3.1. Sink-float separations were carried out on a coal: the results are shown in columns 2 and 3 of Table E3.1.1. What are the separability curves for the ash and coal components?

Solution. The necessary calculations are indicated under the headings in Table E3.1.1, and the results shown in the relevant columns. The cumulative separability curves, using data from columns 6 and 9, are plotted in Fig. E3.1.1. Thus, a perfect

TABLE E3.1.1: SEPARABILITY CURVES EXAMPLE

Specific Gravity Fractions	Mass	Assay	Ash Distribution			Coal Distribution		
	%	% Ash	Mass	%	Cum. %	Mass	%	Cum. %
(1)	(2)	(3)	$(4)=\frac{(2)\times(3)}{100}$	$(5)=\frac{(4)}{\Sigma(4)}$	$(6)=\Sigma(5)$	$(7)=(2)-(4)$	$(8)=\frac{(7)}{\Sigma(7)}$	$(9)=\Sigma(8)$
Float to 1.30	41.0	2.1	0.86	3.00	3.00	40.14	56.26	56.26
1.30 to 1.35	7.0	5.0	0.35	1.22	4.22	6.65	9.32	65.56
1.35 to 1.40	2.3	12.6	0.29	1.01	5.23	2.01	2.82	68.40
1.40 to 1.45	2.2	19.1	0.42	1.47	6.70	1.78	2.49	70.89
1.45 to 1.50	2.9	24.0	0.70	2.44	9.14	2.20	3.08	73.97
1.50 to 1.60	5.9	30.7	1.81	6.32	15.46	4.09	5.73	79.70
1.60 to 1.70	6.0	38.4	2.30	8.03	23.49	3.70	5.19	84.89
1.70 to 1.80	4.4	47.2	2.08	7.26	30.75	2.32	3.25	88.14
1.80 to Sink	28.3	70.1	19.84	69.25	100.00	8.46	11.86	100.00
			28.65	100.00		71.35	100.00	

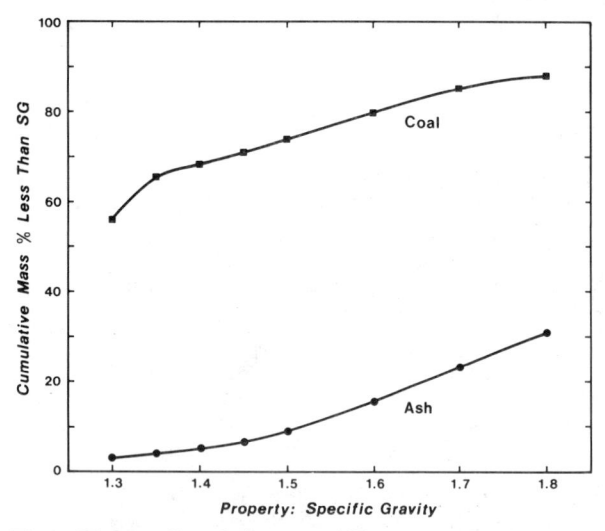

Figure E3.1.1. Cumulative separability curves of coal and ash, from Example 3.1.

separation at SG = 1.50 would recover 73.97% of the coal, together with 9.14% of the ash. The ash content of the coal would be

$$\frac{9.14 \times 28.65}{(9.14 \times 28.65) + (71.35 \times 73.97)} \times 100\%$$
$$= 4.73\%$$

3.1.2. Equipment Limitations

In practice commercial separators seldom achieve the best separation indicated by separability curves because of inherent variations in the separator setting, competing forces, and insufficient time (see Section 3.2 for further discussion of separation time). Although a separator has a set point, a range of values occurs because the set point is subjected to random fluctuations that may arise from variations in flow rates, variations in power supplies, wear in the equipment, and so on. Consequently, if the separator settings illustrated in Fig. 3.4 are thought of as bands rather than lines, the resulting reduction in separation becomes more obvious.

The assumption so far has been that the behavior of a particle is determined solely by the property being exploited. In reality, as well as the separating force, the separator has other forces that act on the particle: gravity is particularly significant (see, e.g., Fig. 14.19). Such additional forces may result in other properties having a secondary but important effect that may either hinder or aid the intended separation. It is possible, for example, that particles of the same true specific gravity may be separated by gravity methods because the shape of one of the sets of particles produces an apparent difference in the bulk specific gravity.[4] In general, with a constant exploiting force, increasing the competing force will increase grade at the expense of recovery (Fig. 14.19).

In most separators some particles leave in the wrong stream because the suspension is subjected to mixing, either natural or deliberate, as well as to a spatial dispersion caused by the occurrence of the entry stream over an area instead of at a point. Furthermore, as the particles become smaller, the separating force becomes progressively less effective and any disturbing forces become more effective, resulting in every separator having a lower size limit beyond which adequate separation cannot be achieved.

3.2. KINETICS: THE RATE OF SEPARATION

Virtually all mineral separations are carried out in continuous separators, which means that any particle has a finite time to response to the separating force. The result is that the maximum separation indicated by the separability curves is seldom achieved. In spite of their complexity, most mineral processes have been found to follow a first-order rate law, so that for particles with identical properties, they can be represented by

$$\frac{dm}{dt} = -km \quad (3.1)$$

(m = mass fraction of particles present at time t, having identical properties; k = rate constant).

For identical particles in a batch or a continuous process with constant residence time, Eq. 3.1 integrates to

$$m = m_I \exp(-kt) \quad (3.2)$$

(m_I = mass fraction at $t = 0$). Realistically, the particles entering a process are not identical, so that k varies from particle to particle, particularly with particle size. Furthermore, the particles are subjected to the separating force for varying times (i.e., the residence time in the separator varies). (Although the latter subject is introduced briefly in Section 3.3, the integration of Eq. 3.1 under these conditions is beyond the scope of this book.)

Kinetics: The Rate of Separation

Alternatively, a number of mineral separating operations, such as flotation, electrostatic separation, and screening, can be considered to be a number of identical stages in series. Under these conditions, the rate of the process can be represented by a simple probability concept[5,6] (see Examples 9.1 and 16.1):

$$m = m_I(1 - p)^n \qquad (3.3)$$

(m_I = initial mass fraction in the feed stream; m = mass fraction remaining in the stream after n stages; p = probability of leaving the stream in any stage). Rearranging,

$$\log \frac{m}{m_I} = n \log(1 - p) \qquad (3.4)$$

Thus a semilog plot of the mass change through a series of operating stages allows p to be evaluated.

Comparison of Eqs. 3.2 and 3.3 clearly shows that p has a similar connotation to k. Note also that there will be occasions where m in Eqs. 3.1 to 3.4 can be replaced by a total mass M.

The response of a set of particles is best illustrated with the frequency separability curve (Fig. 3.3). In Fig. 3.8b the upper curve is the curve of the property values to be exploited; the lower curve is the actual property distribution of the particles recovered in the positive response outlet in time t (it is assumed that no particle with $\mathfrak{P} > \mathfrak{P}^*$ is recovered there). Because the relative area under the two curves in Fig. 3.8b (expressed as a percentage) is the recovery of the mineral in the set, it follows from Eq. 3.2 that any increase in the rate constant or time will raise the lower curve toward the equilibrium one, thus raising the recovery.

In general, as $\mathfrak{P}$ tends to $\mathfrak{P}^*$, the driving force for

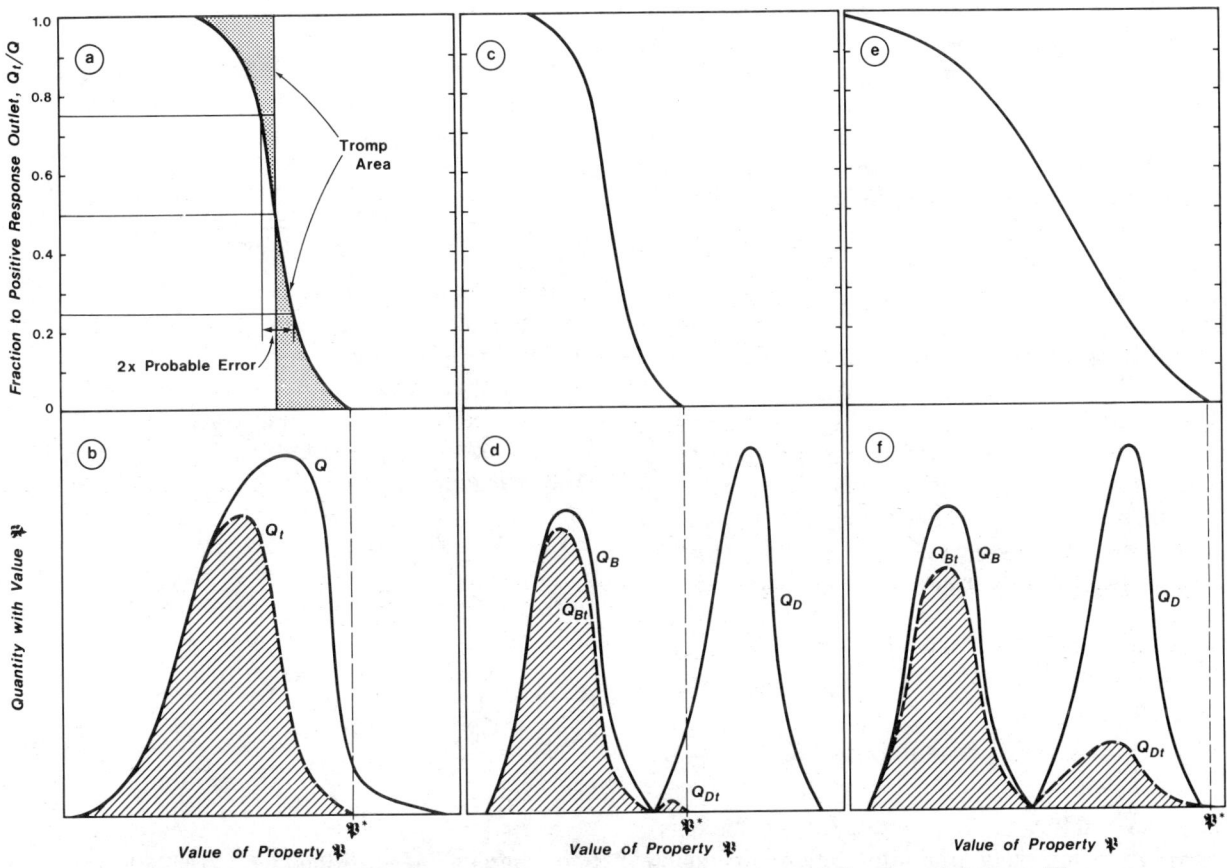

Figure 3.8. (b) (d) and (f) Separability curves Q, and portion of the set Q_t that responds in time t. The ratio of these two quantities at various values of the property $\mathfrak{P}$ forms the separator's performance curves, shown in (a), (c), and (e). A separator with a fast response (d) produces a steep performance curve (c) and can be considered to be efficient, whereas one with a slow response (f) has a dispersed performance curve (e) and is comparatively inefficient.

the separation decreases such that k can be approximated by

$$k = k' (\mathfrak{P}^* - \mathfrak{P}) \tag{3.5}$$

(k' = rate constant independent of $[\mathfrak{P}^* - \mathfrak{P}]$ and is related to the design and operation of the separator). Thus the rate constant k depends on the separator design, its operating conditions, and the mineral properties.

If the separation of B and D in Fig. 3.4 is now reconsidered, two extreme cases can be envisaged. When k' is high and the same for both sets, the separator setting can be set just above that predicted by the separability curves and the separation (Fig. 3.8d) will be close to that predicted before. If however k' is low for both sets, the separator setting has to be well beyond the maximum value of $\mathfrak{P}_B$ to recover an acceptable amount of B, resulting also in an appreciable recovery of D (Fig. 3.8f). Furthermore, since the area under the Q_t curve gives the recovery of each set in the positive response outlet, the ratio of the areas is a measure of the grade of the product. Significantly, in both cases, *increased recovery is obtained at the expense of grade;* a feature typical of any separator. The grade/recovery tradeoff is normally handled by operating separators in series with different separator settings and is the reason for use of rougher, scavenger, and cleaner stages in a concentrating circuit. Such stages may involve recycling and/or property modification such as regrinding.

In many separators the rate constant of small particles is less than that of large particles, and under these conditions it follows from Fig. 3.8d and f that the separator is capable of separating only material that has a limited size distribution.

Although Eq. 3.1 has been suggested for gravity separations such as jigging[4] (Section 5.7), it is more realistic to consider the potential significance of particle acceleration in these separations, in which case the kinetic behavior should be described by a second-order equation of the form

$$\frac{d^2H}{dt^2} + K_1 \frac{dH}{dt} + K_2 \text{fn}(H) + K_3 = 0 \tag{3.6}$$

(H = vertical location parameter; K = coefficients). As yet comparatively little information is available on the applicability of this equation in mineral processing, although it has been applied in other fields,[7] and in the Russian literature.[8]

3.3. MIXING PATTERNS

It was shown in Section 3.2 that an important consideration in any process is the rate. To fully analyze the extent of processing in any continuous process, it is therefore necessary to know how long the material takes to proceed from the inlet(s) to the outlet(s). To estimate these *residence times* it is necessary to know the *flow patterns* (not to be confused with flow type) in the device.[9] The patterns can be determined by injecting an impulse tracer of a conveniently detectable material into the flow device, and following its emergence through the output stream(s). An alternate method is to instantaneously (step) change the concentration of a component in the feed, and then follow the concentration change in the outlet stream. The resulting patterns can be considered in terms of two ideal types; plug flow or perfect mixing.

3.3.1. Plug Flow

The simplest type of flow pattern through a process is when all elements of the feed stream have an equal residence time τ in the device; so called *plug flow*. The residence time is therefore

$$\tau = \frac{\text{volume of device}}{\text{volumetric flow rate}} = \frac{V}{I_V} \tag{3.7}$$

In this situation no mixing can occur in the flow direction, although there may be mixing laterally. If an impulse tracer (or a step change in the composition) is introduced into the feed stream, it will emerge after time $t = \tau$ as illustrated in Fig. 3.9.

It is worth noting at this stage that a plug flow system is equivalent to a batch processing system.

3.3.2. Perfect Mixing

In the case of *perfect mixing*, any input is immediately dispersed uniformly throughout the whole volume of the device. When a tracer impulse is introduced at time $t = 0$, its concentration C in the exit stream at time t is given by

$$C = C_I \exp\left(-\frac{I_V}{V} t\right) \tag{3.8}$$

(C_I = initial concentration at $t = 0$) as illustrated in Fig. 3.9. It can be seen that some of the tracer leaves instantaneously, while some theoretically never leaves, so that there is a distribution of resi-

Mixing Patterns

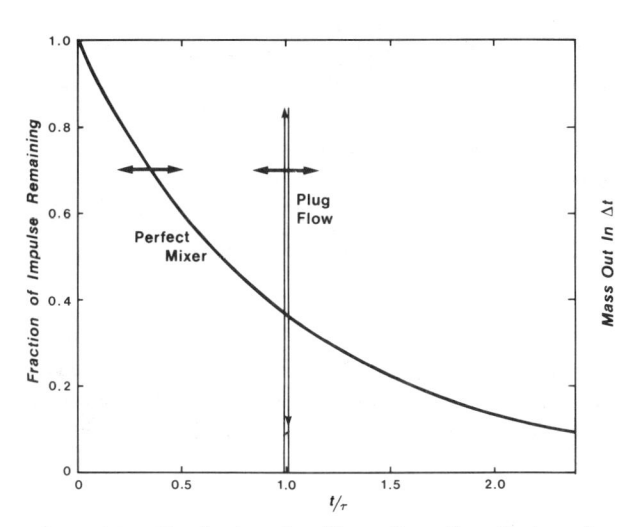

Figure 3.9. Distribution of residence times (impulse tracer).

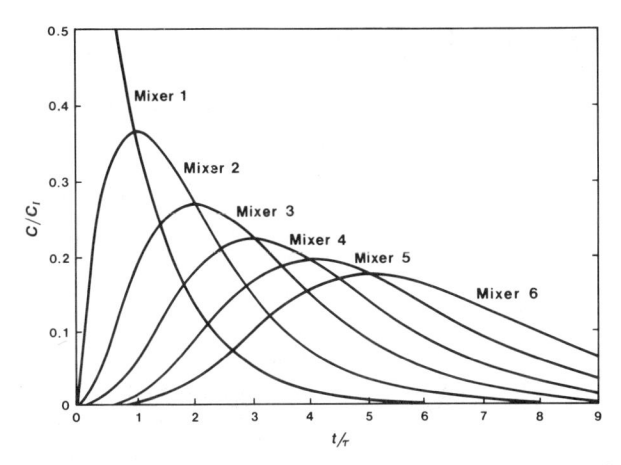

Figure 3.10. The distribution of residence times for a number of perfect mixers in series, predicted by Eq. 3.10 (impulse tracer).

dence time from zero to infinity. A nominal residence time can still be defined by Eq. 3.7, so that Eq. 3.8 may be written

$$C = C_I \exp \frac{-t}{\tau} \quad (3.9)$$

from which it follows that after the nominal residence time the concentration is down to 37% of the initial value when half the impulse has been discharged.

Equation 3.9 may be extended for N perfect mixers in series, resulting in

$$\frac{C_N}{C_I} = \frac{(t/\tau)^{N-1}\exp(-t/\tau)}{(N-1)!} \quad (3.10)$$

(C_N = concentration in the discharge of the Nth device at time t; τ = nominal residence time in each mixer). The theoretical curves obtained from Eq. 3.10 are shown in Fig. 3.10.

3.3.3. Comparison of Plug Flow and Perfect Mixing

It is useful to compare the performance of a series of perfect mixers with a plug flow device. Figure 3.10 shows that a stream will be dispersed in time as it passes through a series of mixers. If, however, the total nominal residence time τ_Σ in the series is held constant while the number of mixers is changed, the results are as shown in Fig. 3.11 for τ_Σ = 8 min.[10] It can be seen that as the number of mixers is increased, the residence time distribution of the material in the series of mixers is more closely distrib-

uted about the nominal value. Since plug flow on this figure would be a vertical line at 8 min, it is clear that an infinite number of mixers in series would represent this situation.

It follows that a small number of continuously agitated devices will give a poorer performance than a plug flow (or batch) device, because some material will not remain in the device long enough, and other material will remain too long. Increased performance can generally be obtained by increasing the number of units in series. In some situations, however, kinetic effects may predominate to such an extent that the number of units is not so significant.[11]

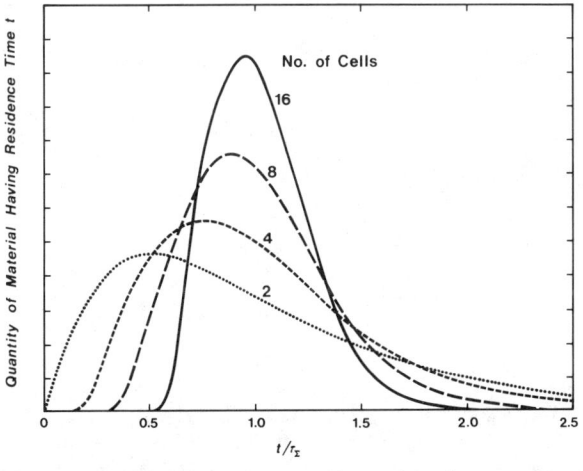

Figure 3.11. The distribution of residence times for a number of perfect mixers in series, when the total (average) residence time remains constant (impulse tracer). (After Roman.[10])

3.3.4. Practical Cases

In practice, plug flow is seldom encountered in processing devices, and even the most vigorously agitated devices seldom behave as perfect mixers. Deviations from Eqs. 3.9 and 3.10 can be explained in terms of:

1. Dead spaces, where the liquid is stagnant.
2. Bypass flow, where a portion of the liquid appears to bypass the mixer.
3. Recycle flow, where a portion of the existing material remixes with the feed stream.

Generally these complicating factors can be overcome by considering the flow in any device to be made up of suitable combinations of apparently perfect mixers in series and parallel. Some examples of practical residence time distributions are shown in Fig. 3.12. The results from a bank of six flotation cells in series shown in Fig. 3.12a can be represented by $4\frac{1}{2}$ perfect mixers in series[12] (cf. Fig. 3.10). Typical data from a ball mill appear in Fig. 3.12b.[13] The distribution shows a delay component followed by perfect mixing and can normally be represented by three perfect mixers in series, with the first two having relatively short residence times.

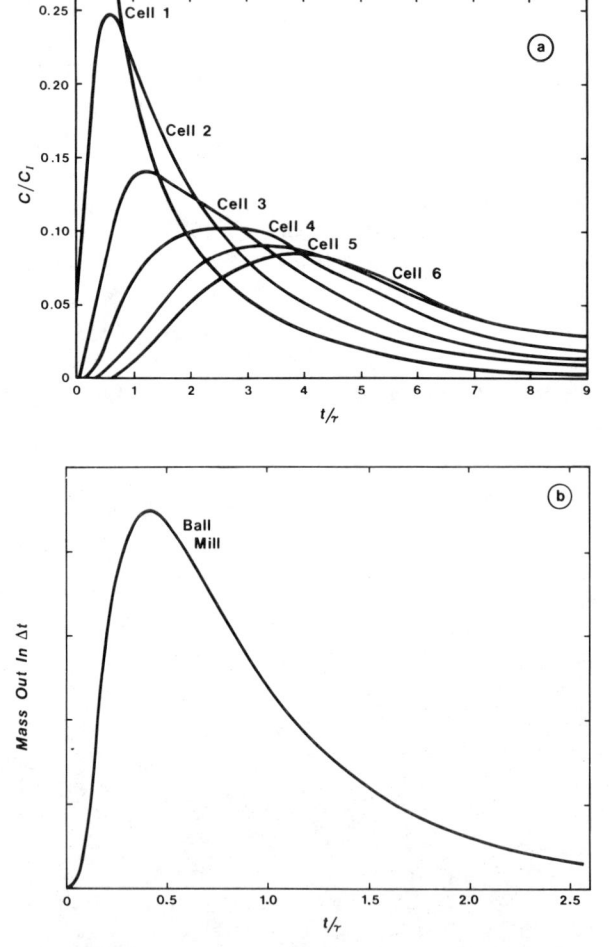

Figure 3.12. Examples of actual residence time distributions. (a) A bank of flotation cells. (After Bull and Spottiswood.[12]) (b) A ball mill. (After Kelsall et al.[13]) (c) A spiral classifier. (d) Arrangement of the mixers that simulate the data in (c). (After Stewart and Restarick.[14]) (a) and (b) should be compared with Figs. 3.10 and 3.9, respectively. (Figures 3.9 and 3.12b have the same vertical scale, and in both cases the area under the curve equals the mass of the impulse sample.)

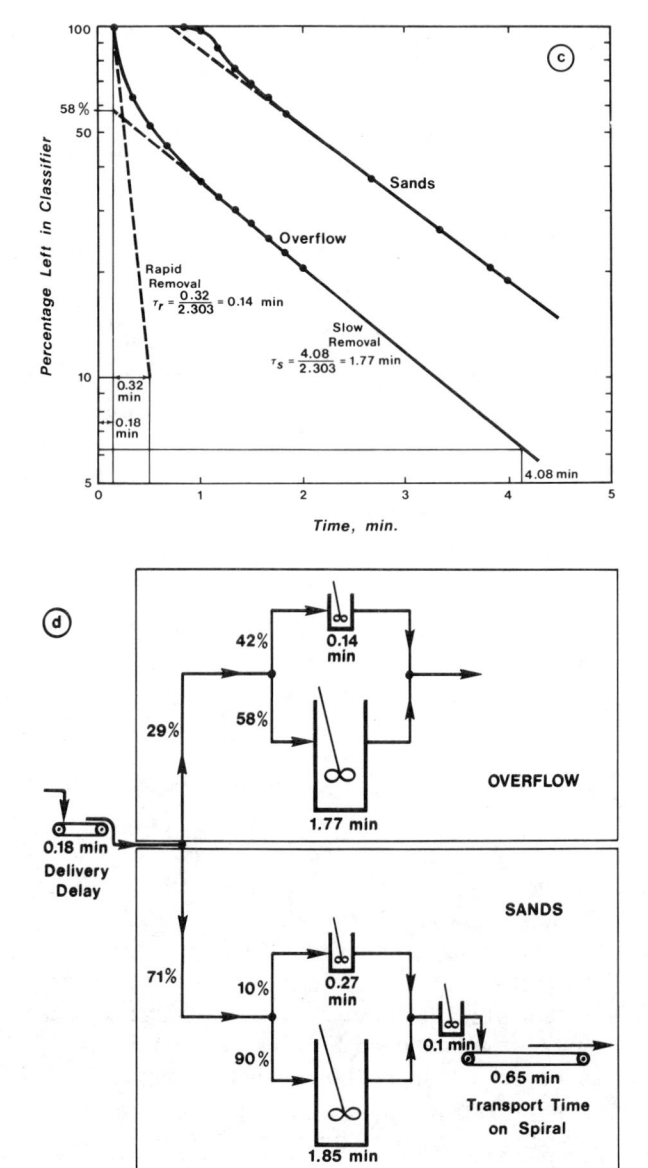

Determination of Separability Curves

The residence time distribution of solids in a spiral classifier is more complex[14] (Fig. 3.12c), but a typical representation is shown in Fig. 3.12d.

3.4. DETERMINATION OF SEPARABILITY CURVES

Unfortunately, separability curves generally cannot be determined rapidly, and an appreciable amount of chemical analysis may be necessary. The data are obtained with laboratory separators having separator settings that can be varied widely up to relatively high levels. A batch separator is preferable, but where this is impractical continuous devices must be used with a very low throughput to eliminate kinetic effects.

3.4.1. Size Separations

Size separations are the most straightforward separability curves to determine, and they may be obtained using any of the laboratory sizing methods described in Chapter 2. It should be remembered that when more than one sizing technique is used, it may be necessary to make corrections if each method measures a different nominal diameter.

3.4.2. Specific Gravity Separations

Specific gravity-based separability curves are normally determined by successive sink-float fractioning of a sample with liquids of differing specific gravities. Some of the liquids that can be used, together with their specific gravities, are listed in Table 3.1. Intermediate values can be obtained by mixing. Standard laboratory equipment such as beakers, test tubes, or separatory flasks are adequate for carrying out separations, although centrifuging may sometimes be necessary to speed up the separation of fine particles. A thorough description of the laboratory methods and problems involved has been presented elsewhere.[1,15,16] Once separated, the individual fractions may be analyzed for the materials of interest, and separability curves prepared.

3.4.3. Magnetic Separation

A number of devices are available for obtaining magnetic separability data and the use of some has been described.[8,17–19] More care is necessary when interpreting the data from these tests, since the results depend on the experimental conditions. This dependence results because all magnetic separations are affected by secondary forces, which means that only *relative* rather than absolute property values are obtained.

3.4.4. Flotation Separation

In the case of flotation separation, the exploited property can be considered to be floatability, although in reality this is virtually impossible to define, let alone measure. Consequently, an indirect measure is necessary. The most widely used one is reagent addition, but it must be appreciated that reagents are themselves very complex sets of variables. The actual test method is comparatively straightforward and is normally conducted in a batch laboratory flotation machine.[20] After floating a number of samples with different reagent levels, the floated and residual products are analyzed for the materials of interest. Figure 3.6 illustrates the type of data that can be obtained. The real difficulty is the vast number of reagents and reagent combinations that have to be considered.

Release analysis has been described as an alternative method for considering flotation separability.[21] In reality, like grade/recovery curves (Section 3.5.4), the analysis combines individual separability curves into one curve independent of property values. The method involves a series of straightforward batch flotation tests that result in four concentrates and a tailings. After assaying these products, the data are treated as illustrated in Example 3.2. This graphic presentation has the advantage that a vector between any two points on the curve

TABLE 3.1: DENSE LIQUIDS

SG	Compound	Formula
5.0	Clerici Solution (aqueous solution of thallium malonate and formate)	$(TlCOOH)_2C/TlCOOH$
3.31	Methylene iodide	CH_2I_2
2.95	Acetylene tetrabromide	$CHBr_2 CHBr_2$
2.89	Bromoform	$CHBr_3$
2.75	Tribromo-flueor-methane	CBr_3F
2.48	Methylene bromide	CH_2Br_2
1.92	Methylene chlorobromide	CH_2BrCl
1.67	Pentachloroethane	$CCl_3\text{-}CHCl_2$
1.61	Perchloroethylene	$CCl_2\text{-}CCl_2$
1.59	Carbon tetrachloride	CCl_4
1.46	Trichloroethylene	$CCl_2\text{-}CHCl$
1.33	Methyl chloroform	$CCl_3\text{-}CH_3$
1.25	Ethylene dichloride	$CH_2Cl\text{-}CH_2Cl$

TABLE E3.2.1: RELEASE ANALYSIS EXAMPLE

	Test Data		Calculated Data				
Sample	Mass	Assay	Mass	Recovery of PbS		Mass of Concentrate per 100 Units of PbS	
	(gms)	% PbS	PbS	%	Cum. %	%	Cum. %
Calculations:	(2)	(3)	$(4) = \frac{(2) \times (3)}{100}$	$(5) = \frac{(4) \times 100}{\Sigma(4)}$	$(6) = \Sigma(5)$	$(7) = \frac{(2) \times 100}{\Sigma(4)}$	$(8) = \Sigma(7)$
Concentrate 1	83.28	95.98	79.93	64.99	64.99	67.71	67.71
Concentrate 2	44.33	49.92	22.13	17.99	82.98	36.04	103.75
Concentrate 3	60.99	18.15	11.07	9.00	91.98	49.58	153.75
Concentrate 4	106.60	4.62	4.92	4.00	95.98	86.67	240.00
Tailing	934.80	0.53	4.95	4.02	100.00	766.00	1000.00
Total	1230.00	10.00	123.00	100.00		1000.00	

represents a possible concentrate whose mass is proportional to the horizontal length, and whose grade is given by the slope. A practical application of this technique is described in Chapter 23.

Example 3.2. A release analysis flotation test was carried out on a sample of the ore described in Example 2.4. The resulting data are presented in Table E3.2.1, in columns 2 and 3. Prepare the release analysis plot.

Solution. The necessary calculations are indicated under the headings in Table E3.2.1, and the results are shown in the relevant columns. The release curve, plotted in Fig. E3.2.1, uses the data in columns 6 and 8. The data are discussed further in Chapter 23 (Section 23.3.2).

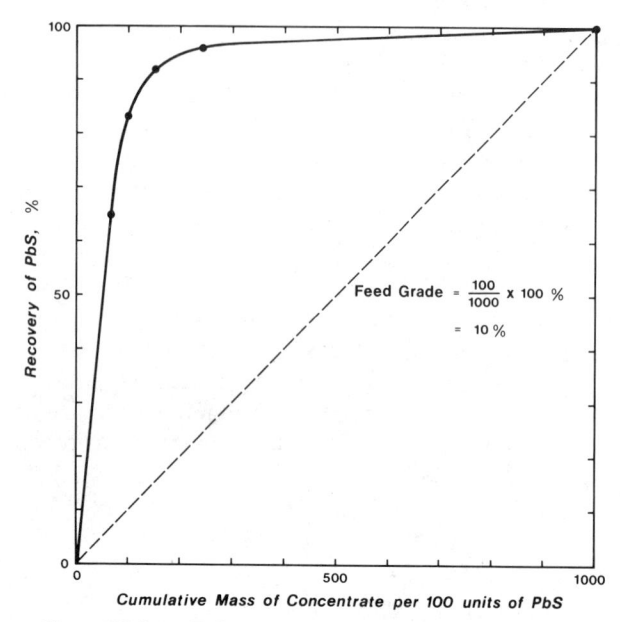

Figure E3.2.1. Release analysis curve for Example 3.2.

3.5. SEPARATION EFFICIENCY

The concept of efficiency in mineral dressing operations must be interpreted with considerable care. Although qualitative use of the term "efficiency" is valid and useful, difficulties arise in expressing it quantitatively because two conflicting criteria, grade and recovery, are normally involved in its evaluation. It is undesirable to express efficiency as a single value, even though this is frequently done in practice, since there are strictly speaking an infinite number of combinations of grade and recovery that can give that one value. Although in a few situations this may not matter, it is strongly recommended that the graphic approaches described below be used to *analyze* "efficiency." Inefficiencies in processes can

Separation Efficiency

be attributed to three broad sources:

Mineral limitations as represented by the separability curves (Section 3.1.1).

Equipment limitations, which prevent the maximum separation indicated by the separability curves [represented by performance curves (Section 3.5.2)].

Human failure to minimize these inefficiencies.

It is essential to know which of the three is causing poor separation, and here the graphic methods are of some help.

3.5.1. Separability Curves

Separability curves (Section 3.1.1) indicate the inefficiencies in a separation that are due to the mineralogical characteristics. However, it must always be remembered that it may be possible to change the shape of the curves. The importance of liberation in this regard has already been emphasized; other factors, such as the use of activators and depressants in flotation, are discussed in subsequent chapters. It is sufficient to point out at this stage that separability (and release analysis) curves are valuable in assessing the worth of different methods of altering the distribution of particle properties.

Frequently, only the curve of the mineral to be removed in the positive response outlet is graphed, it being assumed that maximizing the slope of the curve will lead to the best separation. In many cases this approach may provide sufficient information, but there is always the risk of failing to detect the increasing significance of such secondary properties as particle size.

3.5.2. Performance Curves

The difference between the two curves in Fig. 3.8b provides a method of assessing the effectiveness and efficiency of the separator. In this figure all particles with low property values leave through the positive response outlet. However, as higher property values are considered, increasing fractions are misplaced to the negative response outlet, until at $\mathfrak{P}^*$ all the material is misplaced. If this information is replotted as Q/Q_t for all values of $\mathfrak{P}$, the result is a curve of the form shown in Fig. 3.8a, c, e. Such a curve has variously been called the Tromp, partition, distribution, efficiency, and performance curve.[22] Since the shape depends on the separator rather than the minerals, the title *performance curve* is preferable and is used here.

The value of $\mathfrak{P}$ where particles are split equally between the two outlets is called the cut or *separation point* $\mathfrak{P}_{50}$ (not to be confused with the separator setting, which is higher). If the curve is *normalized* (made dimensionless) by dividing all values of $\mathfrak{P}$ by $\mathfrak{P}_{50}$, the result is the *reduced performance curve*. Clearly, a separator with a steeper reduced performance curve is more efficient and will give a sharper separation on an ore that has separability curves close together. Thus, just as the separability curves show how the mineral properties prevent perfect separation, so the reduced performance curve shows how the separator behavior *further* limits the separation. Note that even if the performance curve could be vertical (i.e., the separator is 100% efficient) the mineral characteristics (separability curves such as A and B in Fig. 3.4) can still prevent a perfect separation.

So far in this section and in Section 3.2 we have implied that given sufficiently high rate constants, or long residence times, a vertical performance curve is possible. In practice, this is not the case. Rather, most separators have a *limiting reduced performance curve*, and although various interpretations can be used, the most convenient approach is to attribute this phenomenon to disturbances that effectively cause some remixing of the separated streams. Put another way, if a separator is fed perfectly separated material, the products would not emerge perfectly separated, but would show contamination. Analogous with interdiffusion theory,[23] it has been shown that separation dynamics can be described by an equation of the form[24a]

$$\frac{dQ}{dt} = \mathscr{D}\frac{d^2Q}{dH^2} - \frac{g}{\mu^*}\frac{d[Q(\mathfrak{P} - \overline{\mathfrak{P}})]}{dH} \quad (3.11)$$

($\mathscr{D}$ = diffusion coefficient; Q = quantity of material with value of property $\mathfrak{P}$ [i.e., essentially the frequency separability data given by the solid line in Fig. 3.3] at a given position H in a stream at time t; $\overline{\mathfrak{P}}$ = average value of the property at position H and time t; μ^* = coefficient of viscous friction). The first term on the right-hand side of Eq. 3.11 can be considered to be the dispersive or mixing effect, with the second representing the separating effect (note that this may be either positive or negative, depending on whether the driving force for separation [$\mathfrak{P} - \overline{\mathfrak{P}}$] is positive or negative). The time constant τ_d in

Eq. 3.11 is of the form

$$\tau_d = \mathscr{D} \left[\frac{\mu^*}{g(\mathscr{P}_{max} - \mathscr{P}_{min})} \right]^2 \quad (3.12)$$

Provided the residence time in the separator is three to four times τ_d, dQ/dt is approximately zero. Under these conditions, the dispersive effect can be considered to balance the separating effect, and Eq. 3.11 can be solved to give a performance curve[24b]

$$\mathscr{R} = \tfrac{1}{2} - \tfrac{1}{2} \operatorname{erf} \left\{ \left[\frac{g\delta(m_{\mathscr{P}l}/\Delta\mathscr{P})}{\mathscr{D}\mu^*} \right]^{1/2} (\mathscr{P} - \mathscr{P}_{50}) \right\} \quad (3.13)$$

(erf = error function; $m_{\mathscr{P}l}$ = mass fraction of material in feed with value of property between $\mathscr{P}$ and $[\mathscr{P} + \Delta\mathscr{P}]$; $\mathscr{R}$ = recovery; δ = thickness of process bed or stream).

At this point, it is worthwhile making some general comments about Eq. 3.13:

1. The variables can be provisionally divided into those related to the mineral properties and those related to the separator operation ($\mathscr{D}$, μ^*).

2. The product $\mathscr{D}\mu^*$ represents the degree of random motion in the separating zone, and is therefore a significant separator parameter. Some separators have adjustments that alter $\mathscr{D}$ and μ^* (e.g., gravity concentrators, Chapter 13), and it follows that to obtain a given sharpness of separation, the separator adjustments will have to be set to suit the feed material. Equally, there may be a variety of combinations of adjustments that can produce the same separation.

3. Clearly, minimizing the product $\mathscr{D}\mu^*$ gives the sharpest possible separation. This tends to substantiate what is found in practice (e.g., with classifiers and gravity concentrators in particular): that some types of separator have a *limiting reduced performance curve*.[25,26]

4. The full significance of mineral properties is difficult to evaluate from Eq. 3.13, since they also affect $\mathscr{D}$ and μ^* in an unpredictable manner. Particle size in particular affects $\mathscr{D}$ and μ^*, and it is well established that separators can have different limiting performance curves for each particle size. When treating a feed with a range of particle sizes, the curves must be combined in appropriate proportions to obtain the net curve.[25] Other property effects are less well established. Hydrocyclones (on which most data are available) are generally accepted as having the curve shown in Fig. 10.9a for example, but some data suggest that the curve is not the same for all minerals (e.g., coal[26]). With the exception of particle size, it appears however that the mineral effects are less significant than the separator effects.

5. It must be remembered that Eq. 3.13 assumes that the separating tendency has reached a balance with the dispersive tendency. If the residence time is too short, for example, when the separator is overloaded, the separation will not be as sharp.

Performance curves provide one of the best measures of separation efficiency. If the separation achieved in practice is less than that given by the separator's limiting performance curve, the separator is not performing as well as it should. This implies inefficiencies that can often be attributed to failure of some part of the equipment, or to human factors such as poor settings or overloading. Sometimes, the inefficiency in the separation is restricted to a small section of the performance curve, but this can often be an aid in locating the source of the problem; such information cannot be obtained from a single value of efficiency, as is frequently used.

In spite of their limitations, single value efficiency terms are still widely used to quantify performance curves (and separations), particularly gravity separations. Two of these, the *Tromp area* and the *probable error* are illustrated in Fig. 3.8a. However, values are normally derived from performance curves, rather than reduced performance curves, and as such are virtually meaningless because their magnitude is a function of $\mathscr{P}_{50}$. Reduced values of Tromp area and probable error are more meaningful, but until recently this difference has not been widely appreciated in the literature.

It has been shown[27] that some reduced performance curves can be approximated by the Weibull (or Rosin-Rammler) equation (Eq. 2.1b in Table 2.3), in which case the exponent is a measure of the sharpness of separation. In this situation, the simplification is somewhat justifiable, in that it provides a convenient mathematical description of the separation, suitable for simulation studies.

Performance curves can be determined by methods similar to those used for separability curves. In this case, the analysis for a particular separator has to be carried out on the products from the positive and negative response outlets.

Separation Efficiency

3.5.3. Prediction of Separation

In addition to their usefulness in assessing efficiency, separability and performance curves are useful in predicting separations. When applied to a separability curve, the performance curve gives a better representation of the likely separation than does the older "±10% curve" method.[1] Release analysis curves have also been shown to have considerable worth in deciding circuit arrangements.[28] These are considered in more detail in Chapter 23.

3.5.4. Cumulative Grade Versus Cumulative Recovery Curves

Another useful representation for considering the efficiency of an operation or process is cumulative grade versus cumulative recovery curve (often called a *grade/recovery curve*). In effect it is an alternative method of presenting separability data, which, although it no longer shows the property values, gives more direct information about the product. A typical grade/recovery curve is illustrated in Fig. 3.13. Ideally at low recoveries the grade should be that of the pure mineral being separated; this eventually falls to the feed grade as the recovery rises to 100%. If the efficiency is expressed as a single value, it will occur on this diagram as a series of isobars. For example, a simple, typical definition of efficiency is:

efficiency = (% recovery of valuable mineral
 in the concentrate)
 × (mass fraction of valuable mineral
 in the concentrate) %   (3.14)

Some values of the efficiency η plotted in Fig. 3.13 and it can be seen that in general any efficiency value occurs twice on a given grade/recovery curve. It may be argued that the best conditions occur at the maximum efficiency that just touches the grade/recovery curve (52% on Fig. 3.13), but this is not necessarily true, since the grade (or recovery) represented by this value may be unacceptable commercially.

The grade/recovery curve can be established in the plant or laboratory by widely varying the separator setting; in the case of the A-B separation in Fig. 3.4, from $\mathcal{P}_1^*$ to $\mathcal{P}_2^*$. Experimental error and random fluctuations make accurate determination of the curve difficult, and in particular data should be collected with constant feed composition and rate. However, once the curve has been established, any

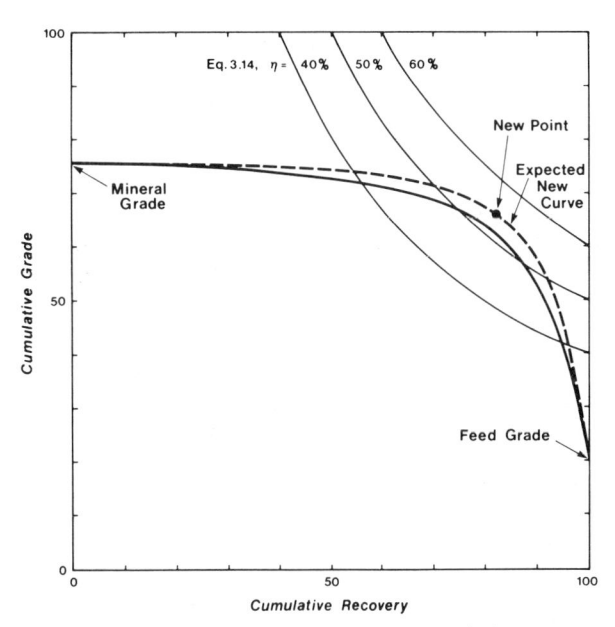

Figure 3.13. The cumulative grade versus cumulative recovery curve.

action (e.g., more grinding to increase liberation) that conclusively generates a point outside the grade/recovery curve can be said to have increased the separation efficiency. This follows because subsequent variations in the separator setting should generate a new grade/recovery curve through this point as illustrated in Fig. 3.13. It is worth noting that the new point could conceivably have caused no change in the efficiency as defined by Eq. 3.14.

Having established the best possible grade/recovery curve, the problem then becomes one of finding the most suitable operating point on that curve. If the concentrate is being sold to an outside smelter, the smelter schedule (Chapter 1) provides a useful method. (If this concentrate is sent to a company smelter, it should still be possible to develop a suitable smelter schedule.) By operating the smelter schedule on the grade/recovery curve, it is possible to determine the maximum profit obtainable per unit mass of ore, as illustrated in the following example.

Example 3.3. Using the release analysis grade/recovery relationship shown in Fig. 23.4 (from Example 3.2), find the optimum economic operating point when lead is worth $1.10/kg and silver is worth $1.23/g. What is the effect of a fall in the silver price to 19¢/g?

Solution. The calculation procedure is essentially that used in Example 1.1. To convert the re-

TABLE E3.3.1: ECONOMIC RETURN EXAMPLE

		Return/tonne Concentrate, $						Return/t Ore, $	
Grade %	Recovery %	Smelting (Cost)	Pb	Ag ($1.23/gm)	Ag (19c/gm)	Net ($1.23 Ag)	Net (19c Ag)	($1.23 Ag)	(19c Ag)
10	100	69.07	65.08	198.40	30.44	194.42	26.46	194.42	26.46
19	99	65.64	135.93	408.48	62.68	478.77	132.46	249.46	69.28
26	98	64.80	191.04	571.87	87.75	698.10	213.98	263.13	80.65
40	96	61.88	301.24	898.65	137.89	1138.01	377.25	273.13	90.54
46	95	61.88	348.47	1038.70	159.38	1325.29	445.97	273.70	92.10
51	94	61.88	387.83	1155.40	177.28	1481.36	503.24	273.03	92.75
60	92	61.88	458.68	1365.48	209.52	1762.28	606.32	270.22	92.97
65	90	61.88	498.04	1482.19	227.43	1918.35	663.59	265.62	91.88
83	80	61.88	639.74	1902.33	291.89	2480.19	869.75	239.05	83.83
93	71	61.88	718.46	2135.75	327.71	2792.32	984.28	213.18	75.14
99	42	61.88	765.69	2275.80	349.20	2979.60	1053.00	126.41	44.67

turn to tonnes of ore instead of tonnes of concentrate, a PbS mass balance can be used thus:

Basis. 1 tonne ore

mass PbS in ore =
mass PbS in concentrate

1 tonne × grade × recovery = mass × grade

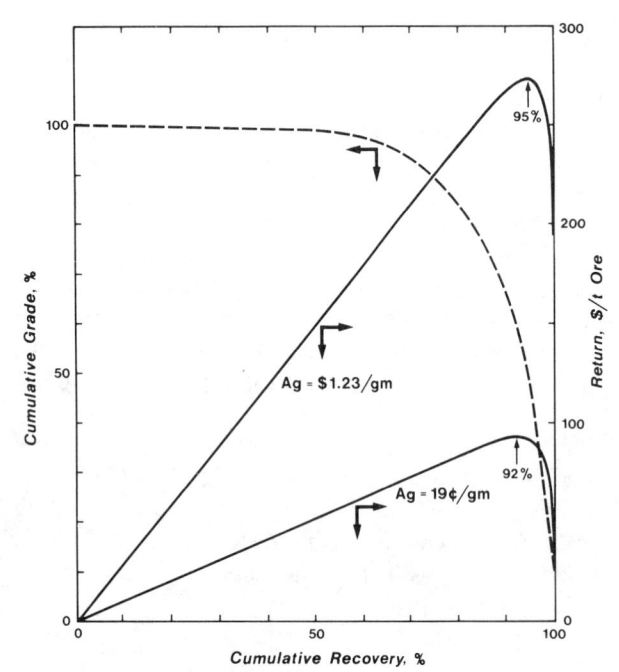

Figure E3.3.1. Graphed data of Example 3.3.

For example, for Example 1.1

concentrate grade = 80%
feed grade = 10%
recovery = 83% (Fig. 23.4)

∴ return/tonne ore = $2386.56 $\frac{10\%}{100} \left| \frac{83\%}{100} \right| \frac{100}{80\%}$

= $247.61/tonne

The resultant data are tabulated in Table E3.3.1, and graphed in Fig. E3.3.1.

Thus, when silver is worth 19¢/g, the optimum point is at 92% recovery, and when silver is worth $1.23/g, the optimum point is at 95% recovery.

Although the variation in the price of silver may appear large, it is representative of the extremes that occurred during 1979-1980. As the silver price rises, the original lead ore becomes a silver ore, and recovery becomes increasingly important. In turn, as the recovery rises slightly, a comparatively large decrease in grade occurs. In general even small shifts in prices may justify changing operating levels, assuming of course that the process can be sufficiently well controlled.

Weiss[29] has suggested that operating costs can be ignored, as was done in this example. This, however, is debatable. An interesting analysis by Steane showed that although higher throughputs may produce significantly larger tailings losses (by lowering the grade/recovery curve), the plant may still give a higher total profit.[30] Clearly ethics then becomes yet another factor in efficiency. It must also be emphasized at this stage that mineral processing is only one stage in the production of a metal, and maximum efficiency in the *total* operation is generally more important than maximum efficiency in the individual stages (cf. Section 25.5).

REFERENCES

1. G. C. Coe, "An Explanation of Washability Curves for the Interpretation of Float-and-Sink Data on Coal," U.S. Bureau of Mines Information Circular IC 7045 (1938).
2. A. P. Prosser, "Analysis of Mineral Separation Processes," *Chem. Proc. Eng.*, **49**, 86–88, 96 (1969).
3. L. Lidström, "Amine Flotation of Ore Minerals and Silicates," *Acta Polytech. Scand., Chem. Metall. Ser., No. 66 (1967).*
4. F. W. Mayer, "Fundamentals of a Potential Theory of the Jigging Process," *Proc. 7th Int. Miner. Process. Congr., 1964*, pp. 75–86, Gordon and Breach (1965).
5. D. F. Kelsall, "Application of Probability in the Assessment of Flotation Systems," *Trans. IMM*, **70**, 191–204 (1961).
6. F. Fraas, "Electrostatic Separation of Granular Materials," U.S. Bureau of Mines Bulletin 603 (1962).
7. P. Harriott, *Process Control*, McGraw-Hill (1964).
8. N. N. Vinogradov et al., "Research on the Separation Kinetics of Gravity Processing in Mineral Suspensions," *Proc. 11th Int. Miner. Process. Congr., Cagliari, 1975*, pp. 319–336, Università di Cagliari (1975).
9. P. V. Danckwerts, "Continuous Flow Systems," *Chem. Eng. Sci.*, **2**, 1–13 (1953).
10. R. J. Roman, "Large Flotation Cells—Selection of the Proper Size and Number," *Min. Congr. J.*, **56**, 56–59 (June 1970).
11. A. Jowett and D. N. Sutherland, "A Simulation Study of the Effect of Cell Size on Flotation Plant Costs," *The Chem. Engr.*, No. 347/8, 603–607 (1979).
12. W. R. Bull and D. J. Spottiswood, "A Study of Mixing Patterns in a Bank of Flotation Cells," *Colorado School Mines Q.*, **69**, Pt. I, 1–26 (1974).
13. D. F. Kelsall, K. J. Reid, and P. S. B. Stewart, "The Study of Grinding Processes by Dynamic Modelling," Paper 2509 presented at the IPAC Symp., Sydney, August 1968; also published in *Elec. Eng. Trans. Inst. Eng. Aust.*, EE5, No. 1, 173–186 (1969).
14. P. S. B. Stewart and C. J. Restarick, "Dynamic Flow Characteristics of a Small Spiral Classifier," *Trans. IMM (C)*, **76**, C225–C230 (1967).
15. J. S. Browning, "Heavy Liquids and Procedures for Laboratory Separation of Minerals," U.S. Bureau of Mines Information Circular IC 8007 (1961).
16. J. D. Bignell, "Prediction and Assessment of Gravity Separator Performance from Heavy Liquid Data," Paper 51, 11th Commonw. Min. and Metall. Congr., Hong Kong (1978).
17. N. F. Schulz, "Determination of the Magnetic Separation Characteristics with the Davis Magnetic Tube," *Trans. AIME/SME*, **229**, 211–216 (June 1964).
18. G. Panou, "Use of Washability Curves in Magnetic Separation," *Proc. 7th Int. Miner. Process. Congr., New York, 1964*, pp. 375–389, Gordon and Breach (1965).
19. E. Occella, "Extension of Beneficiation Curves to the Magnetic Separation of Minerals," *4th Int. Miner. Dress. Congr., Stockholm, 1957*, pp. 407–414, Almqvist & Wiksells (1958).
20. *Flotation Fundamentals and Mining Chemicals*, Dow Chemical Co. (1968).
21. C. C. Dell, "The Analysis of Flotation Test Data," *Colorado School of Mines Q.*, **56**, 115–127 (1961).
22. "New Methods of Computing the Washability of Coals," *Colliery Guardian*, 154, 955–959, 1009 (1937).
23. G. H. Geiger and D. R. Poirier, *Transport Phenomena in Metallurgy*, Ch. 13, Addison-Wesley (1973).
24. (a) O. N. Tikhonov and V. V. Dembovsky, "Automatic Process Control in Ore Treatment and Metallurgy. Part I. Process Dynamics," El-Tabbin Metallurgical Institute for Higher Studies (1973).
    (b) O. N. Tikhonov and V. V. Dembovsky, "Automatic Process Control in Ore Treatment and Metallurgy. Part IV. Industrial Measuring and Control Systems," El-Tabbin Metallurgical Institute for Higher Studies (1973).
25. B. S. Gottfried, "A Generalisation of Distribution Data for Characterizing the Performance of Float-Sink Coal Cleaning Devices," *Int. J. Miner. Process.*, 5, 1–20 (1978).
26. A. J. Lynch and T. C. Rao, "Modeling and Scale-Up of Hydrocyclone Classifiers," *Proc. 11th Int. Miner. Process. Congr., Cagliari, 1975*, pp. 245–269, Università di Cagliari (1975).
27. C. C. Harris, "Graphical Representation of Classifier-Corrected Performance Curves," *Trans. IMM (C)*, **81**, C243–C245 (1972).
28. R. H. Lamb and L. R. Verney, "Investigation into RST Group Concentrator Practices," *Trans. IMM (C)*, **76**, C154–C168 (1967).
29. N. Weiss, "Flotation Economics," Ch. 22 in *Flotation*, D. W. Fuerstenau (Ed.), AIME (1962).
30. H. A. Steane, "Coarser Grind May Mean Lower Metal Recovery but High Profits," *Can. Min. J.*, **97**, 44–47 (May 1976).

Chapter Four

Fluid Dynamics

> *Gardner's Rule:* "*The society which scorns excellence in plumbing because plumbing is a humble activity and tolerates shoddiness in philosophy because it is an exalted activity will have neither good plumbing nor good philosophy. Neither its pipes nor its theories will hold water.*"

Although mineral processing is primarily concerned with solids, interactions between solids and fluids (liquid or gas) are an integral part of any operation. The most important fluid is water, and concentrating devices may require considerable quantities for their successful operation; in some cases water up to 10 times the weight of solid may be needed, although recycling may reduce the net ratio. In wet concentrating plants, water has an important secondary role in that it provides convenient transportation of the solid. By comparison, the interaction between air and solid in dry separations is generally less marked, although there are significant exceptions.

4.1. CHARACTERIZATION OF FLOW BY REYNOLDS NUMBER

Several factors affect the manner in which a fluid flows,[1] particularly the presence of a solid surface, be it a wall, body, or particle. When a fluid is flowing past a solid surface it adheres to the solid at the solid-liquid interface. Well away from the solid, the fluid may be flowing unaffected by the surface. In between these possible extremes there is a transition where the fluid must be in a state of shear, and Prandtl proposed the existence of a boundary layer where such shear would exist.

Viscosity, or resistance to flow, is thus an additional factor in fluid flow, incorporating stress (force per unit area) between adjacent layers of fluid. This resistance to flow increases as the flow rate increases, so viscosity μ is defined as

$$\mu = \frac{F/A}{dv/dy} = \frac{\mathscr{S}}{dv/dy} \quad (4.1)$$

(F/A = force/area = $\mathscr{S}$ = stress; v = velocity in the flow direction; y = distance perpendicular to the flow direction). In practical terms, higher pressure is required to force water rather than air through a pipe, because water has higher viscosity.

When a flowing fluid encounters a surface parallel to it, a boundary layer develops as illustrated in Fig. 4.1. In the boundary layer, velocity increases from zero at the solid surface to the bulk fluid velocity at the boundary layer surface. Provided the flow rate is low enough, shear stress will change linearly across the boundary layer, which will thicken with distance from the leading edge. If the surface is the entrance to a pipe, the boundary layer will reach the pipe center as illustrated in Fig. 4.2 and a parabolic velocity profile will exist (Fig. 4.3). The relationship may be developed from Eq. 4.1 as

$$v_r = \frac{\mathscr{S}_w}{2R\mu}(R^2 - r^2) \quad (4.2)$$

(v_r = velocity in flow direction at radius r; $\mathscr{S}_w$ = wall stress; R = radius of pipe).

It follows that the maximum velocity is

$$v_{\max} = \frac{\mathscr{S}_w R}{2\mu} \quad (4.3)$$

Characterization of Flow by Reynolds Number

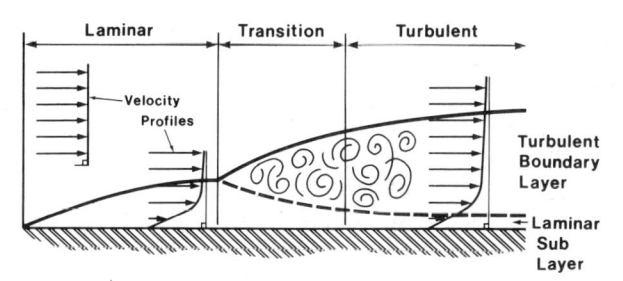

Figure 4.1. Boundary layer buildup on a flat surface (vertical scale greatly magnified).

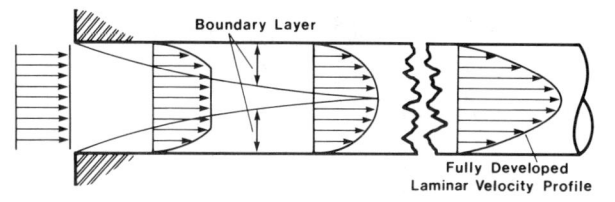

Figure 4.2. Development of a laminar velocity profile at the entrance of a pipe.

and the average velocity $\bar{v}$ may be obtained by integrating Eq. 4.2 over the radius, giving

$$\bar{v} = \frac{\mathscr{S}_w R}{4\mu} = 0.5\, v_{\max} \qquad (4.4)$$

Flow that has a velocity profile given by Eq. 4.2 is described as *laminar* or *viscous*. Should a disturbance occur in this type of flow, it will be damped out by the fluid viscosity and the closeness of the wall. Damping can be considered to be an absorption or transfer of momentum, and for laminar flow it can be characterized by $\mu\bar{v}/D$ (where $D = 2R$). If the flow rate becomes high enough, the laminar flow pattern is no longer maintained and the flow becomes *turbulent*. Under these conditions the movement of any element of fluid is chaotic and can be described only by averages. The transfer of momentum between neighboring elements of fluid is by eddying, and is therefore kinetic in nature. Consequently momentum transfer in this situation can be characterized by $\rho_f \bar{v}^2$ (ρ_f = fluid density). The ratio of momentum transfer by these two mechanisms can be used to characterize the change of flow from laminar to turbulent, and the resulting dimensionless ratio is called the *Reynolds number* **Re**

$$\mathbf{Re} = \frac{\text{momentum transfer by turbulence mechanism}}{\text{momentum transfer by viscous mechanism}} \qquad (4.5)$$

$$= \frac{\rho_f \bar{v}^2}{\mu \bar{v}/D}$$

$$= \frac{D \bar{v} \rho_f}{\mu}$$

Although originally developed for pipe flow, a Reynolds number can be similarly defined for any flowing system by using an appropriate characteristic size, velocity, density, and viscosity. Each system will have its own value of **Re**, which represents a transition from laminar to turbulent conditions.

The occurrence of turbulence in a boundary layer is also illustrated in Fig. 4.1. When the **Re** becomes high enough, laminar flow can no longer be maintained; the turbulent zone forms and increases in thickness. However, there is still a very thin viscous or pseudolaminar layer close to the surface, with a thin buffer layer separating it from the turbulent zone.

In a pipe the turbulent velocity profile is much blunter than that found with laminar flow (Fig. 4.3), and it can be approximated by the empirical relationship

$$\frac{v_r}{v_{\max}} = \left(\frac{R - r}{R}\right)^{1/7} \qquad (4.6)$$

(v_r in this case is the mean velocity with the turbulence fluctuations averaged out). From this it can also be shown that

$$\bar{v} = 0.817\, v_{\max} \qquad (4.7)$$

and $v_r = \bar{v}$ at

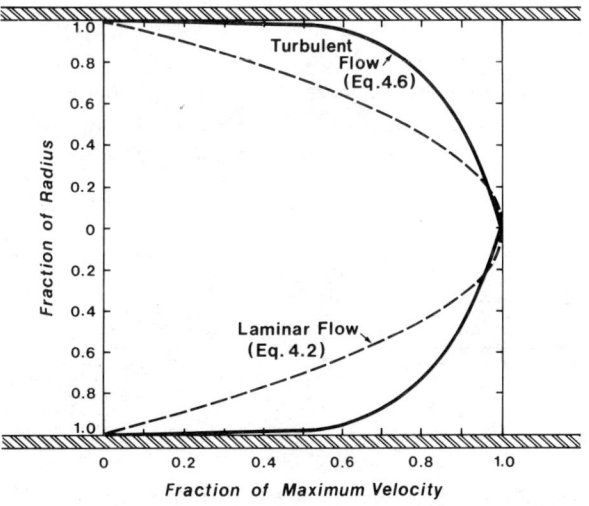

$$r = 0.76 R \qquad (4.8)$$

Figure 4.3. Laminar and turbulent velocity profiles in a pipe.

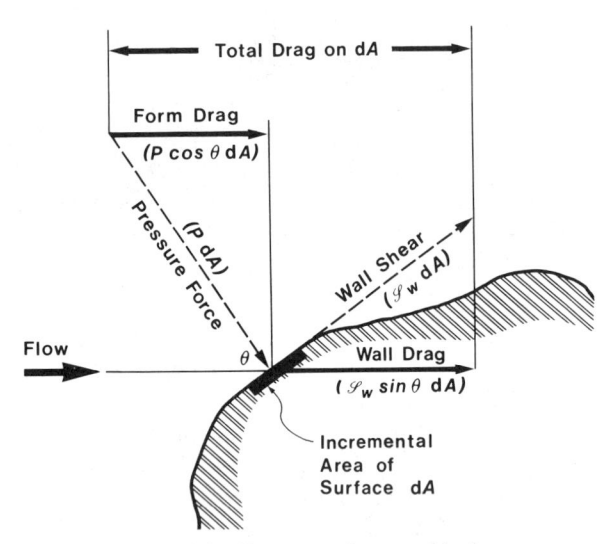

Figure 4.4. Drag on an immersed body.

For **Re** $> 10^5$ the exponent in Eq. 4.6 becomes even smaller. Despite its empirical nature, the equation shows that the velocity gradient is very steep near the wall, that is, in the thin viscous layer.

It is essential to appreciate that in the situations described so far, the shear stress of the fluid represents friction or resistance to flow, and, since it is transmitted to the wall, it is frequently referred to as *wall drag* or skin friction $\mathscr{S}_w$. When flow is turbulent, skin friction is high not only because there is a large amount of stress between the eddies but also because the shear rate is very high near the walls.

So far the discussion has applied only to the situation where the fluid is flowing parallel to the surface. Many situations are not this simple, and the fluid strikes the surface at an angle as illustrated in Fig. 4.4. This results in an additional force on the surface arising from liquid pressure; the component of this force in the flow direction is known as *form drag*. The total drag on the surface is then the sum of wall drag plus form drag in the flow direction.

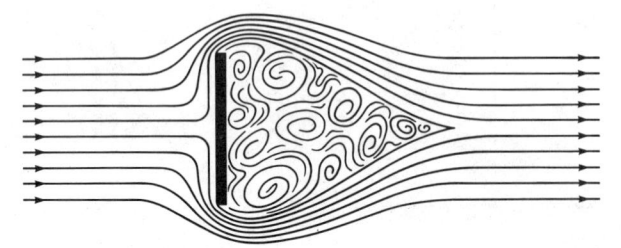

Figure 4.5. Wake formation behind an immersed body, in this case a flat plate normal to the flow direction.

If the body is of a finite size, the momentum of the fluid may prevent it from flowing around behind the body as in Fig. 4.5. The fluid then separates from the body, causing a backwater zone of strongly decelerated fluid in which large eddies, called vortices, are found. This zone is known as the *wake* and is characterized by a large friction loss. It also develops a large form drag; in fact most form drag occurs as the result of wakes.

4.2. FLUID RHEOLOGY

Deformation during flow of a fluid, or rheology, has already been introduced in Eq. 4.1 as the definition of viscosity. According to this equation, a plot of fluid shear stress versus shear rate (the flow curve) will be a straight line through the origin, the slope of the line being viscosity. Fluids that exhibit this behavior are said to be *Newtonian* and are characteristic in that the single parameter, viscosity, is sufficient to relate shear stress and shear rate.

In some cases, the flow curve may no longer be linear through the origin as illustrated in Fig. 4.6. Such behavior is then described as *non-Newtonian* and at least two parameters are necessary to correlate shear stress and shear rate. Non-Newtonian behavior may change with time and can therefore be subdivided into two categories: time independent and time dependent.

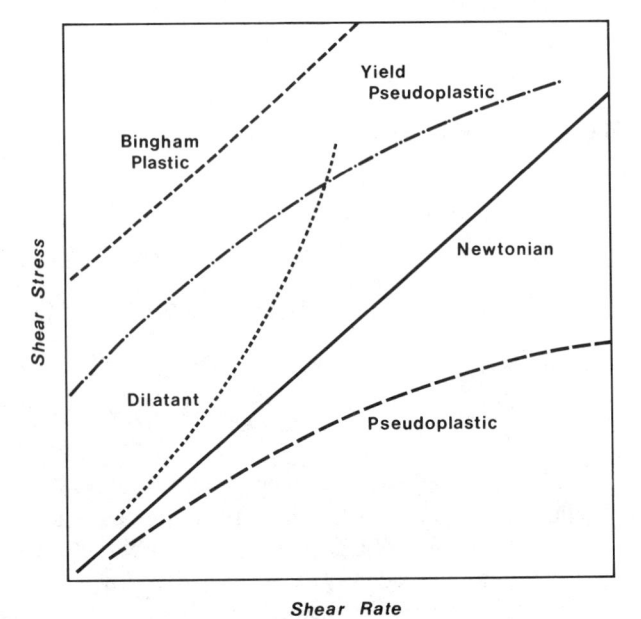

Figure 4.6. Rheological behavior of fluids.

Bingham plastic behavior (Fig. 4.6) is characterized by a straight line flow curve having a yield shear stress $\mathscr{S}_y$, which must be exceeded before flow can commence. Such a phenomenon arises when the fluid at rest has a three-dimensional structure sufficiently rigid to resist any stress less than $\mathscr{S}_y$. Once this stress is exceeded, the material behaves like a Newtonian fluid moving under shear stress $\mathscr{S} - \mathscr{S}_y$. This can therefore be described by an equation such as

$$\mathscr{S} - \mathscr{S}_y = \mu_B \frac{dv}{dy} \quad (4.9)$$

(μ_B = coefficient of rigidity).

A pseudoplastic fluid (Fig. 4.6) has a flow curve whose slope decreases with increasing rate of shear until a limiting slope is reached. Such behavior can be described by the "power law" equation

$$\mathscr{S} = \mu_P \left(\frac{dv}{dy}\right)^n \quad (4.10)$$

(μ_P = consistency; n = flow index). Consequently, when $n = 1.0$, μ_P is equal to the viscosity.

Dilatant fluids are rarely encountered in practice, but they can be described by Eq. 4.10 with $n > 1.0$.

Non-Newtonian behavior consequently introduces a difficulty in defining a suitable viscosity for analytical treatments. One approach is to define an *apparent viscosity* μ_a in a manner similar to Eq. 4.1, that is,

$$\mu_a = \frac{\mathscr{S}_w}{dv/dy} \quad (4.11)$$

The dependence of μ_a on shear rate can be seen by substituting $\mathscr{S}_w$ in Eq. 4.11 for $\mathscr{S}$ in Eqs. 4.9 and 4.10. Thus, for Bingham plastics

$$\mu_a = \frac{\mathscr{S}_y - \mu_B(dv/dy)}{dv/dy} \quad (4.12)$$

and for pseudoplastics

$$\mu_a = \mu_P \left(\frac{dv}{dy}\right)^{n-1} \quad (4.13)$$

Although viscosity is basically a property of the fluid, the addition of solid particles to a fluid normally results in the suspension having a viscosity greater than that of the fluid itself. Whether the suspension behavior is Newtonian or non-Newtonian can depend on the characteristics of the medium or the solids; solids concentration, as well as particle size and shape, are the principal factors affecting viscosity. Generally, suspensions of rounded particles over 50 μm tend to have Newtonian characteristics and the viscosity is a function of the volumetric solids concentration and can be corrected by an empirical equation.[2,3]

Slurries of irregularly shaped mineral particles tend to exhibit Bingham plastic non-Newtonian behavior at even larger particle sizes.[4,5] Heiskanen and Laapas[4] presented

$$\frac{\mu}{\mu_{H_2O}} = 1 + 2.5\mathfrak{F}_{V,sl} + 14.1\mathfrak{F}_{V,sl}^2 \\ + 0.00273 \exp(16.0\mathfrak{F}_{V,sl}) \quad (4.14)$$

($\mathfrak{F}_{V,sl}$ = volumetric fraction of solids in slurry), which is a modification of the equation for spheres.[2,3] However, their data also showed a very significant dependence on specific surface (i.e., particle size), which is not incorporated in Eq. 4.14. They also showed that the yield stress could be described by an equation of the form

$$\mathscr{S}_y = K_1 \exp(K_2 \rho_{sl}) \quad (4.15)$$

but again, the parameters (K_1, K_2) were dependent on specific surface. Thus, even though viscosity can be adequately described, experimental data are still essential.

It is also worthwhile noting at this point that viscosity is a parameter that is particularly susceptible to variations in temperature.

4.3. FRICTION FACTORS

Many flow problems in mineral processing fall into one of two broad categories: flow of bodies through fluids, and the flow of fluids in conduits. Examples of the former include classification and sedimentation, while examples of the latter are the flow of a liquid through a filter or the pumping of mineral slurries through a pipe.

Problems of the flow of bodies through fluids are generally concerned with the relationship between the drag force on the body and the relative solid-fluid velocity. When considering the problem of flow in conduits on the other hand, one is generally interested in the relationship between pressure drop along the conduit and the flow rate. Except for laminar flow and a few simple turbulent situations,[6] relationships cannot conveniently be predicted, and it is necessary to resort to constructed charts of correlations derived from experimental data. These charts are most conveniently constructed in terms of dimensionless quantities and because of its importance in characterizing flow, one of the quantities is logically **Re**. The forces of drag or pressure drop are considered in another dimensionless number that is called a *friction factor*. This can be defined generally as

$$\mathbf{f} = \frac{\mathscr{G}}{E} = \frac{F}{AE} \quad (4.16)$$

(F = force [drag force or pressure drop]; A = characteristic area [on which the force is considered to act]; E = characteristic energy per unit volume). Like **Re**, **f** represents a ratio of momentum transfers; in this case

$$\mathbf{f} = \frac{\text{total momentum transfer}}{\text{momentum transfer by turbulence}}$$

In general terms, **f** decreases as the turbulence increases (i.e., **Re** increases); but because any flow system always has some laminar boundary layer associated with it, **f** never reaches zero, but rather tends to a minimum value.

A number of friction factor charts are available, and some of those relevant to mineral processing are shown in Figs. 4.7, 4.8, 4.10, 4.14, and 4.18. These are discussed in more detail below, but it must always be remembered *that they represent data that have been determined under conditions that may not match the application being considered. Even so, they are still more accurate than some of the theoretical and empirical correlations presented in subsequent sections. Separate friction factor charts are necessary for non-Newtonian fluids.*[7]

4.4. FLOW IN PIPES

The flow of fluids in pipes is a special case of flow in a conduit. In such cases, E in Eq. 4.16 is taken as $\frac{1}{2}\rho_f \bar{v}^2$ (where $\bar{v}$ is the average fluid velocity) and A is taken as the wetted surface. Thus for a pipe of diameter D and length L, $A = \pi DL$. Generally F is measured in terms of the total pressure drop along the pipe length and it consists of two components: the pressure loss due to fluid friction (the stress of the fluid on the wall), and the difference in elevation, that is,

$$F = [(P_I - P_O) + \rho_f g(H_I - H_O)] \frac{\pi D^2}{4} \quad (4.17)$$
$$= \frac{\Delta P \pi D^2}{4}$$

(ΔP = pressure drop; P_I = inlet pressure; P_O = outlet pressure; H_I = inlet height; H_O = outlet height). Substituting in Eq. 4.16

$$\mathbf{f}_p = \frac{\Delta P \pi D^2}{4} \cdot \frac{1}{\pi DL} \cdot \frac{1}{\frac{1}{2}\rho_f \bar{v}^2}$$

$$\mathbf{f}_p = \frac{1}{2} \left(\frac{D}{L}\right) \frac{\Delta P}{\rho_f \bar{v}^2} \quad (4.18)$$

This pipe friction factor $\mathbf{f}_p$ is sometimes called the Fanning friction factor. Its definition varies from text to text, so extreme caution should be taken when selecting friction factor data.

A chart of $\mathbf{f}_p$ versus **Re**, where

$$\mathbf{Re} = \frac{D \bar{v} \rho_f}{\mu} \quad (4.19)$$

is shown in Fig. 4.7.[8] The linear section on the left of the graph indicates that laminar flow is stable up to **Re** = 2100.

Since it can be shown[1] for laminar flow in a pipe that

$$P = \frac{4 \mathscr{G}_w L}{D} \quad (4.20)$$

Eqs. 4.4 and 4.18 can be combined to give a relationship between $\mathbf{f}_p$ and **Re**:

$$\mathbf{f}_p = \frac{16}{\mathbf{Re}} \quad (4.21)$$

At values of **Re** > 2100, the flow is usually unstable and turbulent and the transition between the two types of flow is unpredictable, depending significantly on disturbances. This section of the friction factor chart has been established by experimental data.

By assuming that the turbulent stress is transmitted to the wall across the viscous layer adjacent to the wall, it is possible to use the turbulent velocity profile in a smooth pipe (Eq. 4.6) to develop a relationship between $\mathbf{f}_p$ and **Re** in the turbulent region[1]:

$$\frac{1}{\sqrt{\mathbf{f}_p}} = 4.06 \log(\mathbf{Re} \sqrt{\mathbf{f}_p}) - 0.6 \quad (4.22)$$

When $2100 < \mathbf{Re} < 10^5$, a simpler empirical form is

$$\mathbf{f}_p = \frac{0.079}{\mathbf{Re}^{0.25}} \quad (4.23)$$

If the pipe has a rough surface, a higher pressure drop is required to produce a given flow in the turbulent regime. Such roughness can be treated by a relative roughness parameter K_R/D as shown in Fig. 4.7. It can be seen that roughness causes $\mathbf{f}_p$ to reach

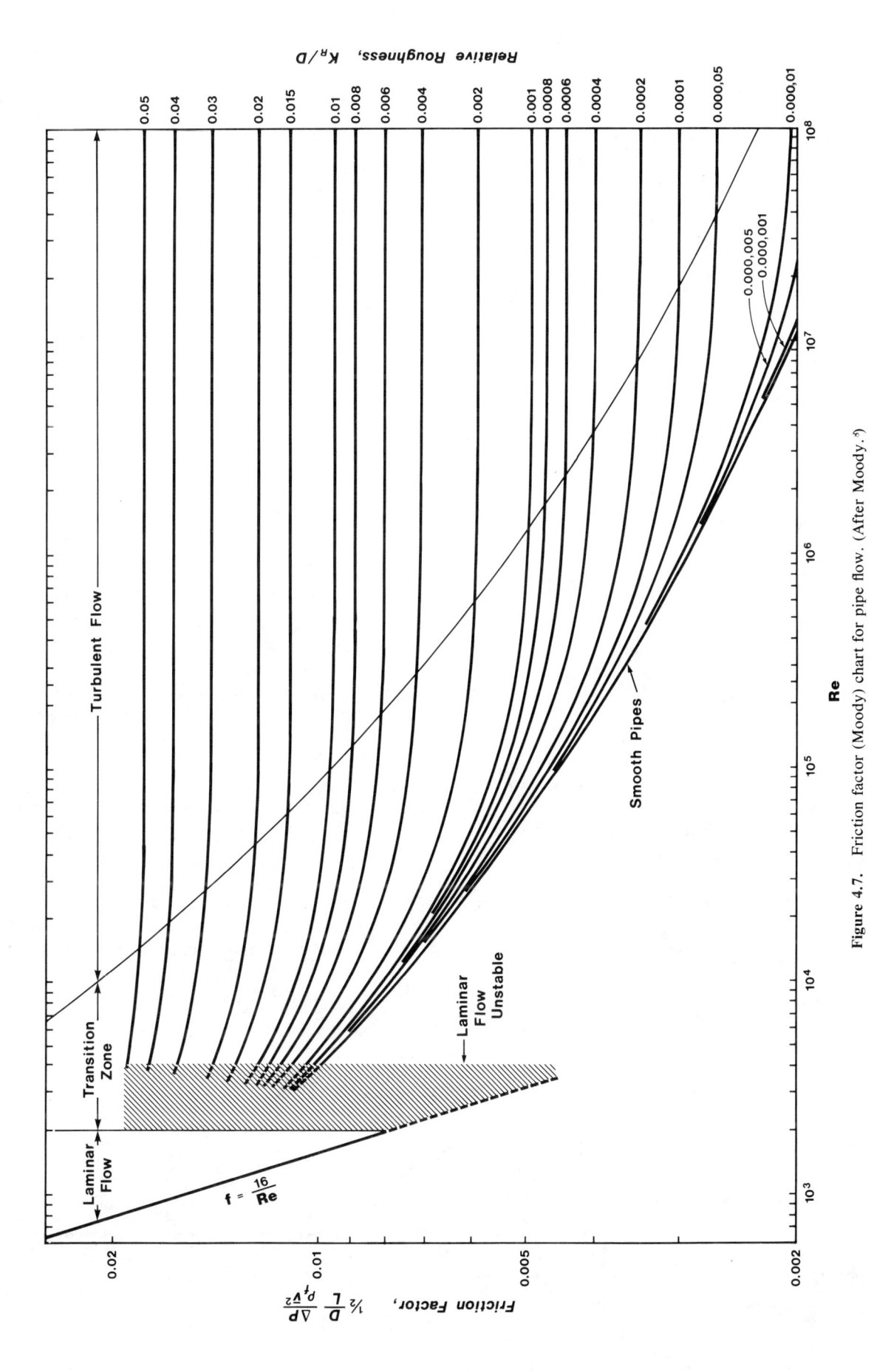

Figure 4.7. Friction factor (Moody) chart for pipe flow. (After Moody.[3])

a higher constant value sooner. Values of the parameter have been published for different surfaces.[1]

For noncircular pipes with turbulent flow, the mean hydraulic radius R_h can be substituted for D according to

$$D = 4R_h = \frac{4A}{Z} \quad (4.24)$$

(A = cross-sectional area; Z = wetted perimeter).

Example 4.1. Water has to be pumped 0.8 km from a dam to a concentrator 20 m higher. A 0.2 m pipe is used, and the flow rate is 2.5 m³/min. If the pump is 65% efficient, what power is required? (Data: $K_R/D = 0.0003$; pipe fittings are equivalent to 25 m of pipe; $\rho_f = 1000$ kg/m³; $\mu = 10^{-3}$ kg/m · sec).

Solution. The pump must provide the power to overcome the static head of 20 m, and the friction loss in the pipe. That is,

$$\Sigma P = \Delta P \text{ (friction)} + \Delta P \text{ (head)}$$

Friction:

$$\bar{v} = \frac{2.5 \text{ m}^3}{\text{min}} \left| \frac{4}{(0.2 \text{ m})^2 \pi} \right| \frac{\text{min}}{60 \text{ sec}}$$
$$= 1.326 \text{ m/sec}$$

$$\therefore \quad \mathbf{Re} = \frac{D\bar{v}\rho}{\mu}$$
$$= \frac{0.2 \times 1.326 \times 1000}{10^{-3}}$$
$$= 2.7 \times 10^5$$

From Fig. 4.7, at $K_R/D = 0.0003$:

$$\mathbf{f}_p = 0.00425$$
$$= \frac{1}{2} \left(\frac{D}{L}\right) \frac{\Delta P}{\rho_f \bar{v}^2}$$

$$\therefore \quad \Delta P = $$
$$\frac{2 \times 0.00425}{} \left| \frac{(800 + 25)\text{m}}{0.2 \text{ m}} \right| \frac{1000 \text{ kg}}{\text{m}^3} \left| \frac{1.326^2 \text{ m}^2}{\text{sec}^2} \right.$$
$$= 6.16 \times 10^4 \text{ kg/m} \cdot \text{sec}^2$$

Static head:

$$\Delta P = \frac{20 \text{ m}}{} \left| \frac{1000 \text{ kg}}{\text{m}^3} \right| \frac{9.81 \text{ m}}{\text{sec}^2}$$
$$= 1.96 \times 10^5 \text{ kg/m} \cdot \text{sec}^2$$

$$\therefore \quad \Sigma(\Delta P) = 19.6 \times 10^4 + 6.2 \times 10^4$$
$$= 25.8 \times 10^4 \text{ kg/m} \cdot \text{sec}^2$$

$$\therefore \quad \text{power} = \frac{2.58 \times 10^5 \text{ kg}}{\text{m} \cdot \text{sec}^2} \left| \frac{2.5 \text{ m}^3}{\text{min}} \right| \frac{\text{min}}{60 \text{ sec}} \left| \frac{\text{kW}}{1000 \text{ W}} \right.$$
$$= 10.7 \text{ kW}$$

$$\therefore \quad \text{pump power} = \frac{10.7}{0.65}$$
$$= 16.5 \text{ kW}$$

4.5. FLOW OF BODIES THROUGH FLUIDS

4.5.1. The Drag Coefficient

When a particle is moving relative to a fluid, the force F in Eq. 4.16 is the drag force F_d, which is a combination of boundary layer wall drag and form drag (Fig. 4.4). The characteristic area A is usually taken to be that obtained by projecting the particle into a plane perpendicular to the direction of relative movement, and the characteristic energy can be taken as $\frac{1}{2}\rho_f v^2$, so that

$$\mathbf{f}_d = \frac{2F_d}{\rho_f v^2 A} \quad (4.25)$$

This friction factor $\mathbf{f}_d$ is normally referred to as the *drag coefficient*. It may be determined from friction factor charts such as Fig. 4.8.[9] Care must be taken when using these charts for irregularly shaped bodies; strictly speaking, a different $\mathbf{f}_d$ versus **Re** relationship exists for each shape and orientation.

4.5.2. Terminal Velocities

The behavior of a particle moving through a fluid can be analyzed by

$$\Sigma F = M\left(\frac{dv}{dt}\right) \quad (4.26)$$

(ΣF = resultant of the forces acting on a body of mass M; dv/dt = resulting acceleration). If the par-

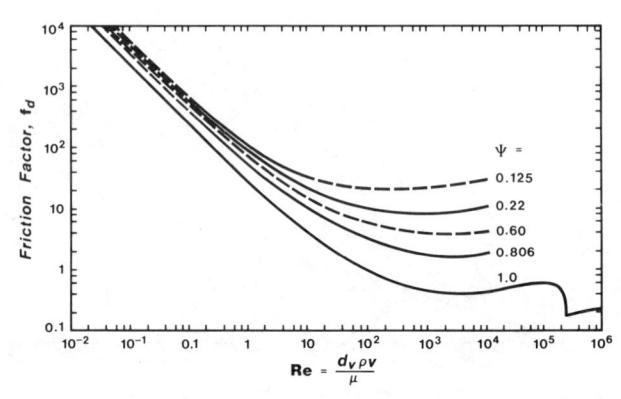

Figure 4.8. Friction factor (drag) chart for particles. (After Waddell.[9])

ticle is moving under an external accelerating force F_x, it will have opposing forces of F_d (the drag force) and F_b (the buoyancy force), that is,

$$F_x - F_d - F_b = M \left(\frac{dv}{dt}\right) \quad (4.27)$$

The external accelerating force may be gravitational or centrifugal and is expressed by

$$F_x = Ma_x = Mg \quad \text{gravitation acceleration} \quad (4.28a)$$
$$= M\omega^2 r \quad \text{centrifugal acceleration} \quad (4.28b)$$

(a_x = acceleration produced by the external force; r = radius of the path; ω = angular velocity in rads/sec). F_d is obtained from Eq. 4.25:

$$F_d = \tfrac{1}{2}\mathbf{f}_d \rho_f v^2 A \quad (4.29)$$

and the buoyancy force from Archimedes' principle:

$$F_b = \frac{M\rho_f}{\rho_s} a_x = \frac{M\rho_f g}{\rho_s} \quad \text{gravitational acceleration} \quad (4.30a)$$
$$= \frac{M\rho_f \omega^2 r}{\rho_s} \quad \text{centrifugal acceleration} \quad (4.30b)$$

Combining Eqs. 4.27–4.30 gives

$$M \frac{dv}{dt} = Mg \left(\frac{\rho_s - \rho_f}{\rho_s}\right) - \tfrac{1}{2}\mathbf{f}_d \rho_f v^2 A \quad (4.31a)$$

for gravitation acceleration, and

$$M \frac{dv}{dt} = M\omega^2 r \left(\frac{\rho_s - \rho_f}{\rho_s}\right) - \tfrac{1}{2}\mathbf{f}_d \rho_f v^2 A \quad (4.31b)$$

for centrifugal acceleration.

If the particle is allowed to accelerate from rest, the second term in the right-hand side of Eq. 4.31 will increase until $dv/dt = 0$, that is, the particle stops accelerating. Thus, for gravitational acceleration

$$Mg \left(\frac{\rho_s - \rho_f}{\rho_s}\right) = \tfrac{1}{2}\mathbf{f}_d \rho_f v_\infty^2 A \quad (4.32)$$

and the maximum velocity is known as the *terminal velocity* v_∞. Subsequently, gravitation acceleration is assumed, but equivalent equations for centrifugal acceleration may be obtained by replacing g with $\omega^2 r$.

4.5.3. Motion of Spherical Particles

For a spherical particle, $A = \pi d_0^2/4$ and $M = \pi d_0^3 \rho_s/6$, so that Eq. 4.32 becomes

$$v_\infty = \left[\frac{4}{3} \frac{g d_0}{\mathbf{f}_d} \left(\frac{\rho_s - \rho_f}{\rho_f}\right)\right]^{1/2} \quad (4.33)$$

Many mineral processing problems are associated with terminal velocity, and this equation may be used to calculate it for a spherical particle in any type of flow, if $\mathbf{f}_d$ is evaluated from Fig. 4.8 at $\Psi = 1.0$.

Stokes[10] assumed that under laminar flow conditions the drag force on a spherical particle was entirely due to viscous forces within the fluid, and he deduced the expression

$$F_d = 3\pi d_0 v \mu \quad (4.34)$$

If instead of Eq. 4.29 this is substituted in Eq. 4.26 with $dv/dt = 0$, the result is

$$v_\infty = \frac{\rho_s - \rho_f}{18\mu} g d_0^2 \quad (4.35)$$

Equations 4.34 and 4.35 are statements of Stokes' law. However, while the latter equation applies to the terminal velocity, Eq. 4.34 is applicable during the initial acceleration of a particle in laminar flow, even though its terminal velocity may not occur in laminar flow.

Combination of Eqs. 4.25 and 4.34 gives an expression for $\mathbf{f}_d$ in the laminar flow region ($\mathbf{Re} < 0.2$)

$$3\pi d_0 v \mu = \tfrac{1}{2}\mathbf{f}_d \rho_f v^2 \left(\frac{\pi d_0^2}{4}\right)$$

that is,

$$\mathbf{f}_d = \frac{24}{\mathbf{Re}_p} \quad (4.36)$$

(cf. Eq. 4.21 for pipes)
where

$$\mathbf{Re}_p = \frac{d_0 v \rho_f}{\mu} \quad (4.37)$$

In this case the Reynolds number of the particle $\mathbf{Re}_p$ is expressed in terms of the diameter of a sphere, but an appropriate nominal diameter may be necessary in other situations.

At high $\mathbf{Re}_p$ (> 500) the flow around the particle shows the maximum turbulence effect and the drag coefficient is almost independent of $\mathbf{Re}_p$:

$$\mathbf{f}_d \sim 0.44 \quad (4.38)$$

Between the laminar and turbulent regions a transition zone occurs, and the drag curve can be approximated by a straight line given by

$$f_d \sim \frac{18.5}{\mathbf{Re}_p^{0.6}} \quad (4.39)$$

Figure 4.8 or Eqs. 4.36 to 4.39 can be used to evaluate a sphere's diameter or terminal velocity, but difficulties arise because the unknown occurs in both f_d and $\mathbf{Re}_p$. One method of avoiding trial and error is to use a modified form of the drag chart.[11] From Eqs. 4.33 and 4.37

$$f_d \mathbf{Re}_p^2 = \frac{4}{3}(\rho_s - \rho_f)\frac{\rho_f g}{\mu^2} d_0^3 \quad (4.40)$$

and

$$\frac{f_d}{\mathbf{Re}_p} = \frac{4}{3}\left(\frac{\rho_s - \rho_f}{\rho_f^2}\right)\frac{\mu g}{v_\infty^3} \quad (4.41)$$

These equations are independent of v and d_0, respectively, and since a unique relationship exists between $\mathbf{Re}_p$ and f_d, a plot of $(f_d \mathbf{Re}_p^2)$ or $(f_d/\mathbf{Re}_p)$ versus $\mathbf{Re}_p$, as shown in Fig. 4.9, allows v_∞ or d_0 to be evaluated.

Example 4.2. What is the terminal velocity of a 150 μm diameter (d_0) particle having $\rho_s = 3145$ kg/m³ settling in (1) water and (2) air? ($\rho_{H_2O} = 1000$ kg/m³; $\rho_{air} = 1.2$ kg/m³; $\mu_{H_2O} = 1 \times 10^{-3}$ kg/m · sec; $\mu_{air} = 17.5 \times 10^{-6}$ kg/m · sec.)

Solution

1. Since $\mathbf{Re}_p$ is unknown, the most convenient solution is by Eq. 4.40 and Fig. 4.9.

$$f_d \mathbf{Re}_p^2 = \frac{4(\rho_s - \rho_f)\rho_f g d_0^3}{3\mu^2}$$

$$= \frac{4(3145 - 1000)1000 \times 9.81(150 \times 10^{-6})^3}{3(10^{-3})^2}$$

$$= 95$$

From Fig. 4.9 ($\Psi = 1.0$)

$$\mathbf{Re}_p = 2.9 = \frac{d_0 v_\infty \rho_f}{\mu}$$

$$\therefore \quad v_\infty = \frac{2.9 \times 10^{-3}}{(150 \times 10^{-6})1000}$$

$$= 0.0193 \text{ m/sec}$$

2. Then we write:

$$f_d \mathbf{Re}_p^2 = \frac{4(3145 - 1.2)1.2 \times 9.81(150 \times 10^{-6})^3}{3(17.5 \times 10^{-6})^2}$$

$$= 5.4 \times 10^2$$

From Fig. 4.9 ($\Psi = 1.0$)

$$\mathbf{Re}_p = 12$$

$$\therefore \quad v_\infty = \frac{12(17.5 \times 10^{-6})}{(150 \times 10^{-6})1.2}$$

$$= 1.17 \text{ m/sec}$$

4.5.4. Calculation of Flow Characteristics When Re Is Unknown

The use of one of the approximate Eqs. 4.36, 4.38, or 4.39 requires knowledge of the flow characteristics. When v or d_0 is unknown, Eqs. 4.40 and 4.41 provide a method for overcoming this difficulty. Although Stokes' law is reliable only for $\mathbf{Re}_p < 0.2$, Eq. 4.36 is still more accurate than Eq. 4.39 over the range $0.2 < \mathbf{Re}_p < 2.0$, that is,

$$\frac{d_0 v_\infty \rho_f}{\mu} < 2.0 \quad (4.42)$$

If v_∞ is unknown, it may be eliminated from this equation by combination with Eq. 4.35, thus

$$\frac{d_0 \rho_f}{\mu} \cdot \frac{g(\rho_s - \rho_f)}{18\mu} d_0^2 < 2.0$$

and when this is combined with Eq. 4.40

$$\tfrac{1}{18} \cdot \tfrac{3}{4} \cdot f_d \mathbf{Re}_p^2 < 2.0$$

that is,

$$f_d \mathbf{Re}_p^2 < 48 \qquad \mathbf{Re}_p < 2.0 \quad (4.43)$$

Similarly, for Eq. 4.38 to be more accurate than Eq. 3.39, $\mathbf{Re} > 500$, which, combined with Eqs. 4.38 and 4.40, gives

$$f_d \mathbf{Re}_p^2 > 1.1 \times 10^5 \qquad \mathbf{Re}_p > 500 \quad (4.44)$$

If d_0 is unknown, a similar treatment based on Eq. 4.41 gives

$$\frac{f_d}{\mathbf{Re}_p} > 6.0 \qquad \mathbf{Re}_p < 2.0 \quad (4.45)$$

and

$$\frac{f_d}{\mathbf{Re}_p} < 8.8 \times 10^{-4} \qquad \mathbf{Re}_p > 500 \quad (4.46)$$

Example 4.3. What is the terminal velocity of the particle in the preceding example settling in water in a 0.5 m radius centrifuge rotating at 2000 rpm?

Solution. On the basis of Part 2 of Example 4.2, it is assumed that the flow is in the transitional re-

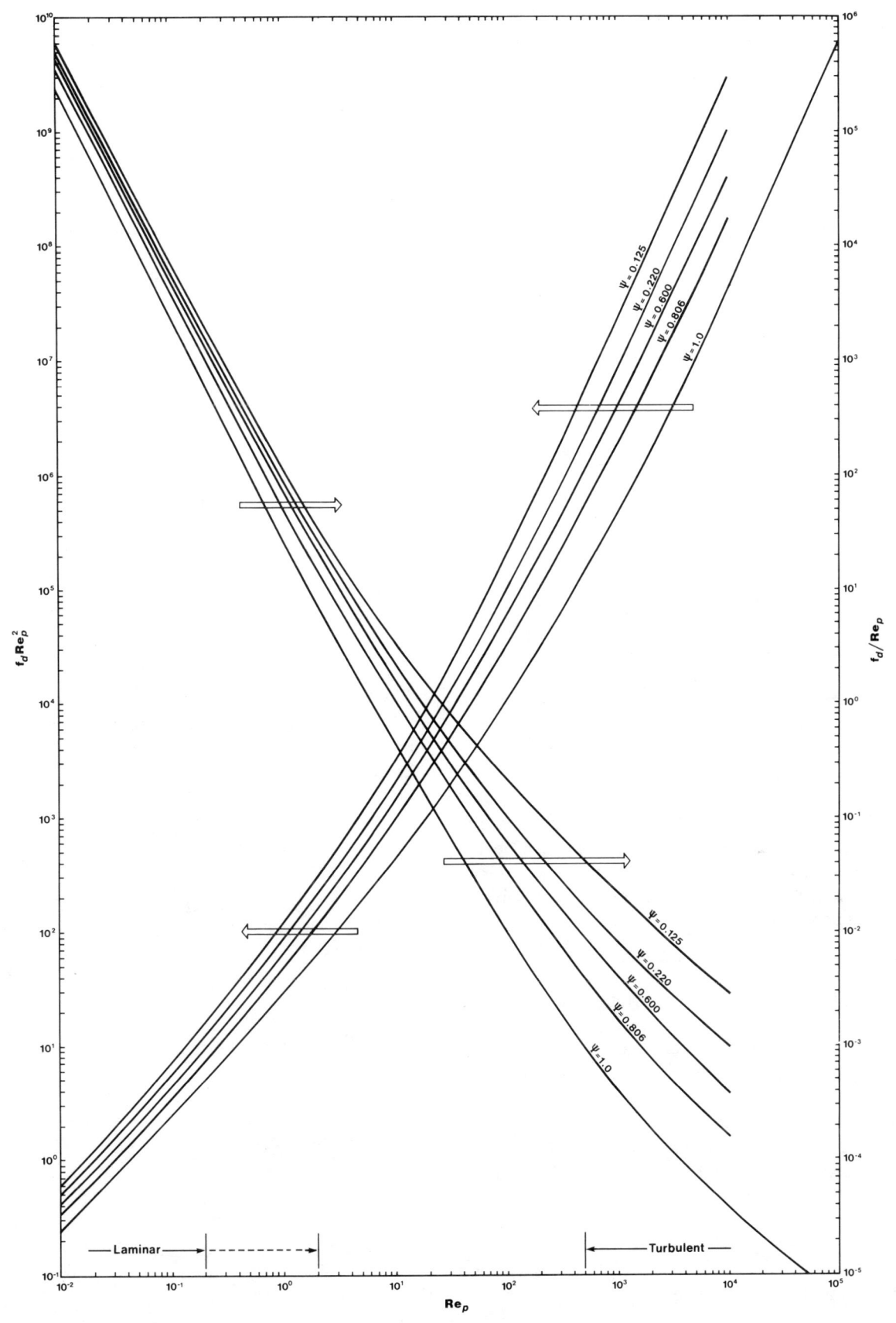

Figure 4.9. Data of Fig. 4.8 replotted as $f_d \, Re_p^2$ and f_d/Re_p versus Re_p.

gion, so that Eq. 4.39 should be valid. Combining Eqs. 4.33 and 4.39 and replacing g by a ($= \omega^2 r$)

$$v_\infty^2 = \frac{4\omega^2 r}{3} d_o \left(\frac{\rho_s - \rho_f}{\rho_f}\right) \frac{1}{18.5} \left(\frac{d_o v_\infty \rho_f}{\mu}\right)^{0.6}$$

$$\omega = \frac{2000 \text{ rev}}{\text{min}} \bigg| \frac{2\pi \text{ rads}}{\text{rev}} \bigg| \frac{\text{min}}{60 \text{ sec}}$$
$$= 209.4 \text{ rads/sec}$$

$$\therefore v_\infty^{1.4} = \frac{4(209.4)^2 \times 0.5(150 \times 10^{-6})(3145 - 1000)}{3 \times 18.5 \qquad\qquad 1000}$$
$$\times \left(\frac{150 \times 10^{-6} \times 1000}{10^{-3}}\right)^{0.6}$$
$$= 10.28$$
$$\therefore v_\infty = 5.28 \text{ m/sec}$$

Check $\mathbf{Re}_p$:

$$\mathbf{Re}_p = \frac{150 \times 10^{-6} \times 5.28 \times 1000}{10^{-3}}$$
$$= 792$$

That is, Eq. 4.39 is *NOT* valid, since $\mathbf{Re}_p > 500$. (This is because the acceleration is now $(209.4)^2 \times 0.5/9.81 = 2235\,g$.)

Using Eq. 4.38 (and Eq. 4.33)

$$v_\infty = \left[\frac{4}{3} \frac{(209.4)^2 \times 0.5(150 \times 10^{-6})}{0.44}\right.$$
$$\left.\left(\frac{3145 - 1000}{1000}\right)\right]^{1/2}$$
$$= 4.62 \text{ m/sec}$$

Although the two answers are comparatively close, this may not always be the case. Consequently the method shown in Example 4.2 is preferable, since it overcomes the problem of selecting the appropriate equation.

4.5.5. Acceleration of Particles

In mineral processing units such as jigs and hydrocyclones, particles may have insufficient time to reach their terminal velocity. An indication of the time required to approach this is therefore desirable and this can be obtained by integrating Eq. 4.31a. For a spherical particle in laminar flow, $M = \pi d_o^3 \rho_s/6$ and $\mathbf{f}_d$ is given by Eq. 4.36 so that

$$\frac{dv}{dt} = g\left(\frac{\rho_s - \rho_f}{\rho_s}\right) - \frac{18\mu v}{d_o^2 \rho_s} \quad (4.47)$$

This equation is of the form

$$dt = \frac{dv}{K_1 - K_2 v}$$

and can therefore be integrated for a particle starting at rest to give

$$v = v_\infty \left[1 - \exp\left(\frac{-18\,\mu t}{d_o^2 \rho_s}\right)\right] \quad (4.48)$$

A completely rigorous solution of Eq. 4.31a is not practical for turbulent conditions because the particle's initial acceleration will be in laminar flow. However, use of Eq. 4.38 for $\mathbf{f}_d$ gives an adequate approximation

$$\frac{dv}{dt} = \frac{g(\rho_s - \rho_f)}{\rho_s} - \frac{3}{4} 0.44 \frac{v^2}{d_o} \frac{\rho_f}{\rho_s} \quad (4.49)$$

which is of the form

$$dt = \frac{dv}{K_1 - K_2 v^2}$$

On integrating for a particle starting at rest, this results in

$$\frac{v_\infty - v}{v_\infty + v} = \exp(-Kt) \quad (4.50)$$

where

$$K^2 = 1.32 \frac{\rho_f(\rho_s - \rho_f)g}{\rho_s^2 d_o}$$

Since distance traveled is $v \cdot dt$, integration of Eq. 4.50 will give the approximate distance traveled during the time the particle is attaining its terminal velocity. Thus

$$H = v_\infty \int_0^t \left[\frac{1 - \exp(-Kt)}{1 + \exp(Kt)}\right] dt$$

which on integration gives

$$H = \left(\frac{2v_\infty}{K}\right) \ln \frac{1}{2}\left[\exp\left(\frac{Kt}{2}\right)\right.$$
$$\left. + \exp\left(\frac{-Kt}{2}\right)\right] \quad (4.51)$$

Again, this equation is approximate because it neglects the nonturbulent behavior occurring during the initial motion.

Example 4.4. Using a long tube, you intend to measure the terminal velocity of a 500 μm (d_o) particle of coal ($\rho_s = 1500$ kg/m^3) in air ($\rho = 1.2$ kg/m^3). Estimate how much height should be reserved to allow the particle to reach its terminal velocity.

Solution. It is first necessary to determine v_∞ for the particle. Using Eq. 4.40 and Fig. 4.9,

$$\mathbf{f}_d \cdot \mathbf{Re}_p^2 = \frac{4(1500 - 1.2)\,1.2 \times 9.81 \times (500 \times 10^{-6})^3}{3(17.5 \times 10^{-6})^2}$$
$$= 9.6 \times 10^3$$

From Fig. 4.9:

$$\mathbf{Re}_p = 88$$

$$v_\infty = \frac{88(17.5 \times 10^{-6})}{500 \times 10^{-6} \times 1.2}$$

$$= 2.57 \text{ m/sec}$$

Because of the difficulty of integrating Eq. 4.31a in the transitional flow region, it is assumed that Eqs. 4.50 and 4.51 give an adequate approximation. Assuming that 98% v_∞ is sufficient accuracy, Eq. 4.50 gives

$$\frac{1 - 0.98}{1 + 0.98}$$

$$= \exp\left\{-t\left[\frac{1.32 \times 1.2 \,(1500 - 1.2)\, 9.81}{1500^2 \times 500 \times 10^{-6}}\right]^{1/2}\right\}$$

$$= \exp(-4.55t)$$

$$\therefore \quad t = 1.01 \text{ sec}$$

From Eq. 4.51:

$$H = 2 \times \frac{2.57 \times 0.98}{4.55} \ln \frac{1}{2}\left[\exp\left(\frac{4.55 \times 1.01}{2}\right) + \exp\left(\frac{-4.55 \times 1.01}{2}\right)\right]$$

$$= 1.79 \text{ m}$$

This is a significant distance, although it tends to err on the high side because turbulent flow was assumed. However, it does illustrate that in some practical situations (generally with relatively light, large particles in air), a significant distance is necessary for the particle to attain its terminal velocity.

4.5.6. Irregularly Shaped Particles

The equations presented in this section so far have been derived in terms of the diameter of a sphere d_0, and appropriate nominal diameters and shape factors (Chapter 2) need to be included before irregular particles can be treated.

Equation 4.33 rearranged immediately defines the free fall diameter d_f for an irregular particle

$$d_f = \frac{3}{4} \mathbf{f}_d \frac{v_\infty^2}{g}\left(\frac{\rho_f}{\rho_s - \rho_f}\right) \qquad (4.33)$$

Furthermore, if d_a is the drag diameter (the projected area diameter of the particle perpendicular to the direction of motion), $A = \pi d_a^2/4$ and $M = \pi d_v^3 \rho_s/6$, so that Eq. 4.32 becomes

$$\frac{\pi d_v^3 \rho_s}{6} \cdot g\left(\frac{\rho_s - \rho_f}{\rho_s}\right) = \frac{\mathbf{f}_d v_\infty^2 \rho_f}{2} \cdot \frac{\pi d_a^2}{4}$$

that is,

$$\frac{d_v^3}{d_a^2} = \frac{3}{4} \mathbf{f}_d \frac{v_\infty^2}{g}\left(\frac{\rho_f}{\rho_s - \rho_f}\right) = d_f = \frac{d_v^3}{d_a^2} \qquad (4.52)$$

During laminar flow, particles fall with a random orientation; hence d_a is the mean projected area diameter d_a in a random orientation. In turbulent flow the particles orient to give maximum resistance so that d_a is the mean d_a in a stable orientation. This means that irregular particles have a greater variation in velocity between laminar and turbulent flow than do spherical particles.

Since the particle volume is equal to $\lambda_{va} d_a^3$ (Eq. 2.10), Eq. 4.32 may also be expressed as

$$\lambda_{va} d_a^3 \rho_s g\left(\frac{\rho_s - \rho_f}{\rho_s}\right) = \frac{\mathbf{f}_d v_\infty^2 \rho_f}{2} \frac{\pi d_a^2}{4}$$

that is,

$$d_a = \frac{\mathbf{f}_d}{8} \cdot \frac{v_\infty^2}{g}\left(\frac{\rho_f}{\rho_s - \rho_f}\right) \cdot \frac{\pi}{\lambda_{va}} = \frac{\pi}{6\lambda_{va}} d_{St} \qquad (4.53)$$

which provides a method of determining λ_{va}.

A number of sets of friction factor data have been published for irregular particles,[9,12–14] although their use is often limited by a nominal diameter that is difficult to measure, or has been inadequately defined. Some of the most carefully defined work is that of Heywood,[12] published in tabular form, in terms of d_a and λ_{va}. Figures 4.8 and 4.9 are in terms of the more widely used sphericity shape factor $\Psi = (d_v/d_s)^2$; here the appropriate nominal diameter is d_v. A more practical method is to have the data determined on actual crushed minerals, sized by sieving (i.e., d_A). Such plots have been published,[13–15] based mainly on the original data of Richards.[16] Unfortunately, although this work is widely assumed[9,13,15] to have been carried out on sieved particles, most of the results appear to have been obtained on particles that were sized by an undefined microscopic method and thus have no known relationship to sieve-sized particles.

Example 4.5. The terminal velocity of a PbS particle ($\rho_s = 7600$ kg/m³) has been measured in water as 0.1 m/sec. Microscopic examination shows that the shape is approximately cubic. Find (1) the free fall diameter d_f, (2) the volume diameter d_v, (3) the surface diameter d_s, and (4) the volume surface diameter d_{vs}.

Solution. Since the diameter is unknown, Eq. 4.41 is used, in conjunction with Fig. 4.9:

$$\frac{\mathbf{f}_d}{\mathbf{Re}_p} = \frac{4}{3} \frac{(7600 - 1000)}{1000^2} \frac{10^{-3} \times 9.81}{(0.1)^3}$$
$$= 8.6 \times 10^{-2}$$

1. In Fig. 4.9, when $\Psi = 1.0$, d_v becomes d_f

   $\therefore \quad \mathbf{Re}_p = 25 = \dfrac{d_f v \rho}{\mu}$

   $\therefore \quad d_f = \dfrac{25 \times 10^{-3}}{0.1 \times 1000}$
   $\quad\quad\quad = 250 \ \mu\text{m}$

2. Figure 4.9 also applies to irregular particles, where the relevant diameter is d_v, used with an appropriate value of Ψ. To determine Ψ, consider a cube with edges of unit length, in which case $V_p = 1$ cubed unit, $S_p = 6$ square units. A sphere of equal volume has

$$d_o = \left(\frac{6}{\pi}\right)^{0.33}$$

and the surface area of a sphere of equal volume is

$$\pi \left[\left(\frac{6}{\pi}\right)^{0.33}\right]^2$$

Substituting in Eq. 2.18,

$$\Psi = \frac{\pi \left[\left(\frac{6}{\pi}\right)^{0.33}\right]^2}{6}$$
$$= 0.806$$

From Fig. 4.9, for $\Psi = 0.806$:

$\mathbf{Re}_p = 50$

$\therefore \quad d_v = \dfrac{50 \times 10^{-3}}{0.1 \times 1000}$
$\quad\quad\quad = 500 \ \mu\text{m}$

3. Since

$$\Psi = \left(\frac{d_v}{d_s}\right)^2$$

$$d_s = \frac{d_v}{\Psi^{1/2}}$$
$$= \frac{500}{(0.806)^{1/2}}$$
$$= 557 \ \mu\text{m}$$

4. To find the volume surface diameter:

$$d_{vs} = \frac{d_v^3}{d_s^2}$$
$$= (500)^3/(557)^2$$
$$= 403 \ \mu\text{m}$$

5. $\quad d_f = \dfrac{d_v^3}{d_a^2} \quad$ for turbulent conditions

   $\therefore \quad d_a = \left(\dfrac{500^3}{250}\right)^{1/2}$
   $\quad\quad\quad = 707 \ \mu\text{m}$

The spread in the d_n values in this example is wider than it should be. This is because the determinations rely on empirical data (Fig. 4.9), which in the case of irregular particles are notoriously unreliable, since they have such erratic behavior.

In particular in this example, d_v could have considerable error, yet it forms the basis of calculating the other nominal diameters. One alternative is to physically measure d_n; however a better approach is to experimentally determine the nominal diameter under conditions that correspond to those that will be used in practice. Then, the fluid mechanics equations can be used to estimate particle behavior under slightly different conditions. For example, the value above of d_f would be satisfactory for particle behavior under transitional and turbulent conditions, but it would not be suitable for predicting behavior under laminar conditions because the particle's orientation in the flow direction would be unpredictable.

4.5.7. Other Correction Factors

A number of other factors may have to be considered when treating the flow of particles through a fluid. The most important in practice is hindered settling, discussed in Section 4.7.

Although generally not as important in full size operations, a wall effect may be significant in laboratory testing. This problem arises because the moving particle pulls fluid along with it, and in the vicinity of the stationary wall, the fluid movement is slowed. The increase in drag can be corrected by a factor of the form $[1 - K_w(d_n/L)]$, where L is the distance of the particle center from the wall. For a single wall $K_w + 0.563$; for two walls $K_w = 1.004$, and $K_w = 2.104$ for a cylinder.[17]

The size of very fine particles can become comparable to the mean free path of the fluid molecules, and the particle then effectively falls partly between the molecules at a rate faster than that deduced by theory. Cunningham[18] introduced a correction factor to Stokes' law, so that Eq. 4.35 becomes

$$v_\infty = \frac{(\rho_s - \rho_f)}{18\mu} g d_n^2 \left(1 + 2K_C \frac{\Lambda}{d_n}\right) \quad (4.54)$$

(K_C = a factor approximately equal to unity; Λ = mean free path of the particles).

4.6. FLOW THROUGH A PACKED BED

Although primarily of importance in chemical engineering,[1] flow of a fluid through a packed bed of particles is relevant here for filtration and can be extended to fluidization and hindered settling.

The resistance to the flow of fluid through voids in a bed of particles is the resultant of the drag on all particles in the bed. Although the bed can be considered to be a collection of submerged objects, a better method of correlating is based on consideration of the total drag of the fluid on the solid boundaries of the tortuous channels through the bed of particles. Thus, it is assumed that the channels can be replaced by parallel pipes of variable cross-section, and that the mean hydraulic radius is able to account for the variations in channel cross-section and shape.

From Eqs. 4.18 and 4.24 the friction factor will therefore be of the form

$$f_b = \frac{R_h}{L} \frac{\Delta P}{2\rho \bar{v}^2} \quad (4.55)$$

The hydraulic radius may be converted to measurable terms as follows:

$$R_h = \frac{\text{average cross-section available for flow}}{\text{wetted perimeter}}$$

$$= \frac{\text{volume available for flow}}{\text{total wetted surface}}$$

$$= \frac{\text{volume of voids/volume of bed}}{\text{total wetted surface/volume of bed}}$$

$$= \frac{\varepsilon}{S_w} \quad (4.56)$$

where ε, the porosity, is the fraction of the bed volume that is voids (bed volume is the volume of particles plus the volume of voids). The term S_w is therefore given by

$$S_w = S_p \frac{(1 - \varepsilon)}{V_p} \quad (4.57)$$

(S_p = particle surface area; V_p = particle volume). Since

$$d_0 = \frac{6}{S_p/V_p} = d_{vs} \quad (4.58)$$

d_{vs} should be used for irregular particles. If a range of particle sizes exists, d_{vs} is replaced by $\bar{d}_{vs32}$. Finally, the velocity of the fluid in the voids can be given by

$$\bar{v} = \frac{v_s}{\varepsilon} \quad (4.59)$$

(v_s = superficial velocity, the average linear velocity the fluid would have if no particles were present).

Since no relationship between the effective channel length and L is possible, Eqs. 4.55 to 4.59 are combined without the constants to give a suitable modified friction factor for a packed bed:

$$f_b = \frac{\Delta P \bar{d}_{vs32}}{L\rho_f v_s^2} \frac{\varepsilon^3}{(1 - \varepsilon)} \quad (4.60)$$

Figure 4.10[13] shows the friction factor chart of f_b versus $\text{Re}_p/(1 - \varepsilon)$ with

$$\text{Re}_p = \frac{\bar{d}_{vs32} v_s \rho_f}{\mu} \quad (4.61)$$

Unlike the previous charts, this entire curve has been determined experimentally. It can be described by the Ergun equation

$$f_b = \frac{150(1 - \varepsilon)}{\text{Re}_p} + 1.75 \quad (4.62)$$

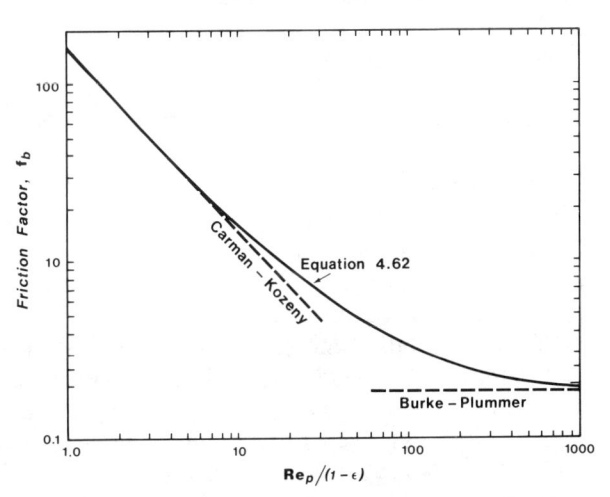

Figure 4.10. Friction factor chart for packed beds. (After Foust et al.[13])

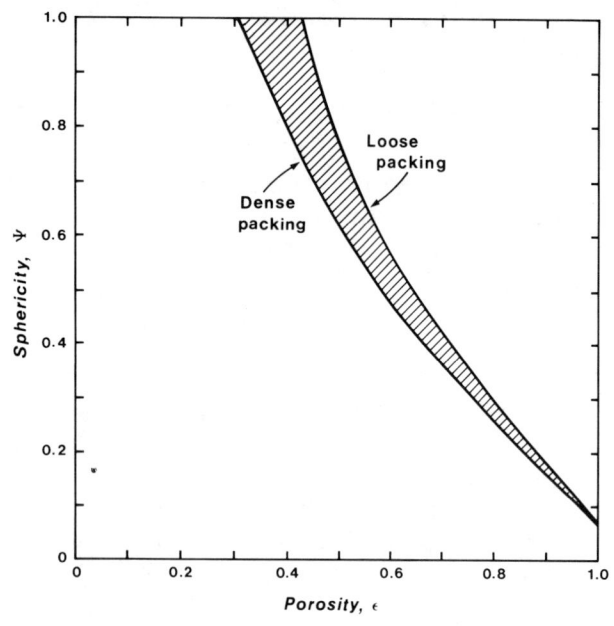

Figure 4.11. Sphericity as a function of porosity for randomly packed beds of particles of uniform size (After Brown et al.[19])

which for laminar flow $[\mathbf{Re}_p/(1 - \varepsilon) < 10]$ reduces to the Carman-Kozeny equation

$$\mathbf{f}_b = \frac{150(1 - \varepsilon)}{\mathbf{Re}_p} \qquad (4.63)$$

and for the fully turbulent flow $[\mathbf{Re}_p/(1 - \varepsilon) > 1000]$ reduces to the Burke-Plummer equation

$$\mathbf{f}_b = 1.75 \qquad (4.64)$$

The porosity should preferably be determined experimentally, although Fig. 4.11, which shows its dependence on Ψ, may be useful for particles of uniform size.[19]

4.7. FLUIDIZATION AND SEDIMENTATION

Fluidization and sedimentation can be treated together as an extension of a packed bed,[20,21] the difference being that in a packed bed the particles are restrained against each other, whereas in fluidization and sedimentation they are not. Hydrodynamically there is little difference between fluidization and sedimentation, in that both can be considered as an expanded packed bed where the particles are free to move relative to one another. Fluidization and sedimentation are distinguished by whether the fluid or the solid moves relative to the containing vessel.

If a fluid is passed up through a bed of unrestrained particles, the pressure drop will increase until a point is reached where the pressure drop and weight just counteract the buoyancy forces. Provided the bed is free flowing, it will start to behave like a boiling liquid; each particle is separated from its neighbor and bubbles of fluid appear at the surface. Such a bed is said to be *fluidized* (or *teetering*). If the fluid velocity is further increased, the bed continues to expand and eventually, as terminal velocity is approached, the particles are entrained in the fluid. The situation is illustrated in Fig. 4.12, where it can be seen that the minimum fluidization velocity $v_{mf} \ll v_\infty$. Above v_{mf}, the pressure drop increase is negligible until the initiation of entrainment, when the pressure drop (due to the resistance of the particles) falls towards zero as the bed disappears. With smooth particles, the point of minimum fluidization is not a sharp feature on the curve in Fig. 4.12, while irregular particles may cause a peak on the graph due to particle interlocking (dashed portion of curve, Fig. 4.12).

Sedimentation or hindered settling essentially represents the opposite of fluidization and is best illustrated by the events that occur in a batch sedimentation test. If a slurry of particles of constant size and density is mixed in a vertical cylinder and allowed to settle, the sedimentation process is as illustrated in Fig. 4.13a. Generally, the particles are small enough to accelerate rapidly and settle at their terminal velocity, producing an upper interface between the clear liquid A and the slurry B. As

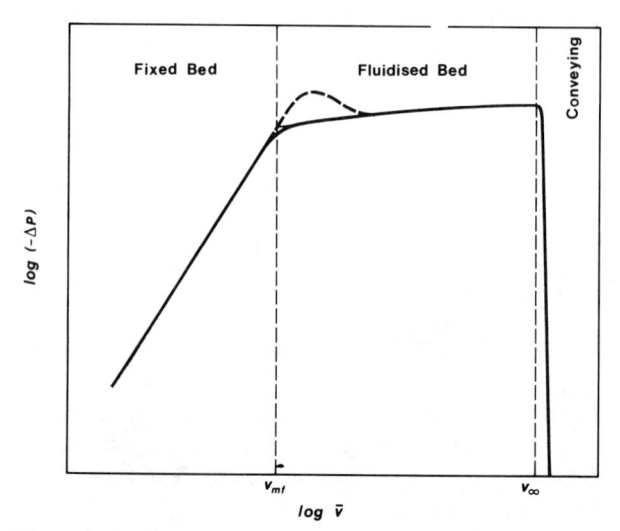

Figure 4.12. Pressure drop versus superficial fluid velocity in a particulate bed.

Fluidization and Sedimentation

(a) Uniform Particle Size and Density

(b) Some Particle Size Distribution

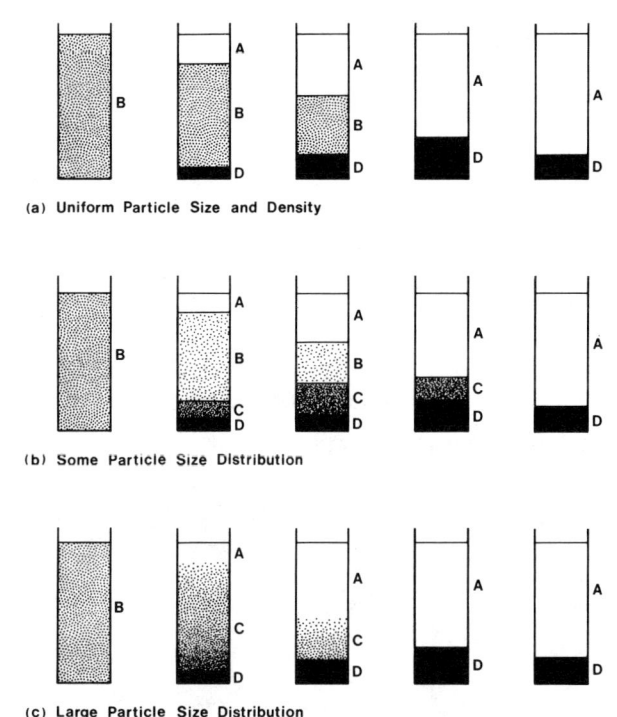

(c) Large Particle Size Distribution

Figure 4.13. Batch sedimentation behavior.

the particles settle on the bottom of the cylinder, an interface between settled and settling particles rises from the bottom. Significantly *the slurry concentration in zone B remains unchanged.* Eventually the two interfaces meet at a critical point, after which zone D goes through a slow process of compression involving very slow settling.

If a small range of particle sizes exists, the process is as depicted in Fig. 4.13b. Zone D has a higher proportion of faster settling particles and zone C is a poorly defined region of variable size distribution and concentration.

When the particle size distribution is extensive, the situation may be as depicted in Fig. 4.13c, where zone C immediately forms instead of zone B. However, at high slurry concentrations, provided they have the same density, the particles can show an apparent interlocking effect[22] such that the spaces between the large particles are too small for the smaller particles to get through, and the slurry behaves like that shown in Fig. 4.13a. Generally, the more irregular the particles, the lower the slurry density at which this occurs. More recent work indicates that this interlocking effect is caused not by mechanical factors, but rather by hydrodynamic ones.[23]

The situation may be further complicated by variations in particle density. If the concentration is high enough, particles that normally settle at equal terminal velocities will show the heavy ones settling preferentially, and conditions may be such that the lighter material cannot sink until the heavier material has settled out, resulting in the formation of separate layers of each material.[24]

The analysis of packed beds may be extended to fluidization and sedimentation as follows. Minimum fluidization velocity v_{mf} occurs when the pressure drop is just capable of supporting the bed, that is,

$$\Delta P_{mf} = \text{weight} - \text{buoyancy}$$
$$\Delta P_{mf} A = A L_{mf}(\rho_s - \rho_{sl})g$$
$$= A L_{mf}(1 - \varepsilon_{mf})(\rho_s - \rho_f)g \quad (4.65)$$

where ε_{mf} and L_{mf} are used because of the slight expansion that occurs just before fluidization. For rounded particles $\varepsilon_{mf} \sim 1.03\varepsilon$ and since the volume of particles is unchanged,

$$L(1 - \varepsilon)A = L_{mf}(1 - \varepsilon_{mf})A \quad (4.66)$$

In principle, these equations can be combined with the Ergun equation (4.62) to eliminate ΔP and give the conditions for minimum fluidization thus:

$$\frac{\varepsilon_{mf}^3 d_{vs}}{(1 - \varepsilon_{mf}) L_{mf} \rho_f v_{mf}^2} \cdot L_{mf}(1 - \varepsilon_{mf})(\rho_s - \rho_f)g$$
$$= \frac{K_1(1 - \varepsilon_{mf})\mu}{v_{mf} d \rho_f} + K_2$$

$$\therefore \quad \mathbf{f}_s = \frac{K_1(1 - \varepsilon_{mf})}{\mathbf{Re}_{mf}} + K_2 \quad (4.67)$$

where

$$\mathbf{f}_s = \frac{d_{vs} g}{v_{mf}^2}\left(\frac{\rho_s - \rho_f}{\rho_f}\right)\varepsilon_{mf}^3 \quad (4.67a)$$

and

$$\mathbf{Re}_{mf} = \frac{d_{vs} \rho_f v_{mf}}{\mu} \quad (4.67b)$$

In practice the constants K_1 and K_2 for mobile beds are different from those for packed beds. However, reliable values are difficult to quote for two reasons: first, Eq. 4.67 is not an exact representation of the experimental data, and second, the values depend on whether the fluid is a liquid or a gas. Figure 4.14 shows the water data of Richardson,[21] which in general agree with those presented by Morse.[25] The data fall below those given by the Ergun equation and can be approximated by

$$\mathbf{f}_s = \frac{120(1 - \varepsilon)}{\mathbf{Re}_p} + 1.0 \quad (4.68)$$

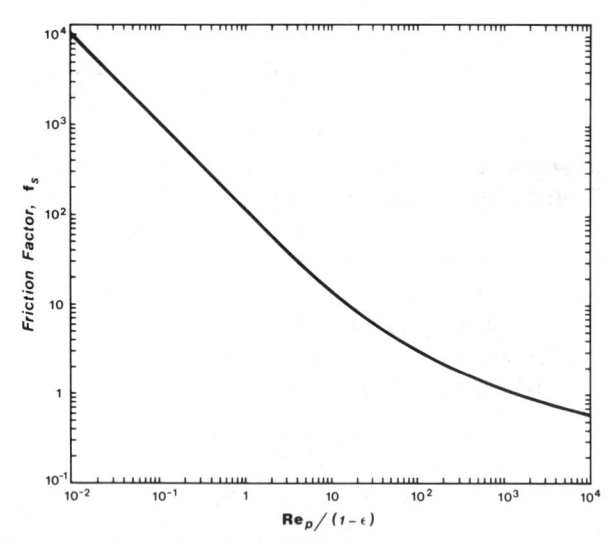

Figure 4.14. Friction factor chart for sedimentation and fluidization. (After Richardson and Meikle.[24])

For gases, the data lie above those given by the Ergun equation and

$$f_s = \frac{200(1-\varepsilon)}{Re_p} + 3.0 \qquad (4.69)$$

In general Eqs. 4.68 and 4.69 give a poor representation of behavior at $Re_p > 1000$.

Example 4.6. What would be the settling velocity of the particle in Example 4.5 when the volumetric solids concentration $\mathfrak{F}_V$ ($= 1 - \varepsilon$) is 0.25?

Solution. Equation 4.69 is expressed in terms of d_{vs}, and this has been calculated in part 4 of Example 4.5. However, as was mentioned, this value (403 μm) incorporates any error inherent in Figs. 4.8 and 4.9. Because Eq. 4.69 was derived from data on spherical particles, it is probably more appropriate to use an experimentally determined value of d that represents d_0: in this case d_f, since this represents *actual* particle behavior.

Thus, the appropriate value of d in Eq. 4.69 is taken as 250 μm (d_f). Substituting

$$(250 \times 10^{-6}) \frac{(1-0.25)^3}{v^2} \left(\frac{7600-1000}{1000} \right) 9.81$$
$$= \frac{120(0.25)10^{-3}}{(250 \times 10^{-6})v\,1000} + 1.0$$

that is,

$$\frac{0.00683}{v^2} = \frac{0.120}{v} + 1.0$$

$$\therefore \quad 1.0v^2 + 0.120v - 0.00683 = 0$$

$$\therefore \quad v^2 = \frac{-0.12 + [(0.12)^2 + 4 \times 0.00683]^{1/2}}{2}$$

$$\therefore \quad v = 0.042 \text{ m/sec}$$

Example 4.7. The particles in Example 4.2 are to be dried in a fluidized bed. (1) At what superficial gas velocity will the bed start to fluidize? (2) What will be the bed expansion if the superficial gas velocity is 50% of the particles' terminal velocity?

1. Assuming the bed is lightly packed, $\varepsilon = 0.42$ (Fig. 4.11, $\Psi = 1.0$). Substituting in Eq. 4.69

$$\frac{150 \times 10^{-6}(0.42)^3(3145-1.2) \times 9.81}{1.2 v_s^2}$$
$$= \frac{200(1-0.42)17.5 \times 10^{-6}}{150 \times 10^{-6} v\, 1.2} + 3.0$$

$$\therefore \quad \frac{0.2856}{v^2} = \frac{11.278}{v} + 3.0$$

That is,

$$3.0v^2 + 11.278v - 0.2856 = 0$$

$$v = \frac{-11.278 \pm (11.278^2 + 4 \times 3.0 \times 0.2856)^{1/2}}{2 \times 3.0}$$
$$= 0.025 \text{ m/sec} \quad \text{(i.e., approx. 2\% of } v_\infty\text{)}$$

2. To find the bed expansion:

$$50\% \, v_\infty = 0.5 \times 1.17 \text{ m/sec} \quad \text{(Example 4.2)}$$
$$= 0.585 \text{ m/sec}$$

From Eq. 4.69:

$$\frac{150 \times 10^{-6} \varepsilon^3 (3145-1.2) \times 9.81}{(0.585)^2 \quad 1.2}$$
$$= \frac{200(1-\varepsilon)\,17.5 \times 10^{-6}}{150 \times 10^{-6} \times 0.486 \times 1.2} + 3.0$$
$$11.26\,\varepsilon^3 = 40.01(1-\varepsilon) + 3.0$$

Try:

	L.H.S.	R.H.S.
$\varepsilon = 0.80$:	5.76	$\neq$ 7.8
$\varepsilon = 0.90$:	8.21	$\neq$ 7.00
$\varepsilon = 0.88$:	7.67	$\neq$ 7.80
$\varepsilon = 0.882$:	7.73	$\sim$ 7.72

$$\therefore \quad \text{bed expansion} = \frac{L}{L_I} = \frac{1-\varepsilon_I}{1-\varepsilon}$$
$$= \frac{1-0.42}{1-0.882} \times 100$$
$$= 492\%$$

An alternative and better approach to the fluidization and sedimentation of uniform spherical particles is to use empirical relationships. One of the most widely quoted is that of Richardson and Zaki[20,24]:

$$\frac{v_h}{v_\infty} = \varepsilon^n \qquad (4.70)$$

(v_h = falling velocity of suspension with respect to a fixed horizontal plane [sedimentation], or superficial velocity [fluidization]; v_∞ = terminal velocity of a single particle at infinite dilution). A satisfactory approximation for n is[26]

$$n = \frac{4.8}{2 + \beta} \qquad (4.71)$$

where β arises from the exponents in Eqs. 4.33 and 4.35 and is therefore the slope of the graph in Fig. 4.8, using

$$\mathbf{Re}_\infty = \mathbf{Re}_p = \frac{d_0 \rho_f v_\infty}{\mu} \qquad (4.72)$$

When $D < 100\, d_0$ wall effects may become significant and more rigorous relationships for n should be used.[20,27]

For laminar flow, an alternative to Eq. 4.70 is that presented by Steinour[28]:

$$\frac{v_h}{v_\infty} = \varepsilon^2 10^{-1.82(1-\varepsilon)} \qquad (4.73)$$

Both equations give similar results, since they compensate for three effects,[29] although these are more clearly seen in the second equation. The first can be considered to be the relative buoyancy, since it can be shown that

$$\varepsilon = \frac{\rho_s - \rho_{sl}}{\rho_s - \rho_l} \qquad (4.74)$$

while the second ε is due to the rising liquid displaced by the settling particles. The term $10^{-1.82(1-\varepsilon)}$ has been shown to be the relative viscosity. This implies that Eq. 4.73 is simply Eq. 4.35 with the fluid density and viscosity terms replaced by the *apparent density* and *apparent viscosity*. On this basis, it has also been suggested that Eq. 4.33 can be modified in the same way.[29]

Information concerning irregular particles and size distributions is still limited. For uniform non-spherical particles at $\mathbf{Re}_\infty > 200$, the value of the exponent in Eq. 4.70 is[21]:

$$n = 2.7 \lambda_{va}^{0.16} \qquad (4.75)$$

TABLE 4.1: REPORTED VALUES OF n FOR EQ. 4.70

Material (shape)	Size (d_A), (μm)		n
Coal[31]*	−355	+300	6.48
	−250	+212	6.9
	−180	+150	7.07
	−125	+106	6.89
	−90	+75	7.5
Ground Methyl-methacrylate[30]	−212	+180	6.9
	−150	+125	7.5
	−106	+90	8.3
	−75	+53	9.5
Alumina powder[24]	−7	+4	10.5
Salt[26] (cubic)	340 (Re = 0.03)		5.5
ABS[26] (cubic)	2900 (Re = 0.06)		5.5
Sugar[26] (cubic)	1200 (Re = 0.03)		5.7
Silicate crystals[26]	400 (Re = 0.06)		5.8

(* References at end of chapter)

where λ_{va} is obtained from Eq. 2.11. Equation 4.75 shows that shape has a relatively slight effect as full turbulence develops. Such is not the case for laminar flow where it appears that n increases as the size of an irregular particle decreases. Some of the few reported values of n are listed in Table 4.1.[24,27,30,31] These can be taken only as a guide; realistically n should be determined experimentally for the material of interest. It has been shown that the increase in n can be attributed to liquid in the surface roughness of the irregular particle. By considering the consequent "hydraulic size," Whitmore[30] was able to correlate particle size effects (including size distributions) with apparent viscosity. Even so, the data do not correlate with those for spherical particles, an effect attributed to the hydraulic film not being rigid. Furthermore, the analysis is unable to explain the fraction of the velocity drop due to non-viscosity effects. His data can, however, be explained using Steinour's modified equation[32]:

$$\frac{v_h}{v_\infty} = \left(\frac{\varepsilon - \delta}{1 - \delta}\right)^2 10^{-1.82(1-\varepsilon)/(1-\delta)} \qquad (4.76)$$

(δ = a dimensionless constant that compensates for what is considered to be stagnant liquid surrounding the particle). Rao and Sirois[33] have shown that hindered settling rates of fine particles also depend on pH, since this affects the zeta potential (Chapter 6), and thus the closeness with which particles can approach each other. Kirchberg et al.[5] have shown (Fig. 4.15) that further complications arise because

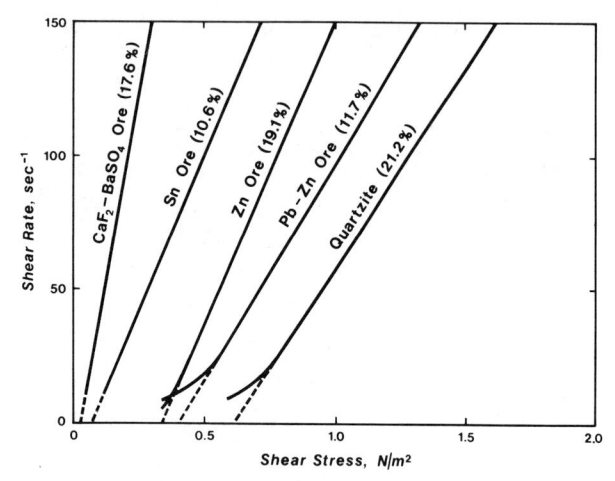

Figure 4.15. Flow curves of selected classifier overflow slurries. Numbers in parentheses are volume % solids. (After Kirchberg et al.[5])

mineral slurries behave as Bingham plastics (Eq. 4.9). This implies that a slurry of fine particles can be essentially stable, because such particles may not have enough mass to overcome the yield stress (in reality, they are actually subject to a very high viscosity, and so settle extremely slowly). Also significant is the point that the yield stress can be provided by a number of other means, such as mechanical agitation, or even by larger particles settling through the fine particles.

Work by Lockett and Al-Habbooby[23,34] has shown that Eq. 4.70 can be extended to heterogeneous systems of particles having two different settling rates. This necessitates the use of the total solids concentration; in addition, values of n and v_∞ must be determined for and applied to each component. Thus, for a mixture of components 1 and 2,

$$v_{h,1} + v_\varepsilon = v_{\infty,1}(1 - \mathfrak{F}_{V,1} - \mathfrak{F}_{V,2})^{n_1-1} \quad (4.77a)$$

$$v_{h,2} + v_\varepsilon = v_{\infty,2}(1 - \mathfrak{F}_{V,1} - \mathfrak{F}_{V,2})^{n_2-1} \quad (4.77b)$$

where

$$v_\varepsilon = \frac{v_{h,1}\mathfrak{F}_{V,1} + v_{h,2}\mathfrak{F}_{V,2}}{1 - \mathfrak{F}_{V,1} - \mathfrak{F}_{V,2}} \quad (4.77c)$$

(v_ε = upward interstitial velocity; $\mathfrak{F}_V$ = volume fraction of solid per unit volume of slurry). A theoretical treatment has been developed for considering the voidage between spherical particles,[35] but in spite of vast amounts of computational time, it is barely as accurate as the empirical methods.

In the past it has frequently been assumed[1,36,37] that, despite a lack of published experimental verification, hindered settling rates can be calculated using apparent density and viscosity in Eqs. 4.33 and 4.35. It follows from this discussion that such a method is probably valid only occasionally, for example, with spherical particles or with irregular particles having significantly different sizes (say > 100×) such as occurs in heavy media separations. Where it is desirable to compare the settling behavior of particles having settling rates of a similar order of magnitude, such as occurs in classifiers and gravity separators, it appears that more reliable results could be obtained by using equations of the form of Eq. 4.77. The calculations are more involved, but this approach has the advantage of being able to predict the negative settling rates that can occur[34] (i.e., $v_\varepsilon > v_h$). Furthermore, the effects of non-Newtonian behavior are minimized by incorporating actual experimental data.

So far the discussion has assumed that an increased solids concentration decreases the settling rate. It has been shown that in dilute suspensions (< 3% volumetric) particles of uniform size may tend to cluster and under these conditions the average drag is reduced, so that the settling rate increases.[27,38] Because these conditions are seldom encountered in commercial mineral processing (they may exist in laboratory sizing), they need be considered no further.

Example 4.8. Rework Example 4.6 using Eq. 4.70.

Solution. Equation 4.71 is used to estimate n. Since $\mathbf{Re}_p = 25$,

$$\beta = -0.55 \text{ (from Fig. 4.8)}$$

$$\therefore \quad n = \frac{4.8}{(2 - 0.55)}$$

$$= 3.31$$

(It must be assumed that the shape effect is negligible because no information is available on the effect of shape when $0.2 < \mathbf{Re}_p < 500$.)

From Eq. 4.70

$$v_h = 0.1(1 - 0.25)^{3.31}$$

$$= 0.039 \text{ m/sec}$$

This is a satisfactory agreement with Eq. 4.69 (note that if Eq. 4.69 is solved with $d_{vs} = 403\ \mu m$, $v_h = 0.073$, a substantially different result).

Fluidization and Sedimentation

4.7.1. Flux Curves

In some situations, instead of considering the settling rate v, it is more useful to consider the flow rate of particles per unit area, the *solids flux* ψ, which is given by

$$\psi = Cv \quad (4.78)$$

(ψ = mass flux of particles; C = concentration, mass per unit volume). Since $1 - C/\rho_s = \varepsilon$ and v can be represented by an equation such as Eq. 4.70, it follows that

$$\psi = v_\infty \varepsilon^n (1 - \varepsilon) \rho_s \quad (4.79)$$

which results in the characteristic flux curve for sedimentation shown in Fig. 4.16. It can be seen that a low solids flux occurs at low solids concentration (when few particles exist) *and* at high concentrations (where settling is severely reduced by hindrance effects).

4.7.2. Settling Pools

A number of mineral processing units behave like an ideal settling pool or pond. Essentially this is a pool that has a slurry fed in at one end and discharges (by overflow) at the far end a slurry of much lower solids content. As illustrated in Fig. 4.17, liquid enters the pool and flows through in plug flow. Although solids can be considered to have a horizontal velocity component equal to the horizontal liquid velocity, they are also subjected to gravitational settling forces and thus have a vertical velocity component. The limiting size of particle is

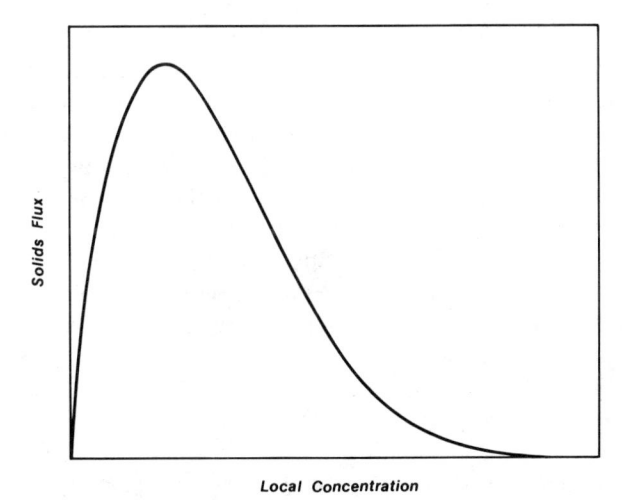

Figure 4.16. The characteristic sedimentation flux curve.

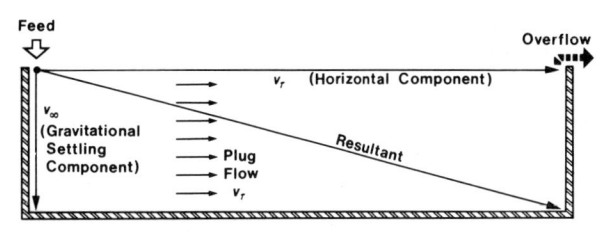

Figure 4.17. Velocity components of the limiting size particle settling in an ideal pool.

that which can just settle the depth of the pool during its passage, as illustrated vectorially in Fig. 4.17. The time t_H to settle the depth H of the pool is

$$t_H = \frac{H}{v_\infty} \quad (4.80)$$

which is also equal to the residence time τ, the time to travel the length L of the pool, given by

$$\tau = \frac{L}{v_\tau} \quad (4.81)$$

But

$$v_\tau = \frac{I_V}{BH} \quad (4.82)$$

(v_τ = horizontal velocity of slurry; I_V = volumetric flow rate through pool; B = breadth or width of pool), which leads to

$$v_\infty = \frac{I_V}{BL} = \frac{I_V}{A_T} \quad (4.83)$$

(A_T = surface area of pool).

This shows a significant fact about the ideal pool: that its ability to capture a particle is independent of pool depth, a feature known as the *area principle*. Superficially it appears from Eq. 4.83 that the liquid is flowing upward, but this is not the case. It should also be noted that the same result can be obtained by considering the pool during filling. A practical application of this principle is that batch settling tests can be used to determine data for continuous settling in ideal pools. In practice, concentration gradients, flow profiles, and eddies exist in real pools, and this results in the required area being greater than the theoretical area. It is therefore possible to define the areal efficiency η_A as

$$\eta_A = \frac{\text{theoretical surface area}}{\text{actual surface area required}} \times 100 \quad (4.84)$$

Typical values of η_A may be in the range 50–100%, but at high concentrations and flow rates the value may be considerably lower.

4.8. AGITATION, MIXING, AND PARTICLE SUSPENSION

Although mineral processing is primarily concerned with the separation of materials, in many situations the opposite, mixing or the prevention of separation,[1,7,39] is an essential part of the process. Practical cases involve the mixing of solids and liquids as in the conditioning of mineral slurries with reagents before flotation, and agitation in the flotation cell itself. Alternatively, the pumping of slurries in pipes requires maintenance of a suspension.

4.8.1. Agitation in Tanks

Mixing is normally carried out in a suitable tank, using an agitator (although on some occasions air may be used to provide agitation). As with other flow systems a Reynolds number can be used to characterize the flow type: in this case,

$$\mathbf{Re}_a = \frac{D_a^2 \mathcal{N} \rho_f}{\mu} \qquad (4.85)$$

($\mathbf{Re}_a$ = Reynolds number of the impeller or agitator; D_a = agitator diameter, $\mathcal{N}$ = agitator speed in revolutions per unit time). Similarly the power requirements can be expressed in the form of a friction factor. In this case it is effectively a drag coefficient for the agitator, called the power number $\mathbf{Po}$, given by

$$\mathbf{Po} = \frac{\mathcal{P}}{\rho_f \mathcal{N}^3 D_a^5} \qquad (4.86)$$

($\mathcal{P}$ = power). As would be expected, this correlates with $\mathbf{Re}_a$. However, for agitated tanks the situation is made far more complex by the large number of variations possible in the shape of tanks and agitators. This generally necessitates a separate friction factor chart of $\mathbf{Po}$ versus $\mathbf{Re}_a$ for each system, although it has been shown[40] that some of the agitator variables can be incorporated by means of modified $\mathbf{Po}$ and $\mathbf{Re}_a$. A further correction is necessary if the tank has no baffles, since at $\mathbf{Re}_a > 300$ the fluid may form a vortex. Under these conditions a further dimensionless number, the Froude number $\mathbf{Fr}$, must be introduced: $\mathbf{Fr}$ is the ratio of the inertia force to the gravitational force (the weight), and for agitation it is given by

$$\mathbf{Fr} = \frac{\mathcal{N}^2 D_a}{g} \qquad (4.87)$$

Thus, the friction factor chart is frequently presented as a plot of $\mathbf{Po}/(\mathbf{Fr})^n$ versus $\mathbf{Re}_a$ (Fig.

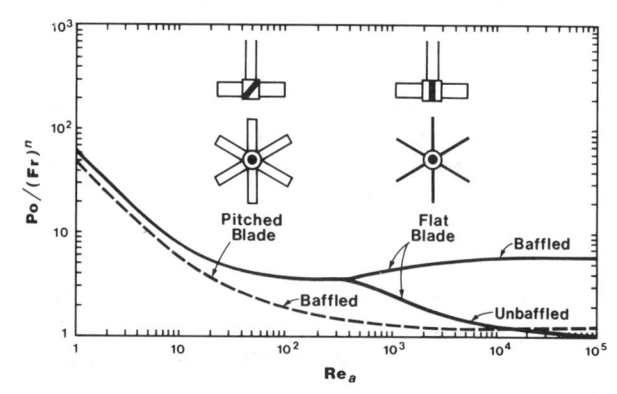

Figure 4.18. Friction factor chart for six-blade (ratio of width to diameter = 1 : 8) turbine agitator. (Data from Rushton et al.[41] and Bates et al.[42])

4.18[41,42]): in baffled tanks $\mathbf{Fr} = 1.0$; otherwise appropriate values of n must be selected.

With tank agitation, one is also concerned with the bulk movement of fluid in the tank; it is essential that the velocity of the stream leaving the impeller be sufficient to carry the currents to the remotest part of the tank. This effect can be covered by the flow or pumping number $\mathbf{Fl}$

$$\mathbf{Fl} = \frac{I_{V,a}}{\mathcal{N} D_a^3} \qquad (4.88)$$

($I_{V,a}$ = volumetric flow rate through the impeller). Experience has shown that each system requires a minimum value of $\mathbf{Fl}$ for satisfactory performance. As distinct from $\mathbf{Fl}$, $I_{V,a}$ is the rate at which fluid is moved by the impeller and is a measure of the intensity of agitation; although strictly speaking the intensity of agitation should be expressed by $I_{V,a}/A_T$ (A_T = cross-sectional area of tank).

A flotation cell is a solid/liquid/gas system; it must have desirable suspension characteristics,[43] and air distribution, also, is important.[44] It has been shown that a specific air flow number $\mathbf{Fl}_g$ is required for each type of cell and operation: $\mathbf{Fl}_g$ is still defined by Eq. 4.88 except that $I_{V,a}$ is the air flow rate to the cell $I_{V,g}$. In this situation however, as well as a minimum value of $\mathbf{Fl}_g$, there is also a maximum value, since excess air reduces the slurry volume and also causes a sweeping of undesirable minerals to the surface.

4.8.2. The Suspension of Particles

The two relevant components of particle suspension are: the force that counteracts the tendency of the particle to settle, and the degree of dispersion of the particles through the fluid (i.e., the solids concentration gradient).

Conventional practice is to retain particles in suspension by turbulence; which can be expressed in terms of an eddy diffusion coefficient thus[45]:

$$\mathscr{D}\frac{dC}{dH} = -Cv_h \qquad (4.89)$$

(C = particle concentration at height H). (This equation also serves as a definition of $\mathscr{D}$.) Even though the particles are not actually settling in the accepted sense, Eq. 4.70 can be used to correlate v_h.

A limitation of Eq. 4.89 is that it implies a transfer process only in the direction of the settling tendency. In many instances (e.g., flow through a pipe or tank) the solids concentration can vary with vertical and horizontal position. Under these conditions[46]:

$$\frac{\partial}{\partial H}\left[\mathscr{D}\frac{\partial C}{\partial H}\right] + \mathscr{D}\frac{\partial^2 C}{\partial L^2} = v_\tau \frac{\partial C}{\partial L} - v_h \frac{\partial C}{\partial H} \qquad (4.90)$$

(L = horizontal distance; v_τ = local time-averaged horizontal velocity component) (i.e., Eq. 3.11, applied at steady state in two directions).

Depending on the intensity of agitation, varying degrees of dispersion are possible. Even when all the particles are just suspended, an interface between clear liquid and slurry may exist. Increasing agitation eventually raises this interface to the top of the liquid, but still leaves a slurry density gradient from top to bottom, which can be eliminated only by even more intense agitation.

In practice, the difficulty in predicting $\mathscr{D}$ restricts the usefulness of Eqs. 4.89 and 4.90. Some workers have found them useful in describing processes where actual settling occurs[45,47] (e.g., wet classification and clarification, Chapter 10); however empirical methods have proved to be more suitable for calculating suspension conditions in mixing tanks[48-50] and pipes.[2,51]

For complete dispersion in a tank, it is generally assumed that the power per unit volume is constant. Weisman and Efferding[48] found that

$$\frac{\mathscr{P}}{V_T} = 0.092\, g v_\infty \frac{D_T}{D_a}\left(\frac{1-\varepsilon}{\varepsilon}\right)^{1/2}$$
$$\exp\left(5.3\frac{H_a}{D_T}\right)(\rho_s - \rho_l) \qquad (4.91)$$

(V_T = volume of tank; v_∞ = slip velocity, taken as the fully turbulent settling velocity; D_T = diameter of tank; H_a = height of agitator from bottom). This equation normally leads to excessively conservative design when applied to large tanks, because it allows for completely uniform dispersion, which is seldom necessary in practice. Methods for calculating power requirements for nonuniform dispersions have been described by Gates et al.[49] and are discussed further in Section 21.2.

The conveyance of slurries in horizontal pipes can conveniently be divided into two cases; homogeneous flow and heterogeneous flow.[2] In the former a negligible slurry concentration gradient exists from the top to the bottom of the pipe, and the slurry can be treated as a homogeneous fluid (e.g., Fig. 4.7). This situation normally occurs with fine particles, but the flow behavior of the slurry may no longer be Newtonian.

Heterogeneous flow occurs with larger particle sizes, and an appreciable concentration gradient exists up the pipe diameter. It is normally assumed that an essential requirement of this operation is turbulent flow, since laminar flow would allow the particles to settle. In that it is the ratio of inertia to gravitational forces, the Froude number can be expected to be an appropriate indication of minimum flow velocity. Durand[51] has found that

$$\mathrm{Fr}_{sl} = \frac{v_d}{(2gD)^{1/2}}\cdot\left(\frac{\rho_f}{\rho_s - \rho_f}\right)^{1/2} \qquad (4.92)$$

(Fr_{sl} = modified Froude number for slurry; v_d = deposition velocity, the average liquid velocity in pipes at which the particle will just settle out) gives sufficiently conservative values for v_d. It can be seen that Fr_{sl} is a modified Fr and has values from about 0.7 to 1.5 depending on particle size and concentration. More rigorous correlations have been developed[2] but are not discussed here.

The presence of particles in the liquid increases the pipe friction factor, and this increase can be determined by equations of the form[13,52]

$$\mathbf{f}_{sl} - \mathbf{f}_p = K(1-\varepsilon)^{n_1}(\mathbf{f}_p)^{n_2}(\mathbf{f}_d)^{n_3}\left[\frac{\bar{v}^2}{Dg(\rho_s/\rho_f - 1)}\right]^{n_4} \qquad (4.93)$$

($\mathbf{f}_{sl}$ = pipe friction factor with slurry; K, n_1 to n_4 = constants). The specific values of K and n_1 to n_4 depend on the type of particle dispersion that occurs in the slurry (homogeneous, heterogeneous, saltation; Fig. 21.11), and this can be determined by trial and error use of a generalized form of Eq. 4.92.[13,52] The calculation of $\mathbf{f}_{sl}$ is discussed further in Section 21.3.

Particles that have been subjected to size reduction may result in a slurry consisting of a heterogeneous slurry of large particles suspended in a homogeneous, normally non-Newtonian, slurry vehicle. Methods for treating this situation have been described in the literature[2] but are beyond the scope of this book.

The discussion so far has considered that particles either remain in suspension because of fluid turbulence or, in the case of laminar flow, have a negligible settling rate. Bagnold[53] carried out studies that showed that those particles which would otherwise have significant settling rates can be maintained in suspension even under laminar flow conditions, provided the concentration and shear rate are sufficiently high. This occurs because in concentrated slurries subjected to shear (Section 5.5) dispersive stresses are set up normal to the shear plane. The Bagnold forces are considered more fully in Chapter 5, but it is worth noting here that despite the limited amount of experimental data available,[53-56] the mechanism has proved to be capable of explaining a number of phenomena that previously defied interpretation.[53,57] Nonsettling conditions can also be attained by introducing sufficient quantities of fine particles so that the fluid viscosity becomes non-Newtonian with pseudoplastic or Bingham plastic properties.[5,58] It follows that slurry transportation does not have to be carried out using turbulent suspension. Rather, with sufficiently high concentrations laminar flow conditions can be maintained and power requirements per tonne of solid conveyed can be reduced.[58,59]

In vertical pipes deposition is not a problem, provided the flow is higher than the particles' terminal velocity.

4.9. FLOW OF LIQUID IN LAYERS

4.9.1. Straight Flow

A number of mineral processing devices involve the flow of a thin layer of water down an inclined surface under the force of gravity. Such flow is often assumed to be laminar and analyzed by assuming that flow is steady state, has constant thickness, and contains no ripples.

Assuming further that there is no shear on the upper surface (between the air and water), the shear force on the lower surface is balanced by the gravitational forces component and it can be shown that[1]

$$v_\delta = \frac{\rho_f g \sin \theta (\delta^2 - \delta_H^2)}{2\mu} \qquad (4.94)$$

($\sin \theta$ = the inclination of the plane from the horizontal; δ = thickness of the layer; δ_H = depth from the top surface at which velocity v_δ occurs). This shows that laminar flow down an inclined surface exhibits a parabolic velocity profile, with zero liquid velocity at the liquid-solid interface as illustrated in Fig. 4.19.

Integration of the flow through a cross-section of the liquid layer gives an expression for the liquid loading $\mathcal{L}$ in kilograms per second per meter of width:

$$\mathcal{L} = \frac{\delta^3 \rho_f^2 g \sin \theta}{3\mu} \qquad (4.95)$$

Provided $\delta \gg D$, these equations also apply to the laminar flow of a layer of liquid down the inside of a pipe or a surface such as occurs in a spiral concentrator. The Reynolds number in these cases is given by

$$\mathbf{Re} = \frac{v \rho_f 4 R_h}{\mu} = \frac{4\mathcal{L}}{\mu} \qquad (4.96)$$

As in the case of a pipe, turbulent flow occurs when $\mathbf{Re} > 2100$: such conditions will generally occur in a number of gravity concentrators, such as sluices and cones.

4.9.2. Spiral Flow

When a liquid is confined to a circular flow path, the centrifugal force causes transverse radial currents

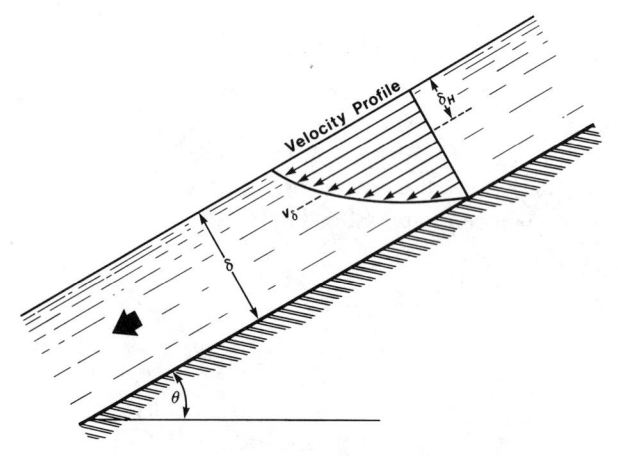

Figure 4.19. Flow of a falling film having laminar behavior.

to be superimposed on the main flow. This "river bend" action involves a radial flow inward along the bottom, an upward flow toward the inner radius, a radial outward flow along the surface of the liquid, with a downward return flow at the outer radius. Burch[60] presented a simplified analysis of such behavior and derived expressions for the transverse flow directions, the shape of the bottom surface to provide a constant upward velocity from the inward to the outward flow, and the effect of variables on this upward velocity.

REFERENCES

1. W. L. McCabe and J. C. Smith, *Unit Operations of Chemical Engineering*, 3rd ed., McGraw-Hill (1976).
2. E. J. Wasp, *Solid-Liquid Flow: Slurry Pipeline Transportation*, Trans Tech Publications (1977).
3. D. G. Thomas, "Transport Characteristics of Suspensions. VIII. A Note on the Viscosity of Newtonian Suspensions of Uniform Spherical Particles," *J. Colloid Sci.*, **20**, 267–277 (1965).
4. K. Heiskanen and H. Laapas, "On the Effects of the Fluid Rheological and Flow Properties in the Wet Gravitational Classification," *Preprints of 13th Int. Miner. Process. Congr., Warsaw, 1979*, Vol. 1, pp. 183–204, Polish Scientific Pub. (1979).
5. H. Kirchberg, E. Töpfer, and W. Scheibe, "The Effect of Suspension Properties on Separating Efficiency of Mechanical Classifiers," *Proc. 11th Int. Miner. Process. Congr., Cagliari, 1975*, pp. 219–243, Università di Cagliari (1975).
6. J. T. Davies, *Turbulence Phenomena*, Academic Press (1972).
7. A. H. P. Skelland, *Non-Newtonian Flow and Heat Transfer*, Wiley (1967).
8. L. F. Moody, "Friction Factors for Pipe Flow," *Trans. ASME*, **66**, 671–684 (1944).
9. H. Wadell, "The Coefficient of Resistance as a Function of Reynolds Number for Solids of Various Shapes," *J. Franklin Inst.*, **217**, 459–490 (1934).
10. G. G. Stokes, "Mathematical and Physical Papers" (1901), *Trans. Cambridge Phil. Soc.*, **9**, Part II, 51ff (1851).
11. W. H. Walker et al., *Principles of Chemical Engineering*, McGraw-Hill (1937).
12. H. Heywood, "Uniform and Non-Uniform Motion of Particles in Fluids," *Symp. on the Interaction Between Fluids and Particles*, Inst. Chem. Eng., **1**, (1962).
13. A. S. Foust et al., *Principles of Unit Operations*, 2nd ed., Wiley (1980).
14. M. L. Albertson, "Effect of Shape on the Fall Velocity of Gravel Particles," *Proc. Fifth Hydraulics Conf., 1952*, J. S. McNown and M. C. Boyer, (Eds.), pp. 243–261, State University of Iowa (1953).
15. W. H. Graf and E. R. Acaroglu, "Settling Velocities of Natural Grains," *Bull. Int. Assoc. Sci. Hydrol.*, **11**, No. 4, 27–43 (1966).
16. R. H. Richards, "Velocity of Galena and Quartz Falling in Water," *Trans. AIME*, **38**, 210–235 (1907).
17. H. Lorentz, *Abh. Theor. Phys.*, **82**, 541ff. (1906).
18. E. Cunningham, "Terminal Velocity of Spheres Falling in a Fluid," *Proc. R. Soc., Ser. A*, **83**, 357–365 (1910).
19. G. G. Brown et al., *Unit Operations*, Wiley (1950).
20. J. F. Richardson and W. N. Zaki, "Sedimentation and Fluidisation: Part 1," *Trans. Inst. Chem. Eng.*, **32**, 35–53 (1954).
21. J. F. Richardson, "Incipient Fluidisation and Particulate Systems," Ch. 2 in *Fluidisation*, J. F. Davidson and D. Harrison (Eds.), Academic Press (1971).
22. R. Davies, "The Experimental Study of the Differential Settling of Particles in Suspensions at High Concentrations," *Powder Technol.*, **2**, 43–51 (1968–1969).
23. M. J. Lockett and H. M. Al-Habbooby, "Relative Particle Velocities in Two-Species Settling," *Powder Technol.*, **10**, 67–71 (1974).
24. J. F. Richardson and R. A. Meikle, "Sedimentation and Fluidisation. Part III," *Trans. Inst. Chem. Eng.*, **39**, 348–356 (1961).
25. R. D. Morse, "Fluidisation of Granular Solids," *Ind. Eng. Chem.*, **41**, 1117–1124 (1949).
26. A. D. Maude and R. L. Whitmore, "A General Theory of Sedimentation," *Br. J. Appl. Phys.*, **9**, 477–482 (1958).
27. Y. S. Chong, D. A. Ratkowsky, and N. Epstein, "Effect of Particle Shape on Hindered Settling in Creeping Flow," *Powder Technol.*, **23**, 55–66 (1979).
28. H. H. Steinour, "Rate of Sedimentation: Non-Flocculated Suspensions of Uniform Spheres," *Ind. Eng. Chem.*, **36**, 618–624 (1944).
29. N. Zuber, "On the Dispersed Two-Phase Flow in the Laminar Flow Region," *Chem. Eng. Sci.*, **19**, 897–917 (1964).
30. R. L. Whitmore, "The Relationship of the Viscosity to the Settling Rate of Slurries," *J. Inst. Fuel*, **30**, 238–242 (1957).
31. C. Moreland, "Settling Velocities of Coal Particles," *Can. J. Chem. Eng.*, **41**, 108–110 (1963).
32. H. H. Steinour, "Rate of Sedimentation: Suspensions of Uniform-Sized Angular Particles," *Ind. Eng. Chem.*, **36**, 840–847 (1944).
33. S. R. Rao and L. L. Sirois, "Study of Surface Chemical Characteristics in Gravity Separation," *CIM Bull.*, **67**, 78–93 (June 1974).
34. M. J. Lockett and H. M. Al-Habbooby, "Differential Settling by Size of Two Particle Species in a Liquid," *Trans. Inst. Chem. Eng.*, **51**, 281–291 (1973).
35. C. R. Phillips and T. N. Smith, "Random Three-Dimensional Continuum Models for Two-Species Sedimentation," *Chem. Eng. Sci.*, **24**, 1321–1335 (1969).
36. G. A. Vissac, "Coal Preparation with the Modern Feldspar Jig," *Min. Eng. (Trans. AIME/SME)*, 649–655 (July 1955).
37. B. Fitch, "Why Particles Separate in Sedimentation Processes," *Ind. Eng. Chem.*, **54**, 44–51 (1962).
38. B. H. Kaye and R. P. Boardman, "Cluster Formation in Dilute Suspensions," *Symp. on Interaction Between Fluids and Particles*, pp. 17–21, Institution of Chemical Engineers (1962).

39. Shinji Nagata, *Mixing: Principles and Applications*, Kodansha (1975); Halsted Press (1975).
40. P. H. Calderbank and M. B. Moo-Young, "The Power Characteristics of Agitators for the Mixing of Newtonian and Non-Newtonian Fluids," *Trans. Inst. Chem. Eng.*, **39**, 337–347 (1961).
41. J. H. Rushton, E. W. Costich, and H. J. Everett, "Power Characteristics of Mixing Impellers. Parts I and II," *Chem. Eng. Prog.*, **46**, 395–404, 467, 476 (1950).
42. R. L. Bates, P. L. Fondy, and R. R. Corpstein, "An Examination of Some Geometric Parameters of Impeller Power," *Ind. Eng. Chem. Process Des. & Dev.*, **2**, 310–314 (1963).
43. K. Fallenius, "The Suspension of Solid Particles in Mixers and Flotation Cells," *Acta Polytech. Scand., Chem. Incl. Met. Ser.*, No. 138, Helsinki (1977).
44. N. Arbiter, C. C. Harris, and R. F. Yap, "Hydrodynamics of Flotation Cells," *Trans. AIME/SME*, **244**, 134–148 (1969).
45. H. Schubert and T. Neesse, "The Role of Turbulence in Wet Classification," *Proc. 10th Int. Miner. Process. Congr., London, 1973*, pp. 213–239, IMM (1974).
46. W. E. Dobbins, "Effect of Turbulence on Sedimentation," *Trans. ASCE*, **109**, 629–656 (1944).
47. H. Z. Sarikaya, "Numerical Model for Discrete Settling," *J. Hydrol. Div. ASCE*, **103**, 865–876 (August 1977).
48. J. Weisman and L. E. Efferding, "Suspension of Slurries by Mechanical Mixers," *AIChEJ.*, **6**, 419–426 (1960).
49. L. E. Gates, J. R. Morton, and P. L. Fondy, "Selecting Agitator Systems to Suspend Solids in Liquids," *Chem. Eng.*, **83**, 144–150 (May 24, 1976).
50. D. S. Dickey, R. W. Hicks et al., "C.E. Refresher Reprint on Liquid Agitation," published in *Chem. Eng.* (1975–1976).
51. R. Durand, "The Hydraulic Transportation of Coal and Other Minerals in Pipes," Colloq. Nat. Coal Board, London (November 1952).
52. R. M. Turian and T. Yuan, "Flow of Slurries in Pipelines," *AIChE J.*, **23**, 232–243 (1977).
53. R. A. Bagnold, "Experiments on a Gravity-Free Dispersion of Large Solids Spheres in a Newtonian Fluid Under Shear," *Proc. R. Soc., Ser. A*, **225**, 49–53 (1954).
54. R. A. Bagnold, "Some Flume Experiments on Large Grains," *Proc. Inst. Civ. Eng.*, **4**, Pt. III, Paper 6041, 174–205 (1955).
55. C. A. Shook and S. M. Daniel, "Flow of Suspensions of Solid in Pipelines. Part I. Flow with a Stable Stationary Deposit." *Can. J. Chem. Eng.*, **43**, 56–61 (1965).
56. C. A. Shook et al., "Flow of Suspensions in Pipelines. Part II. Two Mechanisms of Particle Suspension," *Can. J. Chem. Eng.*, **46**, 238–244 (1968).
57. R. A. Bagnold, "The Flow of Cohesionless Grains in Fluids," *Phil. Trans. R. Soc., Ser. A*, **249**, 235–297 (1956–1957).
58. M. E. Charles and R. A. Charles, "The Use of Heavy Media in the Pipeline Transport of Particulate Solids," Paper 12 in *Advances in Solid-Liquid Flow in Pipes and Its Application*, I. Zandi (Ed.), Pergamon (1971).
59. A. J. Carlton, "Hydraulic Conveying of Aggregates at High Solids Concentrations," *Mine and Quarry*, **7**, 52 and 55 (March 1978).
60. C. R. Burch, written contribution to the paper "Performance of a Shaken Helicoid as a Gravity Concentrator," *Trans. IMM*, **71**, 406–415 (1961–1962).

Chapter Five

Mechanisms and Processes of Particulate Separation

Law of Probable Dispersal:
"Whatever hits the fan will not be evenly distributed."

In reviewing the literature it is obvious, over and above the conflicting explanations, that no single mechanism explains the behavior in any given separator.[1-3] It is also apparent that certain mechanisms are common to devices that at first sight appear to be distinctly different. To emphasize these common features, this chapter briefly reviews in isolation some of the basic mechanisms and processes that have been proposed as those bringing about segregation. Relevance and validity are considered in later chapters.

5.1. HINDERED SETTLING

Hindered settling, or more accurately differential hindered settling, is the mechanism most often cited as producing gravity concentration.[1-3] An indication of the maximum possible separation is usually obtained by considering equisettling particles. Equations 4.33 and 4.35 provide a suitable starting point; both show that terminal velocity is a function of a density difference term and a particle size term. Thus, the conditions for equisettling particles can be expressed by

$$\frac{d_1}{d_2} = \left(\frac{\rho_{s,2} - \rho_f}{\rho_{s,1} - \rho_f}\right)^{(2+\beta)/2} \quad (5.1)$$

(β comes from the exponent to d in Eqs. 4.33 and 4.35, and is the slope of the line for $\Psi = 1.0$ in Fig. 4.8).

Typically any concentrator (separator) has a range of particle sizes fed to it and according to Eq. 5.1, this limits the separation possible. Clearly, to minimize the effect of particle size (or to increase the range of particle sizes that can be separated by density) ρ_f should be as close as possible to the less dense mineral (the case where ρ_f is between that of the two minerals is considered in Chapter 12). Equation 5.1 also implies that high values of $\mathbf{Re}_p$ (>2.0) will reduce the effect of particle size.

In practice of course, the free settling equations are unrealistic because most separators operate at comparatively high slurry concentrations where terminal velocities may be reduced by orders of magnitude. The usual method[1,2] of compensating for slurry concentration is to replace ρ_f in Eq. 5.1 by the apparent slurry density ρ_{sl}, thus

$$\frac{d_1}{d_2} = \left(\frac{\rho_{s,2} - \rho_{sl}}{\rho_{s,1} - \rho_{sl}}\right)^{(2+\beta)/2} \quad (5.2)$$

Although this approximation may be satisfactory where particle sizes or relative quantities are vastly different, it is considered that a more realistic measure of relative behavior can be obtained using Eq. 4.77, which, for particles settling with the same velocity v_h, gives

$$\frac{v_{\infty,1}}{v_{\infty,2}} = \varepsilon^{n_2 - n_1} \quad (5.3)$$

which combined with Eq. 4.35 gives, for laminar flow,

$$\frac{d_1}{d_2} = \left(\frac{\rho_{s,2} - \rho_f}{\rho_{s,1} - \rho_f}\right)^{1/2} \varepsilon^{(n_2 - n_1)/2} \quad (5.4)$$

or combined with Eq. 4.33, for turbulent conditions,

$$\frac{d_1}{d_2} = \left(\frac{\rho_{s,2} - \rho_f}{\rho_{s,1} - \rho_f}\right) \varepsilon^{(n_2-n_1)} \quad (5.5)$$

In the latter case $n_2 \simeq n_1$ even for irregular particles, so that Eq. 5.5 reduces to Eq. 5.1. With laminar flow the respective values of n depend on particle size and shape (Section 4.7). Thus, if the values of n are equal, comparison of Eqs. 5.2 and 5.4 shows that the frequently used equation (5.2) gives an unrealistically high diametric ratio. On the other hand, if the values of n are different, Eq. 5.2 may give either excessively low or high values. Put another way, the effect on size and shape of variations in n can negate or magnify the density difference. An effect worth remembering is that Eq. 4.77 shows that under certain conditions it is possible for a denser particle to rise rather than settle, when in the presence of larger, light particles.

5.2. DIFFERENTIAL ACCELERATION

In most gravity concentrators, a particle can move for only a short distance or time before its motion is stopped or significantly altered by a surface or by another particle. Thus particles are subjected to repeated accelerations (and decelerations) and under certain conditions these acceleration periods may occupy a significant proportion of the period the particle is in motion. The importance of this may be seen by considering Eq. 4.31

$$M\frac{dv}{dt} = Mg\,\frac{\rho_s - \rho_f}{\rho_s} - \tfrac{1}{2}f_d\rho_f v^2 A \quad (4.31a)$$

Initial acceleration occurs with $v = 0$, so that

$$\left(\frac{dv}{dt}\right)_I = \left(\frac{\rho_s - \rho_f}{\rho_s}\right) g = g\left(1 - \frac{\rho_f}{\rho_s}\right) \quad (5.6)$$

that is, the initial acceleration depends on the ratio of solid and liquid densities and is independent of particle size.

Equations such as Eqs. 4.48, 4.50, and 4.51 can be used to obtain estimates of the time required to approach terminal velocity, and the distance traveled during this period. Figure 5.1 shows time versus distance plots for quartz and galena in water, and it is apparent that cyclic motion starting from rest can bring about separations over wide size ranges, the range becoming wider as the cycle time decreases.

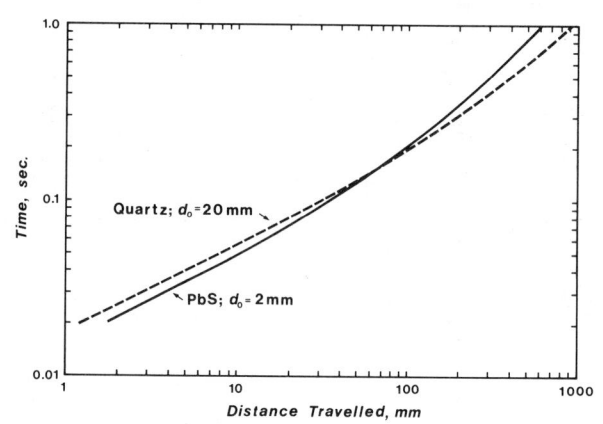

Figure 5.1. Distance versus time, for particles starting from rest (assuming Eqs. 4.38 and 4.51).

It must be emphasized that this analysis gives only an indication of the maximum separation that can be achieved. In reality hindered settling effects may severely alter the terminal velocity.

5.3. TRICKLING

Gaudin originally proposed a trickling mechanism for jigging which he called consolidation trickling.[1] Briefly, this mechanism consists of a bed that has small particles on top of large particles, with the latter in some areas resting on each other in such a manner that passageways exist between them, these passages being large enough to allow the descent of small particles. Consolidation trickling must be noticeably restricted in many normal situations where a gradual size distribution occurs. On the other hand, it is possible to imagine a bed that is in a slightly expanded state, which allows the trickling of small dense particles.

5.4. DIFFERENTIAL VELOCITY IN A FLOWING FILM

Section 4.9 described how a liquid flowing down an inclined surface has a velocity profile. From Eqs. 4.94 and 4.95 it can be concluded that the viscous drag forces on any particle in the film depend on its size, position in the film, and film thickness. In addition, each particle is subjected to gravitational forces, and a frictional force may occur if the particle is in contact with the solid surface.

Differential Velocity in a Flowing Film

Extensive analytical studies of this situation have also been presented by Gaudin.[1] Using a force balance on a particle at the liquid/solid interface

$$M \frac{dv}{dt} = F_{gravity} + F_{drag} - F_{friction} \quad (5.7)$$

and it can be shown that

$$\frac{dv}{dt} = \left(\frac{\rho_s - \rho_f}{\rho_s}\right) g(\sin\theta - \Phi \cos\theta) - \frac{18\mu v}{\rho_s d^2 \Psi} + 9g \frac{\sin\theta}{\Psi} \frac{\rho_f}{\rho_s} \left(\frac{\delta}{d} - \frac{3}{8}\right) \quad (5.8)$$

(v = velocity of particle relative to the solid surface; δ = thickness of the liquid layer; θ = inclination of solid surface to horizontal; Φ = coefficient of friction; Ψ = coefficient [near unity] to allow for nonsphericity of particle). Some conclusions that have been drawn from this equation are as follows.[1] First, there is a critical slope θ_{crit} below which sliding of the particle will not occur. In such a situation, acceleration and velocity are zero, so that the equation reduces to

$$\tan \theta_{crit} = \frac{\Phi_{crit}}{\left(\frac{\rho_f}{\rho_s - \rho_f}\right)\left(\frac{9\delta}{d} - \frac{27}{8}\right) \frac{1}{\Psi} + 1} \quad (5.9)$$

Table 5.1 gives some data derived from this equation to show the relative significance of θ and ρ_s. Thus, to hold different minerals on a sloping surface, the necessary slope depends on the mineral system.

Once a particle starts to slide, it can in principle accelerate to its terminal velocity (it actually decelerates with respect to the liquid), when dv/dt approaches zero. In this situation, Eq. 5.8 gives

$$v_{max} = \frac{d^2 g \sin\theta}{\mu} \left[\frac{(\rho_s - \rho_f)(1 - \Phi \cot\theta)\Psi}{18} - \frac{3\rho_f}{16}\right] + \frac{dg\rho_f \delta \sin\theta}{2\mu} \quad (5.10)$$

TABLE 5.1: CRITICAL SLOPES (EQ. 5.9*)

Mineral SG		Film Thickness / Particle Diameter			
		1	2	5	10
1.3	(bituminous coal)	0°44'	0°18'	0°6'	0°3'
2.65	(quartz)	3°26'	1°30'	0°33'	0°16'
4.0	(sphalerite)	5°26'	2°33'	0°59'	0°29'
7.5	(galena)	8°48'	4°46'	2°0'	1°1'
19	(gold)	13°14'	9°5'	4°40'	2°35'

*$\Phi = 0.33$; $\Psi = 0.75$ (After Gaudin.[1])

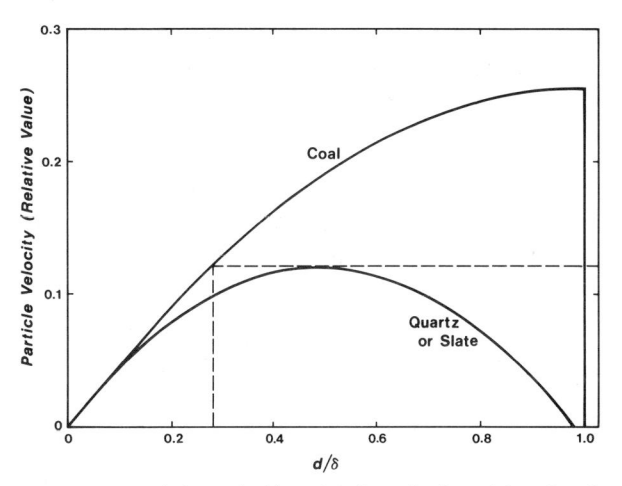

Figure 5.2. Relative velocities of various sized particles of coal, quartz, and slate on a table (Eq. 5.12; $\theta = 2°$, $\psi = 0.75$, $\Phi = 0.2$, $\rho_f = 1.0$). This implies that for a separation between coal and quartz, the feed should also be restricted to particles in the size range 0.28–1.0δ. (From A. M. Gaudin, *Principles of Mineral Dressing*. Copyright 1939, McGraw-Hill. Used with the permission of McGraw-Hill Book Company.)

(v_{max} = maximum velocity of particle relative to solid surface = velocity of liquid relative to solid surface minus the particle's terminal velocity in the liquid). It can be seen that v_{max} is the sum of two terms, one in d^2 and one in d, with the former negative and the latter positive in most practical situations. Thus, depending on the parameters, the velocity of the particle can pass through a maximum with increasing size, as shown in Fig. 5.2. In this ideal example, particles of quartz having diameters zero to δ travel down the slope more slowly than any coal particle between 0.28δ and δ, thus providing a potential separation. Again particle density and size are involved, so that for a given pair of minerals the adjustable parameters (slope and δ) should be chosen to give the maximum possible ratio of terminal velocities. It should be noted that by itself a differential velocity in a continuously fed flow down a slope does not produce a separation. To achieve this, some additional motion, such as a horizontal movement at right angles to the flow, is necessary.

The roughness of the sloping surface is too complex to analyze. If the roughness is of a magnitude considerably less than the particle size, it will of course affect the friction coefficient Φ. If it is of a magnitude greater than the particle size, the latter can be shielded or wedged by the roughness.

As the particle shape becomes more spherical, rolling occurs instead of sliding. With typically expected values of Φ (0.2–0.5), it appears that particles that can be approximated by solid figures having six or fewer faces are less likely to roll. It would therefore appear that only a relatively few materials, such as cubelike galena, or platelike mica and shale, are more inclined to slide than to roll. It follows that in many situations Φ may be essentially zero, either because the particles are rolling or because they are no longer in contact with the surface. Then

$$v_{\max} = \frac{dg \sin \theta}{18\mu} [d\Psi(\rho_s - \rho_f) - 3.4 d\rho_f + 9\delta\rho_f] \quad (5.11)$$

and the velocity increases with increasing ρ_s, the opposite effect to that found with sliding particles. Consequently, for differential velocity to be a potential separating mechanism, either the heavy mineral or both minerals must have an effective friction coefficient on the sloping surface. To what extent this coefficient could result from the presence of particles already resting on the surface is unknown.

5.5. SHEAR

When a concentrated slurry is subjected to a shear force, a dispersive force can be set up between the particles, resulting from collision between them. This situation has been studied by Bagnold,[4,5] who recognized two limiting situations: one where particle inertia dominates and a second where fluid viscosity dominates. The inertia situation can be illustrated by reference to Fig. 5.3. As the upper layer overtakes the slower one below, a succession of glancing collisions causes oscillations in the particles. It can be shown that a repulsive pressure $P_{y,i}$ should exist between the two layers, given by

$$P_{y,i} = K_B \rho_s C_L \, \text{fn}(C_L) d^2 \left(\frac{dv}{dy}\right)^2 \cos \theta_B \quad (5.12)$$

where C_L = linear concentration

$$\left[\left(\frac{C_{V,\max}}{C_V}\right)^{0.33} - 1\right]^{-1} \quad (5.12a)$$

(C_V = volumetric solids concentration; $C_{V,\max}$ = maximum volumetric solids concentration = 0.74 for spheres; K_B = coefficient; y = direction perpendicular to the layers; v = velocity of layer; θ_B = "collision" angle, Fig. 5.3). As well, there will be a proportional particle shear stress $\mathscr{S}_B$ in the shear

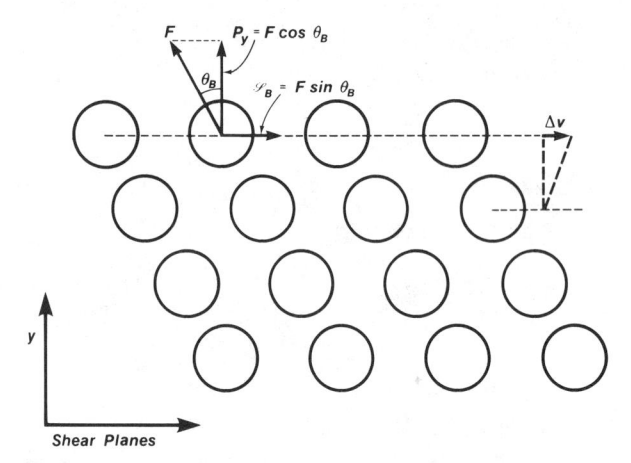

Figure 5.3. Two-dimensional schematic representation of possible statistically preferred particle arrangement (nonequidistant) that could produce a dispersive pressure proportional to shear stress in a viscous fluid. (After Bagnold.[4])

plane given by

$$\mathscr{S}_B = P_y \tan \theta_B \quad (5.13)$$

These equations assume random particle movement. The actual angle θ_B depends on collision conditions such as packing density and difficult-to-measure factors such as the amount of spin the particle has at the time of impact. Experimental results[4] showed that $\text{fn}(C_L) \sim C_L$, and for $C_L < 12$, $K_B \sim 0.042$ while $\tan \theta_B \sim 0.32$. Thus an empirical form of Eq. 5.12 is

$$P_{y,i} = 0.04 \rho_s (C_L d)^2 \left(\frac{dv}{dy}\right)^2 \quad (5.14)$$

When the fluid viscosity predominates (the second limiting case), particles in an overtaking layer are temporarily moved out of their average position as they pass by particles in the layer below. If the velocity fluctuation is assumed to be harmonic, the mean shear stress $\overline{\mathscr{S}}_{B,v}$ is expressed by

$$\overline{\mathscr{S}}_{B,v} = \mu(1 + C_L)\left[1 + \frac{\text{fn}(C_L)}{2}\right]\frac{dv}{dy} \quad (5.15)$$

This gives reasonably similar results to the experimentally determined expression

$$\mathscr{S}_{B,v} = 2.25 \mu C_L^{3/2} \frac{dv}{dy} \quad (5.16)$$

Experimental results again show $\tan \theta_B$ to have a limiting value (0.75) so that from Eqs. 5.13 and 5.16

$$P_{y,v} = 3.0 \mu C_L^{3/2} \frac{dv}{dy} \quad (5.17)$$

The transition from inertial to viscous conditions can be estimated from the dimensionless shear strain $\Re_B$ given by

$$\Re_B = \frac{\text{inertial stress}}{\text{viscous stress}} = \sqrt{C_L}\,\frac{\rho_s d^2}{\mu}\,\frac{dv}{dy} \quad (5.18)$$

Inertial conditions exist when $\Re_B > 450$ and viscous conditions when $\Re_B < 40$; transitional conditions occur between these limits.

These equations may not be rigorously applicable to mineral particles, because they were obtained under restricted conditions involving particles that were:

Possibly more elastic than mineral particles.[6]
Of the same density as the liquid.
Uniformly sized spheres.
Of one size only.

However, the relatively sparse information available suggests that they are widely applicable, since Bagnold has been able to interpret a number of phenomena that previously could not be explained.[4,5,7]

If a particle does not rise or fall under conditions of shear, it must be in equilibrium with the dispersive pressure P_y (or more generally, the vertical component thereof) balanced by the net downward pressure due to the weight less the buoyancy:

$$P_y = \frac{2(\rho_s - \rho_f)gd}{3} \quad (5.19)$$

with inertia conditions ($\Re_B > 450$), Eqs. 5.14 and 5.19 give

$$C_L^2 d \left(\frac{dv}{dy}\right)^2 = \frac{(\rho_s - \rho_f)g}{0.06\rho_s} \quad (5.20)$$

while with viscous conditions ($\Re_B < 40$), from Eqs. 5.17 and 5.19,

$$\frac{C_L^{3/2}}{d}\left(\frac{dv}{dy}\right) = \frac{2(\rho_s - \rho_f)g}{9\mu} \quad (5.21)$$

Obviously, the interpretation of particle behavior under shear conditions is complex. Not only are ρ_s, d, C_L, and dv/dy potential variables, but the transitional conditions are likely to be encountered in a number of gravity concentrating devices.

Even so, it is worth making some general comments about likely particle behavior in flowing liquid films. In a homogeneous mineral system (ρ_s, d constant), the porosity can be expected to increase toward the solid/liquid interface where the shear is highest. If a particle size distribution occurs, under inertia conditions the particles will segregate with the largest particles on top, since $d \propto 1/(dv/dy)^2$. Furthermore, with inertia conditions,

$$d \propto \frac{\rho_s - \rho_f}{\rho_s} \quad (5.22)$$

whereas with viscous conditions

$$d \propto \frac{1}{\rho_s - \rho_f} \quad (5.23)$$

Thus when inertia conditions predominate in a flowing film, segregation tends to be by particle size (with the larger particles on top), whereas under viscous conditions segregation is by density (with the denser material near the solid/liquid interface).

Equation 5.18 can be used to deduce another relevant factor. As the solids concentration increases, $\Re_B$ decreases (C_L decreases), implying that there is a tendency to suppress turbulence, although in some instances the effect may be to make the suspension more uniform.[8] Either way, it follows that in a gravity concentrator there is an optimum solids concentration (normally relatively high) at which the voidage is low enough to minimize turbulence and eliminate a concentration gradient, yet leave sufficient space between the particles to allow them movement for reasonably rapid segregation.

Bagnold also considered the conditions whereby a bed of particles can remain stationary on a sloping surface.[7] For such a situation to exist, the load of the bed on the surface has to counteract the tangential stress, and under these conditions the critical inclination of the slope θ_{crit} is given by

$$\tan\theta_{\text{crit}} = \frac{C_V \tan\theta_B}{[\rho_f/(\rho_s - \rho_f)] + C_V} \quad (5.24)$$

This should be compared with Eq. 5.9 derived by Gaudin.

5.6. SEGREGATION OF DRY PARTICLES

An increasing amount of literature on the segregation of dry particles is becoming available,[9,10] mainly because of its significance in containers such as hoppers or railway wagons. The phenomenon also is of interest in screening in particular,[11] and possibly concentration. Much of the information has been presented only in qualitative terms, such as segregation by percolation of fines through gaps between large particles, or vibration causing large particles to move up through a mass of finer particles. The former situation is of course similar to the trickling mechanisms previously considered. That large particles can rise through fine particles has been attributed to a wedging mechanism whereby the mass of the large particle, aided by friction, wedges fine particles under itself as it vibrates, and thus rises. It must also be pointed out that Bagnold's inertia shear mechanism is likely to apply in air/solid systems so that the nature of the shear force could determine whether this mechanism, or wedging, caused a large particle to rise through fines. It is likely that a lubricating effect would make the wedging noticeably less important in a solid/water system.

In general, segregation of dry powders is primarily affected by size, density having a negligible affect, although the wedging mechanism allows higher density particles to rise rather than sink, and to rise at a rate that may increase as the density difference increases.

In one of the few quantitative studies on the subject, Olsen et al.[6,12] showed that simple two-component segregation can be described by a first-order rate process (Section 3.2), and that the rate is significantly reduced by the presence of particles with intermediate properties. They also showed that segregation can be severely restricted when a significant proportion of the material (e.g., one component) has slightly elastic behavior (i.e., an ability to absorb energy). Such a situation could occur during the screening of slightly damp materials, where the dampness not only produces stickiness, but causes energy to be absorbed, thus further hindering stratification and preventing undersize particles from reaching an aperture.

5.7. CENTER OF GRAVITY

The importance of the instantaneous potential energy of a slurry system appears to have been recognized first by Mayer when he applied it to jigging separations.[13] The situation can be described with reference to Fig. 5.4. Initially, a system containing a dense and a light mineral mixed together has a center of gravity at $H/2$ as shown, and the potential energy E_I is

$$E_I = \tfrac{1}{2} A g H (H_l \rho_{bl} + H_h \rho_{bh}) \quad (5.25)$$

(A = area of bed; H_l, H_h = heights, or bulk volume fractions, of light and dense minerals respectively; ρ_{bl}, ρ_{bh} = bulk densities [including void fluid] of light and dense minerals respectively). If the bed is

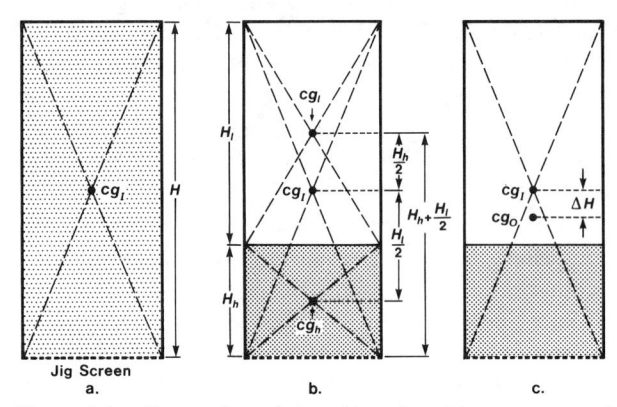

Figure 5.4. Center-of-gravity positions in a binary system of light mineral and dense mineral before (a) and after (b) stratification. The center-of-gravity lowering is shown in (c). Subscripts to cg: I = initial average; 0 = final average; l = light; h = dense fraction. (After Mayer.[13])

brought to the state shown in Fig. 5.4b with the dense mineral on the bottom and the light mineral on top, the potential energy of the segregated system E_s has been lowered to

$$E_s = \tfrac{1}{2} A g [\rho_{bl} H_l (H_l + 2 H_h) + \rho_{bh} H_h^2] \quad (5.26)$$

Thus, the energy available for segregation ΔE is

$$E_I - E_s = \Delta E = \tfrac{1}{2} A g H_l H_h (\rho_{bh} - \rho_{bl}) \quad (5.27)$$

and the center of gravity lowering ΔH is given by

$$\Delta H = \frac{1}{2} \left(\frac{H_h H_l (\rho_{bh} - \rho_{bl})}{\rho_{bh} H_h + \rho_{bl} H_l} \right) \quad (5.28)$$

For any binary mixture ΔH has a maximum value given by

$$\Delta H_{\max} = 50 - \frac{100}{(\rho_{bh}/\rho_{bl})^{1/2} + 1} \quad (5.29)$$

From this, it follows that after a separation that involves this maximum lowering of the center of gravity, the center of gravity will lie in the boundary between the two materials. Whether the maximum value can be achieved therefore will depend on whether the materials are present in the correct proportions.

Even if the proportions are not correct for a maximum value, ΔH (Eq. 5.28) is still a true measure of the ease with which a binary mixture can be separated, since it represents the potential energy per unit mass available for bringing about a separation. Because it was originally applied to jigging, ΔH was referred to as the jiggability of the material; but this is equivalent to the widely used more general term "washability" (used for coal cleaning), or even as a general concept of separability.

Mayer proposed that the rate of segregation would be first order, that is (from Eq. 3.1),

$$\frac{d\,\Delta E}{dt} = -k\,\Delta E \qquad (5.30)$$

(ΔE = potential energy reduction still possible), from which it can be shown that

$$H_t = \Delta H \exp(-kt) \qquad (5.31)$$

(H_t = height of the center of gravity, at time t, above its position when a perfect separation exists). The rate constant k could be expected to depend on ΔH, the resistance to particle movement (friction and drag), and bed action. Thus, ΔH is actually a measure of both the quality of separation possible *and* how rapidly it can be achieved.

Another significant aspect of this analysis, the use of bulk densities rather than mineral densities, recognizes the fact that particle shape and packing can effect separations, in that a denser material can actually have a lower bulk density than a less dense material and so apparently give a reverse separation. Again, because the bulk density is important, the particle size distribution becomes significant, since this markedly affects the space filling ability of a sample.[14]

This concept is also useful in explaining the difficulty of separating streams containing large proportions of middlings, since their center of gravity is the same whether they are evenly distributed or in a layer between the dense and light mineral. Thus, their presence simply lowers the total possible potential energy change of the system and makes the separation more difficult. This can be seen by making the denominator $\Sigma \rho_b H$ in the right-hand side of Eq. 5.28.

5.8. SEPARATION LIMITS AND MIXING

In practice a number of factors prevent the achievement of perfect separations. Some of these have been considered here and in other chapters. First, there is always a kinetic limitation; insufficient time is available for the particles to separate (Section 3.2). Even if sufficient time is available, the separation can be limited by the presence of middling particles, or by particles of different minerals having the same performance under the conditions used. Competing forces are also a factor to be considered and in some situations they may not be obvious. For example, difficulties may be encountered when separating fine particles in gravity concentrators because the surface charge on the particles starts to become a significant force.[15,16]

But beyond these factors, there is the aspect that can be considered to be mixing by the separator. This can be attributed in many cases to hydrodynamic factors,[17] such as non-plug flow, eddying resulting from equipment design, and random agitation of particles; or the forces necessary to separate or move the heavier particles may be sufficient to remix the lighter particles. Rigorous analysis of these factors is obviously not possible, but a number of workers have analyzed separators by assuming that mixing effects limit a separation and have used empirical diffusion coefficients to account for this[17-19] (Sections 3.5.2, 4.8.2, 10.3.2, and 13.3).

5.9. GENERAL COMMENTS

In reviewing these aspects of separation, it is possible to consider them in terms of two concepts: thermodynamics and kinetics. The thermodynamics concepts simply show what is possible, or consider the cause, without presenting any information on the actual mechanism whereby particle separation occurs, or how rapidly the separation is achieved. The center-of-gravity and shear processes fall into this category. These concepts are useful in that they indicate the *driving force* for the separation, and also give some indication of the degree of separation possible.

Most of the other segregation processes discussed involve the older hydrodynamic concepts, which attempt to analyze each particle's performance individually, using a sort of mixed thermodynamic/kinetic analysis. Although these methods can reasonably predict a single particle's behavior, and thus indicate the *mechanisms* whereby particles come to a segregated state, they are quite inadequate for dealing with the variable, concentrated suspensions encountered in practice. One approach used to simplify the problem is to consider that the separation process occurs at a constant rate: the kinetic approach, using first-order equations (Section 3.2). This technique is most suitable for separations that can be conveniently subdivided into a number of incomplete stages (e.g., flotation, screening), and as such, it is less suitable for classification and gravity separations (although

Mayer has suggested its application to jigging,[13] Eq. 5.31).

In most gravity separations, the process rate is not constant; rather, acceleration is a significant factor. This implies that a second-order equation is necessary to describe the process (Eq. 3.6). Some Russian work has shown that consideration of a total energy balance in a gravity separator, such as a jig, can lead to a second-order equation that satisfactorily describes the separation.[20]

REFERENCES

1. A. M. Gaudin, *Principles of Mineral Dressing*, McGraw-Hill (1939).
2. A. F. Taggart, *Elements of Ore Dressing*, Wiley (1951).
3. E. J. Pryor, *Mineral Dressing*, 3rd ed., Elsevier (1965).
4. R. A. Bagnold, "Experiments on a Gravity-Free Dispersion of Large Solids Spheres in a Newtonian Fluid Under Shear," *Proc. R. Soc., Ser. A* **225**, 49–63 (1954).
5. R. A. Bagnold, "Some Flume Experiments on Large Grains," *Proc. Inst. Civ. Eng.*, **4** Pt. III, 174–205 (1955).
6. E. G. Rippie, J. L. Olsen, and M. D. Faiman, "Segregation Kinetics of Particulate Solids Systems. II. Particle Density-Size Interactions and Wall Effects," *J. Pharm. Sci.*, **53**, 1360–1363 (1964).
7. R. A. Bagnold, "The Flow of Cohesionless Grains in Fluids," *Phil. Trans R. Soc., Ser. A,* **249**, 235–297 (1956–1957).
8. C. A. Shook et al., "Flow of Suspensions in Pipelines. Part II, Two Mechanisms of Particle Suspension," *Can. J. Chem. Eng.*, **46**, 238–244 (1968).
9. J. C. Williams, "The Segregation of Particulate Materials: A Review," *Powder Technol.*, **15**, 245–251 (1976).
10. M. H. Cooke, D. J. Stephens, and J. Bridgewater, "Powder Mixing: A Literature Review," *Powder Technol.*, **15**, 1–20 (1976).
11. J. C. Williams and G. Shields, "The Segregation of Granules in a Vibrated Bed," *Powder Technol.*, **1**, 134–142 (1967).
12. J. L. Olsen and E. G. Rippie, "Segregation Kinetics of Particulate Solids Systems. I. Influence of Particle Size and Particle Size Distribution," *J. Pharm. Sci.*, **53**, 147–150 (1964).
13. F. W. Mayer, "Fundamentals of a Potential Theory of the Jigging Process," *Proc. 7th Int. Miner. Process. Congr., New York, 1964,* Vol. I, pp. 75–97, Gordon and Breach (1965).
14. C. W. J. Van Koppen, "A Contribution to the Fundamentals of the Jigging Process," *5th Int. Coal Prep. Congr., Pittsburgh,* Paper B3, pp. 85–97 (1966).
15. S. R. Rao and L. L. Sirois, "Study of Surface Chemical Characteristics in Gravity Separation," *CIM Bull.*, **67**, 78–93 (June 1974).
16. R. O. Burt, "A Study of the Effect of Deck Surface and Pulp pH on the Performance of a Fine Gravity Concentrator," *Int. J. Miner. Process.*, **5**, 39–44 (1978).
17. V. I. Revnivtsev et al., "Hydrodynamic Research into Gravity Concentration Processes and Methods for Their Improvement," *Proc. 10th Int. Miner. Process. Congr., London, 1973,* pp. 293–310, IMM (1974).
18. H. Schubert and T. Neesse, "The Role of Turbulence in Wet Classification," *Proc. 10th Int. Miner. Process. Congr., London, 1973,* pp. 213–239, IMM (1974).
19. O. N. Tikhonov and V. V. Dembovsky, "Automation Process Control in Ore Treatment and Metallurgy. Part IV. Industrial Measuring and Control Systems," Ch. 11, El-Tabbin Metallurgical Institute for Higher Studies (1973).
20. N. N. Vinogradov et al., "Research on the Separation Kinetics of Gravity Processing in Mineral Suspensions," *Proc. 11th Int. Miner. Process. Congr., Cagliari, 1975,* pp. 319–336, Università di Cagliari (1975).

Chapter Six

Surfaces and Interfaces

Anderson's Law: "I have yet to see any problem, however complicated, that, when looked at in the right way, did not become still more complicated."

To gain an understanding of the processes involved in mineral concentration, it is necessary to study the chemical and physical properties of mineral surfaces. It is also necessary to understand the relationship between the bulk solid, liquid, and gaseous phases, and the interfaces between them.

Surface properties are important with both wet and dry processes. Surface conductivity is the basis for separation in electrostatic separations (dry). The chemistry of mineral surfaces (more correctly, the interfacial chemistry at the mineral surface) is the basis for the flotation, oil extraction, flocculation, and agglomeration processes, and it is a factor to be considered in magnetic and gravity separations, thickening and filtration, grinding and classification.

The literature on both the fundamentals of interfacial chemistry and their application in mineral processing is extensive.[1-9] It is clear, however, that the task of relating fundamental studies to commercial mineral processing operations is not a simple one. Most basic work has been carried out on particles of pure minerals in a well-controlled environment. Ore particles, however, are not pure. The mineralogical heterogeneity of the particles of an ore can have a significant effect on the particles' surface properties.[10] Also, minerals differ from one ore to another (even within an ore body) because of the presence of trace elements, and these trace elements sometimes concentrate at grain surfaces and have influence far out of proportion to their concentration in the mineral. Particle shape and surface topography (Fig. 6.1), which can vary from ore to ore, also affect surface properties.

The crystalline structure of a mineral particle terminates abruptly at the particle surface. Thus, the atoms at the surface of the crystal structure must come to equilibrium with their surroundings.[11] In an aqueous system, the surface atoms should be in equilibrium with the ions existing in solution, and since these ions can be controlled, it is possible to control the chemical nature of the mineral surface (Fig. 6.2). Clearly, the interface between the bulk of the mineral particle and the bulk solution is complex.

The mineral-solution interface is the most important in mineral processing. However, many separation processes require the establishment of other interfaces, such as between the mineral and air, or between the mineral and another liquid.

6.1. THERMODYNAMICS OF INTERFACES

The thermodynamics of interfaces has been discussed in great detail in the technical literature.[5-7,12-14] The aspects of most relevance to the mineral processing engineer have been discussed by Gaudin[5] and Klassen and Mokrousov,[6] and summarized in recent reviews.[13,14] In this section, two aspects are considered; the thermodynamics of adsorption on mineral surfaces and at air-water interfaces, and the thermodynamics that control the attachment of mineral particles to air bubbles.

The enthalpy of a unit area of any surface, which is equal to the total surface energy E_S, is given by

$$E_S = \gamma + T\mathfrak{E}_S \qquad (6.1)$$

(γ = *surface free energy* per unit area; $\mathfrak{E}_S$ = surface entropy per unit area; T = absolute temperature). Any interface is the site of a tensile force in the

Figure 6.1. Photomicrographs showing variations in particle size, fine particle adsorption, and morphology.

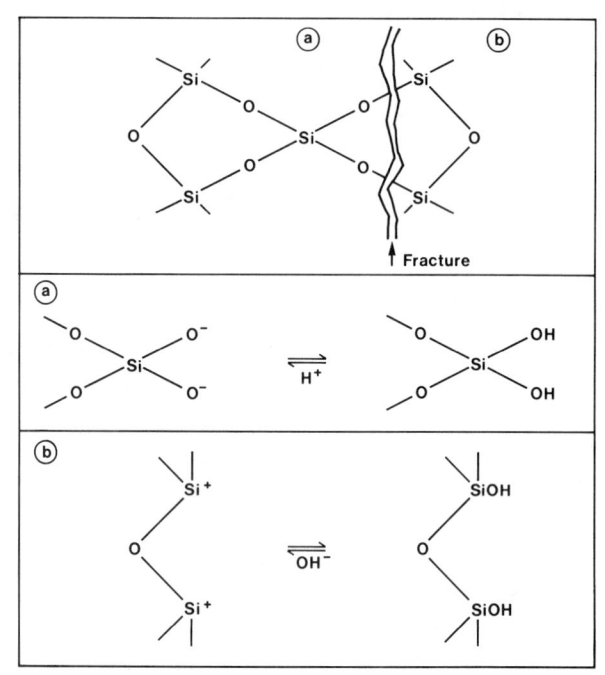

Figure 6.2. Example mechanism showing the origin of the electrical charge on a mineral in aqueous solution. (After Gaudin and Fuerstenau.[16])

plane of the interface and characteristic of the interface. This tensile force is known as the *interfacial tension* or *surface tension* and can be considered to be equal to the surface free energy (the interfacial tension with units of newtons per meter is numerically equal to the surface free energy with units of joules per square meter). The surface entropy at constant pressure is given by

$$\mathfrak{E}_s = -\frac{d\gamma}{dT} \qquad (6.2)$$

so that the total surface energy can be related to the interfacial tension by

$$E_s = \gamma - T\frac{d\gamma}{dT} \qquad (6.3)$$

For water and aqueous solutions (and in fact for most liquids) in contact with a gas, the surface tension decreases linearly with temperature.[7] Water in contact with air at 20°C has a surface tension of 0.073 N/m and $d\gamma/dT$ equal to -1.6×10^{-4} N/m · °C. The total surface energy of the water is thus 0.12 N/m. For solid-liquid and solid-gas interfaces the evaluation of surface energies is complex[7] and is not considered here.

6.1.1. Adsorption at Interfaces

Adsorption at an interface is described by the *Gibbs adsorption equation*. This equation is found in various forms, but the most useful form for mineral processing purposes is[12,13]:

$$d\gamma = -\mathfrak{E}_s dT - \sum_{i=1}^{i} \Gamma_i d\mathcal{U}_i \qquad (6.4)$$

Thermodynamics of Interfaces

This equation relates the interfacial tension at any temperature T to the chemical potential $\mathcal{U}_i$ of the various species in the bulk and the *surface excess* or *adsorption density* Γ_i of those species. The surface excess is the excess concentration at the interface over that which would be expected if the solution phase were of uniform concentration up to the dividing line between the two phases. By defining Γ_i such that there is no surface excess of water, and considering that temperature is maintained constant, Eq. 6.4 can be written as

$$d\gamma = -\sum_{i=1}^{i} \Gamma_i d\mathcal{U}_i \quad (6.5)$$

Now the chemical potential $\mathcal{U}_i$ of a species in solution is related to the activity $\mathcal{A}_i$ of that species by

$$\mathcal{U}_i = \mathcal{U}_i^0 + RT \ln \mathcal{A}_i \quad (6.6)$$

($\mathcal{U}_i^0$ = standard chemical potential of species i in solution). The change in chemical potential is then given by

$$d\mathcal{U}_i = RT\, d(\ln \mathcal{A}_i) \quad (6.7)$$

Thus Eq. 6.5 can be written as

$$\Gamma_i = -\frac{1}{RT} \frac{\partial \gamma}{\partial \ln \mathcal{A}_i} \quad (6.8)$$

with the term $\partial \gamma / (\partial \ln \mathcal{A}_i)$ evaluated at constant temperature and with the chemical potential of all other species constant. For the adsorption at an interface of a single species and at constant temperature, this equation is most useful in the form

$$\Gamma_i = -\frac{1}{RT} \frac{d\gamma}{d \ln C_i} \quad (6.9)$$

with the bulk concentration C_i being a good approximation of the bulk activity $\mathcal{A}_i$ at the low concentration involved.

The surface excess or adsorption density as determined by the equations above can be either positive or negative. A negative value corresponds to a depletion at the surface in the concentration of the species being considered. This is shown in Fig. 6.3. In fact, all inorganic ions fall in this category when in an aqueous solution in contact with air. If an ion or molecule is adsorbed at the air-water interface (i.e., it has a positive adsorption density), it is said to be *hydrophobic*. If the ion or molecule is not adsorbed at the interface (i.e., has a negative adsorption density), it is said to *hydrophilic*.

The two cases of most interest are the adsorption at the air-water interface, and the adsorption at the surface of minerals in water. These are now considered.

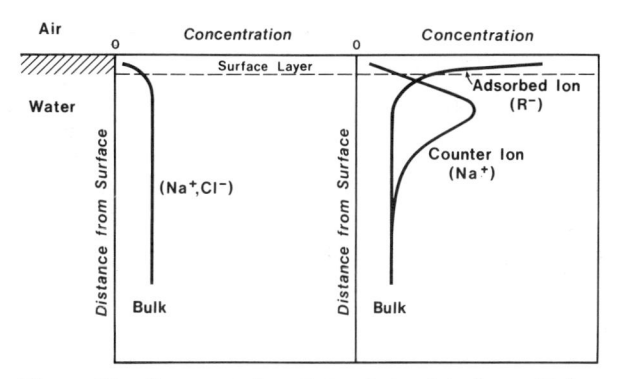

Figure 6.3. Representation of air-solution interface with inorganic and organic ions present. (After Gaudin.[5])

6.1.2. The Air-Water Interface

Adsorption at the air-water interface is of particular interest in the flotation process (Chapter 16). Frothers, many of which are alcohols, are surface-active agents that are used to control the froth in flotation. The curve shown in Fig. 6.4 is typical of the way the addition of an alcohol reduces the surface tension. The surface tension of the solution is reduced from that of water (0.073 N/m) to close to that of the alcohol, although the alcohol is only a small fraction of the solution. This indicates that a large fraction of the surface of the alcohol-in-water solution must be the alcohol.

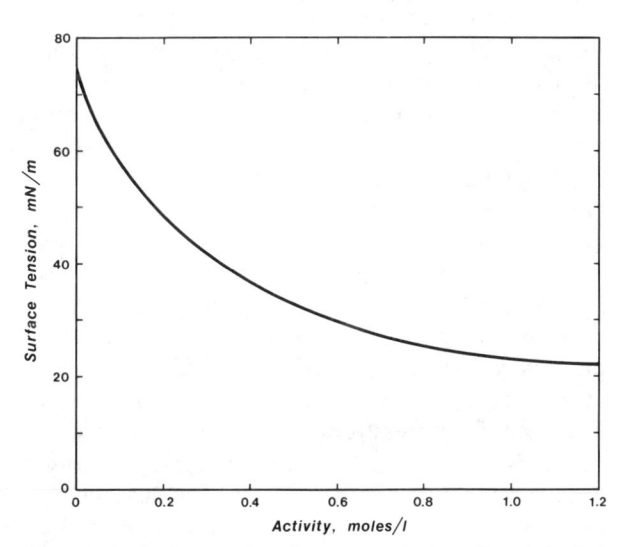

Figure 6.4. Surface tension of aqueous solution of butyl alcohol at 20°C. (After de Bruyn and Agar.[15])

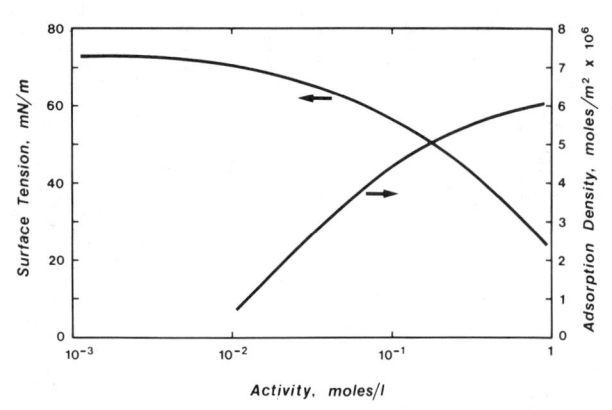

Figure 6.5. Surface tension and corresponding adsorption density curves for aqueous butyl alcohol solutions at 20°C. (After de Bruyn and Agar.[15])

To quantitatively determine the extent of the adsorption of the alcohol at the air-water interface, the Gibbs adsorption equation (Eq. 6.8 or 6.9) can be used. A common and convenient practice is to plot the experimental data shown in Fig. 6.4 in the form shown in Fig. 6.5, that is, surface tension against the logarithm of the bulk activity or concentration. The slope of this curve then gives the adsorption density directly. The adsorption density curve is also shown in Fig. 6.5. When plotted for a constant temperature, this curve is known as the *adsorption isotherm*. From the adsorption isotherm for butyl alcohol at 20°C, it can be seen that, for an activity of 0.712 moles/l, the adsorption density is 6.03×10^{-6} moles/m^2, or 3.65×10^{18} molecules/m^2. The area per adsorbed butyl alcohol molecule is the reciprocal of the adsorption density, and is equal to 0.274 nm^2. Since the area per molecule for close-packed films of long-chain carboxylic acids and alcohols is about 0.216 nm^2, it would appear that the adsorbed butyl alcohol at the water surface is monomolecular.[13]

In flotation, collectors are also used. Collectors are heteropolar organic compounds that are required to adsorb at both the mineral-water interface and the air-water interface. Their function in flotation is to selectively attach mineral particles to air bubbles. Their adsorption at mineral-water interfaces is considered in Section 6.1.4. The adsorption at the air-water interface is discussed briefly here.

Collectors differ from frothers in that they are generally electrolytes, so the adsorption of ionic species at the air-water interface must also be considered. The adsorption of the collector sodium dodecyl sulfate was analyzed by deBruyn and Agar.[15] For this system, the Gibbs adsorption equation (Eq. 6.5) may be written

$$d\gamma = -\Gamma_{R^-} d\mathcal{U}_{R^-} - \Gamma_{Na^+} d\mathcal{U}_{Na^+} \quad (6.10)$$

where R$^-$ refers to the dodecyl sulfate anion. The dodecyl sulfate ion will be adsorbed right at the air-water interface, whereas the sodium ion will be adsorbed as a counterion adjacent to the interface but within the aqueous phase to provide for electrical neutrality (Fig. 6.3). If no other electrolytes are present and R$^-$ and Na$^+$ are the only species adsorbing at the interface, Eq. 6.10 becomes

$$d\gamma = -2\mathbf{R}T \, \Gamma_R d(\ln \mathcal{A}_{NaR}) \quad (6.11)$$

If the collector is a weak electrolyte, such as the salt of a fatty acid, the adsorption of the associated acid must also be considered. This means that the pH of a solution can have a marked effect on the surface tension.

This discussion has considered only the interface between an aqueous solution and air. It should be clear, however, that the same principles apply when a different third phase is used, either another gas (such as nitrogen) or a liquid, such as in the oil extraction process.

6.1.3. The Electrical Double Layer

Adsorption at the surface of a mineral particle is largely controlled by the electrical nature of that surface. From Fig. 6.3, it is seen that if an ionic species is adsorbed at an interface, then for electroneutrality *counterions* must also adsorb.

Figure 6.6 represents a mineral surface and the electrical double layer.[16,17] This is known as the *Stern model* of the electrical double layer.

Those ions that are *chemisorbed* on the mineral surface establish the *surface charge* and are termed *potential-determining ions*. The potential-determining ions may be ions of which the mineral is composed, hydrogen or hydroxyl ions, collector ions that form insoluble salts with ions in the mineral surface, or ions that form complex ions with the ions in the mineral surface. The surface charge on a mineral is determined by the adsorption density of the potential-determining ions on the mineral surface. For a univalent salt, the surface charge σ_s is given by

$$\sigma_s = \mathscr{F}(\Gamma_{M^+} - \Gamma_{A^-}) \quad (6.12)$$

(Γ_{M^+}, Γ_{A^-} = adsorption densities of the potential-determining cation and anion respectively; $\mathscr{F}$ = Faraday constant). For many minerals, such as

Thermodynamics of Interfaces

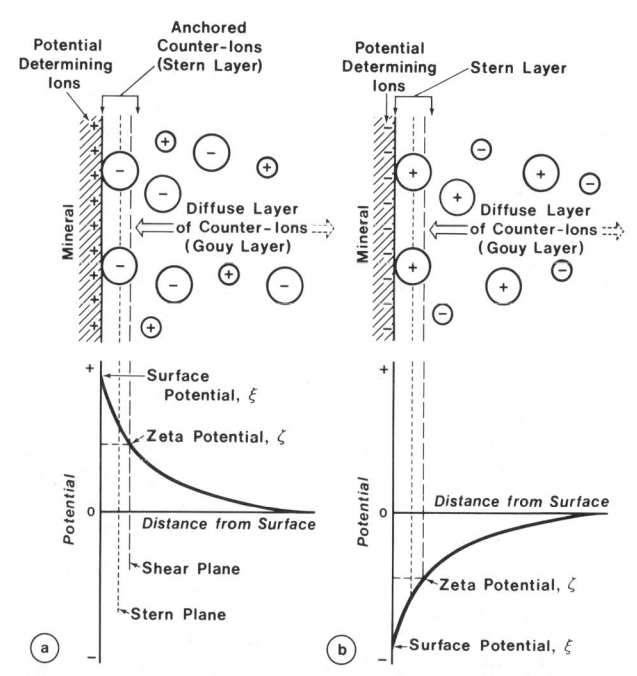

Figure 6.6. The electrical double layer at a mineral surface in aqueous solution. (After Gaudin and Fuerstenau[16]; Kruyt.[17])

Each ion in solution is in equilibrium with the mineral surface, thus

$$M^+_{aq} \rightleftharpoons M^+_s$$

$$A^-_{aq} \rightleftharpoons A^-_s$$

Considering only the cation adsorption, the associated free energy change can be written as

$$\Delta \gamma_{ads} = \Delta \gamma^\circ_{ads} + RT \ln \left[\frac{\mathscr{A}^s_{M+}}{\mathscr{A}_{M+}} \right] \quad (6.15)$$

($\Delta \gamma_{ads}$ = free energy change on adsorption; $\Delta \gamma^\circ_{ads}$ = free energy change at standard state; $\mathscr{A}^s_{M+}$ = activity of cation at surface).

At equilibrium, chemical work must be balanced by the electrical work, so

$$\Delta \gamma_{ads} = -\mathscr{F} \xi \quad (6.16)$$

(ξ = total potential drop across interface [= surface potential]).

At the PZC, $\xi = 0$, therefore

$$\Delta \gamma^\circ_{ads} = -RT \ln \left[\frac{\mathscr{A}^{so}_{M+}}{\mathscr{A}^0_{M+}} \right] \quad (6.17)$$

($\mathscr{A}^{so}_{M+}$, $\mathscr{A}^0_{M+}$ = activity of M^+ at surface and in bulk solution at PZC). From Eqs. 6.15–6.17,

$$\xi = \frac{RT}{\mathscr{F}} \ln \left[\frac{\mathscr{A}_{M+} \mathscr{A}^{so}_{M+}}{\mathscr{A}^s_{M+} \mathscr{A}^0_{M+}} \right] \quad (6.18)$$

oxides, the potential-determining ions are H^+ and OH^- ions, so Eq. 6.12 can be written as

$$\sigma_s = \mathscr{F}(\Gamma_{H^+} - \Gamma_{OH^-}) \quad (6.13)$$

The conditions at which the surface charge becomes zero are of particular importance. The activity of the potential-determining ions at which the surface charge is zero is called the *point of zero charge* (PZC) of the mineral. When the H^+ and OH^- ions are potential determining, the PZC is expressed in terms of pH. Figure 6.7 represents the relationship between surface potential and the activity of the potential-determining ion at a mineral surface. Note that in region A in Fig. 6.7, the surface is as shown in Fig. 6.6a, and in region B the surface is as shown in Fig. 6.6b.

Although the surface charge cannot be measured, it is possible to determine the potential difference between the surface and the bulk solution, that is, the surface potential. Consider a sparingly soluble salt-type mineral (M^+A^-) in equilibrium with an aqueous solution. The concentration of the ions in solution is given by

$$K_{sp} = \mathscr{A}_{M+} \mathscr{A}_{A-} \quad (6.14)$$

(K_{sp} = solubility product; $\mathscr{A}_{M+}$, $\mathscr{A}_{A-}$ = activities of cation and anion, respectively).

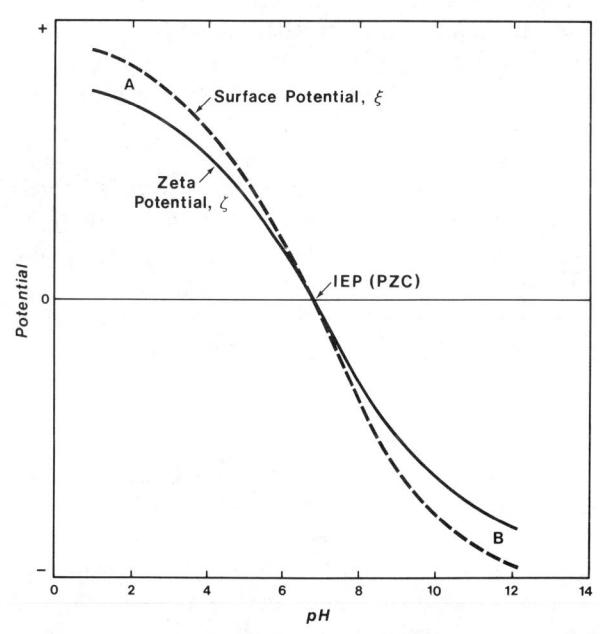

Figure 6.7. Representation of the effect of the activity of the potential-determining ion on the surface potential and zeta potential at a mineral-solution interface.

Since M^+ is a constituent of the mineral, its activity at the surface can be assumed constant, so:

$$\mathscr{A}^s_{M^+} = \mathscr{A}^{so}_{M^+} \qquad (6.19)$$

Thus

$$\xi = \frac{RT}{\mathscr{F}} \ln \left[\frac{\mathscr{A}_{M^+}}{\mathscr{A}^o_{M^+}} \right] \qquad (6.20a)$$

This equation is frequently presented in the general form[17]:

$$\xi = \frac{RT}{N\mathscr{F}} \ln \left[\frac{\mathscr{A}}{\mathscr{A}_{PZC}} \right] \qquad (6.20b)$$

($\mathscr{A}$ = activity of potential-determining ion; $\mathscr{A}_{PZC}$ = activity of potential-determining ion at PZC; N = valence of potential-determining ion).

The PZC is important because the sign of the surface charge is of major significance. Counterions are adsorbed by Coulombic or electrostatic attraction at the mineral surface bearing the potential-determining ions. The counterions may be any ions from the solution of opposite charge to that of the mineral surface after chemisorption of the potential-determining ions. If the counterions are adsorbed by electrostatic attraction alone, their source in solution is called an *indifferent electrolyte*. If the counterions possess a special affinity for the mineral surface but are not chemisorbed, they are known as *specifically adsorbed ions*. Specifically adsorbed ions adsorb strongly at the surface because of such phenomena as covalent bond formation and solvation effects. Examples of specifically adsorbed ions include hydrated polyvalent metal cations and their hydroxy complexes, and also certain flotation collectors.

Counterions approach the mineral surface only as close as their size or sheath of hydration will allow. The plane through the center of charge of the counter ions adjacent to the surface is called the *Stern plane* (Fig. 6.6). This layer of counterions is known as the *Stern layer*. Adjacent to the Stern layer is the *diffuse layer of counterions* known as the *Gouy layer*. Figure 6.6 also schematically represents the change in potential through the double layer, from the surface potential at the mineral surface to zero at an "infinite" distance from the surface, that is, in the bulk of the solution. Equations describing the charge distribution in the double layer (e.g., the Gouy-Chapman equation), and the thickness of the double layer, are available in the literature[13,17,18] and are not discussed here. It should be noted that Grahame[18] has modified the Stern model of the double layer to include two planes of closest approach for counterions, the "inner Helmholtz plane" for nonhydrated ions and the "outer Helmholtz plane" for hydrated ions. Although this in fact might well be a more exact representation of the actual situation, it does not add significantly to the understanding of mineral processing systems.

Electrokinetic Phenomena. When there is motion of the solution relative to the mineral surface, shear occurs at a plane close to the boundary between the Stern and Gouy layers, although not necessarily coincident with it,[16-19] as shown in Fig. 6.6. This means that the ions in the Stern layer remain "anchored" to the mineral surface, while the ions in the diffuse Gouy layer are carried with the solution. The electrical potential at this shear plane is known as the *zeta potential*, ζ. The zeta potential can be measured readily by a number of methods as indicated in Section 6.1.7. The plane of shear is shown in Fig. 6.6. Also shown is a plot of potential as a function of the distance from the mineral surface through the double layer. As can be seen from this diagram, the magnitude of the zeta potential is always less than that of the surface potential. In Fig. 6.7, the zeta potential is plotted against the solution pH. As mentioned earlier when discussing the surface potential, the nature of the mineral-solution interface corresponding to points A and B on this figure is as shown in Fig. 6.6.

A knowledge of how the zeta potential of a mineral varies with changes in solution conditions is of great value to the mineral engineer. By assuming that the shear plane and the Stern plane coincide, a measure of the zeta potential can provide an understanding of the adsorption phenomena occurring. This assumption is possible because any potential difference between the shear plane and the Stern plane is small compared to the total potential difference across the double layer. It is an assumption very widely used in the literature on the surface chemistry of mineral systems and on the surface phenomena involved in flotation, flocculation, and agglomeration.

Just as the point at which the surface charge became zero (the PZC) was shown to be important, so the point at which the zeta potential becomes zero has special significance. This point is known as the *isoelectric point* (IEP). There has been considerable confusion in the use of the terms "point of zero charge," "isoelectric point," and a third term, the "zero point of charge," which has been used for

Thermodynamics of Interfaces

both the PZC and the IEP. Our usage follows that proposed by Somasundaran,[20] which appears to have become accepted as the standard usage in the mineral processing literature. It should be noted, however, that in many cases the PZC and the IEP coincide or are extremely close. But this is not always the case, especially in the presence of specifically adsorbing species, such as polyvalent and surfactant ions.[20] The IEP is shown in Fig. 6.7, and in some cases this point may also be the PZC.

To illustrate the importance of the zeta potential and the value of identifying the isoelectric point, the results of flotation tests on goethite by Iwasaki et al.[21] are presented in Fig. 6.8. Figure 6.8a shows the zeta potential as a function of pH for three different levels of NaCl. The IEP is independent of the NaCl level at a pH of 6.7, so the NaCl is acting as an indifferent electrolyte. Figure 6.8b shows the recovery by flotation of the goethite using both anionic and cationic collectors. When the zeta potential is positive, that is, for pH values less than the IEP, the anionic collector will adsorb on the goethite surface and flotation results. The anionic collector does not adsorb on the goethite surface when the surface

TABLE 6.1: PZC OR IEP VALUES FOR SELECTED MINERALS

Mineral	PZC or IEP
Barite	3.4 (pBa 3.9 - 7.0)
Calcite	8.2
Cassiterite	4.5
Chalcopyrite	2.0 - 3.0
Chromite	5.6 - 7.2
Corundum	9.0 - 9.4
Fluorite	6.2 (pCa 2.6 - 7.7)
Galena	2.4 - 3.0
Goethite	6.7 - 6.8
Hematite	4.8 - 6.7
Kaolinite	3.4
Magnetite	6.5
Molybdenite	1.0 - 3.0
Pyrite	6.2 - 6.9
Pyrrhotite	3.0
Quartz	2.0 - 3.7
Rutile	5.8 - 6.7
Scheelite	10.2(pCa 4.0 - 4.8)
Sphalerite	2.0 - 7.5
Zircon	4.0 - 5.8

(ζ) is negative; only the cationic collector adsorbs on the goethite surface when the surface (ζ) is negative.

Values of the PZC or IEP have been determined for many minerals; Table 6.1 lists values for selected minerals. Values for other minerals are available in the literature.[13,22-25] It should be noted that the value of the PZC or IEP for a particular mineral is affected by the presence of trace elements in the mineral structure and by the previous history of the mineral particle. As an example, values of the IEP for quartz are reported from a pH of 2 to a pH of 3.7.[16,26]

6.1.4. The Mineral-Water Interface

In all mineral processing operations that are carried out in water, the nature of the mineral-water interface is of interest. In flotation, flocculation, and agglomeration processes the importance of this interface is readily apparent. However, it is also important in gravity separation,[27,28] in wet classifying and screening, in filtering, and in grinding.[29]

The adsorption of ions from solution at the mineral-water interface is dependent on the chemical composition and structure of the mineral surface and the electrical double layer at the interface. This adsorption can be due to chemical reaction between the adsorbing species and the ions comprising the mineral surface, that is, *chemisorption*, or it can be *physical adsorption* of counterions in the electrical double layer as discussed in the preceding section. There is no clear division between the two modes of adsorption. Chemisorption is frequently arbitrarily

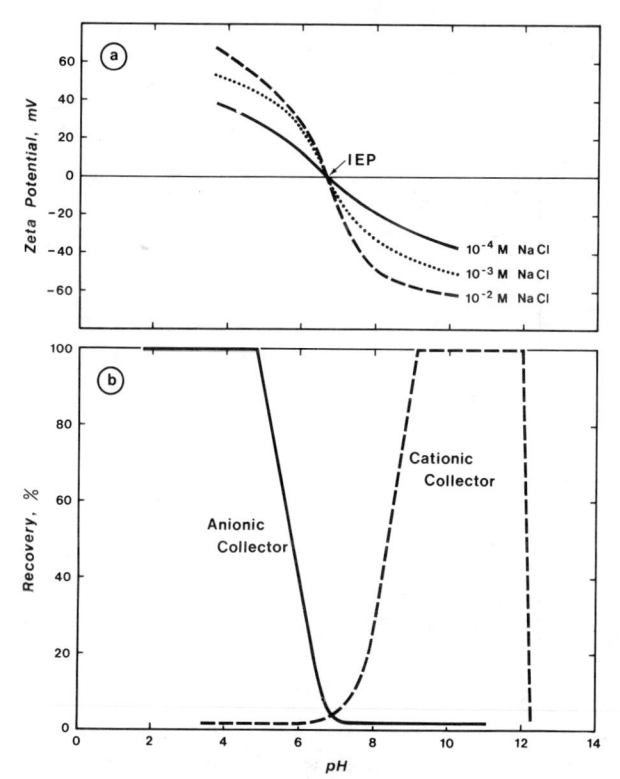

Figure 6.8. The effect of zeta potential and the IEP on the flotation of goethite. (After Iwasaki et al.[21])

defined as an adsorption process in which the free energy of adsorption ≥ 40 kJ/mole.

Adsorption isotherms can be determined experimentally and used to explain the adsorption occurring at the mineral-water interface. The Stern model of the electrical double layer has been used to explain the adsorption of both anions[16,30–32] and cations[31,33,34] as counterions at the mineral surface.

In a heterogeneous system, equilibrium is attained when the chemical potential of all species i is equal in all phases. Thus the chemical potential as given by Eq. 6.6 must be the same at the surface as in solution when the interface is at equilibrium. That is,

$$\mathcal{U}_i = \mathcal{U}_i^s = (\mathcal{U}_i^\circ)^s + RT \ln \mathcal{A}_i^s \qquad (6.21)$$

where the superscript s indicates the surface. The standard free energy of adsorption, $\Delta\gamma_{ads}^\circ$, is defined as

$$\Delta\gamma_{ads}^\circ = (\mathcal{U}_i^\circ)^s - \mathcal{U}_i^\circ \qquad (6.22)$$

and by substituting $\Gamma_i/2R_i$ for $\mathcal{A}_i$, where R_i is the effective radius of the adsorbed ion and Γ_i is the adsorption density of species i at the Stern plane, the Stern-Grahame equation is obtained.[18]

$$\Gamma_i = 2R_i\mathcal{A}_i \exp\left(\frac{-\Delta\gamma_{ads}^\circ}{RT}\right) \qquad (6.23)$$

The standard free energy of adsorption $\Delta\gamma_{ads}^\circ$ is the driving force for adsorption. It may be considered to be made up of a number of terms, each term representing a type of interaction that is contributing to the adsorption. Thus[13]

$$\Delta\gamma_{ads}^\circ = \Delta\gamma_{elect}^\circ + \Delta\gamma_{chem}^\circ + \Delta\gamma_{CH_2}^\circ + \Delta\gamma_{H_2O}^\circ + \ldots \qquad (6.24a)$$

where $\Delta\gamma_{elect}^\circ$ is the electrostatic contribution to the standard free energy of adsorption, $\Delta\gamma_{chem}^\circ$ is the standard free energy due to the formation of chemical (covalent) bonds with the mineral surface, $\Delta\gamma_{CH_2}^\circ$ is the standard free energy due to the association of hydrocarbon chains of the adsorbed species (i.e., hemi-micelle formation), and $\Delta\gamma_{H_2O}^\circ$ is the standard free energy due to solvation of the adsorbing ion and the mineral surface. It is often convenient to simplify Eq. 6.24a by combining all terms that represent specific adsorption, either chemisorption or physical adsorption, to give

$$\Delta\gamma_{ads}^\circ = \Delta\gamma_{elect}^\circ + \Delta\gamma_{spec}^\circ \qquad (6.24b)$$

where $\Delta\gamma_{spec}^\circ$ is the standard free energy due to specific interaction given by

$$\Delta\gamma_{spec}^\circ = \Delta\gamma_{chem}^\circ + \Delta\gamma_{CH_2}^\circ + \Delta\gamma_{H_2O}^\circ + \ldots \qquad (6.25)$$

If the adsorption is by electrostatic attraction alone, the standard free energy of adsorption is given by

$$\Delta\gamma_{ads}^\circ = \Delta\gamma_{elect}^\circ = N\mathcal{F}\xi_S \qquad (6.26)$$

(N = valence of the adsorbed ion, including the sign; ξ_S is the potential at the Stern plane). And if the zeta potential and the potential at the Stern plane are assumed equal as discussed earlier, Eq. 6.26 can be written as

$$\Delta\gamma_{ads}^\circ = N\mathcal{F}\zeta \qquad (6.27)$$

Thus the Stern-Grahame equation (6.23) can be written as

$$\Gamma_i = 2R_i\mathcal{A}_i \exp\left(\frac{N\mathcal{F}\zeta}{RT}\right) \qquad (6.28)$$

In a recent study of the adsorption of the collector sodium dodecyl sulfonate on alumina,[30] the adsorption isotherm and corresponding zeta potential data shown in Fig. 6.9 were reported. The adsorption isotherm has three distinct regions. Where the

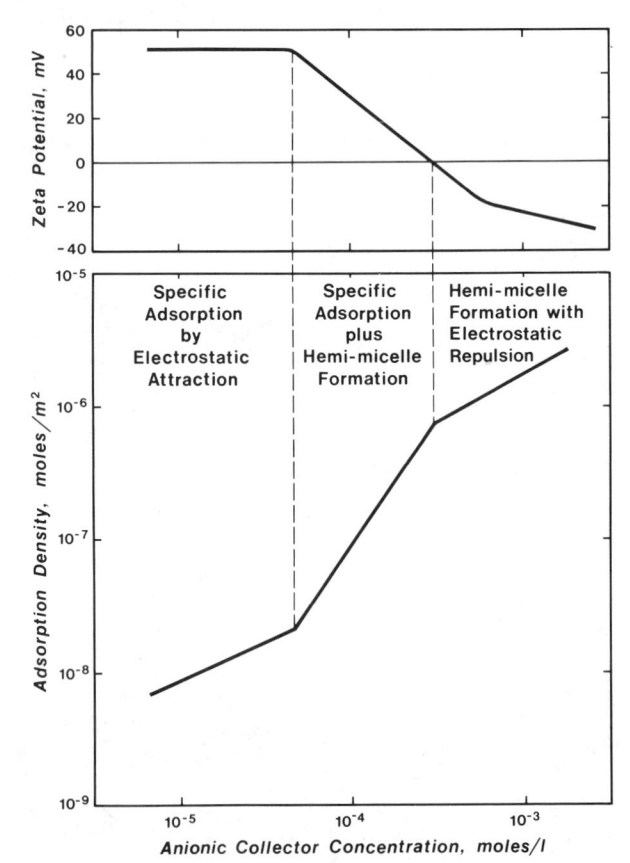

Figure 6.9. Adsorption isotherm and corresponding zeta potential data for the adsorption of sodium dodecyl sulfonate on alumina. (After Wakamatsu and Fuerstenau.[30])

concentration of the collector is low, individual dodecyl sulfonate anions adsorb by electrostatic attraction at the positively charged alumina surface. This is shown schematically[21,24,33] in Fig. 6.10a, and the adsorption density under these conditions is given by Eq. 6.28 with $N = 1$. When the concentration of the anions is increased, the adsorption density becomes sufficiently high for interaction between the hydrocarbon chains to occur. This results in *hemi-micelle* formation.[33,34] Once the ions of a collector reach a concentration at the mineral-water interface of about the same level as the *critical micelle concentration* (CMC) of the bulk solution, the adsorbed ions associate into patches on the mineral surface in much the same way as micelles form in the bulk solution. This is shown in Fig. 6.10b. When hemi-micelle formation occurs, the adsorption density is given by

$$\Gamma_i = 2R_i\mathcal{A}_i \exp\left(\frac{N\mathcal{F}\zeta + \Delta\gamma°_{CH_2}}{RT}\right) \quad (6.29)$$

with $\Delta\gamma°_{CH_2}$ having a value of 2.6 kJ/mole per CH_2 group.

When the concentration of the collector anion in Fig. 6.9 is increased again, the sign of the zeta potential changes from positive to negative and the electrostatic interaction opposes the specific adsorption effects, so the slope of the adsorption isotherm decreases. It should be noted, however, that adsorption continues to increase.

For adsorption beyond a complete *monolayer*, *multilayer* adsorption as shown in Fig. 6.10c can occur.[33] This is clearly undesirable when a hydrophobic surface is required.

Hemi-micelles, and also micelles, are limited to a maximum number of surfactant ions because of the electrostatic repulsion between the charged "heads" of the ions. The presence of neutral molecules, such as long-chain alcohols, reduces the repulsive forces between the charged heads and so lowers the CMC.[22,24] Figure 6.10d represents the lowering of the CMC at a mineral surface.

6.1.5. Wetting and Three-Phase Contact

Adsorption at the air-water and mineral-water interfaces was treated in earlier sections. The air-mineral or oil-water interface is now discussed briefly. This interface must be established for a particle to be collected in the flotation and oil extraction processes. For more detailed explanations the reader is referred to the excellent review papers of Laskowski[35] and Finch and Smith.[14]

The classical illustration of three-phase contact between water, an air bubble, and a smooth ideal mineral surface is shown in Fig. 6.11. At equilibrium, the three interfacial free energies γ are related by the Young equation

$$\gamma_{ma} - \gamma_{mw} = \gamma_{wa} \cos \theta_c \quad (6.30)$$

with θ_c being the *contact angle* and γ_{wa}, γ_{ma}, and γ_{mw} being the interfacial energies at the water-air, mineral-air, and mineral-water interfaces, respectively.

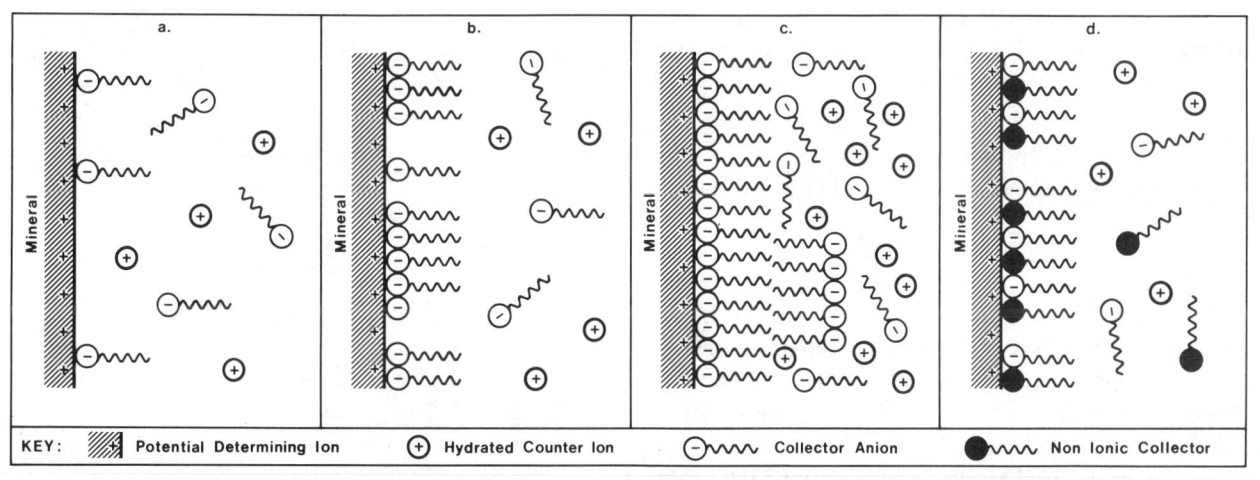

Figure 6.10. Schematic representation of the mineral-solution interface in the presence of an anionic collector. (*a*) Single ion adsorption at low concentrations. (*b*) Hemi-micelle formation at higher concentrations. (*c*) Multilayer adsorption. (*d*) Coadsorption of neutral molecules. (After Gaudin and Fuerstenau.[33])

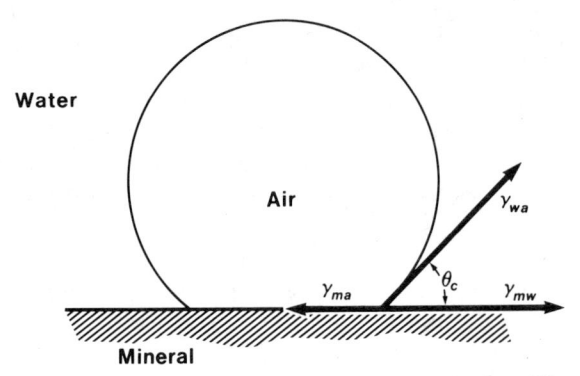

Figure 6.11. Idealized representation of three-phase equilibrium contact between air, water, and mineral surface.

It is obvious that the situation shown in Fig. 6.11 is unrealistic, because in flotation particles attach to relatively small bubbles. However, when a large spherical bubble and a small smooth particle surface are considered, the same equation (Eq. 6.30) is obtained.[7] This is still an idealization of the real situation, since mineral surfaces are frequently far from smooth (Fig. 6.1), and other forces act on the bubble. However, it has been widely accepted in the literature that the Young equation provides an adequate basis for the thermodynamic analysis of three-phase systems. As can be seen in Fig. 6.12, the contact angle can be related to the adsorption density, zeta potential, and also the recovery obtained in a flotation process.

For flotation to occur, a mineral-air interface must be created with the simultaneous destruction of water-air and mineral-water interfaces of equal area. Thus for a bubble-mineral particle attachment to take place,

$$\gamma_{ma} - \gamma_{mw} < \gamma_{wa}. \qquad (6.31)$$

The change in free energy $\Delta\gamma$ associated with the creation of the mineral-air interface is given by

$$\Delta\gamma = \gamma_{ma} - (\gamma_{wa} + \gamma_{mw}) \qquad (6.32)$$

$\Delta\gamma$ is sometimes referred to as the *work of adhesion*[4,5] between air and the mineral surface (bubble and particle) or the *tenacity of adhesion*.[4] It has also been called the *spreading coefficient*.[36] For flotation to be possible, that is, for a mineral-air interface to be created, the change in free energy $\Delta\gamma$ as given by Eq. 6.32 must be negative.

When Eq. 6.32 is combined with the Young equation (Eq. 6.30), the following expression for the change in free energy is obtained:

$$\Delta\gamma = \gamma_{wa}(\cos\theta_c - 1) \qquad (6.33)$$

This equation has frequently been used to analyze the flotation process: the more negative the value of $\Delta\gamma$, the greater the probability of particle-bubble attachment, and hence flotation. Such a thermodynamic treatment, however, describes only the overall free energy change occurring and does not take into account the intermediate stages in the bubble-particle attachment. Leja and Poling[37] have shown that the energy required to deform the air-water interface is a significant fraction of the interfacial energy available, and this energy requirement is affected by both the gravitational and kinetic energy. Thus the contact angle is affected by the motion of both particles and bubbles. It is thus clear that the application of Eq. 6.33 is limited.[35]

Influence of Reagents. In their natural state in water, most minerals are hydrophilic. By the addition of flotation collectors, the value of $(\gamma_{ma} - \gamma_{mw})$ is

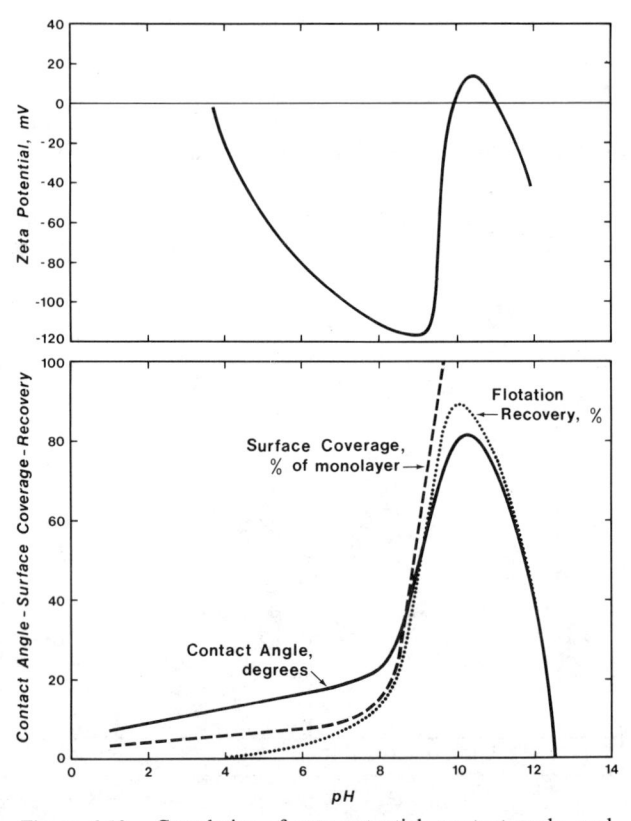

Figure 6.12. Correlation of zeta potential, contact angle, and collector adsorption with the flotation of quartz using a cationic collector. (After Fuerstenau.[22])

decreased, hence the contact angle is increased (Eq. 6.30). DeBruyn et al.[15,38] combined the Gibbs equation (Eq. 6.5) with the Young equation (Eq. 6.30) to show that

$$\Gamma_{ma} > \Gamma_{mw} \qquad (6.34)$$

where Γ_{ma} is the adsorption density of the collector at the mineral-air interface and Γ_{mw} is the adsorption density of the collector at the mineral-water interface. This means that in the collector solution any increase in the contact angle has to be explained by the decrease in the mineral-air interfacial energy γ_{ma}, a decrease that has to be larger than the decrease in γ_{mw}. This conclusion is in agreement with the penetration theory of Leja and Schulman,[39,40] in which it was suggested that the frother layer on an air bubble penetrates into the collector layer on the mineral to form a mixed surfactant layer at the mineral-air interface. The relationship of Eq. 6.34 has been confirmed experimentally by Somasundaran[41] (Fig. 6.13).

Bubble-Particle Contact. The attachment of a particle to a bubble in flotation can be considered in stages:

1. The approach of the bubble and the particle.
2. The thinning of the water film between the bubble and particle until rupture occurs.
3. The establishment of equilibrium contact.

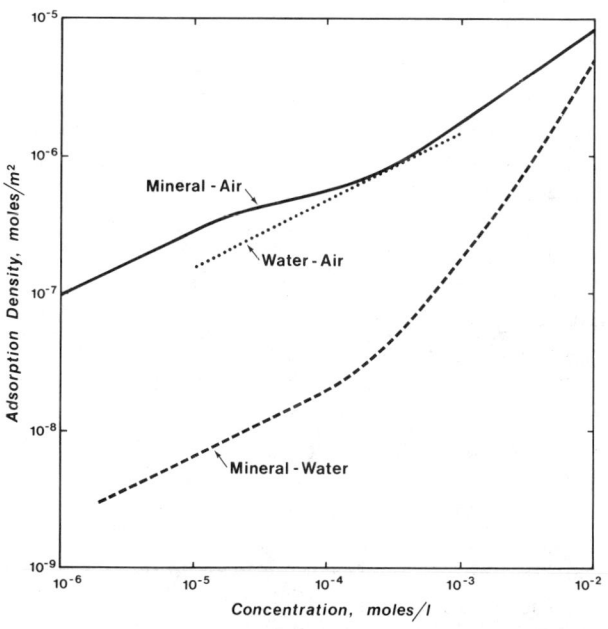

Figure 6.13. Comparison of the adsorption of dodecylammonium acetate at different interfaces. (After Somasundaran.[41])

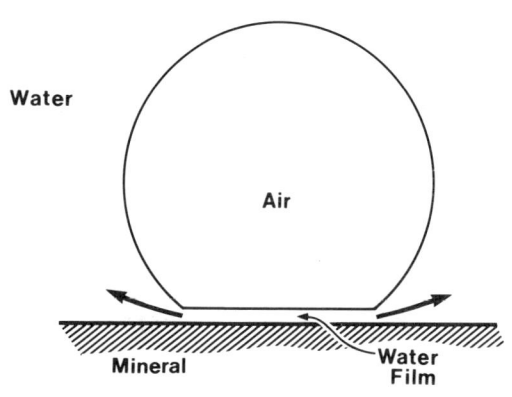

Figure 6.14. Representation of the thinning of the water film during bubble-particle contact.

The first stage is governed by the hydrodynamics of the flotation process. The third stage is defined to some extent by the contact angle as discussed above. The second stage, which is claimed by some researchers[14,35,42] to be the most important, is now considered.

When the bubble-particle separation is greater than about 1000 nm, hydrodynamic forces are dominant. Below this separation distance, molecular forces become important in the water film. Three types of molecular force have been recognized as important.[42] They are:

1. Van der Waals forces of attraction.
2. Electrical forces arising from interacting double layers in the water and around the particles.
3. Hydration of any hydrophilic groups on the surfaces of the particles.

In the simplified view of the water film shown in Fig. 6.14, a bubble is pressed against a flat mineral surface. The water film behaves as though an excess pressure P_d were acting normal to the film and opposing the thinning of the film. Derjaguin termed this excess pressure a *disjoining pressure*.[43] When the film is sufficiently thin, this disjoining pressure arises as a result of van der Waals forces, hydrogen bonding, and the deformation of the electrical double layers. The films are stable if $dP_d/d\delta$ is less than zero and unstable if $dP_d/d\delta$ exceeds zero, where δ is the film thickness.[35] This is shown[44] in Fig. 6.15. For values of δ greater than about 20 nm, the most important contribution to the disjoining pressure arises from the compression of the diffuse double layers.[45] At reduced thicknesses, dispersion forces and hy-

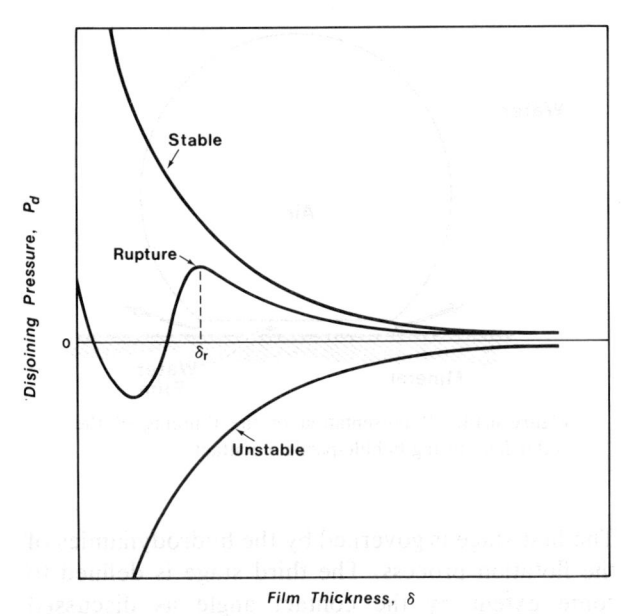

Figure 6.15. Variation of disjoining pressure with film thickness. (After Laskowski and Iskra.[44])

drogen bonding become important. It is believed[46,47] that hydration of mineral surfaces is the major reason for their being hydrophilic, and by removing the hydrogen bonding junctions between mineral and water by the adsorption of an appropriate organic species, the mineral can be transformed into a hydrophobic one. It should be noted that at the reagent concentrations and film thicknesses present in some flotation systems, the disjoining pressure is not so significant.[48]

The thinning of the wetting film essentially depends on the van der Waals constants (which determine the van der Waals forces) for the attraction of water to the mineral surface and of water to itself. The van der Waals forces that oppose the thinning of the wetting film also oppose the rupture of the film. Derjaguin has shown that the condition for the disappearance of the "force barrier," and hence mineral-bubble attachment, is that[43]

$$\frac{\delta_D K_\varepsilon \zeta^2}{K_A} > 3 \qquad (6.35)$$

(δ_D = Debye thickness of the ionic atmosphere; K_ε = dielectric constant of the solution; $K_A = K_{A11} - K_{A12}$ = Hamaker constant,[50] which is the difference between the Hamaker constants for the attraction of water to the mineral and of water to itself, denoted by K_{A11} and K_{A12}, respectively). This "criterion for floatability" based on the Hamaker constants and given by Eq. 6.34 has been discussed by Rao.[42] He points out that if K_{A11} exceeds K_{A12}, the mineral is intrinsically hydrophobic and should be floatable if the value of ζ is not too far from the IEP of the mineral. However, this criterion cannot be applied to many minerals at the present time.

Figure 6.15 includes a third curve that is most typical of practical flotation systems. The film is stable until a critical thickness, δ_r, is reached. At this thickness the film becomes unstable and so will spontaneously rupture.

The film life, defined as the time required for the disjoining film to drain to such a thickness that rupture takes place (i.e., δ_r), is called the *induction time* (Section 6.2.1). Induction times have been measured by a number of researchers[51,52] and found to be of the same order of magnitude as the time of contact of a particle and bubble under the hydrodynamic conditions found in a flotation machine.

Clearly, if the induction time exceeds the contact time, flotation would be impossible even if the equilibrium contact angle were found to be large. Rendering a mineral floatable by collector addition can be interpreted as increasing the rate of film thinning (i.e., reducing the induction time) or increasing the film thickness at which rupture occurs.

This section's discussion has been totally in terms of air as the third phase. The same principles, however, apply when the third phase is another liquid, such as in the oil extraction process.[53–56]

6.1.6. Interparticle Contact

Particles interact directly with each other in a number of processes. The interaction can be the attachment of very fine particles onto a large particle, or the clustering of particles as in flocculation and agglomeration. It can also be the interaction of fine particles undergoing sedimentation, filtration, or balling.

In each case, the particle in water solution will have an electrical double layer surrounding it (Section 6.1.3); when particles are in close proximity, their double layers can be considered to interact. If particles are of opposite charge, clearly electrostatic attraction can result in the attachment or "adsorption" of a fine particle onto the surface of a larger particle. This frequently occurs in flotation systems; it is a cause of reduced selectivity when fines are present, and it is the basis of carrier flotation (both discussed in Chapter 16).

When the sign of the charge on particles is the same, the situation is more complex. Very fine particles have repulsive forces arising from the presence of the electrical double layers, but also attractive van der Waals forces. A stable situation can result, explained by the DLVO (for Derjaguin, Landau, Verwey, and Overbeek) theory of colloid stability.[57,58]

Using this theory, the surface conditions (zeta potential) have been determined[59-61] to give interparticle energies represented by the curves of Fig. 6.16. Clearly, it is possible to promote or prevent aggregation by controlling the charge on the particles so that attractive energy between them will exceed the repulsive energy. The DLVO theory has been successfully applied in explaining the formation of stable aggregates in agglomeration and flocculation[61] and it has also been used to explain sedimentation occurring under hindered settling conditions.[62]

6.1.7. Determination of Interfacial Properties

Two properties of major importance in the study of interfaces are interfacial free energy and zeta potential.

The interfacial free energy (or surface tension) between two liquids or a liquid and gas can readily be measured using a variety of methods, described elsewhere.[4-7] The interfacial free energy between liquid or air and the mineral surface requires measurement of the contact angle, and again the methods for this measurement are described elsewhere.[4-7] Contact angles can be predicted using the concept of a critical surface tension for wetting of a solid surface.[63]

A number of methods are available for the determination of the zeta potential of a mineral particle in contact with a solution.[64] Two widely used techniques are the measurement of *streaming potential* and the measurement of *electrophoretic mobility*.

The streaming potential method[64-67] involves passing a stream of solution through a porous plug of particles (relatively coarse) mounted between electrodes. The pressure drop across the plug and the potential difference between the electrodes are measured. The zeta potential can then be calculated by the Helmholtz-Smoluchowski equation

$$\zeta = \frac{4\pi\mu\mathfrak{C}\mathscr{E}_a}{K_\varepsilon P} \quad (6.36a)$$

(μ, K_ε, $\mathfrak{C}$ = viscosity, dielectric constant, and specific conductivity, respectively; $\mathscr{E}_a$ = streaming potential; P = driving pressure).

In the electrophoresis method,[64] individual fine particles moving through a solution between two electrodes are observed through a microscope. A modified version of Eq. 6.36a is used to calculate the zeta potential:

$$\zeta = K_Z \frac{\pi\mu L v_e}{K_\varepsilon \mathscr{E}_a} \quad (6.36b)$$

(K_Z = constant [= $\frac{1}{6}$ for particles small compared to the double layer thickness; $\frac{1}{4}$ for large particles]; v_e = observed electrophoretic velocity; L = distance between electrodes).

The IEP for a mineral in solution can of course be determined by measurements of the zeta potential. However, it can also be measured by a number of other techniques that are often more convenient.[17,68] Also, methods have been developed to predict the IEP or PZC for minerals based on their chemical composition and crystalline structure.[69,70]

6.2. MECHANISMS OF INTERFACIAL ATTACHMENT

The thermodynamics of bubble-particle and interparticle attachment has been treated in Section 6.1. The mechanisms by which such attachment occurs are now discussed. Clearly, both aspects of attachment must be considered.

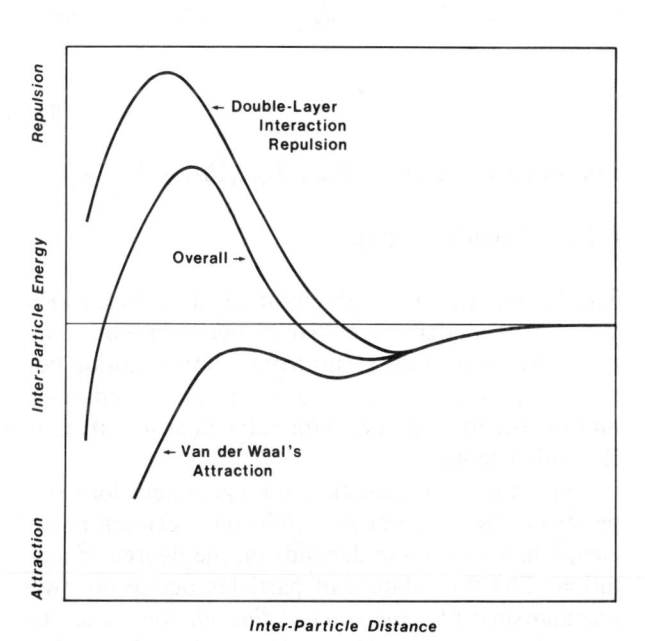

Figure 6.16. Illustration of the variation in interparticle energy with distance between particles.

6.2.1. Bubble-Particle Attachment

The approach of a particle to a bubble and the subsequent bubble-particle attachment have been examined by many workers, both experimentally and theoretically.[71-75] The rate of flotation is considered to equal the product of three factors[35,72,74]:

1. The rate of collision between particles and bubbles.
2. The probability of adhesion.
3. The probability that the adhering particle will not be subsequently detached because of turbulence.

The probability of flotation of a particle thus can be represented by

$$p = p_c p_a p_s \qquad (6.37)$$

(p = probability of flotation of a particle; p_c, p_a, p_s = probability of collision between particle and bubble, probability of adhesion following collision, and probability of formation of stable bubble-particle attachment, respectively). Expressions have been derived for each probability factor (or corresponding rate constant, Eqs. 3.2 and 3.3) and overall equations obtained. Sutherland[72] derived an equation that can be written as

$$p = [3\pi R_b R_p v_{bp} C_{Nb}] \left[\text{sech}^2 \left(\frac{3v_{bp}t_I}{4R_b} \right) \right] p_s \qquad (6.38)$$

(R_b, R_p = radius of bubble and particle, respectively; v_{bp} = relative bubble-particle velocity; C_{Nb} = bubble concentration [number per unit volume]; t_I = induction time). In this equation, the first term in brackets corresponds to p_c and the second to p_a. Though having limitations,[76] this equation has been used to predict differences in flotation rates with variation in particle size.[71]

The induction time (Section 6.1.5) in Eq. 6.38 can be calculated by an equation such as that of Scheludko,[77] presented by Jowett[71] for a small particle in contact with a large bubble as

$$t_I = K_1 \frac{d_v^3}{P_d} \qquad (6.39)$$

(K_1 = constant; P_d = disjoining pressure).
The relative bubble-particle velocity in a flotation system is determined primarily by the bubble velocity because of the comparatively large bubble terminal velocity. However, the driving pressure on collision of a particle with a bubble results primarily from the momentum of the particle traveling at its terminal velocity, and the resulting deceleration. Thus, Eq. 6.39 can be modified to

$$t_I = K_2 d_v^n \qquad (6.40)$$

(K_2 = constant; $0 \leq n \leq 1.5$, where $n = 0$ for laminar flow conditions, $n = 1.5$ for turbulent conditions, and $n = 1$ [approx.] in the transition region). This theoretical equation gives a good representation of experimental results[71] and clearly indicates that at fine particle sizes induction time is independent of particle size, whereas at coarser sizes the induction time is dependent on particle size.

Equation 6.38 appears to predict a decrease in flotation rate with reduced bubble size. However, when bubble size is changed in a real flotation system, a corresponding change occurs in both the concentration of bubbles and the average bubble velocity. When these factors are taken into account, Eq. 6.38 is found[71] to predict the increased flotation rate when bubble size is reduced, as found in practice.

The one remaining factor to consider in Eq. 6.38 is the detachment of a particle. This factor relates to the equilibrium attachment conditions (represented by a contact angle) and the hydrodynamic conditions. A number of methods for calculating p_s have been proposed[5,74,78] and critical particle sizes for stability determined.[5,71] It has been shown that if the maximum particle size to remain attached under prevailing conditions is $d_{v,\text{max}}$, then for all smaller particles[74]

$$p_s = 1 - \left(\frac{d_v}{d_{v,\text{max}}} \right)^{1.5} \qquad (6.41)$$

For all particles larger than $d_{v,\text{max}}$, $p_s = 1$.

6.2.2. Particle Aggregation

The flocculation or agglomeration of particles in a liquid is dependent on collisions between those particles. The collisions result from relative motion between the particles caused by external agitation, motion due to an external force (such as gravity), or Brownian motion.

The rate of flocculation (or agglomeration) depends on the frequency of collisions between particles, which of course depends on the degree of agitation. The flocculation of particles occurs by two mechanisms[17,61]: *perikinetic flocculation* due to Brownian motion, and *orthokinetic flocculation* due to more vigorous agitation.

When there is no surface repulsion between the particles (Fig. 6.16), the following equation for the rate of flocculation can be derived for perikinetic flocculation[61]:

$$\frac{dC_{Np}}{dt} = -4\pi \mathscr{D} d_v C_{Np}^2 \quad (6.42a)$$

(C_{Np} = number of particles per unit volume; $\mathscr{D}$ = diffusion coefficient; d_v = particle size). Using the Stokes-Einstein equation, the diffusion coefficient in Eq. 6.42a can be replaced to give

$$\frac{dC_{Np}}{dt} = -\frac{4K_{Bo}T C_{Np}^2}{3\mu} \quad (6.42b)$$

(K_{Bo} = Boltzmann's constant; T = absolute temperature; μ = viscosity). This equation shows that in perikinetic flocculation, the rate is independent of particle size.

Again, when there is no surface repulsion between the particles, the rate of flocculation for orthokinetic flocculation is[61]:

$$\frac{dC_{Np}}{dt} = -\frac{2}{3}\left(\frac{dv}{dz}\right) d_v^3 \sum_{i=1}^{N} (C_{Np,i}^2 \cdot i) \quad (6.43)$$

($C_{Np,i}$ = number of aggregates of type i [a type i aggregate has diameter $d_{v,i}$ given by $i \times (\pi/6)d_v^3 = (\pi/6)d_{v,i}^3$]; (dv/dz) = velocity gradient; N represents maximum stable aggregate size). As with perikinetic flocculation, the rate is second order with respect to particle concentration. However, the orthokinetic rate is highly dependent on the initial particle size. It is also dependent on the velocity gradient, which explains why gentle agitation has been observed to promote flocculation.

It is of interest to compare the rates of perikinetic flocculation and orthokinetic flocculation. When the rates are equal

$$d_v = \left[\frac{2K_{Bo}T}{(dv/dz)\mu}\right]^{1/3} \quad (6.44)$$

Using this equation, it has been shown[61] that, even with agitation, the perikinetic flocculation rate is greater for particles smaller than 1 μm and the orthokinetic rate greater for particles larger than 1 μm. However, it should be assumed that both mechanisms are always occurring to some extent.

In establishing Eqs. 6.42 and 6.43, it was assumed that every collision resulted in aggregation. Although this is true when there is no repulsive force between the particles (the particles have been completely destabilized), there also exists a situation (Fig. 6.16) characterized by some repulsive force (stable suspension) and only a fraction of the collisions are effective. This can be accounted for in Eqs. 6.42 and 6.43 by including a factor that represents the fraction of collisions that are successful.

When polymers are used as flocculating agents, heteroflocculation can be considered to exist. In this case, the rate of adsorption of the polymer, assuming each "collision" between polymer and particle leads to attachment, is given by

$$\frac{dC_{Nf}}{dt} = -kC_{Np}C_{Nf} \quad (6.45)$$

(C_{Nf} = number of polymer molecules per unit volume; k = rate constant for collisions between polymer molecules and particles).

In flocculation with polymers, the floc initially formed is loose and bulky. By applying local and uneven fluctuating forces over the surface of the floc, a densification occurs by exudation of the dispersion medium.[79] This additional mechanism of flocculation is called *mechanical syneresis*.

A qualitative model has been used to describe the reactions occurring in flocculation with polymers.[80] This "bridging theory" includes seven separate reactions as indicated in the seven parts of Fig. 6.17:

a. Initial adsorption of the polymer (at optimum dosage).
b. Perikinetic or orthokinetic flocculation.
c. Secondary adsorption of polymer (no contact with available sites on another particle).
d. Initial adsorption of excess polymer (> optimum dosage).
e. Rupture of floc due to intense or prolonged agitation.
f. Secondary adsorption of polymer.
g. Mechanical syneresis due to uneven forces on floc.

Hindered Settling. When flocculation or agglomeration occurs (either naturally or controlled), settling rates are changed. The Richardson-Zaki equation, widely used to obtain the settling velocities of suspensions of nonflocculated particles, is:

$$\frac{v_h}{v_\infty} = \varepsilon^n \quad (4.70)$$

(v_h, v_∞ = suspension velocity and terminal velocity of particles, respectively; ε = porosity; n = exponent). This equation has been adapted for floccu-

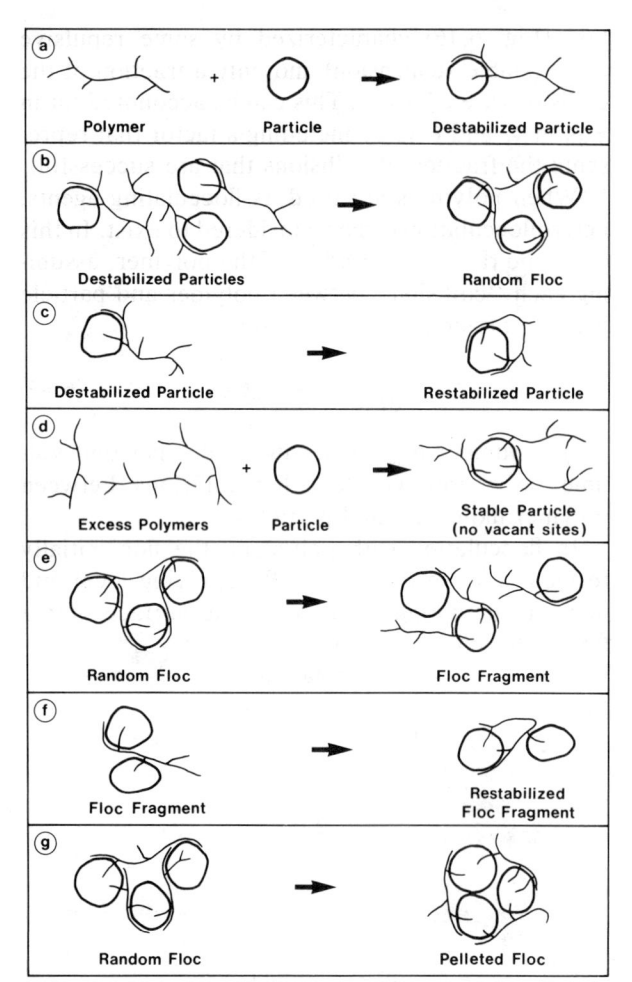

Figure 6.17. Representation of flocculation mechanisms. (After Yusa.[79])

lated suspensions by replacing the porosity with an "effective porosity" such that[81]

$$\frac{v_h}{v_{\infty,f}} = (1 - K_f C_V)^n \qquad (6.46)$$

($v_{\infty,f}$ = terminal velocity of a single floc; C_V = volume concentration of particles $[1 - \varepsilon]$; K_f = volume of floc per unit volume of particles in the floc). Although the flocs are not smooth spheres, the value of the exponent used is 4.65 and the values of K_f and $v_{\infty,f}$ selected to fit the data.[81,82] Alternatively, a value of 4.8 could be used (Eq. 4.71).

REFERENCES

1. P. Somasundaran (Ed.), *Fine Particles Processing*, AIME (1980).
2. M. C. Fuerstenau (Ed.), *Flotation—A.M. Gaudin Memorial Volume*, AIME/SME (1976).
3. D. W. Fuerstenau (Ed.), *Froth Flotation—50th Anniversary Volume*, AIME/SME (1962).
4. K. L. Sutherland and I. W. Wark, *Principles of Flotation*, Australasian IMM (1955).
5. A. M. Gaudin, *Flotation*, 2nd ed., McGraw-Hill (1957).
6. V. I. Klassen and V. A. Mokrousov, *An Introduction to the Theory of Flotation*, Butterworths (1963).
7. A. W. Adamson, *Physical Chemistry of Surfaces*, 3rd ed., Wiley (1976).
8. K. V. S. Sastry (Ed.), *Agglomeration 77*, AIME/SME (1977).
9. E. Matijevic (Ed.), *Surface and Colloid Science*, Vols. 1–12, Wiley (1969–1981).
10. R. D. Kulkarni and P. Somasundaran, "Mineralogical Heterogeneity of Ore Particles and Its Effects on Their Interfacial Characteristics," *Powder Technol.*, **14**, 279–285 (1976).
11. J. D. Miller and J. V. Calara, "Analysis of the Surface Potential Developed by Non-Reactive Ionic Solids," in *Flotation—A. M. Gaudin Memorial Volume*, M. C. Fuerstenau (Ed.), pp. 66–86, AIME/SME (1976).
12. R. Defay and I. Prigogine, *Surface Tension and Adsorption*, Wiley (1966).
13. D. W. Fuerstenau and S. Raghavan, "Some Aspects of the Thermodynamics of Flotation," in *Flotation—A. M. Gaudin Memorial Volume*, M. C. Fuerstenau (Ed.), pp. 21–65, AIME/SME (1976).
14. J. A. Finch and G. W. Smith, "Contact Angle and Wetting," *Miner. Sci. Eng.*, **11**, 36–63 (1979).
15. P. L. de Bruyn and G. E. Agar, "Surface Chemistry of Flotation," in *Froth Flotation—50th Anniversary Volume*, D. W. Fuerstenau (Ed.), pp. 91–138, AIME/SME (1962).
16. A. M. Gaudin and D. W. Fuerstenau, "Quartz Flotation with Anionic Collectors," *Trans. AIME/SME*, **202**, 66–72 (1955).
17. H. R. Kruyt (Ed.), *Colloid Science*, Vol. I, Elsevier (1952).
18. D. C. Grahame, "The Electrical Double Layer and the Theory of Electrocapillarity," *Chem. Rev.*, **411**, 441–501 (1947).
19. S. S. Dukhin and B. V. Derjaguin, "Electrokinetic Phenomena," Vol. 7 in *Surface and Colloid Science*, E. Matijevic (Ed.), Wiley (1974).
20. P. Somasundaran, "Zeta-Potential of Apatite in Aqueous Solutions and Its Change During Equilibration," *J. Colloid Interface Sci.*, **27**, 659–666 (1968).
21. I. Iwasaki, S. R. B. Cooke, and A. F. Colombo, "Flotation Characteristics of Goethite," U.S. Bureau of Mines Report of Investigations RI 5593 (1960).
22. F. F. Aplan and D. W. Fuerstenau, "Principles of Non-Metallic Mineral Flotation," in *Froth Flotation—50th Anniversary Volume*, D. W. Fuerstenau (Ed.), pp. 170–214, AIME/SME (1962).
23. T. W. Healy and M. S. Moignard, "A Review of Electrokinetic Studies of Metal Sulfides," in *Flotation—A. M. Gaudin Memorial Volume*, M. C. Fuerstenau (Ed.), pp. 275–297, AIME/SME (1976).

References

24. M. C. Fuerstenau and B. R. Palmer, "Anionic Flotation of Oxides and Silicates," in *Flotation—A. M. Gaudin Memorial Volume*, M. C. Fuerstenau (Ed.), pp. 148–196, AIME/SME (1976).
25. H. S. Hanna and P. Somasundaran, "Flotation of Salt-Type Minerals," in *Flotation—A. M. Gaudin Memorial Volume*, M. C. Fuerstenau (Ed.), pp. 197–272, AIME/SME (1976).
26. I. Iwasaki, S. R. B. Cooke, and H. S. Choi, "Flotation of Cummingtonite," *Trans. AIME/SME*, **220**, 394–395 (1961).
27. R. O. Burt, "A Study of the Effect of Deck Surface and Pulp pH on the Performance of a Fine Gravity Concentrator," *Int. J. Miner. Process.* **5**, 39–44 (1978).
28. S. R. Rao and L. L. Sirois, "Study of Surface Chemical Characteristics in Gravity Separation," *CIM Bull.*, **67**, 78–83 (1974).
29. R. R. Klimpel and W. Manfroy, "Chemical Grinding Aids for Increasing the Throughput in the Wet Grinding of Ores," *Ind. Eng. Chem., Process Des. Dev.*, **17**, 518–523 (October 1978).
30. T. Wakamatsu and D. W. Fuerstenau, "Effect of Alkyl Sulfonates on the Wettability of Alumina," *Trans. AIME/SME*, **254**, 123–126 (1973).
31. D. W. Fuerstenau, "Interfacial Processes in Mineral/Water Systems," *Pure Appl. Chem.*, **24**, 135–164 (1970).
32. P. Somasundaran and D. W. Fuerstenau, "Mechanisms of Alkyl Sulfonate Adsorption at the Alumina/Water Interface," *J. Phys. Chem.*, **70**, 90–96 (1966).
33. A. M. Gaudin and D. W. Fuerstenau, "Quartz Flotation with Cationic Collectors," *Trans. AIME/SME*, **202**, 958–962 (1955).
34. P. Somasundaran, T. W. Healy, and D. W. Fuerstenau, "Surfactant Adsorption at the Solid/Liquid Interface—Dependence of Mechanism on Chain Length," *J. Phys. Chem.*, **68**, 3562–3566 (1964).
35. J. Laskowski, "Particle-Bubble Attachment in Flotation," *Miner. Sci. Eng.*, **6**, 223–235 (1974).
36. W. D. Harkins, *The Physical Chemistry of Surface Films*, Reinhold (1952).
37. J. Leja and G. W. Poling, "On the Interpretation of Contact Angle," *5th Int. Miner. Process. Congr., London*, pp. 325–341; IMM (1960).
38. P. L. de Bruyn, J. T. G. Overbeek, and R. Schuhmann, "Flotation and the Gibbs Adsorption Equation," *Min. Eng.*, **6**, 519–523 (1954).
39. J. Leja and J. H. Schulman, "Flotation Theory: Molecular Interactions Between Frothers and Collectors at Solid/Liquid/Air Interfaces," *Trans. AIME/SME*, **199**, 221–228 (1954).
40. J. Leja, "Mechanisms of Collector Adsorption and Dynamic Attachment of Particles to Air Bubbles as Derived from Surface Chemical Studies," *Trans. IMM*, **66**, 425–437 (1957).
41. P. Somasundaran, "The Relationship Between Adsorption at Different Interfaces and Flotation Behavior," *Trans. AIME/SME*, **241**, 105–108 (1968).
42. S. R. Rao, "Surface Forces in Flotation," *Miner. Sci. Eng.*, **6**, 45–53 (1974).
43. B. V. Derjaguin and Z. M. Zorin, "Optical Study of the Adsorption and Surface Condensation of Vapors on a Smooth Surface," in *Second International Congress on Surface Activity*, Vol. 2, pp. 145–152, Butterworths (1957).
44. J. Laskowski and J. Iskra, "Role of Capillarity Effects in Bubble-Particle Collision in Flotation," *Trans. IMM (C)*, **79**, C6–C10 (1970).
45. A. D. Read and J. A. Kitchener, "Wetting Films on Silica," *J. Colloid Interface Sci.*, **30**, 391–398 (1969).
46. P. A. Rehbinder, *General Course in Colloid Chemistry*, Moscow University (1949).
47. J. Laskowski and J. A. Kitchener, "The Hydrophylic-Hydrophobic Transition on Silica," *J. Colloid Interface Sci.*, **29**, 670–679 (1969).
48. K. Mushiake, T. Imaizumi, and T. Inoue, "Thinning and Rupture of Liquid Film on Polarized Mercury—An Experimental Approach to the Theory of Flotation," *Proc. 11th Int. Miner. Process. Congr., Cagliari, 1975*, pp. 405–425, Università di Cagliari (1975).
49. B. V. Derjaguin and N. D. Shukakidse, "Dependability of the Floatability of Antimonite on the Value of the Zeta-Potential," *Trans. IMM*, **70**, 569–574 (1961).
50. H. C. Hamaker, "The London-van der Waals Attraction Between Spherical Particles," *Physica*, **4**, 1058–1072 (1937).
51. L. F. Evans, "Bubble-Mineral Attachment in Flotation," *Ind. Eng. Chem.*, **46**, 2420–2424 (1954).
52. M. A. Eigeles and M. L. Volova, "On the Mechanism of Activating and Depressant Action on Soap Flotation," *Proc. 7th Int. Miner. Process. Congr., New York, 1964*, pp. 269–277, Gordon and Breach (1965).
53. J. M. W. MacKenzie, "Interaction Between Oil Drops and Mineral Surfaces," *Trans. AIME/SME*, **247**, 202–208 (1970).
54. H. L. Shergold and O. Mellgren, "Concentration of Minerals at the Oil Water Interface: Hematite/Iso-octane/Water System in the Presence of Sodium Dodecyl Sulfonate," *Trans. IMM (C)*, **78**, C121–C131 (1969).
55. G. Zambrana et al., "Recovery of Minus Ten Micron Cassiterite by Liquid-Liquid Extraction," *Int. J. Miner. Process.*, **1**, 335–345 (1974).
56. R. W. M. Lai and D. W. Fuerstenau, "Liquid-Liquid Extraction of Ultrafine Particles," *Trans. AIME/SME*, **241**, 549–555 (1968).
57. B. V. Derjaguin and L. Landau, "A Theory of the Stability of Strongly Charged Lyophobic Sols and of the Adhesion of Particles in Solutions of Electrolytes," *Acta Phys. Chem. URSS*, **14**, 633–662 (1941).
58. E. J. W. Verwey and J. T. G. Overbeek, *Theory of the Stability of Lyophobic Colloids*, Elsevier (1948).
59. G. R. Wiese and T. W. Healy, "Effect of Particle Size on Colloid Stability," *Trans. Faraday Soc.*, **66**, 490–499 (1970).
60. T. W. Healy, "Principles of Dispersion and Aggregation in Mineral Fines," in *Beneficiation of Mineral Fines*, P. Somasundaran and N. Arbiter (Eds.), pp. 161–178, AIME/SME (1979).
61. K. J. Ives (Ed.), *The Scientific Basis of Flocculation*, Sijthoff and Noordhoff (1978).

62. Z. Sadowski, J. Mager, and J. Laskowski, "Hindered Settling in Coagulating Suspensions," *Powder Technol.*, **21**, 73–79 (1978).
63. W. A. Zisman (Ed.), "Contact Angle, Wettability, and Adhesion," Vol. 43 in Advances in Chemistry Series, ACS (1964).
64. J. M. W. MacKenzie, "Zeta-Potential Studies in Mineral Processing: Measurement, Techniques, and Applications," *Miner. Sci. Eng.*, **3**, 25–43 (1971).
65. D. W. Fuerstenau, "Measuring Zeta-Potentials by Streaming Potential Techniques," *Trans. AIME/SME*, **205**, 834–835 (1956).
66. P. Somasundaran and R. D. Kulkarni, "A New Streaming Potential Apparatus and Study of Temperature Effects Using It," *J. Colloid Interface Sci.*, **45**, 591–600 (1973).
67. B. Ball and D. W. Fuerstenau, "A Review of the Measurement of Streaming Potentials," *Miner. Sci. Eng.*, **5**, 267–277 (1973).
68. A. L. Mular and R. B. Roberts, "A Simplified Method to Determine Isoelectric Points of Oxides," *CIM Bull.*, **59**, 1329–1330 (1966).
69. G. A. Parks, "The Isoelectric Points of Solid Oxides, Solid Hydroxides and Aqueous Hydroxo Complex Systems," *Chem. Rev.*, **65**, 177–198 (1965).
70. R. H. Yoon, T. Salman, and G. Donnay, "Predicting Points of Zero Charge of Oxides and Hydroxides," *J. Colloid Interface Sci.*, **70**, 483–493 (1979).
71. A. Jowett, "Formation and Disruption of Particle-Bubble Aggregates in Flotation," in *Fine Particle Processing*, P. Somasundaran (Ed.), pp. 720–754, AIME (1980).
72. K. L. Sutherland, "Physical Chemistry of Flotation. XI. Kinetics of the Flotation Process," *J. Phys. Colloid Chem.*, **52**, 394–425 (1948).
73. B. V. Derjaguin and S. S. Dukhin, "Theory of Flotation of Small and Medium-Sized Particles," *Trans. IMM*, **70**, 221–246 (1960–1961).
74. E. T. Woodburn, R. P. King, and R. P. Colburn, "The Effect of Particle Size Distribution on the Performance of a Phosphate Flotation Process," *Met. Trans.*, **2**, 3163–3174 (1971).
75. L. R. Flint and W. J. Howarth, "The Collision Efficiency of Small Particles with Spherical Air Bubbles," *Chem. Eng. Sci.*, **26**, 1155–1168 (1971).
76. W. J. Trahar and L. J. Warren, "The Floatability of Very Fine Particles—A Review," *Int. J. Miner. Process.*, **3**, 103–131 (1976).
77. A. Scheludko, "Thin Liquid Films," in *Advances in Colloid and Interfacial Science*, Vol. 1, pp. 391–464 (1967).
78. H. J. Schultze, "New Theoretical and Experimental Investigations on Stability of Bubble/Particle Aggregates in Flotation: A Theory on the Upper Particle Size of Floatability," *Int. J. Miner. Process.*, **4**, 241–259 (1977).
79. M. Yusa, "Mechanisms of Pelleting Flocculation," *Int. J. Miner. Process.*, **4**, 293–305 (1977).
80. V. K. LaMer and T. W. Healy, "Adsorption-Flocculation Reaction of Macromolecules at the Solid-Liquid Interface," *Rev. Pure Appl. Chem.*, **13**, 112–133 (1963).
81. A. S. Michaels and J. C. Bolger, "Settling Rates and Sediment Volumes of Flocculated Kaolin Suspensions," *Ind. Eng. Chem. Fundam.*, **1**, 25–33 (1962).
82. D. C. Dixon, "Comments on *Hindered Settling Theories*," *Powder Technol.*, **17**, 147–149 (1977).

Chapter Seven

Mechanisms of Fracture

Jake's Law: "*Anything hit with a big enough hammer will fall apart.*"

Size reduction is of vital importance in the processing of minerals. An ore must be reduced in size until the valuable minerals are liberated from the host rock or exposed for chemical action. Sometimes, as in aggregate production, size reduction is required to produce a marketable product. Size reduction may also be required before a subsequent processing step such as pelletizing. In all cases, the breakage of the rock required to achieve this size reduction is an energy intensive and energy inefficient operation.

It is often stated that the efficiency of energy utilization during the fragmentation of solid particles is only about 1% with respect to the new surface created.[1,2] It represents a major cost in any mineral processing operation.

The literature on the breakage of materials is extensive. That directly relating to the breakage of mineral particles has been the subject of controversy over many years. In this chapter, the breakage of particles is considered in terms of both single particle fracture and the breakage of distributions of particles, and the literature is reviewed accordingly.

7.1. SINGLE PARTICLE FRACTURE

In any industrial size reduction operation, the breakage of any individual particle is occurring simultaneously with that of many other particles. The breakage products from any particles are intimately mixed with those of other particles and are indistinguishable from them. Thus, an industrial size reduction operation can be analyzed only in terms of a distribution of feed particles being reduced to a distribution of product particles. However, each individual particle breaks as a result of the stresses applied to it and it alone, so it is of value to investigate how a single particle fractures.

7.1.1. Mechanics of Particle Fracture

In seeking to understand the fundamental mechanisms by which ore particles break, a number of investigators over many years have attempted to apply the concepts of "fracture physics" and "fracture mechanics" as used in the materials science and rock mechanics disciplines. Ore particles are heterogeneous, normally flawed on both a macro- and a microscale, and do not always behave like brittle materials. However, by examining well-defined materials and then considering the effects of the heterogeneous nature of ore particles, substantial progress has been made toward describing the breakage process. Major contributions have been made by Rumpf[3] and Schonert,[4] and by Oka and Majima.[5] Discussion of other work is available in the literature.[6,7]

Except at very small sizes, an ore particle may be considered to be a brittle material; that is, the strain is proportional to the applied stress until the point at which fracture occurs. The fracture of brittle materials was analyzed by Griffith,[8] and this has formed the basis for much subsequent work. Griffith observed that under stress, the presence of flaws in a material could lead to a stress concentration in a solid. The "tip" or leading edge of a flaw or crack is shown in Fig. 7.1. Chemical bonds at the crack tip range from unstressed to fully strained at the point of breaking. However, although virtually no stress is required to bring about bond breakage, stress is

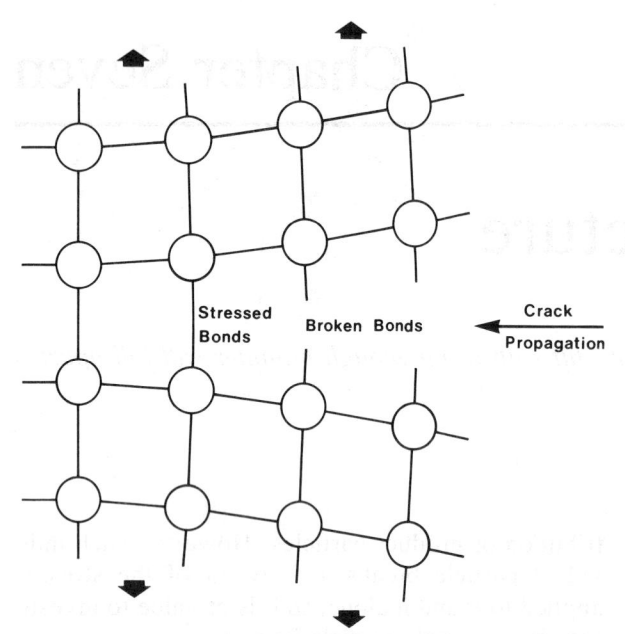

Figure 7.1. Propagation of a crack by rupture of chemical bonds under external stress. (After Somasundaran and Lin.[37])

required to provide the energy necessary for crack propagation and the consequent production of new surface. Thus, fracture is assumed to occur when the local strain energy at the crack tip is just sufficient to provide the surface energy of the two new surfaces produced by the fracture. This critical applied tensile stress normal to the crack, $\mathcal{S}_G$, also known as the Griffith stress, is given by

$$\mathcal{S}_G \cong \left(\frac{2\gamma\mathcal{Y}}{L_{cr}}\right)^{1/2} \quad (7.1)$$

($\mathcal{Y}$ = Young's modulus for the material; γ = the surface free energy per unit area of crack surface; L_{cr} = the crack length). Equation 7.1 is known as the *Griffith criterion* for fracture. The concept of cracks propagating from tiny flaws in a material is widely accepted, although modifications to the Griffith theory have been made to account for such aspects as the kinetic energy associated with the stress field propagating at the tip of the crack, and the localized plastic behavior of a material at the crack tip.[7]

It should be noted that although it is not necessary to provide enough energy to strain all bonds to the point of breaking (because of the presence of flaws), more energy is required than that which is just sufficient to provide the free energy of the new surfaces, because bonds away from the eventual fracture surfaces also become strained, hence absorb energy.

The Griffith theory requires that a tensile stress exists across a crack to open it further. A uniform compressive loading can only close up a crack. However, nonuniform compressive loading leads to localized tensile stresses; hence *in comminution particles normally break in tension, not in compression.*

The Griffith theory has been applied by Gilvarry[9,10] to the breakage of ore particles. He considers single particle breakage, assuming that flaws are randomly distributed throughout the particle and that each new fracture surface will contain flaws at which new fractures may originate. From this basis a distribution of fracture products is derived that has been found to agree well with experiment (see Eq. 7.17).

In the work of Rumpf[3] and his group on the fracture of single particles, the stress-strain relationships for various mineral particles have been examined and related to the Griffith crack theory. They have shown that for smaller particles having fewer flaws, the applied stress at which fracture occurs is greater. They also note that irrespective of the distribution and density of flaws, a greater stress is required to fracture a smaller particle. Since the energy absorbed by a crack is proportional to an area, whereas the strain energy is proportional to volume, the amount of energy available at a given stress condition decreases as the particle size decreases. This means that a greater energy density, hence a greater applied stress, is necessary to produce a continuous crack in a smaller particle, irrespective of the crack initiation or the material strength.

The fracture of single particles at very small sizes has also been investigated.[3] In this situation plastic deformation becomes a factor, and when significant deformation can occur without fracture, the *limit of grindability* has been reached. This limit is the smallest particle size that can be fractured for a given material, and should not be confused with the smallest product particle size (which can be smaller than the limit of grindability). Rumpf[3] reports the limits of grindability for quartz particles as 1 μm and for limestone particles as 3–5 μm.

Oka and Majima[5] have theoretically analyzed fracture as it occurs in comminution. They consider what happens when the breakage of a particle is due to the tensile stresses generated by a pair of concentrated loads F directed in compression as shown in Fig. 7.2. The stress and deformation characteristics

Single Particle Fracture

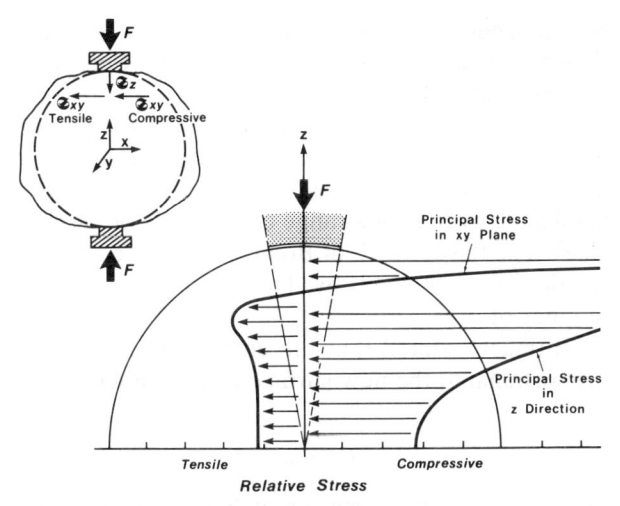

Figure 7.2. Distribution of principal stresses in a particle under localized compressive loading. (After Oka and Majima.[5])

for an irregular particle are shown to be similar to those of a sphere under the same conditions, and so the analysis is based on a sphere. The stress distributions on the axis of the sphere (particle) are shown in Fig. 7.2. Note that although the principal stress in the z-direction is a compressive stress throughout the particle, the principal stress in the x-y plane is a compressive stress adjacent to the load points but a tensile stress within the particle. Since mineral particles are much weaker in tension than in compression (i.e., their tensile strength is much lower than their compressive strength), the fracture of a particle occurs primarily because of the tensile stress. Subjecting a particle to a concentrated load as in Fig. 7.2 has resulted in breakage into a small number of large pieces due to tensile fracture, plus a large number of small pieces due to compression adjacent to the loading points. The equations describing the stress distribution in the particle are complex[5] and are not presented here. It can be noted, however, that for any given Young's modulus and Poisson's ratio, the stress is proportional to the load and inversely proportional to the square of the particle diameter, whereas the deformation of the particle is proportional to the load, inversely proportional to the square of the particle diameter, and also proportional to the distance from the center of the particle along the z-axis.

The energy E_p acting on the particle shown in Fig. 7.2 is the product of the loading compression F and the deformation from the loading point z_d, that is,

$$E_p = F \cdot z_d \quad (7.2)$$

The deformation at this point is

$$z_d = \frac{F}{d\mathcal{Y}} \cdot K_\nu \quad (7.3)$$

(K_ν = a constant dependent only on the Poisson's ratio of the rock particle; d = particle diameter). Using elasticity theory, Oka and Majima[5] show and confirm experimentally that the tensile stress at fracture (or the tensile strength) $\mathfrak{S}_f$ that is generated in a particle can be approximated by

$$\mathfrak{S}_f = 0.9 \frac{F_0}{d^2} \quad (7.4)$$

(F_0 = value of the applied load at fracture). By substituting Eqs. 7.4 and 7.3 into Eq. 7.2, the energy required to break a particle can be expressed in terms of the particle size and the properties of the material of the particle as

$$E_p = 1.23 K_\nu \frac{\mathfrak{S}_f^2 d^3}{\mathcal{Y}} \quad (7.5)$$

with K_ν being essentially constant.[5]

From the Griffith theory of fracture,[8] discussed above, it is known that (1) the tensile strength of a particle is dependent on the cracks or flaws in the particle, and (2) the probability of cracks being present in a particle decreases as the particle size decreases. Oka and Majima[5] show that the crack strength in a particle may be represented by a Weibull probability density function (as in Table 2.3), and that assuming a constant number of cracks per unit volume, the relationship between the crack strength, hence the particle strength, and particle volume is therefore given by

$$\mathfrak{S}_f \propto V^{-1/s} \quad (7.6)$$

(s = index in the Weibull function). For a constant rock particle volume, the variance in the crack strength decreases when s increases, and so s is known as the coefficient of uniformity of a rock.

Equation 7.6 can be written in terms of particle size as

$$\mathfrak{S}_f = \mathfrak{S}_{fr} \left(\frac{d}{d_r}\right)^{-3/s} \quad (7.7)$$

($\mathfrak{S}_{fr}$ = strength of reference particle of size d_r). By substituting Eq. 7.7 into Eq. 7.5, the following is obtained.

$$E_p = 1.23 K_\nu \frac{\mathfrak{S}_{fr}^2}{d_r^{-6/s} \mathcal{Y}} d^{3(1-2/s)} \quad (7.8)$$

Thus the energy required for single particle breakage is proportional to the 3(1 − 2/s)th order of the particle size d, that is,

$$E_p = K_1 d^{3(1-2/s)} \quad (7.9)$$

(K_1 = constant dependent on particle properties).

7.1.2. Energy-Size Reduction Relationships

An *empirical relationship* between energy and size reduction that has been proposed[11] and generally accepted is

$$dE_0 = -K\, dd/d^{\text{fn}(d)} \quad (7.10a)$$

(E_0 = specific energy necessary to supply the surface energy of the new surface; d = particle size). This equation has also been widely reported[12] with fn(d) replaced by a constant n, that is,

$$dE_0 = -K\, dd/d^n \quad (7.10b)$$

Figure 7.3 shows the general form of the energy-size reduction relationship, and clearly n is not a constant over the whole range of particle sizes. However, within limited size ranges a constant value for n can be used. This empirical equation (Eq. 7.10b) is then the generalized form of the theoretical and empirical energy-size reduction equations of Rittinger,[13] Kick,[14] Bond,[15,16] Holmes,[17] Charles,[18] Svensson and Murkes,[19] and others.

The equations of Rittinger, Kick, and Bond have been referred to as "laws." Although the first two have some theoretical basis, they are not generally applicable over a wide range of materials or particle sizes. The third "law" of Bond is a purely empirical relationship obtained by batch grinding a large number of ores.

Rittinger[13] stated that the energy of breakage is proportional to the new surface area produced, that is,

$$E_0 = K_1(S_{oO} - S_{oI}) \quad (7.11a)$$

(S_{oI}, S_{oO} = specific surface area before and after the breakage event, respectively). This equation can be obtained by integrating Eq. 7.10b if $n = 2$ and assuming that $S_0 \propto 1/d$. Equation 7.11a can also be written as

$$E_0 = K_2 \left(\frac{1}{d_O} - \frac{1}{d_I} \right) \quad (7.11b)$$

(d_I, d_O = particle size before and after breakage event, respectively). It should be noted that for a distribution of particle sizes, d should be replaced by a representative particle size such as a mean. The definition of particle size used must be the same before and after breakage.

Kick[14] proposed the theory that equivalent geometrical changes in the sizes of particles require equal energy. The resulting equation is

$$E_0 = K_3 \ln \left(\frac{d_I}{d_O} \right) \quad (7.12)$$

This equation can also be obtained by integration of Eq. 7.10b, this time with $n = 1$.

In an extensive series of tests covering a wide range of ores, Bond[15,16] found that an equation of the following form was widely applicable:

$$E_0 = K_4 \left(\frac{1}{d_O^{1/2}} - \frac{1}{d_I^{1/2}} \right) \quad (7.13a)$$

The particular form of the equation used by Bond was

$$E_0 = K_4 \left(\frac{1}{d_{80,O}^{1/2}} - \frac{1}{d_{80,I}^{1/2}} \right) \quad (7.13b)$$

($d_{80,I}$, $d_{80,O}$ = 80% passing size before and after the breakage event, respectively). Equations 7.13a and 7.13b can be obtained by integration of Eq. 7.10b with $n = 1.5$.

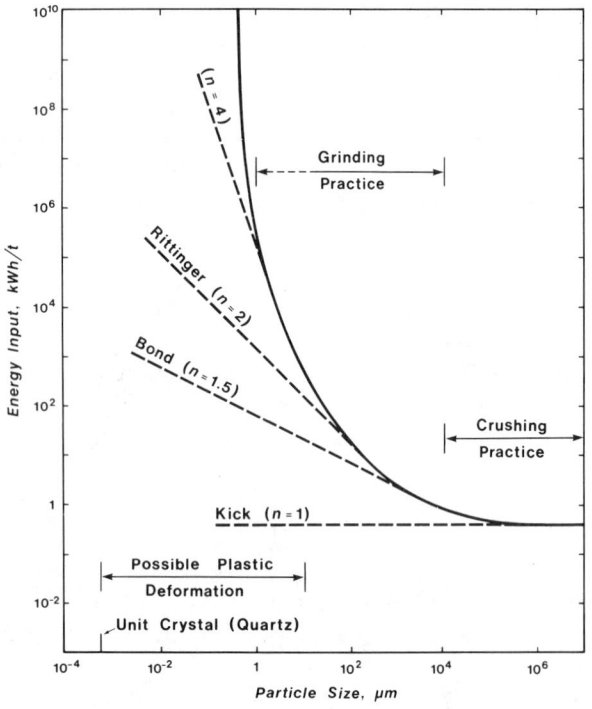

Figure 7.3. General form of the energy-size reduction relationship (Eq. 7.10). (After Hukki.[11])

Single Particle Fracture

Holmes,[17] Charles,[18] and others[19,20] have used a size distribution in the energy-size reduction equation. Their equations can be generalized as

$$E_0 = K_5(d^*)^{-n_1} \qquad (7.14)$$

(d^* = size modulus as described in the Gaudin-Schuhmann equation [Table 2.3]; n_1 = constant). The equations of this form have been discussed widely in the literature.[6,20,21]

The equation obtained by Oka and Majima (Eq. 7.9) for breakage of a single particle can be adapted for multiple particle breakage[5] to give

$$dE_0 = -K_6 \, dd/d^{(1+6/s)} \qquad (7.15)$$

which on integration becomes

$$E_0 = K_7(d_0^{-6/s} - d_I^{-6/s}) \qquad (7.16)$$

except when $s = \infty$. Clearly when $s = \infty$, Eq. 7.15 on integration becomes the Kick equation (Eq. 7.12). When $s = 6$, Eq. 7.16 becomes the Rittinger equation (Eq. 7.11b), and when $s = 12$, it becomes the Bond equation (Eq. 7.13a).

Similarity Equations. In an entirely different approach to the development of an energy–size reduction relationship, Rumpf and his co-workers[3] used dimensional analysis to develop a series of similarity equations for fracture under various conditions. They have shown that this approach is useful in describing the fracture that occurs on the basis of material properties and loading conditions.

7.1.3. Mechanisms of Fracture

For a particle to fracture, a stress high enough to exceed the fracture strength of the particle is required. The manner in which the particle fractures depends on the nature of the particle, and on the manner in which the force on the particle is applied.[22,23] The force on the particle can be a compressive force as shown in Fig. 7.2, causing the particle to fracture in tension. This force could be applied at either a fast or a slow rate, and the rate affects the nature of the fracture. Also, instead of a compressive force, the particle could be subject to a shear force, such as is exerted when two particles rub against each other.

Many terms have been used to describe the various mechanisms of single particle fracture, and there is little consistency in their usage. The terms used here are as shown in Fig. 7.4: *abrasion, cleavage,* and *shatter*. They are interpreted as follows.

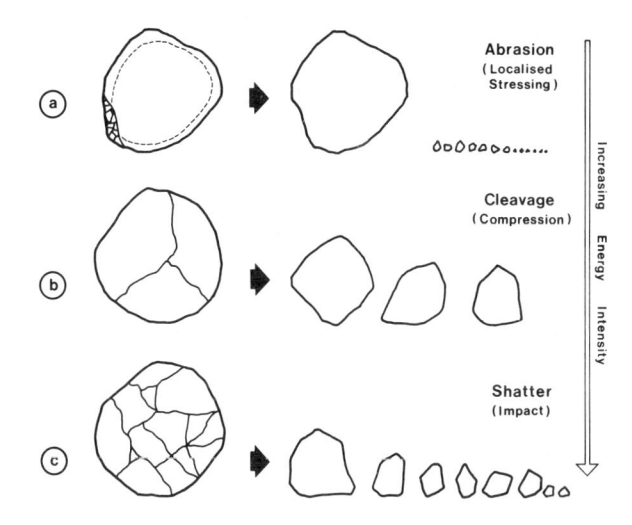

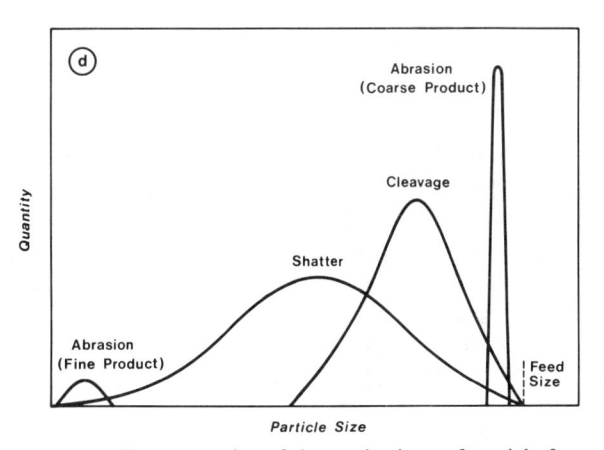

Figure 7.4. Representation of the mechanisms of particle fracture and the resulting product size distributions.

Abrasion fracture (Fig. 7.4a) occurs when insufficient energy is applied to cause significant fracture of the particle. Rather, localized stressing occurs and a small area is fractured to give a distribution of very fine particles (effectively localized shatter fracture). Fracture by *cleavage* (Fig. 7.4b) occurs when the energy applied is just sufficient to load comparatively few regions of the particle to the fracture point, and only a few particles result. Their size is comparatively close to the original particle size. Typically this situation occurs under conditions of slow compression where the fracture immediately relieves the loading on the particle. Fracture by *shatter* (Fig. 7.4c) occurs when the applied energy is well in excess of that required for fracture; under these conditions many areas in the particle are overloaded and the result is a comparatively large number of particles with a wide spectrum of sizes. This

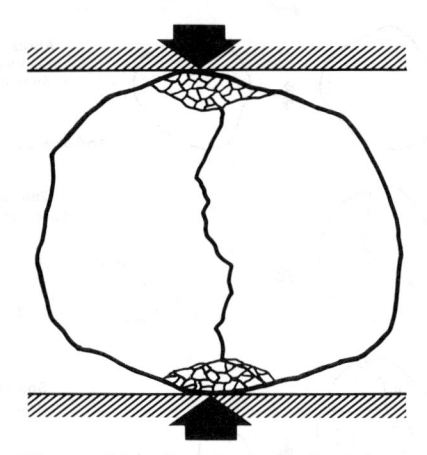

Figure 7.5. Combination of fracture mechanisms, as occurs in practice.

occurs under conditions of rapid loading such as in a high velocity impact.

In practice of course, these events do not occur in isolation. Particularly significant is the situation illustrated in Fig. 7.5 (which commonly occurs in crushers), where attrition occurs at the loading points. Fracture due to a high velocity impact in which there is insufficient energy to result in shatter can yet cause attrition of the particle surface; the resultant fracture is essentially abrasion.

If very small fractions of a particle are broken off because of a shear force applied at the surface, the cumulative effect is abrasion. Strictly speaking, of course, this is not a single fracture event.

Another term used to describe a mechanism of fracture is chipping. However, chipping, which is the breaking of edges or corners of a particle, may be considered to be a special case of cleavage.

7.1.4. Fracture Product Size Distributions

The size distribution of the products resulting from each fracture mechanism is shown in Fig. 7.4d. Many attempts have been made to mathematically describe these size distributions. For single particle shatter breakage as a result of impact, continuous functions have been derived.

Gilvarry[9] used the Griffith crack theory approach (Section 7.1.1) to show that

$$Y^- = 1 - \exp\left[-\left(\frac{d}{K_1}\right) - \left(\frac{d}{K_2}\right)^2 - \left(\frac{d}{K_3}\right)^3\right] \quad (7.17)$$

(Y^- = cumulative mass fraction finer than size d; K_1, K_2, K_3 = constants dependent on the activated edge, area, and volume flaw densities, respectively).

When the assumption is made that edge flaws are predominent, and for small particles, Eq. 7.17 reduces to the Rosin-Rammler equation

$$Y^- = 1 - \exp\left[-\left(\frac{d}{d^*}\right)^s\right] \quad (2.1b)$$

(d^* = a reference size; s = index of the Weibull or Rosin-Rammler distribution function) given in Table 2.3.

Using a statistical approach, Gaudin and Meloy[24,25] derived an equation

$$Y^- = 1 - \left(1 - \frac{d}{d^*}\right)^n \quad (2.1e)$$

(d^* = a reference size, in this case, the initial particle size; n = distribution modulus) that is applicable to the larger fragments of a single breakage event. Combining the approach of Gilvarry and the statistical approach, Klimpel and Austin[26] derived a general equation

$$Y^- = 1 - \left[1 - \left(\frac{d}{d^*}\right)\right]^{n_1}\left[1 - \left(\frac{d}{d^*}\right)^2\right]^{n_2}\left[1 - \left(\frac{d}{d^*}\right)^3\right]^{n_3} \quad (7.18)$$

(n_1, n_2, n_3 = constants dependent on the edge, area, and volume flaw densities, respectively), which for coarse distributions in which a few large particles are produced reduces to

$$Y^- = 1 - \left[1 - \left(\frac{d}{d^*}\right)^3\right]^{n_4} \quad (7.19)$$

Klimpel and Austin[26] also show that the other distributions reported, including Eq. 7.17, are either approximations or special cases of Eq. 7.18.

For fine particles, and a slope near unity, Eqs. 7.17–7.19 all approximate the Gaudin-Schuhmann equation[6,7,20]

$$Y^- = \left(\frac{d}{d^*}\right)^n \quad (2.1c)$$

(d^* = size modulus; n = distribution modulus).

Broadbent and Callcott[27,28] used another product size distribution

$$Y^- = \frac{1 - \exp\left[-(d/d^*)^n\right]}{1 - \exp(-1)} \quad (2.1d)$$

to calculate values of Y^- in a geometric series of particle sizes. Although this distribution has no ap-

Single Particle Fracture

parent theoretical basis (except as a modification of the Rosin-Rammler equation), the discrete value form of it (i.e., a matrix of values) has been widely used in the mathematical analysis of size reduction operations.[28,29]

7.1.5. The Breakage Function

The preceding section presented a number of distributions that have been used to describe the products of single breakage events. In industrial size reduction operations, a large number of single breakage events are occurring simultaneously, and the products of each breakage event in turn undergo breakage. To mathematically describe the breakage occurring in such size reduction operations, it has been found useful to consider the breakage in terms of a series of single particle breakage events, with each breakage event resulting in a distribution of product particles. This distribution of breakage products is known as the *breakage function*; it is most commonly used as a *cumulative breakage function*. The mass fractions comprising the breakage functions are called *breakage parameters*.

There are two major schools of thought on the concept of a breakage function. One is to consider the breakage function independent of initial particle size. The other requires the determination of a breakage function for each initial particle size. Both approaches are widely used and have been shown to provide useful descriptions of the breakage process.

The breakage function has been determined in a number of ways. One approach has been to use a breakage distribution equation. Equation 2.1d (Table 2.3: with $n = 1$) has been widely used,[28,29] with the cumulative breakage function being Y^-, the cumulative mass fraction passing. Kelsall[30] found that the equation

$$Y^- = \left(\frac{d}{d^*}\right)^n \qquad (2.1c)$$

satisfactorily described the breakage function for initial particle sizes less than 590 μm, with n having values from 0.90 to 0.95.

However, it is now generally accepted that it is preferable to use an experimentally determined breakage function and that no advantage is gained by attempting to use a distribution equation. Kelsall and Reid[31] report the breakage function shown in Fig. 7.6 for particle sizes defined by a $1/\sqrt{2}$ sieve series. Note that the breakage function is independent of initial particle size. The values of the break-

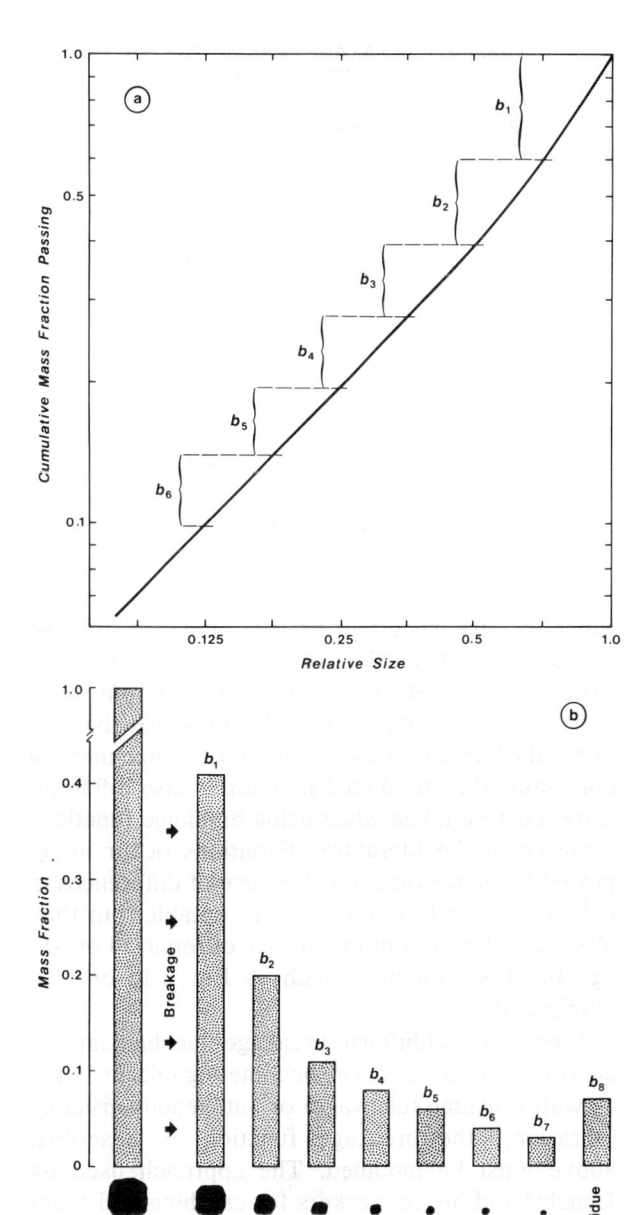

Figure 7.6. The breakage function for particle sizes defined by a $1/\sqrt{2}$ sieve series. (a) The breakage function. (b) Illustration of the breakage function. (After Kelsall and Reid.[31])

age parameters b_i, corresponding to the breakage function of Fig. 7.6, are shown in Table 7.1. Such data can be described by the general equation[32]:

$$Y^- = K_b \Re_b^{n_1} + (1 - K_b)\Re_b^{n_2} \qquad (7.20)$$

(K_b = coefficient such that $0 \leq K_b \leq 1$; $\Re_b = d/d_I$; d_I = initial particle size; $n_2 \geq n_1 \geq 0$).

The breakage function of Fig. 7.6 is based on the size intervals in a $1/\sqrt{2}$ sieve series. Other breakage

TABLE 7.1: BREAKAGE PARAMETERS FOR $1/\sqrt{2}$ SIEVE SERIES

Breakage Parameter	$1/\sqrt{2}$ Series
b_1	0.410
b_2	0.200
b_3	0.114
b_4	0.081
b_5	0.057
b_6	0.040
b_7	0.029
b_8	0.021
b_9	0.015
b_{10}	0.010
b_{11}	0.007
b_n (balance)	0.016

functions are reported for $1/2$ and $1/\sqrt[4]{2}$ sieve series.[29,32] It should be noted that the relation between the breakage functions for the various sieve series is not a simple one. Not only are the size intervals different in each case, but the size interval containing the initial feed material is also different. Care must be taken when using breakage functions reported in the literature. Variations occur in reported breakage functions because of differences in definition, which is of course acceptable, but they also occur because of the misuse of reported breakage function equations such as Eq. 2.1c or 2.1d (Table 2.3).

When the additional breakage mechanisms of cleavage and/or abrasion become significant, such as with crushing (cleavage) or autogenous grinding (abrasion), the breakage function as described above must be modified. The approach used by Lynch[29] and his co-workers for crushing and Stanley[33] for autogenous grinding has been to determine the breakage function corresponding to each breakage mechanism, then to combine them as follows:

$$b_i = \mathfrak{F}_{b1} b_{i1} + (1 - \mathfrak{F}_{b1}) b_{i2} \quad (7.21)$$

(b_{i1}, b_{i2} = breakage parameters corresponding to each breakage mechanism; $\mathfrak{F}_{b1}$ = fraction of material broken by mechanism 1). The nature of these particular breakage functions is discussed further in Chapter 8.

7.1.6. Breakage as a Rate Process

The rate at which particles break in a size reduction process can be represented by a first-order equation (Eq. 3.1). Thus,

$$\text{absolute rate of breakage of size } i = k_i m_i \quad (7.22a)$$

or

$$\frac{dm_i}{dt} = -k_i m_i \quad (7.22b)$$

(k_i = *rate of breakage* out of size fraction i [the first-order rate constant]; m_i = mass fraction in size i). This equation has been found to be generally applicable to breakage processes.[34] Using the concept of a breakage function introduced in the preceding section and the rate of breakage as described by Eq. 7.22, mathematical descriptions of size reduction processes have been developed (see Chapter 8).

The rate of breakage of particles has been related to their material properties. Everell[35,36] measured both the tensile strength and the uniaxial compressive strength of limestone particles (approx. 5 mm), and also determined the rate of breakage of these particles in a laboratory ball mill (Fig. 7.7).

There is an alternative approach that has been extensively reported in the literature.[27–29] A particle in a size reduction process has some *probability of being broken*. In any size fraction, a proportion of the particles are *selected for breakage* and the remainder pass through the process unbroken. Using this concept, (Fig. 7.8), a *selection function* has been defined that is the probability of breakage in each size fraction. As with the rate of breakage, the selection function can be used in combination with the breakage function to mathematically describe a size reduction process.

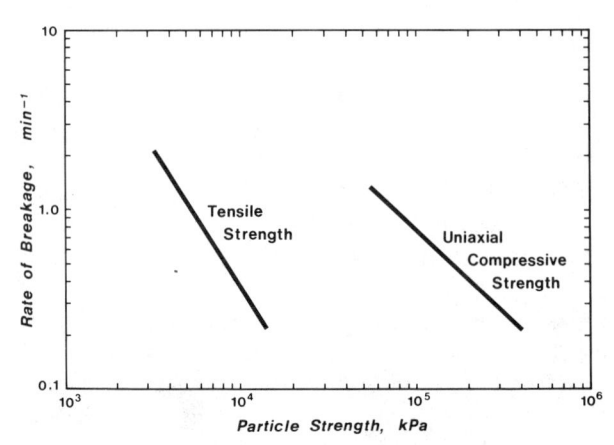

Figure 7.7. The effect of particle strength on rate of breakage. (After Everell.[35])

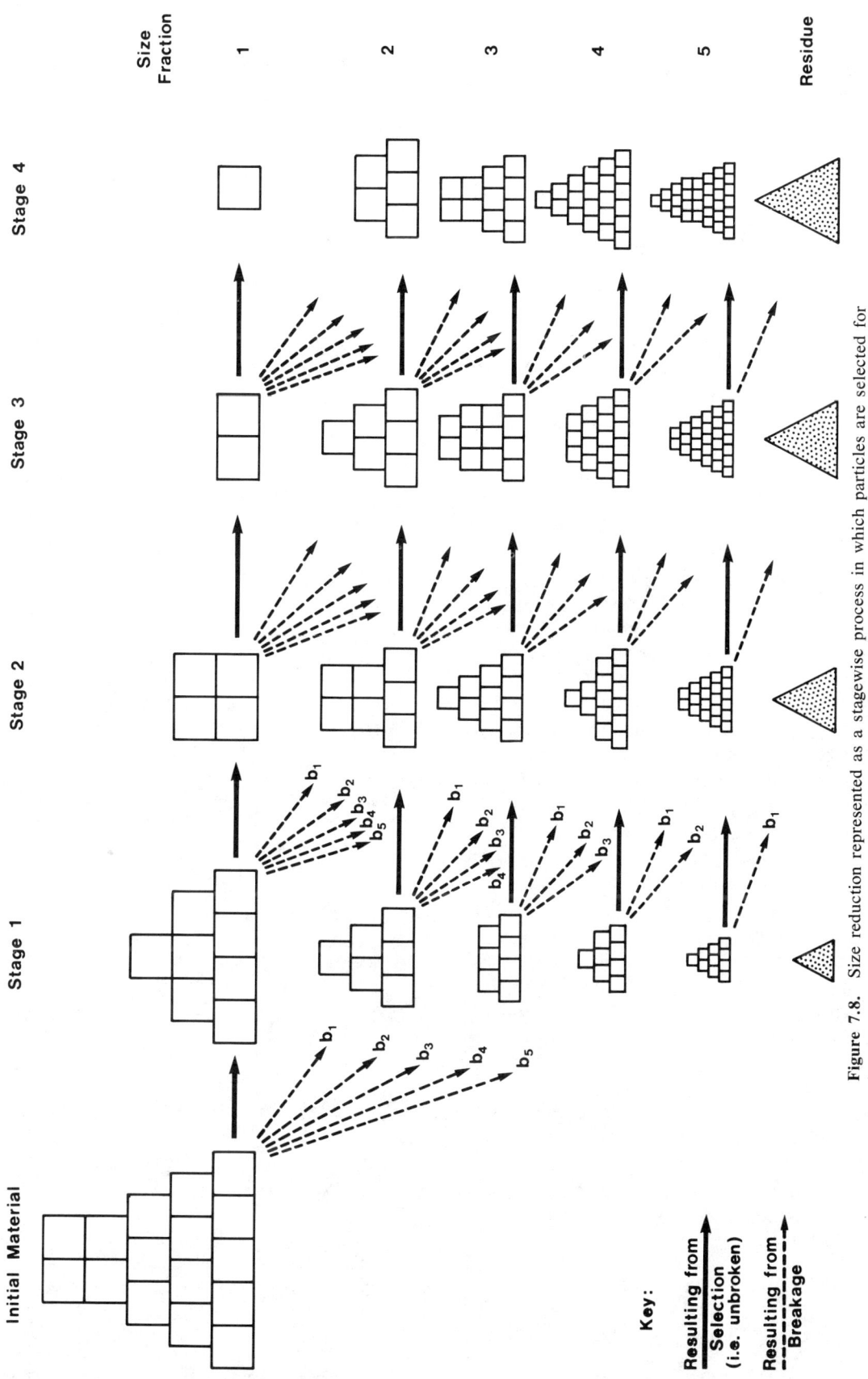

Figure 7.8. Size reduction represented as a stagewise process in which particles are selected for breakage and break according to their breakage function.

In Chapter 3, the use of rate and probability approaches to the description of concentration processes was discussed. There is a clear relation between the use in size reduction and in concentration.

7.2. THE FRACTURE ENVIRONMENT

Size reduction involves the rupturing of chemical bonds to produce new surfaces (Fig. 7.1). Any phenomenon that enhances this rupturing is of value. The chemical nature of a surface can be expected to affect the fracture energy. As discussed in Chapter 6, a surface exists in equilibrium with its surroundings, and the electrical and chemical nature of the surface is dependent on those surroundings.

Somasundaran and Lin[37] have provided a detailed review of the effect of environment on size reduction. The most common effect is the presence of water, which often results in improved breakage efficiency.[38] However, the addition of inorganic ions and organic surface-active agents can also have a major effect. In general, the rate of breakage is enhanced by the addition of moderate amounts of surfactants[37,39] (see Fig. 8.26f), although where surfactant is opposite in charge to the mineral surface, the effect can be to decrease the breakage rate. One common operating practice is to add flotation reagents at the wet grinding stage; recently it has been shown that in some circumstances this can result in reduced grinding rates.[40] In dry grinding, organic additives have been extensively used.[41]

Two mechanisms have been proposed to explain the increase in breakage rates or the decrease in effective hardness of rocks due to surfactant additions (known as the "Rehbinder effect"). Rehbinder[42,43] attributed this phenomenon to the reduction in surface energy on adsorption, leading to a lower energy requirement for the production of new surfaces. This requires that the surfactant penetrate to some extent into a crack before rupture. An alternative explanation is offered by Westwood[44]: the mobility of dislocations near the surface (which result in crack formation) is related to electrical charges at the surface resulting from adsorption.

REFERENCES

1. J. Gross and S. R. Zimmerley, "Crushing and Grinding. III. Relation of Work Input to Surface Produced in Crushing Quartz," *Trans. AIME/SME*, **87**, 35–50 (1930).
2. R. T. Hukki, "The Principles of Comminution: An Analytical Summary," *Eng. Min. J.*, **176**, 106–110 (1975).
3. H. Rumpf, "Physical Aspects of Comminution and New Formulation of a Law of Comminution," *Powder Technol.*, **7**, 145–159 (1973).
4. K. Schonert, H. Umhauer, and W. Klemm, "The Influence of Temperature and Environment on the Slow Crack Propagation in Glass," *Proc. 2nd Int. Conf. on Fracture*, Chapman and Hall (1969).
5. Y. Oka and W. Majima, "A Theory of Size Reduction Involving Fracture Mechanics," *Can. Met. Q.*, **9**, 429–439 (1970).
6. C. C. Harris, "On the Role of Energy in Comminution: A Review of Physical and Mathematical Principles," *Trans. IMM (C)*, **75**, C37–C56 (1966).
7. L. G. Austin and R. R. Klimpel, "The Theory of Grinding Operations," *Ind. Eng. Chem.*, **56**, 18–29 (November 1964).
8. A. A. Griffith, "The Phenomena of Rupture and Flow in Solids," *Phil. Trans. R. Soc., Ser. A*, **221**, 163–198 (1920–1921).
9. J. J. Gilvarry, "Fracture of Brittle Solids. I. Distribution Function for Fragment Size in Single Fracture (Theoretical)," *J. Appl. Phys.*, **32**, 391–399 (1961).
10. J. J. Gilvarry, "Fracture and Comminution of Brittle Solids (Theory and Experiment)," *Trans. AIME/SME*, **220**, 380–389 (1961).
11. R. T. Hukki, "Proposal for a Solomonic Settlement Between the Theories of von Rittinger, Kick, and Bond," *Trans. AIME/SME*, **220**, 403–408 (1961).
12. W. H. Walker et al., *Principles of Chemical Engineering*, McGraw-Hill, p. 255 (1937).
13. P. R. von Rittinger, *Lehrbuch der Aufbereitungskonde*, Berlin (1857).
14. F. Kick, *Das Gesetz der Proportionalen Widerstande und seine Anwendung*, Leipzig (1885).
15. F. C. Bond, "The Third Theory of Comminution," *Trans. AIME/SME*, **193**, 484–494 (1952).
16. F. C. Bond, "Crushing and Grinding Calculations," *Br. Chem. Eng.*, **6**, 378–385, 543–548 (1961).
17. J. A. Holmes, "A Contribution to the Study of Comminution—A Modified Form of Kick's Law," *Trans. Inst. Chem. Eng.*, **35**, 125–141 (1957).
18. R. J. Charles, "Energy-Size Reduction Relationships in Comminution," *Trans. AIME/SME*, **208**, 80–88 (1957).
19. J. Svensson and J. Murkes, "An Empirical Relationship Between Work Input and Particle Size Distribution Before and After Grinding," *Int. Miner. Dressing Congr.*, Stockholm, pp. 37–57, Almkuist & Wiksell (1957).
20. A. L. Mular, "Comminution in Tumbling Mills–A Review," *Can. Met. Q.*, **4**, 31–74 (1965).
21. R. Schuhmann, "Energy Input and Size Distribution in Comminution," *Trans. AIME/SME*, **217**, 22–25 (1960).
22. D. D. Crabtree et al., "Mechanisms of Size Reduction in Comminution Systems. Part I. Impact, Abrasion, and Chipping Grinding," *Trans. AIME/SME*, **229**, 201–206 (1964).
23. R. S. Kinasevich et al., "Mechanisms of Size Reduction in

References

Comminution Systems. Part II. Interpreting Size Distribution Curves and the Comminution Event Hypothesis," *Trans. AIME/SME*, **229**, 207–210 (1964).

24. A. M. Gaudin and T. P. Meloy, "Model and a Comminution Distribution Equation for Single Fracture," *Trans. AIME/SME*, **223**, 40–43 (1962).

25. T. P. Meloy, "A Three-Dimensional Derivation of the Gaudin Size Distribution Equation," *Trans. AIME/SME*, **226**, 447–448 (1963).

26. R. R. Klimpel and L. G. Austin, "The Statistical Theory of Primary Breakage Distributions for Brittle Materials," *Trans. AIME/SME*, **232**, 88–94 (1965).

27. S. R. Broadbent and T. G. Callcott, "A Matrix Analysis of Processes Involving Particle Assemblies," *Phil. Trans. R. Soc., Ser. A*, **249**, 99–123 (1956).

28. S. R. Broadbent and T. G. Callcott, "Coal Breakage Processes," *J. Inst. Fuel*, **29**, 524–539 (1956); **30**, 13–25 (1957).

29. A. J. Lynch, *Mineral Crushing and Grinding Circuits*, Elsevier (1977).

30. D. F. Kelsall, "A Study of Breakage in a Small Continuous Open Circuit Wet Ball Mill," *Proc. 7th Int. Miner. Process. Congr., New York, 1964*, pp. 33–42 Gordon and Breach (1965).

31. D. F. Kelsall and K. J. Reid, "The Derivation of a Mathematical Model for Breakage in a Small, Continuous, Wet Ball Mill," *AIChE Inst. Chem. Eng. Symp. Ser.* No. 4, 14–20 (1965).

32. L. G. Austin and P. T. Luckie, "The Estimation of Non-Normalized Breakage Distribution Parameters from Batch Grinding Tests," *Powder Technol.*, **5**, 267–271 (1971–1972).

33. G. G. Stanley, "Mechanisms in the Autogenous Mill and Their Mathematical Representation," *J. S. Afr. IMM*, **75**, 77–97 (1974).

34. L. G. Austin, "A Review Introduction to the Mathematical Description of Grinding as a Rate Process," *Powder Technol.*, **5**, 1–17 (1971–1972).

35. M. D. Everell, "Empirical Relations Between Grinding Selection Functions and Physical Properties of Rocks," *Trans. AIME/SME*, **252**, 300–306 (1972).

36. M. D. Everell, D. E. Gill, and L. L. Sirois, "Relation of Grinding Selection Functions to Physicomechanical Properties of Rocks," *Proc. 6th Can. Rock Mech. Symp.* (1970).

37. P. Somasundaran and I. J. Lin, "Effect of the Nature of Environment on Comminution Processes," *Ind. Eng. Chem. Process Des. Dev.*, **11**, 321–331 (1972).

38. A. F. Taggart, *Handbook of Mineral Dressing*, pp. 6–15, Wiley (1927).

39. R. R. Klimpel and W. Manfroy, "Chemical Grinding Aids for Increasing Throughput in the Wet Grinding of Ores," *Ind. Eng. Chem. Process Des. Dev.*, **17**, 518–523 (1978).

40. A. Ryncarz and J. Laskowski, "Influence of Flotation Reagents on the Wet Grinding of Quartz," *Powder Technol.*, **18**, 179–185 (1977).

41. R. H. Snow, "Size Reduction," *Ind. Eng. Chem.*, **62**, 36–43 (November 1970).

42. P. A. Rehbinder, "On the Effect of Surface Energy Changes on Cohesion, Hardness, and Other Properties of Crystals," *Proc. 6th Phys. Congr.*, State Press, Moscow (1928).

43. P. A. Rehbinder, L. A. Schreiner, and K. F. Zhigach, "Hardness Reducers in Drilling," *Trans. Counc. Sci. Ind. Res., Melbourne*, pp. 163ff. (1948).

44. A. R. C. Westwood, "Environment Sensitive Mechanical Behavior, Status and Problems," in *Environment-Sensitive Mechanical Behavior*, A. R. C. Westwood and N. S. Stoloff (Eds.), Gordon and Breach (1966).

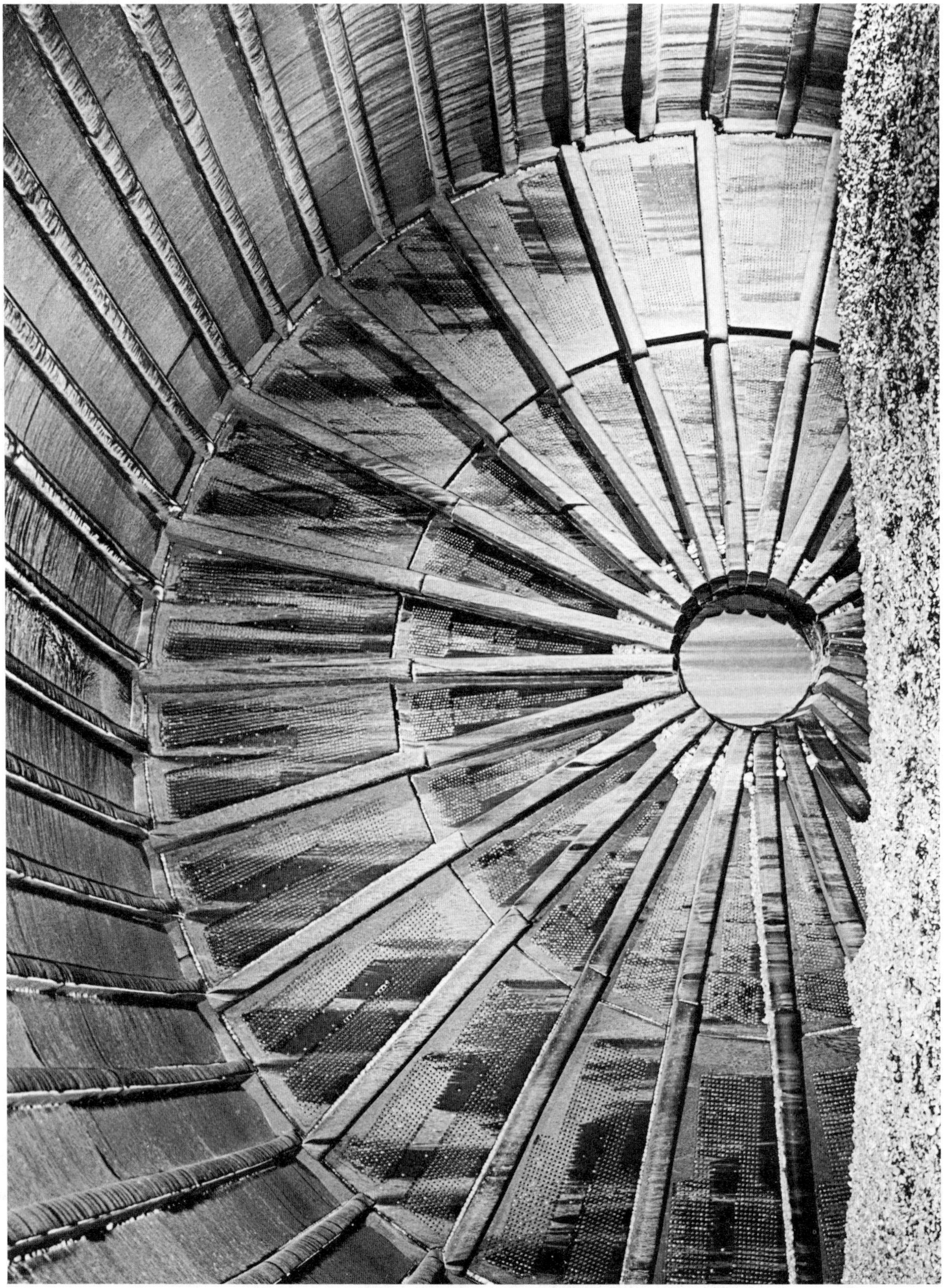

Part II

Size Reduction

Size reduction, or comminution, is an important step in the processing of most minerals, in that it may be used:

To produce particles of the required size and shape.

To liberate valuable minerals from gangue so that they can be concentrated.

To increase the surface area available for chemical reaction.

A variety of equipment is available, but individually each is restricted in its application (Fig. II.1). In the minerals industry, most initial size reduction (crushing) is carried out by compression crushers, with tumbling mills used for subsequent fine size reductions (grinding).

Over recent years, the most noticeable development has been the increasing size of equipment, as ore grades have decreased and mine sizes have increased. Indeed, it is now normal to find very large gyratory crushers treating run-of-mine ore, whereas in the past jaw crushers had more than sufficient capacity.

Conventional rod and ball mills have also undergone substantial size increases. Much of the incentive for this development lay in the knowledge that larger mills are more energy efficient. Even when it was found that ball mill diameters greater than 4 m gave no further energy savings, the savings in capital costs still justified increased size. However at present mills appear to be limited to about 5 m, because at this size the grinding action starts to deteriorate.

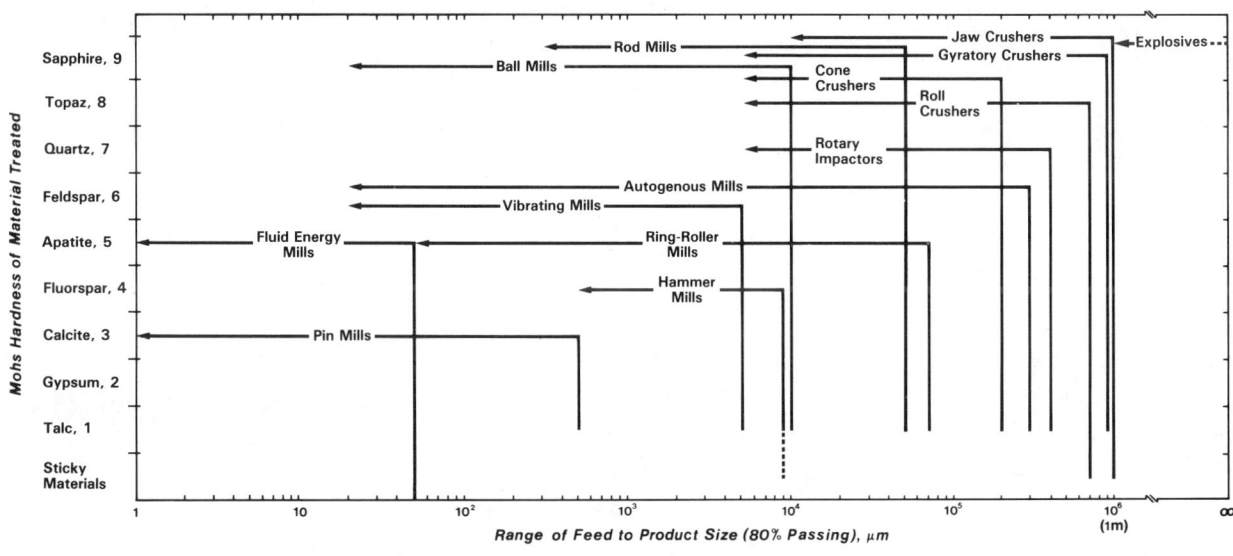

Figure II.1. Applicability of size reduction equipment.

For a period of time, very large autogenous grinding mills, with their lower capital investment requirement, appeared to be the grinding mills of the future. Though still widely used, particularly as semi-autogenous mills, their ascent has been limited for two reasons. First, because early design knowledge was inadequate, a number of mills failed to perform as well as expected. It was, however, the rapid increase in energy costs during the 1970's that had the most significant effect. In an already energy inefficient operation such as size reduction, autogenous mills do not look quite so attractive.

Higher energy costs are focusing attention on alternative equipment. Ring-roller mills have long been used in Europe for dry grinding, and along with vibrating mills, are receiving attention in the United States. Chemical and electrical methods of size reductions have also been suggested.

In recent years considerable progress has been made in developing methods for theoretical analysis of size reduction. Unfortunately, the widespread use of these methods is restricted by the scarcity of scale-up data, and although some data are already available in some manufacturing companies, it appears that it will be some time before it is available in the literature.

Chapter Eight

Crushing and Grinding

Anthony's Law of Force: "Don't force it, get a larger hammer."

The methods of size reduction may be grouped in a number of ways, but since reduction occurs in stages, particle size provides the primary method of grouping. If the ore body is a massive one, mining is actually the first stage of size reduction and is generally carried out with explosives, although mechanical means may be used on softer ores. The term *crushing* is applied to subsequent size reductions down to about 25 mm, with finer reductions considered to be *grinding*. Both crushing and grinding can be further subdivided into primary, secondary, tertiary, and sometimes even quaternary, stages. Since these stages relate to the machinery used, the boundaries are not rigid, and in any given operation not all may be required. Grinding can be yet further subdivided by the type of mill, the type of grinding media, and whether the grinding is carried out wet or dry.

The extent of the size reduction achieved by any machine is described by the *reduction ratio*, which can be broadly defined as the feed size divided by the product size. In reality, both these sizes must be defined, and although a number of definitions are possible, the most widely used is simply the 80% passing size of the cumulative size distribution.

8.1. EQUIPMENT

The design requirements of size reduction machines change markedly as the particle size changes. In virtually all machines, the breakage forces are applied either by compression or impact (Section 7.1). The products in each case are similar and the difference between machines is associated mainly with the mechanical aspects of applying the force to the various sizes of particles. When the particle is large, the energy to fracture each particle is high, even though the energy per unit mass is low (Fig. 7.3). As the particle size decreases, the energy to fracture each particle decreases but the energy per unit mass rises more rapidly. Consequently, crushers have to be massive and structurally strong, whereas grinding machines (*mills*) must be capable of dispersing energy over a large area.

The various types of size reduction equipment commonly used in mineral processing are illustrated in Tables 8.1 and 8.2, and Figs. 8.1–8.9; the tables also summarize basic information about the equipment. More detailed information is available elsewhere in the literature[1-9] and in manufacturers' catalogues.[10]

8.1.1. Jaw and Gyratory-Type Crushers

The jaw and gyratory-type crushers (e.g., Figs. 8.1 and 8.2) cause fracture by compression, since this is the most practical method of applying a fracture force to large particles. This in turn means that the machine must be constructed in such a way that the openings impose limitations on the feed and product size: capacity therefore becomes dependent on the size of the discharge opening and the machine speed. Furthermore, the fixed size of the input and output openings (defined in Fig. 8.3) results in these crushers having a limited reduction ratio.

A notable characteristic of gyratory-type crushers is that as the particle size decreases, the outer crushing surface changes to an upright cone and becomes more parallel to the mantle surface.

In addition to the basic machines illustrated in Table 8.1, minor variations occur, particularly with respect to the crushing chamber.[1-4] One such variation is the reaction to unbreakable objects or over-

TABLE 8.1: THE MAJOR TYPES OF CRUSHERS

			Size (mm)	Power (kW)	Speed (r.p.m.)	Reduction Ratio	Characteristics and Applications
ELECTRO-ENERGETIC			to 250		2 - 3 min for 5 - 10 tonne boulder		Primarily used to break oversize rocks before a primary crusher.
JAW CRUSHERS	Blake (Double-Toggle)		125 (gape) x 150 (width) to 1600 x 2100	2.25 to 225**	300 to 100	Average 7:1 Range 4:1 to 9:1	Originally the standard jaw crusher used for primary and secondary crushing of hard, tough abrasive rocks. Also for sticky feeds. Relatively coarse slabby product, with minimum fines. Flywheel evens power draft.
	Overhead Pivot (Double Toggle)		180 x 305 to 1220 x 1525	11 to 150**	390 to 257	Average 7:1 Range 4:1 to 9:1	Similar applications to Blake. Overhead pivot; reduces rubbing on crusher faces, reduces choking, allows higher speeds and therefore higher capacities. Energy efficiency higher because jaw and charge not lifted during cycle.
	Overhead Eccentric (Single-Toggle)		125 x 150 to 1600 x 2100	2.25 to 400**	300 to 120	Average 7:1 Range 4:1 to 9:1	Originally restricted to smaller sizes by structural limitations. Now in same sizes as Blake, which it has tended to supersede, because overhead eccentric encourages feed and discharge, allowing higher speeds and capacity, but with higher wear and more attrition breakage and slightly lower energy efficiency. Unsuitable for very hard, tough abrasive rock. Sometimes made with twin swing jaws.
	Dodge		100 x 150 to 280 x 380	2.25 to 11**	300 to 250	Average 7:1 Range 4:1 to 9:1	Bottom pivot gives closer sized product than Blake, but Dodge is difficult to build in large sizes, and is prone to choking. Generally restricted to laboratory use.
GYRATORY CRUSHERS	True Gyratory		760 (opening width) x 1400 (mantle maximum diameter) to 2135 x 3300	5 to 750	450 to 110	Average 8:1 Range 3:1 to 10:1	True gyratory crushers characterised by diverging crushing surfaces (outer surface or bowl has inward slope towards bottom). Used for primary and secondary rock, with minimum fines. Taller, higher capacity, and more suitable for slabby feeds than jaw crusher.
	Cone	Short Head	600 (cone diameter) to 3050	22 to 600	290 to 220	Secondary crushing 6:1 to 8:1 Tertiary crushing 4:1 to 6:1	Cone gyratories are characterised by converging crushing surfaces (outer surface tends to parallel mantle surface). Used for secondary and tertiary crushing. Generally as the particle size decreases (e.g. tertiary crushing) the outer crushing surface is made straighter and more parallel to a steeper mantle (often called a "Short Head" crusher). Tertiary crushers are often choke fed.

* Data intended only to indicate capacity. Manufacturers' catalogs and Bond's Law should be consulted for reliable information.
** For very hard rock, power may be up to 50% higher provided machine is strengthened.
® Reg. Trade Mark, Rexnord Inc.

TABLE 8.1: (Continued)

			Size (mm)	Power (kW)	Speed (r.p.m.)	Capacity* (t/hr)	Reduction Ratio	Characteristics and Applications
GYRATORY CRUSHERS	Gyradisc®		900 (mantle diameter) to 2100	100 to 400	325 to 260		2:1 to 4:1	For very fine or quaternary crushing. Choke feeding and low cone angle cause fracture between particle layers, reduce wear, give more cubic particle shape. Used to produce agregate (instead of rod mill), or to ensure uniform sized rod mill feed. Unsuitable for sticky feeds.
ROLL CRUSHERS	Single Rolls		500 (diameter) x 450 (width) to 1500 x 2100	15 to 300	60 to 23	20 to 1500	to 7:1	Basically primary or secondary crusher, suitable for softer, friable, non-abrasive materials such as coal or limestone. Better than jaw and gyratory crushers on wet and sticky materials.
	Double Rolls		750 (diameter) x 350 (width) to 1800 x 900 or 860 x 2100	27 to 112	150 to 50	20 to 2000	3:1	At low reduction ratios, product is comparatively low in fines. Still in use in some plants as tertiary (rock) crushers, but largely superceded by cone crushers. Toothed rolls (with width ~ twice dia.) are used for coal crushing.
ROTARY BREAKERS			2100 (diameter) x 3650 (length) to 4300 x 9750	7 to 112	18 to 12	400 to 2000	Run-of-mine coal to product of 40 to 150mm	Breaks run-of-mine coal to a predetermined top size (with minimum of fines) as well as removing coarse refuse.
IMPACT CRUSHERS	Hammer Mills		Feed opening 160 x 230 to 640 x 1470	11 to 375	1800 to 600	to 2500	to 20:1, open circuit to 40:1, closed circuit	Characterised by bar screen across outlet. Many forms; reversible/non-reversible, adjustable/non-adjustable cage, non-clog, ring granulator. Most breakage by impact, some by attrition. Used for primary, secondary, tertiary crushing, for cubic shape, maximum fines. Feed not hard or abrasive.
	Impactors		Feed opening to 1400 x 2300	to 450	to 900	to 1200	to 40:1 closed circuit	Characterised by breaker plates and open discharge. For primary, secondary, tertiary crushing of soft, friable materials. Recommended when large reduction ratio, high capacity, cubic shaped, well-graded product and minimum fines are required. Higher speeds can be used to give more fines.
	Cage Disintegrators		750 (diameter) to 1300	22 to 260	1500 to 480	5 to 80		May have 1, 2, 4 or 6 cages mounted concentrically. Feed enters center of inner cage, and centrifuges outwards, being subjected to increasingly higher impact forces at each stage. Otherwise similar to impactors.
	Vertical Spindle		685 (rotor diameter) to 990	55 to 150	2300 to 1400	200 to 100	2:1	Feed centrifuged by rotor. Breaker ring protected from wear by persistent layer of product. Essentially a tertiary crusher for very hard rock. Less wear and more cubic product relative to hammer mills.

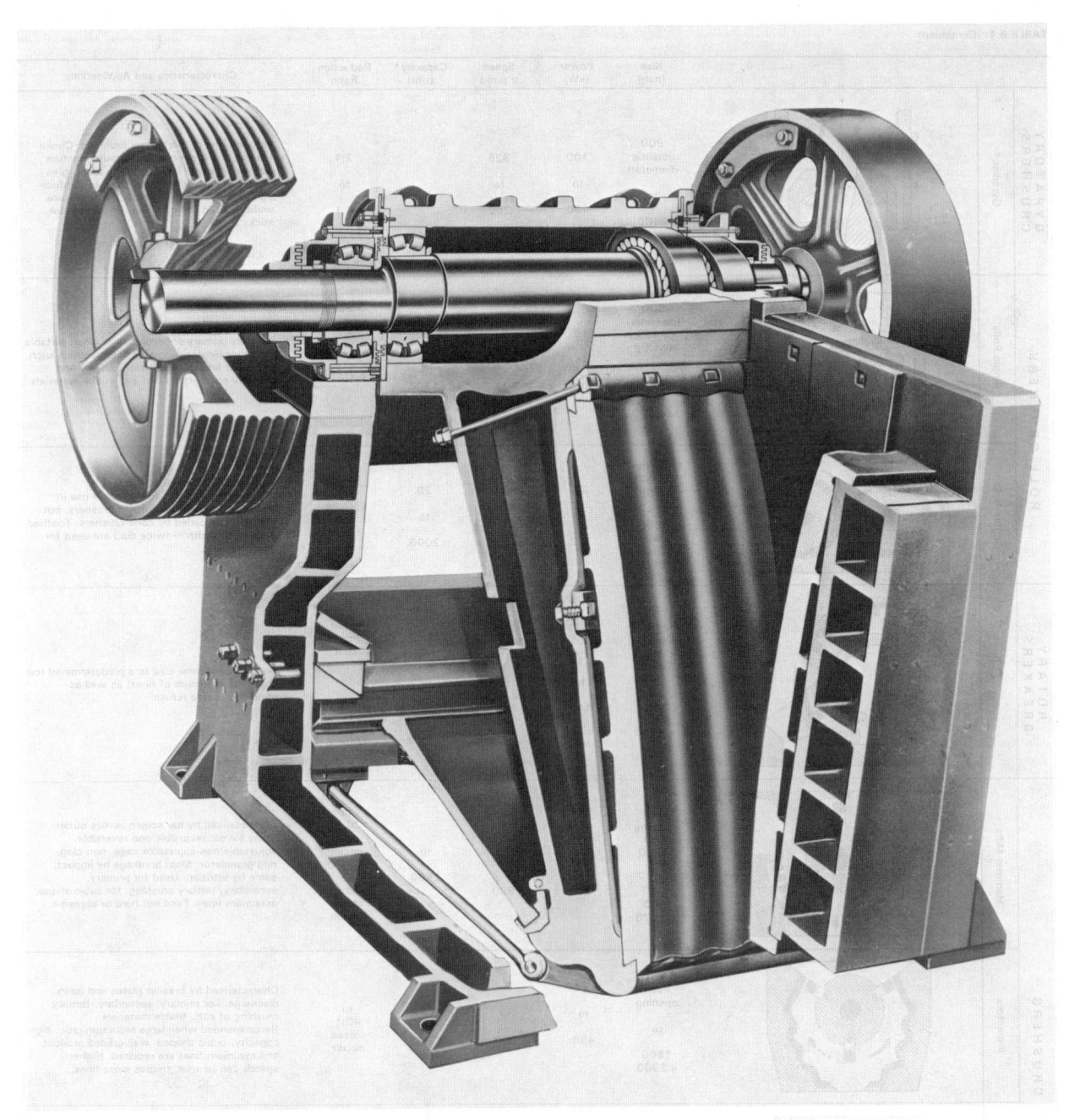

Figure 8.1. Overhead eccentric jaw crusher. (Courtesy Portec Inc.)

load. Jaw crushers are generally built with the toggle as the weakest part so that it can break without causing too much damage. Gyratory-type crushers may have the outer crushing surface spring loaded or the mantle height hydraulically adjustable.

The actual shape of the crushing surfaces is also important, and two basic types can be distinguished.[4] The standard types have straight faces and these give the maximum feed opening, the choke point (point of minimum capacity) at the discharge opening, and a slightly lower reduction ratio. The curve profile (e.g., Fig. 8.1) on the other hand:

Permits the use of a smaller discharge setting for a given crusher.

Gives a slightly higher reduction ratio.

Equipment

TABLE 8.2 APPLICATIONS OF GRINDING MILLS

APPLICATIONS	AUTOGENOUS Prim.	AUTOGENOUS Int.	AUTOGENOUS Sec.	ROD Overflow	ROD Peripheral Discharge	BALL Overflow	BALL Diaphragm	MULTI-COMPARTMENT	SPECIAL (Pebble)	VBM	RING-ROLLER MILL
Ores (Ferrous and Non-ferrous)											
Preparation of fine aggregate											
Talc and ceramic materials											
Cement raw materials											
Cement clinker											
Coal, oil shale and petroleum coke											
Silica, ceramics, etc. (must be free of iron)											
Production of a specific mesh size											
Production of a specific surface area											
Wet grinding											
Dry grinding											
Damp feed (1% to 15% moisture)											
Large feed (minus 10″)											
Large feed (minus 4″)											
Large feed (minus 1″)											
Intermediate size feed (minus ½″)											
Fine feed (minus 14 mesh)											
Coarse product (minus 6 mesh)											
Fine product (minus 35 mesh)											
Maximum production of fines											
Minimum production of fines											
Production of cubical particles											
Primary mill of two-stage circuit											
Secondary mill of two-stage circuit											
Operation in open or closed circuit											
Operation in closed circuit only											

Gives a higher capacity.

Distributes wear on the surfaces more evenly.

Gives a more uniform product.

Results in lower power consumption.

Reduces the feed opening size.

Crusher surfaces are normally made of tough manganese steel, and are replaced either with new sections or by building up new surface with suitable weld deposit.

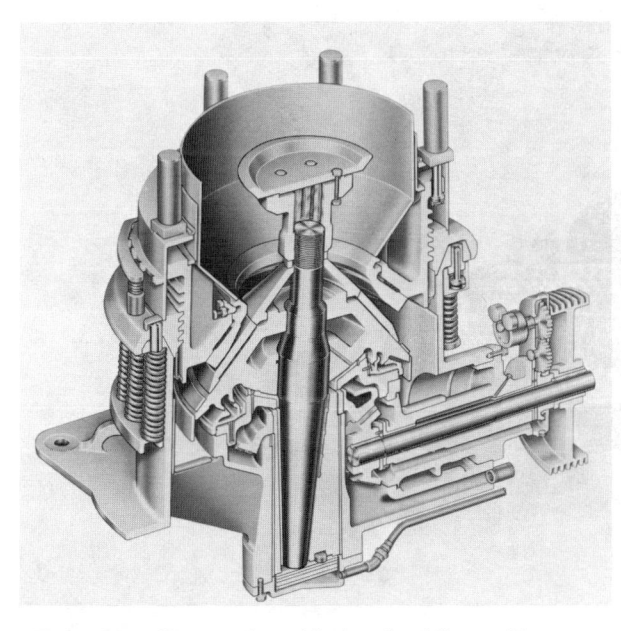

Figure 8.2. Cone crusher with short head liners. (Courtesy Rexnord Inc.)

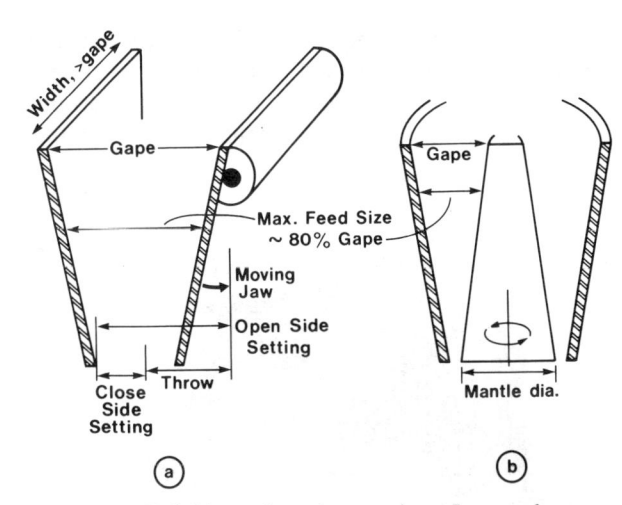

Figure 8.3. Definitions of crusher openings. Jaw crushers are normally specified by the gape × width, gyratory crushers by gape × mantle diameter, and cone crushers by the diameter of the feed opening (approx. 2 × gape).

Primary crushing is the first mechanical size reduction after mining, and the alternatives are jaw or gyratory crushers. Since the latter have become large enough to take feed sizes that previously could be handled only by jaw crushers, the selection has become simpler. Provided the feed size opening can be decided (a factor that is determined largely by the mining operation) selection between the two types is essentially based on throughput, gyratories having three to four times the capacity for a given feed size opening. However, other factors such as mineral characteristics and the lower head room requirements of the jaw crusher should be considered.[11]

Cone-type gyratory crushers are generally used for secondary and subsequent stages of hard rock crushing.

8.1.2. Roll and Impact Crushers

The essential features of roll and impact crushers are also described in Table 8.1 and two are illustrated in Fig. 8.4. There are in fact a large number of variations on the basic types,[1-5] but because of the limited use of these machines they are not considered in detail here. Roll crushers may have one roll, which has teeth, or two rolls, in which case teeth are optional. Some fracture may occur by cleavage in toothed machines, but most is by shatter.

Impact crushers vary mainly in the design of the impactors (which may be built onto a roll or as a series of hammers) and in the nature of the crushing chamber (which consists of plates, bars, or screens). When there is a screen across the outlet the product cannot leave until it is smaller than the screen aperture, and this may result in some fracture by abrasion. Some impact machines are reversible; that is, both sides of the machine are identical, which permits the impactor to be run in either direction to even out wear.

8.1.3. Tumbling Grinding Mills

The tumbling grinding mill[1-4] (Figs. 8.5 and 8.6) is the solution to the problem of applying a small fracture force to a large number of particles; the effect achieved by using a grinding media to produce predominantly shatter fracture. Grinding media are either *rods* (steel), *balls* (steel or ceramic), or particles of the ore itself; the latter process is called *autogenous* grinding.[12] As originally conceived, autogenous grinding included all size reduction beyond primary crushing and as such a certain amount of the size reduction of the larger particles occurs by abrasion and cleavage, although most size reduction in these mills is by shatter. However, relatively few ores have suitable fracture characteristics for fully autogenous operation, and an alternative is to carrying out final grinding in a *pebble mill* using se-

Figure 8.4. (*a*) Impact crusher. (Courtesy Allis-Chalmers.) (*b*) Rotary breaker. (Courtesy Pennsylvania Crusher Corp.)

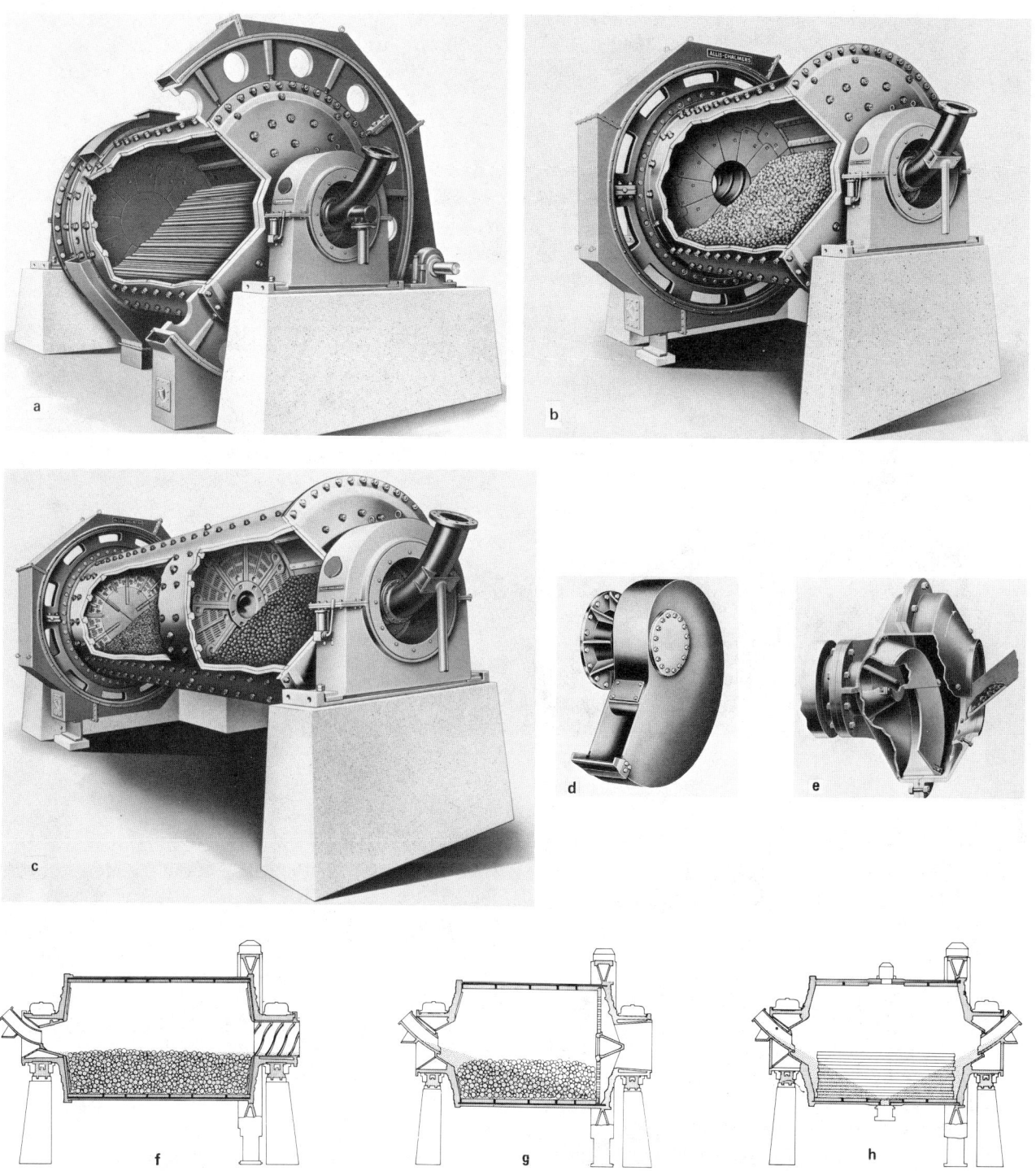

Figure 8.5. Conventional tumbling mills. (*a*) End peripheral discharge rod mill, with spout feeder. (*b*) Overflow ball mill with spout feeder. (*c*) Diaphragm (grate) discharge multicompartment mill, with spout feeder. (*d*) Scoop feeder. (*e*) Conical feeder. (*f*) Slurry gradient in overflow mill. (*g*) Slurry gradient with diaphram (and end peripheral) discharge. (*h*) Slurry gradient with center peripheral discharge. [(*a*)–(*c*) Courtesy Allis-Chalmers; (*d*), (*e*) Courtesy Koppers Co. Inc., (*f*)–(*h*) Courtesy Rexnord, Inc.]

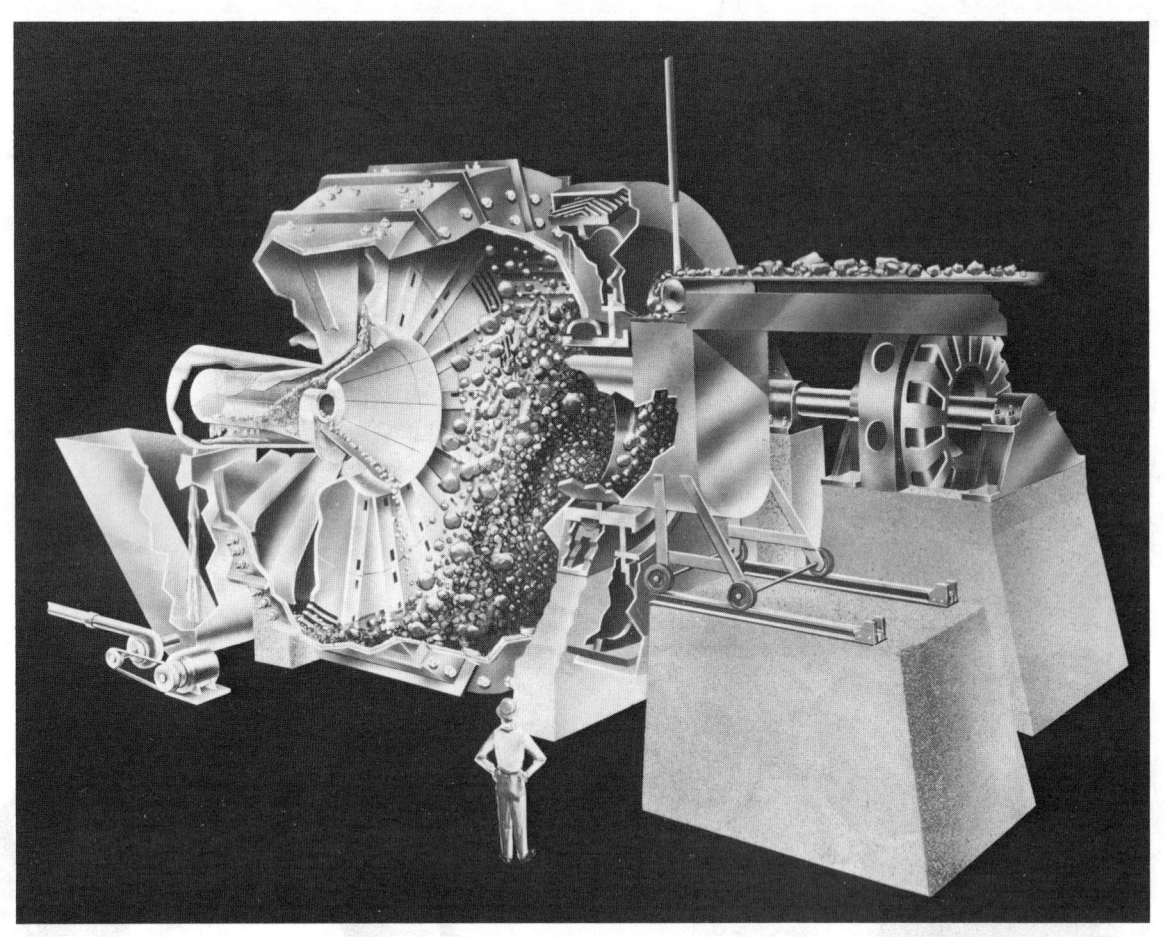

Figure 8.6. Wet autogenous mill. (Courtesy Koppers Co. Inc.)

lected pieces of ore as the pebbles. A further variation on autogenous grinding *semi-autogenous* grinding; the use of a small amount of large steel balls to break down certain intermediate sizes of particles that otherwise would build up in the mill. In other instances, the addition of balls can be avoided by diverting this critical size of particle through a small crusher.

As well as being grouped by grinding media, tumbling mills may be further subdivided by the method of feeding and discharging (Fig. 8.5), or whether they are operated wet or dry.

There are two significant differences between compression crushers and tumbling mills. The first is that tumbling mills impose comparatively little restriction on the feed and product size, since the openings are normally orders of magnitude larger than the mineral particles. However, some size control on the feed to a tumbling mill is generally provided by the preceding stage of size reduction. A diaphragm or grate discharge can provide some classifying effect on the product, but the method has limited use because the aperture size has to be relatively large. As a consequence, ball mills are generally operated in closed circuit with an external classifier. Rod mills exert a limited amount of control on the product size because large particles are selectively broken while fine particles are selectively passed by the mechanism illustrated in Fig. 8.7. Thus for the most part, the transport of solids through these mills is dependent on flow characteristics, as distinct from the physical constraints imposed by the discharge opening of compression crushers.

The second difference concerns the application of energy. Jaw and gyratory crushers can be considered as applying the energy directly to the particles. With tumbling mills, the energy is consumed in keeping the mill shell, the media, and the mineral charge in motion; fracture occurs as a by-product of

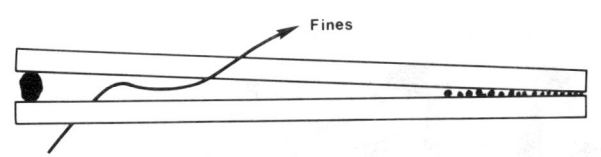

Figure 8.7. Schematic illustration of selective grinding of large particles in rod mill.

passage through the mill and is a statistical process. It follows that for a given size, pebble and autogenous mills will draw less power than rod or ball mills because of the lighter load. (This is *not* to say that they will be more energy efficient: see Section 8.4.3.)

8.1.4. Other Equipment

A common problem in any crushing operation is the treatment of rocks too large for the primary crusher. Ideally such material is sorted out before being transported to the crusher, and it is generally broken down with explosives. Occasionally, some oversize rocks are fed to the crusher, where they jam. Pneumatic jackhammers (Fig. 8.8) and Electro-Breakers[6] can provide quick and economical solutions in both of the above situations.

Historically, ring-roller mills have not been widely used for grinding in mineral operations because they have a high capital cost and operate dry. On the other hand, these mills are comparatively energy efficient, and consequently they have been fairly widely used for dry grinding where energy costs are high[7] (e.g., in Europe for cement manufacture). A notable feature of these mills is that they often contain an air classifier as an integral part of the machine (Fig. 8.9).

Vibrating mills (Fig. 8.10) are also relatively energy efficient, but because of their comparatively low capacity, they are restricted to smaller, specialized applications, particularly where grinding has to be carried out under a special or inert atmosphere.[8]

It has been claimed that a smaller amount of fines would be produced if coal were fractured by chemical methods. This could be achieved by using low molecular weight compounds such as ammonia to penetrate the coal and disrupt the bonding at natural interfaces.[9]

8.2. CHARACTERISTICS OF TUMBLING MILLS

8.2.1. Wet or Dry Grinding

There are a number of factors that influence the decision to grind wet or dry:

Whether subsequent processing is wet or dry.
The availability of water.
Wet grinding requires less power per tonne of material.
Wet classification requires less space than does dry classification.
Wet grinding does not need dust control equipment.
A low moisture content is essential for dry grinding, and thus may necessitate drying.
Wet grinding uses more steel grinding media and mill lining material, per tonne of product, because of corrosion.
Dry grinding eliminates the need for product dewatering.
Whether the material reacts with water (e.g., cement clinker).

Figure 8.8. Pneumatic breaker for breaking oversize rock jammed in gyratory crusher. Note the weld resurfaced mantle of the crusher. (Courtesy Kent Air Tool Co.)

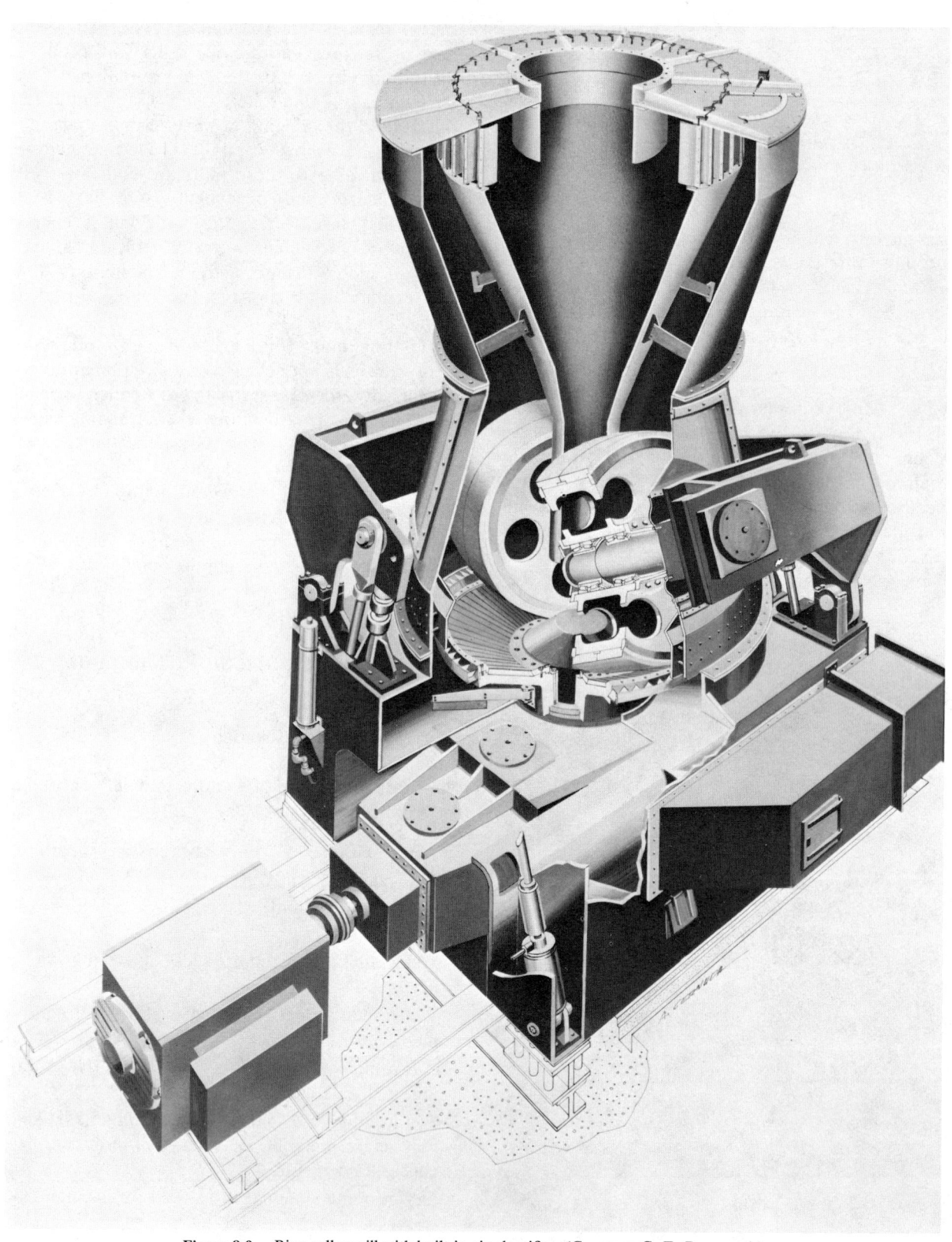

Figure 8.9. Ring-roller mill with built-in air classifier. (Courtesy C. E. Raymond.)

Characteristics of Tumbling Mills

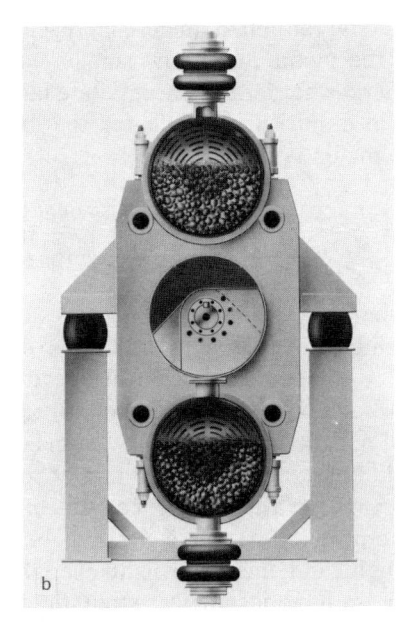

Figure 8.10. Vibrating mill. (Courtesy KHD Humboldt Wedag AG.)

When subsequent processing is to be carried out wet (e.g., flotation or gravity concentration), wet grinding is the logical choice, and as a consequence, most minerals are ground wet.

8.2.2. Charge Volume

The charge volume of a tumbling mill is the percentage of the mill interior filled with grinding media and includes the void spaces between the media. It can be approximated by the relationship[13]:

$$\text{charge volume (\%)} = 113 - 126 \frac{H_c}{D_M} \quad (8.1)$$

(H_c = inside distance from top of mill to top of stationary charge; D_M = mill diameter inside lining). Figure 8.11 indicates how the charge volume affects the power drawn by the mill. This graph is only an approximation, since power draft is affected by charge slippage, which is in turn dependent on the liners and the mill speed. Only grate discharge mills can draw maximum power (by operating with a 50% charge), since the grate holds the balls in. Overflow ball mills are restricted to less than 45% charge to prevent balls discharging, and rod mills are restricted to 32–40% because the charge is swollen to 40–50% by the feed particles.

8.2.3. Size of Grinding Media

As a result of wear, all tumbling mills have a range of media sizes known as a *seasoned charge*. Make-up media (of the largest size only) must be added regularly, and the need is determined by a falloff in the power draft to the mill.[14] The size of the make-up media determines the average size of the charge, and in turn this is determined by the average size of the particles being ground.

The selection of media size is a compromise between two conflicting factors. As the relative size of the media decreases, the surface area for grinding increases, giving a higher capacity. On the other hand, as the media size increases, the force between the grinding surfaces increases so that larger particles can be broken. Several other factors also determine whether the grinding media has sufficient energy to fracture a given particle. Both higher mill

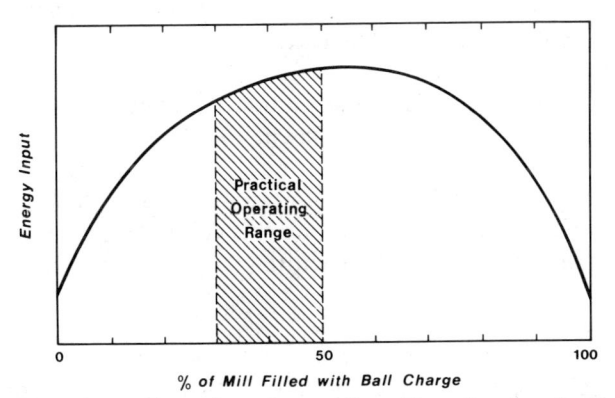

Figure 8.11. Power drawn by tumbling mill as a function of mill filling.

speeds and larger mill diameters increase the energy input to the media and allow smaller media to be used, whereas harder ores necessitate larger media. A suitable empirical formula for the size of new make-up media is[15]

$$D_m = \left[\frac{d_{80,I} W_i}{K D_M^{0.25}} \left(\frac{\rho_s}{\rho_f} \right)^{0.5} \left(\frac{\mathcal{N}_c}{\mathcal{N}} \right) \right]^{0.5} \left(\frac{7800}{\rho_m} \right)^{0.33} \quad (8.2)$$

(D_M = diameter of media, m; $d_{80,I}$ = 80% passing size of feed, m; K = 0.46 for ball mills, 0.69 for rod mills; $\mathcal{N}$ = speed; $\mathcal{N}_c$ = critical speed [Eq. 8.4]; W_i = work index, kWh/t [Section 8.6]; ρ_m = density of media; ρ_s = density of feed). Seldom is it economical to add balls smaller than 25 mm or rods smaller than 60 mm. As a consequence, rod mills are restricted to primary or coarse grinding, because rods thin enough, and therefore light enough, to grind fine particles have insufficient strength to resist the tumbling action.

The media-particle size relationship also implies that it would be desirable to have the ball size decrease toward the discharge end. This is generally not considered to be important in closed circuit grinding, where high recycle leads to short residence times, but it can be important in open circuit operation, particularly with long mill lengths. Three methods are used to segregate balls. In one of the earliest methods developed, the second half of the mill is conical (Hardinge mill); this results in the smaller balls segregating toward the smaller diameter. This method appears to have been restricted to relatively small mills and so is not widely used today. A second approach is the compartment mill (Fig. 8.5c), with each compartment having a different media or media size. A more recent method is to build a spiral profile into the mill lining,[16,17] which again results in size segregation. This method is particularly convenient with rubber liners, and simply involves the lifter bar pattern having a slight twist along its length (Fig. 8.12).

An appropriate distribution of grinding media must be used when starting new mills. Although it is not essential with pebble mills, it is advisable with ball mills, and is vital with rod mills, since the rods wear relatively evenly, and this can result in catastrophic tangling of the entire charge.

8.2.4. Liners

The purpose of the mill liner is twofold; to protect the mill shell from wear, and to reduce slip between the shell and the grinding media. Because they wear away, liners are made replaceable, and although their wear rate is less than that of the media, their effect on operating cost is comparatively high because production has to be stopped for relining. Even so, mills typically can achieve 99% operating time.

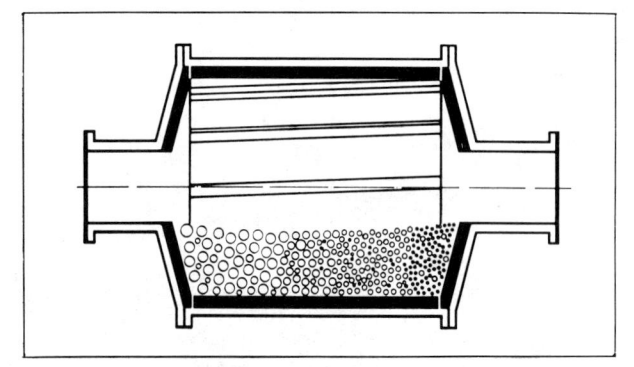

Figure 8.12. Spiral twists in liner profile cause size segregation of the grinding media. (Courtesy Trelleborg AB.)

Liners can be considered in terms of material, profile, and mounting method. Ceramics are occasionally used, but by far the most commonly used materials are cast or rolled steel, and rubber. Although the application of rubber has increased rapidly since its introduction in the 1960s, it still cannot replace steel in many situations. In general, rubber is more suitable with smaller ball sizes (i.e., finer feeds) and harder ores[18] (Fig. 8.13). Where applicable, rubber liners give longer life, reduced noise, with lighter and safer liner replacement; generally at a slightly higher initial cost.

Some liner profiles are illustrated in Fig. 8.14. The smoother liners result in more slip between charge and liners, so to achieve the same power draft higher speeds are necessary. Both the wave form and the similar double wave form wear more slowly than shiplap while giving similar lift. The wave forms are commonly used with steel linings because they tend to maximize metal utilization. The rubber linings that feature a smooth plate with a rib similar to a lifter bar are the preferred types, since the flexure in the rib virtually eliminates wear on the plate. Rib spacing is critical for maximum life, which can be further extended by reversing the rib.[16]

Liners are bolted to the shell, and two methods are illustrated in Fig. 8.15. Often, an epoxy filler is used behind metal liners.

Characteristics of Tumbling Mills

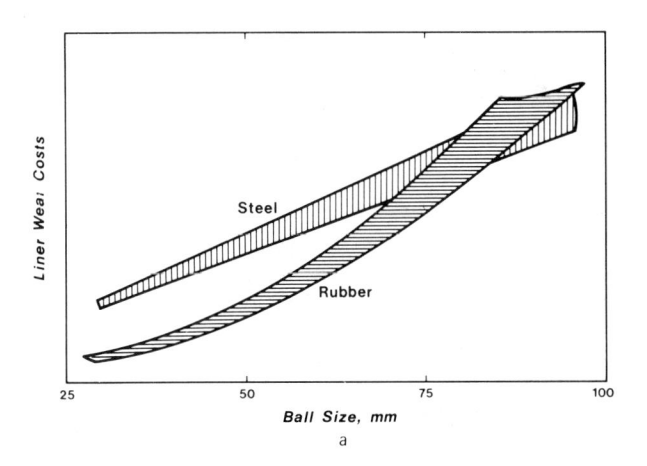

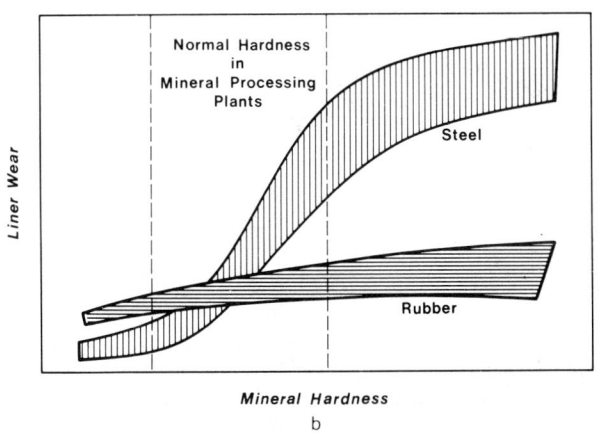

Figure 8.13. Comparison of steel and rubber liners. (Courtesy Skega AB.)

8.2.5. Media and Liner Wear

Some typical wear rates are given in Table 8.3. Such wear can be considered to result from three mechanisms: corrosion, impact, and abrasion. With wet grinding, corrosion must be considered the major contributor to wear,[19] which explains why dry grinding and rubber linings can have lower wear rates. It is for this reason also that the hardest grinding balls are seldom justified in wet grinding because they are being consumed by corrosion rather than wear.

For steels, impact and abrasion resistance are conflicting properties. Low impact energies are absorbed by elastic deformation without damage, but if the impact is severe, large amounts of either elastic deformation or plastic deformation must be tolerable. Unfortunately the latter property is in conflict with wear resistance. For example, manganese steel is resistant to impact but has poor abrasion

TABLE 8.3: METAL CONSUMPTION

	Media	Lining
Ball Mill (wet), kg/kWh	0.31	0.044
(dry), kg/kWh	0.04	0.006
Rod Mill (wet), kg/kWh	0.46	0.057
(dry), kg/kWh	0.07	0.008
Jaw/Gyratory		
Crusher, kg/t		0.035

resistance, whereas Ni-hard has the opposite properties. The impact resistance of rubber can be improved by bonding it to metal, since this gives a composite material of higher elastic modulus; this technique is applied to lifter bars. Any factor that increases the energy of the individual pieces of media also results in higher wear. There are indications that wear is at least proportional to the square of mill speed; Fig. 8.13a shows the effect of increasing media size. Tests measuring the actual impact force applied by grinding media suggest that the mineral slurry almost halves the impact force.[20] An important implication of this is the potential for disastrous liner damage from operating the mill, even briefly, without ore.

Abrasion resistance of a metal is essentially a function of its hardness, but the ability of rubber to tolerate large deformations makes it a superior material for use with harder minerals (Fig. 8.13b). An exception to this appears to occur when the mineral has a sharp cutting edge, in which case the rubber cuts and wears excessively.[18]

8.2.6. Media Action

Media action is best considered initially in terms of a smooth lining, and by considering the action of a single ball.[16] The changing forces on such a ball are illustrated in Fig. 8.16. At point 1 the force holding the ball against the shell is the normal component of the weight $M_b g \cos \theta$, plus the centrifugal force,

$$F_c = \frac{2 M_b v_m^2}{D_M} \quad (8.3)$$

(M_b = mass of ball; v_m = tangential velocity of ball). This total force rises to a maximum at point 2 and returns to the initial value at point 3. Still traveling in a circular path, the force decreases until at point 4 the component of the weight just balances the centrifugal force. At this point the ball is thrown from

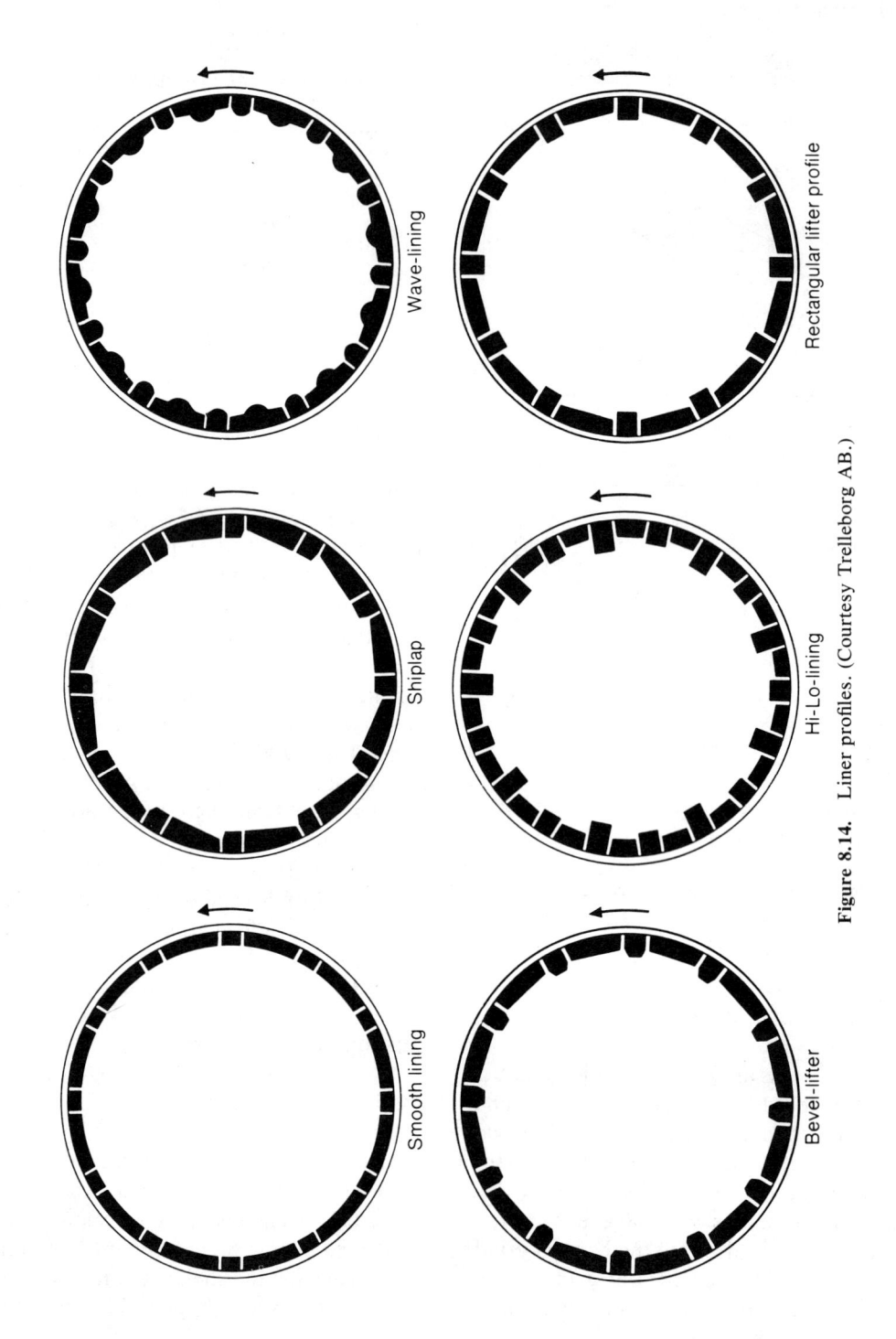

Figure 8.14. Liner profiles. (Courtesy Trelleborg AB.)

Characteristics of Tumbling Mills

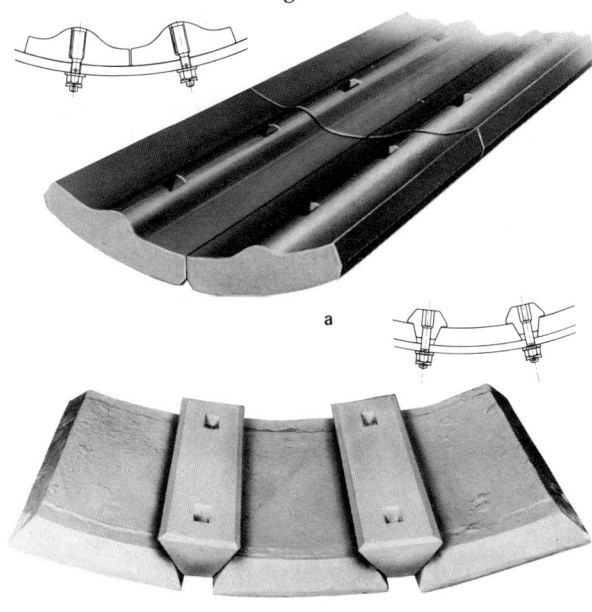

Figure 8.15. Liner mountings. (*a*) Wave type. (*b*) Wedge bar. (Photographs Courtesy Koppers Co. Inc.; diagrams courtesy U.S. Steel.)

the shell and returns to point 1 along a parabolic trajectory. It is common to define the *critical speed* as that at which the single ball will just hold against the shell for the full cycle. Thus for $D_M \gg D_m$

$$M_b g = \frac{2 M_b v_m^2}{D_M} \tag{8.4}$$

or

$$\mathcal{N}_c = \frac{42.3}{\sqrt{D_M}} \tag{8.5}$$

($\mathcal{N}_c$ = critical speed, rpm).

When the mill has a normal charge of balls, the situation is slightly different, as illustrated in Fig. 8.17. In zone A, the balls are moving over one another in concentric layers and breakage is by shatter and possibly some cleavage. The upper boundary of this zone results from the ball weight being neutralized by the centrifugal force. This boundary is higher than was suggested by the analysis in Fig. 8.16 basically because of interball action. Balls farther from the shell (zone B) break away and start to roll back sooner, and produce the region of most *vigorous* grinding action; that is, all breakage is by shatter. Zone C is a small region where shatter breakage also occurs, in this case as a result of the cataracting balls rejoining the charge. Note that in the cataracting zone itself no breakage occurs and consequently this should be kept to a minimum. It is for these reasons that mills operate within a comparatively narrow range of about 65–82% of critical

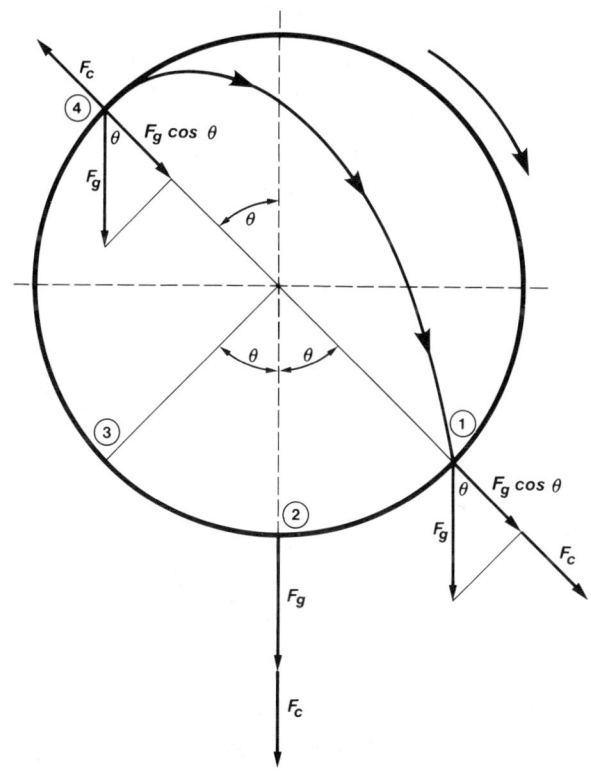

Figure 8.16. Forces on a single ball in a ball mill.

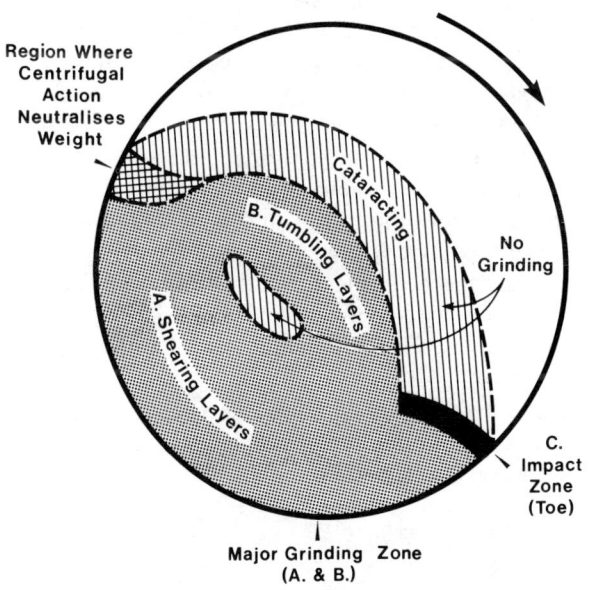

Figure 8.17. Behavior of full ball charge.

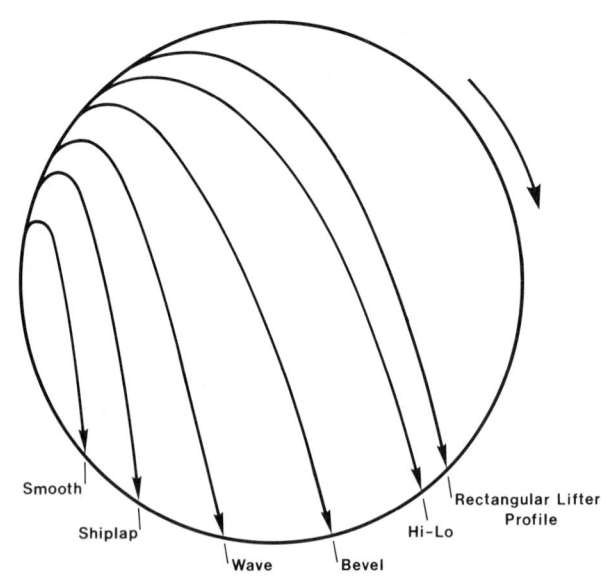

Figure 8.18. The fall parabola for various linear profiles (Fig. 8.14) at a given mill speed and load factor. (Courtesy Trelleborg AB.)

speed. Higher speeds have an additional disadvantage in that they cause the balls to strike the liner above the toe and so increase liner and ball wear. Although there has been some debate on the relative importance of the different fracture zones in a mill, the nature of the product and its rate of production suggest that the breakage products from zones A, B, and C are virtually indistinguishable. However, it is possible that some abrasion breakage occurs between the balls and the liners.

It must be appreciated that the foregoing concepts are generalized and that other factors are involved. These to a large extent relate to the amount of slippage between charge and liners. As well as charge volume, slippage is affected by liner profile; the effect of the various profiles shown in Fig. 8.14 is illustrated in Fig. 8.18 (for a given mill speed and charge volume). Because of these variations in behavior, it has been suggested that mill speeds should be quoted as a percentage of the actual critical speed instead of the theoretical critical speed (Eq. 8.5).

Although this discussion has been presented in terms of ball mills, the principles also apply to rod, autogenous, and pebble mills.

8.2.7. Power Draft

Power supplied to a mill is consumed essentially in tumbling the media (and to a lesser extent the ore), but some power is also used in turning the shell (basically a flywheel) and overcoming drive friction. Five main parameters determine the power drawn by the mill: diameter, length, charge volume, speed, and mill type.

Theoretically, the power drawn $\mathscr{P}$ is proportional to the length L, the charge mass ($\propto D_M^2$), the length of the torque arm ($\propto D_M$), and the angular velocity ($\propto 1/\sqrt{D_M}$); thus

$$\mathscr{P} \propto L D_M^{2.5} \tag{8.6}$$

In practice, reported exponents to D_M have ranged from 2.3 to 3.0. The low exponents presented by Bond[13] (2.4) and Rowland[21] (2.3) have been fairly widely described as valid for rod and ball mills and have been attributed to lower efficiencies in smaller mills. Gow et al.[22] found an exponent of 2.6 in their ball mill study, and although Rowland[21] quoted 2.3 for autogenous grinding, most workers suggest that short autogenous mills ($D_M > 2L$) have above-theoretical exponents[23–25] (2.6–2.85). It is likely that figures exceeding 2.5 are caused by an end-wall effect that enhances the lift, and thus increases the power draft, so that a more realistic relationship is of the form

$$\mathscr{P} \propto [L + (KD_M)]D_M^{2.5} \tag{8.7}$$

This exponent 2.5 is in fact supported by the data given in most manufacturers' catalogues.

As illustrated in Fig. 8.19, the power draft initially increases linearly with speed, but as slippage increases the rate of increase decreases, and after passing through a maximum, falls toward zero as the charge centrifuges. It is worth noting that the theoretical critical speed is less than the actual point at which the complete charge centrifuges, because

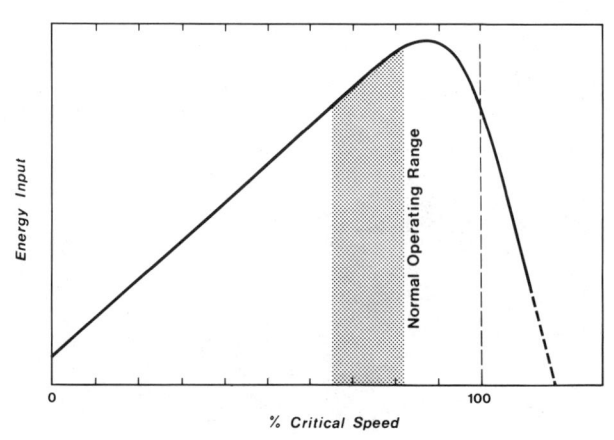

Figure 8.19. The effect of mill speed on power draft.

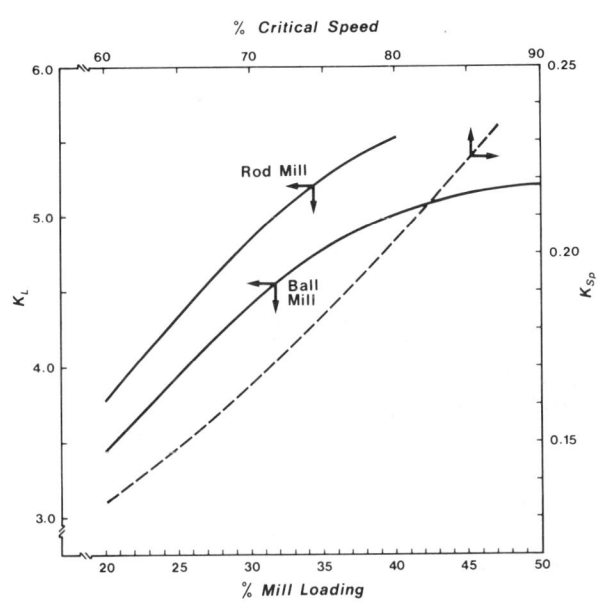

Figure 8.20. Factors for calculating power draft.[15]

of slip and the smaller circumferential path of the inner media. The effect of mill charge has already been illustrated in Fig. 8.11.

A typical equation for calculating power draft[15] is:

$$\mathscr{P} = 8.44 D_M^{2.5} L K_{Mt} K_L K_{Sp} \quad (8.8)$$

(D_M = mill, inside diameter, m; K_{Mt} = mill type factor [1.0 for wet overflow mills, 1.13 for wet diaphragm ball mills and wet peripheral rod mills, 1.25 for dry diaphragm ball mills and dry peripheral rod mills]; K_L = loading factor [Fig. 8.20]; K_{Sp} = speed factor [Fig. 8.20]; L = mill length, m; $\mathscr{P}$ = power draft, kW).

Example 8.1. What size of wet overflow discharge ball mill is required to draw 2100 kW?

Solution. A variety of combinations of diameter, length, speed, and charge loading will give the required power draft. As an illustration, assume $D \sim L$; a 40% charge loading (typical range 35–45% for overflow mills), and 74% of critical speed (typically 65–82%, depending on the liner shape). Thus, from Fig. 8.20,

$$K_L = 5.02$$
$$K_{Sp} = 0.18$$
$$K_{Mt} = 1.0$$

Substituting in Eq. 8.8:

$$2100 = 8.44 \times D_M^{2.5} \times D_M \times 1.0 \times 5.02 \times 0.18$$
$$\therefore D_M^{3.5} = 275$$
$$\therefore D_M = 4.98 \text{ m}$$

that is, a mill 4.98 × 4.98 m. Allowing for liners about 75 mm thick, the actual mill diameter would be 5.13 m. By using a higher speed, the mill size could be reduced (lower capital cost), but the wear on the liners and media would be higher. Hence this calculation is suitable only for preliminary planning.

8.2.8. Feed and Discharge Methods

A number of feed and discharge methods are possible with tumbling mills and the common methods are illustrated in Fig. 8.5. Peripheral discharge mills minimize overgrinding, supposedly because a low pulp line (Fig. 8.5g) produces a rapid discharge.[26] More recent information suggests that the low pulp line may not occur, and that more rapid grinding occurs when the media void space is incompletely filled.[27]

8.3. CIRCUITS

The three basic types of size reduction circuit used to produce a fine product are shown in Fig. 8.21. The notable feature of the conventional circuit (Fig. 8.21a) is that size reduction is carried out in a number of stages. There are three major factors contributing to this. First, most size reduction equipment performs more efficiently when the reduction ratio is limited. This is especially true for the compression crushers where the crushing chamber is designed to handle the larger particles rather than the smaller. Hence, once a particle has been broken once, twice, or three times, it is too small to be efficiently broken further in the same chamber. The relationship between media and particle size in tumbling mills has already been considered, and this again limits the reduction ratio that should be attempted in a given mill. In practice, ball mills in particular can have large reduction ratios, but to achieve this most efficiently requires decreasing media size along the mill.

The second reason for limiting the reduction in a given machine applies mainly to the final stages of grinding and concerns the desirability of preventing

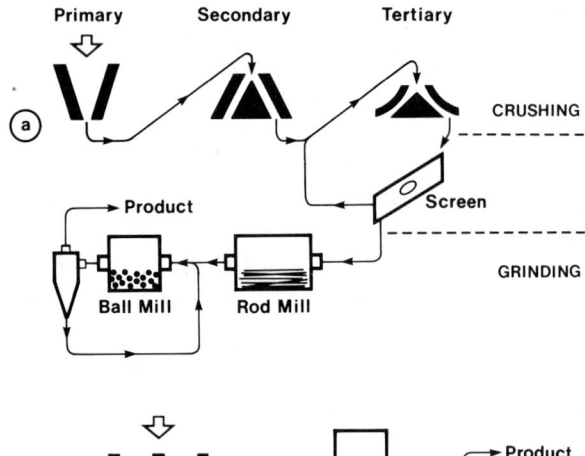

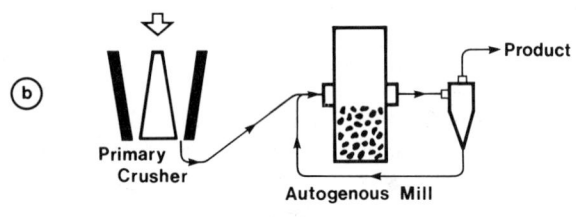

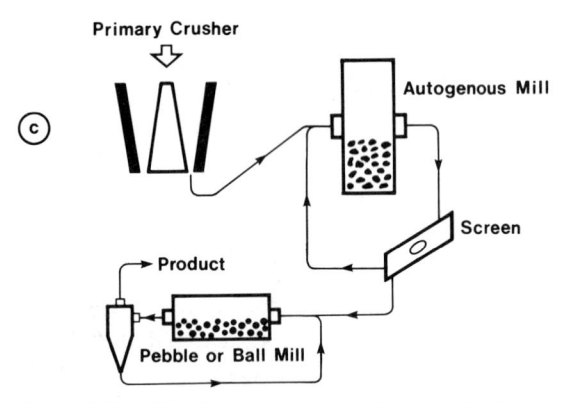

Figure 8.21. The three basic types of size reduction circuit. (*a*) Conventional. (*b*) Autogenous. (*c*) Autogenous with separate fine grinding.

overgrinding. Such an aim is achieved by operating the final stages in closed circuit, with comparatively high circulating loads, so that the material has little chance of being broken a second time before it is removed from the circuit by a classifier.

Rod mills are normally operated in open circuit. No one factor is responsible for this, but there are several contributing factors: (1) the feed size is normally well controlled by the preceding stage of reduction, (2) the mill has some internal classification effect (Fig. 8.7), (3) the reduction ratio is relatively large, and (4) control over the product size is seldom critical because the product passes either to a classifier or to a closed circuit ball mill.

Examination of the crushing section of the conventional circuit (Fig. 8.21a) shows that sizing separations are also used between crushers. This is because any breakage produces a range of product sizes, and when the reduction ratio is comparatively low, some of the feed is already as fine as the product. By screening this material out and bypassing the next reduction stage, the size of machine can be reduced because throughput is lower. Furthermore, the removal of the finer particles will make the crusher more efficient by allowing a freer flow of material and reduced cushioning effects.

Obviously, autogenous grinding (Fig. 8.21b and c) refutes many of the principles described above, and in fact such grinding is less energy efficient primarily because the mill no longer balances media energy with the fracture energy of a given particle. Fortunately, autogenous grinding has compensating features that can give a lower cost per tonne of product, such as a reduction in the amount of equipment (crushing, grinding, classifying, and conveying), as well as savings in grinding media.

8.4. THEORY

Much of the early literature on the theory of crushing and grinding centers around the so-called laws of size reduction, which were discussed in Chapter 7. This section considers the more recent mass-size balance equations for analyzing size reduction. Although insufficient information has been published to allow this method to be used for reliable scale-up and design, some manufacturers are using proprietary information to design new mills on the basis of this approach. Hopefully, this information will soon become generally available. That this approach is emphasized here is not to belittle the Bond method, which has been the main-stay of size reduction design since it was introduced in the 1950s. Furthermore, it remains useful, and will continue to be a valuable aid, particularly in making straightforward comparisons and in measuring plant efficiency.

8.4.1. Mass-Size Balance Equations

Up to four functions can be necessary to analyze size reduction by this method: a breakage rate, a breakage function, a distribution of residence time, and a classification function. The *rate of breakage* (Section 7.1.6) is the fractional rate at which a given size of particle disappears, having been broken into

Theory

smaller particles. The resultant cumulative distribution of the daughter fragments is described as the *breakage function* (Section 7.1.5). A completely rigorous analysis of any mill also requires a distribution of residence times, a representation that allows for the fact that particles may spend different times in the device. As Chapter 3 explained, this is achieved by representing the flow as a suitable combination of mixers. As described above, compression crushers, grate discharge mills, and rod mills either will not release a particle until it is below a certain size, or will selectively break larger particles. In these cases, it is necessary to incorporate a classification/discharge function, such as that shown in Fig. 8.22.

A complete analysis therefore becomes complex and can be handled conveniently only by computer; as such it is beyond the scope of this book. However, some of the factors that affect the four basic functions are reviewed, and as well, consideration is given to some simplified approaches, to familiarize the reader with the treatment.

8.4.2. Breakage Function

In essence, this function is the cumulative size distribution resulting from an average single fracture event. By comparison with the other functions, it shows relatively little variation.[28–35] For example, in ball mills it is often independent of mill conditions, the material being broken, and the initial particle size, and is represented by the data in Fig. 7.6 and Table 7.1.

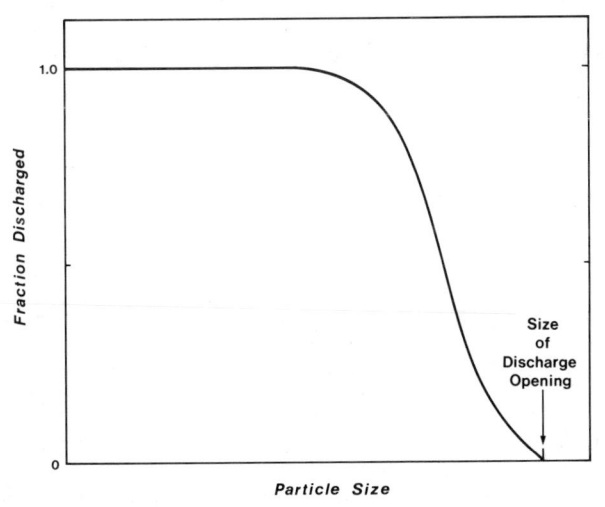

Figure 8.22. Classification/discharge function.

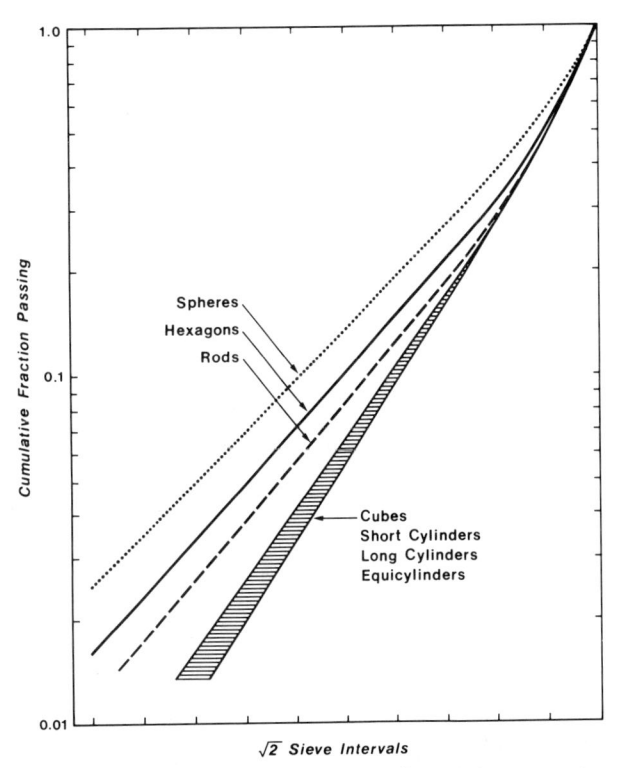

Figure 8.23. The effect of grinding media shape on the cumulative breakage function. (After Kelsall et al.[32,33])

Where variations have been found in the breakage function, they have been attributed mainly to the way the fracture force is applied[31,32] and to the material being broken.[34] Figure 8.23 shows the differing size distributions produced by various media shapes in a tumbling mill.[32,33] It can be seen that the point contact produced by a sphere produces the greatest amount of fines.

A few materials have been found to have breakage functions that cannot be normalized; that is, they are functions of the initial particle size.[36–40] When this occurs, the spread in the product size distribution decreases as the initial particle size increases (i.e., larger particles produce a smaller amount of fines, Fig. 8.24). Equation 7.21 then has to include a correction factor for K_b

$$K_b = K_1 \left(\frac{d_i}{d_1}\right)^{-n_3} \qquad (8.9)$$

(d_1 = size at which K_1 is determined). As yet, no explanation for this effect has been found, although it is worth noting that the narrower size distribution suggests a trend toward a cleavage/abrasion fracture mechanism.

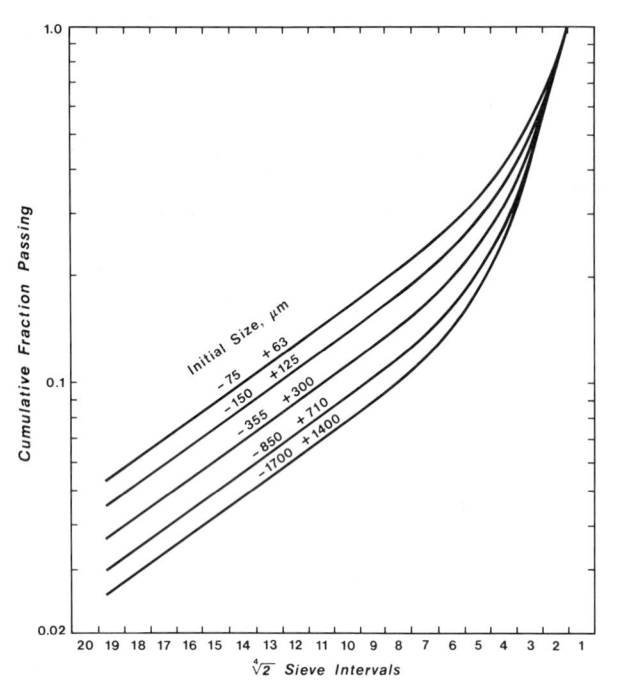

Figure 8.24. Nonnormalizable cumulative breakage function. (After Austin and Luckie.[38])

Comparatively few data have been published for crushers. Lynch found that the breakage function of a cone crusher had two components: one that was attributed to large particles breaking between the opposing faces of the crusher (essentially cleavage fracture), the other to breakage adjacent to the crusher faces (Fig. 7.5) or to smaller particles being crushed between other particles.[41] The former resulted in a small number of large particles and the latter produced a spread of sizes more typical of shatter fracture, and accounted for 5–20% of the product size distribution. The actual percentage correlated with the close-side setting of the crusher.

In autogenous grinding, multiple fracture processes are even more apparent in that shatter, abrasion, and cleavage (chipping) fracture occurs. Consequently autogenous grinding is difficult to analyze. Stanley[42] showed that two mechanisms were sufficient to describe the breakage function: the first was shatter (which applied basically to the lower part of the size range), and the second abrasion (which applied basically to the larger particles). Since the larger particles essentially maintain their original size after breakage, their product was represented by a normal shatter breakage function, displaced a number of screen sizes below the initial particle's size. In their analysis, Austin et al.[43] found it neces-

sary to attribute additional breakage to the presence of imaginary steel balls.

8.4.3. Breakage Rate

As noted first in Section 7.1.5, size reduction can be represented by a first-order rate process, that is,

$$\frac{dm_i}{dt} = -k_i m_i \qquad (7.22)$$

The first-order rate constant k_i gives the fractional rate of breakage out of the initial size i. The breakage rate is the relationship between k_i and the initial size, and Fig. 8.25 shows the typical form of the function for a ball mill. This can be described by an equation of the form[36]:

$$k_i = K d_i^n \, \text{fn(LP)} \qquad (8.10)$$

(K = constant; fn(LP) = log probability function to account for the peak in the curve). The exponent n is in the range 0.5–1.5 for tumbling mills, but may be as high as 3.0 in a hammer mill.

As might be expected, the breakage rate depends markedly on the hardness of the material, as well as the equipment, in particular the number of fracture events generated per unit time.[27–36,39,40,44–48] The effects of some variables are illustrated in Figs. 8.26

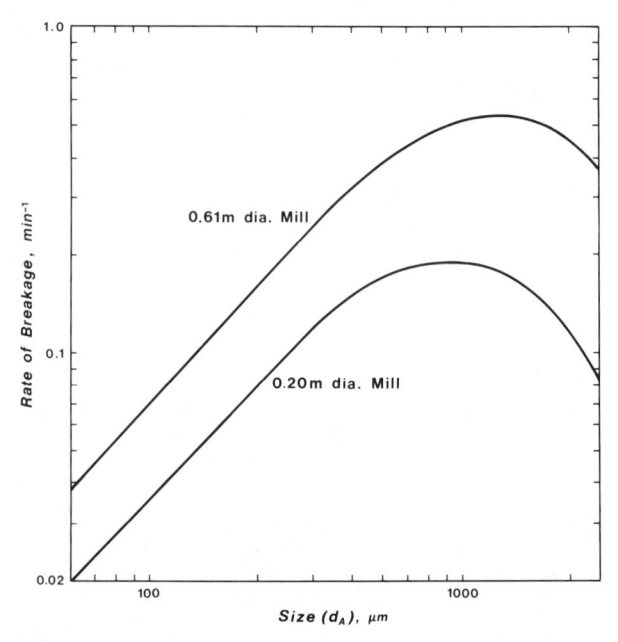

Figure 8.25. Breakage rates for dry ball milling of anthracite. [Reprinted with permission from L. G. Austin, *Ind. Eng. Chem. Process Des. Dev.*, **12**, 121–129 (1973). Copyright 1973 American Chemical Society.]

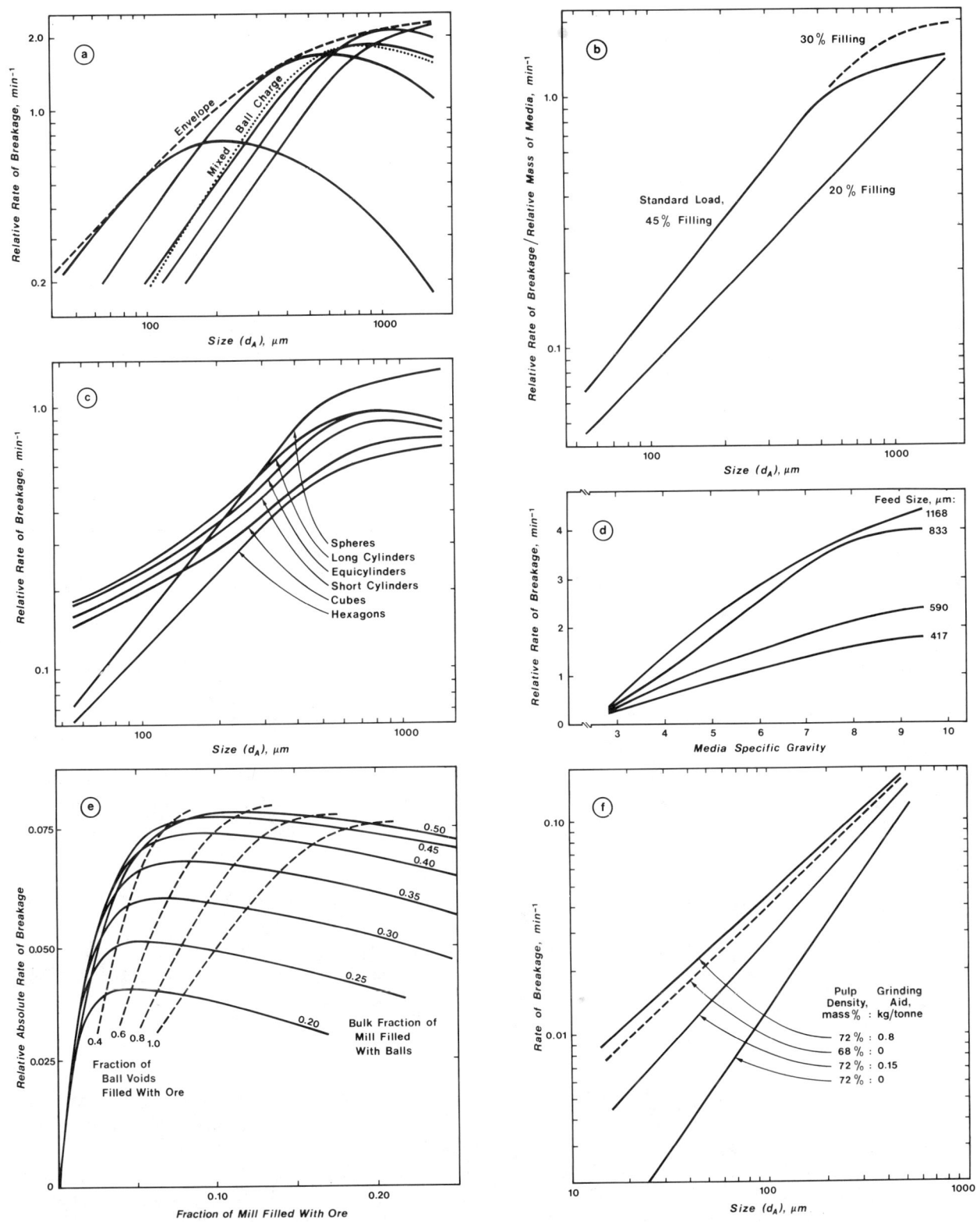

Figure 8.26. The effect of some variables on breakage rates in ball mills. (*a*) Ball size. (After Kelsall et al.[28]) (*b*) Media load. (After Kelsall et al.[31]) (*c*) Media shape. (After Kelsall et al.[32]) (*d*) Media density. (After Kelsall et al.[31]) (*e*) Ball and powder loading. (After Shoji et al.[27]) (*f*) Slurry viscosity. (After Klimpel and Manfroy.[52])

and 8.27. For any given grinding conditions, the rate of breakage is characteristic of the material being ground; in rod mills up to twofold variations have been found in k_i at a given particle size (Fig. 8.27a).

The occurrence of a maximum and falloff in k_i with increasing particle size is characteristic of equipment whose structure imposes no significant control on through flow, such as hammer mills and tumbling mills. Such a shape results because the grinding media (or impactor) have insufficient energy to cause fracture with every impact, the effect being most clearly seen with differing ball sizes (Fig. 8.26a).

In reality the situation is slightly more complex. When a maximum occurs, the initial fracture is apparently no longer first order. Austin has shown that although this can be interpreted as particles having a range of strengths, or the media having a range of energies, the breakage can be analyzed by assuming large particles to be made up of two components, one hard (low breakage rate) and one soft (high breakage rate).[36,46,49,50] Data such as those in Fig. 8.26a for particle sizes above the maximum represent an average value of k_i derived assuming first-order breakage. This concept of two strengths is also useful for treating the fracture of heterogeneous materials.[51]

Austin's analysis of the effect of ball size indicates that the maximum in k_i, $k_{i,\max}$, relates to the diameter of the ball D_m and the mill D_M according to[39,46]:

$$k_{i,\max} \propto D_M^n D_m^2 \qquad (8.11)$$

($n = 0.1$–0.2). Note that although the maximum in Fig. 8.26a *decreases* with decreasing ball size, provided the balls still have sufficient energy for fracture (i.e., to the left of the maximum) the specific rate of breakage of a given particle size *increases* as the ball size decreases, because in a given mill there are more fracture events due to the greater number of balls. For a mixture of ball sizes, the breakage rate is the weighted mean of the breakage rates of the individual sizes.[28]

Close to the optimum ball loading, the grinding rate is proportional to the mass of balls in the mill; at lower loadings the rate falls off[31] (20%, Fig. 8.26b). A reduction in the specific gravity of the grinding media causes a disproportionate decrease in k_i for the coarse particles compared to the fines, and as the specific gravity of the media approaches that of the slurry, the rate decreases to zero[31] (Fig. 8.26d).

The effect of media shape can be seen in Figs. 8.26c and 8.27b.[32,33] Compared with balls, rods break large particles at higher rates and finer particles at lower rates. This selectivity is in addition to the fact that the product from each rod-induced fracture has a narrower size distribution (Fig. 8.23), and partly explains why a rod mill seldom needs to be operated in closed circuit. This overall effect is shown more clearly in Fig. 8.28, where the rod mill

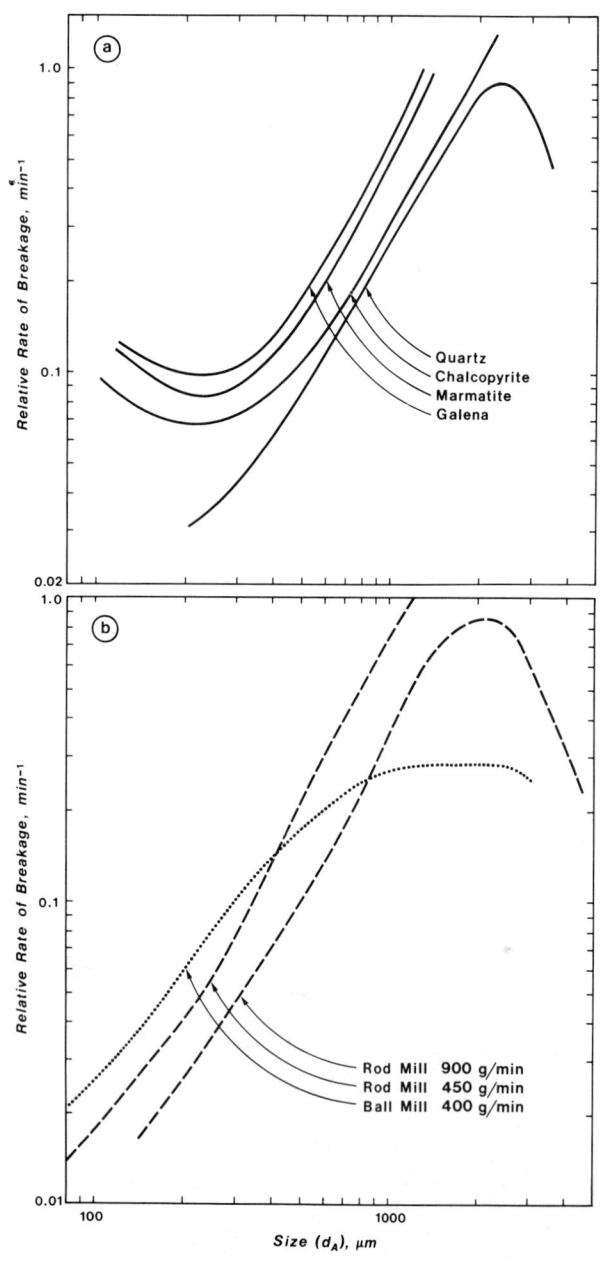

Figure 8.27. The effect of some variables on breakage rates in rod mills. (*a*) Ore minerals. (After Heyes et al.[34]) (*b*) Feed rate. (After Heyes et al.[33])

Theory

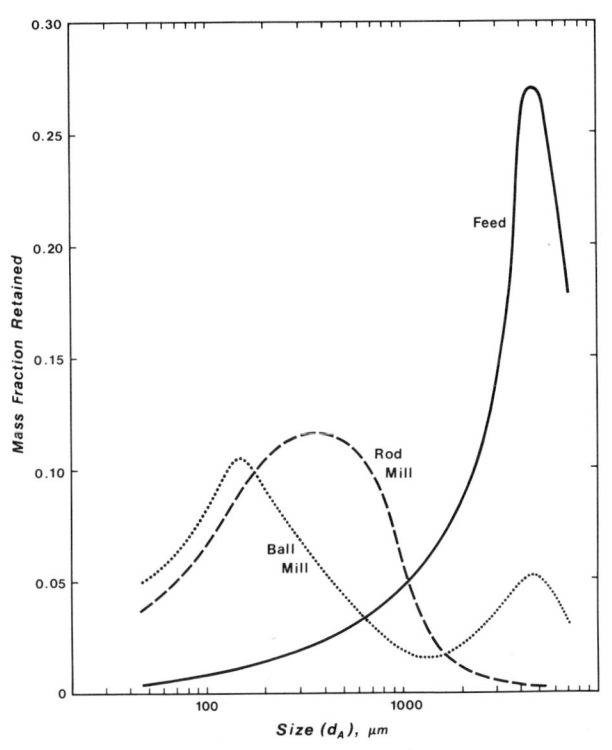

Figure 8.28. Comparison of steady state size distributions in rod and ball mills. (After Heyes et al.[33])

shows a unimodal product distribution on single size feed, whereas a ball mill shows a bimodal distribution, one peak representing unbroken feed, and the other product. The effect of other media shapes is complex, although spherical media apparently give the higher overall throughput with the narrowest size distribution (Fig. 8.29).

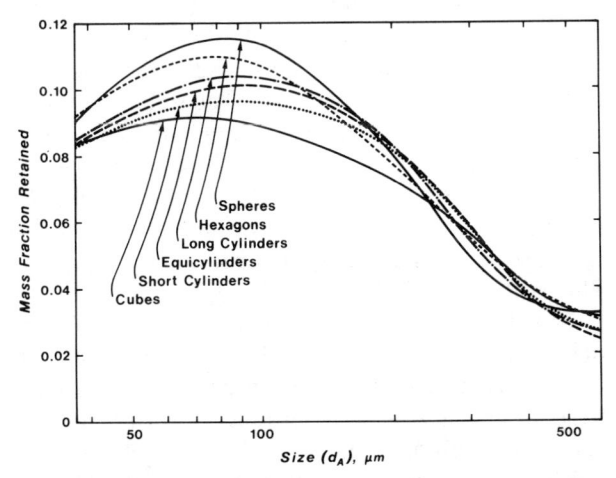

Figure 8.29. The effect of media shape on product size distribution. (After Kelsall et al.[32])

In their study of the effect of the amount of powder in the mill, Kelsall et al.[32] found that the breakage rate remains constant as the void space between the media fills, so that the absolute rate of breakage is given by the product of k_i and the hold-up in the mill M_h. When the void space overfills, the breakage rate decreases in proportion to the amount of powder present, so that the product $k_i M_h$ remains approximately constant. LeHouiller and Marchand[47] presented a more rigorous relationship of the form

$$k_i = K_1 + \frac{K_2}{\sqrt{M_h}} \qquad (8.12)$$

The constants K_1 and K_2 are different for the starved region (< 100% media void filling) and the highly saturated region (100–250% void filling), and in the latter situation the initial breakage is not first order. In an even more detailed study, Shoji et al.[27] found that abnormal breakage also occured at low void fillings (< 40%). As well, they presented a correlation for compensating k_i for different levels of void filling and grinding media (Fig. 8.26e). This correlation shows that there is a gradual decrease in the relative breakage rate as the media void space filling rises from 50 to 100%, which possibly explains why grate discharge mills are more efficient than overflow mills.

The nature of the mill contents also has an effect on breakage rates. A finer average size leads to higher breakage rates, as does a decrease in viscosity; the latter can be attained either by dilution of the slurry, or by chemical additives[52] (Fig. 8.26f).

One of the most important factors (for design) is the effect of mill diameter. Relatively few data have as yet been published, but Fig. 8.25 compares the rates for two different sizes of mills, and for the linear sections of the curves[53]:

$$\frac{k_{i,l}}{k_{i,sm}} = \left(\frac{D_l}{D_{sm}}\right)^{0.6} \qquad (8.13)$$

(D_l, D_{sm} = inside diameter of large and small mills, respectively; $k_{i,l}$, $k_{i,sm}$ = breakage rate of size i in large and small mills, respectively), which is the same as that implicit in Bond's analysis.[13,53]

Figure 8.30 illustrates the form of the breakage rate found in autogenous grinding.[42] The left peak is comparable with the curve of a ball mill (Fig. 8.26a), the difference being that its right-hand slope continues to decrease as the particle size increases. Characteristic of autogenous grinding is the second peak on the right, which represents the rate of breakage of material disappearing as a result of

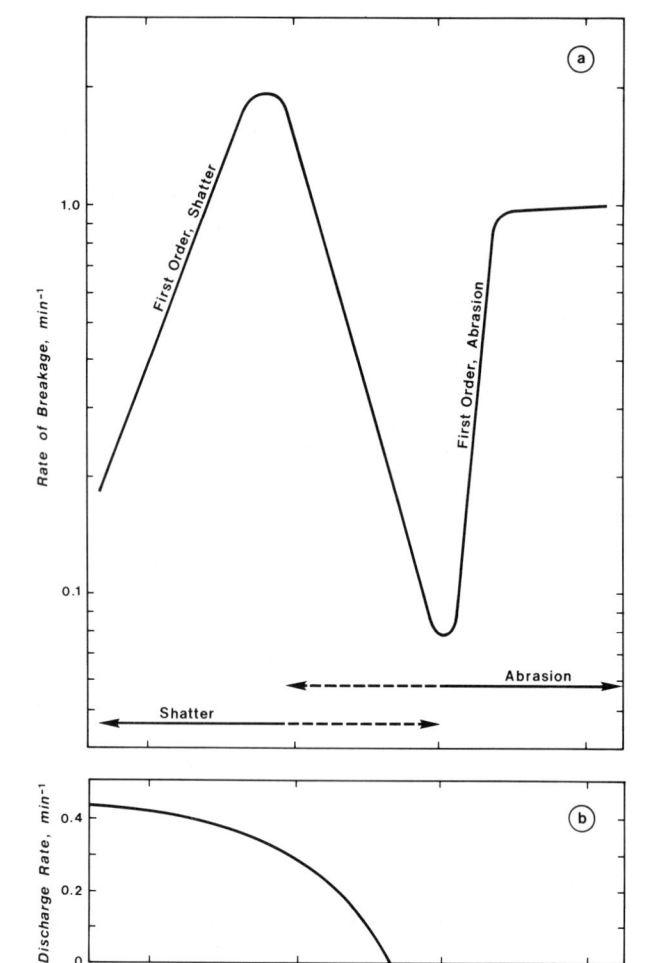

Figure 8.30. Breakage rates in autogenous grinding. (After Stanley.⁴²)

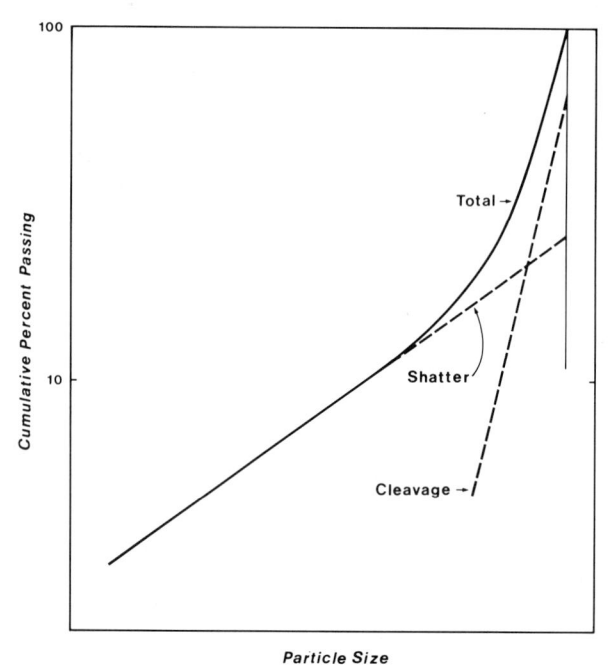

Figure 8.31. Schematic illustration showing that the breakage function can be considered as being made up of two size distributions.

abrasion. The minimum at the intermediate sizes indicates one of the difficulties in operating an autogenous mill—material in this size range can tend to build up in the mill. Solutions to this problem are the addition of a few large grinding balls (semi-autogenous grinding), or crushing these size fractions in the recycle circuit.

8.4.4. Summary of Breakage Processes

In view of the discussion above, the breakage processes in tumbling mills can be interpreted as follows. Each of the two components of the breakage function (Eq. 7.25) can be taken to represent a different fracture mechanism (Fig. 8.31). In the finer sizes where the media has more than sufficient energy to break any particle, fracture is essentially all shatter. With some materials, as the particle size increases, the fracture can tend to approach that represented in Fig. 7.5; that is, cleavage fracture exists because the energy applied is no longer sufficient for shatter. Where such a transition in the fracture mechanism occurs, the result is a nonnormalizable breakage function (Fig. 8.24).

At some stage, the particle size will become so large that a single collision event may have insufficient energy for fracture. The number of collisions then necessary for fracture will obviously increase with increasing particle size. Under these conditions the particles can be considered to have a range of strengths (as distinct from a range of media energies); and the breakage of any particle size can be described by two rate constants: one for fast breakage (soft material, or high energy media) and one for slow breakage (hard material, or low energy media). This is illustrated in Fig. 8.32.

Eventually (e.g., in autogenous grinding) the particle becomes so large that few cleavage fracture events occur. However, these collision events do have sufficient energy to produce localized breakage, that is, abrasion fracture.

Thus, the sequence abrasion, cleavage, shatter represents increasing energy intensity (Fig. 7.5).

Theory

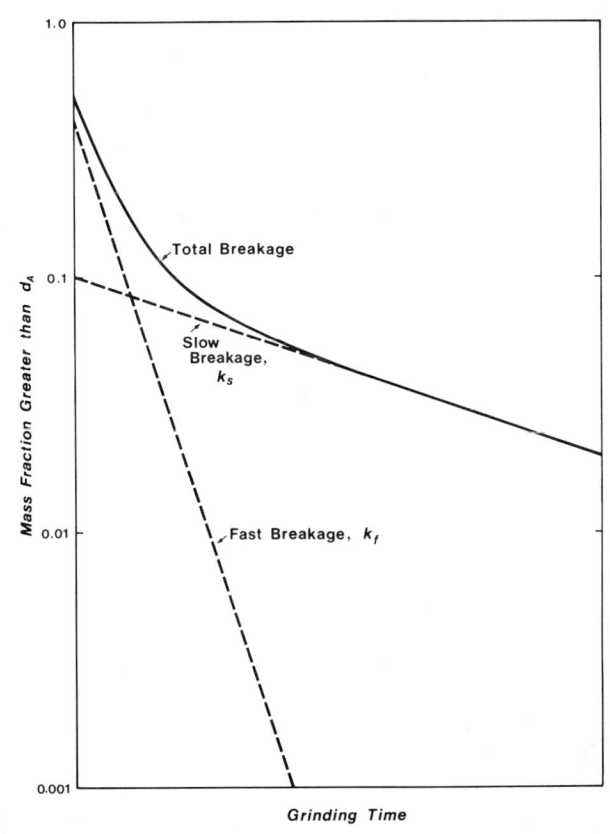

Figure 8.32. Use of two breakage rates to represent non-first-order breakage.

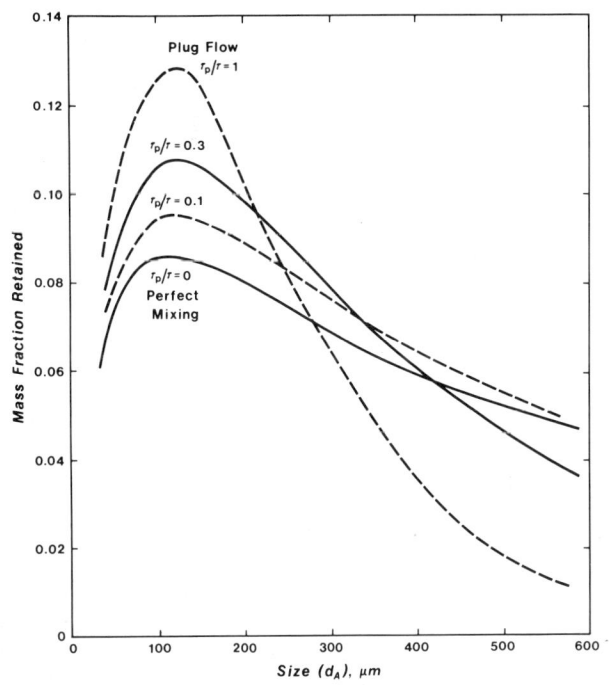

Figure 8.33. Simulated ball mill product size distributions, to show the effect of flow behavior. (After Kelsall et al.[30])

8.4.5. The Distribution of Residence Times

The distribution of residence time for a mill can be determined by adding an impulse tracer to the mill and measuring its recovery in the output.[54,55] Two extreme examples of flow can be considered: plug flow and perfect mixing (Chapter 3). With plug flow, all the sample leaves at a time equal to the residence time, whereas for perfect mixing some leaves instantly, but about 15% is retained after a time equal to twice the average residence time. Ball mills show a behavior between these two extremes (Fig. 3.12b) with closed circuit mills tending more toward perfect mixing and very long open circuit mills tending toward plug flow.

The significance of the flow behavior is readily apparent in Fig. 8.33, where it can be seen that plug flow produces a noticeably narrower size distribution than perfect mixing.

Practical distributions of residence times can be described by dividing the average residence time τ into a delay component τ_p (representing the plug flow component) and the balance τ_m (representing the perfect mixing component). One of the more thorough sets of information on distributions of residence times has been published by Kelsall et al.[28-31] These studies on ball and rod mills show that as a first approximation particles of different sizes have the same flow characteristics within the mill and these are similar to that of the water. Major changes in operating conditions, such as flow rate, ball charge, and ball size, primarily effect the average residence time.[30] Increasing pulp density, particle composition, and media shape also have some effect[27] on the ratio of τ_p to τ_m (Table 8.4 and Fig. 8.34).

8.4.6. Discharge Conditions

As has been described (Section 8.1), a number of size reduction devices impose a classifying effect on the discharge. For example, compression crushers and grate discharge mills have an actual physical barrier, whereas larger particles in rod mills show above average residence times. Both types of behavior can be described by including a classification/discharge function of the form shown in Fig. 8.22.[41,56,57]

TABLE 8.4: DISTRIBUTION OF RESIDENCE TIME IN BALL MILL

Variable	Value	τ_p/τ	τ_m/τ
Pulp density,	0.32	0.3	0.7
(Vol. Fraction	0.43	0.22	0.78
Solids)	0.53	0.18	0.82
Mill L/D	0.5	0.24	0.76
Ratio	1.0	0.225	0.775
	1.5	0.26	0.74
Ball Load	1	0.23	0.77
(Relative)	0.66	0.25	0.75
	0.45	0.45	0.55
Mill Speed	50	0.23	0.77
(r.p.m.)	60	0.26	0.74
	70	0.26	0.74
Media Shape	Spheres	0.25	0.75
	Cubes	0.23	0.77
	Short Cyl.	0.27	0.73
	Equi. Cyl.	0.27	0.73
	Long Cyl.	0.31	0.69
	Hexagonal	0.30	0.70
Feed rate	200	0.16	0.84
(g/min)	400	0.23	0.77
	1330	0.30	0.70
Mill Holdup	1	0.22	0.78
(Relative)	1.31	0.19	0.81
	1.61	0.15	0.85
	5.65	0.09	0.91
Rod Mill	450 g/min	0.072	0.938
	900 g/min	0.127	0.873

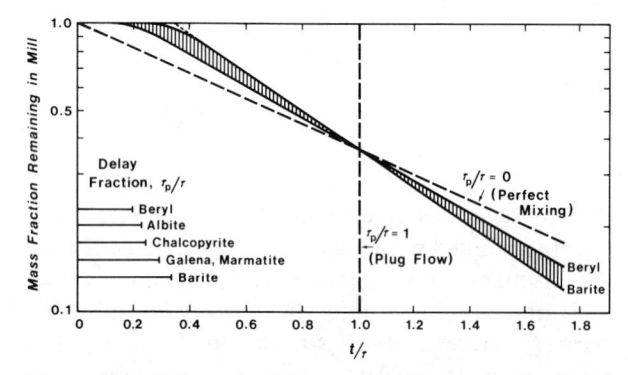

Figure 8.34. Effect of particle composition on distribution of residence time in a ball mill. This should be compared with Figs. 3.9 and 3.12b, noting that on a logarithmic scale the perfect mixing line is straight. (After Kelsall et al.[30])

8.4.7. Plug Flow and Batch Grinding

In batch grinding all the material remains in the mill for the required time. Plug flow through a continuous mill is in fact equivalent to batch grinding, with the average residence time equal to the batch grinding time. Under these conditions part of the material in any size fraction is disappearing because of breakage, while other material is entering the size fraction as a result of breakage of particles in larger size fractions. Thus, for the top size fraction[58]:

$$\frac{dm_1}{dt} = -k_1 m_1 \qquad (8.14a)$$

for the second size fraction:

$$\frac{dm_2}{dt} = -k_2 m_2 + b_1 k_1 m_1 \qquad (8.14b)$$

for the third size fraction:

$$\frac{dm_3}{dt} = -k_3 m_3 + b_1 k_2 m_2 + b_2 k_1 m_1 \qquad (8.14c)$$

and in general:

$$\frac{dm_i}{dt} = -k_i m_i + \sum_{j=1}^{i-1} b_{i-j} k_j m_j \qquad (8.15)$$

that is,

accumulation = breakage + appearance in size i
　　　　　　　　out　　　　by fracture of parti-
　　　　　　　　　　　　　cles in size j

In most cases the solution of this equation requires a numerical approximation method in which the changes in all fractures Δm during successive short time intervals Δt are calculated by the approximation

$$\frac{dm_i}{dt} = \frac{\Delta m_i}{\Delta t} \qquad (8.16)$$

The calculations are normally lengthy and require a computer; however the following example illustrates the general method.

Example 8.2. A 3.35 m × 3.35 m ball mill has 800 t/hr fed through it. The feed size distributions I_i, breakage rates k_i (min^{-1}), and breakage functions b_{i-j} are shown in Table E8.2.1.

What is the product size distribution?

[Assume plug flow through the mill, 45% charge filling, and $\rho_s = 3.145$ t/m³ (i.e., use the same feed material as that used in Example 10.1, but for

Theory

TABLE E8.2.1: PLUG FLOW GRINDING EXAMPLE

i	Size (μm)		I_i (%)	b_{i-j}	k_i	$M_{i,t}$					
						$t=1$	$t=2$	$t=3$	$t=4$	$t=5$	$t=30$
1		+2360	2.4	0.410	9.6	2.0160	1.6934	1.4225	1.1949	1.0037	0.0128
2	-2360	+1700	3.1	0.200	6.8	2.9061	2.7090	2.5131	2.3216	2.1368	0.1573
3	-1700	+1180	4.0	0.114	4.8	3.9008	3.7883	3.6653	3.5344	3.3978	0.7465
4	-1180	+850	5.0	0.081	3.4	4.9619	4.9113	4.8496	4.7779	4.6975	2.0343
5	-850	+600	6.6	0.057	2.4	6.5873	6.5652	6.5343	6.4951	6.4483	4.1144
6	-600	+425	9.1	0.040	1.7	9.0939	9.0811	9.0620	9.0369	9.0059	7.0661
7	-425	+300	13.1	0.029	1.2	13.0901	13.0757	13.0568	13.0336	13.0063	11.3434
8	-300	+212	16.4	0.021	0.85	16.4231	16.4425	16.4581	16.4701	16.4786	15.7618
9	-212	+150	12.7	0.015	0.60	12.8186	12.9338	13.0454	13.1536	13.2585	14.9249
10	-150	+106	7.6	0.010	0.42	7.7449	7.8876	8.0280	8.1663	8.3024	11.0610
11	-106	+75	4.9	0.007	0.30	5.0197	5.1383	5.2558	5.3721	5.4874	8.0216
12	-75	+53	3.7	0.005	0.21	3.7909	3.8812	3.9710	4.0602	4.1489	6.1871
13	-53	+38	2.8	0.004	0.15	2.8681	2.9360	3.0035	3.0708	3.1378	4.7189
14	-38		8.6	0.007	0	8.7786	8.9566	9.1346	9.3125	9.4901	13.8499
			100.0	1.000							

simplicity, assume that both minerals have the same breakage parameters.)]

Solution. To estimate the residence time in the mill, assume that the charge has 32% porosity, and that the feed is 50% solids by volume. Thus, the hold-up is given by:

$$\text{hold-up} = (3.35)^2 \left(\frac{\pi}{4}\right)$$

$$3.35 \times 0.45 \times 0.32 \times 0.5 \times 3.145$$

$$= 6.69 \text{ tonnes}$$

$$\therefore \quad \tau = \text{hold-up/flow rate}$$

$$= \frac{6.69 \text{ t}}{800 \text{ t}} \bigg| \frac{\text{hr}}{\text{hr}} \bigg| \frac{60 \text{ min}}{}$$

$$= 0.50 \text{ min}$$

Equation 8.15 can be rewritten as

$$\frac{\Delta m_i}{\Delta t} = -k_i m_i + \sum_{j=1}^{i-1} b_{i-j} k_j m_j$$

If Δm_i is replaced by $(m_{i,t+1} - m_{i,t})$, then provided Δt is small enough

$$m_{i,t+1} = m_{i,t}(1 - k_i \Delta t) + \Delta t \sum_{j=1}^{i-1} b_{i-j} k_j m_{j,t}$$

In this example, Δt is 1 sec, and I_i is used for m_i. Thus, for the top size fraction ($i = 1$)

$$m_{1,t+1} = m_{1,t}(1 - k_1 1.0) + 1.0 \sum_{j=1}^{i-1} b_{1-1} k_j m_{j,t}$$

After 1 sec, since $b_{1-1} = 0$,

$$m_{1,1} = 2.4\left(1 - \frac{9.6}{60} 1.0\right)$$

$$= 2.0160$$

and after 2 sec

$$m_{1,2} = 2.0160\left(1 - \frac{9.6}{60} 1.0\right) + 0$$

$$= 1.6934$$

and so on.

For the second size fraction, after 1 sec

$$m_{2,1} = m_{2,0}(1 - k_2 1.0) + 1.0 \times b_1 k_1 m_{1,0}$$

$$= 3.1\left(1 - \frac{6.8}{60} 1.0\right) + 1.0 \times 0.41 \frac{9.6}{60} 2.4$$

$$= 2.9061$$

and after 2 sec

$$m_{2,2} = 2.9061\left(1 - \frac{6.8}{60} 1.0\right) +$$

$$1.0 \times 0.41 \frac{9.6}{60} 2.0160$$

$$= 2.7090$$

and so on.

For the fifth size fraction, after 4 sec,

$$m_{5,4} = 6.5343\left(1 - \frac{2.4}{60} 1.0\right) + \overbrace{0.081(9.6/60)1.4225}^{b_{i,j} \times k_j \times m_{j,3}}$$
$$+ 0.114(6.8/60)2.5131$$
$$+ 0.200(4.8/60)3.6653$$
$$+ 0.410(3.4/60)4.8496$$

$$= 6.4951$$

The fourteenth size fractions (i.e., the subsieve fractions) are obtained by difference, since no breakage occurs from this size fraction.

Clearly, such calculations become very tedious as the number of size fractions and the number of time intervals increase. The more extensive results in Table E8.2.1 were generated by digital computer.

An alternative approach to this problem is the use of matrix algebra,[41,59] in which case the rate of breakage is replaced by a sequence of breakage events, with each size essentially having a certain probability of breaking during each event. From a mass balance of a breakage process, it can be shown that

$$[O] = [X]^n \cdot [I] \quad (8.17)$$

($[I]$, $[O]$ = $N \times 1$ matrices of input (feed) and output (product) size distributions, respectively; N = number of size intervals; n = number of breakage events). To describe $[X]$, it is necessary to consider it in terms of its two components: material left after a breakage event, plus material appearing from breakage of coarser sizes. During the breakage event, a particle in any size fraction i has a certain probability p_i of being broken (i.e., is selected for breakage), and for all particles the probability is $[p]$. Since the unbroken particles pass through, they are given by $([1] - [p])$; (where $[1]$ = the unit matrix). When any particle breaks, it will form the familiar range of products that can be described by a breakage parameter matrix $[b]$, so that the fraction appearing by breakage of coarser sizes can be given by $[b] \cdot [p]$. Thus, summing these two components

$$[X] = [b] \cdot [p] + ([1] - [p]) \quad (8.18)$$

and Eq. 8.17 becomes

$$[O] = \{[b] \cdot [p] + ([1] - [p])\}^n [I] \quad (8.19)$$

When the device imposes a classification/discharge function, this can be incorporated.

8.4.8. Perfect Mixing Mill

The performance of a short tumbling mill can sometimes be approximated by assuming the contents to be perfectly mixed.[57] Under these conditions, a mass balance on the mill is

accumulation = (feed + appearance from breakage of larger sizes)
− (breakage + discharge) (8.20)

which for the ith size fraction under steady state conditions gives

$$\left(I_i + \sum_{j=1}^{i-1} b_{i-j} k_j m_j \tau \right) - (k_i m_i \tau + O_i) = 0 \quad (8.21)$$

Use of this equation is illustrated with an example. (The method used is essentially that described by Whiten,[57] which allows the use of a small calculator.)

Example 8.3. Repeat Example 8.2, assuming perfect mixing.

Solution. Equation 8.21 can be put in the form

$$O_i = \frac{I_i + \sum_{j=1}^{i-1} b_{i-j}(\Sigma O)_j}{k_i \tau + 1} \quad (E8.3.1)$$

To use this equation, two intermediate calculations are performed: first,

$$K_{B,i} = \sum_{j=1}^{i-1} b_{i-j}(\Sigma O)_j \quad (E8.3.2)$$

(i.e., material appearing in size i by breakage of particles greater than size i), and second,

$$(\Sigma O)_j = O_j k_j \tau \quad (E8.3.3)$$

(i.e., an intermediate calculation to carry the quantity broken in size j down to smaller sizes i).

Basis. 100 units of mill input/min

Starting with the first size fraction (i.e., $N = i = 1$):

STEP 1. Material appearing from larger sizes (Eq. E8.3.2):

$$K_{B,1} = \sum_{j=1}^{i-1} b_{i-j}(\Sigma O)_j$$
$$= 0 \times 0$$

STEP 2. Evaluate Eq. E8.3.1:

$$O_1 = \frac{I_1 + K_{B,1}}{k_1 \tau + 1}$$
$$= \frac{2.4 + 0}{9.6 \times 0.5 + 1}$$
$$= 0.4138$$

(N.B. Answers are given to four decimal places to eliminate confusion over rounding off: this does not imply that such accuracy is normally justified.)

Theory

STEP 3. Evaluate carry-on breakage (Eq. E8.3.3):

$$(\Sigma O)_1 = 0.4138 \times 9.6 \times 0.5$$
$$= 1.9862$$

Repeating for the second size fraction:

STEP 1

$$K_{B,2} = b_1(\Sigma O)_1$$
$$= 0.41 \times 1.9862$$
$$= 0.8143$$

STEP 2

$$O_2 = \frac{I_2 + K_{B,2}}{k_2\tau + 1}$$
$$= \frac{3.1 + 0.8143}{6.8 \times 0.5 + 1}$$
$$= 0.8896$$

STEP 3

$$(\Sigma O)_2 = 0.8896 \times 6.8 \times 0.5$$
$$= 3.0247$$

Repeating for the third size fraction:

STEP 1

$$K_{B,3} = b_1(\Sigma O)_2 + b_2(\Sigma O)_1$$
$$= 0.41 \times 3.0247 + 0.20(1.9862)$$
$$= 1.6374$$

STEP 2

$$O_3 = \frac{4.0 + 1.6374}{4.8 \times 0.5 + 1}$$
$$= 1.6580$$

STEP 3

$$(\Sigma O)_3 = 1.6580 \times 4.8 \times 0.5$$
$$= 3.9793$$

Repeat for the fourth size fraction:

STEP 1

$$K_{B,4} = b_1(\Sigma O)_3 + b_2(\Sigma O)_2 + b_3(\Sigma O)_1$$
$$= 0.41 \times 3.9793 + 0.20 \times 3.0247$$
$$+ 0.114 \times 1.9862$$
$$= 2.4629$$

and so on.

TABLE E8.3.1: PERFECT MIXING GRINDING EXAMPLE

i	Size (μm)	$K_{B,i}$	O_i	$(\Sigma O)_i$
1	+2360	0.0	0.4138	1.9862
2	+1700	0.8143	0.8896	3.0247
3	+1180	1.6374	1.6580	3.9793
4	+850	2.4629	2.7640	4.6989
5	+600	3.2281	4.4673	5.3608
6	+425	3.9496	7.0538	5.9957
7	+300	4.6403	11.0877	6.6526
8	+212	5.3239	15.2448	6.4790
9	+150	5.6611	14.1239	4.2372
10	+106	4.9793	10.3961	2.1832
11	+75	3.8613	7.6185	1.1428
12	+53	2.8900	5.9638	0.6262
13	+38	2.1406	4.5959	0.3447
14	+0		(13.7228)	
			100.00	

The complete product size analysis is shown in Table E8.3.1. Again, the subsieve size ($i = 14$) is determined by difference.

With care, some streamlining of the calculations is possible. For example, the results $K_{B,i}$, O_i, and $(\Sigma O)_i$ can be set out in tabular form as follows:

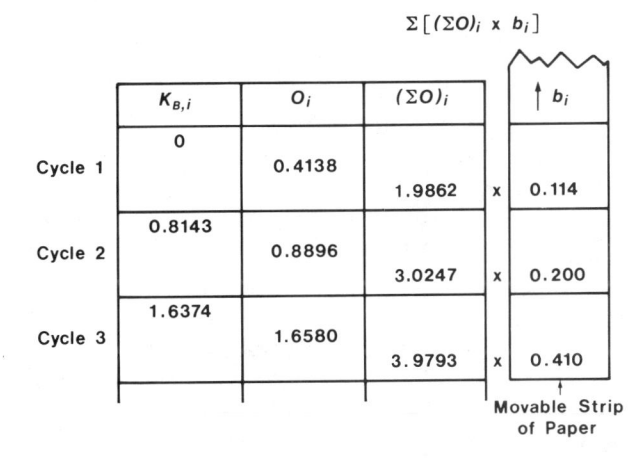

$\Sigma[(\Sigma O)_i \times b_i]$

	$K_{B,i}$	O_i	$(\Sigma O)_i$		b_i
Cycle 1	0	0.4138			
			1.9862	×	0.114
Cycle 2	0.8143	0.8896			
			3.0247	×	0.200
Cycle 3	1.6374	1.6580			
			3.9793	×	0.410

Movable Strip of Paper

This movable slip of paper makes evaluation of $K_{B,i}$ much simpler, although it must be noted that the alignment illustrated is being used to calculate $K_{B,4}$ (i.e., K_B for cycle 4).

Programs have been published that allow this analysis to be carried out on small programmable calculators.[57]

8.4.9. Complex Situations

It must be emphasized that the two examples considered so far are only approximations, and they neglect flow behavior and the classifying effect that occurs in crushers and certain tumbling mills. This latter situation requires the inclusion of data such as that of Fig. 8.22, and has been described elsewhere.[27,41,57] The incorporation of realistic flow patterns has been extensively described in the literature,[28-34,60] the most common method being the representation of the mill flow by a number of mixers in series. Equations 8.14 and 8.15 therefore become:

Top size fraction:

$$\text{Mixer A} \quad \frac{dm_{1A}}{dt} = \frac{m_{I1}}{\tau_A} - k_1 m_{1A} - \frac{m_{1A}}{\tau_A} \quad (8.22a)$$

$$\text{Mixer B} \quad \frac{dm_{1B}}{dt} = \frac{m_{1A}}{\tau_B} - k_1 m_{1B} - \frac{m_{1B}}{\tau_B} \quad (8.22b)$$

$$\text{Mixer C} \quad \frac{dm_{1C}}{dt} = \frac{m_{1B}}{\tau_C} - k_1 m_{1C} - \frac{m_{1C}}{\tau_C} \quad (8.22c)$$

Second size fraction:

$$\text{Mixer A} \quad \frac{dm_{1A}}{dt} = \frac{m_{I2}}{\tau_A} + b_1 k_1 m_{1A}$$
$$- k_2 m_{2A} - \frac{m_{2A}}{\tau_A} \quad (8.22d)$$

and so on.

($m_{I,i}$ = mass fraction of size i in the input [feed]; τ_A, τ_B, τ_C = residence time in given mixer).

Kelsall et al.[28-34,56,60] have shown that three mixers in series are sufficient in most cases, the residence times in each of A, B, and C (Eq. 8.22) being typically 0.15τ, 0.15τ, and 0.7τ, respectively (τ = overall average residence time in the mill).

8.4.10. Applications of Mass-Size Balance Equations

Although mass-size balance equations have been shown to be capable of simulating breakage in a wide variety of size reduction operations,[36,41-44,51,53,57,61,62] they are at present capable of only limited application to the design of new size reduction circuits.[40,53,63,64] Furthermore, despite claims to the contrary, their predictive abilities are limited because the effects of many variables are still not known. However, there is a growing amount of literature describing how the method has been used to simulate circuits to analyze their effectiveness,[39,40,60,65,66] to determine their susceptibility to various control mechanisms,[60,67] and to highlight the areas where modifications could be worthwhile.[65,66]

8.4.11. Measurement of Mass-Size Balance Parameters

A variety of methods for determining breakage parameters have been described.[37,38,54-56,58,68-73] Determination of the breakage function in a compression crusher is comparatively straightforward in principle, since it is possible to adjust particle size and machine settings so that only primary fracture occurs. Crushing a number of equal sized particles allows the breakage function to be obtained by sieve analysis of the product.

The determination of the breakage function is more difficult with tumbling mills because of the possibility of secondary fracture.[72] In principle, grinding a single particle size for a short time is a realistic approach, but in practice a correction is necessary to compensate for sieving errors. A more versatile method has been described to handle longer grinding times by making appropriate corrections for secondary breakage, but the method requires estimated values of the breakage rate.

The use of tracers has also been described. Ra-

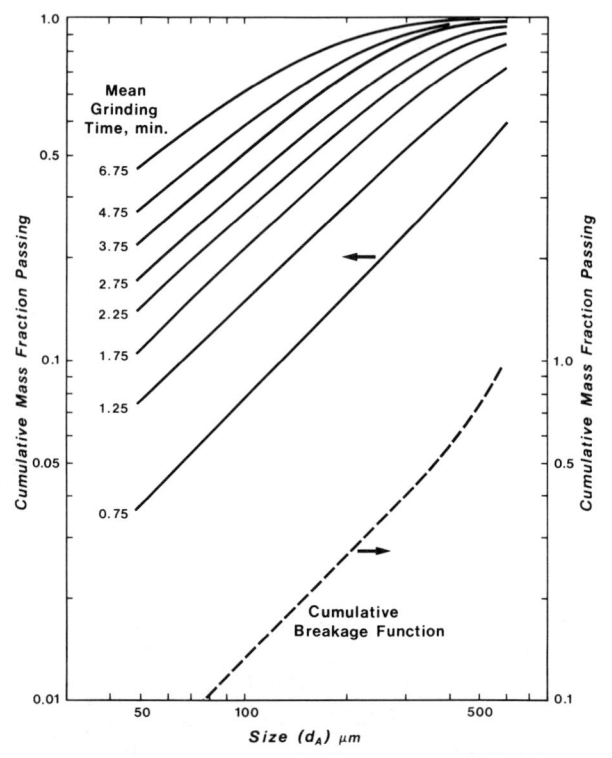

Figure 8.35. Changing size distribution of a $-833 +590$ μm quartz impulse in a small ball mill grinding calcite. Correction of the early size distributions for flow effects gives the breakage function. (After Kelsall et al.[28])

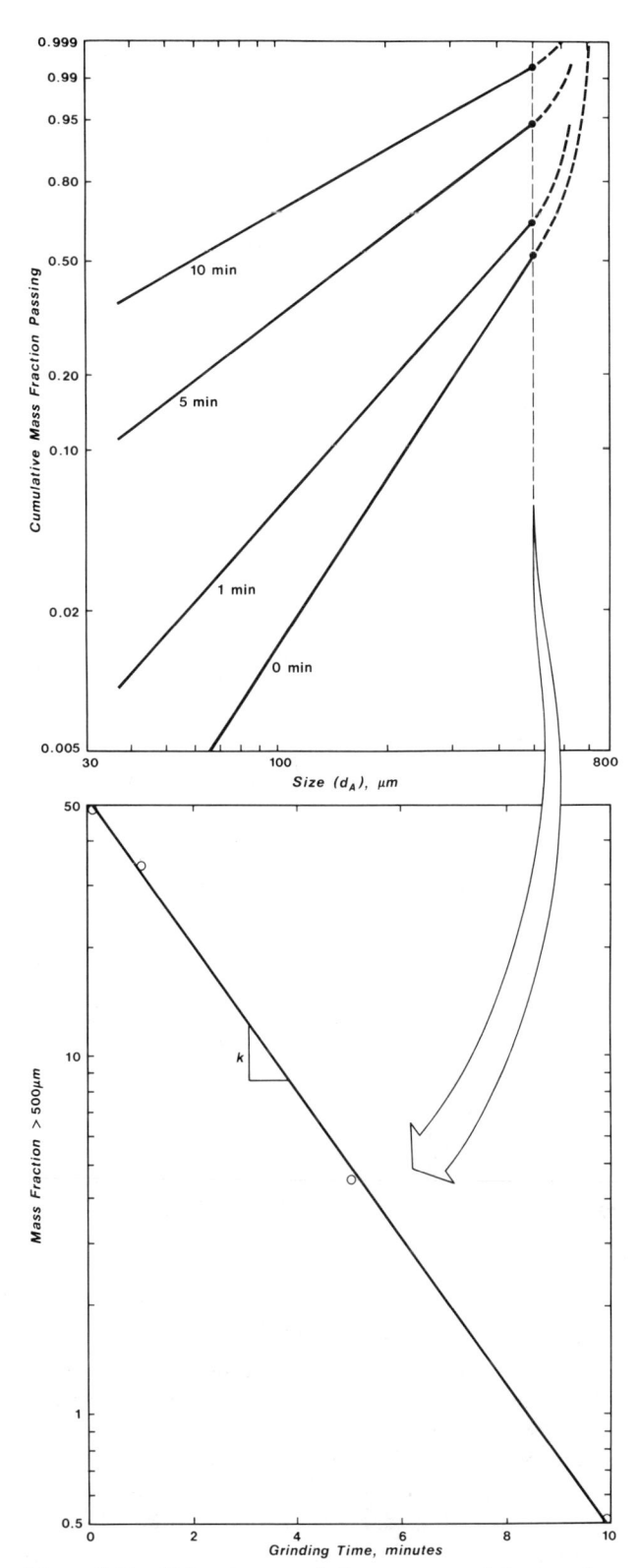

Figure 8.36. Using the rate of disappearance of the top size fraction to determine the breakage rate. [Reprinted with permission from L. G. Austin, *Ind. Eng. Chem. Process Des. Dev.*, **12**, 121–129 (1973). Copyright 1973 American Chemical Society.]

dioactive tracers were used by earlier workers, but the method is experimentally unattractive.[68] Kelsall and his co-workers derived their data using mineral tracers, using the size distribution of the tracer after very short grind times to derive the breakage function[54,55] (Fig. 8.35).

Breakage rates in tumbling mills can be obtained by grinding a size distribution in a mill and following the rate of disappearance of the top size fraction as illustrated in Fig. 8.36.[53]

Klimpel and Austin have shown that the breakage rates and the breakage function can be back-calculated from batch grinding test data using statistical methods. Experimentally, the method is rather straightforward, and although the computer calculations are somewhat extensive, the method can accommodate nonnormalized breakage functions.[72] Kelsall and his co-workers also use a back-calculation approach; their analysis was done by trial and error and the possibility of nonnormalized breakage functions was not considered.[55,73]

Distributions of residence times are determined by dye, chemical, or mineral tracers, the latter having the advantage of allowing mill classification/discharge effects to be detected.[55,56,65] Some experimental difficulties occur when the mill is in closed circuit, because recycle tracer soon appears in the mill discharge.[74] To date, there is no practical means of predicting mean residence times, or distributions of residence times, for new systems.

8.5. EQUIPMENT CAPACITIES

Few attempts have been made to theoretically describe the capacity of compression crushers, and most capacity determinations are based on empirical data available in manufacturers' catalogues. Some typical data are shown in Fig. 8.37.

In essence, the capacity of a compression crusher is determined by the physical size of the passageways through the device (provided of course sufficient power is available to break the feed). The throughput of a jaw crusher has been analyzed by Rose and English.[75] They assumed the throughput to be limited by the amount that could fall under gravity as the jaws opened. At the slower speeds capacity should increase linearly with speed $\mathcal{N}$, because there is sufficient time for the particles to fall to the crushing surface before it starts its return stroke. At higher speeds the particles fall against the closing crushing surface, and the

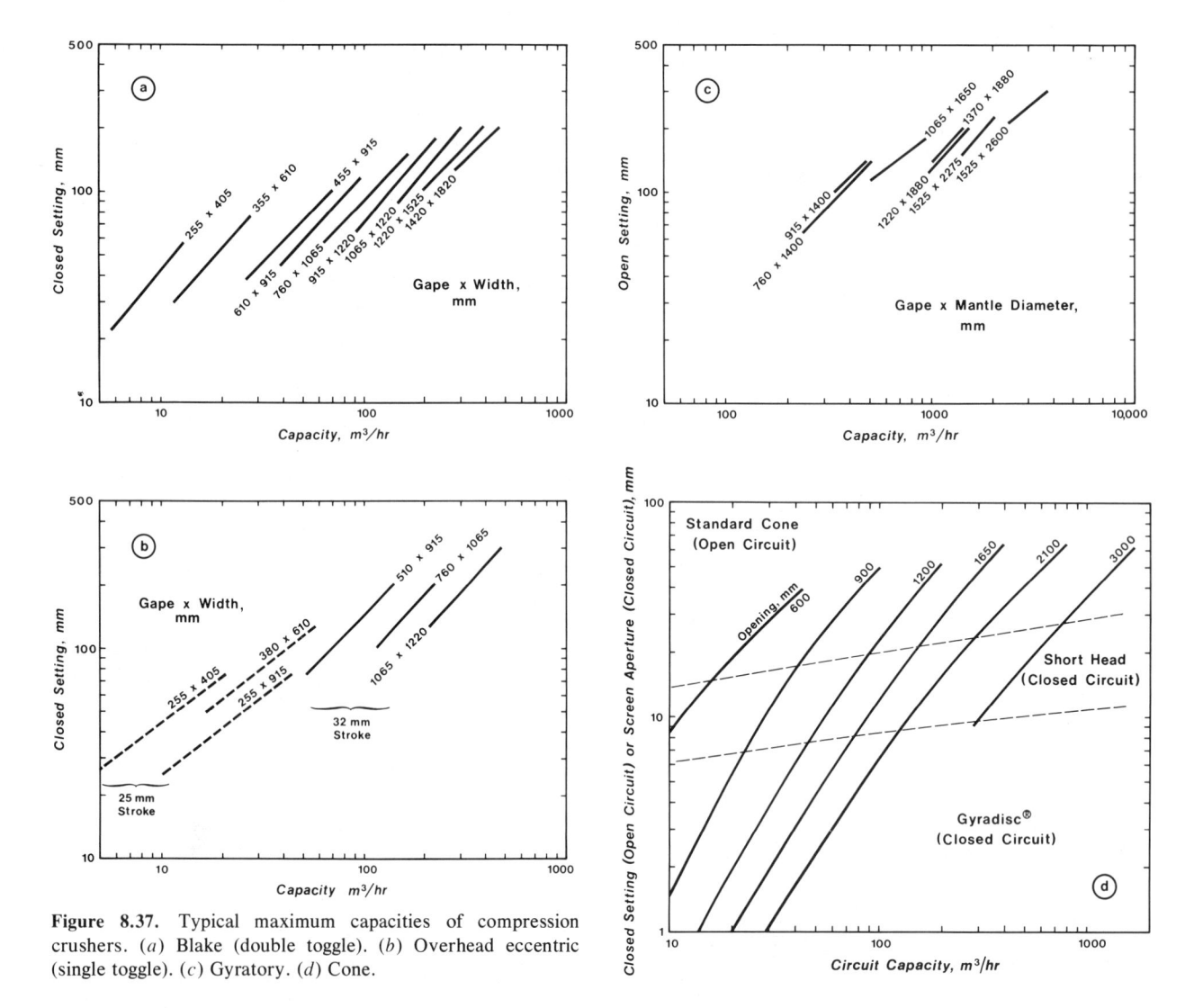

Figure 8.37. Typical maximum capacities of compression crushers. (*a*) Blake (double toggle). (*b*) Overhead eccentric (single toggle). (*c*) Gyratory. (*d*) Cone.

throughput becomes proportional to $1/\mathcal{N}$. Based on these assumptions Rose and English derived a correlation for jaw crusher throughput. The general effect is illustrated in Fig. 8.38, showing that a jaw crusher has a critical speed that produces the maximum throughput. Theoretically, this critical speed $\mathcal{N}_c$ (rpm) is given by

$$\mathcal{N}_c = 47 \left(\frac{\mathfrak{R} - 1}{\mathfrak{R}_r x} \right)^{1/2} \qquad (8.23)$$

($\mathfrak{R}_r$ = reduction ratio for crusher, gape/set; x = throw of crusher, m).

Comparison of this theory with manufacturers' data suggests that recommended speeds are lower than $\mathcal{N}_c$ for the smallest and largest gapes, and that manufacturers' data can be described by the empirical equation

$$\mathcal{N}_{op} = 280 \exp(-0.212 G^3) \qquad (8.24)$$

($\mathcal{N}_{op}$ = operating speed; G = gape setting, m).

Their study also suggested that jaw crushers are underpowered, and that they normally operate below the maximum throughput possible at the actual operating speed.

The work of Babu and Leonard[76] indicates that water flushing during crushing can increase the throughput, while decreasing the amount of fines produced.

Cone crusher capacity can supposedly be calculated by a "bouncing ball" treatment, but no useful details have been given.[77]

Equipment Capacities

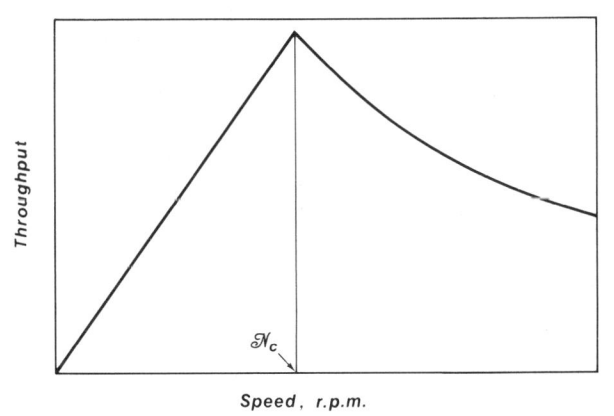

Figure 8.38. Theoretical throughput of a jaw crusher, as a function of speed. (After Rose and English.[75])

The capacity of tumbling mills, for a given material and size distribution, is described by

$$\text{capacity} \propto D_M^n L \quad (8.25)$$

Bond[13] claims that the theoretical value of n is 3.0, (such a value is apparently valid for some grate discharge mills, and is used by one manufacturer for overflow mills[2]). He found $D_M^{2.6}$ to be more realistic in practice, and this is the most widely accepted value. It is worth noting that the 0.6 in the 2.6 corresponds to the value of 0.6 in Eq. 8.13; in other words the value of 2.6 implies that the scale-up of breakage rates is independent of particle size. As was described in Section 8.4.3, this may not be the case with the larger particles (see Fig. 8.25); thus relationships such as Eq. 8.25 must be used with care. The difference between exponents in Eqs. 8.25 and 8.6 leads to a size advantage for larger mills,[53] but the uncertainty of the exponents poses some doubt on the true difference, although it is normally taken[13] as 0.2 (i.e., 2.6 − 2.4). Part of this uncertainty must be attributed to the size advantage not being constant; there is clear evidence that the advantage is markedly decreased above 4.1 m,[21] to the extent that it becomes a disadvantage above 5 m.[78] The cause of the inefficiency of very large diameter mills is unknown (although some suggestions have been made[78]), but Fig. 8.25 suggests that the size advantage below this limit is related to the increasing particle size at which the maximum in the breakage rate occurs with increasing mill diameter.

8.5.1. Bond's Method

At present, insufficient information has been published to permit the mass-size balance method to be used for sizing new size reduction equipment and designing circuits, although some preliminary approaches have been described.[36,63,64] The difficulty is not that the method itself is inadequate; to the contrary, the more sophisticated equations have been shown capable of reliably predicting complete size distributions, a feature the Bond approach lacks. Ironically, the inadequacies of the method correspond to the main criticisms of the Bond method, that is, a lack of empirical scale-up correlations. Some manufacturers are using mass-size balance methods for designing new circuits, but since the vital information is not being published, the method cannot yet be considered to be a substitute for Bond's.

In spite of its empirical nature, the Bond method is still the most widely used for designing size reduction circuits. Furthermore, it can be expected to be in use for some time, the reasons being essentially threefold. First, there is a vast amount of published data available for *industrial equipment*, and it is continually increasing.[13,21,79−82] Second, the method is quite satisfactory for initial calculations: even when a circuit is being designed, feed, and product specifications especially, seldom can be given with sufficient reliability to justify more sophisticated methods. Finally, the method, and adaptions of it,[83−86] provide simple methods for measuring mill efficiency.

On the basis of extensive experimental work, Bond[13] found that "the total work useful in breakage which has been applied to a stated weight of homogeneous broken material is inversely proportional to the square root of the diameter of the product particles." (The particle diameter is defined as the 80% cumulative passing size $d_{A,80}$, expressed in micrometers). Thus

$$W = \frac{K}{\sqrt{d}} \quad (8.26)$$

(W = work) so that for a reduction from size d_I to d_O, (strictly $d_{A,3,80,I}$ to $d_{A,3,80,O}$) the total work W is

$$W = W_O - W_I = \frac{K}{\sqrt{d_O}} - \frac{K}{\sqrt{d_I}} \quad (8.27)$$

To eliminate the constant K, the *work index* W_i is defined as the total work to reduce a particle from infinite size to 100 μm, and substituting these values in Eq. 8.27:

$$K = W_i \sqrt{100} \quad (8.28)$$

Combining Eqs. 8.27 and 8.28 gives the basic equation

$$W = 10 W_i \left(\frac{1}{\sqrt{d_o}} - \frac{1}{\sqrt{d_I}} \right) \quad (8.29)$$

(cf. Eq. 7.13b). The work index is basically a measure of the hardness of the material, but it also includes the mechanical efficiency of the machine (in that $\approx 1\%$ of the machine power is converted to useful new surface, it is remarkable that this method works as well as it does). Some typical values of W_i are shown in Table 8.5.[13]

The work index can be evaluated by standard laboratory tests that *have been empirically correlated with full-scale equipment.* The crushing test involves measuring the crushing strength K_{cs} (in m · kg/m) of 10 pieces of rock, from which

$$W_i = \frac{476 K_{cs}}{\rho_s} \quad (8.30)$$

Rod mill grindability G_r, as net grams of undersize generated per revolution, is obtained in a standard rod mill test, and thence W_i is found from

$$W_i = \frac{6.83}{(d_t)^{0.23}(G_r)^{0.625}(1/\sqrt{d_o} - 1/\sqrt{d_I})} \quad (8.31)$$

(d_t = particle size, μm, at which W_i is determined). This W_i conforms with the motor output to an average overflow rod mill of 2.44 m interior diameter, wet grinding in open circuit. Similarly, *ball mill grindability* G_b is evaluated in a standard test that simulates a closed circuit, and W_i is evaluated from

$$W_i = \frac{4.90}{(d_t)^{0.23}(G_b)^{0.82}(1/\sqrt{d_o} - 1/\sqrt{d_I})} \quad (8.32)$$

This W_i corresponds to the motor output power to an average overflow ball mill of 2.44 m interior diameter wet grinding in closed circuit. Equations 8.31 and 8.32 allow for the fact that W_i may change with particle size. (Other workers[87] have attempted to surmount this problem by using Eq. 8.26 in the form

$$W = \frac{K}{d^n} \quad (8.33)$$

but this considerably extends the amount of test work required.)

A considerable number of correction factors have been presented to account for deviations from the "average" conditions represented by Eqs. 8.31 and 8.32. The main correction factors are listed in Table 8.6 and described in detail elsewhere.[13] Bond also described a variety of empirical correlations for calculating other aspects of size reduction, such as ball sizes, amount of charge, and optimum speeds.[13]

Subsequent publications have updated Bond's data, particularly with respect to larger mills and their limited applicability to autogenous grinding.[21,79-83]

In using Bond's method to determine capacity and power requirements, the following aspects must be emphasized. The capacity of jaw and gyratory crushers is determined by the discharge opening. Consequently, the machine size has a significant

TABLE 8.5: TYPICAL VALUES OF W_i

Material	W_i, kWhr/t
All materials	15.19
Barite	6.86
Basalt	22.45
Cement Clinker	14.84
Clay	7.81
Coal	12.51
Copper Ore	14.44
Dolomite	12.44
Emery	64.00
Feldspar	12.84
Galena	10.68
Glass	3.39
Gold Ore	16.31
Granite	15.83
Iron Ore	16.98
Lead Ore	12.54
Limestone	12.77
Mica	148.00
Oil Shale	19.91
Phospate Rock	11.14
Quartz	14.05
Taconite	16.36

TABLE 8.6: CORRECTIONS TO EQ. 8.29

Dry grinding
Mill diameter
Scalped feed
Oversize feed
Product <70 μm
Non-optimum reduction ratio
Open circuit ball mill

Equipment Capacities

correlation with the feed and product sizes. Thus, for a given size reduction and throughput, the machine is selected for this capacity (using published correlations,[1,75,88,89] or, preferably, reliable manufacturers' catalogues). Equation 8.29 is then used to calculate the motor size necessary to carry out the operation (a larger motor being required for a harder ore, and a smaller motor for a soft ore): that is, *compression crushers match power with ore hardness*, W_i. On the other hand, tumbling mills impose little restriction on throughput, and the power can be considered to be consumed in keeping the mill and contents in motion. Size reduction occurs as a by-product of passage through the mill, and Eq. 8.29 is effectively being used to determine the *extent* of this size reduction: if the ore is soft (W_i low) the ore will come out finer; if it is hard (W_i high) it will come out coarser (i.e., in terms of breakage rates, high and low, respectively). Thus, *the power is being matched to the extent of the size reduction*. Consequently, when Eq. 8.29 is used to size a tumbling mill, once the power is calculated, the mill size is essentially fixed, and in principle a mill that draws that amount of power can be selected from manufacturers' catalogues, or by Eq. 8.8. It must be remembered that this power can be divided among a number of mills and that larger diameter mills tend to be more efficient. Empirical equations[89] can also be used, but the results are likely to be less reliable than those obtained using manufacturers' catalogues. However, the need for caution in using manufacturers' "typical mill capacities" must be emphasized, since the capacity of a given tumbling mill (by Eq. 8.29) actually depends on *three* variables, d_I, d_O, and W_i, which are not directly related to mill size.

Example 8.4. An ore is to be broken from 80% −20 mm to 80% −200 μm, at the rate of 300 t/hr (ρ_b = 1.6 t/m³). What equipment could be used, and what power would be required? (W_i = 14.0 kWh/t for both crushing and grinding)

Solution. One method of achieving this size reduction would be to use a Gyradisc crusher to prepare feed for a ball mill. Both machines would be used in closed circuit (with screen and classifier respectively), the crusher giving a stage reduction ratio of about 5, and the ball mill the balance. [In practice, other equipment combinations are possible (e.g., a rod mill followed by a ball mill), and should be investigated when more detailed information is available.]

The crusher must be selected on the basis of its capacity, or in the case of a closed circuit crusher, the circuit capacity.

$$\text{throughput} = \frac{300 \text{ t}}{\text{hr}} \bigg| \frac{\text{m}^3}{1.6 \text{ t}}$$
$$= 188 \text{ m}^3/\text{hr}$$

Because the machine throughput depends on the circuit screen aperture, it is necessary to estimate this. Assume that $d_A \sim 1.5$ times the 80% passing size of the product.

∴ 80% passing size of product = 20/5
$$= 4 \text{ mm}$$
∴ $\qquad d_A = 1.5 \times 4 \text{ mm}$
$$= 6.0 \text{ mm}$$

From Fig. 8.37d, a 2100 mm Gyradisc crusher has a circuit capacity of 95 m³/hr when in closed circuit with a 6 mm screen. Thus, two 2100 mm crushers are required.

The power necessary can be calculated by Eq. 8.29:

$$P = 10 \times W_i \times I \left(\frac{1}{\sqrt{d_O}} - \frac{1}{\sqrt{d_I}} \right)$$
$$= 10 \times \frac{14 \text{ kWh}}{\text{t}} \bigg| \frac{300 \text{ t}}{\text{hr}} \left(\frac{1}{\sqrt{4000}} - \frac{1}{\sqrt{20,000}} \right)$$
$$= 367 \text{ kW}$$

that is,

$$184 \text{ kW/crusher}$$

The ball mill is selected on the basis of the power required for the given size reduction. Assume that the ball mill grinds wet in closed circuit and that the mill will be over 3.81 m in diameter, so that the maximum diameter efficiency can be used [i.e., $(2.44/D_M)^{0.2}$]:

$$\mathscr{P} = 10$$
$$\times \frac{14 \text{ kWh}}{\text{t}} \bigg| \frac{300 \text{ t}}{\text{hr}} \left(\frac{1}{\sqrt{200}} - \frac{1}{\sqrt{4000}} \right) \left(\frac{2.44}{3.81} \right)^{0.2}$$
$$= 2109 \text{ kW}$$

As shown in Example 8.1, a 4.98 × 4.98 m overflow discharge ball mill with 40% charge loading, operating at 74% of critical speed, should draw this power.

Though presented primarily as a method for calculating the capacity and power requirements of size reduction equipment, Bond's method has wider applications. A number of workers have shown that for *comparative* purposes, the laboratory test work can be considerably reduced by standardising a new simplified laboratory test.[84-86] One particularly useful technique is to use such a test to evaluate an arbitrary W_i that can be continually compared with that obtained on an operating mill, so that decreases in the efficiency of the full size mill can be monitored.

8.5.2. Autogenous Grinding Mills

Autogenous grinding mills are designed based on small or pilot scale testing to determine power requirements; then Eq. 8.8 (with the appropriate exponent of D_M) is applied to determine the mill size. The ore used in such small scale tests should be naturally graded in size (as produced from a primary crusher) with a particle size given by[24]:

$$d_{A,80,I} = 5.35 D_M^{2/3} \qquad (8.34)$$

This gives the maximum power draft.

The ore must also be checked for its competency to form an adequate supply of grinding media; manufacturers have established empirical tests for this purpose.

8.5.3. Other Methods of Predicting Machine Performance

A number of other test methods have been developed for predicting machine performance,[2,88] but they have not been as widely adopted as Bond's method. Some of these other methods have been reviewed by Marshall, who also showed that there is generally a poor correlation between the different methods.[88]

Some workers[1,88] have used manufacturers' catalogue data to derive correlations that can be used for machine scale-up (i.e., to extrapolate data from a smaller to a larger machine). These too have been reviewed by Marshall.[88] However such correlations are often unreliable and may not agree with those of other manufacturers, or with theoretical correlations (see, e.g., the discussion on Eqs. 8.6 and 8.22). There are two main reasons: first, efficiency tends to vary with equipment size, and second, manufacturers' data are empirical, and consequently differ from manufacturer to manufacturer, depending on the firm's experience (and also on whether the catalogue data are optimistic or conservative). Thus, such approaches should be avoided where possible, and until better methods are developed, the broad philosophy presented in Section 8.5.1 should be adhered to.

REFERENCES

1. A. F. Taggart, *Handbook of Mineral Dressing*, Wiley (1945).
2. R. H. Perry and C. H. Chilton, *Chemical Engineers' Handbook*, 5th ed., McGraw-Hill (1973).
3. K. Spink, "Comminution Methods and Machinery," *Min. Miner. Eng.*, **8**, 5–21 (January 1972); 5–25 (February 1972).
4. B. McGrew, "Crushing Practice and Theory," Allis Chalmers Bull. 07R.8073-02, Allis-Chalmers (1950), reprinted from *Rock Prod.*, **53**, 118–120 (June 1950); 95 (September 1950); 116–117 (October 1950); 62–63 (November 1950); 128–129, 152 (December 1950).
5. B. A. Bartley and G. J. Macdonald, "The Macdonald Impactor: Its Development and How it Relates to Other Crushers," *Inst. of Quarrying, N.Z. Branch Conf., Auckland* (June 1977).
6. E. Sarapuu, "Electro-Energetic Rock Breaking Systems," *Min. Congr. J.*, **59**, 44–54 (June 1973).
7. (a) E. J. Klovers, "Energy Crisis Spurs Use of Roller Mills," *Rock Prod.*, **77**, 47–49, 80 (October 1974).
   (b) E. J. Klovers, "Performance of Large Roller Mills," presented to Rock Products 9th Cement Ind. Sem. (1973).
8. H. E. Rose and R. M. E. Sullivan, *Vibration Mills and Vibration Milling*, Constable and Co., London (1961).
9. V. C. Quackenbush, R. R. Maddocks, and G. W. Higginson, "Chemical Comminution: An Improved Route to Clean Coal," Coal and Min. Process., **16**, 68–72 (May 1979).
10. S. Levine, "An Update on Crushing and Grinding Equipment," *Rock Prod.*, **77**, 59–71 (June 1974).
11. L. R. Mabson, "Primary Crusher Selection," *Mine Quarry Eng.*, **27**, 114–121 (March 1961).
12. M. Digre (Ed.), *Proceedings of the Autogenous Grinding Seminar*, Norwegian Technical University, Trondheim (1979).
13. F. C. Bond, "Crushing and Grinding Calculations, Parts I and II," *Br. Chem. Eng.*, **6**, 378–385; 543–548 (1961).
14. A. A. Rauth, "Maintaining an Optimum Grinding Charge," *Trans. AIME/SME*, **244**, 82–88 (March 1969).
15. Anon., "Nordberg Grinding Mills; Ball, Rod and Pebble," Bulletin 315, Nordberg Mfg. Co. (1970).
16. Anon., "Trellex Mill Linings," Bulletin 10979, Trelleborg (1976).

References

17. P. A. Korpi, "Angular Spiral Lining Systems in Wet Grinding Grate Discharge Ball Mills," presented to 107th AIME Annual Meeting, New Orleans, 1979.
18. T. Andrén and G. Nilsson, "Appraisal of the Use of Rubber Linings in Grinding Mills," *Proc. 10th Int. Min. Proc. Congr., London, 1973,* pp. 123–142, IMM (1974).
19. G. R. Hoey, W. Dingley, and C. Freeman, "Corrosive Wear of Grinding Media in Grinding a Complex Zinc-Lead-Copper Sulphide Ore," *Proc. Australas. IMM,* No. 265, 27–32 (March 1978).
20. D. J. Dunn, "Measurement of Impact Forces in Ball Mills," AIME/SME Preprint 77-B-316, AIME (1977).
21. C. A. Rowland, "Grinding Calculations Related to the Application of Large Rod and Ball Mills," presented to Can. Miner. Process., Ottawa, 1972, also published in *Can. Min. J.,* **93,** 48–53 (June 1972).
22. A. M. Gow, A. B. Campbell, and W. H. Coghill, "A Laboratory Investigation of Ball Milling," *Trans. AIME,* **87,** 51–74 (1930).
23. C. A. Rowland and D. M. Kjos, "Autogenous and Semi-Autogenous Mill Selection and Design," *Aust. Min.,* **67,** 21–35, (September 1975).
24. A. R. MacPherson and R. R. Turner, "Autogenous Grinding from Test Work to Purchase of a Commercial Unit," Ch. 13 in *Mineral Processing Plant Design,* AIME/SME (1978).
25. F. Peña and R. Hopple, "Cascade Milling Scale-Up Considerations," *Proc. Autogenous Grinding Seminar,* Norwegian Technical University, Trondheim (1979).
26. C. T. Van Winkle, "Recent Tests of Ball-Mill Crushing," *Trans. AIME,* **59,** 227–248 (1918).
27. K. Shoji, S. Lohrasb, and L. G. Austin, "The Variation of Breakage Parameters with Ball and Powder Loading in Dry Ball Milling," *Powder Technol.,* **25,** 109–114 (1980).
28. D. F. Kelsall, K. J. Reid, and C. J. Restarick, "Continuous Grinding in a Small Wet Ball Mill. Part I. Study of the Influence of Ball Diameter," *Powder Technol.,* **1,** 291–300 (1967–1968).
29. D. F. Kelsall, K. J. Reid, and C. J. Restarick, "Continuous Grinding in a Small Wet Ball Mill. Part II. A Study of the Influence of Hold-Up Weight," *Powder Technol.,* **2,** 162–168 (1967–1968).
30. D. F. Kelsall, K. J. Reid, and C. J. Restarick, "Continuous Grinding in a Small Wet Ball Mill. Part III. A Study of Distribution of Residence Time," *Powder Technol.,* **3,** 170–178 (1969–1970).
31. D. F. Kelsall, P. S. B. Stewart, and K. R. Weller, "Continuous Grinding in a Small Wet Ball Mill. Part IV. A Study of the Influence of Grinding Media Load and Density," *Powder Technol.,* **7,** 293–301 (1973).
32. D. F. Kelsall, P. S. B. Stewart, and K. R. Weller, "Continuous Grinding in a Small Wet Ball Mill. Part V. A Study of the Influence of Media Shape," *Powder Technol.,* **8,** 77–83 (1973).
33. G. W. Heyes, D. F. Kelsall, and P. S. B. Stewart, "Continuous Grinding in a Small Wet Rod Mill. Part I. Comparison with a Small Ball Mill," *Powder Technol.,* **7,** 319–325 (1973).
34. G. W. Heyes, D. F. Kelsall, and P. S. B. Stewart, "Continuous Grinding in a Small Wet Rod Mill. Part II. Breakage of Some Common Ore Minerals," *Powder Technol.,* **7,** 337–341 (1973).
35. J. A. Herbst and D. W. Fuerstenau, "Influence of Mill Speed and Ball Loading on the Parameters of the Batch Grinding Equation," *Trans. AIME/SME,* **252,** 169–176 (1972).
36. L. Austin et al., "Some Results on the Description of Size Reduction as a Rate Process in Various Mills," *Ind. Eng. Chem., Process Des. Dev.,* **15,** 187–196 (1976).
37. L. G. Austin and P. T. Luckie, "Methods for Determination of Breakage Distribution Parameters," *Powder Technol.,* **5,** 215–222 (1971–1972).
38. L. G. Austin and P. T. Luckie, "The Estimation of Non-Normalised Breakage Distribution Parameters from Batch Grinding Tests," *Powder Technol.,* **5,** 267–271 (1971–1972).
39. L. G. Austin, P. T. Luckie, and H. M. von Seebach, "Optimisation of a Cement Milling Circuit with Respect to Particle Size Distribution and Strength Development, by Simulation Models," *4th Eur. Symp. on Size Reduction, Nuremburg, September 1975,* pp. 519–537 (1975).
40. L. G. Austin, P. T. Luckie, and D. Wightman, "Steady-State Simulation of a Cement-Milling Circuit," *Int. J. Miner. Process.,* **2,** 127–150 (1975).
41. A. J. Lynch, *Mineral Crushing and Grinding Circuits,* Elsevier (1977).
42. G. G. Stanley, "Mechanisms in the Autogenous Mill and Their Mathematical Representation," *J. S. Afr. IMM,* **75,** 77–98 (November 1974).
43. L. G. Austin et al., "Preliminary Results on the Modeling of Autogenous Grinding," *14th APCOM Symp., 1976,* pp. 207–226, AIME/SME (1977).
44. L. G. Austin, V. K. Jindal, and C. Gotsis, "A Model for Continuous Grinding in a Laboratory Hammer Mill," *Powder Technol.,* **22,** 199–204 (1979).
45. W. J. Taute, P. H. Meyer, and L. G. Austin, "Comparison of Breakage Parameters in Two Tumbling Ball Mills of Different Dimensions," *Symp. on Automatic Control in Min., Miner. and Met. Process, IFAC, Sydney, August 1973,* pp. 11–19 (1973).
46. L. G. Austin, K. Shoji, and P. T. Luckie, "The Effect of Ball Size on Mill Performance," *Powder Technol.,* **14,** 71–79 (1976).
47. R. LeHouillier and J. C. Marchand, "Empirical Correlation Predicting Particulate Mass Effect on Selection Parameters," *Powder Technol.,* **17,** 101–107 (1977).
48. V. K. Gupta and P. C. Kapur, "Empirical Correlations for the Effects of Particulate Mass and Ball Size on the Selection Parameters in the Discretized Batch Grinding Equation," *Powder Technol.,* **10,** 217–223 (1974).
49. L. G. Austin, K. Shoji, and M. D. Everett, "An Explanation of Abnormal Breakage of Large Particle Sizes in Laboratory Mills," *Powder Technol.,* **7,** 3–7 (1973).
50. R. P. Gardner and L. G. Austin, "The Applicability of the First-Order Grinding Law to Particles Having a Distribution of Strengths," *Powder Technol.,* **12,** 65–69 (1975).
51. V. K. Jindal and L. G. Austin, "The Kinetics of Hammer Milling of Maize," *Powder Technol.,* **14,** 35–39 (1976).

52. R. Klimpel and W. Manfroy, "Chemical Grinding Aids for Increasing Throughput in the Wet Grinding of Ores," AIME/SME Preprint, 77-B-327. Presented at AIME/SME Fall Meeting, 1977.
53. L. G. Austin, "Understanding Ball Mill Sizing," *Ind. Eng. Chem. Process Des. Dev.*, **12**, 121–129 (1973).
54. D. F. Kelsall, "A Study of Breakage in a Small Continuous Open Circuit Wet Ball Mill," *Proc. 7th Int. Miner. Process. Congr., New York, 1964*, pp. 33–42, Gordon and Breach (1965).
55. D. F. Kelsall, P. S. B. Stewart, and K. J. Reid, "Confirmation of a Dynamic Model of Closed-Circuit Grinding with a Wet Ball Mill," *Trans. IMM (C)*, **77**, C120–C127 (1968).
56. P. S. B. Stewart and C. J. Restarick, "A Comparison of the Mechanism of Breakage in Full Scale and Laboratory Scale Grinding Mills," *Proc. Australas. IMM*, No. 239, 81–92 (September 1971).
57. W. J. Whiten, "Ball Mill Simulation Using Small Calculators," *Proc. Australas. IMM*, No. 258, 47–51 (June 1976).
58. K. J. Reid, "A Solution to the Batch Grinding Equation," *Chem. Eng. Sci.*, **20**, 953–963 (1965).
59. S. R. Broadbent and T. G. Callcott, "Coal Breakage Processes, Parts I–V," *J. Inst. Fuel*, **29**, 524–539 (December 1956); **30**, 13–25 (January 1957).
60. D. F. Kelsall, K. J. Reid, and P. S. B. Stewart, "The Study of Grinding Processes by Dynamic Modelling," paper 2509 presented at the IPAC Symp., Sydney, August 1968; also published in *Elec. Eng. Trans. Inst. Eng. Aust.*, **EE5**, No. 1, 173–186 (1969).
61. (a) R. R. Klimpel and L. G. Austin, "The Mathematical Modeling and Optimisation of an Industrial Rotary-Cutter Milling Facility," *Proc. 3rd Eur. Symp. on Comminution, Cannes; Dechema Monogr.*, **69**, 449–473 (1972).
    (b) R. R. Klimpel, "The Use of Mathematical Modeling to Evaluate Operating Alternatives in an Industrial Comminution Facility," *Proc. 1973 Int. Conf. Part. Technol., Chicago, August 1973*, IIT Research Inst. (1973).
62. L. G. Austin, D. R. Van Orden, and J. W. Pérez, "A Preliminary Analysis of Smooth Roll Crushers," *Int. J. Miner. Process.*, **6**, 321–336 (1980).
63. R. H. Snow, "Grinding Mill Simulation and Scale-Up of Ball Mills," *Proc. 1st Int. Conf. Part. Technol., Chicago, August 1973*, pp. 29–38, IIT Research Inst. (1973).
64. J. A. Herbst and D. W. Fuerstenau, "Scale-Up Procedure for Continuous Grinding Mill Design Using Population Balance Models," *Int. J. Miner. Process.*, **7**, 1–31 (1980).
65. A. W. Cameron et al., "A Detailed Assessment of Concentrator Performance at Broken Hill South Ltd.," *Proc. Australas. IMM*, No. 240, 53–67 (December 1971).
66. D. F. Kelsall et al., "The Effects of a Change from Parallel to Series Grinding at Broken Hill South," *Australas. IMM Annual Conference, Newcastle, 1972*, pp. 337–347 (1972).
67. P. S. B. Stewart, "The Control of Wet Grinding Circuits," *Aust. Chem. Process. Eng.*, **23**, 18–21 (July 1970).
68. L. G. Austin and R. R. Klimpel, "The Theory of Grinding Operations," *Ind. Eng. Chem.*, **56**, 18–29 (November 1964).
69. L. G. Austin and V. K. Bhatia, "Experimental Methods for Grinding Studies in Laboratory Mills," *Powder Technol.*, **5**, 261–266 (1971–1972).
70. L. G. Austin and V. K. Bhatia, "Note on Conversion of Discrete Size Interval Values of Breakage Parameters S and B to Point Values and vice versa," *Powder Technol.*, **7**, 107–110 (1973).
71. R. P. Gardner and K. Sukanjnajtee, "A Combined Tracer and Back-Calculation Method for Determining Particle Breakage Functions in Ball Milling, Parts I–III," *Powder Technol.*, **6**, 65–83 (1972); **7**, 169–179 (1973).
72. R. R. Klimpel and L. G. Austin, "The Back-Calculation of Specific Rates of Breakage and Non-Normalised Breakage Distribution Parameters from Batch Grinding Data," *Int. J. Miner. Process.*, **4**, 7–32 (1977).
73. K. J. Reid and P. S. B. Stewart, "An Analogue Model of Batch Grinding and Its Application to the Analysis of Grinding Results," *Proc. Chemeca '70, Session 7*, pp. 87–106, Butterworths (Australia) (1970).
74. L. G. Austin, P. T. Luckie, and B. G. Ateya, "Residence Time Distributions in Mills," *Cement Concrete Res.*, **1**, 241–256 (1971).
75. H. E. Rose and J. E. English, "Theoretical Analysis of the Performance of Jaw Crushers," *Trans. IMM (C)*, **76**, C32–C43 (1967).
76. S. P. Babu and J. W. Leonard, "Water-Flushing of Coal During Crushing," *Trans. AIME/SME*, **262**, 278–282 (1977).
77. R. B. De Diemar, "Predicting Crushing Results—The Bouncing Ball Theory," AIME/SME Preprint 77-B-305. Presented at AIME/SME Fall Meeting, 1977.
78. R. A. Stean and D. A. Hinckfuss, "Selection and Performance of Large-Diameter Ball-Mills at Bougainville Copper Ltd, Papua New Guinea," *Proc. 11th Common. Min. Metall. Congr., Hong Kong, 1978*, pp. 577–584, IMM (1979).
79. C. A. Rowland, "Applying Large Grinding Mills," presented at Pacific SW Min. Conf., San Francisco (May 1970).
80. C. A. Rowland, "Comparison of Work Indices Calculated from Operating Data with Those from Laboratory Test Data," *10th Int. Miner. Process. Congr., London, 1973*, pp. 47–61, IMM (1973).
81. C. A. Rowland and R. C. Nealey, "Experts Look at Dry Grinding Rod Mills," *Min. Eng.*, **20**, 85–90 (December 1968).
82. C. A. Rowland, "The Tools of Power Power," AIME/SME Preprint 76-B-311 (1976). Presented at AIME/SME Fall Meeting, Denver, 1976.
83. A. R. MacPherson, "A Simple Method to Predict the Autogenous Grinding Mill Requirements for Processing Ore from a New Deposit," *Trans. AIME/SME*, **262**, 236–240 (1977).
84. R. W. Smith and K. H. Lee, "A Comparison of Data from Bond Type Simulated Closed-Circuit and Batch Type Grindability Tests," *Trans. AIME/SME*, **241**, 91–101 (March 1968).

References

85. W. E. Horst and J. B. Bassarear, "Use of a Simplified Ore Grindability Technique to Evaluate Plant Performance," *Trans. AIME/SME,* **260,** 348–351 (1976).
86. D. F. Kelsall, "The Relation Between Energy Expended in Comminution and the Reduction in Particle Size Achieved," unpublished lecture notes.
87. J. A. Holmes, "A Contribution to the Study of Comminution—A Modified Form of Kick's Law," *Trans. Inst. Chem. Eng.,* **35,** 125–142 (1957).
88. V. C. Marshall (Ed.), "Comminution," I. Chem. E. (1975).
89. A. M. Gow et al., "Ball Milling," Technical Paper No. 517, *Trans. AIME,* **112,** 24ff. (1934).

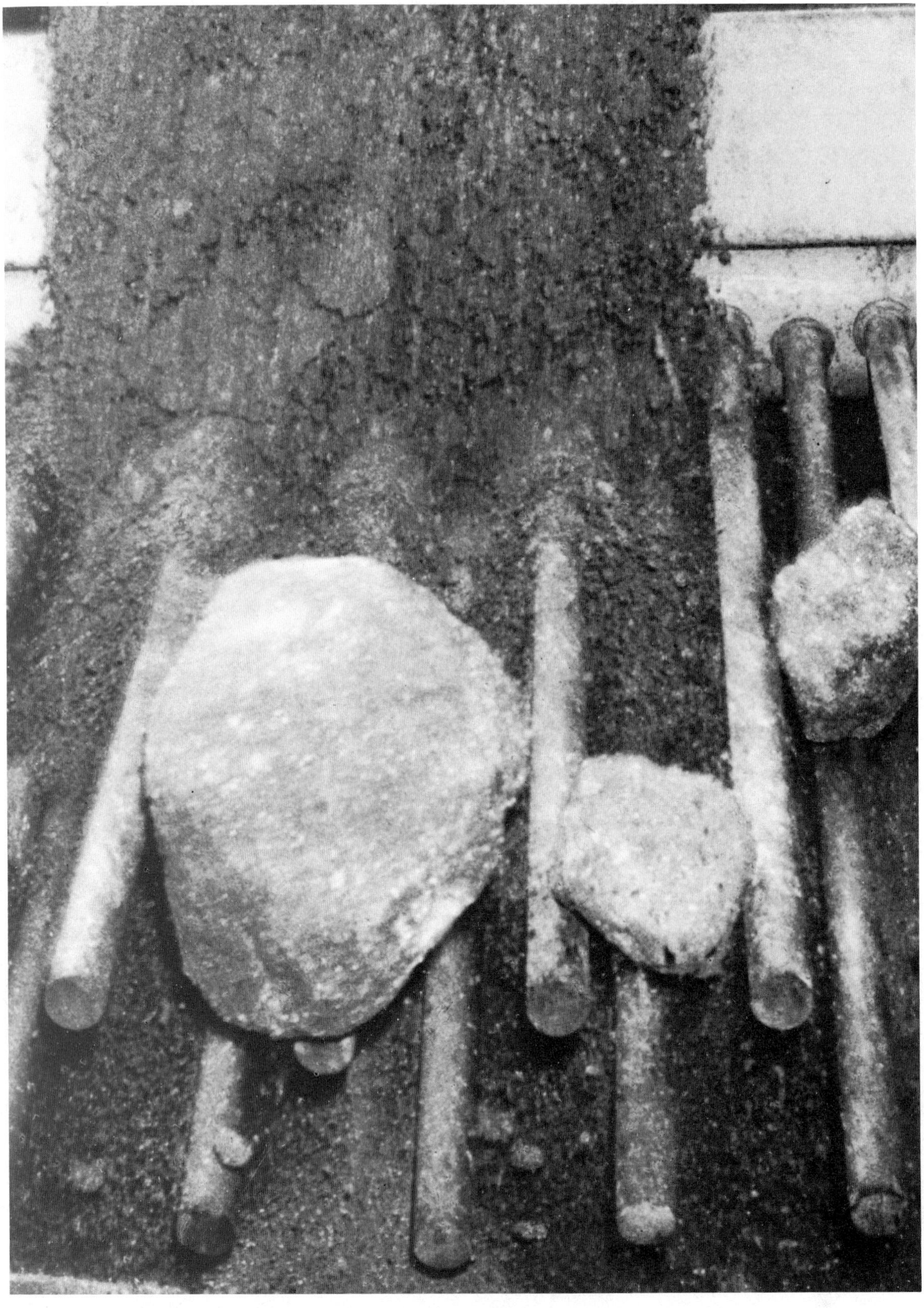

Part III

Sizing Separation

Although the production of a final product having a specific size is sometimes the function of a sizing separator, the most important application is controlling the size of material fed to other equipment. This is because all equipment has an optimum size of material that it can handle most efficiently.

There are two basic types of sizing separator; typically, screens are used for coarser separations, and classifiers for finer separations. Screens are characterized by the use of physical barriers to bring about a separation. Although there is a vast array of equipment available, few significant innovations have occurred since the basic screen was first invented. One of the few innovations is the sieve bend, developed in the 1950s. Although in principle it has advantages that should make it an attractive alternative to a wet classifier, its excessive dewatering tendency appears to have restricted its development. On the other hand, the radically different probability screens have few disadvantages: yet, in spite of their notable advantages over conventional screens, they have not received the recognition they deserve.

Although there are a number of basic types of wet classifier, the comparatively cheap and simple hydrocyclone has become the industry standard for closed circuit grinding, even though it gives relatively inefficient separations in this application. Today, the poor classification in closed circuit grinding is the area of mineral processing most urgently in need of attention. Yet surprisingly, although the benefits of more efficient classification are well established, and the cost of improvement is comparatively small, poor classification is still widely accepted.

No general theories of screening or classification exist. However, there is a growing amount of valuable literature on both subjects and this is reviewed in the two chapters that follow.

Chapter Nine

Screening and Sieving

Clopton's Law: "*For every credibility gap there is a gullibility fill.*"

Screening and sieving are mechanical separations of particles on the basis of size. Such separations are achieved using a uniformly perforated surface that acts as a multiple go/no-go gauge. Ideally, particles larger than the apertures are retained on the surface, while particles smaller (in at least two dimensions) pass through. Material retained on the surface is the *oversize* or *plus* material; that passing through is the *undersize* or *minus* material; and material passing one surface but retained on a subsequent surface is *intermediate* material.

Screening and sieving are distinguished by the fact that screening is a continuous process and is used mainly on an industrial scale, whereas sieving is a batch process used almost exclusively for test purposes. However, both processes are very similar in principle in that particles are repeatedly presented to the apertures of the screening surface until passage occurs or the process is terminated. Both processes are also typical of all separation processes in that a perfect separation is seldom achieved; in particular there are normally some potential undersize particles left in the oversize fraction.

9.1. LABORATORY OR TEST SIEVING

Many product specifications may call for definite sizes of material in terms of a given percentage passing (or greater than) a certain size, the check being determined by standard sieves.[1] These sieves are also used to determine the efficiency of screening equipment and to analyze the performance of crushing and grinding equipment. A laboratory sieve consists of a screening surface, normally of woven wire, which results in square apertures with small tolerances. Large apertures may be obtained with punched plate, and very small apertures (20–90 μm with tolerances of ± 2 μm) can be made from electroformed metal.[2] The sieve surface is mounted in a cylindrical frame that allows the sieves to be stacked on each other, the sieves normally being stacked in series with successively smaller apertures from top to bottom (Fig. 9.1). A lid and bottom pan are used to complete the stack. The sample to be analyzed is placed on the top sieve and the entire stack is shaken. After a period the sample is divided into *size fractions*, each fraction containing material that has passed higher sieves but is unable to pass the sieve it rests on.

Whether a particle passes a given aperture depends on a number of factors, which are discussed more fully in Section 9.6. If the total range of particle sizes is large compared with the difference between two successive sieve apertures, the particles retained on any sieve may be considered to have an average size given by the arithmetic mean of the apertures of the two sieves limiting the fraction. It must always be remembered that this size is purely a nominal diameter if the particles are irregular in shape; even then, it is really meaningful only if all the particles have the same general shape.

The desirability of a geometric progression of aperture sizes has already been considered in Chapter 2, and a number of standard series have been introduced to satisfy this requirement. These series differ mainly in their relationships between aperture and wire size. Although sieves have historically been designated by their *mesh* (the number of openings per linear inch of weave), the different series can be intermixed because apertures can still be selected to give an approximate geometric series. However, because aperture is the significant designation, recent practice is to specify aperture size and de-emphasize mesh designations.

Figure 9.1. The principle of laboratory sieving with a stack of sieves. (Courtesy Tyler Industrial Products).

International Test Sieve Series. The International Standards Organization has recommended an international standard series, and the U.S. series corresponds to this. Consequently sieve analyses intended for international publication should be reported in terms of the apertures of the U.S. series.

Tyler Series. This is one of the original geometric series of sieves and is still widely used. It differs from the U.S. series in that it identifies the sieves by a mesh designation rather than aperture. The series is based on a aperture 0.0029 in. (74 μm) square and a wire diameter of 0.0021 in. (53 μm). Wire diameter plus aperture equals 0.0050 in. (127 μm) so that the sieve has 200 apertures per linear inch and is known as the 200 mesh Tyler sieve. Successive sieves have apertures with a $\sqrt{2}$ ratio, although a "double series" with $\sqrt[4]{2}$ ratios is also used.

British Standard Series. These sieves are based on wire of British Standard Gauge and are adjusted within tolerances to have apertures that are interchangeable with the other series, although again mesh designations are different.

9.1.2. Sieve Shakers

There are a number of machines available for shaking stacks of sieves, and besides taking much of the tedium out of sieving, they give more consistent results. A typical machine is shown in Fig. 9.2. In

9.1.1. Sieve Series

The important sieve series are based as follows. (Actual apertures are given in Appendix C.)

U.S. Sieve Series. This series is based on a sieve having a 1 mm square aperture, with successive sieves now having apertures in a $\sqrt[4]{2}$ ratio. Sieves are designated by the aperture size, apertures over 1 mm being expressed in millimeters, those finer than 1 mm in micrometers (microns). The sieves also have an alternative arbitrary number designation, which although similar to the mesh count is not necessarily the same.

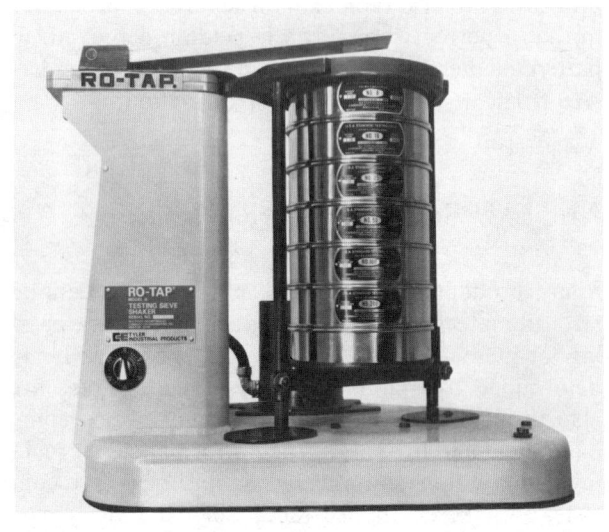

Figure 9.2. A laboratory sieve shaker. (Courtesy Tyler Industrial Products.)

general these machines transmit a consistent circular and oscillating vertical motion to the particles that ideally not only repeatedly exposes the particles to the apertures, but also minimizes *"blinding"* or blockage of the apertures by particles slightly larger than the opening. Even so blinding generally occurs, and for accurate results the amount of material on any sieve at the end of a test should not be more than one or two particles deep. (A quick calculation will show that this is often a surprisingly small amount of material.) The attainment of high accuracy is further complicated by the absence of a real end point to a sieving operation. This problem is generally "solved" by sieving for a fixed length of time, which is assumed to be sufficient. Methods have been developed for estimating true end points[3,4] (e.g., Eq. 9.10), but when this degree of accuracy is required, it may also be desirable to calibrate the apertures of the sieve, since the quoted aperture is really only a nominal value that is subject to manufacturing tolerances.

9.1.3. Hand Sieving

When a sieve shaker is not available, hand sieving must be resorted to. Although the process is very laborious, a regular procedure such as that recommended by the American Society for Testing and Materials (ASTM) should be used if meaningful results are required.[5]

9.1.4. Wet and Dry Sieving

A common problem in sieving is the adherence of very fine particles to larger particles or to each other as a result of electrostatic attraction or surface tension arising from small amounts of moisture. The best solution to this problem is initial wet sieving of a slurried sample on the finest sieve to be used, additional water being used as necessary. Both oversize and undersize fractions are recovered and dried and the former is then sieved in the normal manner.

9.2. SCREENING

Most commonly, screens are used for size separations in conjunction with crushing operations; but there are many other applications and these are described in Table 9.1,[6] with typically used equipment.[6–11] In the mineral industry screens are seldom

TABLE 9.1: TYPES OF SCREENING OPERATIONS

Operation and Description	Type of Screen
Scalping: Strictly, the removal of a small amount of oversize from a feed that is predominantly fines. Typically the removal of oversize from a feed with, approximately, a maximum of 5% oversize, and a minimum of 50% halfsize.	Coarse: grizzly. Intermediate and fine: same as used for separations
Separation, Coarse: Making a size separation at 4.75 mm and larger.	Vibrating screens, horizontal or inclined.
Separation, Intermediate: Making a size separation smaller than 4.75 mm and larger than 425 μm.	Vibrating screens, high-speed, sifter, and centrifugal screens. Static sieves.
Separation, Fine: Making a size separation smaller than 425 μm.	High-speed, sifter, and centrifugal screens. Static sieves.
Dewatering: Removal of free water from a solids-water mixture. Generally limited to 4.75 mm and larger.	Horizontal vibrating, inclined (about 10°), and centrifugal screens. Static sieves.
Trash Removal: Removal of extraneous matter from a processed material. Essentially a form of scalping operation. Screen type will depend on size range of processed material.	Vibrating screens; horizontal or inclined. Sifter and centrifugal screens. Static sieves
Other Applications: Desliming, conveying, media recovery, concentration.	Vibrating screens; horizontal or inclined. Oscillating and centrifugal screens. Static sieves.

used for separations below 0.2 mm because they have inadequate capacity; however there are now plants using sieve bends for separations as low as 50 μm, since these devices give sharper separations than can be achieved with wet classifiers.

Screening equipment can readily be classified (Fig. 9.3) as stationary or dynamic, depending on whether the screening surface is moving. Both stationary and dynamic screens can also be broadly subdivided into conventional or probability screens, depending on whether the particles passage is determined by restriction or statistical principles. Dynamic screens may in turn be further characterized according to the most significant motion imparted to the screening surface. The basic types of screens are described in Table 9.2, and Fig. 9.4 shows the range of particle sizes each is capable of treating. Some of these screens are illustrated in Fig. 9.5.

Vibrating screens are now the most widely used, and this has resulted in such a large variety of designs that this group in particular can be further classified according to:

1. The true vibrating motion of the screen surface.
2. Where the vibrating motion is applied.
3. How the vibrating motion is generated.

TABLE 9.2: THE MAJOR TYPES OF SCREENS

	Distinguishing Characteristics	Classifications	Description	Motion	Speed, Amplitude	Applications	Advantages and Disadvantages
STATIONARY GRIZZLY	Heavy duty surface of fixed bars.	Conventional (Fig. 9.5a)	Heavy bars running in flow direction, sloped to allow gravity transport. Bars may spread along length to minimise blinding.	Stationary surface. (Vibrating grizzlies also available: bar vibrating screens).		Scalping before crushers	Simple, robust. Probability form blinding resistant.
		Probability (Frontispiece Pt III)	Bars divergent in vertical plane.				
ROLL GRIZZLY	Surface of rotating rolls.		Essentially a stationary screen surface, but non-uniform shape of rolls conveys material.			Coarse separations before crushing. Primarily a conveyor.	Conveying action allows near horizontal operation in low head room situations.
SIEVE BENDS	Slurry feed, fixed bar surface.	Straight or curved surface (Fig. 9.5h)	Stationary, parallel bars at right angles to slurry flow. Surface may be straight (with steep incline) or curved to 300°.	Stationary		Separations in range 2 mm to 45 μm, or those too coarse for hydrocyclone, or where density effects make classifier unsuitable. Dewatering.	Relatively high efficiency and capacity. Sharpness of cut less than true screen. Separation slightly affected by mineral density. Excessive dewatering can be a problem.
REVOLVING SCREENS	Screen surface rotating around cylinder axis.	Trommel (Fig. 9.5c)	Slightly inclined cylindrical screen. May have concentric surfaces.	Below critical speed (c.f., ball mill).	15-20 r.p.m.	Wet or dry separations 60 to 10 mm if dry, smaller if wet.	Simple, useful for scrubbing or rough size separations. High wear, low surface utilisation.
		Centrifugal	Vertically mounted cylindrical screen; centrifuges particles through screen.	Operates above critical speed. Also has vertical action of 800-1000 cycles/min.	60-80 r.p.m.	Wet or dry separations 12 mm to 400 μm. Dewatering.	High wear.
		Probability	Particles drop through "surface" formed by bars radiating out like spokes on a wheel.	Radiating bars rotate about vertical axis. Speed of rotation determines cut size.		Developed for separating coal <6 mm.	Relatively high capacity with fine separations. Cut size easily changed and controlled by varying speed.
VIBRATING SCREENS	High speed motion, designed primarily to lift particles off surface.	Inclined (Fig. 9.5d,e,f) (Subclassified by vibrator mechanisms).	Inclined rectangular screening surface which allows material to flow with aid of vibrations.	Mechanical vibrations give circular motion at center; elsewhere it depends on vibrator (Table 9.3). Electro-magnetic vibrators may give linear vibration at center.	600-7000 r.p.m. Low <25 mm	Wide applications, generally down to 200 μm in mineral industry, but down to 38 μm in chemical industry, using the high speeds.	Relatively high efficiency and capacity, but capacity generally inadequate below 200 μm.
		Horizontal	Horizontal rectangular screening surface. Linear vibration must have horizontal component to convey material along screen.	Linear motion, with vertical component to provide lift, and horizontal component for conveying.	600-3000 r.p.m. Low <25 mm	Similar to inclined screens.	Similar to other vibrating screens, but can also be used where head room is restricted.
		Probability (Fig. 9.5g)	Series of relatively small inclined screen surfaces: separates by statistics rather than physical constraint.			Similar to inclined screens	Generally superior to conventional vibrating screen. High capacity and efficiency for given space; low noise, low power. Low efficiency at low loading.
SHAKING SCREENS	Slow linear motion essentially in plane of screen.		Usually slightly inclined. May have several surfaces in series with different apertures.	Linear motion, essentially in plane of screen; particles tend to remain in contact with screen surface.	30-800 r.p.m. 25-1000 mm	Down to 12 mm for coal preparation and non-metallic minerals. Higher speeds may size down to 250 μm.	Low headroom and power requirements. May be used for conveying and sizing. Accurate for large sizes. High maintenance cost, low capacity.
ROTARY SIFTERS	Circular motion applied to screen surface.	Reciprocating	Rectangular screen surface with slight (~5°) incline.	Circular motion is applied at feed end, and produces reciprocating motion at discharge end.	500-600 r.p.m. Low <25 mm	Generally used for finer separations (12 mm to 45 μm, wet or dry) in non-metallurgical industries.	Suitable for finer separations, but with low capacity.
		Gyrating	Circular screen surface.	Circular motion over most of the screen surface.			
		Gyrating	Circular screen surface.	Screen moves with circular motion, but also has oscillating vertical component.			

Screen Surfaces

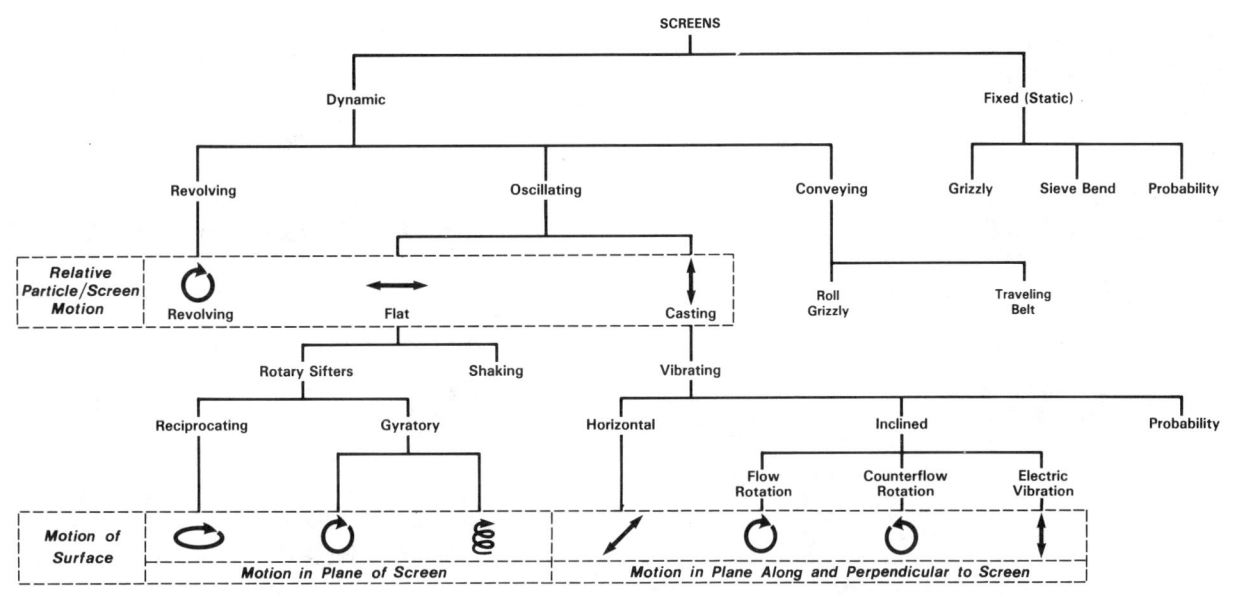

Figure 9.3. The basic screen types and their classification.

4. The nature of the screening surface.
5. How the screen is supported.

Items 1–3, as well as the resulting effects, are summarized in Table 9.3; screening surfaces are discussed in more detail in Section 9.3. Two basic types of support are used: frame and hanging, the former being essential for heavy duty applications. Regardless of the type of support, it must be designed to minimize the transmission of vibration to the building. This requires the use of springs or rubber mounts, although most of the damping is actually achieved by counterbalancing.

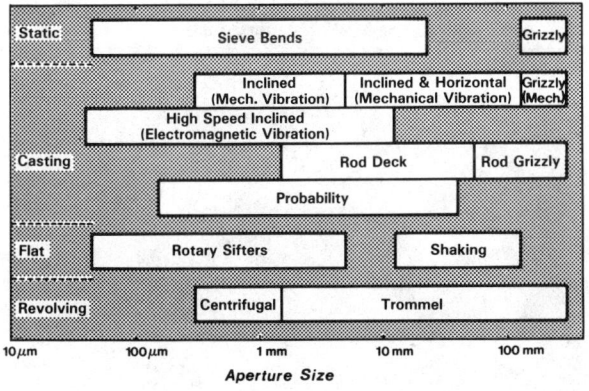

Figure 9.4. Typical separation sizes of the basic screen types.

9.3. SCREEN SURFACES

The screen surface is the medium that contains the apertures for the passage of undersized material. Selection of such a surface appears to be deceptively simple. The surface must be strong enough to support the weight of material being screened, yet flexible enough to yield to the vibrating forces applied to it, and light enough to provide a reasonable percentage of open area to give a practical throughput. The problem is further complicated because screening surfaces are made in a variety of types, shapes, sizes, and materials, often with subtle variations arising from slightly different manufacturing processes. Thus, the catalogues of screen surface manufacturers should be consulted for the vast amount of information they contain. Basically every screening situation is unique, and final selection is always a matter of trial and error.

Three basic types of surface can be distinguished[12–14]: punched or perforated "plate," woven "cloth," and profile "bars." Although there have been many relatively recent developments in these three types, woven cloth surfaces still account for over 75% of all sales.

9.3.1. Perforated Plate

The perforated plate surface is made by punching apertures in steel plate; Table 9.4 illustrates the vara

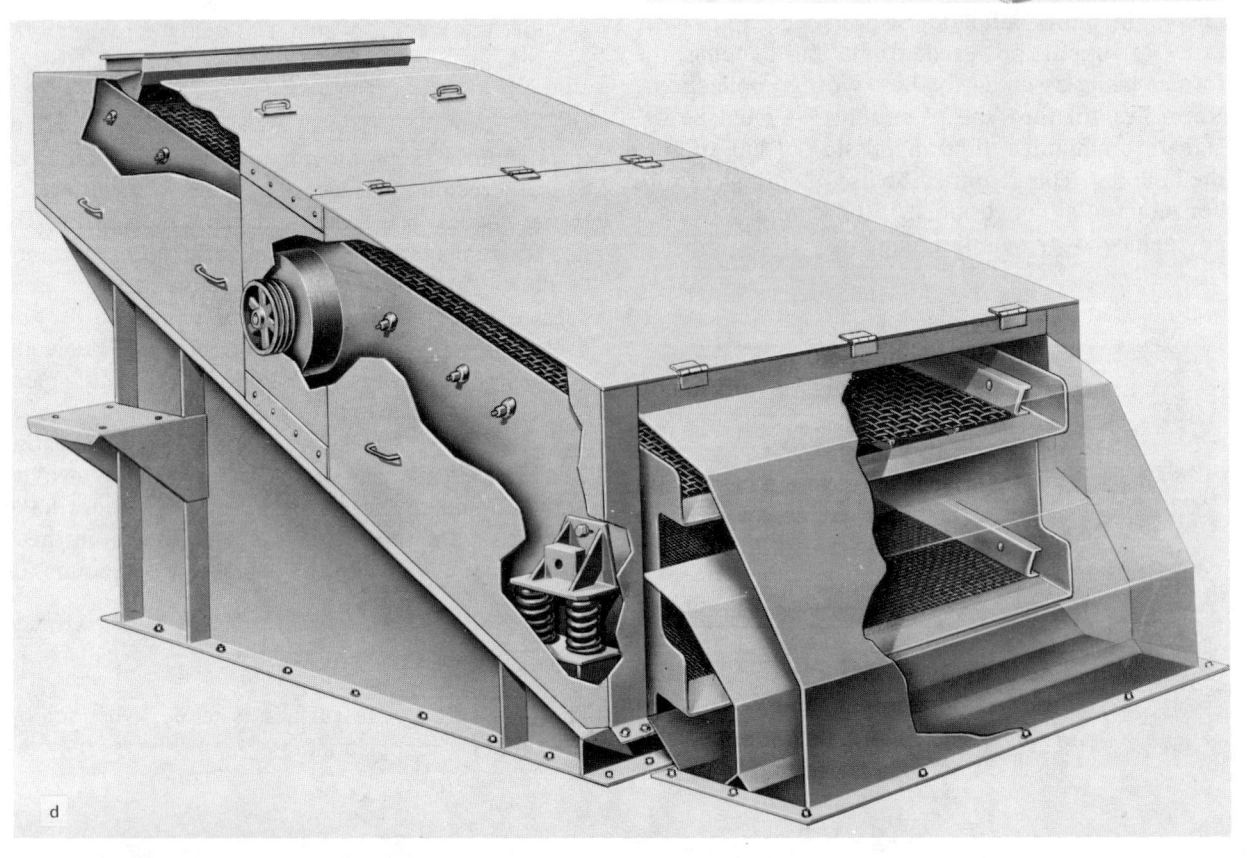

b

c

d

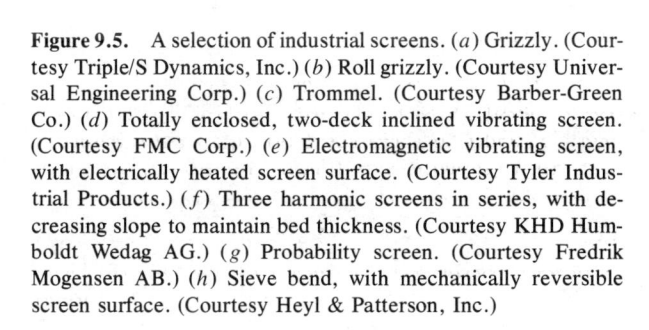

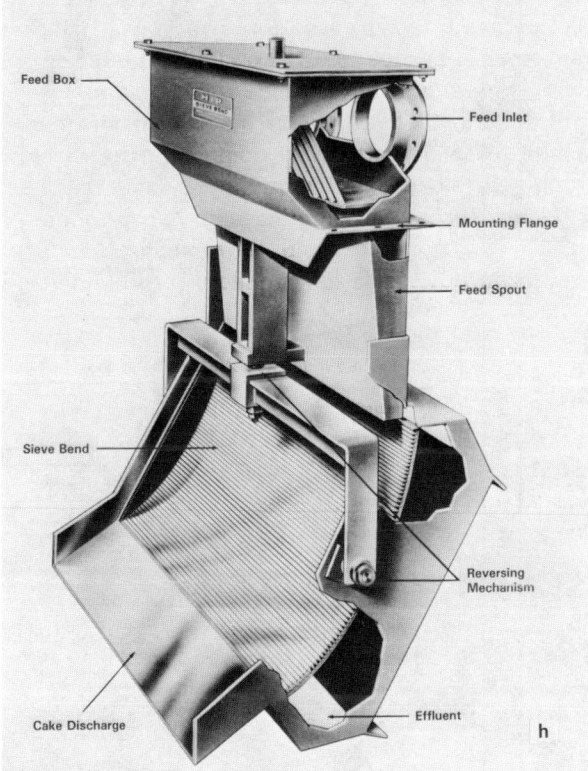

Figure 9.5. A selection of industrial screens. (*a*) Grizzly. (Courtesy Triple/S Dynamics, Inc.) (*b*) Roll grizzly. (Courtesy Universal Engineering Corp.) (*c*) Trommel. (Courtesy Barber-Green Co.) (*d*) Totally enclosed, two-deck inclined vibrating screen. (Courtesy FMC Corp.) (*e*) Electromagnetic vibrating screen, with electrically heated screen surface. (Courtesy Tyler Industrial Products.) (*f*) Three harmonic screens in series, with decreasing slope to maintain bed thickness. (Courtesy KHD Humboldt Wedag AG.) (*g*) Probability screen. (Courtesy Fredrik Mogensen AB.) (*h*) Sieve bend, with mechanically reversible screen surface. (Courtesy Heyl & Patterson, Inc.)

TABLE 9.3: VIBRATING SCREEN MOTIONS

		Characteristics	Applications
VIBRATOR TYPES		Unbalanced pulley type. One concentric shaft with adjustable counterweights and two bearings. Circle-throw motion produces an oscillating vibration. Stroke may be varied by adjusting the counterweights. Frequencies 500-2500 r.p.m.; stroke <10 mm.	Generally used on light duty screens.
		Eccentric shaft type. One eccentric shaft with adjustable counterweights and 2 bearings. (Commonly designated as "2-bearing"). Circle-throw motion produces vibration. Stroke may be varied by adjusting the counterweights. Frequencies 25-500 r.p.m.; stroke, 15-30 mm.	Used on light and heavy-duty inclined vibrating screens.
		"Positive-stroke" or "4-bearing" type. One double-eccentric shaft with 2 sets of bearings; one set supports screen frame the other the shaft. Produces a positive motion that is not dampened by load on screen deck. Shaft is generally on center-of-gravity of screen box. Stroke cannot be varied except by changing shaft.	Used on heavy-duty inclined vibrating screens since it is fairly resistant to overloading.
		Reciprocating type (also commonly designated as "4-bearing"). Two shafts, eccentric or weighted, counter-rotating in phase to produce a positive straight-line motion. Stroke can be inclined by operating slightly out of phase.	Used on horizontal vibrating screens and some conveyors.
		Electromagnetic vibrator. Provides very high oscilation frequencies (1500-7200 r.p.m.)	Used on steeply inclined screens for relatively fine size separations.
VIBRATOR MOUNTING		Center mounting vibrator mounted centrally between side frames produces circular motion. Rotation of vibrator may also be "flow" or "counterflow". Flow gives higher capacity, lower efficiency. Counter flow gives *vice versa*.	Generally used for heavy-duty screens of inclined type.
		Top mounting, flow rotation. Vibrator mounted on top of frame produces elliptical motion at the ends. Flow rotation moves material down the screen faster, increasing capacity and lowering efficiency.	Used for rough screening where high rate of feed is needed.
		Top mounting, counterflow rotation. Vibrator mounted on top of frame produces elliptical motion. Counterflow rotation holds material on screen longer, gives deeper bed, reduces capacity, and raises efficiency.	Used for more efficient size separations.
		Reciprocating, inclined vibrator. Vibrator mounted above (or below) frame with slight inclination of axis using positive straight-line motion to move material along the screen surface.	Used on horizontal screens for; close separation of medium sized material, dewatering, media recovery, limited headroom situations.
		Reciprocating, unphased vibrator. Vibrator mounted above (or below) frame. Straight-line motion is obtained by setting one eccentric to lead the other. Phase adjustment determines the angle of inclination of straight-line force.	
		Resonance vibration. By means of an eccentric drive shaft, screen mountings vibrate between rubber buffers at a resonant frequency, significantly reducing power consumption.	Horizontal screens.

Screen Surfaces

TABLE 9.4: SCREEN SURFACES

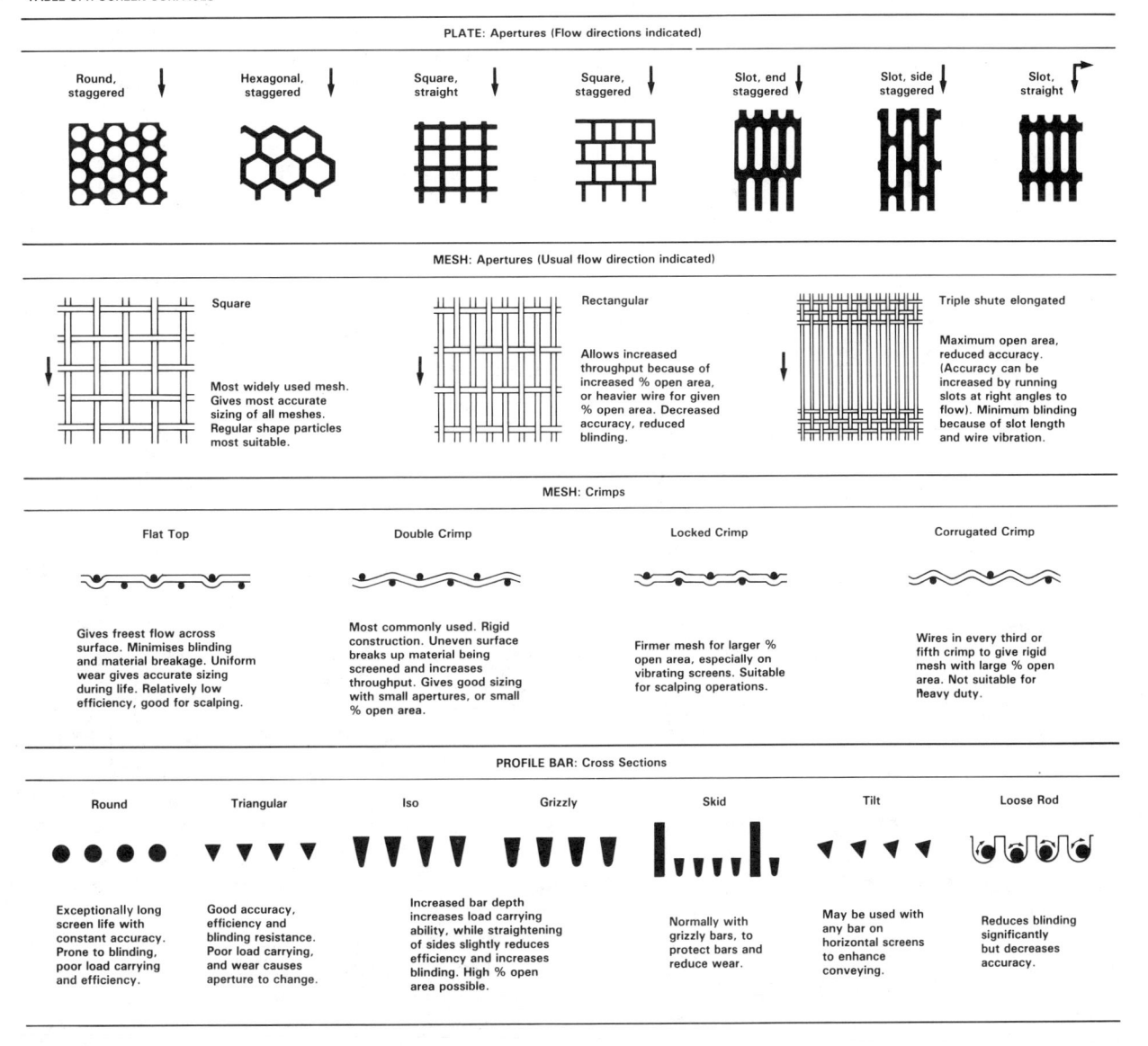

PLATE: Apertures (Flow directions indicated)

Round, staggered	Hexagonal, staggered	Square, straight	Square, staggered	Slot, end staggered	Slot, side staggered	Slot, straight

MESH: Apertures (Usual flow direction indicated)

Square — Most widely used mesh. Gives most accurate sizing of all meshes. Regular shape particles most suitable.

Rectangular — Allows increased throughput because of increased % open area, or heavier wire for given % open area. Decreased accuracy, reduced blinding.

Triple shute elongated — Maximum open area, reduced accuracy. (Accuracy can be increased by running slots at right angles to flow). Minimum blinding because of slot length and wire vibration.

MESH: Crimps

Flat Top — Gives freest flow across surface. Minimises blinding and material breakage. Uniform wear gives accurate sizing during life. Relatively low efficiency, good for scalping.

Double Crimp — Most commonly used. Rigid construction. Uneven surface breaks up material being screened and increases throughput. Gives good sizing with small apertures, or small % open area.

Locked Crimp — Firmer mesh for larger % open area, especially on vibrating screens. Suitable for scalping operations.

Corrugated Crimp — Wires in every third or fifth crimp to give rigid mesh with large % open area. Not suitable for heavy duty.

PROFILE BAR: Cross Sections

Round — Exceptionally long screen life with constant accuracy. Prone to blinding, poor load carrying and efficiency.

Triangular — Good accuracy, efficiency and blinding resistance. Poor load carrying, and wear causes aperture to change.

Iso — Increased bar depth increases load carrying ability, while straightening of sides slightly reduces efficiency and increases blinding. High % open area possible.

Grizzly — (see Iso)

Skid — Normally with grizzly bars, to protect bars and reduce wear.

Tilt — May be used with any bar on horizontal screens to enhance conveying.

Loose Rod — Reduces blinding significantly but decreases accuracy.

---

iety of apertures that are available. Perforated plate generally wears longer than woven wire cloth, and is stronger and more rigid. Since, however, it is heavier and has a lower capacity than woven wire, perforated plate is generally restricted to coarser separations where surface loadings are high.

Perforated plate can be obtained in a variety of materials, including hardened steels, stainless steels, and special corrosion-resistant materials such as Monel.

Rubber surfaces[15-17] offer a number of potential advantages over metal: in particular longer life, because rubber is only temporarily deformed under load; lower blinding, because of the elasticity of the aperture (which may also be tapered outward from the top); significant noise abatement; and protection of soft materials during screening. In some cases, the springier surface may allow a reduction in the oscillation rate, with a consequent improvement in bearing life. Disadvantages are a limited resistance to high temperatures, and a smaller percentage of open area in the finer size ranges. These surfaces were originally made by bonding rubber to the top of steel plate, but modern practice is to use two layers of rubber of differing hardness. The initial cost of rubber may be two to four times that of steel,

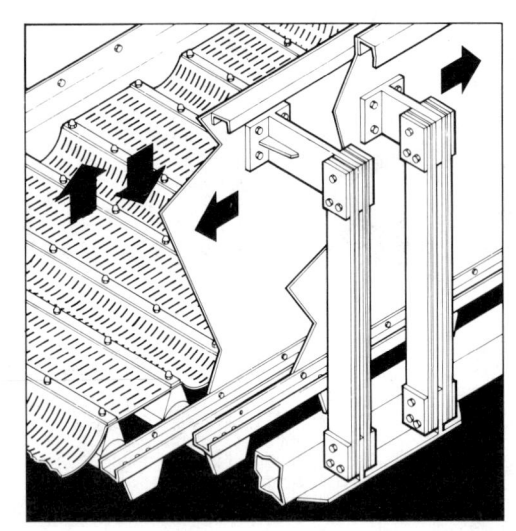

Figure 9.6. Flipflow screen surface. (Courtesy Krebs Engineers.)

but the potentially longer life may result in a significantly lower overall cost.

Perforated plastic is used in the Flipflow screen, which can be used to treat abrasive, wet, or sticky materials[11,13] (Fig. 9.6).

9.3.2. Woven Wire Cloth

Large sheets of woven metal wire are available in thousands of specifications covering various aperture-to-wire ratios, weaves, aperture shapes, wire shapes, wire materials, and edgings.

Weave can be classified as precrimped or plane. In the first type wire is shaped in a crimping machine to produce the desired crimp or bend, then woven into cloth on a loom. Typically, this method is used for apertures greater than 1.5 mm. Three of the most common crimps are described in Table 9.4. To distinguish the two wire directions, the wires running along the screen are termed the *warp* wires; those across the screen are the *shute* or *fill* wires.

Apertures in woven cloth are most commonly square, but rectangular apertures are sometimes used, and some typical forms are described in Table 9.4.

Although woven cloth is normally made from round wire, a limited amount is made in square wire, which is claimed to give increased wear. Such a cloth can be an alternative to punched steel plate in some heavy duty operations.

Three properties are of importance when selecting a mesh material: abrasion, fatigue, and corrosion. High carbon steels give satisfactory performance in many applications particularly with large apertures, although tempered steels may be necessary to counter high abrasion. With some crimps wear is more significant on the upstream side of the wire, and longer life can often be obtained by turning the surface end for end. In cases of extreme impact, manganese steels are superior, but it must be remembered that these steels have to work harden in service to gain maximum benefits. Stainless steel is often preferable in corrosive environments such as wet screening or fine screening below 850 μm. In some situations satisfactory corrosion resistance may be achieved with galvanized steel, but others may require more expensive materials such as phosphor bronze or Monel. Cloths are also available in natural and synthetic fabrics, but these are normally incapable of withstanding the rigors of mineral processing operations. Recently a number of manufacturers have started offering reinforced synthetic cloths of polyurethane rubber.[18] Although initially more expensive, they are guaranteed to give lower overall operating costs than conventional cloths, due to longer life and reduced blinding. Noise levels are also lower.

The type of edge preparation (or hook strip) put on the shute edges of the screen surface to facilitate mounting depends essentially on the diameter of the wire. Cloth having wires smaller than about 2.5 mm will be provided with metal reinforcement on flanges to prevent damage while tensioning during installation. Larger wires between about 2.5 and 6 mm may be hooked only, while the largest wires (>6 mm) may have a steel bar, angle, or hook welded to the straight end of the shute wires.

Finer cloths benefit from the use of additional support underneath, either in the form of cross bars or a coarse mesh screen (preferably smooth top crimp). With any screen surface, it is good practice to have the surface divided into sections, so that only part of the surface has to be replaced when failure occurs.

9.3.3. Profile Bar Surfaces

The profile bar surface is also known as the formed semipermanent, or rod-deck surface.[11-14,19] The multiple terminology is the result of this type of surface being used for both static and dynamic screens. These surfaces are made up of parallel bars anchored to heavier cross members. Some of the large variety of individual surfaces available for dy-

Auxiliary Features and Equipment

namic screens are illustrated in Table 9.4. Profile bar surfaces are characterized by a smooth top and resistance to blinding, so that they are particularly suited to three situations:

1. Washing and/or dewatering of fine sizes.
2. Sizing where blinding is a problem.
3. Treatment of abnormally shaped particles such as slivers or slabs.

The classic example of this type of surface, used primarily for heavy duty scalping operations, is the grizzly (Fig. 9.5a). Originally only a static screen, it is also available today as a vibrating screen. Almost at the other end of the scale is a comparatively recent (dynamic) screen surface made up of stretched high tensile steel piano wire.[13] The sieve bend (Fig. 9.5h) is another special static screen, suitable for separating fine particles that would otherwise blind a conventional screen. Sizing with this device however is determined not so much by the aperture between the bars as by the apparent vertical displacement of bars, which causes a part of the slurry stream to be bled off (Fig. 9.7). (In reality a curved surface is not essential to the operation of these devices, and they can operate with a straight, sloping surface.) Typically, particles passing through the surface are less than half the aperture. By adding rappers[20] or simpler crimps[21] to the back of the screen surface, these surfaces can be used successfully for separations down to 45 μm.

9.3.4. Probability Surfaces

An increasing variety of designs of probability screens is becoming available as the advantages of this approach are appreciated. Although the equipment often appears distinctly different in design (Part III frontispiece; Fig. 9.5g), the operating mechanism is basically the same as that of a conventional screen, and so are the screen surfaces—that is, woven mesh[22] or bars (either fixed[23] or rotating[11,24]).

9.4. AUXILIARY FEATURES AND EQUIPMENT[25]

Feed chutes are an important part of any screening unit. They should be designed and fitted so that the material enters the screen surface along its longitudinal axis; any vertical drops should be kept to the absolute minimum. Furthermore, the material should enter the screen surface evenly spread across the full width. These factors will ensure maximum utilization of the screen surface and action.

True "wet" screening, where the particles are immersed in water to prevent them from sticking together, is seldom practical because screening action is suppressed. Repulping troughs (Fig. 9.8a) are one alternative; a second is to use a series of sprays (Fig. 9.8b), although this method requires considerable amounts of water and may necessitate water recovery systems. These wet screening situations where water is a screening aid, should not be confused with an operation such as the sieve bend, where the water is essential for slurry transport and the sizing action.

If dust is a problem during screening, there are two possible solutions: wet screening or dust covers, the latter being used where wet screening is impractical. Most manufacturers are able to supply fully enclosed screens, which can if necessary be connected to dust extraction systems (Fig. 9.5d).

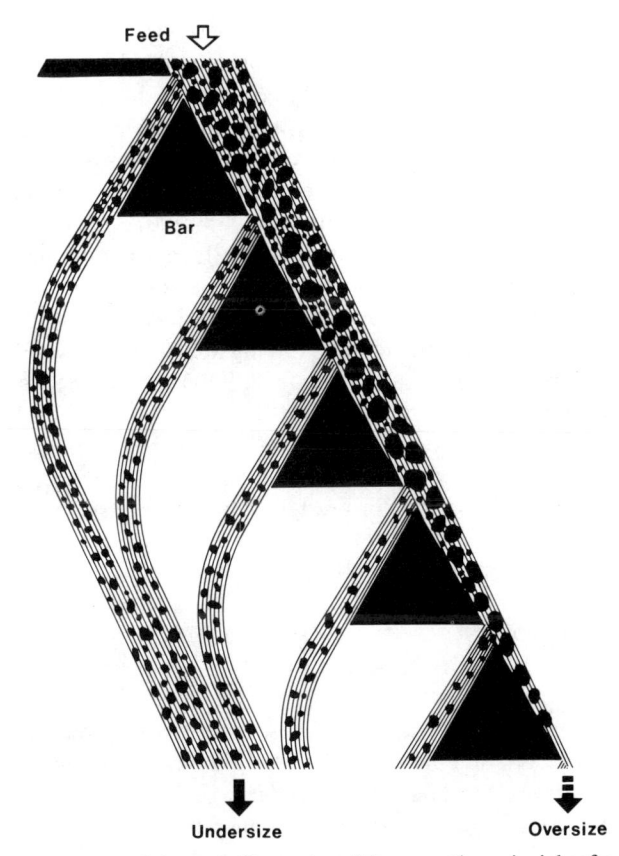

Figure 9.7. Schematic illustration of the operating principle of a sieve bend. (Courtesy KHD Humboldt Wedag AG.)

Figure 9.8. (*a*) Repulping troughs. (Courtesy Tyler Industrial Products.) (*b*) Wet screening sprays. (Courtesy *Pit and Quarry*.)

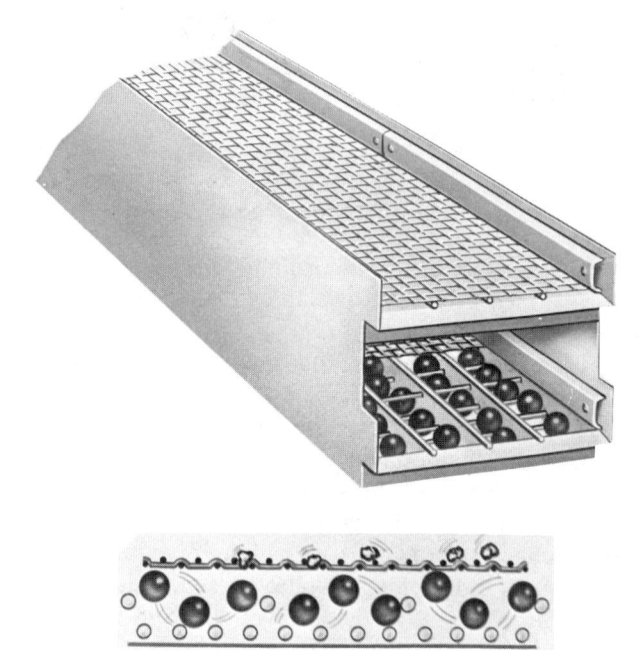

Figure 9.9. The nature and operation of a ball-deck screen. (Courtesy FMC Corp.)

Existing screens can be modified to take sprays, and even simple dust covers can be effective in lowering dust levels: but care must be taken because any additional weight may have a significant effect on the screen motion.

A common problem with screens is the "blinding" or "pegging" of the screen apertures with particles that are just slightly oversize. The problem tends to become more persistent as aperture size decreases, and it can result in a significant reduction in capacity. Blinding can often be minimized by correct screen motion[26] or a suitable surface material, but supplementary methods may be necessary in many cases. If the problem results from small amounts of moisture making the material sticky, electrically heated screen cloths can be employed. This method may double the capital cost of the screen; however, operating costs may decrease because of longer surface life.[27,28] Alternatively, a gas flame underneath and parallel to the screen surface can be used, but there are obvious difficulties with this method. With screens having apertures between 0.5 and 5.0 mm, ball decks (Fig. 9.9) are suitable for cases of severe blinding.

Although the foregoing methods increase capacity, they generally do not raise the efficiency as such because fine particles are left sticking to larger particles.

9.5. MEASUREMENT OF SCREEN PERFORMANCE

Two criteria are used to assess screen performance: capacity and efficiency. Capacity is easy to define; it is simply the quantity of material fed to the screen per unit time (strictly, per unit area of screen surface). In reality "capacity" is a meaningless parameter unless efficiency is also stated. As is typical of separation processes, capacity and efficiency are generally conflicting requirements. Any screen can have its capacity increased, but this is likely to be achieved at the expense of efficiency.

Although one intuitively expects a screen to produce a perfect separation, this is not the case (as explained further in Section 9.7). It therefore becomes necessary to express the efficiency of a screening separation. A number of methods can be used, but the best is the use of the performance curves described in Chapter 3. Because the separability curves of oversize and undersize sets (Fig. 3.5) have no overlap (which may not be strictly true if, e.g., the material is heterogeneous in composition or shape), the performance curve is essentially a measure of machine performance only. Unlike the reduced performance curves of classifiers and some gravity separation devices, the reduced performance curves of screens *are* dependent on machine operation as well as on design.

Because performance curves are so dependent on operating conditions, simpler (but far less informative) measures of efficiency are commonly used in industry. Two methods are available, depending on whether one is interesting in removing oversize or undersize material. For the latter, efficiency $\eta_{u/s}$ is defined as

$$\eta_{u/s} = \frac{\text{mass \% of feed reporting to the undersize product}}{\text{mass fraction of true undersize in feed (determined by sieving)}} \quad (9.1)$$

More commonly, the oversize is important, as for example in closed circuit crushing, and in this case the efficiency $\eta_{o/s}$ can be defined as

$$\eta_{o/s} = \frac{\text{mass \% of true oversize in feed (determined by sieving)}}{\text{mass fraction of feed left in the oversize product}} \quad (9.2)$$

These conditions are based essentially on the amount of *misplaced* material that, provided the screen surface has no breaks in it, should be only undersize particles. The limitation of these definitions is that they give no consideration to the size of the misplaced particles relative to the aperture size. This is significant because it is the particles whose size is close to that of the aperture (*near-mesh particles*) that cause the problems in screening, by being difficult to pass or by blinding the screen surface.

9.6. FACTORS AFFECTING SCREENING

A number of factors determine the rate at which particles pass through a screen surface, and as shown in Table 9.5 they can be divided into two groups: those related to particle properties, and those dependent on the machine and its operation.[29,30] In reality, many of the factors are interdependent and cannot really be considered in isolation. For example, the rate at which material travels over the screen surface is also important, since this determines the bed depth and the residence time. It in fact depends on machine characteristics such as angle of inclination, amplitude, frequency, and type of screen surface, as well as such particle properties as shape and feed size distribution.

Much of the published information on sieve and screen performance is empirical, and although data are sometimes conflicting, the relative significance of most parameters is indicated.

9.6.1. Relative Size of Particle and Aperture

One of the most important factors in screening and sieving is the size of particles relative to the aperture. Whether the particle will pass through the aperture is primarily a statistical problem. In the simplest situation the probability of passage in one attempt is the ratio of the area available for passage to the total area on which the particle can fall. For a spherical particle falling vertically to a square aperture, Gaudin[31] showed that the probability of passage p is given by

$$p = \left(\frac{D_A - d_0}{D_A + D_W}\right)^2 \quad (9.3)$$

(D_A = length of side of aperture; D_W = wire diameter). If account is taken of the fact that the particles

TABLE 9.5: FACTORS AFFECTING SCREENING

Material Factors	Machine Factors
Bulk density	Screen surface: area
	percentage of open area
Shape of size distribution curve	size of aperture
	shape of aperture
	thickness
Shape of particles	
	Vibration: amplitude
Surface moisture	frequency
	direction
	Angle of inclination
	Method of feeding screen

can pass after deflection off a round wire, then

$$p = \frac{[(D_A + D_W) - (D_W + d_0)\cos\theta_p]^2}{(D_A + D_W)^2} \quad (9.4)$$

[$\cos\theta_p$ = the angle above the horizontal at which the particle deflects from the wire and is deflected against the equator of the wire opposite and is given by

$$\cos\theta_p = \frac{1 + \sqrt{1 + 8K^2}}{4K}$$

where

$$K = \frac{2D_A + D_W - d_0}{D_W + d_0} \quad]$$

Mogensen[22] has produced a relationship to allow for the inclination θ of the screen surface:

$$p = \frac{(D_A + \phi D_W - d_0)[(D_A + D_W)\cos\theta - (1 - \phi)D_W - d_0]}{(D_A + D_W)[(D_A + D_W)\cos\theta]} \quad (9.5)$$

(ϕ = a function that decreases with increasing d_0/D_A; as an approximation it may be assumed to be 0.2, 0.15, 0.10, and 0.05 when d_0/D_A is 0.3, 0.4, 0.6, and 0.8 respectively). However, even such rigorous relationships as this do not include other practical aspects, such as the inability of particles to act independently (Section 9.7).

In spite of their limitations, these equations can be used to illustrate two significant features. Consider what happens when the probability of passage is the same for *all* attempts at passage for *all* particles. The total mass of undersize M can be substituted for m in Eq. 3.3, and the fraction of undersize remaining M/M_I on the screen after n attempts is given by

$$\frac{M}{M_I} = (1 - p)^n \quad (9.6)$$

This equation can be combined with Eq. 9.4 (or Eq. 9.3). Table 9.6 shows some representative figures.[31] The case of 10 attempts can be considered to represent industrial screening, and that of 1000 attempts test sieving. The data illustrate the two significant features: that in practice any separation is unlikely to be complete, and that the likelihood of passage decreases rapidly as the particle size approaches the aperture size.

It follows that the particle size distribution is also significant, since the number median size (d_{A1m}) and the range of size determines the number of near-submesh particles. If this number is high, a long time will be required for a significant proportion of them to pass through the surface. Hence a rule of thumb is not to try to screen near the median size. When it is necessary, oversize apertures are particularly advisable.

TABLE 9.6: PROBABILITY OF SPHERICAL PARTICLE PASSING THROUGH A SQUARE APERTURE

	Probability of Passage, %			
	10 attempts		1,000 attempts	
$\frac{d_0}{D_A}$	$D_W = D_A$	$D_W = \frac{D_A}{4}$	$D_W = D_A$	$D_W = \frac{D_A}{4}$
0.0	99.0	100.0	100.0	100.0
0.1	97.5	100.0	100.0	100.0
0.2	94.8	99.9	100.0	100.0
0.4	83.9	98.8	100.0	100.0
0.6	57.0	86.2	100.0	100.0
0.8	20.8	43.0	100.0	100.0
0.9	6.3	14.5	99.8	100.0
0.95	1.8	4.1	84.3	98.5
0.99	0.1	0.2	7.2	16.5
1.00	0	0	0	0

(After Gaudin.[31])

9.6.2. Screen Surface

Percentage of Open Area. The basis for choosing a specific aperture size is obviously the particle size that needs to be separated. More difficult is the selection of wire diameter (or bar size). The smaller this diameter, the greater will be the percentage of the surface available for particle passage and thus the greater will be the capacity per unit area. Furthermore, the particles will have a greater chance of passage at each presentation, resulting in higher efficiency. A thicker wire on the other hand is more resistant to breakage and stretching, and therefore will give longer life. In turn the larger wire surface also is more susceptible to the buildup of wet sticky material or fines. This and the more rigid mesh can lead to more blinding, with a further reduction in open area. Again, the size distribution is also important, since this determines the proportion of material which is 1–1½ times the aperture size, the primary cause of blinding.

It follows that the selection of a surface is essentially a matter of the surface that produces a given product at the cheapest overall cost. In practice, open area can range from 20 to 80%.

Aperture Shape. Although square apertures are the most widely used because of their ease of construction, other shapes are sometimes justified. All types of screen surface can be obtained with rectangular apertures or slots, which offer the advantage of increased open area, less susceptibility to blinding, and easier passage of elongated particles. Furthermore, the probability of passage through a rectangular aperture is higher. These advantages carry the penalty of slightly less accurate sizing.

The orientation of rectangular slots is relevant. Slots running parallel to the axis show the greatest wear but maximize capacity and the passage of elongated particles, which naturally orient themselves along the screen axis. On the other hand, slots running across the screen still give higher capacity than square apertures while giving a better balance between capacity and wear.

With punched plate it is possible to use circular apertures. These provide the most accurate sizing, although to pass a given size of particle a round aperture has to be 10–40% larger (linearly) than a square aperture. As well as providing regular opportunity for passage because the particle cannot ride the bar, staggering of round apertures also results in the highest fraction of open area. The concept of staggered apertures is practical and beneficial with any punched surface.

Aperture and Wire Diameter. Since the capacity is approximately proportional to aperture, the separating size determines the capacity of the screen. This relationship normally results in oscillating screens with apertures less than 0.2 mm having insufficient capacity to be used industrially, although fixed screens such as sieve bends are successfully used below this size, particularly for coarse closed circuit grinding.

Increased wire diameter or bar depth means that the particles have further to fall through the screen surface, and this also lowers the capacity.

Size of the Screen Surface. Capacity is almost directly proportional to screen width. Increased length provides more chance for passage, primarily increasing the efficiency but only slightly increasing the capacity,[32] as illustrated in Fig. 9.10. Experience shows that screen length should be two to three times the width, but departures from this are used in special situations such as restricted space.

Angle of Inclination. As the slope of a screen is increased, the aperture is effectively reduced by the cosine of the angle of inclination. At the same time,

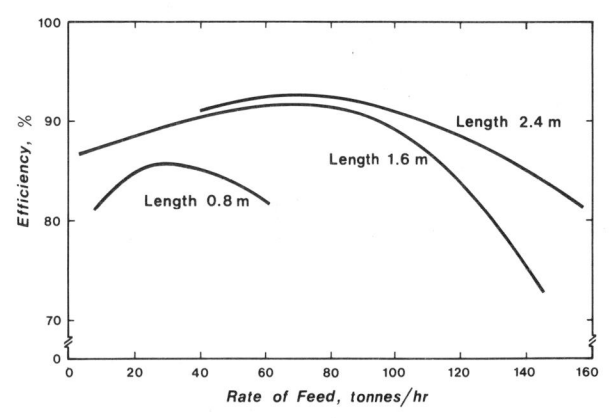

Figure 9.10. The effect of feed rate and length on screening efficiency (as given by Eq. 9.1 or 9.2). (After Gluck.[32])

the material moves across the screen faster and there are indications that stratification is also more rapid.[33] Initially, therefore, the bed is thin, causing both efficiency and capacity to rise; but eventually the decreasing aperture and low residence time result in a decrease in efficiency. Most screens therefore operate best at angles of 12–18°, although some screens are designed to operate at very low angles and some electromagnetically vibrated screens may have inclinations up to 35° (Fig. 9.5e).

9.6.3. Screen Movement

The primary purpose of screen movement is to repeatedly present the particles to the apertures for passage; but either directly, or with the aid of gravity, the screen movement also affects the transport of material along the surface. Screens can be classified, by their movement, as casting, flat, or revolving screens (Fig. 9.3).

With casting screens, a number of factors contribute to maximum efficiency[34]: the particles should be ejected from and return to the screen surface as steeply as possible; they move forward one-half to one aperture per vibration; the aperture and frequency must be high enough to prevent blinding, but not so high that the number of presentations is decreased; and the maximum height of the particle trajectory should occur when the screen surface is at its lowest point. From this it follows that there is an optimum frequency and amplitude of vibration[35] (Fig. 9.11). Furthermore, the frequency of vibration must decrease and the amplitude must increase as the aperture size increases (Fig. 9.12).

Movement and slope interact to affect the transportation of material along, or the capacity of, a

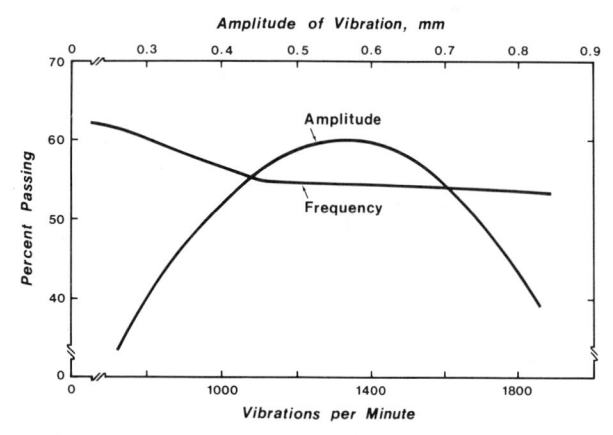

Figure 9.11. Dependence of screening efficiency on the amplitude and frequency of vibration. (After Dalla Valle[35]).

casting screen. When movement is linear, a component exists down the slope, and the screen can have a relatively low slope. In the case of circular vibrations, the slope has to be steeper. By having the circular motion counterrotating, even steeper slopes are required to prevent blinding but the particles are retained on the screen longer, thus raising efficiency at the expense of capacity.

Although it is often desirable to change the load, frequency, or amplitude of a vibrating screen, this has a significant effect on bearing life, which is proportional to

$$\left(\frac{1}{\text{rpm}}\right)^{7.66} \times \left(\frac{1}{\text{load}}\right)^{3.33} \times \left(\frac{1}{\text{amplitude}}\right)^{3.33} \quad (9.7)$$

With flat screens there is a minimum relative acceleration [amplitude × (frequency)^2/g] before relative motion commences between the particles and the surface. A further increase in acceleration leads to a rapid increase in the rate of screening to a maximum, which is followed by a falloff. At low speeds screening efficiency is high, but blinding is severe. Blinding decreases at high speeds, but efficiency is low because the particles have difficulty falling through the rapidly moving apertures.[34]

Revolving screens show, up to a limiting velocity, increased capacity with increased rotational speed: the limit depending on D_A/d_p and D_A. Beyond it, capacity falls off, because of the onset of blinding.

9.6.4. Depth of Bed

Although an exact depth is hard to specify, it is apparent that there is an optimum bed depth that produces the maximum screening rate (of passage). It is generally accepted that there are three distinct flow regions as the particle loading is increased.[34] Region I is characterized by insufficient particles to form a monolayer on the screen. This results in a low flow rate because the particles tend to have excessive unrestrained motion, and because the full screen surface is not being fully utilized. In Region II, there is at least a monolayer of particles over the surface and the maximum flow rate occurs. Under these conditions the closeness of the particles tends to restrain the bouncing action and potential passing material has maximum exposure to the screen apertures. The lowest flow rate occurs in region III, that is, when the bed is excessively deep and presentation for passage is severely restricted, not only because the quantity of oversize material is high, but because of insufficient segregation. Segregation or stratification is a phenomenon whereby vibration in particular causes small particles to work their way to the bottom of the bed while large particles rise to the top (Section 5.6). It is an especially important aspect in industrial screening, where the screen movement must encourage it.[33] (One of the major drawbacks of the slower moving shaking screens is that they have insufficient vibration to promote good stratification.) The development of these regions on an industrial screen is illustrated in Fig. 9.13. As a general indication, the feed thickness should not exceed four times the size of the aperture for material having a bulk density of 1600 kg/m^3 (100 lb/ft^3), or 2½–3 times the aperture for material with a bulk density of 800 kg/m^3 (50 lb/ft^3).

Because the most efficient screening occurs in region II, it is desirable to maintain this region over as much of the surface as is possible. One way of achieving this is to decrease the slope along the screen surface[36] (Fig. 9.5f). This reduces the flow

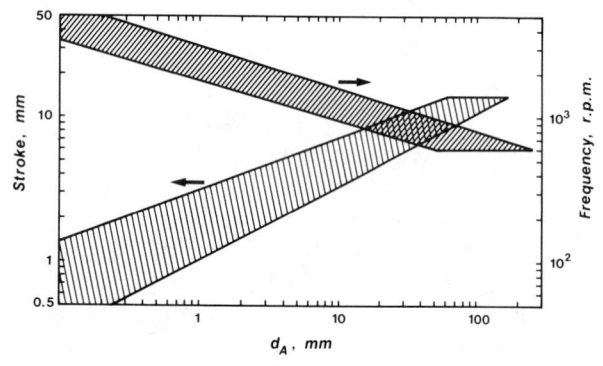

Figure 9.12. Recommended strokes and frequencies for vibrating screens.

Theory of Conventional Screening and Sieving

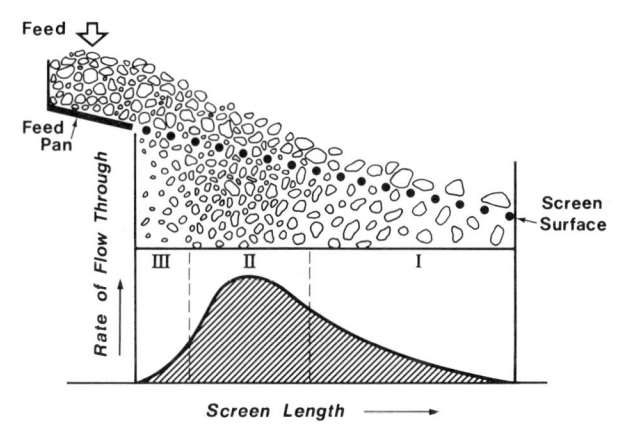

Figure 9.13. The three major regions occurring along a screening surface.

rate along the screen, and maintains or even increases the bed depth. Such an approach takes advantage of stratification that has already occurred and prevents the formation of region I (unrestrained particle motion). It is claimed that this method can reduce area requirements by up to 50%.

Because bed depth depends on flow rate along the screen, it is impossible to calculate. However, under a given set of conditions it will be proportional to the feed rate. Consequently, as the feed rate rises, the load increases and as illustrated in Fig. 9.14, the amplitude must be raised to maintain movement and so keep efficiency high. Even so, efficiency generally falls off as capacity rises. It is also worthy of note that efficiency can fall off because of insufficient loading on the screen, for the reasons discussed above (Fig. 9.27).

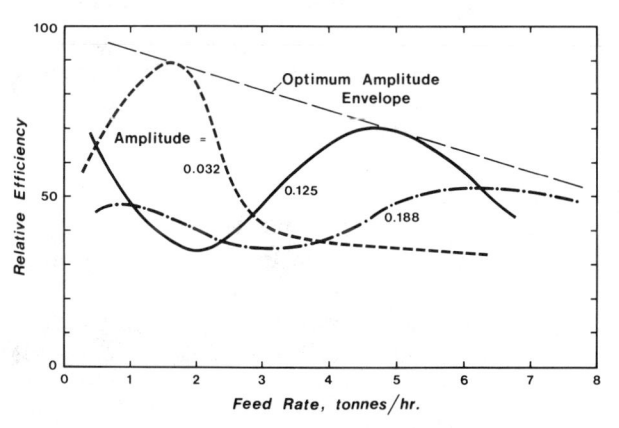

Figure 9.14. As the load on a screen surface increases, amplitude must be increased to maintain movement of the heavier bed: copper slag on an inclined vibrating screen, at 1640 rpm, 600 μm aperture. (Courtesy Triple/S Dynamics, Inc.)

In the case of sieving, surface loading is also important; the mechanisms discussed above are generally applicable, but blinding plays a much more significant role.

9.6.5. Moisture

Completely dry or completely wet materials are comparatively easy to screen, but materials with as little as 1% *surface* moisture can cause severe problems.[30] In general the finer the size distribution, the less moisture is needed to impair the screening action, and some materials such as clays, fertilizers, and chemicals have very low moisture tolerances.

Impaired performance can result from either of two main phenomena: *plastic* blinding, caused by fine moist particles such as clay clinging to the screen wires and gradually plastering over areas of the screen with tightly adhering material, and *blanketing*, caused by particles sticking together either by cementing or surface tension effects.

Mechanical deblinding devices such as ball trays offer protection against either cause. Electric screen heaters can be used on vibrating screens to control plastic blinding, but they are ineffective against blanketing. Drying, or wet screening, can be used to treat both effects. Drying has the disadvantage of expense, possible dust problems, and the possibility of cementing particles together. Wet screening has the advantage that it encourages the passage of small particles, but that surface tension effects hinder the passage of near mesh particles.

9.7. THEORY OF CONVENTIONAL SCREENING AND SIEVING

At present there is no completely satisfactory theory of sieving or screening. A number of studies have been described in the literature, but many of these are of limited use because of the narrow range of conditions studied. Some of the studies have shown that under certain circumstances screening and sieving can be represented by a single rate process, normally assumed to be first order.[3,37–42] However, the typical screening illustrated in Fig. 9.13 suggests that in practical situations a number of overlapping processes must occur. Despite an increasing amount of evidence that commercial screening can often be adequately described by two limiting processes[43,44] (crowded screening and dispersed screening), we give a wider review here.

9.7.1. Single Process Treatments

In most single process treatments, first-order kinetics is assumed, that is, represented by Eq. 3.2 or 3.3

$$m = m_I \exp(-kt) \quad (3.2)$$

$$m = m_I(1 - p)^n \quad (3.3)$$

For example, the true end point of sieving can be determined as follows.[3] Assuming that Eqs. 3.2 and 3.3 are equivalent

$$\exp(-kt) = (1 - p)^n \quad (9.8)$$

n will be the number of presentations to passage. If the number per unit time n/t is constant then

$$-k = \left(\frac{n}{t}\right) \ln(1 - p) \quad (9.9)$$

During the latter stages of sieving, the particles passing will be essentially all near-mesh particles of the same size, so that m can be replaced by M and p will be constant. Furthermore the mass of material on the sieve will be virtually constant, so that provided a sieve shaker is used, n/t can be considered to be constant. Thus, as the end point is approached, k is constant, and Eq. 3.2 gives

$$\ln(M_t - M_\infty) = kt + C \quad (9.10)$$

(M_t = mass of sample retained by the sieve at time t; M_∞ = mass of sample retained at infinite time, i.e., the true end point). Hence, when log $(M_t - M_\infty)$ is plotted versus sieving time, using estimated values of M_∞, the correct value of M_∞ will be that one which produces a straight line.

Miwa used Eq. 3.3 to develop a useful screening index.[42a] He assumed that on a screen the number of presentations per unit length N_L would be constant and that p could be approximated by Eq. 9.3. Substituting (Eqs. 3.3, 9.3)

$$\frac{M}{M_I} = \left[1 - \left(\frac{D_e - d}{D_e + D_W}\right)^2\right]^{N_L L} \quad (9.11)$$

(D_e = effective aperture of the screen; L = length of screen). If d is taken as the size at which 50% of the material is still retained on the screen (i.e., 50% is passed), then $d = d_{50}$ and $M/M_I = 0.5$, so that Eq. 9.11 becomes, after taking logarithms,

$$\ln 0.5 = N_L L \ln \left[1 - \left(\frac{D_e - d_{50}}{D_e + D_W}\right)^2\right] \quad (9.12)$$

Since the term in the inner brackets of this equation must be less than unity, the logarithmic term on the right (when expanded in a series from which all terms of the fourth power or higher are dropped) can be shown to be approximately equal to $-(D_e - d_{50})^2/(D_e + D_W)^2$, so that Eq. 9.12 becomes

$$\ln 0.5 \simeq -N_L L \left(\frac{D_e - d_{50}}{D_e + D_W}\right)^2$$

that is,

$$d_{50} = D_e - \frac{0.833(D_e + D_W)}{\sqrt{L} \times \sqrt{N_L}} \quad (9.13)$$

Thus, a plot of d_{50} versus $1/\sqrt{L}$ from screening data obtained by collecting undersize fractions at various distances along a screen, and determining d_{50} for successive cumulative lengths along a screen, should give a straight line from which values of D_e and N_L can be determined. Under these conditions N_L can be regarded as a screening index that characterizes the frequency and amplitude of vibration, the inclination of the screen, and bed characteristics such as the feed rate. D_e is the effective aperture size. To determine these parameters it is necessary to estimate a value for D_W. This can be taken as the measured value, or as a realistic fraction of D_e. Once D_e and N_L are known, the product size distribution can be calculated for a feed of another size distribution by means of Eq. 9.11.

Even though the inclusion of Eq. 9.3 is an approximation this method can be useful,[41,42] although different values of D_e and N_L may occur at different points along the screen as the bed characteristics change (Section 9.7.3, Fig. 9.17a).

Example 9.1. Material is fed across a vibrating screen at the rate of 1 t/hr. Laboratory sieving has shown that 50% of the feed is potential undersize, yet only 45% of the feed occurs in the undersize from the screen. If half of the potential undersize consists of near-mesh particles with a probability of passage (per presentation) of 0.05, and the other half is finer particles with an average probability of passage (per presentation) of 0.1, how many presentations to passage occur as the particles pass along the screen?

If the undersize recovery is to be increased to 47.5% of the feed, what is the maximum flow rate the screen could handle?

Solution. Using Eq. 3.3, the recovery in the undersize of the near mesh is

$$[1 - (1 - 0.05)^n]0.25$$

while the recovery of the fines is

$$[1 - (1 - 0.1)^n]0.25$$

Together these equal 45% of the feed, that is,

$$0.45 = 0.25\{[1 - (1 - 0.05)^n] + [1 - (1 - 0.1)^n]\}$$

that is,

$$0.2 = (0.95)^n + (0.9)^n$$

By trial and error, $n \sim 34$; that is, the number of attempts at passage is 34.

To obtain 47.5% recovery, we write:

$$0.475 = 0.25\{[1 - (1 - 0.05)^n] + [1 - (1 - 0.1)^n]\}$$

which requires approximately 46.5 attempts at passage.

Assuming that the flow rate is inversely proportional to the number of attempts at passage,

$$I = \frac{34}{46.5} \times 1.0 \; \frac{t}{hr}$$
$$= 0.73 \; t/hr$$

9.7.2. Two Process Treatments

A number of authors have pointed out that during screening, particles may be subjected to two distinctly different types of condition, crowded or separated, and this leads to two different rate processes.[43,44]

Crowded Screening. Crowded screening occurs when the flow rate is above a critical level I_{crit} such that the material forms a bed so thick that only particles in the layer immediately in contact with the screen are capable of passage. As long as the upper layers are capable of replenishing the contact layer, the rate of passage will remain constant and will be given by

$$-\frac{dI_L}{dL} = k_c \qquad (9.14)$$

(I_L = mass flow rate on screen per unit width, at distance L from the feed point; k_c = rate constant for crowded conditions). In practical screening a range of particle sizes will occur and a separate equation applies for each size i: a suitable approximation of Eq. 9.14 is then[44]:

$$-\frac{d(I_L m_{iL})}{dL} = k_{ci} m_{iL} \qquad (9.15)$$

(m_{iL} = mass fraction of particles in size class i in the bed at distance L from feed point). Because I and m_{iL} are functions of L, this becomes, on integration (between 0 and L)

$$-\ln\left(\frac{I_L m_{iL}}{I_i m_i}\right) = k_{ci} \int_0^L \frac{dL}{I_L} \qquad (9.16)$$

(I = mass flow rate fed to screen, per unit width; m_i = mass fraction of particles in size class i in feed). This applies only for $0 < L < L_{crit}$, where L_{crit} is the critical length at which the flow rate drops below I_{crit} necessary to maintain crowded screening. Arbitrarily setting the underflow as the positive response stream, Eq. 9.16 can also be written as

$$-\ln(1 - \mathcal{R}_{(+)i}) = k_{ci} \int_0^L \frac{dL}{I_L} \qquad (9.17)$$

($\mathcal{R}_{(+)i}$ = mass fraction of size i reporting to the undersize stream). Figure 9.15a shows how the performance curve changes with screen length and presents results produced on a screen divided into four sections.[44] As the screen length increases, the cumulative-length d_{50} increases and tends to approach the mesh value. This results from each successive section having a steeper performance curve because of the early passage of the finest particles.

Separated Screening. If the flow of material across the screen is less than I_{crit}, the particles behave as if they were isolated and do not interfere with one another. Under these conditions, the quantity of particles dI that passes through the screen in the length dL is proportional to the rate I at which particles enter dL, and to dL itself. Thus

$$-\frac{dI_L}{dL} = k_s I_L \qquad (9.18)$$

that is, a first-order rate equation (cf. Eq. 3.2). Again, the treatment of real materials requires that the equation be applied to each size class i, so that Eq. 9.18 becomes

$$\frac{-d(I_L m_{iL})}{dL} = k_{si} I_L m_{iL} \qquad (9.19)$$

On integration between the limits 0 and L (or L_{crit} and L if crowded screening occurs initially),

$$-\ln\left(\frac{I_L m_{iL}}{I m_i}\right) = k_s L \qquad (9.20)$$

or

$$-\ln(1 - \mathcal{R}_{(+)i}) = k_{si} L \qquad (9.21)$$

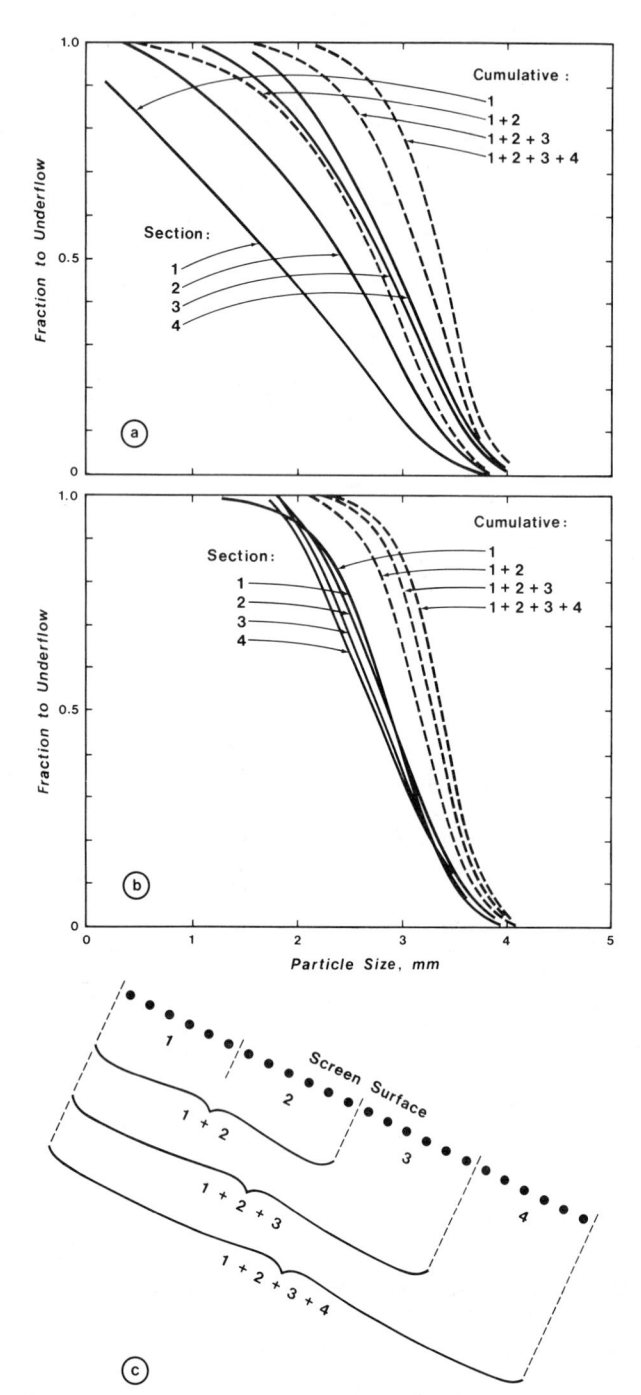

Figure 9.15. Performance curves from tests on a vibrating screen. (*a*) High feed rate, producing crowded screening. (*b*) Low feed rate, producing separated screening. (*c*) Key to the performance curves. (After Ferrara and Preti.[44])

Figure 9.15*b* shows how the performance curve changes with increasing screen length under separated conditions.[44] While the cumulative length performance curve gradually steepens, the curve of each individual section remains essentially constant.

Rate Constants. The crowded rate constant k_c is a function of the individual pass probability p of the particles, the number of presentations per unit length N_L, and the critical flow rate I_{crit}, that is[44]:

$$k_{ci} = I_{\text{crit}} N_L p_i \qquad (9.22)$$

For separated behavior k_s is given by the cumulative pass probability of an isolated particle. From Eq. 3.3, replacing n by $N_L L$

$$\frac{m_i}{m_{iI}} = (1 - \mathscr{R}_{(+)i}) = (1 - p_i)^{N_L L} \qquad (9.23)$$

which combined with Eq. 9.21 gives

$$k_{si} = -N_L \ln(1 - p_i) \qquad (9.24)$$

Both Eqs. 9.22 and 9.24 show that the rate constant is a function of the probability of passage (i.e., the screen surface) and the number of presentations per unit length (i.e., essentially the vibration characteristics). Ferrara and Preti have described one of the more thorough investigations of these rate constants.[44] They found that for crowded screening p_i could be approximated by Eq. 9.3. Because N_L is difficult to measure, they introduced the reduced coefficient

$$\chi_{i,1/2} = \frac{k_{ci}}{k_{c,1/2}} \qquad (9.25)$$

($k_{c,1/2}$ = rate constant for the size i corresponding to $d/D_A = 0.5$). According to Eq. 9.3

$$\chi_{i,1/2} = 4\left(1 - \frac{d}{D_A}\right)^2 \qquad (9.26)$$

that is, a log-log plot of $\chi_{i,1/2}$ versus $(1 - d/D_A)$ should be a straight line of slope 2. Experimental data give straight line plots, but of slightly different slopes, so that

$$\chi_{i,1/2} = 2^{n_c}\left(1 - \frac{d}{D_A}\right)^{n_c} \qquad (9.27)$$

(n_c = slope of lines). Variations in the feed rate and the vibration direction gave n_c values in the range 1.61–2.18. It follows that the two parameters $k_{c,1/2}$ and n_c are sufficient to define the kinetic characteristics of the crowded screening process. Of the two parameters, $k_{c,1/2}$ is the more significant: it represents the mass rate at which particles having half the mesh aperture pass through unit area of screen. Although n_c may be approximated by the theoretical

Theory of Conventional Screening and Sieving

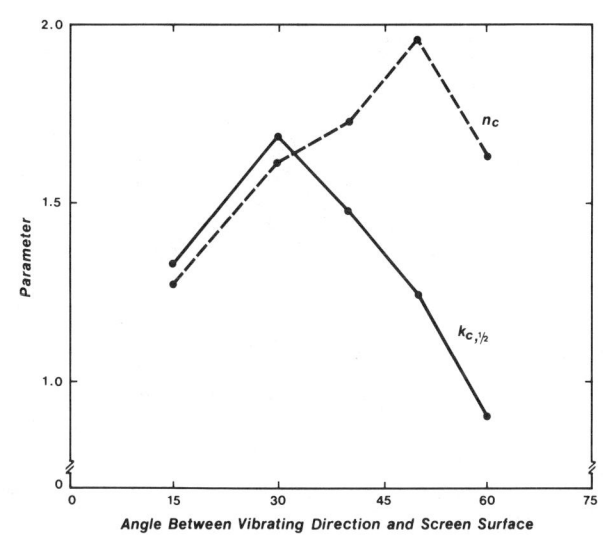

Figure 9.16. Values of $k_{c,1/2}$ and n_c as functions of the angle between the vibrating direction and the screen surface. (After Ferrara and Preti.[44])

value of 2, the actual value has a precise significance: it indicates whether the operating conditions are more favorable for the passage of fine particles (high n_c) or near-mesh particles (low n_c). Figure 9.16 illustrates the effect of one variable (the angle between the vibrating direction and the screen surface) on the two parameters: there is a clear maximum in the rate of passage, and extreme inclinations facilitate the passage of near-mesh particles *relative* to fine particles (but at the expense of lower total flow rate). Thus the two parameters can be very useful in comparing screening operations. Alternatively, combined with Eq. 9.17, they allow screening performance to be predicted.

9.7.3. Multiprocess Treatments

Section 9.7.2 described crowded screening by a single (zero-order) rate process. In reality the eventual passage of a particle under crowded conditions depends on two distinct statistical phenomena: the probability of a particle reaching the screening surface, and the subsequent probability of passage through the surface. Equation 9.14 effectively assumes that the former is much higher than the latter, that is, the bed is perfectly mixed. However, the rate of bed penetration (or bed stratification) may be rate determining, and in fact, this was assumed when discussing Fig. 9.13. Figure 9.17a shows some actual experimental data[43] plotted according to Eq. 9.13. It can be seen that up to three approximately linear sections occur at most flow rates, from which it can be concluded that the entire screening process can be represented by three first-order rate processes (since Eq. 9.13 is derived from Eq. 3.3). That

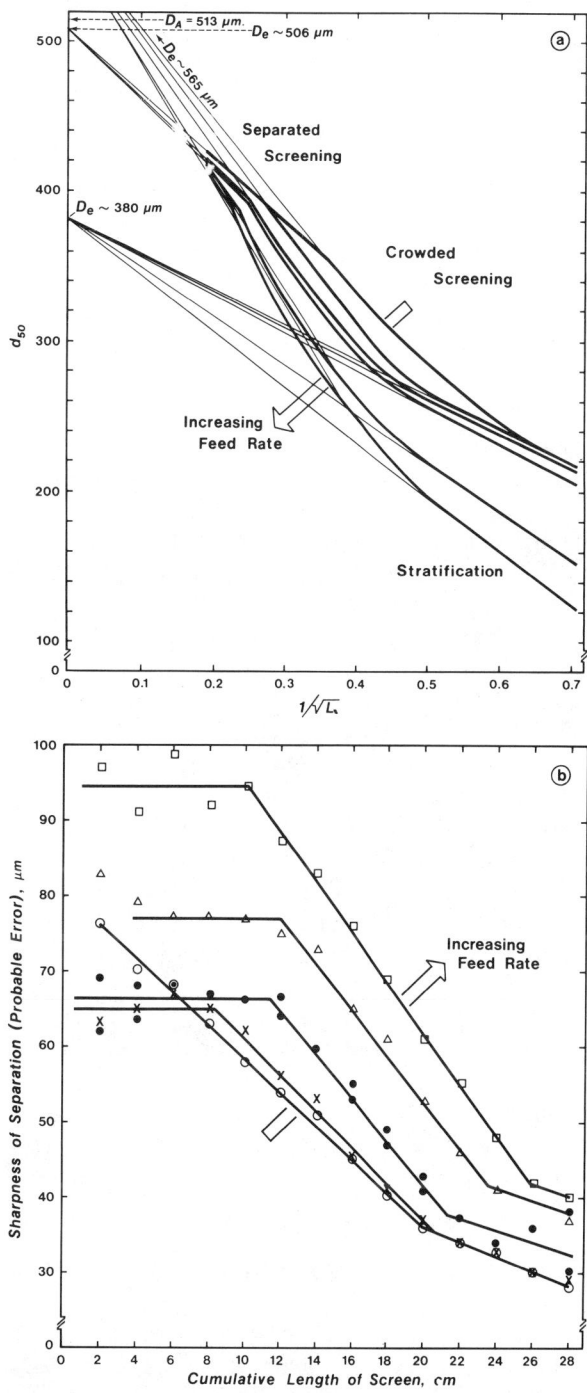

Figure 9.17. Data of Brereton and Dymott[43] (*a*) plotted according to Eq. 9.13 and (*b*) showing how the performance curve steepens with increasing length of screen.

the third (final) stage of screening is first-order is in agreement with Eq. 9.9. The initial stage (if it occurs) can be attributed to stratification being rate controlling. Meinel and Schubert[45,46] have shown that penetration of the bed by a particle can be considered to be a combination of a diffusion process and a segregation process, both of which can be characterized by coefficients that can be related to system variables. However, the results of Olsen et al.[40] suggest that segregation (or stratification) may be represented by a first-order rate process, and this is supported by the initial behavior in Fig. 9.17a. It is also interesting to note that during the initial screening at high feed rates the data indicate a value of D_e that is about $0.75D_A$: a feature that can be interpreted as the effective pore size of the bed.

That the middle stage of screening can be approximated by a first-order rate process does not rule out constant rate screening (Eq. 9.15). In fact, there are two features that suggest that crowded screening, as discussed in the Section 9.7.2, does actually occur. First, close examination of Fig. 9.17a shows that at the higher feed rates (when I_L should be greater than I_{crit}) the plot is actually curved in the center, as expected. Second, for the middle stage, the section performance curve (i.e., for the material presented to the small section of screen dL) is steepening as L increases (Fig. 9.17b), a feature that is characteristic of constant rate screening (Fig. 9.15a). On the other hand, during the initial and final stages of screening, the section performance curves essentially retain the same slope (characteristic of separated behavior: Fig. 9.15b).

English[38] has shown that there are actually three rate constants of significance when sieving near-mesh particles: a rate constant for passage k_p, a rate constant k_b for the rate of blinding of the screen, and a rate constant k_{ub} for the unblinding or clearing of the apertures. Although the blinding and unblinding rates can have significant effects in sieving, they normally are in balance during screening and can be accounted for by a permanent reduction in the available screen area.

9.8. THEORY OF PROBABILITY SCREENING

Probability or statistical screening is so named because the separation is based on the probability of a particle passing through the screen, rather than by physical constraint, as in conventional screening (Fig. 9.18). Essentially the principle is that the smaller the particle, the sooner it will pass through a

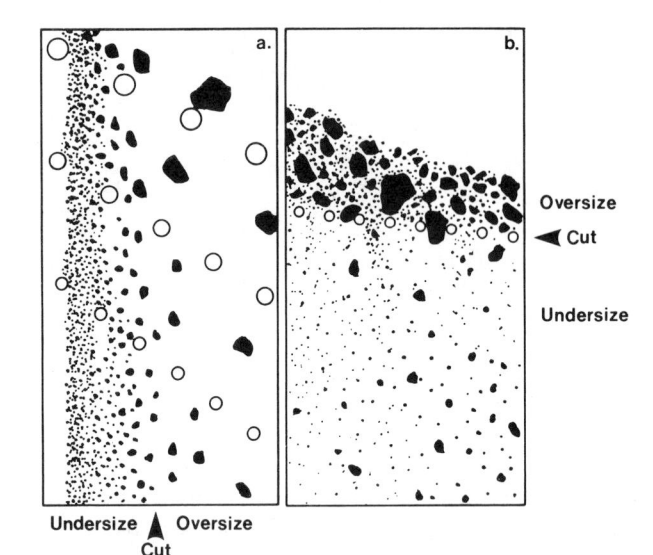

Figure 9.18. Comparison of probability screening (a) and conventional screening (b). (Courtesy Fredrik Mogensen AB.)

mesh, because its probability of passage is higher. A further spread is achieved by stacking layers of screen surfaces.

The attainment of a separation based on these concepts has been illustrated as follows.[22,47] If the probability of passage for a given size particle is p, then a fraction p passes at the first attempt and a fraction $(1 - p)$ is reflected to the next attempt. It follows that the fraction of sample passing the nth opening (aperture and attempt can be treated as the same thing) is

$$\frac{M_n}{M_I} = p(1 - p)^{n-1} \quad (9.28)$$

If there are N layers of screen, the general case is

$$\frac{M_{n,N}}{M_I} = \frac{(n + N - 2)!}{(n - 1)!(N - 1)!} p^N (1 - p)^{n-1} \quad (9.29)$$

Equation 9.29 is illustrated in Fig. 9.19. Since N is fixed for any screen, the probability of N passages in $(n - N)$ attempts is the cumulative fraction passing given by

$$\sum \left(\frac{M_{n,N}}{M_I}\right)_N = Y_{\bar{N}}$$

$$= \sum_{n=1}^{n=n} \frac{(n + N - 2)!}{(n - 1)!(N - 1)!} p^N (1 - p)^{n-1} \quad (9.30)$$

where

$$\sum_{n=1}^{n=\infty} \left(\frac{(n + N - 2)!}{(n - 1)!(N - 1)!}\right) p^N (1 - p)^{n-1} = 1$$

Theory of Probability Screening

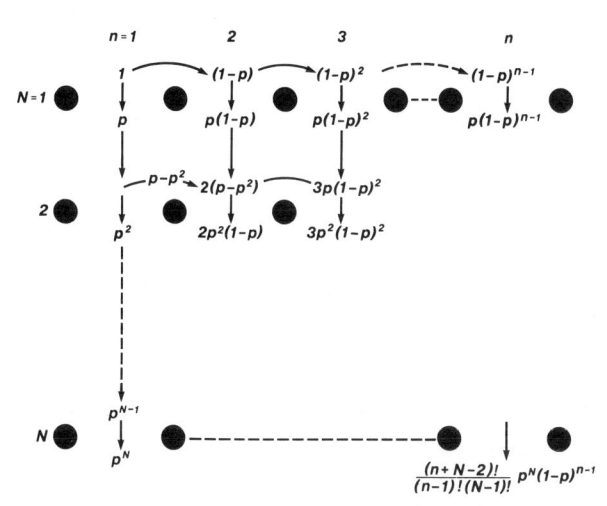

Figure 9.19. Idealized cross-section of a probability screen. (After Schultz and Tippin.[47])

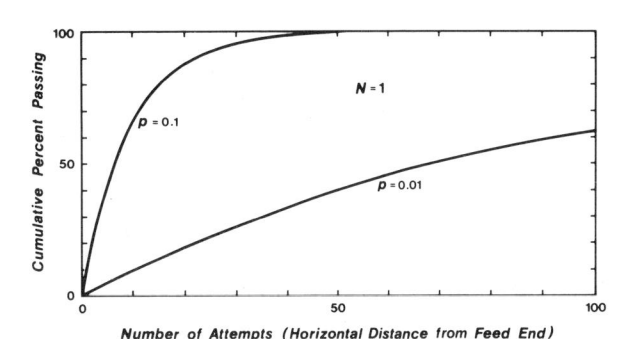

Figure 9.21. Predicted passage of two particles of differing size: differing probabilities of passage. (After Schultz and Tippin.[47])

The extent of the separation achieved can be made clearer by showing the effect of the parameters, p and N graphically. Figure 9.20 shows the effect of the number of screen layers on the cumulative passage of material having a passage probability of 0.1. It can be seen that as the number of screens is increased, the particles are spread out, the number of attempts (i.e., the length of screen) necessary to pass 50% of the material being approximately proportional to the number of screens.

The distribution on a single screen of two sizes of particles having probabilities of passage of 0.1 and 0.01 is shown in Fig. 9.21. This illustrates the significant difference between conventional and probability screening. In the latter, the aperture is selected so that the material presented has a significant probability of passing, nominally $p > 0.1$. In conventional screening the problem is essentially one concerned with the near-mesh particles, those with low probabilities of passage ($p < 0.1$); this clearly requires a far longer screen length.

Figure 9.22 shows the effect of using a stack of screens, in this case nine. If the separation is to be achieved at $p = 0.2$ (which, from Eq. 9.3, $d_p \sim 0.7 D_A$), then 50% passage of this probability (size) occurs at $n = 35$. A vertical line at this point gives the fraction of other probabilities (sizes) separated by this length of screen: in this case the undersize will contain all material with $p > 0.4$, ~90% with $p = 0.3$ and ~3% with $p = 0.1$.

In practice, a further sharpening of the separation is achieved by two modifications. First, by giving successive screens down the stack higher slopes, the effective aperture is reduced and the probability of passage decreases (Eq. 9.5) as the particle falls through the stack. Figure 9.23 illustrates the improved separation achieved between particles of $p = 0.2$ and $p = 0.4$ by using 5 screens with increasing slopes, compared to 10 parallel screens. Second, successively lower screens are given smaller apertures to accentuate this phenomenon.

Although the probability screen gives slightly lower efficiencies than conventional vibrating screens at low loadings, the efficiency remains rela-

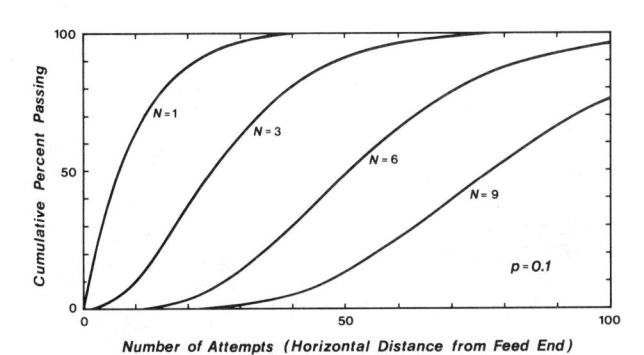

Figure 9.20. The effect of the number of screens on the distribution: Eq. 9.30, $p = 0.1$. (After Schultz and Tippin.[47])

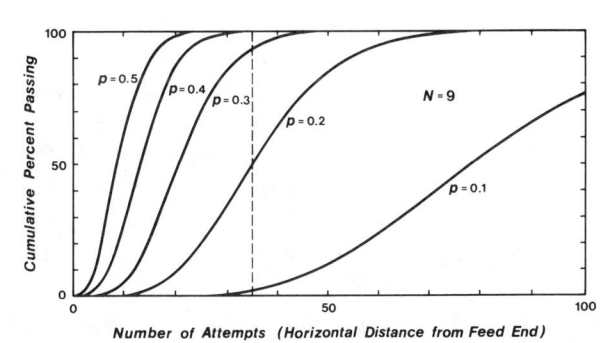

Figure 9.22. Prediction of the separation obtained with a nine-deck screen. (After Schultz and Tippin.[47])

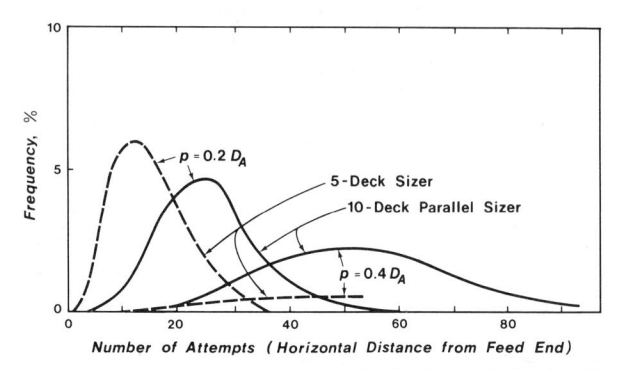

Figure 9.23. Comparison of the distributions obtained with decks of 10 parallel screens and 5 different-slope screens. (After Mogensen.[22])

tively constant as the load increases.[48] Consequently, these screens are capable of higher capacity with reduced space requirements, less power, less wear, and lower noise.

9.9. THEORY OF SIEVE BENDS

The sieve bend is normally a curved surface of bar screen that has a slurry fed tangentially over the surface. Very thin layers of slurry are peeled off as the mixture passes over the surfaces, as illustrated schematically in Fig. 9.7.

The parameters of interest in the sieve bend are[49] d_{50}/D_A and the fraction of the feed stream reporting to the undersize stream $O_{(+)}/I$. Ideally d_{50}/D_A should be low to minimize blinding, while $O_{(+)}/I$ is a measure of the separating size: high values of $O_{(+)}/I$ implying large separating sizes and low values small sizes. According to Fontein, the factors that contribute to the thickness of the layer peeled off, and thus d_{50}/D_A and $O_{(+)}/I$, are:

1. The quantity of the stream that hits the next bar, since in principle it travels in a straight line across the slot (occurring with concave surfaces only).
2. The centrifugal force pressing the slurry against the surface (occurring with concave surfaces only).
3. The gravitational force pulling the slurry through the slot.
4. The surface tension of the slurry on the metal.

Analysis of these factors gives the following qualitative equations:

$$\frac{d_{50}}{D_A} = \frac{K_1 D_A}{R} + \frac{K_2}{\text{fn}(\mathbf{Re}_s)}$$
$$\left(\frac{\delta_I}{R} + \frac{\delta_I g \sin\theta}{v_I^2} + \frac{K_5 \sigma_s}{\rho_{sl} v_I^2 D_A}\right)^{1/2} \quad (9.31)$$

$$\frac{O_{(+)}}{I} = \frac{K_3 D_A^2 N}{\delta_I R} \text{fn}(\mathbf{Re}_s) + \frac{K_4 D_A N}{\delta_I}$$
$$\left(\frac{\delta_I}{R} + \frac{\delta_I g \sin\theta}{v_I^2} + \frac{K_5 \sigma_s}{\rho_{sl} v_I^2 D_A}\right)^{1/2} \quad (9.32)$$

(K = constants; N = number of slots; R = radius of curvature of screen surfaces; $\mathbf{Re}_s$ = Reynolds number of slot, $D_A v \rho_{sl}/\mu$; v_I = feed velocity; δ_I = thickness of feed layer; θ = angle of arc of surface; σ_s = surface tension). Both equations indicate the trends found in practice.

Figure 9.24 makes the significance of the two parameters more obvious. It can be seen that provided the flow rate is high enough (high $\mathbf{Re}_s$), d_{50}/D_A falls to a constant value of about 0.5. Furthermore, this coincides with the slurry split leveling off at a constant, relatively high value. Such dewatering characteristics are typical of screens (and are thus beneficial when a screen is used for dewatering), but they also constitute the factor that can make sieve bends difficult to use in closed circuit grinding, because a water balance can be difficult to control when the feed contains large quantities of oversize.[50]

Although not readily apparent from Eqs. 9.31 and 9.32, two other characteristics are important. First, the sieve bend tends to concentrate the denser mineral in the *undersize* fraction,[20] unlike wet classifiers, which concentrate it in the oversize. Consequently,

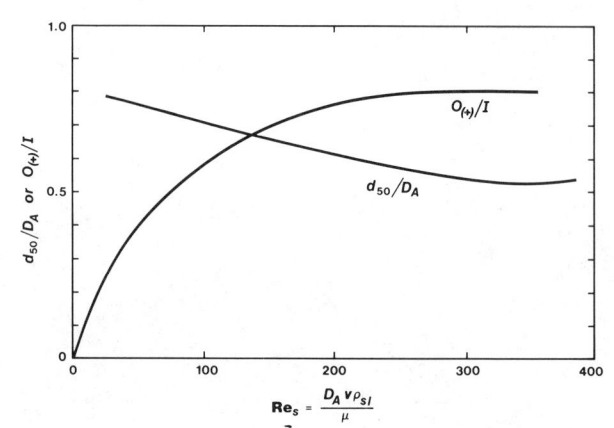

Figure 9.24. Values of $O_{(+)}/I$ and d_{50}/D_A as a function of $\mathbf{Re}_s$, for a sieve bend. (After Fontein.[49])

Screen Sizing

sieve bends are an attractive alternative to wet classifiers, where it is desirable to minimize overgrinding of a dense mineral being ground in closed circuit. The second feature concerns the effect of screen wear. Because wear occurs at the leading edge of the bar, the thickness of the peeled-off layer decreases with time, so that the cut size *decreases*. This effect is the opposite to that occurring with conventional screening and because of the relatively severe wear conditions, means that the surfaces need regular reversing. On some units, this can be done automatically (Fig. 9.5h).

9.10. SCREEN SIZING

There are two basic approaches to screen size calculations: essentially empirical methods based on tabulated or graphical data, and methods based on tests on the actual material.[30,51-55] The former group can be classified according to whether the basis of calculation is the capacity of material fed to, passing over, or passing through the screen. Such methods are of limited reliability and different manufacturers' data give different results. Consequently, these methods should be used only to obtain an initial estimate. One method is described here to illustrate the approach.

More reliable estimates of screen requirements can be obtained by using techniques that employ tests on representative samples of the actual material to be screened.[30] The major limitation of these methods is the more extensive test equipment required.

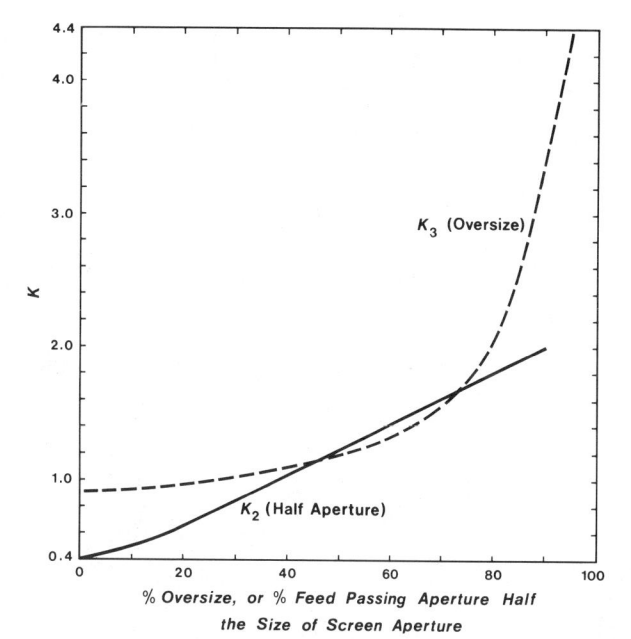

Figure 9.26. Typical values of the oversize K_3 and half-aperture K_2 screening factors.

9.10.1. Empirical Factor Method

A basic formula for calculating screen area is

$$A = \frac{I}{(I_u/\rho_b)\rho_b \cdot K_\Sigma} \quad (9.33)$$

(A = area of screening surface; I = feed mass flow rate; I_u = unit capacity; ρ_b = bulk density of feed; K_Σ = the product of various correction factors). Figures 9.25–9.27 and Tables 9.7 and 9.8[54] show typical values of I_u and K_Σ. Various sources of data give different emphasis to some factors or introduce others, thereby resulting in different answers; differences that in most cases are hardly significant in terms of the limited accuracy of these methods.

TABLE 9.7: FACTOR METHOD

I_u	=	Unit capacity; (Fig. 9.25)
K_Σ	=	$K_1 \times K_2 \times K_3 \times K_4 \times K_5 \times K_6 \times K_7 \times K_8 \times K_9 \times K_{10}$
K_1	=	Open area factor: (% open area of screen)/100
K_2	=	Half size factor: to correct for the percentage of feed which passes through an aperture half the size of the screen aperture (Fig. 9.26)
K_3	=	Oversize factor: a factor to correct for the percentage of oversize in the feed (Fig. 9.26)
K_4	=	Screening efficiency factor: (Fig. 9.27)
K_5	=	Deck factor: to correct for the reduced effective length of lower decks (Table 9.8)
K_6	=	Screen angle factor: to correct for non-optimum screen inclination (Table 9.8)
K_7	=	Wet screening factor: (Table 9.8)
K_8	=	Aperture shape factor: (Table 9.8)
K_9	=	Particle shape factor: (Table 9.8)
K_{10}	=	Tenacity or surface moisture condition: (Table 9.8)

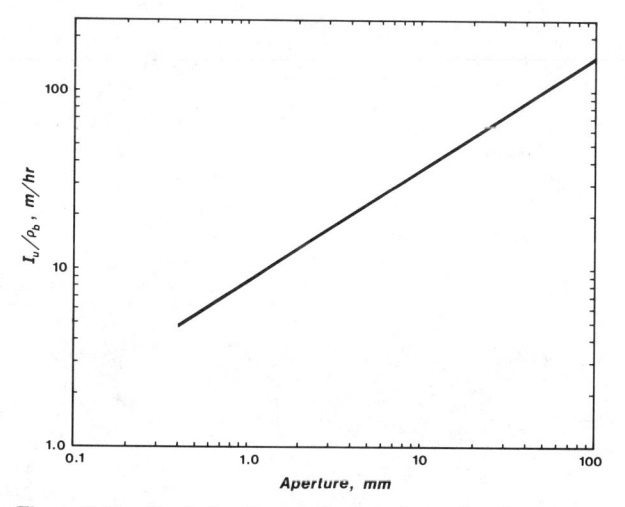

Screen should have length to width ratio of 1.5 to 2.0:1.
Effective width of screens is 150 mm less than true width.

Figure 9.25. Typical unit capacity data for a vibrating screen.

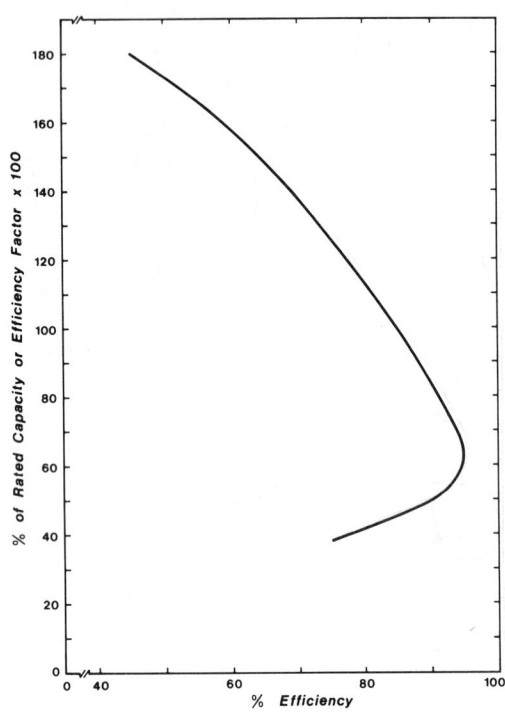

Figure 9.27. Typical values of the efficiency factor K_4.

Example 9.2. Estimate the screen area necessary to screen at 6.67 mm 400 t/hr of moist coal, having the size distribution shown in Fig. E9.2.1. Bulk density is 835 kg/m³, and screen open area is 64%. Required efficiency is 80%.

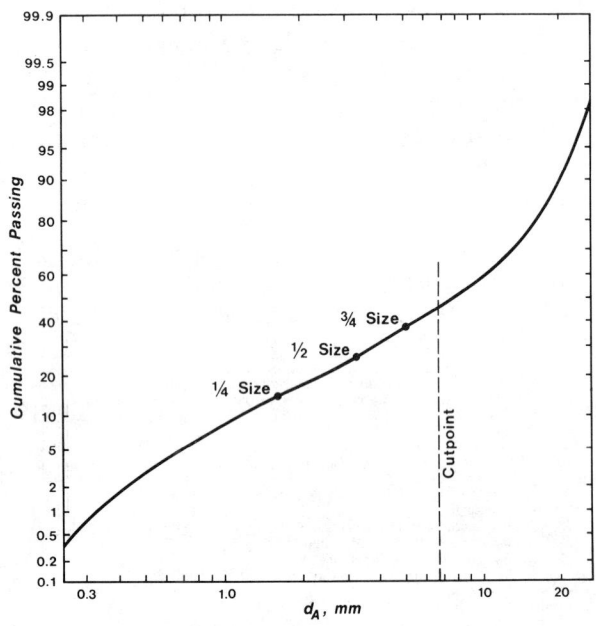

Figure E9.2.1. Size distribution of wet coal fines. (Courtesy Triple/S Dynamics, Inc.)

TABLE 9.8: FACTOR VALUES

K_5: Deck Number

Deck	K_5
Top	1.00
Second	0.90
Third	0.80
Fourth	0.70

K_6: Screen Angle

Angle of Incline (Deg.)	K_6
Horizontal	1.20
5	1.15
10	1.05
15	1.00
20	0.95

K_7: Wet Screening

Size of Opening, (mm)	K_7	Size of Opening, (mm)	K_7
0.8	1.25	9.5	2.25
1.6	1.5	12.7	2.5
3.2	1.75	19.0	2.71
4.75	1.9	25.4	2.9
7.9	2.1		

K_7 is used when water is added to the material at a rate of 1 to 2.5 vol. %. Note: for feed sizes larger than 25 mm, wet screening becomes much less effective. Below 850 μm, wet screening poses problems.

K_8: Aperture Shape

Shape	K_8
Square opening	1.0
Slot length 6 or more times width	1.60
Slot length 3 to 6 times width	1.40
Slot length 2 to 3 times width	1.10
Round openings	0.80

K_9: Particle Shape

Elongated Particles, % *	K_9	Elongated Particles, %	K_9
5	1.00	40	0.75
10	0.95	50	0.70
15	0.90	60	0.65
20	0.85	70	0.60
30	0.80	80	0.55

*The percentage of elongated feed particles that have more than a 3-to-1 ratio of length to width, and that have a width greater than half the aperture width but smaller than one and a half times the aperture width.
Note: factors K_8 and K_9 are closely interdependent.

K_{10}: Moisture

Tenacity and/or Surface Moisture Condition	K_{10}
Wet, muddy, or otherwise sticky rock; gypsum, phosphate rock, etc.	0.75
Surface-wet quarried or mined material; material from stockpiles with surface moisture greater than 14 vol. % but nonhygroscopic	0.85
Dry pit-run material; dry, lumpy, crushed manufactured chemicals. Surface moisture less than 10 vol. %.	1.00
Naturally dry material, uncrushed; materials that have been dried prior to screening; or materials screened while hot.	1.25

Solution. From Fig. 9.25:

$$\frac{I_u}{\rho_b} \text{ at } 6.67 \text{ mm} = 28 \text{ m/hr}$$

$$K_1 = \frac{64\%}{100}$$

From Fig. 9.26:

$$K_2 = 0.7$$

From Fig. 9.26:

$$K_3 = 1.25$$

From Fig. 9.27:

$$K_4 = 1.2$$

Screen Sizing

From Table 9.8:

$$K_{10} = 0.85$$

$$\therefore K_\Sigma = 0.64 \times 0.7 \times 1.25 \times 1.2 \times 0.85 = 0.57$$

$$\therefore A = \frac{400\ t}{hr} \left| \frac{hr}{28\ m} \right| \frac{m^3}{0.835\ t} \left| \frac{1}{0.57} \right. = 30\ m^2$$

In keeping with the trend of developing empirical regression equations for separations, Karra has published equations derived from a commercial vibrating screen installation.[55] Effectively these are mathematical forms of the tabular or graphical data described above. One particularly valuable contribution is the presentation of a correlation for d_{50}. However, the paper also presents a reduced performance curve for the system studied. This concept may be adequate for predicting the results of slight variations in operating conditions, but such curves must be used with considerable caution, because in theory the performance curve of a screen cannot be reduced. This is primarily because the curve is constrained at the d_{100} size by the aperture.

9.10.2. Screenability Characteristics Method

The second screen sizing method[30] involves using a laboratory sieve to determine the screening characteristics of the material in question, with a single closed cycle test on a small industrial screen to generate a scale-up factor to account for the two different screening actions.

Initial test work involves the determination of the material's bulk density and a full size analysis so that screening efficiencies can be calculated using Eq. 9.2. This size analysis is also useful in giving an initial indication of the difficulty of the screening problem.

A series of tests is then carried out with a single sieve using different loadings (i.e., initial bed depths). These tests must be performed with only one aperture size, that which is to be used in practice. By operating the sieve on a shaker for various times the increasing accumulation of undersize in the pan allows one to obtain plots of efficiency versus initial bed depth. Illustrative results are shown in Figs. 9.28 and 9.29, these sets of curves representing the *screenability characteristics* of the material under test.

The closed cycle test must be conducted on a small industrial machine of the type to be used in practice, again using the same aperture size. Such a test produces a single efficiency value for that type of machine for a given bed depth and flow rate across the screen. The screenability characteristics can then be used to correct the efficiency on this machine to that required in practice. The actual mechanics of the method are most easily illustrated by the following example.

Example 9.3. Recalculate Example 9.2, using the screenability characteristics shown in Figs. 9.28 and 9.29. A closed cycle test with a 0.6 m × 1.2 m screen of the recommended type had 74% efficiency at 18.3 t/hr. The average rate of travel over the screen was 15.2 m/min.

Solution. It is first necessary to determine the bed depth and residence time of the closed cycle

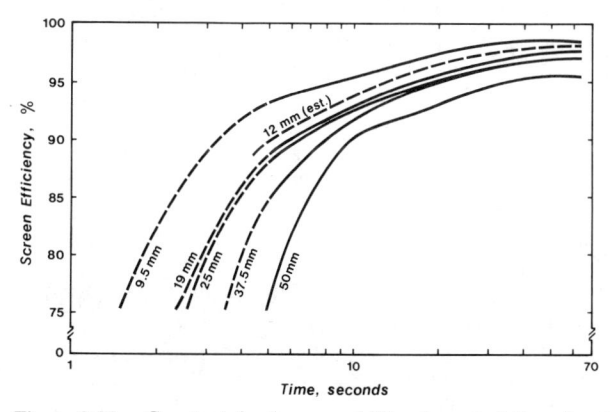

Figure 9.28. Constant depth screenability characteristics of wet coal fines. (Courtesy Triple/S Dynamics, Inc.)

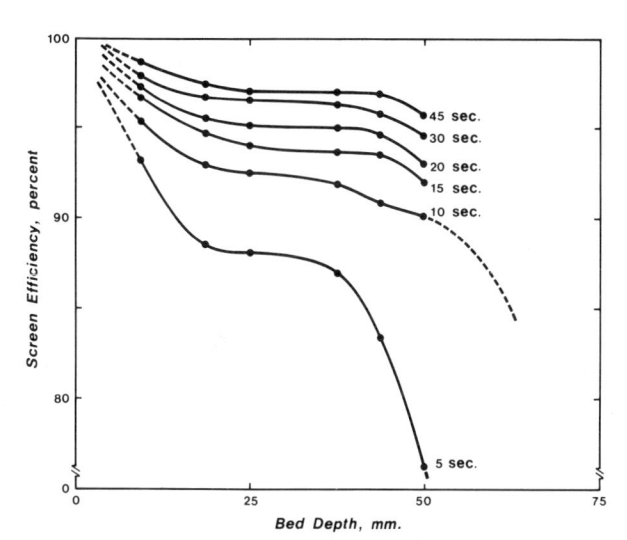

Figure 9.29. Constant time screenability characteristics of wet coal fines. (Courtesy of Triple/S Dynamics, Inc.)

test. The bed depth may be obtained from the formula[30]:

$$\delta_I = \frac{I}{\rho_b v B} \quad (E9.3.1)$$

(v = travel rate along screen; δ_I = initial bed depth; B = width of screen).

$$\therefore \delta_I = \frac{18.3 \text{ t}}{\text{hr}} \left| \frac{\text{m}^3}{0.835 \text{ t}} \right| \frac{\text{min}}{15.2 \text{ m}} \left| \frac{\text{hr}}{60 \text{ min}} \right| \frac{1}{0.6 \text{ m}} \left| \frac{10^3 \text{ mm}}{\text{m}} \right.$$
$$= 40 \text{ mm}$$

Residence time, τ, on the 1.2 m long closed cycle test screen is

$$\tau = \frac{1.2 \text{ m}}{\left| 15.2 \text{ m} \right.} \left| \frac{60 \text{ sec}}{\text{min}} \right. = 4.7 \text{ sec}$$

That is, the closed cycle test gives 74% efficiency with $\delta_I = 40$ mm and $\tau = 4.7$ sec. Figure 9.28 shows that the equivalent reference point ($\delta_I = 40$ mm, $\tau = 4.7$ sec) for the test sieves produces 84% efficiency. Thus to obtain 80% efficiency on an industrial screen, the corresponding efficiency on a test sieve is

$$\frac{80\%}{74\%} \times 84\% = 90.8\%$$

Although a number of conditions can produce this result, examination of Fig. 9.29 shows that bed depths between 20 and 40 mm have little effect on efficiency. Thus select 40 mm bed depth, and to obtain an efficiency of 90.8 from Fig. 9.28, $\tau = 8.8$ sec.

Using a travel rate of 15.2 m/min

$$L = \frac{8.8 \text{ sec}}{\left| \text{min} \right.} \left| \frac{15.2 \text{ m}}{\text{min}} \right| \frac{\text{min}}{60 \text{ sec}} = 2.23 \text{ m}$$

say a 2.44 meter screen.
The total screen width, from Eq. E9.3.1., is

$$B = \frac{400 \text{ t}}{\text{hr}} \left| \frac{\text{m}^3}{0.835 \text{ t}} \right| \frac{\text{min}}{15.2 \text{ m}} \left| \frac{\text{hr}}{60 \text{ min}} \right| \frac{1}{40 \text{ mm}} \left| \frac{10^3 \text{ mm}}{\text{m}} \right.$$
$$= 13.1 \text{ m}.$$
$$A = 13.1 \times 2.23 = 29.2 \text{ m}^2$$

Using screens 1.5 m × 2.4 m, nine screens would be required. This could be achieved with three triple deck screens or some other suitable combination.

Alternatively, using a 50 mm bed depth, $\tau = 11.5$ sec.

$$L = \frac{11.5 \text{ sec}}{\left| \text{min} \right.} \left| \frac{15.2 \text{ m}}{\text{min}} \right| \frac{\text{min}}{60 \text{ sec}} = 2.9 \text{ m}$$

$$B = \frac{400 \text{ t}}{\text{hr}} \left| \frac{\text{m}^3}{0.835 \text{ t}} \right| \frac{\text{min}}{15.2 \text{ m}} \left| \frac{\text{hr}}{60 \text{ min}} \right| \frac{1}{50 \text{ mm}} \left| \frac{10^3 \text{ mm}}{\text{m}} \right.$$
$$= 10.5 \text{ m} = \text{six 1.8 m screens}$$

$$A = 10.5 \text{ m} \times 2.9 \text{ m} = 30.5 \text{ m}^2$$

which is similar to that calculated above. However, this screening could be satisfied by a six-screen deck of 2.9 × 10.5 m screens, a cheaper solution.

A word of caution: that the two methods gave the same solution in this problem should not be taken as a vindication of the first method. Substantial divergence is far more likely.

It should be remembered that the main difficulty in screening is the passage of near-submesh particles. In many cases and in particular those where a significant proportion of the undersize is near-mesh, significant reductions in screen area requirements (for a given efficiency) can be achieved by using screen apertures larger than the desired cut point. Under these conditions the loss of product quality or recovery may be slight by comparison with a significant reduction in capital investment.

REFERENCES

1. Anon., *Testing Sieves and Their Uses*, Handbook 53, W. S. Tyler, Mentor (1973).
2. T. Allen, *Particle Size Measurement*, 2nd ed., Ch. 5, Chapman & Hall (1974).
3. B. H. Kaye, "Investigation into the Possibilities of Developing a Rate Method for Sieve Analysis," *Powder Metall.*, **5**, 199–217 (1962).
4. K. T. Whitby, "The Mechanics of Fine Sieving," Symp. Particle Size Measurement, ASTM Spec. Tech. Publ. No. 234, pp. 3–23 (1958).
5. ANSI/ASTM E11-70 (reapproved 1977), "Standard Specification for Wire Cloth Sieves for Testing Purposes," in *Annual Book of ASTM Standards*, Part 41, ASTM (Issued Annually).
6. C. W. Matthews, "What You Should Know About Screening," *Rock Prod.*, **72**, 44–51 (August 1969).
7. C. W. Matthews, "Buyers' Guide to Screens," *Eng. Min. J.*, **172**, 53–65 (Dec. 1971).
8. C. W. Matthews, "Special Screens Solve a Wide Variety of Screening Problems," *Rock Prod.*, **72**, 61–70, 99–101 (October 1969).

References

9. Anon., "Screens," *Coal Min. Process.*, **10**, 40–46 (October 1973).
10. D. J. Bevan and T. Martyn, "The Dry Screening Centrifuge," *Mine Quarry*, **1**, 57–63 (May–June 1972).
11. M. P. Armstrong et al., "The Dry Extraction of Small Coal and Fines from Moist Raw Coal," *Mine Quarry*, **2**, 41–57 (June 1973).
12. C. W. Matthews, "The Screening Media Dilemma: Perforated Plate, Wire Cloth, or Profile Wire," *Rock Prod.*, **72**, 90–96, 114–115 (November 1969).
13. C. W. Matthews, "The New Look in Screening Media," *Rock Prod.*, **73**, 75–82, 90 (February 1970).
14. Anon., "The Screening Surface: Which is the Right One?", *Coal Min. Process.*, **10**, 46–51 (October 1973).
15. Anon., "Screening Developments," *Min. Miner. Eng.*, **4**, 38–40 (August 1968).
16. B. L. A. Dehlen, "Rubber in Vibrating Screens—An Efficient Weapon Against Wear, Noise and Dust," *Screening and Grading of Bulk Materials*, pp. 5–12, Institution of Mechanical Engineers, London (1975).
17. H. J. Ephithite, "The Applications of Rubber in the Screening of Bulk Materials," *Screening and Grading of Bulk Materials*, pp. 13–21, Institution of Mechanical Engineers, London (1975).
18. Anon., "New Wear-Resistant Surface from U.S. Company," *Min. J.*, **288**, 29 (Jan. 14, 1977).
19. O. Jones and R. F. Swift, "The Application of the Loose Rod Deck in the Mining Industry," *Mine and Quarry*, **1**, 25 (July 1972).
20. J. H. Healy et al., "Erie Mining Co. Patents New Method for Screening Ores in Very Fine Size Ranges," *Min. Eng.*, **19**, 65–70 (April 1967).
21. R. O. Burt, "Fine Sizing of Minerals," *Min. Mag.*, **128**, 463–465, (June 1973).
22. F. Mogensen, "A New Method of Screening Granular Materials," *Quarry Managers' J.*, **49**, 409–414 (October 1965).
23. L. A. Adorján, "Mineral Processing," *Min. Annu. Rev.*, 233–268 (1978).
24. L. A. Adorján, "Mineral Processing," *Min. Annu. Rev.*, 233–269 (1979).
25. C. W. Matthews, "Screen Installations to Meet Maintenance and Environmental Requirements," *Rock Prod.*, **73**, 82–84, 87 (April 1970).
26. J. J. Bandholz, "Horizontal Screen Blinding—Cause and Cures," *Rock Prod.*, **72**, 96–97 (February 1969).
27. M. Jean, "Use of Heated Screens in the Sarée-et-Moselle Group of the Houillères du Bassin de Lorraine," 3rd Int. Coal Prep. Congr., Paper C6, pp. 354–360, *Ann. Mines Belg.* (1958).
28. E. Burstlein, "Some Advances in Direct Screening of Moist Materials," *Iron Coal Trades Rev.*, **178**, 791–798 (Apr. 3, 1959).
29. S. E. Gluck, "Some Technological Factors Affecting the Economics of Screening," *J. Met.*, **18**, 373–377 (March 1966).
30. J. F. Sullivan, *Screening Technology Handbook*, Triple/S Dynamics Inc. (1975).
31. A. M. Gaudin, *Principles of Mineral Dressing*, McGraw-Hill (1939).
32. S. E. Gluck, "Vibrating Screens," *Chem. Eng.*, **75**, 151–168 (Feb. 15, 1968).
33. J. C. Williams and G. Shields, "The Segregation of Granules in a Vibrated Bed," *Powder Technol.*, **1**, 134–142 (1967).
34. M. L. Jansen and J. R. Glastonbury, "The Size Separation of Particles by Screening," *Powder Technol.*, **1**, 334–343 (1967–1968).
35. J. M. Dalla Valle, *Micromeritics*, Pitman Publishing Corp. (1948).
36. Anon., "High Efficiency Iron Ore Screening Plant," *Min. Mag.*, **129**, 529–535 (December 1973).
37. R. Feller and A. Foux, "Screening Duration and Size Distribution Effects on Sizing Efficiency," *J. Agric. Eng. Res.*, **21**, 347–353 (1976).
38. J. E. English, "A New Approach to the Theoretical Treatment of the Mechanics of Sieving and Screening," *Filtr. Sep.*, **11**, 195–203 (1974).
39. H. E. Rose, "Mechanics of Sieving and Screening," *Trans. IMM (C)*, **86**, C101–C114 (September 1977).
40. (a) J. L. Olsen and E. G. Rippie, "Segregation Kinetics of Particulate Solid Systems. Part I," *J. Pharm. Sci.*, **53**, 147–150 (February 1964).
    (b) E. G. Rippie, J. L. Olsen, and M. D. Faiman, "Segregation Kinetics of Particulate Solid Systems. Part II," *J. Pharm. Sci.*, **53**, 1360–1363 (November 1964).
41. W. J. Whiten, "The Simulation of Crushing Plants with Models Developed Using Multiple Spline Regression," *J. S. Afr. IMM*, 257–264 (May 1972).
42. (a) S. Miwa, "Proposal of a New Index for Expressing the Performance of Screens," *Chem. Eng. (Japan)*, **24**, 150–155 (1960).
    (b) C. Orr, *Particulate Technology*, Macmillan (1966).
43. T. Brereton and K. R. Dymott, "Some Factors Which Influence Screen Performance," *Proc. 10th Int. Miner. Process. Congr., London, 1973*, pp. 181–194, IMM (1974).
44. G. Ferrara and U. Preti, "A Contribution to Screening Kinetics," *Proc. 11th Int. Miner. Process. Cong., Cagliari, 1975*, pp. 183–217, Università da Cagliari (1975).
45. A. Meinel and H. Schubert, "Über einige zusammenhänge Zwischen der Einzelkorndynamik und der stochastischen sieb Theorie bei der Klassierung auf Stösselschwingsiebmaschinen," *Aufbereitungs-Techn.*, **7**, 408–416 (1972).
46. A. Meinel and H. Schubert, "Zu den Grundlagen der Feinsiebung," *Aufbereitungs-Techn.*, **3**, 128–133 (1971).
47. C. W. Schultz and R. B. Tippin, "Fundamentals of Statistical Screening," *Trans. AIME/SME*, **247**, 314–316 (December 1970).
48. C. W. Hoffman and W. R. Hinken, "Probability Sizing—Principles, Problems and Development in the Mining Industry," *Trans. AIME/SME*, **244**, 149–153 (June 1969).
49. F. J. Fontein, "Some Variables Influencing Sieve-Bend Performance," *AIChE–Inst. Chem. Eng. Symp. Ser.* No. 1, London, pp. 1:123–1:130 (1965).
50. L. D. Keller, "Fine Screening: The Current State of the Art," AIME/SME Preprint 71-B-25, AIME (1971).

51. "Product Facts," Bulletin No. 2100.0601, Hewitt-Robins (1968).
52. "Selection of Vibrating Machines," Bulletin No. PM 1.1, Allis-Chalmers (1963).
53. "Nordberg Process Machinery: Reference Manual," Form No. 415, Rexnord (1976).
54. S. E. Gluck, "Vibrating Screens: Surface Selection and Capacity Calculations," *Chem. Eng.*, **72**, 179–184 (March 15, 1965).
55. V. K. Karra, "Development of a Model for Predicting the Screening Performance of a Vibrating Screen," *CIM Bull.*, **72**, 167–171 (April 1979).

Chapter Ten

Classification

Blaauw's Law: *"Established technology tends to persist in the face of new technology."*

Classification is the separation of particles according to their settling rate in a fluid. Because water is the fluid most commonly used in mineral processing, this chapter considers primarily wet classification, but it should be appreciated that in most cases the principles can be extended to other fluids, particularly air.

As with most separating devices, classifiers typically produce two products. The positive response stream contains the faster settling particles and is called the *sands, underflow,* or *oversize*. This stream is generally low in water, which necessitates the removal of the stream from the device by gravity, mechanical means, or induced pressure. The negative response stream is called the *overflow* or *slimes*. It contains the slower settling particles in the remaining water, and essentially discharges as a result of displacement by the incoming feed. By varying the effective separating force, classification may also be carried out in stages to produce intermediate products or size fractions.

Although classification generally aims to separate particles by size, particle density and other factors also have a significant effect and the operation is more realistically viewed as one of sorting rather than sizing. As a consequence, classifications may be applied in a number of situations:

1. Separation into relatively coarse and relatively fine size fractions, typically for separations too fine to be economically screened.
2. To effect a concentration of smaller heavier particles from larger lighter particles.
3. To split a long-range size distribution into fractions.
4. To restrict the property distribution of particles entering a concentration process.
5. To control closed circuit grinding.

10.1. EQUIPMENT

Classifiers are characterized by two overlapping features: the method of discharging the sands product, and the reference point for relative water/particle motion.

Mechanical classifiers use some mechanical means for removing the sands products from the device: generally this mechanism operates against gravity. *Nonmechanical classifiers* rely for sands discharge on the flow properties of the sands stream, aided by gravitational or centrifugal forces. In *sedimentation classifiers* the faster settling particles settle through a pool of water that is formed out of the feed stream (Fig. 10.1a). *Hydraulic* or *fluidized bed classifiers* require the particle to settle against an upward flowing stream of supplementary hydraulic water (Fig. 10.1b). In general, the latter situation results in a separation under hindered settling conditions with comparatively high solids concentrations.

A wide range of equipment is available[1-6] and some of the more common types are described in Table 10.1, which also discusses their suitability and applications. Figures 10.2–10.7 show a selection of wet classifiers; a typical air classifier (separator) is shown in Fig. 10.8. Other classifiers have been described in the literature, but they have not as yet found widespread acceptance.[7-9]

Selection of a classifier for a particular applica-

TABLE 10.1: THE MAJOR TYPES OF CLASSIFIERS

CLASSIFIER	(Type*)	DESCRIPTION	SIZE (m) - Width - Diameter - Max. Length	LIMITING SIZE (Max. Feed Size)	FEED RATE (t/hr)	VOL. % SOLIDS - Feed - Overflow - Sands	POWER (kW)	SUITABILITY AND APPLICATIONS
Sloping Tank Classifier (spiral, rake, drag)	(M-S)	Classification occurs near deep end of sloping, elongated pool. Spiral (Fig. 10.4), rake or drag mechanism lifts sands from pool.	0.3 to 7.0 2.4 (spiral) 14	1 mm to 45 μm (25 mm)	5 to 850	Not critical 2 to 20 45 to 65	0.4 to 110	Used for closed circuit grinding, washing and dewatering, desliming; particularly where clean dry sands are important. (Drag classifier sands not so clean). In closed circuit grinding discharge mechanism (spirals especially) may give enough lift to eliminate pump.
Log Washer	(M-S)	Essentially a spiral classifier with paddles replacing the spiral. (Fig. 10.3).	0.8 to 2.6 0.6 to 1.1 4.6 to 11	(100 mm)	40 to 450		7.5 to 60	Used for rough separations such as removing trash, clay from sand. Also to remove or break down agglomerates.
Bowl Classifier	(M-S)	Extension of sloping tank classifiers, with settling occuring in large circular pool, which has rotating mechanism to scrape sands inwards (outwards in Bowl Desiltor) to discharge rake or spiral.	0.5 to 6.0 1.2 to 15 12	150 μm to 45 μm (12 mm)	5 to 225	Not critical 0.4 to 8 50 to 60 (15 to 25 in Bowl Desiltor)	Bowl: 0.75 to 7.5 Rake: 0.75 to 20	Used for closed circuit grinding (particularly regrind circuits) where clean sands are necessary. Larger pool allows finer separations. Bowl Desiltor has larger pools (and capacities). Relatively expensive.
Hydraulic Bowl Classifier	(M-F)	Basically a hydraulic bowl classifier. Vibrating plate replaces rotating mechanism in pool. Hydraulic water passes through perforations in plate and fluidises sands.	1.2 to 3.7 1.2 to 4.3 12	1 mm to 100 μm (12 mm)	5 to 225	Not critical 2 to 15 50 to 65	Vib: 2.2 to 7.5 Rake: 3.7 to 15	Gives very clean sands and has relatively low hydraulic water requirements (0.5 t/t sand). One of the most efficient single stage classifiers available for closed circuit grinding and washing. Relatively expensive.
Cylindrical Tank Classifier	(M-S)	Effectively an overloaded thickener. Rotating rake feeds sands to central underflow.	- 3 to 45 -	150 μm to 45μm (6 mm)	5 to 625	Not critical 0.4 to 8 15 to 25	0.75 to 11	Simple, but gives relatively inefficient separation. Used for primary dewatering where the separations involve large feed volumes, and sand drainage is not critical.
Hydraulic Cylindrical Tank Classifier	(M-F)	Hydraulic form of overloaded thickener. Siphon-Sizer (N-F) uses siphon to discharge sands instead of rotating rake.	- 1.0 to 40 -	1.4 mm to 45 μm (25 mm)	1 to 150	Not critical 0.4 to 15 20 to 35	0.75 to 11	Two-product device giving very clean sands. Requires relatively little hydraulic water (2 t/t sand). Used for washing, desliming, and closed circuit grinding.
Cone Classifier	(N-S)	Similar to cylindrical tank classifier, except tank is conical to eliminate need for rake.	- 0.6 to 3.7 -	600 μm to 45 μm (6 mm)	2 to 100	Not critical 5 to 30 35 to 60	None	Low cost (simple enough to be made locally), and simplicity can justify relatively inefficient separation. Used for desliming and primary dewatering. Solids build-up can be a problem.
Hydraulic Cone Classifier	(M-F)	Open cylindrical upper section with conical lower section containing slowly rotating mechanism.	- 0.6 to 1.6 -	400 μm to 100μm (6 mm)	10 to 120	Not critical 2 to 15 30 to 50	3 to 7.5	Used primarily in closed circuit grinding to reclassify hydrocyclone underflow.

TABLE 10.1: (Continued)

CLASSIFIER	(Type*)	DESCRIPTION	SIZE (m) - Width - Diameter - Max. Length	LIMITING SIZE (Max. Feed Size)	FEED Rate (t/hr)	VOL. % SOLIDS - Feed - Overflow - Sands	POWER (kW)	SUITABILITY AND APPLICATIONS
Hydrocyclone	(N-S)	(Pumped) pressure feed generates centrifugal action to give high separating forces, and discharge (Fig. 10.6)	– 0.01 to 1.2 –	300 μm to 5 μm (1400 μm to 45 μm)	to 20 m³/min	4 to 35 2 to 15 30 to 50	35 to 400 kN/m² pressure head	Small cheap device, widely used for closed circuit grinding. Gives relatively efficient separations of fine particles in dilute suspensions.
Air Separator	(N-S)	Similar shape to hydrocyclone, but higher included angle. Internal impellor induces recycle within classifier (Fig. 10.8)	– 0.5 to 7.5 –	2 mm to 38 μm	to 2100		4 to 500	Used where solids must be kept dry, such as cement grinding. Air classifiers may be integrated into grinding mill structure. (Fig. 8.9).
Solid Bowl Centrifuge	(M-S)	Power generates high settling forces. Slurry centrifuged against rotating bowl, and removed by slower rotating helical screw conveyor within bowl.	– 0.3 to 1.4 1.8	74 μm to 1 μm (6 mm)	0.04 to 2.5 m³/min	2 to 25 0.4 to 20 5 to 50	11 to 110	Relatively expensive, but high capacity for a given floor space; used for finer separations.
Scrubber	(M-S)	Essentially a rotating drum mounted on slight incline. (Fig. 10.2)	– 1.5 to 3.5 3 to 10	(450 mm)	to 700		1 to 55	Similar applications to log washer, but lighter action. Tumbling (85% critical speed) provides attrition to remove clay from sand. Also removes trash.
Counter Current Classifier	(M-F)	One form based on scrubber, another on spiral classifier. They have wash water added to flow essentially horizontally in opposite direction to sands which are conveyed and resuspended by some form of spiral.	– 0.5 to 3.3 (spiral type) 12 (spiral type)	2 mm to 40 μm	3 to 600	Not critical 2 to 15 50 to 65	0.2 to 19	Very clean sands product, but relatively low capacity for a given size.
Elutriator	(N-F)	Basically a tube with hydraulic water fed near bottom to produce hindered settling. Sands withdrawn through valve at base. Column may be filled with network to even out flow. (Fig. 10.7)	– 1.2 to 4.3 –	2.4 mm to 100 μm (7.5 mm)	4 to 120	15 to 35 0.4 to 5 20 to 35	0.75 for valves	Simple and relatively efficient separation. Normally a two product device but may be operated in series to give a range of size fractions.
Pocket Classifier	(N-F)	A series of classification pockets, with decreasing quantities of hydraulic water in each, producing a range of product sizes. (Fig. 10.5)	0.5 to 6.0 – 12	2.4 mm to 100 μm (10 mm)	4 to 120	15 to 35 0.4 to 5 20 to 35		Efficient separations, but requires 3 t hydraulic water/t sand. Used to produce exceptionally clean sands fractioned into narrow size ranges.

*M: Mechanical transport of sands to discharge
N: Non-mechanical (gravity or pressure) discharge of sands
S: Sedimentation classifier
F: Fluidised bed classifier

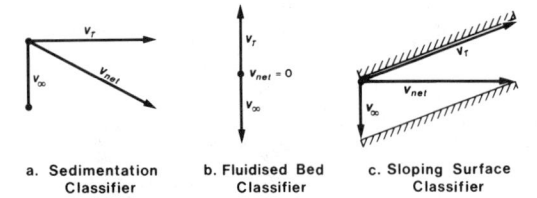

Figure 10.1. Characterization of classifiers on the basis of operating principle.

- a. Sedimentation Classifier
- b. Fluidised Bed Classifier
- c. Sloping Surface Classifier

tion can be narrowed down fairly easily, and there normally is a single machine which is best for a given situation. The most common need for a classifier is in closed circuit grinding, and since its introduction in the 1950s, the Hydrocyclone has rapidly become the industry standard, mainly because of its mechanical simplicity, low capital cost, and small space requirements.

10.2. CLASSIFIER PERFORMANCE

No universally acceptable theory exists for classifiers. Thus this chapter reviews a number of

Figure 10.3. Log washer. (Courtesy McLanahan Corp.)

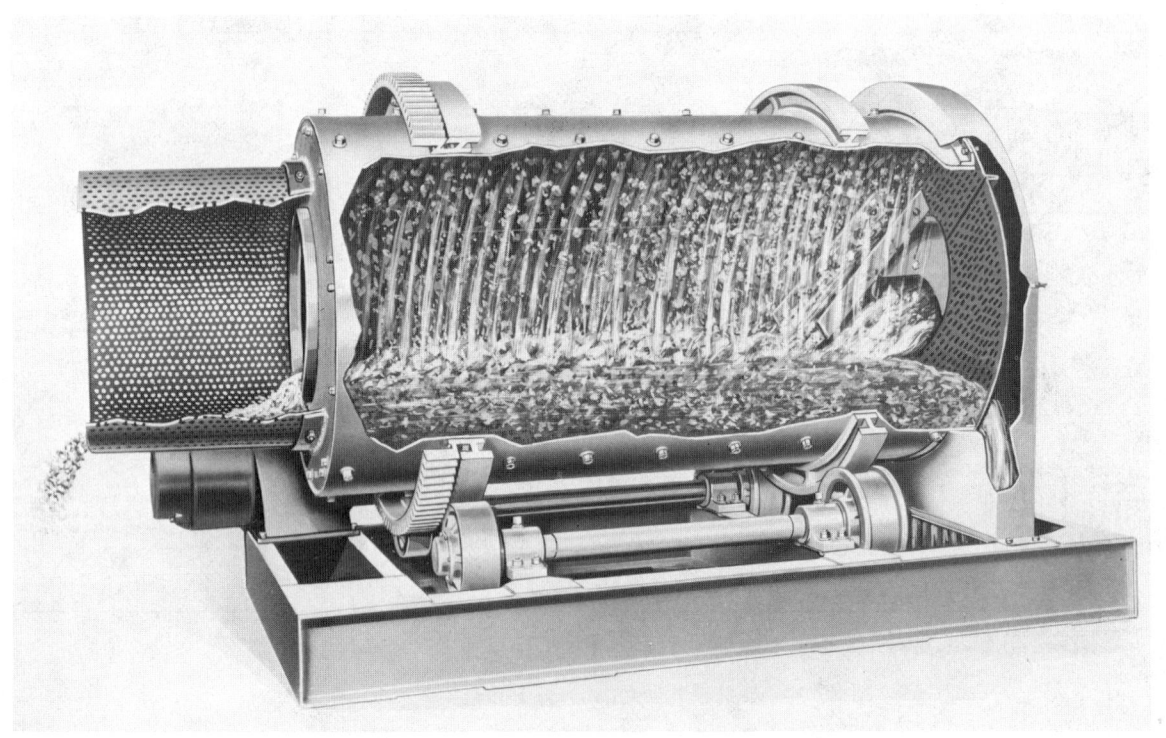

Figure 10.2. Drum washer. (Courtesy Barber-Greene Co.)

Classifier Performance

Figure 10.4. Spiral classifier. (Courtesy Mine and Smelter Division of Barber-Green Co.)

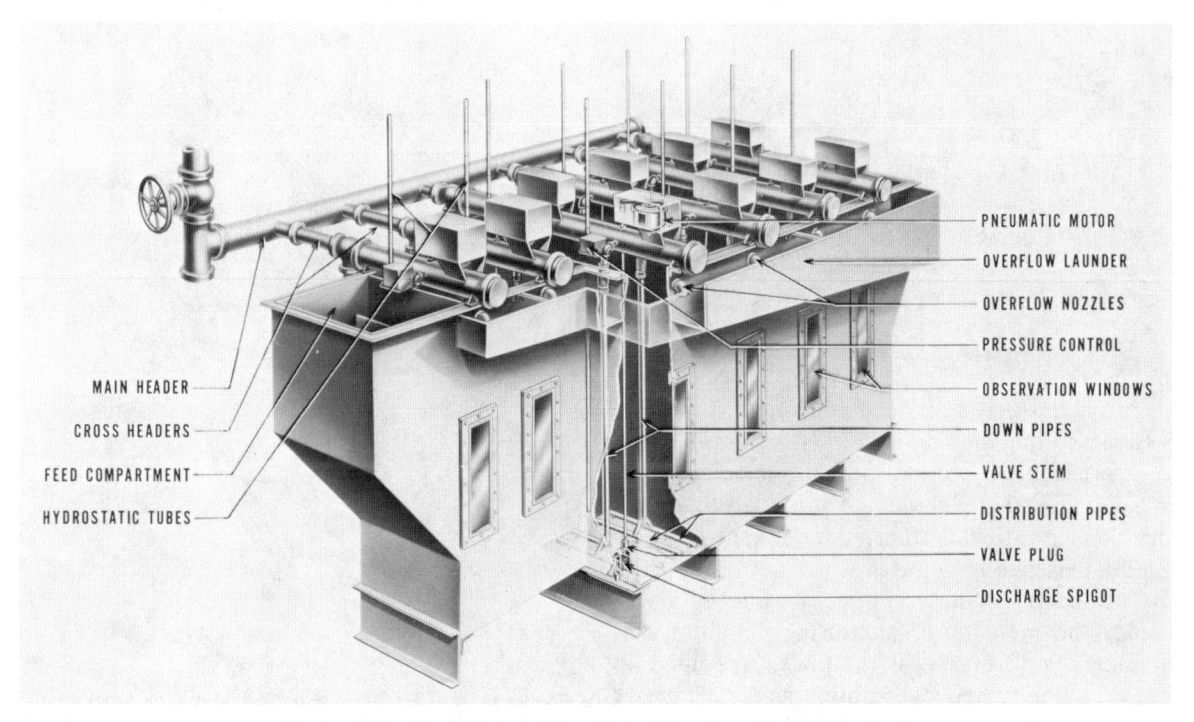

Figure 10.5. Pocket sizer: $9\frac{1}{2}$ pockets. (Courtesy Dorr-Oliver Inc.)

time, although both solid and liquid components must be specified to define it fully. The size of a separation usually quoted, especially by equipment manufacturers, is the size of screen that will retain a certain percentage of solids in the overflow. Generally, this percentage is in the range 1–3% and in keeping with this notation the separating sizes in Table 10.1 are quoted for the $1\frac{1}{2}\%$ level. Realistically, though, this definition of separation size is inadequate and we term it the *limiting size*.

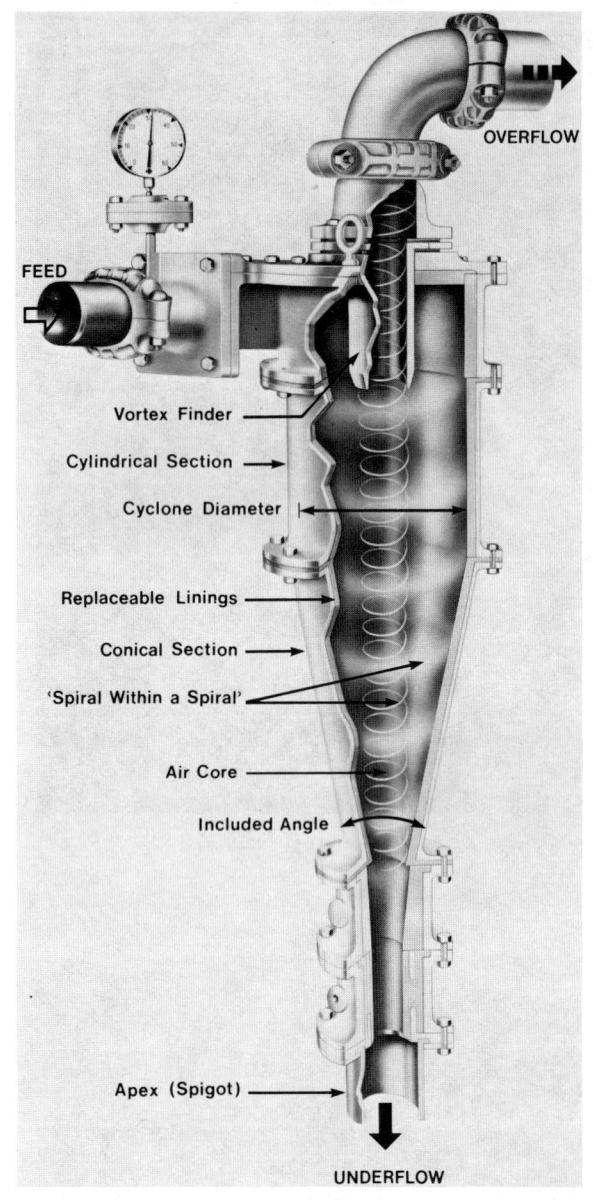

Figure 10.6. Hydrocyclone. (Courtesy Krebs Engineers.)

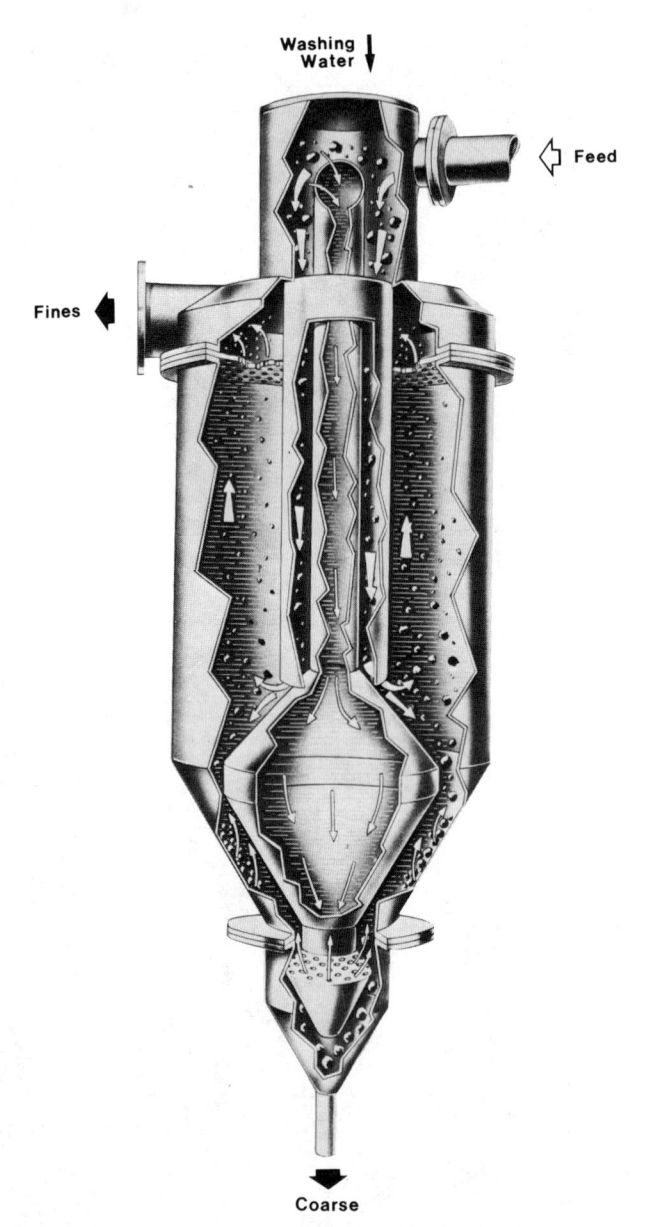

Figure 10.7. Elutriator. (Courtesy Alsthom Atlantique Co.—France.

the approaches that have been published. In spite of the coverage included here, it must be emphasized that at present it is still not possible to reliably design a classifier that will operate at moderate to high slurry concentrations; rather, a manufacturer's empirical experience is required.

Before reviewing any theories, it is necessary to consider the methods of measuring classifier performance. Performance is normally specified in terms of capacity and "separation size." Capacity is obviously the quantity of material fed per unit

Classifier Performance

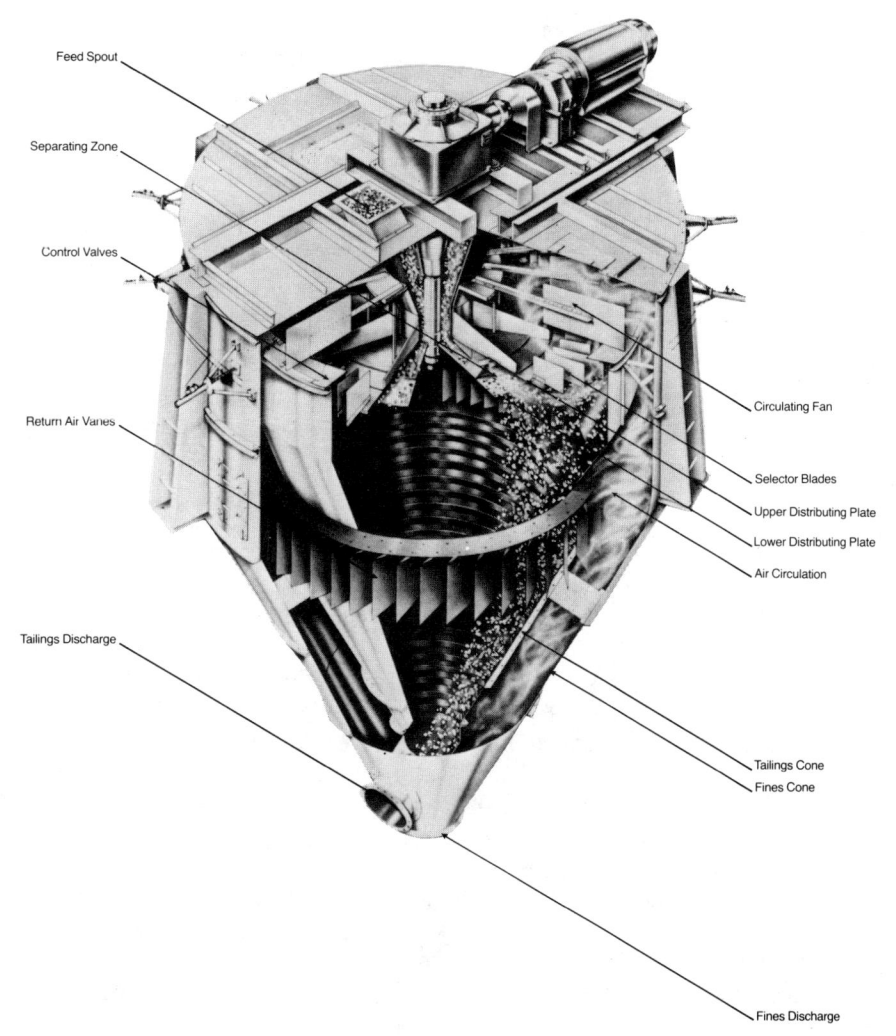

Figure 10.8. Air separator. (Courtesy Sturtevant Mill Co.)

A far better description of the separation is given by the performance curves introduced in Chapter 3. Since a separation is commonly described in terms of particle size, the performance curve is presented on a plot of mass fraction of size d_n in the feed passing to the underflow versus size d_n. A typical curve is illustrated in Fig. 10.9a. It must be emphasized that although the particle diameter scale is usually in terms of aperture diameter, this is a different nominal diameter from that actually "seen" by the classifier, which is a nominal diameter based on settling conditions inside the classifier. The performance curve for perfect classification is a vertical line about the "cut size." In practice the curve is spread along the size axis, an effect that can arise from the initial dispersion of particles across the feed inlet, and from dispersal due to disturbances during passage through the classification pool. The curve may also show two other imperfections, indicated by y_1 and y_2 in Fig. 10.9a. These are effectively caused by part of the inlet stream passing out of the classifier without being classified (Fig. 10.9b). That denoted by y_1 is the more common, and is due to slimes in the underflow. It can reliably be estimated by assuming that the finest particles behave like water and therefore the fraction of these particles y_1 in the underflow will be the same as the fraction of the feed water leaving through the underflow. The imperfection y_2 is due to feed material bypassing to the overflow. Generally it is negligible, although it may appear to exist when the particles have a range of densities (Section 10.7.2).

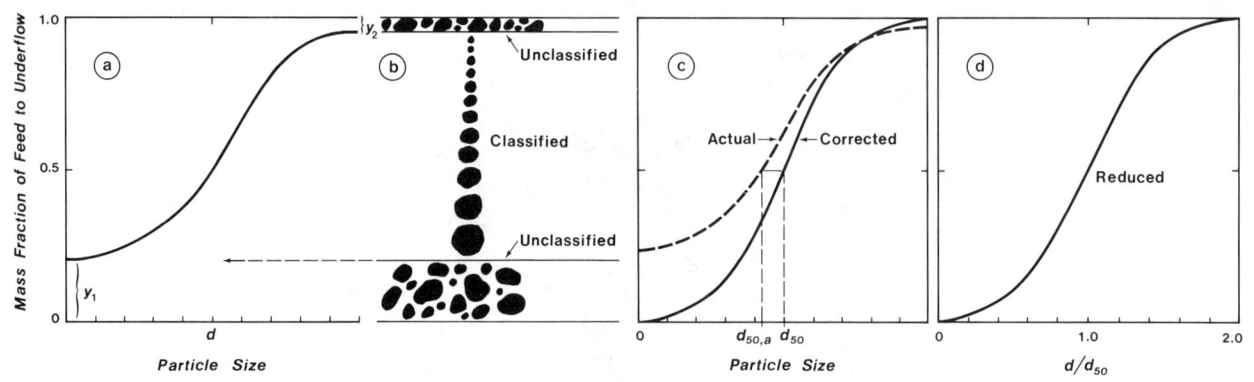

Figure 10.9. (a) Characteristic form of a sedimentation classifier performance curve. (b) Relationship of the performance curve to particle classification. (c) The corrected performance curve. (d) The reduced performance curve.

The possibility of y_2 occurring also shows the danger of the conventional measure of separation size (limiting size): clearly the steep section of the performance curve is the true measure of the separation size, whereas the 1–3% size of the overflow tends to be near the upper section of the performance curve, which is nearly horizontal and therefore prone to error.

A clearer representation of the actual classification can be obtained from the *corrected performance curve*, shown in Fig. 10.9c. It is obtained by applying the following relationship to each size:

corrected fraction of size d_n to underflow
$$= \frac{\text{actual fraction of size } d_n \text{ to underflow} - y_1}{1 - y_1 - y_2}$$

that is,

$$\mathcal{R}_{(+)} = \frac{\mathcal{R}_{(+)a} - y_1}{1 - y_1 - y_2} \quad (10.1)$$

($\mathcal{R}_{(+)}$ = corrected recovery in positive response outlet, i.e., the underflow; $\mathcal{R}_{(+)a}$ = actual or uncorrected recovery in positive response outlet).

A significant parameter of this curve is the d_{50} size, the size of particle that as a result of the *classifying* action, has an equal chance of reporting to the underflow or to the overflow. (It should be noted that d_{50} must be obtained from the corrected performance curve; the d_{50} on the performance curve is only an *apparent* $d_{50,a}$.) By virtue of this representation and its position on the corrected curve, d_{50} is the real measure of the *separation* or *cut size* and consequently it is defined here as the separation size, rather than the less definite 1–3% size of the overflow (the limiting size).

In turn the corrected performance curve can be normalized (made dimensionless) by dividing the size scale by d_{50}. The resulting curve, illustrated in Fig. 10.9d, is referred to as the *reduced performance curve* (Section 3.5.2). It is now widely accepted that the shape of this curve is largely independent of the nature of the solid particles and is characteristic of the type of classifier, within a reasonable range of design. However, data are still insufficient to permit us to identify conclusively the characteristic curves of the different classifiers. Even so, the curves can be powerful tools in assessing classifier efficiency.

Figure 10.10 shows the range of reduced performance curves commonly found in practice. The sharp separation (curve A) is obtainable in hydro-

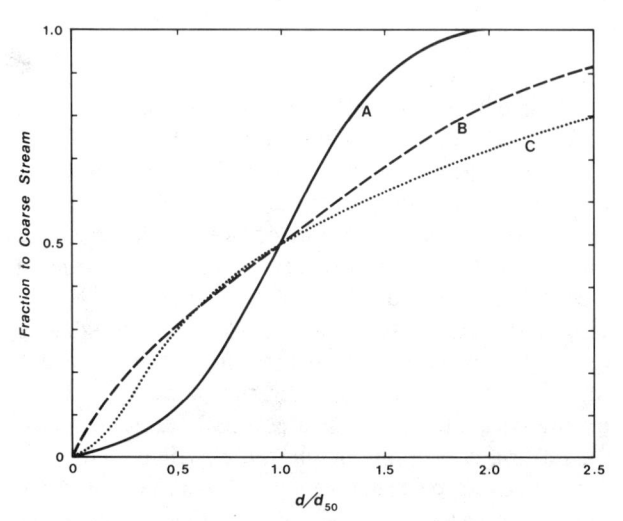

Figure 10.10. In practice, commercial sedimentation classifiers show a range of reduced performance curves, from comparatively efficient (curve A) to low efficiency (curves B and C).

Sedimentation Classifiers

cyclones, up to moderate slurry concentrations, when the feed contains only one mineral.[10] Sharp separations are also achieved by mechanical classifiers (spiral, rake) discharging relatively low tonnages of solid to the overflow (expressed as tonnes of solid per meter of weir per minute), and by fluidized bed classifiers with small proportions of hydraulic water. On the other hand, the dispersed separation (curves B and C) is more characteristic of overloading, such as occurs when hydrocyclones are used for closed circuit grinding.[11] It also occurs when a mechanical classifier discharges a relatively high tonnage of solids to the overflow.[12]

A variety of equations have been presented to describe the reduced performance curves of classifiers. Most equations have two parameters, one to represent the separation size (generally d_{50}) and a second to represent the sharpness of classification. A few equations have been presented with a third parameter (for the shape of the curve). This gives a better description, but the additional complexity is seldom justified.

To estimate the performance of a classifier it is necessary to know the effect of variables on the performance curve. A considerable amount of work has been carried out in this respect on hydrocyclones, but with only limited success. Because of their decreasing importance, other classifiers have received less attention, with the result that they are even less well understood. Although the separation in both sedimentation and fluidized bed classifiers is based on the volumetric flow rate per unit area of pool (Eq. 4.83), the way this is achieved in each case is different, and consequently the two types must be considered separately. Hydrocyclones are essentially sedimentation-type classifiers, but in view of their importance they are treated further in Section 10.4.

10.3. SEDIMENTATION CLASSIFIERS

10.3.1. Classification in the Ideal Pool

The performance of sedimentation classifiers can be developed from the concept of the ideal settling pool introduced in Section 4.7.2. Figure 10.11 is an extension of this ideal pool to include distribution walls at each end, between which plug flow occurs. In the ideal pool, the limiting size particle is that particle which has a settling rate just sufficient to allow it to settle the depth of the pool during its

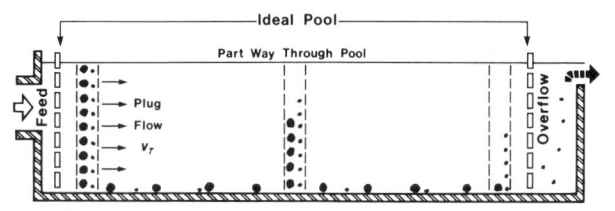

Figure 10.11. Extension of the ideal settling pool (Fig. 4.17) to include the imaginary distribution walls at each end. Also illustrated is the behavior of two sizes of particle: those close to the limiting size and those smaller. [Reprinted with permission from B. Finch, *Ind. Eng. Chem.*, **54**, 44–51 (October 1962). Copyright 1962 American Chemical Society.]

residency. Obviously all particles with higher settling rates will be collected. The d_{50} or separation size will be that particle which, if it enters the pool halfway down, just settles to the bottom, since half the particles of this size will initially be lower down and reach the bottom, and the other half will start above and therefore be carried out in the overflow. Extending this concept, it can be seen that the ideal performance curve can be constructed using the relationship

$$\mathcal{R}_{(+)} = \frac{\text{settling rate of particle of size } d}{\text{settling rate of limiting size particle}} = \frac{v_\infty}{v_{\infty l}} \quad (10.2)$$

($\mathcal{R}_{(+)}$ = [for a classifier] recovery of d-sized particle in the positive response outlet, i.e., the underflow). We return to the significance of particle size later (Section 10.6.2); sufficient to note at this stage that in principle the settling rate can be determined, using equations such as Eqs. 4.33 and 4.35 (and strictly speaking Eq. 4.70). In the case of laminar flow, the reduced performance curve resulting from Eq. 10.2 is as shown in Fig. 10.12.

The extent to which this concept can be applied to practical classifiers is open to debate. Roberts and Fitch[13,14] have shown that it can be applied to rake classifiers in particular. They argue that significant disturbance damping effects result from the fact that the slurry density increases with depth in the pool, and that even though flow patterns are nothing like horizontal, a series of flow lines exist: both features allowing the ideal pool concepts to remain valid. These conclusions of course imply that all overflow particles have the same residence time. In reality they do not; even the simplest of settling pools has a distribution of residence times, resulting from velocity profiles (from wall drag) and/or turbulence effects (from the high **Re**, corners,

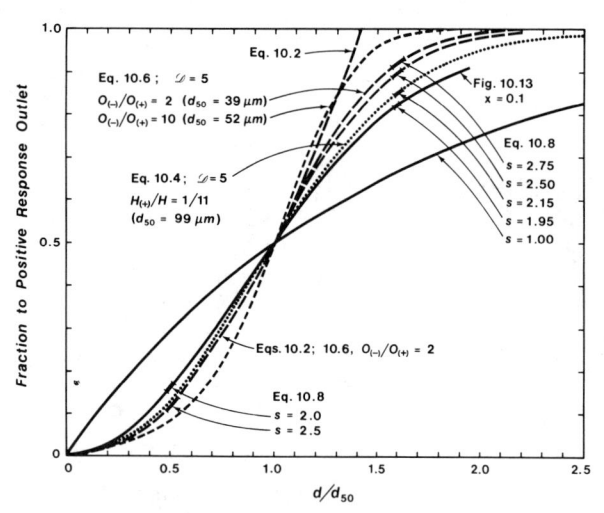

Figure 10.12. Theoretical reduced performance curves, derived from various equations in the text (assuming laminar flow).

or agitation generated by the discharge mechanism). This distribution is apparent in that most operating classifiers have a characteristically S-shaped performance curve, often with considerable tailing (Fig. 10.10), a feature not predicted by Eq. 10.2.

10.3.2. General Classifier Theories

A number of studies have extended the ideal pool concept to include dispersion effects, by introducing a diffusion parameter. The dispersion equation for the steady state transport of suspended particles in a uniform two-dimensional flow is given by[15]:

$$v_\tau \frac{\partial C}{\partial L} - v_\infty \frac{\partial C}{\partial H} = \frac{\partial}{\partial H}\left[\mathscr{D}\frac{\partial C}{\partial H}\right] + \mathscr{D}\frac{\partial^2 C}{\partial L^2} \quad (4.90)$$

(v_τ = local time-average horizontal velocity component; C = concentration of particles; $\mathscr{D}$ = diffusion coefficient; L = length along pool; H = height above the bottom of the pool). Solutions to this equation require v_τ and $\mathscr{D}$ to be expressed as functions of L, although it is generally assumed that the last term on the right-hand side is zero. The simplest solution assumes that v_τ and $\mathscr{D}$ are constant, and the results are shown graphically in Fig. 10.13.[16] In turn Fig. 10.12 shows the reduced performance curve derived from this solution, and it can be seen that it has the S shape characteristic of some classifiers, particularly hydrocyclones operating on dilute slurries. (It is worth noting that by extending the imagination, the hydrocyclone can be thought of as a very long settling pool, in spiral form.) In reality, most pools

have a high enough **Re** to imply turbulent flow, and as a consequence the true velocity profile is a logarithmic function, while $\mathscr{D}$ is a parabolic function. Such functions make Eq. 4.90 difficult, but not impossible, to solve; and although other approximations can be made to facilitate the computations, these more valid solutions[17] do not in fact produce performance curves significantly different from those given by Fig. 10.13.

Schubert and Neesse[18] also used a diffusion coefficient approach, by proposing that in an equilibrium dispersion of particles, the settling is opposed by turbulent diffusion, that is,

$$\mathscr{D}\frac{\partial C}{\partial H} = -v_\infty C \quad (4.89)$$

They consider that classifiers can be represented by one of two concepts, illustrated in Fig. 10.14. In the *partition* concept, the slurry is simply divided at a particular height (strictly by volume), so that for each particle size

$$\mathscr{R}_{(+)a} = \frac{H_{(+)}}{H} \cdot \frac{C_{(+)}}{C_I} \quad (10.3)$$

($C_{(+)}$ = concentration of a given particle size in the positive response stream; C_I = concentration of given particle size in the feed; $H_{(+)}$ = height [strictly, volume fraction] of positive response cut; H = height of slurry in classifier; $\mathscr{R}_{(+)a}$ = actual or

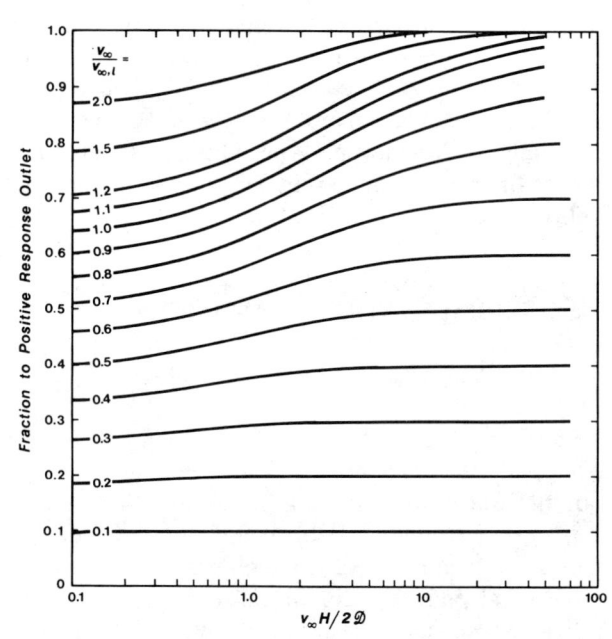

Figure 10.13. Camp's solution of Eq. 4.90. (After Camp.[16])

Sedimentation Classifiers

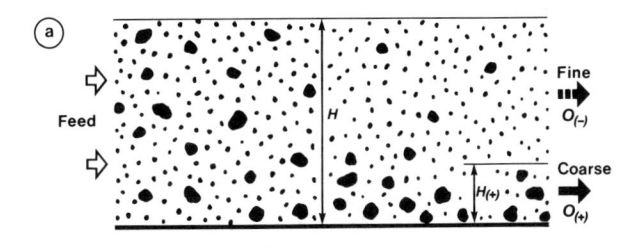

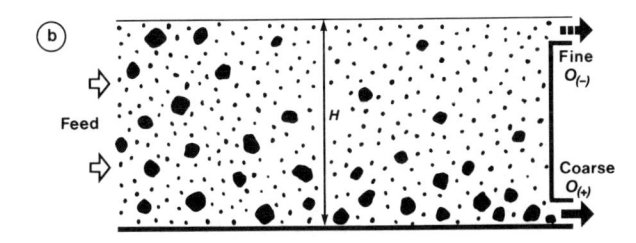

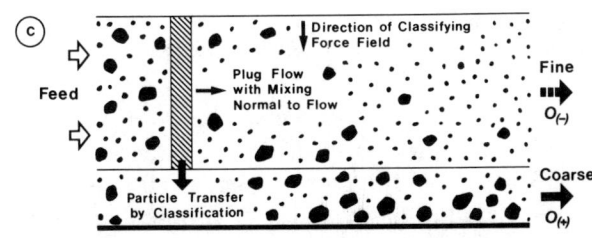

Figure 10.14. (*a*) Partition representation of a classifier. (After Schubert and Neesse.[18]) (*b*) Tapping representation of a classifier. (After Schubert and Neesse.[18]) (*c*) Reid representation of a classifier. (After Reid.[19])

uncorrected $\mathscr{R}_{(+)}$). From Eqs. 4.89 and 10.3 it can be shown that for each particle size

$$\mathscr{R}_{(+)a} = \frac{1 - \exp(-v_\infty H_{(+)}/\mathscr{D})}{1 - \exp(-v_\infty H/\mathscr{D})} \quad (10.4)$$

In the *tapping* concept, a fine and a coarse product are drawn from the top and bottom of the classifier, respectively (Fig. 10.14*b*):

$$\mathscr{R}_{(+)a} = \frac{C_{(+)}O_{(+)}}{C_{(+)}O_{(+)} + C_{(-)}O_{(-)}} \quad (10.5)$$

($C_{(-)}$ = concentration of given size particle in the negative response stream, i.e., the overflow; $O_{(+)}$ = flow rate in positive response stream; $O_{(-)}$ = flow rate in negative response stream), which when combined with Eq. 4.89 leads to

$$\mathscr{R}_{(+)a} = \frac{1}{1 + \dfrac{O_{(-)}}{O_{(+)}} \exp\left(\dfrac{-v_\infty H}{\mathscr{D}}\right)} \quad (10.6)$$

Equations 10.4 and 10.6 also give rise to S-shaped performance curves. For a given set of parameters, the latter equation gives a considerably steeper performance curve than the former, but in terms of reduced performance curves, the difference is significantly less (Fig. 10.12). It is of course debatable which concept is more valid. The authors show that although theoretically the tapping concept represents the better performance, operating classifiers cannot achieve this level. In fact, rake and spiral classifiers, which appear to fit the tapping concept best, generally have a tailing performance curve more characteristic of the partition concept.

There are two major restrictions on these diffusional treatments: not only are $\mathscr{D}$ and v_∞ (or more correctly the hindered settling terminal velocity v_h) difficult to predict, but also they may change throughout the classifier (because of variations in the slurry concentration). On the other hand, the inclusion of a meaningful parameter such as $\mathscr{D}$ means that some indication of the effect of variables, such as the speed of the discharge mechanism, can be obtained.

In a simpler approach Reid[19] assumed that a classifier could be represented by two flowing streams, initially at the feed composition (Fig. 10.14*c*). These travel in plug flow with intense radial mixing within the overflow stream, because of turbulence. Classifying forces move the faster settling particles toward the underflow stream, but the intense mixing in the fine product stream allows particle transfer only from near the interface. By assuming that the probability p of a particle moving from the shaded area (Fig. 10.14*c*) can be related to the particle diameter by

$$p = Kd^s \quad (10.7)$$

it can be shown that

$$\mathscr{R}_{(+)} = 1 - \exp\left[-0.6931\left(\frac{d}{d_{50}}\right)^s\right] \quad (10.8)$$

(i.e., a Weibull or Rosin-Rammler relationship[20]). In contrast to the diffusion coefficient in earlier equations, the parameter s is empirical and without physical meaning, although it does represent the sharpness of separation. Published[21] values of s are in the range 1–3.8.

In essence, Eqs. 4.90 and 10.3–10.8 allow for the distribution of residence times occurring in a classifier. In many instances this distribution is a dispersion about a single mean and can therefore be represented by a number of perfect mixers in series.

Some data indicate that the process may be as simple as one or two mixers in series,[18] but other results suggest a more complex process.[22]

In a study of a small spiral classifier, Stewart and Restarick[22] showed that four main regions can be recognized (Fig. 10.15). Region A is that region above the spiral, where the slurry density is low and particles report mainly to the overflow. The spiral rotates in the main body of the pool (region B) from which particles may leave through the overflow or the sands discharge. In region C virtually all particles pass into the sands, while the final region D is at the bottom of the pool where the spiral conveys sands at high pulp density to the discharge. Thus, as is also shown in Fig. 10.15, discharge can be considered to occur along four main routes. Overflow material leaves by routes 1 and 2. Via route 1, particles are carried through region A almost directly to the overflow. Typically their residence time is 0.1–0.3 of the average for all overflow particles, smaller particles having the higher values. About 40–50% of sub-d_{50} particles leave by this route, again the higher values applying to the smallest particles. Routes 2 and 3 pass through the well-mixed region B. All particles passing along route 2 to the overflow have similar residence times—about 1.6 times the average for all overflow particles. Route 3 carries the majority of particles leaving in the sands, although the residence time decreases as the particle size increases. Even so, these residence times are five to six times as long as those of particles that leave via route 4. The net residence time of all sands particles is further increased by spiral transportation time. That the overflow particles effectively have two distinct residence times implies that the overall performance curve is based on two separate curves, one for each stream, as illustrated in Fig. 10.16 (neglecting the small amount of the sands that has a short residence time).

From this discussion, it follows that although the

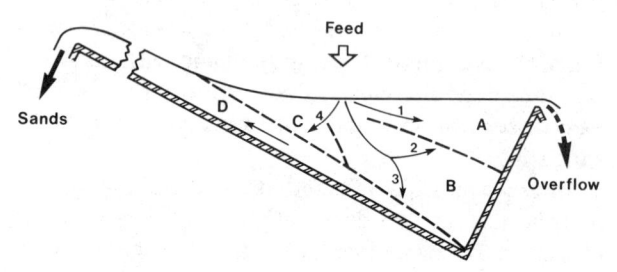

Figure 10.15. Major regions in a spiral classifier, and the main routes taken by the solids. (After Stewart and Restarick.[22])

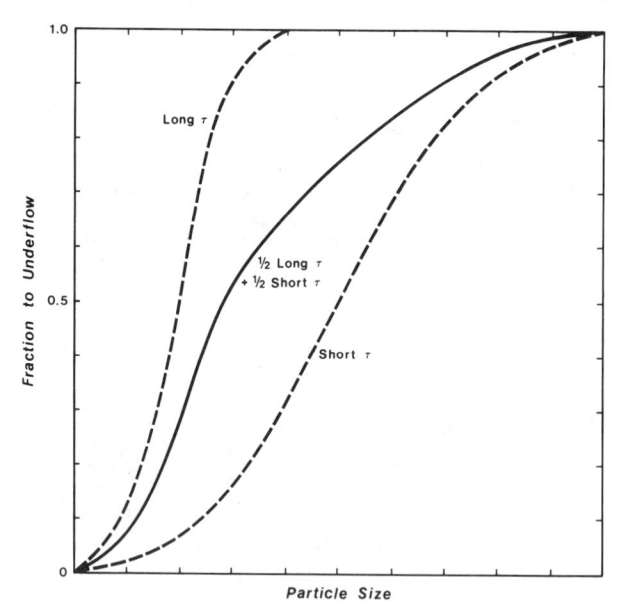

Figure 10.16. The resultant performance curve of a classifier having two major flows through it, one with a short residence time and one with a comparatively long residence time.

performance curve is primarily an extension of the ideal settling pool, it can be significantly dispersed by the classifier design and can also be affected by operating conditions. Clearly even the best designed and operated sedimentation classifier is incapable of giving a perfect separation; but in principle *comparison* of ideal and actual *reduced* performance curves allows the efficiency of any given classifier to be assessed, or in principle allows comparisons to be made between different types and makes of classifiers. Unfortunately, these comparisons are somewhat restricted because of the limited information available from manufacturers.

Example 10.1. The product from the ball mill in Example 8.3 is classified by a hydrocyclone. If the hydrocyclone gives $d_{50} = 150$ μm, what fraction of the feed leaves in each product stream, and what are the size distributions of the two products? (Assume the solids to be homogeneous with $\rho_s = 3.145$ t/m³, 30% of the water recycled in the underflow, and efficient classification.)

Solution. To obtain the split and size distributions, it is necessary to apply a suitable performance curve to the feed size distribution. We consider that Eq. 10.8 is a suitable description of a hydrocyclone reduced performance curve, and efficient classification is represented by $s = 2.5$. Thus, the mass frac-

Sedimentation Classifiers

TABLE E10.1.1: CLASSIFICATION SIZE DISTRIBUTION EXAMPLE

Sieve Apertures	Feed: Mass % in Size i	Mass Fraction to Underflow	Underflow: Mass in Size i	Underflow: Mass % in Size i	Overflow: Mass in Size i	Overflow: Mass % in Size i
(1)	(2)	(3)	(4)=(2)×(3)	(5)=$\frac{(4)}{\Sigma(4)}$	(6)=(2)−(4)	(7)=$\frac{(6)}{\Sigma(6)}$
+2360	0.41	1.000	0.41	0.58	0.0	0.0
−2360 +1700	0.90	1.000	0.90	1.28	0.0	0.0
−1700 +1180	1.66	1.000	1.66	2.35	0.0	0.0
−1180 +850	2.76	1.000	2.76	3.91	0.0	0.0
−850 +600	4.47	1.000	4.47	6.34	0.0	0.0
−600 +425	7.05	1.000	7.05	10.00	0.0	0.0
−425 +300	11.09	.999	11.08	15.71	0.01	0.03
−300 +212	15.24	.950	14.48	20.54	0.76	2.58
−212 +150	14.12	.769	10.86	15.40	3.26	11.05
−150 +106	10.40	.561	5.83	8.27	4.57	15.50
−106 +75	7.62	.425	3.24	4.60	4.38	14.85
−75 +53	5.96	.355	2.12	3.01	3.84	13.02
−53 +38	4.60	.324	1.49	2.10	3.11	10.55
−38	13.72	.303	4.16	5.90	9.56	32.49
	100.00		70.51	100.00	29.49	100.00

tion (classified) in the underflow $\mathscr{R}_{(+)}$ is given by

$$\mathscr{R}_{(+)} = 1 - \exp\left[-0.6931\left(\frac{d}{150}\right)^{2.5}\right] \quad (10.8)$$

where $d = (d_{A,1} + d_{A,2})/2$. To obtain the actual fractional split to the underflow, Eqs. 10.8 and 10.1 must be combined, so that

$$\mathscr{R}_{(+)a} = \left\{1 - \exp\left[-0.6931\left(\frac{d}{150}\right)^{2.5}\right]\right\}(1-y_1) + y_1$$

(assume $y_2 \sim 0$)

Basis. 100 tonnes total feed

Using the size fraction −212 +150 μm as an example

$$\mathscr{R}_{(+)a} = \left\{1 - \exp\left[-0.6931\left(\frac{212+150}{2\times 150}\right)^{2.5}\right]\right\}(1-0.3) + 0.3$$

$$= 0.769$$

Since the feed contains 14.12 tonnes of −212 +150 μm material, the underflow will contain

= 14.12 × 0.769
= 10.86 t

while the overflow will contain

= 14.12 × (1 − 0.769)
= 3.26 t

The procedure is repeated for all size fractions, and the results are shown in Table E10.1.1. Summation of columns 4 and 6 gives 29.5% of the solids in the overflow and 70.5% in the underflow, respectively. The percentage size distribution of the two streams is then obtained by dividing each size fraction by the total (solids) mass of the stream (columns 5 and 7).

10.3.3. Correlations for d_{50}

Thus far we have been primarily concerned with the shape of the performance curve. Also required for classifier design is a correlation for evaluating d_{50}, an even more difficult problem. In principle d_{50} can be calculated from Eqs. 4.33 or 4.35, 4.70, 4.83, and 10.2 (at $\mathscr{R}_{(+)} = 0.5$). In reality, the problem is immediately complicated because the concentration in most classifying pools changes in an unknown manner, so that the appropriate value of ε in Eq. 4.70 is difficult to assess.

As a first approximation Eq. 4.83 can be applied using the overflow concentration $C_{(-)}$, but it must be appreciated that this gives only a minimum area.[23] The error increases markedly as the solids concentration is increased, since this causes an increasing solids concentration down the pool, with a consequent unpredictable decrease in v_h with depth. Corrections are therefore made using the concept of areal efficiency (Eq. 4.84). Typically these can range from 100% in the tranquil pools of a hydrosizer to 30% in rake classifiers set to give maximum agitation (although it appears that even lower values may occur with coarser separations). One major manufacturer of spiral classifiers uses a value of 50% for all separation sizes.

Example 10.2. What pool area would be required in a spiral classifier to carry out the classification specified in Example 10.1? (Assume that the experiments have shown that the settling velocity of the solids can be described by Eq. 4.70 with $n = 6.0$.)

Solution. In principle, the required area can be found from Eq. 4.83, remembering that v_∞ in this equation applies to the limiting particle size.

The terminal velocity of the d_{50} particle $v_{\infty,50}$ was calculated in Example 4.2 as 0.0193 m/sec. From Eq. 10.1

$$0.5 = \frac{v_{\infty,50}}{v_{\infty l}}$$

$$v_{\infty l} = \frac{0.0193}{0.5}$$

$$= 0.0386 \text{ m/sec}$$

This velocity must be corrected for the solids concentration in the overflow. Assuming that the underflow can be withdrawn at about 50% solids by volume, from Example 1.5, the overflow will be 15.3% solids by volume. Thus

$$v_h = v_{\infty l}(\varepsilon)^6$$
$$= 0.0386 (1 - 0.153)^6$$
$$= 0.0143 \text{ m/sec}$$

Substituting in Eq. 4.83:

$$A = \frac{I_V}{v_h}$$
$$= \frac{800 \text{ t}}{\text{hr}} \left|\frac{\text{m}^3}{3.145 \text{ t}}\right| \frac{\text{hr}}{3600 \text{ sec}} \left|\frac{1.0}{0.3}\right| \frac{\text{sec}}{0.0143 \text{ m}}$$
$$= 16.5 \text{ m}^2$$

Allow 50% areal efficiency

$$A = 33 \text{ m}^2$$

Such an area can be obtained with a 1.8 m duplex spiral classifier.

When areal efficiencies are unknown, some estimate of d_{50} (and the whole performance curve) can be obtained by using published values of diffusion coefficients[18] (Fig. 10.17) in Eq. 10.4 or 10.6.

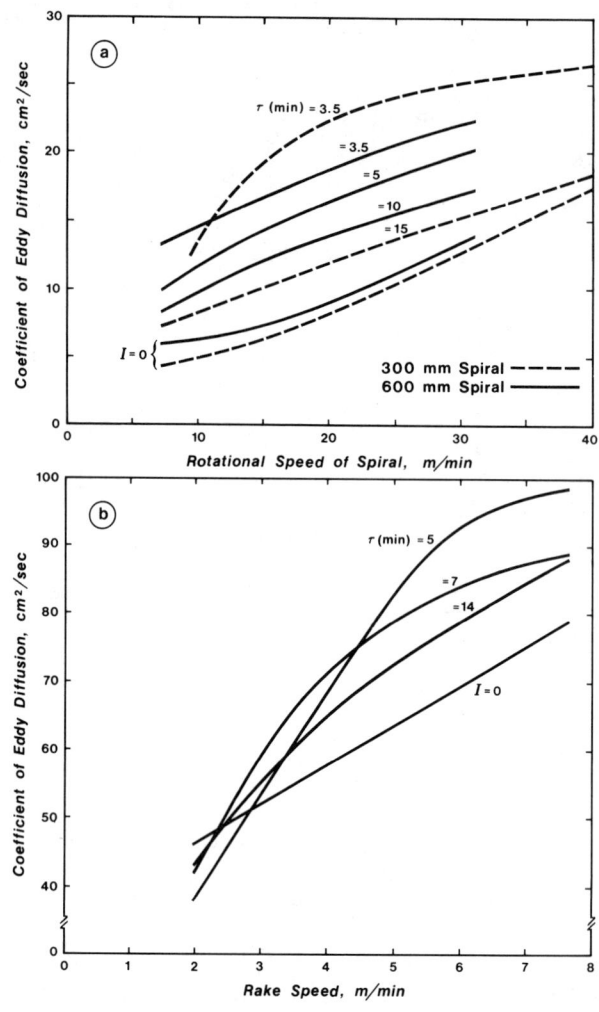

Figure 10.17. Eddy diffusion coefficient in (a) a spiral classifier and (b) a rake classifier. (After Schubert and Neesse.[18])

Example 10.3. If the spiral in the classifier in Example 10.2 was rotating at 10 m/min, what would be the effect of increasing the speed by 50% (to 15 m/min)?

Solution. Unless more specific information can be obtained, a reliable prediction is not possible. However, Fig. 10.17 gives an indication of the effect of rotational speed on $\mathcal{D}$, and this can be used to obtain an indication of the change in the classification behavior.

To estimate the residence time, assume that the average pool depth is 1 m, in which case the volume is about 33 m³. The average throughput is

$$= \frac{800 \text{ t}}{\text{hr}} \left|\frac{\text{m}^3}{3.145 \text{ t}}\right| \frac{1.0}{0.3} \left|\frac{\text{hr}}{60 \text{ min}}\right.$$
$$= 14.1 \text{ m}^3/\text{min}$$

Hydrocyclone Classifiers

$$\therefore \quad \tau \sim \frac{33}{14.1}$$
$$\sim 2.3 \text{ min}$$

Since Fig. 10.17 has no data for this time, the line for $\tau = 3.5$ (600 mm spiral) is used. (Since it can be seen that shorter residence times raise $\mathcal{D}$, while larger spiral diameters lower $\mathcal{D}$, this is an adequate approximation.) Thus, raising the speed by 50% raises $\mathcal{D}$ by about 17/14 (i.e., ~20%).

In principle, either Eq. 10.4 or Eq. 10.6 could be used, the latter is more convenient to manipulate, however. If the flow split is assumed to be unchanged (not altogether true, because the size distribution and the circulating load change), then from Eq. 10.6

$$\frac{v_\infty}{\mathcal{D}} = \text{const}$$

that is,

$$\left(\frac{v_\infty}{\mathcal{D}}\right)_{10} = \left(\frac{v_\infty}{\mathcal{D}}\right)_{15}$$

Because Example 4.2 indicates that the flow behavior is in the transitional region ($\mathbf{Re}_p > 2.0$), we can assume from Eq. 4.39 that $v_\infty \propto d_{50}^{1.14}$. Thus

$$\left(\frac{d_{50}^{1.14}}{\mathcal{D}}\right)_{10} = \left(\frac{d_{50}^{1.14}}{\mathcal{D}}\right)_{15}$$

$$\left(\frac{150^{1.14}}{1.0}\right)_{10} = \left(\frac{d_{50}^{1.14}}{1.2}\right)_{15}$$

$$\therefore \quad d_{50} = 176 \ \mu\text{m}$$

That is, the separation size is raised by about 17%.

Although the necessary settling velocity can be calculated from Eqs. 4.33, 4.35, and 4.70, batch settling tests (e.g., the "long tube" test) can be used to obtain more reliable data based on the similarity between the ideal pool and a batch sedimentation test.[24,25] If the latter is carried out for a time equal to the residence time in the actual pool and then the top liquid is decanted off so that the slurry is split in the same proportions as the overflow and underflow of the pool, the same conditions are achieved. However, to produce a performance curve, the usual test must be extended to include a size analysis of the solids in both fractions. Thus the batch test can be used to obtain basic data, although an areal efficiency correction is still necessary.

It should also be pointed out, that clarification (as defined in Chapter 17), is actually a classification process and consequently, the concepts described in this section should be used to analyze clarification. Normally, clarifier performance is described in terms of the total mass of solids per unit volume of overflow.[24,25] In reality these solids are the range of particles represented by the integral of the area above a performance curve.

10.4. HYDROCYCLONE CLASSIFIERS

A considerable amount of research has been carried out on the hydrocyclone since its introduction.[26,27] Unfortunately, most of the earlier work used dilute slurries, which are hardly typical of those occurring in mineral concentration circuits. This research did lead to a general understanding of the hydrocyclone; it also showed the complexity of the problem, making it apparent that a completely rigorous analysis may be impossible. In recent years, with the emphasis on grinding circuit process analysis, a large amount of empirical information has become available, suitable for describing operating hydrocyclones and (provided the design is flexible enough) adequate for selecting hydrocyclones for new applications.[10,28]

10.4.1. Flow Patterns

Reynolds Number. A number of comments will be made as to whether conditions in the hydrocyclone are laminar or turbulent. Supposedly, the nature of flow can be characterized by the Reynolds number (Section 4.1). Unfortunately, the problem with the hydrocyclone is that the **Re** is hard to define and it is apparent that in fact two Reynolds numbers must be considered. The first is that of the particle, $\mathbf{Re}_p$, and correlations for d_{50} tend to suggest that laminar settling behavior occurs. Clearly there is also an **Re** for the hydrocyclone itself, $\mathbf{Re}_c$, that includes a characteristic dimension of the device and a characteristic velocity. These are obviously difficult to select, but the relationship normally adopted is:

$$\mathbf{Re}_c = \frac{D_c v_I \rho_l}{\mu} \quad (10.9)$$

On this basis, typical values would be in the range 10^4–10^6, which would imply turbulent flow in the inlet pipe. This does not, however, mean that turbulence continues within the hydrocyclone body, since it is known that laminar conditions can be maintained to higher **Re** in curved channels. It has been found, in fact, that there is an optimum Re_c (about 10^4), below which Reynolds number effects predominate, and above which centrifugal effects predominate.[29] This phenomenon appears to be important only in very small cyclones with high liquid viscosity, and is therefore not important in most mineral operations.

General Flows. The most significant flow pattern in a hydrocyclone is the "spiral within a spiral," illustrated in Fig. 10.6. These spirals are generated by the tangential feed and revolve in the same direction; the reversal in velocity applies only to the vertical component. Figure 10.18 shows two addition flows.[29] The short-circuit flow against the roof is due to obstruction of the tangential velocity, and the main purpose of the vortex finder is to minimize this flow. Because the overflow opening cannot handle the natural upflowing vortex, eddy flows also exist in the upper section of the hydrocyclone.

Two further features of hydrocyclone flow are illustrated in Fig. 10.19.[30] The first is the locus of zero vertical velocity and the second, rising from the apex and passing out through the vortex finder, is the air core, the presence of which indicates vortex stability.

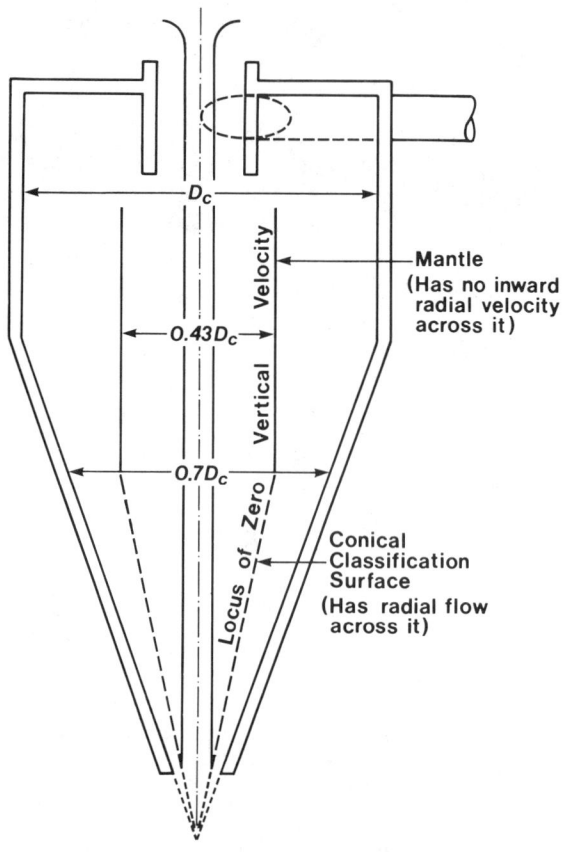

Figure 10.19. The locus of zero vertical velocity in a hydrocyclone. (After Bradley and Pulling.[30])

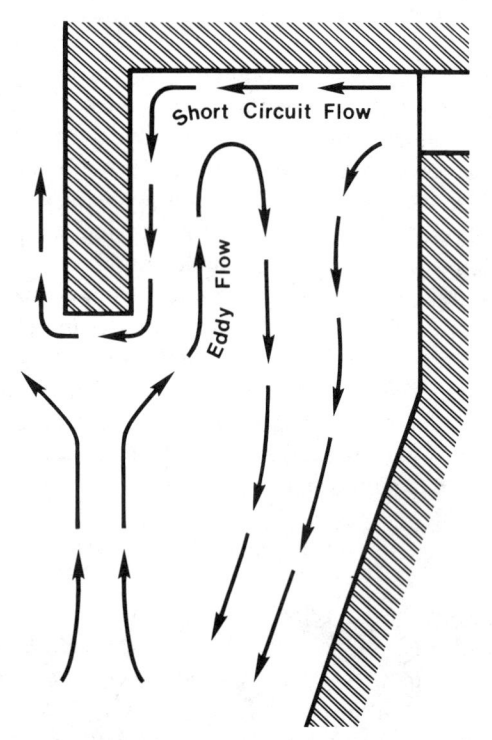

Figure 10.18. Flow patterns in the upper regions of a hydrocyclone. (After Bradley and Pulling.[30])

The velocity of liquid or solid at any point may be resolved into three components; the vertical or axial component $v_{(v)}$, the radial component $v_{(r)}$, and the tangential component $v_{(t)}$. No completely satisfactory measure of these components has been made. The results shown in Fig. 10.20, which were produced by Kelsall in an optical study using a very dilute suspension of fine particles,[31] are considered to be the most reliable indication of liquid flow patterns. Little information is available on the behavior of solid particles at practical slurry concentrations, although the results of Renner and Cohen[32] at least indicate the spatial distribution of different particle size fractions in the hydrocyclone.

Liquid Vertical Velocity. The vertical velocity component of the liquid flow indicates the magnitude of the two spirals, and therefore relates to the volumetric distribution of product between the underflow and overflow. The most noticeable feature of the vertical component shown in Fig. 10.20a

Hydrocyclone Classifiers

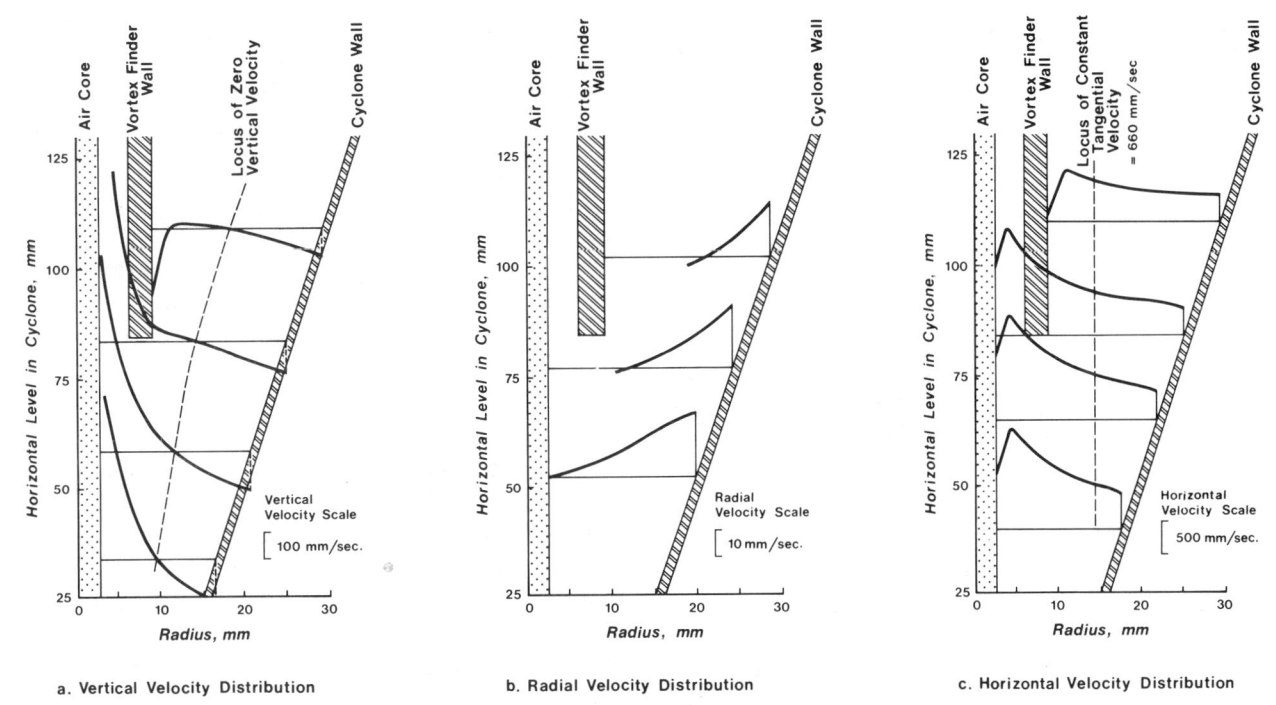

Figure 10.20. The three velocity components in a hydrocyclone. (After Kelsall.[31])

a. Vertical Velocity Distribution

b. Radial Velocity Distribution

c. Horizontal Velocity Distribution

is the locus or envelope of zero vertical velocity. Below the vortex finder, all liquid moves upward inside the envelope, and downward outside. In each case the velocity component increases with distance from the envelope, and the maximum upward component is appreciably higher than the maximum downward component.

Another significant feature not readily apparent in Fig. 10.20a is the existence of a second imaginary envelope between the first one and the outer wall. This envelope represents the division of downward flowing liquid into two parts: an outer layer that is able to leave through the underflow, and the excess, which becomes the innermost layer that is the upward flowing liquid. Figure 10.20a also indicates that above the vortex finder a second locus of vertical velocity exists because of the short circuit flow.

Liquid Radial Velocity. The radial velocity component is the liquid "current" against which the particles must settle because of centrifugal force if they are to be removed in the underflow. The general form of these components is shown in Fig. 10.20b and it must be noted that this component is normally inward. It is the smallest of the three components and increases to a maximum at the wall.

The radial velocities in the region above the base of the vortex finder are not fully shown in Fig. 10.20a; in practice there may be strong inward flows along the roof resulting from short-circuit flows, and below this, the component may be outward because of the eddy flow[30] (Fig. 10.18).

Liquid Tangential Velocity. Below the vortex finder, envelopes of constant tangential velocity are cylinders coaxial within the hydrocyclone (Fig. 10.20c) and the velocity is inversely related to the radius according to

$$v_{(t)}r^n = \text{constant} \qquad (10.10)$$

where n is less than 1.0. It appears that n is more strongly dependent on design variables than operating variables (Section 10.4.2). Values of n are typically 0.4–0.9. This should be compared to the outer region of a free vortex where angular momentum is conserved and the relationship is

$$v_{(t)}r = \text{constant} \qquad (10.11)$$

that is, the hydrocyclone tends to approximate a free vortex as n tends to unity. On the other hand, in a centrifuge the liquid rotates as a solid body (i.e., with constant angular velocity) and

$$v_{(t)}r^{-1} = \text{constant} \qquad (10.12)$$

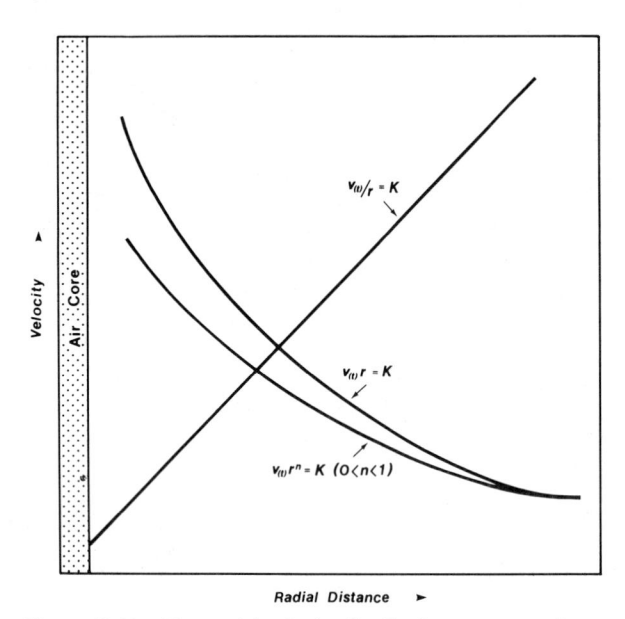

Figure 10.21. Tangential velocity distributions corresponding to various relationships. (After Bradley.[26])

These three relationships are shown in Fig. 10.21. However, while Eqs. 10.10 and 10.11 indicate that $v_{(t)}$ tends to infinity at the center, Fig. 10.20c shows that this does not occur in the hydrocyclone; rather there is an area around the air core where the behavior changes toward constant angular velocity (i.e., solid body rotation).

Practical Liquid Flows. Because of the long vortex finder used in the experimental hydrocyclones,[31] Fig. 10.20a is probably not a completely accurate representation of the locus of zero vertical velocity and Fig. 10.19 is considered to be more representative of practical hydrocyclones.[30] With more typical vortex finders, the shape of the upper locus is cylindrical (termed the mantle) rather than conical, and has no radial flow across it. The lower conical section has radial flow across it because excess downward flowing liquid moves across to the upward flowing stream. Figure 10.19 also shows the approximate position of the locus envelope.

Behavior of Solid Particles. Because little information is available on the behavior of solid particles in a hydrocyclone, only probable behavior can be described. In general, a particle anywhere in the hydrocyclone is subjected to two opposing forces: an outward centrifugal force F_c, and an inward drag force F_d. The former has the form

$$F_c = \frac{\pi d_v^3}{6} \frac{(\rho_s - \rho_l)}{r} v_{(t)}^2 \qquad (10.13)$$

(d_v = particle diameter; r = the instantaneous distance of the particle from the center of the hydrocyclone). Although the drag force is a function of $v_{(r)}$, the exact relationship depends on whether the flow is laminar or turbulent. Indications are that laminar conditions are approximated, giving (from Eq. 4.34):

$$F_d = 3\pi d_0 \mu v_{(r)} \qquad (10.14)$$

Even so, the probability of intermediate or turbulent conditions cannot be ruled out.

When F_c exceeds F_d, the particle moves outward, and may leave through the underflow; if F_c is less than F_d, the particle moves inward and may leave through the overflow. The split of particles is therefore dependent on the relative values of $v_{(r)}$ and $v_{(t)}$.

In reality, a number of factors complicate this simplistic approach. Some of these have already been described in Section 10.3.1, relative to the ideal pool concept. As in the case of the spiral classifier, other factors become apparent when the characteristic regions of the hydrocyclone are determined. In their studies, Cohen et al.[32,33] found that the interior of the hydrocyclone could be divided into four regions (Fig. 10.22). The narrow region A, against the roof and the cylinder wall, contains particles having the feed size distribution (i.e., unclassified feed). Most of the conical section is filled by region B, a region where the particles have essentially the coarse product size distribution. Surrounding and extending below the vortex finder is a further narrow region C, which contains essentially the fine product size distribution. The fourth region D is an elongated toroid lying between C and A–B. Here the size distribution is higher in the intermediate sized particles than either the feed or the contents as a whole. These intermediate sized particles tend to have above average residence times (the discrepancy maximizing at about the d_{70} size), implying that such particles tend to accumulate in this region until displaced by lack of room. This indicates that this region is the locus of active classification. Poor classification can be expected to result when region D fails to form well, because of poor design or poor operation such as would be caused by excessive slurry density overloading the region.

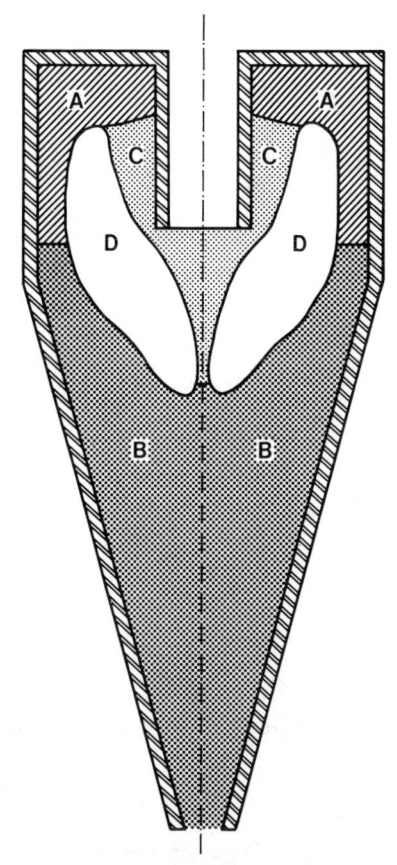

Figure 10.22. Regions of similar size distribution in a hydrocyclone. (After Renner and Cohen.[32])

10.4.2. Hydrocyclone Correlations

The variables that affect hydrocyclone performance can be divided into two groups: those that are dependent on hydrocyclone size and proportions (design variables) and those that are independent of size and proportions (operating variables). The design variables are effectively illustrated in Fig. 10.6, and include feed, overflow and underflow opening sizes, the hydrocyclone size, and its shape. Operating variables depend on the feed stream and include:

Pressure drop (and feed flow rate).
Solids concentration (or liquid-liquid concentration).
Solids size and shape.
Solids densities (or liquid densities).
Liquid medium density.
Liquid medium viscosity.

Because many of these variables interact, it is not possible to consider them individually. Instead, hydrocyclone performance is evaluated in terms of correlations using three criteria: the performance curve (frequently as a correlation for d_{50}), the ratio of overflow to underflow, and pressure drop.

Theoretical Correlations for d_{50}. Since the d_{50} is the most important parameter for describing the performance of the hydrocyclone, being equivalent to the aperture of a screen, attempts have been made to correlate it with operating and design variables.

Many of the theoretical relationships use the equilibrium orbit hypothesis[26]; that is, they balance the outward acting centrifugal force against the drag force due to the inward flow of liquid (Eqs. 10.13 and 10.14), resulting in the intermediate equation:

$$d_{50} = \left[\frac{18\mu \cdot v_{(r)}r}{(\rho_s - \rho_l)v_{(t)}^2}\right]^{1/2} \quad (10.15)$$

The final equations derived from this differ in their assumptions of the locus of zero velocity. One of the more rigorous correlations (by Bradley and Pulling[30]) is based on the locus illustrated in Fig. 10.19 and incorporates two parameters n (Eq. 10.1) and K_c.

$$d_{50} = 3(0.38)^n \frac{D_I^2}{K_c} \left[\frac{\mu}{D_c I_V} \frac{(1 - \mathcal{R}_{(+)l})}{(\rho_s - \rho_l)} \tan\frac{\theta}{2}\right]^{1/2} \quad (10.16)$$

(D_c = inside diameter of hydrocyclone; D_I = inside diameter of feed inlet; I_V = input volumetric flow rate; K_c = a parameter, dependent on design, fluid properties, and I_V; $\mathcal{R}_{(+)l}$ = mass fraction of feed liquid leaving through underflow; θ = hydrocyclone cone angle).

Rietema[29] has claimed that the equilibrium orbit approach is invalid because it assumes that particles have sufficient time to achieve equilibrium velocities and that this may not be true with respect to the short residence times occurring in hydrocyclones. His approach was to assume that the d_{50} particle is that particle which starts at the center of the inlet, and just succeeds in reaching the wall at the apex. The correlation

$$d_{50}^2 \frac{(\rho_s - \rho_l)}{\rho_l} \frac{L_c P}{\mu I_l} = \frac{36}{\pi} \frac{v_{(v)}}{v_{(t)}} \frac{D_c}{D_I} \quad (10.17)$$

was obtained by considering the movement of the d_{50} particle over the distance $\frac{1}{2}D_I$ in the residence time available, assuming laminar conditions. The right-hand side of Eq. 10.17 is approximately constant and is called the Cyclone Characteristic number, Cy_{50}. In practice it may be difficult to evaluate, but it appears to have a minimum value about 3.5, and it can be considered that an "optimum" hydrocyclone will achieve this figure. In reality, the $\frac{1}{2}D_I$ movement of the equilibrium particle is justified only if the underflow discharge is small, the slurry concentration is low, and negligible turbulence occurs. The actual reduced performance curve that can be derived with Eq. 10.17 is steeper than that obtained experimentally, indicating that some turbulence must occur in the cylindrical section of the hydrocyclone.

Fahlstrom,[34] in a third approach, proposed that the apex restricts the flow of solids, and that under the effect of centrifugal force the probability of a particle leaving through the apex is determined by its mass, so that the coarsest and heaviest particles discharge first and the discharge of the smaller and lighter particles becomes progressively more difficult. Thus:

$$\mathcal{R}_{(+)} = \text{fn}(Y_u^+) \tag{10.18}$$

where Y_u^+ is a correlation for the size distribution. For a Rosin-Rammler size distribution:

$$Y_u^+ = \exp - \left(\frac{d}{d^*}\right)^s \tag{2.1b}$$

and when substituted in Eq. 10.18 with $d = d_{50}$, results in

$$d_{50} = d^*(-\ln \mathcal{R}_{(+)})^{1/s} \tag{10.19}$$

This relationship might be expected to be more suitable for concentrated slurries; because of the unclassified material that always occurs in the underflow stream, however, this is not necessarily the case. Thus the d_{50} in Eq. 10.19 will be closer to the $d_{50,a}$ than the true d_{50}.

The diffusion coefficient concept (Section 10.3.2) has also been applied to hydrocyclones.[18]

In practice, none of the theoretical approaches is entirely satisfactory, except under fairly restricted conditions. This is probably because in most situations the ideal conditions assumed by the different approaches do not occur in isolation from one another.

Empirical Correlations for d_{50}. Of the many empirical correlations for d_{50}, that of Dahlstrom[35,36] is the most widely accepted:

$$d_{50} = \frac{3 \times 10^{-3}(D_{(-)}D_I)^{0.68}}{(I_V)^{0.53}(\rho_s - \rho_l)^{0.5}} \tag{10.20}$$

(S.I. units, m, kg, sec; $D_{(-)}$ = inside diameter of vortex finder).

Like most of the empirical correlations, this equation is restricted to low feed concentrations and low underflow splits, conditions hardly typical of closed circuit grinding where hydrocyclones are most widely used. Also, in common with virtually all d_{50} correlations (theoretical and empirical), Eq. 10.20 contains the term $(\rho_s - \rho_l)^{-0.5}$, indicating laminar flow conditions, and if this is the case, it follows that d_{50} is proportional to $\sqrt{\mu}$. There is a limited amount of data[37] that shows that the equation can be extended to slurry densities of about 35% by volume by using apparent slurry density ρ_{sl} and viscosity μ_a, thus

$$d_{50} = \frac{3 \times 10^{-3}(D_{(-)}D_I)^{0.68}}{(I_V)^{0.53}(\rho_s - \rho_{sl})^{0.5}} \left[\frac{\mu_a}{\mu_{H_2O}}\right]^{0.5} \tag{10.21}$$

Work by Marlow[10,38] at high slurry densities has suggested that

$$d_{50} \tilde{\propto} (\rho_s - \rho_l)^{-1.0} \tag{10.22}$$

which implies that full turbulence occurs. However, the interpretation of his results apparently fails to consider the buildup of heavy minerals that occurs in closed circuit grinding,[39] and in view of the large amount of prior evidence supporting laminar flow and the small amount of evidence for even transition flow, this suggestion of complete turbulence is of questionable validity. In fact, most of his data can be interpreted as supporting Eq. 10.21. The failure of the lighter gangue material to support Eq. 10.21 could be due to its larger average size, putting its settling characteristics into the transition region. Equation 10.21 clearly becomes invalid as the slurry density approaches or even exceeds the density of the lightest mineral present, and it is likely that under these conditions the separation is based more on heavy media concepts (Chapter 12). This is supported by results from a circuit grinding a galena ore (ρ_s = 7600 kg/m³) where the d_{50} of the galena was considerably finer than could be accounted for by any of the settling concepts above.[39] However, it is likely that particle behavior is affected by hindered settling conditions, which can be described by Eq. 4.70. If this is the case, the analytical difficulties could then be attributed to the minerals having significantly different values of n in Eq. 4.70.

A number of generalized d_{50} correlations have been published for commercially available classifying hydrocyclones. The most extensive results have been presented by Lynch and Rao[10] and Plitt,[28] who subjected their data to conventional regressional

Hydrocyclone Classifiers

analysis techniques. Typical is the correlation by Plitt:

$$d_{50} = \frac{2.69 \times 10^{-3} D_c^{0.46} D_I^{0.6} D_{(-)}^{1.21} \exp[3.9 \mathfrak{F}_{V,sl}/(d_{Am3})^{0.052}]}{D_{(+)}^{0.71} H_v^{0.38} I_V^{0.45} (\rho_s - \rho_l)^{0.5}} \quad (10.23)$$

(S.I. units, m, kg, sec; $D_{(+)}$ = inside diameter of apex; d_{Am3} = mass median diameter of feed particles; $\mathfrak{F}_{V,sl}$ = volume fraction of solids in slurry; H_v = height of free vortex in hydrocyclone, i.e., height from apex to bottom of vortex finder).

Tarr[40] presented data in a graphical form based on a "typical" hydrocyclone operating under "typical" conditions (Fig. 10.23). Graphic correction factors can be applied for nontypical conditions.

The Sharpness of Classification. The factors that affect the exponent s in Eq. 10.8 are uncertain. It is generally assumed that s is a characteristic of a given type of classifier, and otherwise is a function only of solid properties such as specific gravity and shape. However, it is well established that smaller hydrocyclones are less capable of giving as sharp a separation as larger hydrocyclones (although because the latter are often used with higher pulp densities. this may not be apparent). Plitt attributed the decreasing sharpness of smaller hydrocyclones to lower residence times and found that an increasing $D_{(+)}/D_{(-)}$ ratio also reduced the sharpness.[28]

Flow Split Correlations. Because the performance curve of the hydrocyclone shows the effect of the fraction of liquid leaving through the underflow, the volume split between the two exits must be known before the liquid balance around the hydrocyclone can be evaluated. The flow split is generally defined by

$$\mathfrak{F}_{(+/-)} = \frac{\text{underflow volumetric flow rate}}{\text{overflow volumetric flow rate}} \quad (10.24)$$

In practice the split is controlled by adjusting the size of the underflow outlet to give the desired operating conditions.

Although no satisfactory theoretical relationships have been derived for $\mathfrak{F}_{(+/-)}$, a number of empirical ones are available. According to Plitt,[28] for a free discharge of the spray or vortex type (Section 10.4.3),

$$\mathfrak{F}_{(+/-)} = \left[34.3 \left(\frac{D_{(+)}}{D_{(-)}} \right)^{3.31} H_v^{0.54} (D_{(+)}^2 + D_{(-)}^2)^{0.36} \exp(0.54 \mathfrak{F}_{V,sl}) \right] / \left[\Delta P^{0.24} D_c^{1.11} \right] \quad (10.25)$$

(S.I. units, m, kg, sec)

Pressure Drop Correlation. Knowledge of the pressure drop across a hydrocyclone is necessary to calculate the pumping requirements. The pressure applied is used essentially to produce the centrifugal action that develops the separation in the hydrocyclone. In reality, the total pressure drop contains three items: the inlet velocity head ($\frac{1}{2} \rho v_I$), the wall friction losses in the hydrocyclone, and the centrifugal head, which brings about the actual classification. The latter item is theoretically obtained by integrating between appropriate limits the radial pressure gradient caused by the tangential velocity.

Since the radial pressure gradient is given by

$$\frac{dP}{dr} = \frac{\rho v_{(t)}^2}{r} \quad (10.26)$$

Bradley[26] derived a theoretical correlation by assuming that Eq. 10.10 was valid and that other losses were negligible. This resulted in

$$\frac{\Delta P/\rho}{v_I^2/2} = \frac{K_c^2}{n} \left[\left(\frac{D_c}{D_{(-)}} \right)^{2n} - 1 \right] \quad (10.27)$$

As with his correlation for d_{50} (Eq. 10.16) the parameters K_c and n need to be known, limiting the usefulness of this equation.

Dahlstrohm[35,36] has produced an empirical correlation for pressure drop, given by:

$$\frac{I_V}{(\Delta P)^{1/2}} = 7.4 \times 10^{-6} \left(\frac{D_{(-)}}{D_I} \right)^{0.9} \quad (10.28)$$

which, in spite of its simplicity, appears to be as reliable as most.

For completeness, the empirical correlation of Plitt[28] is given:

$$P = \frac{1.3 \times 10^5 I_V^{1.78} \exp[0.55 \mathfrak{F}_{V,sl}]}{D_c^{0.37} D_I^{0.94} H_c^{0.28} (D_{(+)}^2 + D_{(-)}^2)^{0.87}} \quad (10.29)$$

(S.I. units, m, kg, sec)

The Significance of Hydrocyclone Correlations. Behavior within the hydrocyclone is obviously very complex, despite the mechanical simplicity of the device. In general, one would expect the theoretical relationships to be inferior to empirical ones, because of the simplifying assumptions that must be made. With dilute slurries, reasonable predictions of hydrocyclone behavior can be made using empirical correlations, particularly if the hydrocyclone is to be operated under conditions similar to those used to derive the correlation.

The apparent all-encompassing nature of Plitt and Lynch's correlations (e.g., Eqs. 10.23, 10.25, and 10.29) must not be allowed to deceive one. These

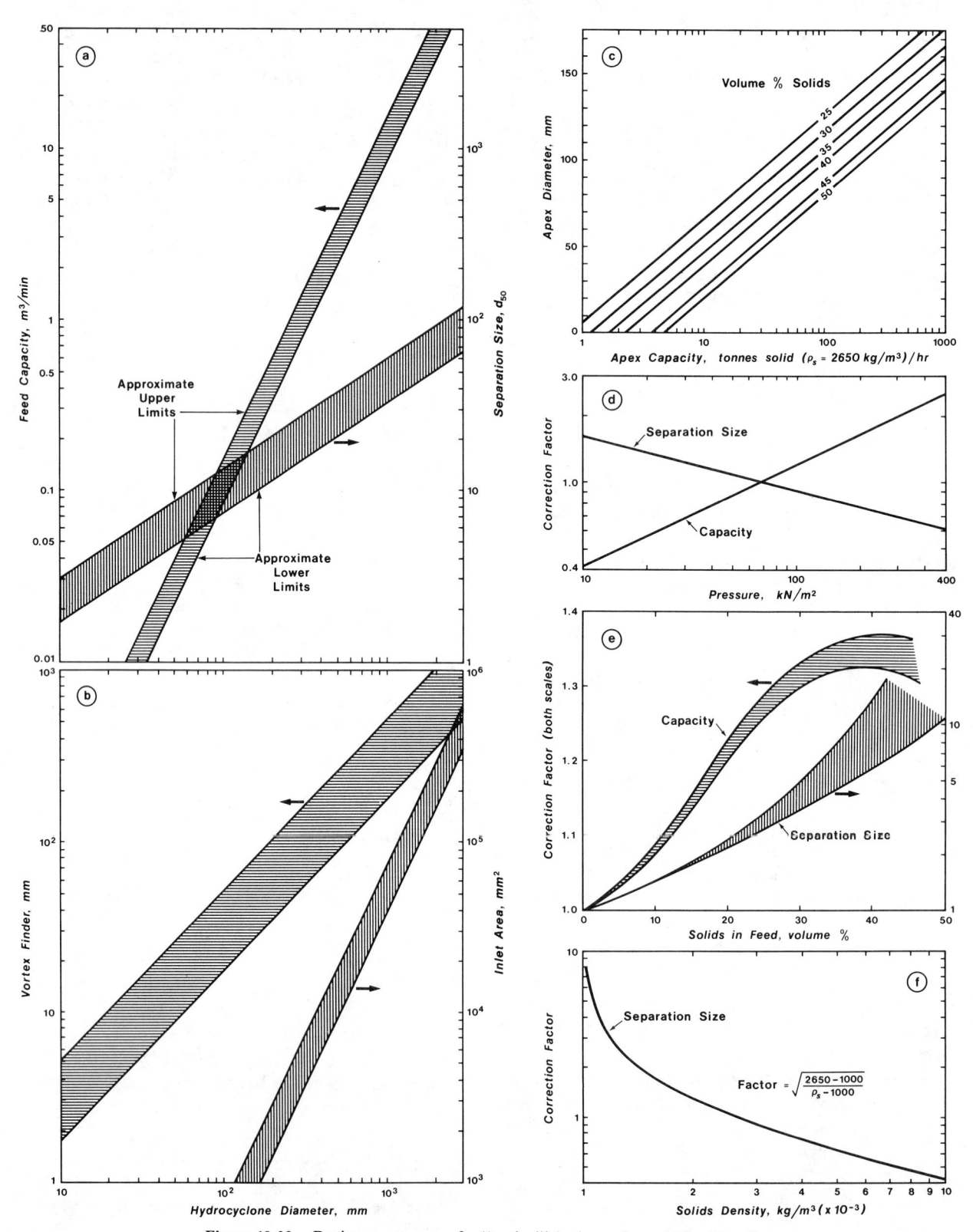

Figure 10.23. Design parameters of a "typical" hydrocyclone. (After Tarr.[40])

have been derived over a wide range of conditions, which, because of the large number of interrelating variables involved, inherently imply considerable error. In itself, each dependent variable may correlate well with the independent variable, but the form of the intercorrelation may not be general to all hydrocyclones. Furthermore, any correlation includes only significant variables, so that omitted terms and accumulated error mean that results may have up to ±50% error. Errors can be expected to arise, particularly under extreme operating conditions, such as the high slurry densities encountered in closed circuit grinding. Although Plitt and Lynch include a term for the solids concentration, it basically allows only for the crowding effect and fails to include the effect of significant factors such as shape and roughness, which become increasingly important as the concentration rises (see Section 4.7). As discussed below, hydrocyclones have fairly constant proportions, so that most of the dimension terms in the correlations can be expressed as direct proportions of D_c. This means that Eq. 10.23 can be reduced to a form similar to Eq. 10.20 and that the apparently more sophisticated correlation may not be as reliable as the simple correlation (Eq. 10.20), which necessitates the inclusion of actual experimental data (μ_a).

In spite of these limitations, the more complex correlations have a place. They may be the only starting point available and they can be useful in *indicating* the effect of changes in any variable (although again reliability may be doubtful). In fact if reliable information is required on an operating hydrocyclone, it would always be desirable to determine an entirely new correlation on that particular unit.[10]

Example 10.4. What hydrocyclones would be required to carry out the classification described in Example 10.1?

Solution. Because the hydrocyclone is operated at such a high pulp density, no reliable solution can be obtained. An approximate solution can be obtained from the data in Fig. 10.23. The procedure requires first, determination of the hydrocyclone size to give the required separation size, and the selection of an appropriate number of hydrocyclones to carry the required flow rate. For the hydrocyclone size:

actual separation size = d_{50} for "typical" hydrocyclone × correction factors

correction factors:

pressure (Fig. 10.23d) = 1.0 (assume "typical" pressure, 70 kN/m²)

feed concentration (Fig. 10.23e) = 4.5 (30% solids by volume)

solids density (Fig. 10.23f) = 0.88 (ρ_s = 3.145 t/m³)

$$\therefore d_{50} \text{ ("typical" hydrocyclone)} = \frac{150}{1.0 \times 4.5 \times 0.88}$$
$$= 38 \ \mu m$$

From Fig. 10.23a, a 38 μm separation size requires an 800 mm hydrocyclone. The nearest commercial size is 762 mm, and from Fig. 10.23a, this has a throughput of about 6.5 m³/min. Figure 10.23e shows that with a 30% solids feed concentration, the capacity increases by a factor of 1.3. Thus, the throughput will be 1.3 × 6.5 ~ 8.5 m³/min:

$$\text{circuit capacity} = \frac{800 \text{ t}}{\text{hr}} \bigg| \frac{\text{m}^3}{3.145 \text{ t}} \bigg| \frac{1.0}{0.3} \bigg| \frac{\text{hr}}{60 \text{ min}}$$
$$= 14.1 \text{ m}^3/\text{min}$$

Thus, two 762 mm hydrocyclones should be used.

Initial inlet, vortex finder, and apex sizes can also be found from Fig. 10.23, but the final selection would have to be found by trial and error during actual operation.

10.4.3. Hydrocyclone Geometry

The effect of the major design variables was considered in the preceding section, and this discussion treats in general terms their relationships to each other and to the flow patterns.

It has been mentioned that the main purpose of the vortex finder is to minimize short-circuit flow. To achieve this, the vortex finder must be larger than the locus of maximum tangential velocity so that large particles can be swept outward. At the other extreme, the vortex finder must not lie outside the mantle section of the locus of zero vertical velocity, since some particles could then be subjected to an inward radial velocity in the cylindrical section, which could sweep them into the overflow stream.

Analysis of the inlet diameter is less amenable to logic, particularly because it affects both d_{50} and the pressure drop. Rietema[29] has suggested that D_I (and other proportions) can be selected for a particular purpose on the basis of minimizing his Cy_{50}.

In general the hydrocyclone is relatively insensitive to variation in proportions, and a configuration of the form $D_I = D_c/7$; $D_{(-)} = D_c/5$; $D_{(+)} = D_c/15$; $\theta = 10°–30°$; $L_c/D_c = 3$; $L_v/D_c = 0.4$ (L_v = length of vortex finder) is satisfactory for classification.[26] Larger cone angles are better for classification, but there is an upper limit because they cause d_{50} to increase with a decrease in throughput for a given pressure drop. The effect can be partly counteracted by lengthening the cylindrical section.

If the hydrocyclone is to be used for dewatering operations, slightly different proportions are desirable[29]:

$$D_I = \frac{D_c}{4}$$

$$D_{(-)} = \frac{D_c}{3}$$

$$\frac{L_c}{D_c} = 5.0$$

$$\frac{L_v}{D_c} = 0.4$$

In practice these proportions are also frequently used in closed circuit grinding classification, because the larger openings reduce the pressure drop. An upper limit in opening size occurs as the inlet stream starts to impinge on the vortex finder, that is, $2D_I + D_{(-)} < D_c$.

The inlet should not be below the center of the vortex finder, and a rectangular involute entry form (Fig. 10.24) against the roof has advantages over a circular entry, although the additional complications in manufacturing and fitting may not be justified. In some cases inlet and overflow delivery pipe diameters should be larger than the size of the actual hydrocyclone opening and in the case of the latter pipe, care should be taken to prevent syphoning effects that could destroy the flow patterns within the hydrocyclone.

The underflow is normally a free discharge and the size of the opening is generally adjusted to suit the application. The appearance of the discharge can show three forms:

1. "Vortex" or "spray," where solid and liquid discharge in a violent spray in the shape of a hollow cone. This is used for the maximum removal of solids.
2. "Sausage" or "rope," where the discharge is a rotating solid spiral, representing minimum liquid content.
3. "Overload," where the discharge is a straight lazy stream with no spiral motion.

For a given hydrocyclone and duty there is only one size for $D_{(+)}$. If the size is too small, complete discharge of the oversize material is not possible, with the result that some will be carried out the overflow. If the size is too large, the underflow will contain too much water, and consequently, carry away too many fine particles.

This discussion is only an introduction to hydrocyclone geometry. Recent studies indicate that currently available commercial hydrocyclones do not necessarily represent an optimum design and significant improvements are still possible.[41]

10.4.4. Relative Merits of the Hydrocyclone

The hydrocyclone is extremely versatile in that it can be used as a thickener, a classifier, or a concentrator. In a relatively short time it has become established in a large number of applications. In virtually all cases it has disadvantages, but its adoption has resulted from the clear strength of its advantages.

Advantages. The major advantages of a hydrocyclone center around its simplicity, smallness, and low cost in comparison with competitive equipment.

The simplicity is due to its straightforward shape and lack of moving parts. Effective hydrocyclones can if necessary be made from old drums, with the range of construction extending through to extremely well-shaped hydrocyclones with replaceable interiors to counteract high wear. Being small, the hydrocyclone requires little in the way of major supports, and the strong separating forces inside

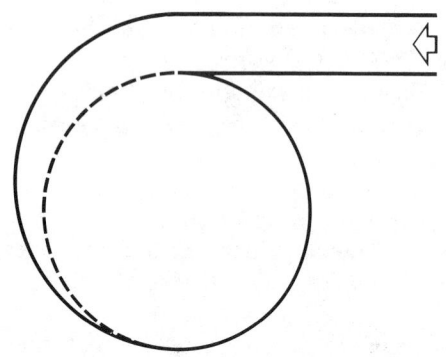

Figure 10.24. Involute hydrocyclone entry.

allow it to be operated in any position, although it is normally mounted with the apex down to allow free underflow discharge. Space savings in the case of classification may be of minor significance, but can be of the order of 10^4 for thickening operations.

The lower capital costs that result from this simplicity may be somewhat offset by the need for a *larger* pumping unit to overcome the additional pressure loss caused by the hydrocyclone. As a result, the hydrocyclone uses more energy than other classifiers and thickeners (except centrifuges), although the difference may not be as large as expected, since most classifiers require a pump to transport either the feed or the product.

Because the hydrocyclone produces a high shear action, it can be superior in treating flocculated material and suspensions with high apparent viscosity. This has been attributed to a dependence on liquid viscosity rather than slurry viscosity, but in view of more recent evidence,[42] it appears that it may be due to the hydrocyclone's ability to overcome the critical shear stress occurring with Bingham plastic-like slurries.

A final factor, which is of increasing importance in the control of grinding circuits, is the very short residence times that occur in hydrocyclones. This has a significance effect on the speed with which control action can be effected.

Disadvantages. The main disadvantages of the hydrocyclone are clearly portrayed by the performance curve: the lack of sharpness of separation, and the amount of material that passes through the device without classification. This is not to say, however, that an alternative piece of equipment is necessarily capable of achieving a better separation, since the drawbacks mentioned are characteristic of all sedimentation classifiers.

As distinct from the sharpness of separation, there is also a limit on the degree of separation; a hydrocyclone can obtain only one relatively pure product—for example, the overflow has some fines in a degritting operation, or the underflow has some water in a dewatering operation.

Although the hydrocyclone has a potentially wide range of separating sizes, it is generally impractical to operate with cut points below about 2 μm, because of excessive energy requirements. Insufficient feed material normally provides an upper cut limit.

On the whole a hydrocyclone is inflexible once in operation, since its continued satisfactory performance requires stable feed behavior. In some cases such as closed circuit grinding, a disturbance in the feed may produce a behavior change that compounds the harmful effect of the feed disturbance.

Finally, there is the problem of erosion. This can be alleviated by the use of rubber or replaceable liners or components.

10.5. FLUIDIZED BED (HYDRAULIC) CLASSIFIERS

The basic principle of the fluidized bed classifier is illustrated in Fig. 10.1b, which shows that the settling velocity of a particle is balanced by a fluid flowing upward at an equal velocity. Ideally any particle with a lower settling velocity is carried out in the overflow, and any particle with a higher settling velocity falls to the underflow.

In contrast to sedimentation classifiers, under these conditions the d_{50} can be determined with some precision; but the performance curve cannot even be estimated. It is apparent that the d_{50} size will still belong to the particle that neither settles nor rises in the fluidized bed. As fresh feed enters the fluidized bed, it will eventually build up a bed of equisized particles (strictly, equisettling particle if the material is not homogeneous) so that eventually these particles will range from the underflow to the overflow and will have an equal chance of leaving through either outlet. Since this situation now represents a conventional fluidized bed, Eq. 4.70 can be used to calculate d_{50}, but in calculating the liquid velocity allowance must be made for the negative effects of the solid and liquid discharge through the underflow. Although reasonable estimates of settling velocities can be made, it is preferable to determine them experimentally, since this is a comparatively simple matter.[43]

Theoretically the fluidized bed classifier should give an essentially perfect separation. In practice this is not the case, because the liquid velocity across the column is not constant (Fig. 4.3), first because of the primary velocity profile and second because of eddying. The existence of velocity profiles in tubes due to wall drag has already been described (Section 4.1). In true fluidized beds this profile is severely flattened by the presence of particles and, if one exists at the liquid inlet, by the porous disperser across the column. By itself a stable velocity profile should not be harmful, in that the separation size should be determined by the maximum velocity in the profile, since differential pressure forces on the particles tend to mix them into the fastest moving liquid.[44] Herein lies some of the problem, because a continuous classifier does not allow sufficient time for all particles to be sorted by the maximum velocity.[45] More serious however is the effect of eddies and other massive flow disturbances. To some extent these are a characteristic of all fluidized beds, but their occurrence may be initiated or accentuated by factors such as poor dispersion of the hydraulic water, or by the energy of the feed stream. To minimize these effects, some classifiers, and elutriators in particular, often have extensive gridworks in the actual classification zone to ensure an even flow across the whole column.[4]

Some of the fluidized bed classifiers (e.g., the Hydroscillator) are essentially sedimentation classifiers using only small amounts of hydraulic water, primarily to wash the pore liquid from the sands. As a result, they have performance curves typical of efficient sedimentation classifiers with few fines in the sands (i.e., the performance curve is virtually a corrected performance curve). Even so, the hydraulic water noticeably raises the limiting particle size. Probably one of the best documented examples of this behavior is comprised of the data on the water injection hydrocyclone described in the next section.[46]

Some air classifiers use an internal fan to recycle

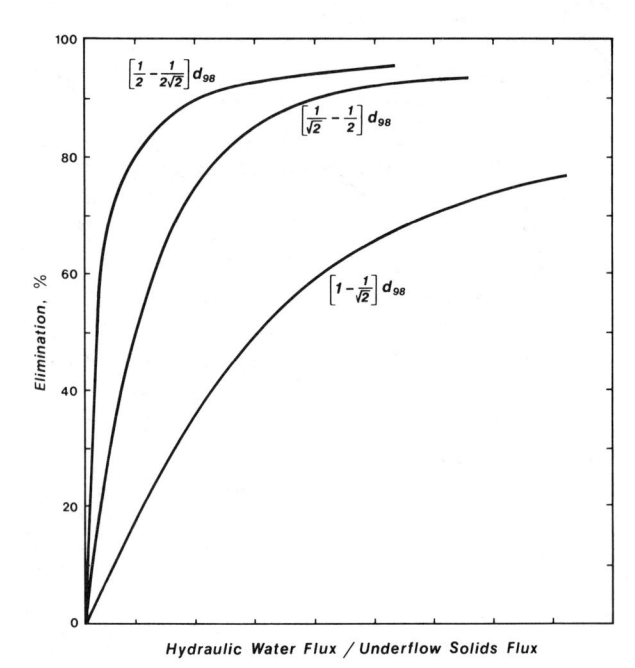

Figure 10.26. Typical elimination curves for a fluidized bed classifier. [Reprinted with permission from B. Fitch, *Ind. Eng. Chem.*, **54**, 44–51 (October 1962). Copyright 1962 American Chemical Society.]

air and thus generate the hydraulic fluid flow. Under these conditions, the classifier is effectively two classifiers in series, each with its own performance curve. When these two curves are summed together, the net performance curve may be fishhook shaped[47] (Fig. 10.25).

It follows that the performance curve of a well-designed fluidized bed classifier is determined primarily by operating conditions, but because the disturbing effects are unpredictable, actual design methodology involves the use of plots of average elimination as a function of relative solid-liquid velocities[14] (Fig. 10.26). In the zig-zag air classifier, particles are exposed to multiple classification and under these conditions, stochastic methods are useful in analyzing the performance.[48]

In general, a fluidized bed classifier is capable of a sharper separation than a sedimentation classifier, but this is achieved at the expense of capacity.

Example 10.5. A continuous commercial elutriator is to be used to classify a slurry. The following information is available:

Elutriator diameter: 1.0 m
Separation size (i.e., d_{50} drag diameter): 150 μm
Solids density: 3145 kg/m³

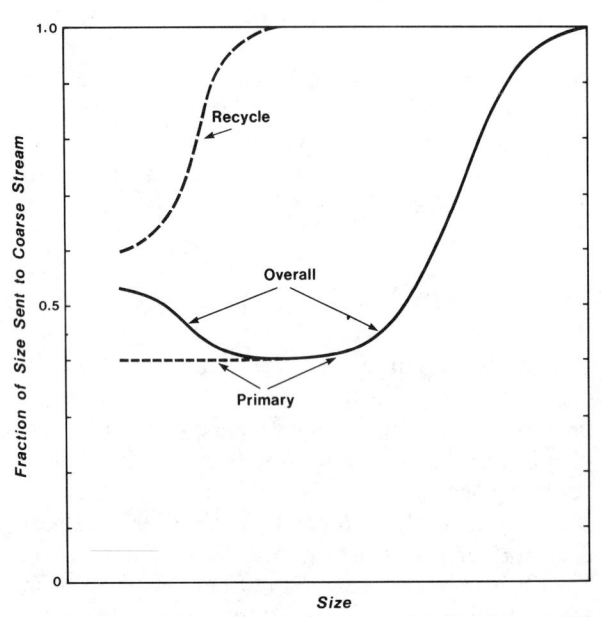

Figure 10.25. Derivation of "fishhook" performance curve from two classification actions in series. (After Luckie and Austin.[47])

Slurry concentration in elutriation zone controlled at 1400 kg/m³

What flow rate is required?

Solution. Because the elutriator operates continuously, the residence time will be comparatively short, so that some particles will be classified by the maximum fluid velocity (in the center of the flow path), while others will be classified by lower velocities nearer the walls. Thus, the appropriate velocity for the d_{50} particle will be the *average* velocity in the column. From Example 4.2, for a 150 μm particle ($\rho_s = 3145$ kg/m³):

$$v_\infty = 0.0193 \text{ m/sec}$$

Because the slurry concentration in the classification zone is controlled at 1400 kg/m³, the particles will be subjected to hindered settling conditions (i.e., $\bar{v} = v_h$). Thus

$$\bar{v} = v_h = v_\infty(\varepsilon)^n$$

where

$$n = \frac{4.8}{(2 + \beta)} \quad \text{(Eq. 4.71)}$$

From Fig. 4.8, the slope of the $\Psi = 1.0$ curve at $\text{Re}_p = 2.9$ (from Example 4.2) gives

$$\beta = -0.85$$

∴ $$n = 4.17$$

Also

$$1400 = 3145(1 - \varepsilon) + 1000\varepsilon$$

∴ $$\varepsilon = 0.814$$

∴ $$\bar{v} = 0.0193(0.814)^{4.17}$$
$$= 0.0082 \text{ m/sec}$$

∴ volumetric rate =
$$\frac{0.0082 \text{ m}}{\text{sec}} \bigg| \frac{\pi \times (1.0)^2 \text{ m}^2}{4} \bigg| \frac{60 \text{ sec}}{\text{min}}$$
$$= 0.39 \text{ m}^3/\text{min}$$

Example 10.6. The size distribution of a powder is to be determined in a laboratory elutriator. One of the tubes is 3 cm in diameter, and the flow rate is to be set to give a separation at 150 μm in this tube. What flow rate should be used? ($\rho_s = 3145$ kg/m³)

Solution. In this case, because the elutriator operates on a batch basis, the residence time should be long enough for the particles to be classified by the *maximum* velocity in the tube. (If the test period is too short, the classifying velocity will be indeterminant, between the average and maximum values).

Assuming laminar flow in the tube, from Eq. 4.4:

$$v_{\max} = 2\bar{v} = v_\infty$$

Again, from Example 4.2

$$2\bar{v} = v_\infty = 0.0193 \text{ m/sec}$$

∴ $$\bar{v} = 0.00965 \text{ m/sec}$$

∴ volumetric rate =
$$\frac{0.00965 \text{ m}}{\text{sec}} \bigg| \frac{60 \text{ sec}}{\text{min}} \bigg| \frac{\pi(0.03)^2 \text{ m}^2}{4} \bigg| \frac{10^3 \text{ l}}{\text{m}^3}$$
$$= 0.41 \text{ l/min}$$

Check tube **Re**:

$$\mathbf{Re} = \frac{Dv\rho}{\mu}$$
$$= \frac{0.03 \times 0.00965 \times 1000}{10^{-3}}$$
$$= 290$$

That is, flow is laminar, and Eq. 4.4 is valid.

Assume that at the end of the test the quantity of powder is measured and found to occupy 2% of the tube volume. Is the foregoing answer valid?

Check for hindered settling behavior:

$$\varepsilon = 0.98$$

∴ $$v_h = v_\infty(0.98)^{4.17}$$
$$= 0.92 v_\infty$$

Thus, even this comparatively low solids concentration has an effect on the result, and the flow rate should be reduced accordingly. Furthermore, since the column probably had a much higher concentration at the start of the test, it would be advisable to start the test at an even lower flow rate, and gradually come up to the required rate to compensate for the *decreasing* hindered settling effect.

10.6. CAPACITIES

The discussion above shows some of the possibilities and limitations in calculating classifier capacities and performance. Substantiation of the methods is difficult because of the scarcity of data, especially since most of the available data are expressed as separations based on the limiting size. Manufacturers' catalogues are of some use, but unfortunately they seldom give any indication of performance curves or d_{50} values. Furthermore the capacities are often based on overflow conditions, which in reality are dependent on input conditions.

A further factor to be considered with respect to mechanical classifiers is the capacity of the mechanical discharge unit. These must have sufficient power and action to handle the sands load.[49] Handbooks[12] and manufacturers' catalogues are useful for these calculations.

10.7. CLOSED CIRCUIT GRINDING

In the past, mechanical sedimentation classifiers (spiral or rake) were used for closed circuit grind-

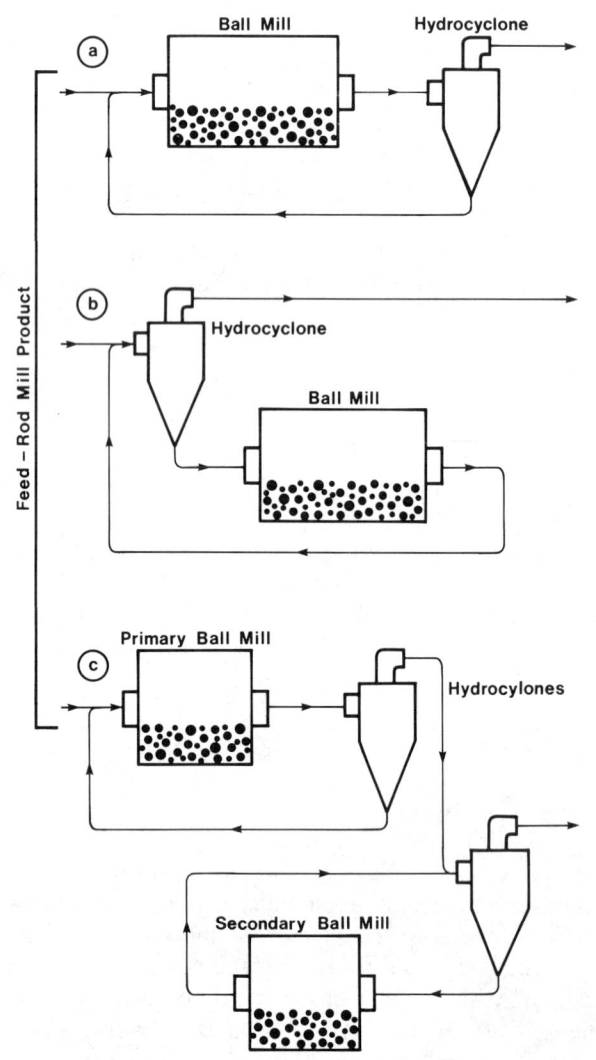

Figure 10.27. Typical mill-classifier arrangements used for closed circuit grinding.

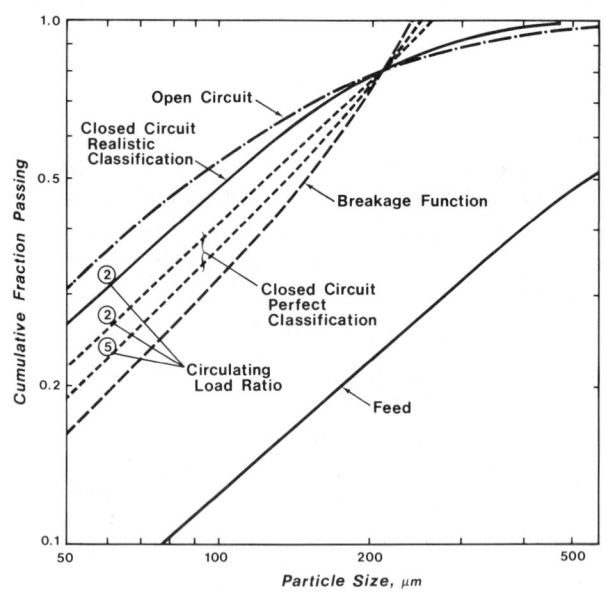

Figure 10.28. Simulated data to show the effect of classification on closed circuit grinding. (After Kelsall, Reid, and Stewart.[50])

ing, but their comparatively poor separating performance eventually led to the frequent use of the physically similar (but more expensive) hydraulic types (e.g., the Hydroscillator). Even though the classifying performance of hydrocyclones generally is little better than that of the old sedimentation classifiers, their simplicity and low cost have resulted in their being adopted as the standard closed circuit classifier today. Some typical circuits are shown in Fig. 10.27; whether the hydrocyclone is placed before or after the mill depends primarily on the liberation characteristics of the ore.

10.7.1. Improving Classifier Performance

There are a number of theoretical analyses available to show the considerable benefits that can be derived from improved classification[50-52] (Fig. 10.28): lower circulating loads reduce grinding costs and/or increase capacity, as well as reduce over- and undergrinding, which minimizes the difficulties of concentrating fine particles.[39]

There are two main factors contributing to inefficient classification. The first is the inherent spread of the performance curve of sedimentation classifiers (arising from the ideal pool basis), and the second is the unclassified material in the underflow (Fig. 10.9b), which at the high pulp densities frequently encountered in closed circuit grinding may be as high as 50% of the feed.

Closed Circuit Grinding

Because of the widespread use of hydrocyclones, particular attention has been given to them. Kelsall and Holmes[46] demonstrated that significant decreases in the amount of unclassified material could be achieved by using the solution originally applied to mechanical sedimentation classifiers, that is, the addition of hydraulic water (Fig. 10.29). Applied to a 36 cm hydrocyclone in a circuit with a 2.7 m × 2.4 m ball mill, the circulating load was reduced from 350 to 160%, and amount of -75 μm material recycled to the mill was reduced from 8.2 to 1.8 t/hr for a fresh feed rate of 22.4 t/hr. The improved separation is illustrated in the performance curves shown in Fig. 10.30.

Although this solution has been adopted by the

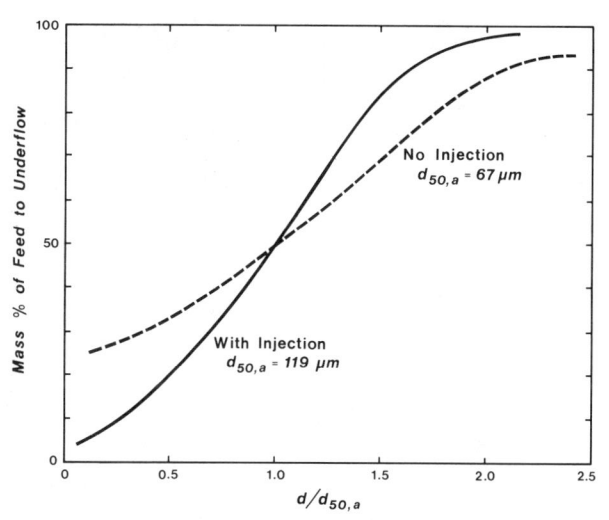

Figure 10.30. The effect of water injection on the performance of a hydrocyclone. (After Kelsall and Holmes.[46])

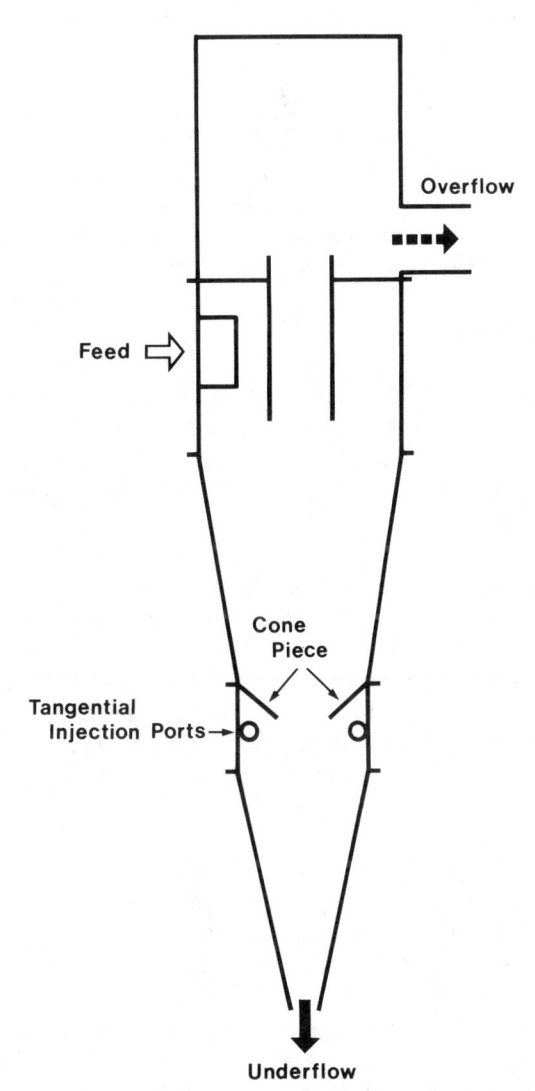

Figure 10.29. Schematic diagram of a water injected hydrocyclone. (After Kelsall and Holmes.[46])

pulp and paper industry, the mineral industry has been very slow to follow suit. It has been claimed that the excess water is undesirable in the circuit, but it is possible to design the circuit to minimize this problem.[53]

The separation limitations imposed by the ideal pool concept can be overcome by successive reclassification of the sands product. Each successive reclassification results in a steeper *overall* performance curve (the stage curve remains unchanged) so that, theoretically with a large number of stages, a perfect separation can be obtained.[54] Water requirements can be minimized by counter-current operation, but the substantial cost of equipment is seldom justified. At present the most attractive approach appears to be the use of two sedimentation classifiers in series. This actually gives benefits by two mechanisms: first, a steeper overall performance curve as a result of double classification, and second, a marked reduction of the fines in the sands product arising from the washing effect resulting from repulping of the sands before the second classification.

When two hydrocyclones are run in series, it may be better to operate them with different discharge characteristics to minimize the amount of unclassified material[55] (Fig. 10.31). This method has the disadvantage of needing two pumping systems, but so far, efforts to eliminate one pump and have the two hydrocyclones directly connected have yet to produce the same benefits.[56] Alternatively, hydrocyclone sands can be treated in some other type

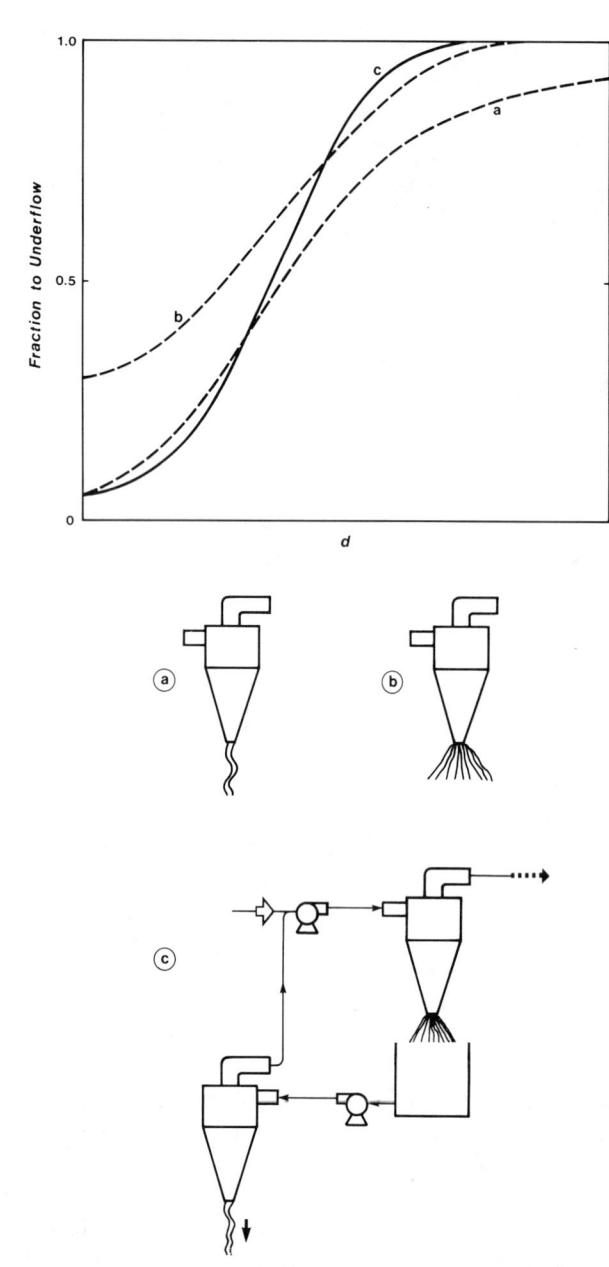

Figure 10.31. Two-stage hydrocyclone classification (one with spray discharge and one with rope discharge), not only reduces the amount of unclassified material but also steepens the overall performance curve. (After Trawinski.[55])

of classifier. Increasingly, the Hukki cone classifier is being employed for this purpose.[57]

A method of increasing classifier capacity is to reduce the distance a particle has to travel before it is effectively out of the flowing slurry. This can be achieved by introducing into the classifier parallel inclined surfaces for the particles to settle against[52,58] (Fig. 10.1c). This concept can be applied to both sedimentation and fluidized bed classifiers, and since the velocity components now have features of each basic type (Fig. 10.1a and b), it can itself be considered as a third type. This method is finding increased use in clarifiers,[58] but, as yet, is seldom used in classifiers.[52]

10.7.2. Classification of Heterogeneous Solids

Thus far the treatment has implied homogeneous solids with settling rates essentially determined by the particle size alone. In many practical situations the feed consists of more than one mineral, either as liberated or as middling particles, with the result that a number of different sized particles can have the same settling rate. Frequently it is desirable to know the relative sizes of these *equisettling* particles. Equation 5.2 is widely quoted for this purpose, but in view of the discussion in Sections 4.7 and 5.1, Eq. 5.3 probably gives a far better indication of the relative behavior of irregular particles.

The significant point that follows from this equation is that in a given classifier, each material will have its own performance curve. This is due primarily to differing densities (and this includes middlings particles), but it may also result from shape or roughness. In turn this has important implications in closed circuit grinding. In most metalliferous mineral systems, the density of the valuable mineral is higher than that of the gangue mineral and consequently the former has a stronger tendency to report to the sands product. Thus, the heavy mineral builds up in the circulating load and eventually leaves the circuit with a markedly smaller average size than the gangue minerals. In some cases this may not be a problem—for example, when the liberation size is much larger than the recovery size of the concentrating device, as is the case in the Zinc Corp. flotation circuit.[11] In other instances it can be a serious problem—for example, with scheelite, which tends to fracture more readily than the gangue material anyway and has fine particles that are very difficult to concentrate.[59] This buildup of heavy minerals in classifiers is one of the reasons for the increasing attention, particularly for coarser sizing, that sieve bends are receiving, since they tend to concentrate the heavy particles in the undersize fraction.[60,61]

Since different minerals may have different performance curves, the net performance curve of the ore is spread out, with the result that the 1–3% of the overflow measure of separation is even less meaningful than usual, because this range may represent the particle size of the wrong mineral.[59]

Example 10.7. If the solids in the slurry fed to the hydrocyclone in Example 10.1 are 90% (vol) quartz ($\rho_Q = 2650$ kg/m³) and 10% (vol) galena, ($\rho_{PbS} = 7600$ kg/m³), what is the d_{50} separation size for these two minerals? (Assume in Eq. 4.70, $n_{PbS} = 6.0$, $n_Q = n_{slurry} = 8.0$).

Solution. A solution can be obtained by Eq. 4.77, by setting v_h for quartz and galena equal to v_h for the bulk slurry. Using v_∞ for the "average" slurry particle (calculated in Example 4.2), then:

For Bulk Slurry

$$v_h + v_\varepsilon = 0.0193(1 - 0.27 - 0.03)^{8-1} \quad (E10.7.1)$$

For Quartz

$$v_{h,Q} + v_\varepsilon = v_{\infty,Q}(1 - 0.27 - 0.03)^{8-1} \quad (E10.7.2)$$

For Galena

$$v_{h,PbS} + v_\varepsilon = v_{\infty,PbS}(1 - 0.27 - 0.03)^{6-1} \quad (E10.7.3)$$

To find $d_{50,Q}$, combining Eqs. E10.7.1 and E10.7.2:

$$0.0193(0.7)^7 = v_{\infty,Q}(0.7)^7$$

$$\therefore \quad v_{\infty,Q} = 0.0193 \text{ m/sec}$$

Substituting in Eq. 4.41:

$$\frac{f_d}{Re_p} = \frac{4}{3} \frac{(2650 - 1000)}{1000^2} \frac{10^{-3} \times 9.81}{(0.0193)^3}$$
$$= 3.0$$

From Fig. 4.9:

$$Re_p = 3.0 = \frac{d_{50,Q} v \rho}{\mu}$$

$$\therefore \quad d_{50,Q} = \frac{3.0 \times 10^{-3}}{1000 \times 0.0193}$$
$$= 155 \text{ μm}$$

To find $d_{50,PbS}$, combining Eqs. E10.7.1 and E10.7.3:

$$0.0193(0.7)^7 = v_{\infty,PbS}(0.7)^5$$

$$\therefore \quad v_{\infty,PbS} = 0.0193(0.7)^{7/5}$$
$$= 0.0117 \text{ m/sec}$$

Substituting in Eq. 4.41:

$$\frac{f_d}{Re_p} = \frac{4}{3} \frac{(7600 - 1000)}{1000^2} \frac{10^{-3} \times 9.81}{(0.0117)^3}$$
$$= 54$$

From Fig. 4.9:

$$Re_p = 0.7$$

$$\therefore \quad d_{50,PbS} = \frac{0.70 \times 10^{-3}}{1000 \times 0.0117}$$
$$= 60 \text{ μm}$$

In practice, the dependence of the d_{50} on ρ_s would have important consequences. The circulating load of PbS in the grinding circuit would be higher, the liberation would be changed, and since the flotation probabilities are normally a function of particle size, the flotation behavior would be changed. Once again, a rigorous analysis would require iteration of the relevant examples.

10.8. CLASSIFIER OPERATION

Two major problems are apparent in classifier design. The difficulties of predicting settling rates have already been considered. No so apparent is the very real problem of knowing the actual cut size required. Thus, classifiers with considerable flexibility must be selected. To this end, one of the major operating variables frequently used is the slurry concentration[12] (i.e., water addition). This can be explained as follows. The capacity design criterion I_V/A, which is determined by the settling velocity, is in fact the volumetric settling flux in the classifier. From Eqs. 4.70 and 4.78

$$\psi_V = vC_V = v_\infty \varepsilon^n(1 - \varepsilon) = \frac{I_V}{A} \quad (10.30)$$

(ψ_V = volumetric solids flux; C_V = volumetric solids concentration). Differentiating yields

$$\frac{d(I_V/A)}{d\varepsilon} = v_\infty[n\varepsilon^{n-1} - (n + 1)\varepsilon^n] \quad (10.31)$$

from which it follows that the capacity per unit area passes through a maximum when

$$\varepsilon = \frac{n}{n + 1} \quad (10.32)$$

Consequently, provided the classifier is designed to operate with a slurry density where $\varepsilon > n/(n + 1)$, the capacity (i.e., total amount of solid settled) can be *increased* by diluting the slurry. (Alternatively, if the capacity is not changed, the d_{50} will increase.)

A further consequence of Eq. 10.31 is that, if the classifier operates in the region where $\varepsilon \sim n/(n + 1)$, it should be relatively insensitive to changes in the slurry concentration. Such behavior has been noted in the Broken Hill South Ltd. concentrator.[39] Operating at this point also means that the capacity can be lowered by either raising *or* lowering the slurry concentration.

There is a limit to the extent to which the slurry concentration can be raised, since a point is eventually reached where classification ceases and only thickening occurs[62] (Chapter 17). The available information on viscosity effects suggests that in the case of mechanical classifiers, this point may be affected by the speed of the sands discharge mechanism.[42]

Analysis of Eq. 10.4 or 10.6 explains why the speed of the mechanical action is also a major operating variable in such classifiers. Since the diffusion coefficient $\mathcal{D}$ is approximately proportional to speed (Fig. 10.17), the speed raises the cut size (or decreases the capacity). The exact effect is more difficult to predict because the higher speed will also raise the slurry concentration and thus reduce the settling rate of any given particle. Note that according to Eqs. 10.4 and 10.6 the reduced performance curve does not change; that is, the efficiency of the equipment does not change.

Hydraulic classifiers with bottom sands discharge are frequently controlled by measuring the pressure drop between the top and bottom of a length of the fluidized bed, a method that is limited by the fact that it is actually measuring the average terminal velocity within what has to be a relatively large length.[63] A more satisfactory method, also used, is to measure the slurry density at a given point in the bed, since, as Eqs. 4.70 and 4.74 show, v is primarily a function of slurry concentration (i.e., the slurry density) for a given particle. Thus, when faster settling particles fall into the measuring zone, they descend until eventually they build up and raise the slurry density. This is then used to activate the bottom discharge mechanism.

In this chapter, we have emphasized classifier performance curves because they provide the most realistic assessment of separation. Because sedimentation classifiers are incapable of perfect separations, their performance must be measured against realistic standards, and it is generally accepted that curve A in Fig. 10.10 represents a suitable "ideal." Whether, as is often claimed, this ideal curve is totally unrealistic at the high slurry concentrations often encountered in closed circuit grinding has yet to be established. The literature abounds with operating data, but even where it is possible to derive performance curves, it is normally impossible to determine whether the deviations are due to high slurry concentrations, flow patterns and their subsequent effect on residence time distributions, variable mineral densities (Section 10.7.2), poor classifier design, or poor classifier operation.

At this stage, is is also impossible to state which of Eqs. 10.3–10.10 is the better theoretical representation of a classifier, because, by selecting suitable parameters, most of the equations can be fitted to a given set of data [particularly when allowance is made for residence times (Sections 10.3.3 and 10.7.2)]. In selecting between equations therefore, it is usually best to aim for mathematical simplicity with a realistic description of the physical system.

REFERENCES

1. R. H. Perry and C. H. Chilton (Eds.), *Chemical Engineers' Handbook*, 5th Edition, Section 21, McGraw-Hill (1973).
2. *Kirk-Othmer Encyclopedia of Chemical Technology*, 2nd ed., Vol. 18, *Size Separation*, Wiley-Interscience (1969).
3. Anon., "Extraction and Processing of a Silty Sand and Gravel Deposit," *Cement Lime Gravel*, **39**, 157–164 (May 1964).
4. Anon., "Classifying Sand by Elutriation," *Cement Lime Gravel*, **40**, 123–134 (April 1965).
5. Anon., "The Bathmos Classifier," *Phosphorous and Potassium* (September–October 1972).
6. R. T. Hukki, "Hydraulic Classification in Gravitational and Centrifugal Fields," Paper C.1, presented at 8th Int. Miner. Process. Congr., Leningrad, 1968.
7. Anon., "Novel Device for Three Way Solid Particle Separation," *World Min.*, **22**, 67 (March 1969).
8. R. C. Fisher, "New Equipment for Mineral Processing. Part 2," *Aust. Min.*, 24–29 (May 15, 1967).
9. Z. Yoshino et al., "Studies on Wet Classification," *Int. Chem. Eng.*, **11**, 547–554 (July 1971).
10. A. J. Lynch and T. C. Rao, "Modelling and Scale-Up of Hydrocyclone Classifiers," *Proc. 11th Int. Miner. Process. Congr., Cagliari, 1975*, pp. 245–269, Università di Cagliari (1975).
11. A. W. Cameron et al., "A Detailed Assessment of Concentrator Performance at Broken Hill South Ltd," *Proc. Australas. IMM*, No. 240, 53–67 (December 1971).
12. A. F. Taggart, *Handbook of Mineral Processing* (Tabulated Data, Section 8), Wiley (1945).
13. E. J. Roberts and E. B. Fitch, "Predicting Size Distributions in Classifier Products," *Trans. AIME/SME*, **205**, 1113–1120 (November 1956).
14. B. Fitch, "Why Particles Separate in Sedimentation Processes," *Ind. Eng. Chem.*, **54**, 44–51 (October 1962).

References

15. W. E. Dobbins, "Effect of Turbulence on Sedimentation," *Trans. ASCE*, **109**, 629–656 (1944).
16. T. R. Camp, "Sedimentation and the Design of Settling Tanks," *Trans. ASCE*, **111**, Paper 2285, 895–958 (1946).
17. H. Z. Sarikaya, "Numerical Model for Discrete Settling," *J. Hydrol. Div. ASCE*, **103**, 865–876 (August 1977).
18. H. Schubert and T. Neesse, "The Role of Turbulence in Wet Classification," *10th Int. Miner. Process. Congr., London, 1973*, pp. 213–239, IMM (1974).
19. K. J. Reid, "Derivation of an Equation for Classifier-Reduced Performance Curves," *Can. Metall. Q.*, **10**, 253–254 (1971).
20. C. C. Harris, "Graphical Representation of Classifier-Corrected Performance Curves," *Trans. IMM (C)*, **81**, C243–C245 (1972).
21. L. R. Plitt, "The Analysis of Solid-Solid Separations in Classifiers," *CIM Bull.*, **64**, 42–47 (April 1971).
22. P. S. B. Stewart and C. J. Restarick, "Dynamic Flow Characteristics of a Small Spiral Classifier," *Trans. IMM (C)*, **76**, C225–C230 (December 1967).
23. A. F. Taggart, *Elements of Ore Dressing*, Wiley (1951).
24. R. H. Perry and C. H. Chilton (Eds.), *Chemical Engineers' Handbook*, 5th ed., Section 19, McGraw-Hill (1973).
25. E. B. Fitch and D. G. Stevenson, "Gravity Separation Equipment," Ch. 4 in *Solid/Liquid Separation Equipment Scale-Up*, pp. 81–153, Uplands Press (1977).
26. D. Bradley, *The Hydrocyclone*, Pergamon Press (1965).
27. D. F. Kelsall, "The Theory and Applications of the Hydrocyclone," Ch. 5 in *Solid-Liquid Separations; A Review and Bibliography*, J. B. Poole and D. Doyle (Eds.), HMSO (1966).
28. L. R. Plitt, "A Mathematical Model of the Hydrocyclone Classifier," *CIM Bull.*, **69**, 114–123 (December 1976).
29. K. Rietema, "Performance and Design of Hydrocyclones," *Chem. Eng. Sci.*, **15**, 298–325 (1961).
30. D. Bradley and D. J. Pulling, "Flow Patterns in the Hydraulic Cyclone and Their Interpretation in Terms of Performance," *Trans. Inst. Chem. Eng.*, **37**, 34–45 (1959).
31. D. F. Kelsall, "A Study of the Motion of Solid Particles in a Hydraulic Cyclone," *Trans. Inst. Chem. Eng.*, **30**, 87–104 (1952).
32. V. G. Renner and H. E. Cohen, "Measurement and Interpretation of Size Distribution of Particles Within a Hydrocyclone," *Trans. IMM (C)*, **87**, C139–C145 (June 1978).
33. E. Cohen et al., "The Residence Time of Mineral Particles in Hydrocyclones," *Trans. IMM (C)*, **75**, C129–C138 (June 1966).
34. P. H. Fahlstrom, "Studies of the Hydrocyclone as a Classifier," in *Mineral Processing, Proc. 6th Int. Miner. Process. Congr., Cannes, 1963*, pp. 87–109, A. Roberts (Ed.), Pergamon (1965).
35. D. A. Dahlstrom, "Cyclone Operating Factors and Capacities on Coal Refuse Slurries," *Trans. AIME*, **184**, 331–344 (1949).
36. D. A. Dahlstrom," Fundamentals and Applications of the Liquid Cyclone," *Chem. Eng. Prog. Symp. Ser.* No. 15, Vol. 50, pp. 41–61 (1954).
37. R. Vajarapongse, "The Investigation of Pulp Density and Viscosity Effects in a Wet Cyclone," M. Eng. Sci. thesis (unpublished), University of Melbourne (1963).
38. D. R. Marlow, "A Mathematical Analysis of Hydrocyclones Data," M.Sc. thesis (unpublished), University of Queensland (1973).
39. D. F. Kelsall et al., "The Effects of a Change from Parallel to Series Grinding at Broken Hill South Ltd," *Aust. IMM Conf. Newcastle*, pp. 337–347 (May–June 1972).
40. D. T. Tarr, "Theory of Hydrocyclones," presented at IADC Conference on Hydrocyclones, Dallas, May 1976.
41. D. T. Tarr, personal communication (1979).
42. H. Kirchberg, E. Töpfer, and W. Scheibe, "The Effects of Suspension Properties on Separating Efficiency of Mechanical Classifiers," *Proc. 11th Int. Miner. Process. Congr., Cagliari, 1975*, pp. 219–244, Università di Cagliari (1975).
43. J. F. Richardson and W. N. Zaki, "Sedimentation and Fluidisation, Part I," *Trans. Inst. Chem. Eng.*, **32**, 35–53 (1954).
44. T. Allen, *Particle Size Measurement*, 2nd ed., Ch. 11, Chapman and Hall (1974).
45. K. Heishanen and H. Laapas, "On the Effects of the Fluid Rheological and Flow Properties in the Wet Gravitational Classification," *Preprints, 13th Int. Miner. Process. Congr., Warsaw, 1979*, Vol. 1, pp. 183–204, Polish Scientific Pub. (1979).
46. D. F. Kelsall and J. A. Holmes, "Improvement of Classification Efficiency in Hydraulic Cyclones by Water Injection," Paper 9, *Proc. 5th Int. Miner. Process. Congr., London*, pp. 159–170, IMM (1960).
47. P. T. Luckie and L. G. Austin, "Mathematioal Analysis of Mechanical Air Separator Selectivity Curves," *Trans. IMM (C)*, **84**, C253–C255 (1975).
48. M. M. G. Senden and M. T. Tels, "A Stochastic Model of Multistage Separation in Zig Zag Air Classifiers," *J. Powder Bulk Solid Technol.*, **2**, 16–35 (Spring 1978).
49. R. E. Riethmann and B. M. Bunnell, "Application and Selection of Spiral Classifiers," Ch. 16 in *Mineral Processing Plant Design*, pp. 362–375, AIME (1978).
50. D. F. Kelsall, K. J. Reid, and P. S. B. Stewart, "The Study of Grinding Processes by Dynamic Modelling," *Elec. Eng. Trans. Inst. Eng. Aust.*, **EE5**, 1, 173–186 (March 1969).
51. S. K. de Kok, Discussion in *J. S. Afr. IMM*, **75**, 280–287 (1975).
52. R. T. Hukki, "A New Way to Grind and Recover Minerals," *Eng. Min. J.*, **178**, 66–74 (April 1977).
53. D. F. Kelsall, "Some Applications of Hydraulic Cyclones in Hydrometallurgical Processes," *Trans. AIME/SME*, **226**, 2–8 (1963).
54. D. F. Kelsall and J. C. H. McAdam, "Design and Operating Characteristics of a Hydraulic Cyclone Elutriator," *Trans. Inst. Chem. Eng.*, **41**, 84–95 (1963).
55. H. Trawinski, "Theory, Applications, and Practical Operation of Hydrocyclones," *Eng. Min. J.*, **177**, 115–127 (September 1976).
56. D. F. Kelsall, P. S. B. Stewart, and C. J. Restarick, "A Practical Multiple Cyclone Arrangement for Improving Classification," Paper E5, *1st Eur. Conf. Mixing and Centrifugal Separation*, E5 [83–93], (September 1974).

57. (a) K. Heiskanen, "The Influence of Two-Stage Classification on Rod Mill/Pebble Mill Grinding Circuit," *Larox News,* **2/1978,** 4–6 (1978).
    (b) K. Heiskanen, "Two Stage Classification," *World Min.,* **32,** 44–46 (June 1979).
58. Kuan M. Yao, "Design of High-Rate Settlers," *J. Environ. Eng. Div. Proc. ASCE,* **99,** 621–637 (October 1973).
59. R. D. Carpenter, "Preparation of Flotation Plant Feed," Ch. 19-1 in *Froth Flotation—50th Anniversary Volume,* D. W. Fuerstenau, Ed., AIME (1962).
60. J. H. Healy et al., "Erie Mining Co. Patents New Method For Screening Ores in Very Fine Size Ranges," *Min. Eng.,* **19,** 65–70 (April 1967).
61. L. D. Keller, "Fine Screening . . . The Current State of the Art," AIME/SME Preprint 71-B-25. Presented at AIME Centennial Meeting, New York, February–March 1971.
62. M. J. Lockett and H. M. Al-Habbooby, "Relative Velocities in Two-Species Settling," *Powder Technol.,* **10,** 67–71 (1974).
63. J. R. Himsworth, "Hindered Settling Classifiers; The By-Pass Particle Size Controller," Powder Technol., **10,** 181–187 (1974).

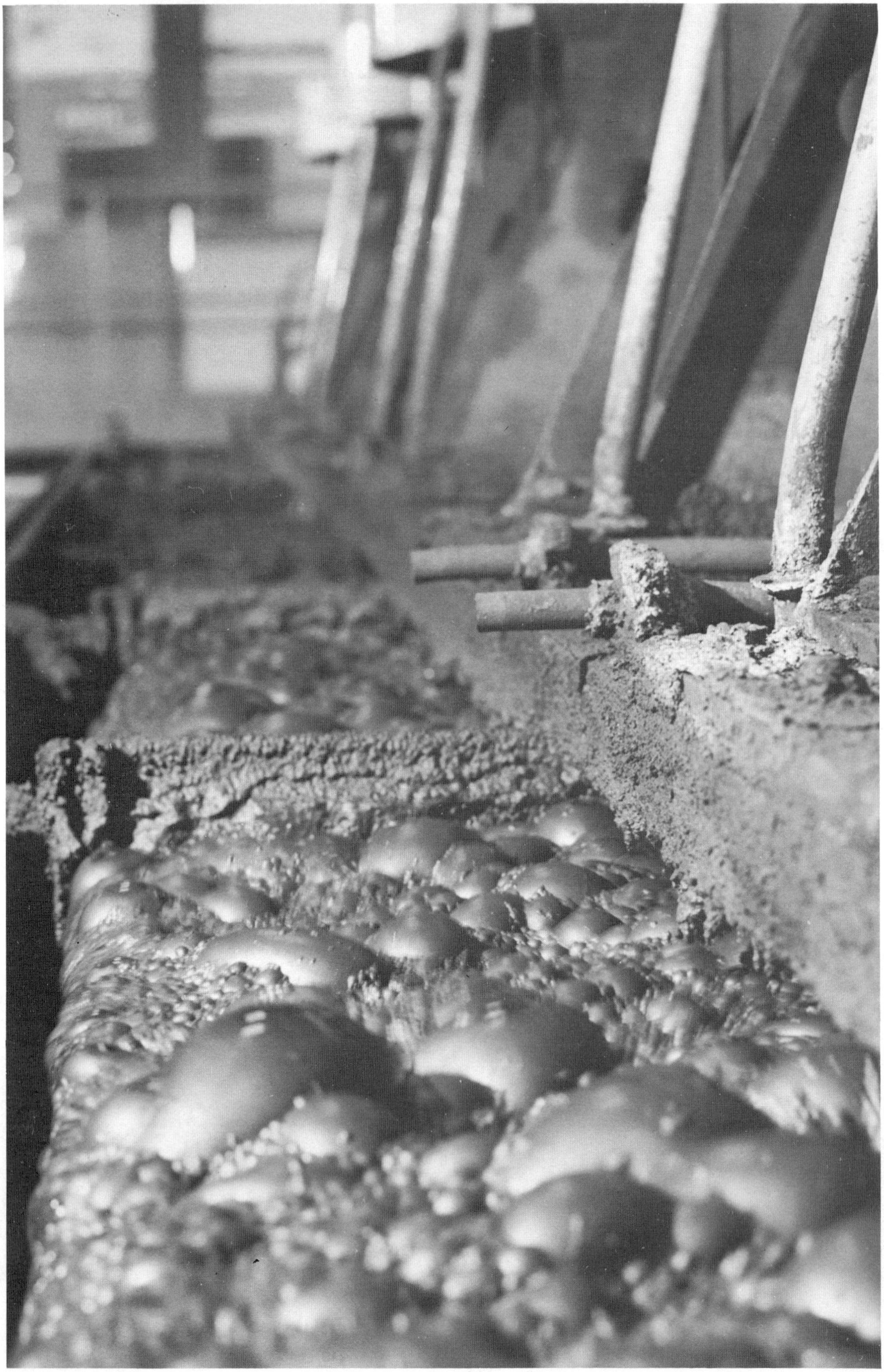

Part IV

Concentration Separation

Although there are exceptions, the reason for processing most ores is to obtain the valuable minerals in a more *concentrated* form. Part IV analyzes the principles and techniques of concentration processes based on the fundamental principles introduced in Part I.

The concentration method selected for a particular concentration separation depends on the nature of the ore and the properties of the minerals to be separated (and the differences between those properties). For separation to be possible, the ore particles must be reduced in size until at least some degree of liberation has been achieved. Since particle size affects the efficiency of all concentration separations, the size at which liberation occurs may influence the choice of concentration process (Fig. IV.1). If more than one separation method is available, the selection must be based on economics.

Minerals can be separated based on both physical and chemical properties. Physical properties utilized in concentration include color, surface reflectance, radioactivity level, magnetic susceptibility, and conductivity. Concentration separation methods using differences in these properties are treated in Chapters 11–15. Chemical properties are also utilized for concentration; however, the concentration processes that are considered to be "mineral processing" are those that utilize *surface* chemical properties only (Chapter 16).

A wide range of processes and equipment have been developed for concentration separations. Sorting by hand and gravity concentration have been

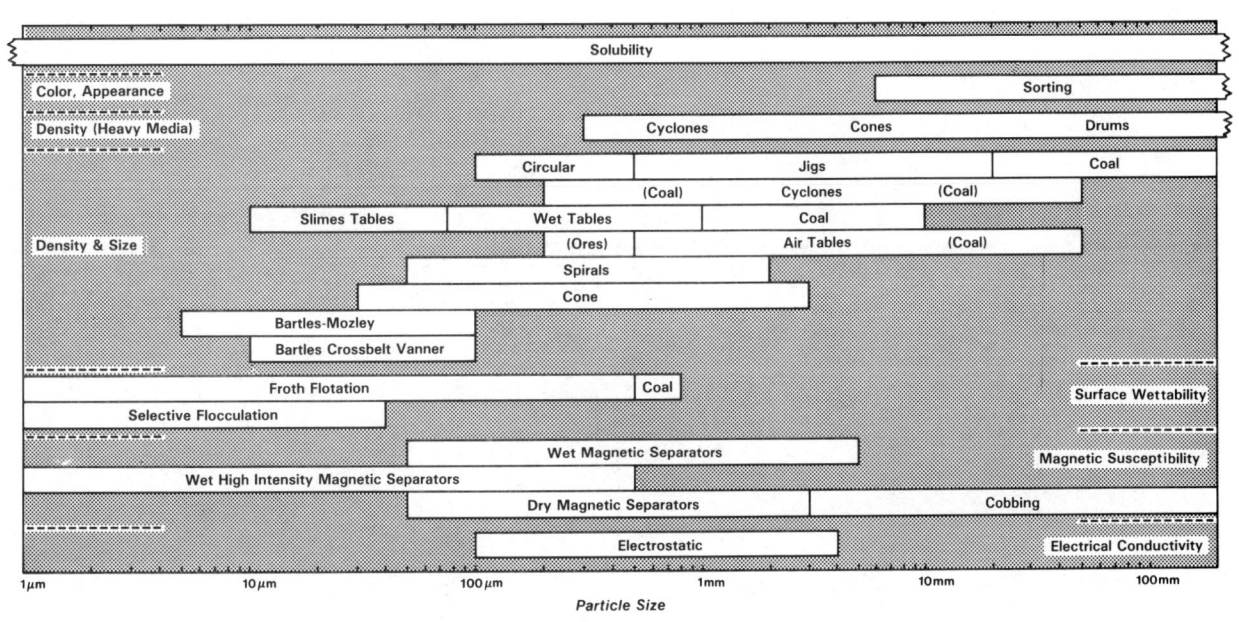

Figure IV.1. Particle size ranges for concentrating equipment.

used for many centuries. By the end of the nineteenth century, these processes had been improved and crude magnetic and electrostatic separation machines introduced. However, it was the development of flotation early in this century that provided a dramatic technological change; a wider range of minerals and lower grade ores could be concentrated, and higher concentrate grades could be obtained. Since that time, improvements have been steadily made in the selectivity of flotation reagents and in the equipment available for concentration separations.

Staged size reduction and concentration is widely used when one mineral liberates at a relatively coarse size. The increases in energy costs of recent years can be expected to result in the reintroduction of a *preconcentration* stage using such low cost techniques as sorting or gravity separation where liberated gangue can be rejected.

Chapter Eleven

Ore Sorting

Baldy's Law: "*Some of it plus the rest of it is all of it.*"

The sorting of an ore as a method of mineral concentration was probably practiced in the very first mining operation. Separation was by hand, after visual inspection and assessment of the value of an individual particle. Sorting today follows the same principles, but there are now a variety of detecting devices and the separation is generally mechanical.

The technique of hand sorting continued to be a major mineral processing tool until early this century, when advances in crushing, grinding, and alternate concentration methods, coupled with the rapidly increasing cost of labor, led to its rapid decline. However, hand sorting is still practiced where labor costs are relatively low and/or the ore is not amenable to other concentration or preconcentration techniques.

The desirability of mechanizing the sorting operation has long been recognized. The early mechanical sorters, such as the Sweet Sorter[1] of 1928, were not commercially successful. Although no operating data are available, it may be assumed that these early devices were inefficient and that losses were even higher than those incurred in hand sorting. This is to be expected when one considers that the human sensory system is very complex yet efficient: a person can simultaneously sense color, luster, shape, and texture, plus the other factors which experience has taught, to distinguish values from waste. In contrast mechanical detectors, no matter how reliable, are single-minded. That is, they can detect only one property, which in turn must be uniformly present to effect separation of the selected particles. The economic importance of achieving high recovery of the valuable mineral species in sorting operations is paramount,[2,3] and it is this high recovery requirement that has limited the success of sorting operations.

The first successful mechanized sorting device was the Lapointe Picker,[4] used to concentrate low grade uranium ores in the late 1940s. Since that time, a number of successful sorting machines have been introduced, taking advantage of the increasing sophistication of sensing devices and electronics.

11.1. SORTING MACHINES AND APPLICATIONS

Ore sorting machines are produced by a number of manufacturers, and though there are certainly differences between them, the major operating principles are common to all (Fig. 11.1).[5,6] The detection system is the heart of the sorting machine, and sorters are usually classified by the detection system used. Optical and photometric sorters are the most widely used; radiometric sorters are used for uranium; X-ray, conductivity, and magnetic sorters have found only limited application.

In sorting machines, each particle is analyzed individually. The particle size to be sorted has a marked effect on the sorter capacity (Fig. 11.2), so sorters are specifically designed to handle a narrow size range. Particles up to 160 mm are sorted, with sorter capacities ranging up to 180 t/hr for the larger particles.

Photometric ore sorting has been used for a range of minerals and ores, and with improved ore sorting equipment becoming available, this range grows wider. These sorters are in use in plants recovering rock salt, magnesite, barite, gypsum, marble, diamond, and other minerals.[7,8] In the Yerankini magnesite operation,[9,10] photometric sorters are being used to separate magnesite from serpentine and talc. The ore is sized, and six separate size

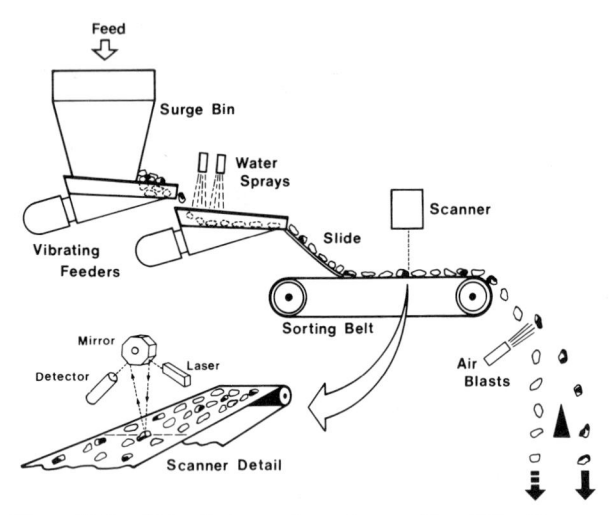

Figure 11.1. Major features of a sorting machine. (After Keys et al.[5])

TABLE 11.1: SORTING OF A MAGNESITE-SERPENTINE ORE

Size Range (mm)	Sorting Technique	Tonnage (t/h)	Sorting per Machine (t/h)	Number of Sorting Machines
200–1600	Hand Sorting	200	—	-
140– 200	Hand Sorting	200	—	-
80– 140	Photometric	160	180	1
45– 80	Photometric	80	100	1
25– 45	Photometric	120	50	4
12– 25	Photometric	100	30	4
–12	(Waste)	340	—	-
		1200		

fractions are sorted (Table 11.1). In the 20–40 mm size range, sorting increases the magnesite grade from 20% to about 92%, with a second pass through the sorter increasing the grade to 98–99% magnesite. Other applications reported include sorting to upgrade aggregate materials and sorting to upgrade asbestos.[10] Of major importance is separation of white mineral-bearing quartz from quartzite of various colors. Gold, cassiterite, and wolframite in quartz are sorted from quartzite,[5,6,10] and in a recent installation white quartz containing tetrahedrite and other minerals (Ag and Cu values) is sorted from quartzite.[11] Scheelite is sensed by its fluorescence under ultraviolet light and so sorted.[12] Asbestos is sorted using infrared radiation.[12,13]

Radiometric ore sorting[6,14] is a valuable preconcentration technique with some uranium ores, since a large fraction of the ore containing less than some specified cutoff grade can be discarded with very little loss of uranium values. Radiometric sorting has also been used for boron and beryllium ores,[15] but in this case the radioactivity is not natural.

Ore sorting based on mineral conductivity has been used with iron ores[16] and with native copper.[3,17] Sorting using other sensing systems has been attempted but has not yet been commercially successful.

11.2. MECHANICS OF SORTING

The modern sorting operation has three distinct stages: singulation, detection, and ejection. Singulation is the control of the flow of feed ore so as to present each particle individually to a detector for inspection. Detection is the sensing of the presence or absence of some desired value in the ore and the electronic evaluation of the signal received. Ejection is the mechanical separation of the "detected" particles from the remainder of the ore. The selection of the technique for carrying out each of these stages can be made independently. The overall result, however, depends on the successful completion of each stage. In general, the singulation and ejection stages are rate limiting, and the detector determines the separation efficiency.

11.2.1. Singulation

The need to examine each particle individually is one of the unique features of ore sorting. Because of this requirement, a high speed operation is necessary to permit an acceptable throughput. In a singu-

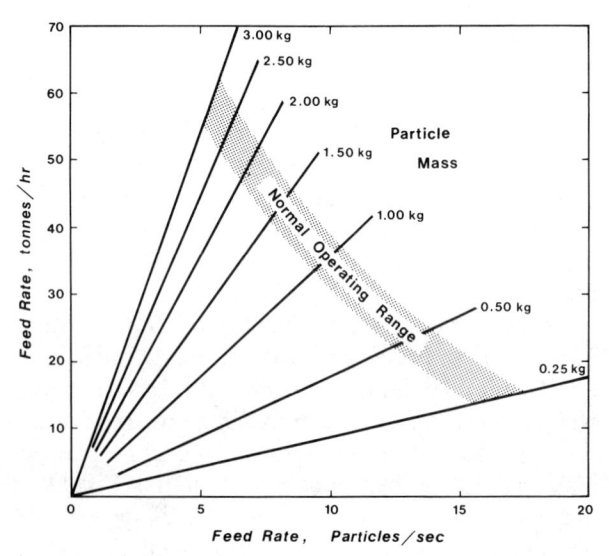

Figure 11.2. Significance of particle size to sorter capacity.

lation operation, the feed must be drawn from some bulk source, then presented before a detector in such a way that each particle can be individually considered. This means that particles must be in a single layer, and sufficiently apart both laterally and in the direction of movement to permit the detector to identify and evaluate each particle.

Most sorting is done either on a conveyor belt or in free fall from a conveyor belt. The design of a method of singulation onto a belt is not particularly difficult, and a number of techniques are in use. The singulation techniques in use fall roughly into two classes: *in-line*, where the particles are made to move in single file, and *single-layer*, where the particles move in a band one layer in depth. The choice of in-line or single-layer singulation is entirely dependent on the subsequent detection system. The single-layer method allows a much greater capacity. However, if the detection system requires scanning from all sides, the in-line method is normally used because it allows for scanning during free fall in air.

An in-line singulation system usually includes acceleration from a hopper to closely spaced, single file delivery with a precisely controlled trajectory through the sensing and separation points. Sorting machines using this singulation method[2,3,7,8] are limited in the particle size range over which they can be operated. A number of techniques are used to obtain an in-line particle presentation. These relatively simple mechanical techniques have been described in detail by Wyman.[14]

Single layer singulation systems involve feeding of the particles from a hopper and even distribution across a conveyor belt, which moves them through the detection and separation points. Greatly increased capacity is obtained by effectively condensing multiple in-line singulation systems into one continuous stream. With single-layer singulation, the particles may be examined from above or below but not from all directions. Figure 11.1 shows a sorter using two vibrating feeders in series to spread the particles laterally, followed by a slide plate to impart acceleration to the particles, hence to increase their spacing. As well as obtaining a single layer of separated particles on a conveyor belt, it is necessary to maintain the relative position of the particles. Once a particle has been scanned, it must maintain its known speed and direction until it reaches the separation zone. To prevent the movement of unstable particles on the belt, a loose belt on a soft pulley can be added above the feed conveyor.[9-11]

11.2.2. Detection

The detection stage of an ore sorting operation has two components: a *sensing* device, and an electronic component.

The sensing method that can be used in a particular situation depends on the ore to be treated. The minerals to be separated must be sufficiently different in some detectable property for a discrimination to be made. The sensing device must produce an electrical signal based on this distinguishing property.

A wide range of mineral properties may be used in sorting. Table 11.2 lists the sensing systems that have been used for ore sorting.[14]

The optical properties of minerals are a common basis for the design of detection systems. In this broad class of properties are surface reflectivity, color, transparency, and fluorescence. Surface reflectivity was the basis of the early sorter of Sweet[1] and is the basis of the present day "photometric" sorters. Reflectivity can be of two types, diffuse or specular, and a sorter for coal was developed based on the ratio of the two.[18] Color sorting is achieved by isolating a narrow spectral range in the reflected light from the desired mineral and then using the intensity of reflectance within that range to effect the separation. Sorting may be achieved by inducing fluorescence on a naturally fluorescent mineral such as scheelite with ultraviolet light.[12,14] Also, when a fluorescent chemical is selectively adsorbed

TABLE 11.2: SENSING SYSTEMS FOR USE IN ORE SORTING

Mineral Property	Detection Device
Optical	
Reflectance	Photomultiplier
General	Photomultiplier
Specific	Photomultiplier
Polarized	Photomultiplier
Photometric	Photomultiplier
Transparency	Photomultiplier
Fluorescence	Photomultiplier
Natural	Photomultiplier
Chemically induced	Photomultiplier
Infrared	Infrared scanner
X-Ray	
Transparency	Scintillation counter and pulse-height analyzer
Fluorescence (visible)	Photomultiplier
Fluorescence (X-ray)	Scintillation counter and pulse-height analyzer
Conductivity	
Low voltage	Resistance
High voltage	Electrical induction or eddy-current type detection
Magnetism	Eddy-current type detection
Radioactivity	
Natural	Scintillation counter and pulse height analyzer
Induced	

onto the surface of one of the minerals in an ore, sorting by fluorescence is possible.[19] This technique adds the adsorption step and a washing step to the sorting flowsheet, but it broadens the applicability of the sorting concept to virtually all minerals.

Closely related to the ultraviolet and visible light properties of minerals are those of the infrared. Differences in thermal conductivity cause minerals to absorb heat at different rates. Thus by heating and subsequently using infrared scanning it is possible to sort asbestos ores.[12,13]

Ore sorting has been investigated using X-rays. The transparency of a mineral to X-rays depends on the density, and this property has been used in the sorting of diamonds. Also, when irradiated with X-rays, diamonds emit visible light, and this can be detected with a photomultiplier tube as used in optical ore sorting.[14] A system using X-ray fluorescence has also been reported.[20] If developed to a commercial level, an X-ray fluorescence unit would be the first that could be made specific to an individual element.

The detection methods described to this point have one common attribute. They examine only the surface of the mineral particles presented to them, and in some machines only a small part of that surface. This imposes certain limitations on their applicability and operation. The major limitation is that the surface must be clean. Dust or slimes on the particle surface would obscure not only the color and luster but the chemical composition as well. Moreover, the sorting operation must be performed in a relatively clean environment to reduce the accumulation of dust on optical surfaces.

Sorting systems utilizing bulk or internal properties of the desired mineral do not require that mineral surfaces be clean and do not need a clean environment. The major consideration, however, is the lack of a requirement for the presence of the desired mineral at the surface of the particle being examined. Therefore, ores in which the values are randomly segregated from the gangue are more readily detected.

Electrical conductivity, magnetic properties, and natural or induced radioactivity are examples of bulk or internal properties that can be used in sorting. Electrical conductivity has been used in a number of devices. In one sorter, the change in inductance due to the presence of a conductor in an electrical field is used.[3,17] Another manufacturer produces a sorter utilizing an eddy current-type sensing system, which can sense differences in electrical conductivity or magnetic properties. In another device, the conductivity between two electrodes is measured;[16] this, of course, is no longer a "bulk property" sorter.

Radioactivity as used in ore sorting may be either natural or induced. Uranium ores emit gamma rays, and these may be detected with a scintillation counter and pulse height analyzer. Thus a radiometric sorter for uranium ores can sort on the basis of uranium grade.[6]

Induced radioactivity may be due to bombardment of the ore particles with either neutrons or gamma rays.[14] In this case, gamma ray bombardment results in characteristic neutron emission, and the emitted neutrons are counted using a scintillation neutron detector.

All detection systems combine the sensors discussed above with some electronic decision making procedure. It is the advances in electronics and electronic decision making hardware and procedures that have permitted the development of the sophisticated ore sorting machines now on the market. Some ore sorters now use a digital computer for the interpretation of the sensor signal,[9,10] and with the ready availability of microprocessors this can be expected to become the standard procedure.

11.2.3. Ejection

Ejection is the simplest of the stages in a sorting system and can be achieved in a number of ways. Early sorters used a solenoid plunger to push detected particles off the conveyor. One machine being developed uses a solenoid activated deflector plate to direct each particle onto the proper waste or concentrate belt.[2] Most sorters, though, use solenoid controlled air valves to eject selected particles while they are in free fall from the end of the sorting conveyor.

There are two major requirements of an ejection system: accurate timing and rapid recovery. In free fall ejection systems using an air valve, the time when the air valve is open must be accurate to within a few milliseconds. There is a time delay while the particle travels from the sensor to the ejector, and this must be accounted for in the detection system electronics. Since the ejector system must act separately on each individual particle, the ejector must return to a null condition between each. In an air ejection system, the air compressor and the delivery system must be designed to deliver the required amount of air at a sufficiently high pressure

Ore Sortability

to be effective. The air requirement is often minimized by designing the sorter to react to the less frequent particle type, whether it be value or waste.

11.3. ORE SORTABILITY

The term "sortability" as used in the sorting of ores appears to have both a qualitative and a quantitative meaning; both meanings are discussed here.

11.3.1. Sortability of Mineral Deposits

To be sortable, the mineral distribution in an ore must meet the requirements of liberation, size, and detectability. Sorting should be applied as soon as sufficient liberation has been achieved; the larger the size of particles that can be discriminated, the greater the value of sorting. There is a minimum particle size below which sorting ceases to be practical.

For a mixture of ore particles to be sortable into two components of different bulk value, the particles must not all have equal value. As is discussed in Chapter 3 and indicated in Fig. 3.4, separation is most readily achieved if the distribution of values is such that all the value of the ore is contained in some small fraction of the total particles (an ore would be sortable if represented by curves A and B in Fig. 3.4; the sorter would be operated with a setting at $\mathfrak{B}_3^*$). Such ores are rarely found and usually middlings particles are present. However, favorable values distributions are frequently found at very coarse size ranges because of the factors discussed below (Fig. 11.3).

Mining Equipment Limitations. Modern high capacity mining operations generally utilize equipment that requires certain minimum mining dimensions. If the orebody has dimensions less than the minimum mining dimensions, it is inevitable that waste rock will be included in the ore shipped to the mill (Fig. 11.3a).

Orebody Meandering. Frequently a tabular or prismatic orebody follows a meandering course not readily tracked by the mining crew. This could result in the inclusion of considerable quantities of waste rock in the ore shipped to the mill if selective mining methods are not practical (Fig. 11.3b).

Orebody Heterogeneity. Most orebodies contain two components broadly classifiable as "ore" and "waste." Clearly the finer the run-of-mine ore is broken, the more it is physically separable into these two basic components. However, in most cases run-of-mine ore contains substantial quantities of rejectable waste at coarse sizes (Fig. 11.3c).

11.3.2. Separability and Sorter Performance

Attempts to quantify the sortability of an ore are limited. Newman and Whelan,[18] working with a barite ore and using the difference between specular and diffuse reflectance as the detected property, measured the frequency of occurrence of both barite and waste over a range of values of the reflectance difference. Their results are shown in the form of separability (sortability) curves in Fig. 11.4.

Although no data are available, a performance curve of the form shown in Fig. 3.8a and 3.8b could be drawn to describe the operation of an ore sorter. The sorter performance curve shows the best separation that can be achieved with a particular machine. The difference between the sortability and sorter performance curve is due to inherent in-

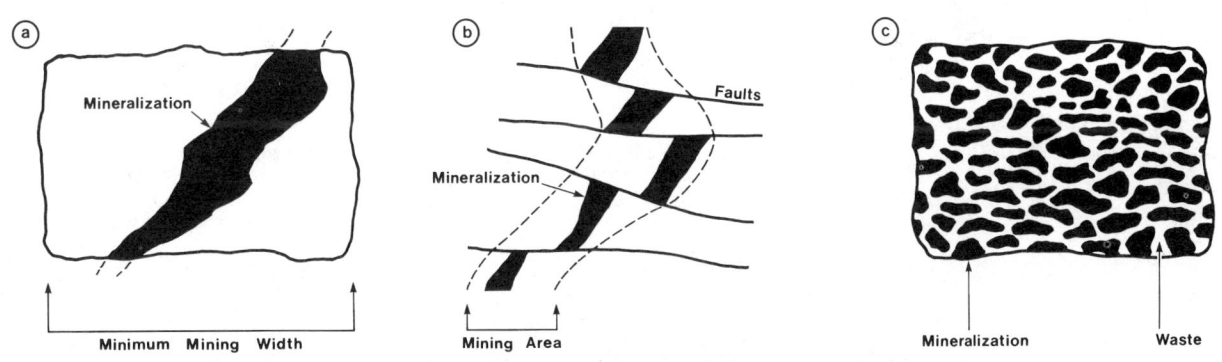

Figure 11.3. Values distribution, hence sortability, are affected by (a) mining equipment requirements, (b) ore body meandering, and (c) ore body heterogeneity.

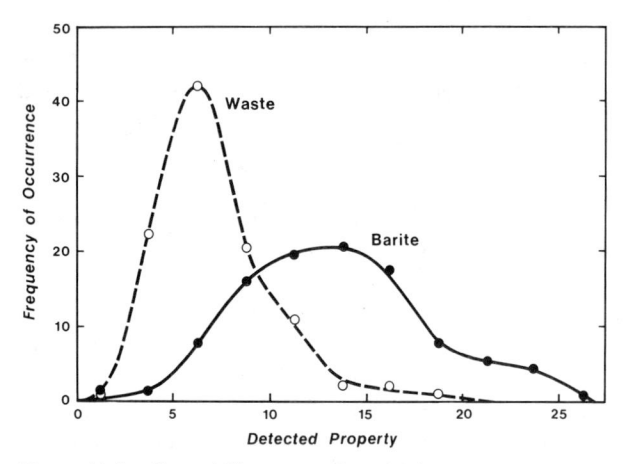

Figure 11.4. Separability curves for a barite ore sorted on the basis of reflectivity differences. (After Newman and Whelan.[18])

efficiencies of the sorter. For example, in a photometric ore sorter, the sensor can detect the presence of the sensed values only if they occur on the face of the particle presented to the sensor. This limitation leads to an inherent inefficiency in this type of sorter. Some effort would be required to prepare both the separability and performance curves for a sorting operation, but they would give a ready indication of the efficiency of operation of the sorter.

Other workers have attempted to quantify sortability,[2,21] but their results have been limited. King[22] has reported a preliminary attempt to model the sorting operation on a theoretical basis, but this approach has not as yet been applied to operating systems.

REFERENCES

1. A. T. Sweet, "Color and Luster as a Basis for Concentration," *Proc. Lake Superior Min. Inst.* (1928).
2. C. W. Schultz, "An Economic Evaluation of Preconcentration in the Mining-Processing System," Ph.D. thesis, University of Minnesota (1975).
3. C. W. Schultz, "The Cost Advantages of Electronic Sorting of Nature Copper Ores," AIME/SME Preprint 75-B-75 (1975).
4. A. H. Betters and C. M. Lapointe, "Electronic Concentration of Low Grade Ores with the Lapointe Picker," Mines Branch Technical Paper No. 10, Canada (1955).
5. N. J. Keys, R. J. Gordon, and N. F. Peverett, "Photometric Sorting of an Ore on a South African Gold Mine," AIME/SME Preprint 74-B-311 (1974).
6. J. R. Goode, "Recent Advances and Applications of Radiometric and Photometric Sorting," 7th Annual Canadian Mineral Processors Meeting, 1975.
7. A. Balint, "Mineral Beneficiation with Optical Separators," AIME/SME Preprint 67-H-305 (1967).
8. R. R. French, "Beneficiation of Low-Grade Gypsum by Electronic Color Sorting," AIME/SME Preprint 67-H-318 (1967).
9. P. J. Barton and H. Schmid, "The Application of Laser/Photometric Sorting Techniques to Ore Sorting Processes," *12th Int. Miner. Process. Congr.* (1977).
10. M. A. Schapper, "Beneficiation at Large Particle Size Using Photometric Sorting Techniques," *Aust. Min.*, **69**, 44–53 (April 1977).
11. D. D. McLaughlin, "Operation of an Optical Ore Sorting System," *Am. Min. Congr.*, Los Angeles (1979).
12. D. Collier, et al., "Ore Sorters for Asbestos and Scheelite, *Proc. 10th Int. Miner. Process. Congr.*, pp. 1007–1022, IMM (1973).
13. "The CSR Asbestos Ore Sorter," CSR Research Laboratories Report, Australia.
14. R. A. Wyman, "Selective Electronic Sorting to 1972," Mines Branch Monograph No. 878, Canada (1972).
15. V. A. Mokrousov et al., "Neutron-Radiometric Processes for Ore Beneficiation," *Proc. 11th Int. Miner. Process. Congr., Cagliari, 1975*, pp. 1249–1290, Università di Cagliari (1975).
16. A. Balint, "Ore Sorting According to Electrical Conductivity," *J. S. Afr. IMM*, **76**, 40–42 (1975).
17. V. R. Miller, R. W. Nash, and A. E. Schwaneke, "Detecting and Sorting Disseminated Native Copper Ores," U.S. Bureau of Mines, Report of Investigations RI7904 (1974).
18. P. C. Newman and P. F. Whelan, "Photometric Separation of Ores in Lump Form," in *Recent Developments in Mineral Dressing*, pp. 359–381, IMM (1953).
19. T. C. Mathews, "New Concepts in Pre-Concentration by Sorting," AIME/SME Preprint 74-B-328 (1974).
20. R. H. Goodman, A. H. Bettens, and C. A. Josling, "Ore Sorting with Radiation," *Miner. Process.*, 10–13 (October 1968).
21. R. M. Doerr, "Radial Distribution Analysis for Sorting," *Trans. AIME/SME*, **255**, 4–8 (1974).
22. R. P. King, "Automatic Sorting of Ores," *Miner. Sci. Eng.*, **10**, 198–207 (1978).

Chapter Twelve

Dense Medium Separation

Murphy's Law, Corollary: "*It is impossible to make things foolproof. Fools are so ingenious.*"

Dense medium separation is used in the concentration of a variety of minerals; its major application, however, is in the cleaning of coal.[1] As a process that is not highly efficient for the separation of fine particles, it is restricted in use to ores in which either the valuable or gangue minerals liberate at a relatively large size. A minor application at present, but one with considerable potential (see Chapter 11), is the preconcentration at coarse sizes of ores that require fine grinding for subsequent concentration by other methods.

Dense medium separation is clearly related to gravity concentration (Chapter 13). The separations discussed in both chapters are based on *differences in mineral density*. The same basic principles apply. The major difference is *the medium in which the separations take place;* in gravity concentration the separation is in water (or air), whereas in dense medium separation, the separation is in a medium of density higher than that of water and between the densities of the minerals to be separated. This medium can be dissolved salts in water, or more commonly a suspension in water of finely divided high density particles.

The process of dense medium separation is referred to by a variety of other names, including heavy media separation and float-sink or sink-float separation. "Medium" or "media" are used interchangeably.

12.1. EQUIPMENT AND APPLICATIONS

The aim of dense medium separation is to produce two products, a *float* product (lower density minerals) and a *sink* product (higher density minerals), and in some instances a third middlings product. A variety of equipment is manufactured for this purpose. This equipment can be considered in two categories, based on the separation forces applied: *gravity* or *static bath separators,* for use with a coarse feed, and *centrifugal separators* for use with finer feeds. A major part of any dense medium separation operation is the *recovery and control of the medium*. Since the method for magnetic medium recovery is not dependent on the category or type of separator in use, it is discussed separately (Section 12.2). The most common medium for separation is a suspension in water of fine magnetite, ferrosilicon (FeSi), or a mixture of the two, depending on the separation density required. Magnetite alone is used when a density is required in the range 1250–2200 kg/m^3, a mixture between 2200 and 2900 kg/m^3, and ferrosilicon alone for a density of 2900–3400 kg/m^3, with even higher densities being possible if spherical particles of ferrosilicon are used.

Magnetite and ferrosilicon are selected as media because they meet a number of essential criteria: they are physically stable and chemically inert, easily removable from the products and readily recovered for reuse (by magnetic separation), and form low viscosity "fluids" over the range of medium densities required.

Other media are also used in coal cleaning,[1] but to a much lesser extent. Dissolved salts are used in the Belknap calcium chloride washer for coal cleaning. In this case, the medium is largely formed from the fine clay particles derived from the coal being recirculated, with the calcium chloride used mainly to stabilize and control the medium density. Sand suspensions are also used, as in the Chance Cone process; in this case the sand suspension is maintained by upward flow of the medium. Sand is removed from the products by screening (with water sprays), and is then recirculated. The separators used with dissolved salts and sand suspensions as media are in the static (or quasi-static) category.

Detailed descriptions of dense medium separation equipment are available elsewhere[1-3] and in manufacturer's literature.

12.1.1. Gravity Dense Medium Separators

In gravity dense medium separators (Figs. 12.1 and 12.2), the feed and medium are introduced to the surface of a large quiescent pool of the medium. The float material overflows or is scraped from the pool surface. The sink material falls to the bottom of the vessel and is removed, the removal mechanisms aimed at minimizing vertical flow of medium.

There are many separators in the gravity or static bath category. In the *drum separator* (Fig. 12.1a), a widely used type, the rotation of the drum lifts the sink product clear of the bath. Drum separators are available in a range of sizes (up to 4.6 m diameter by 7.0 m long, capacities to approximately 800 t/hr) and configurations. They are capable of separating particles in a size range from 6 mm to 20–30 cm.

In the *cone separator* (Fig. 12.1b), also widely used, the sink product is removed from the bottom of the cone either directly or by an airlift in the center of the cone. Cone separators are available in sizes from 0.9 to 6.1 m diameter, with capacities to approximately 450 t/hr. The maximum particle size

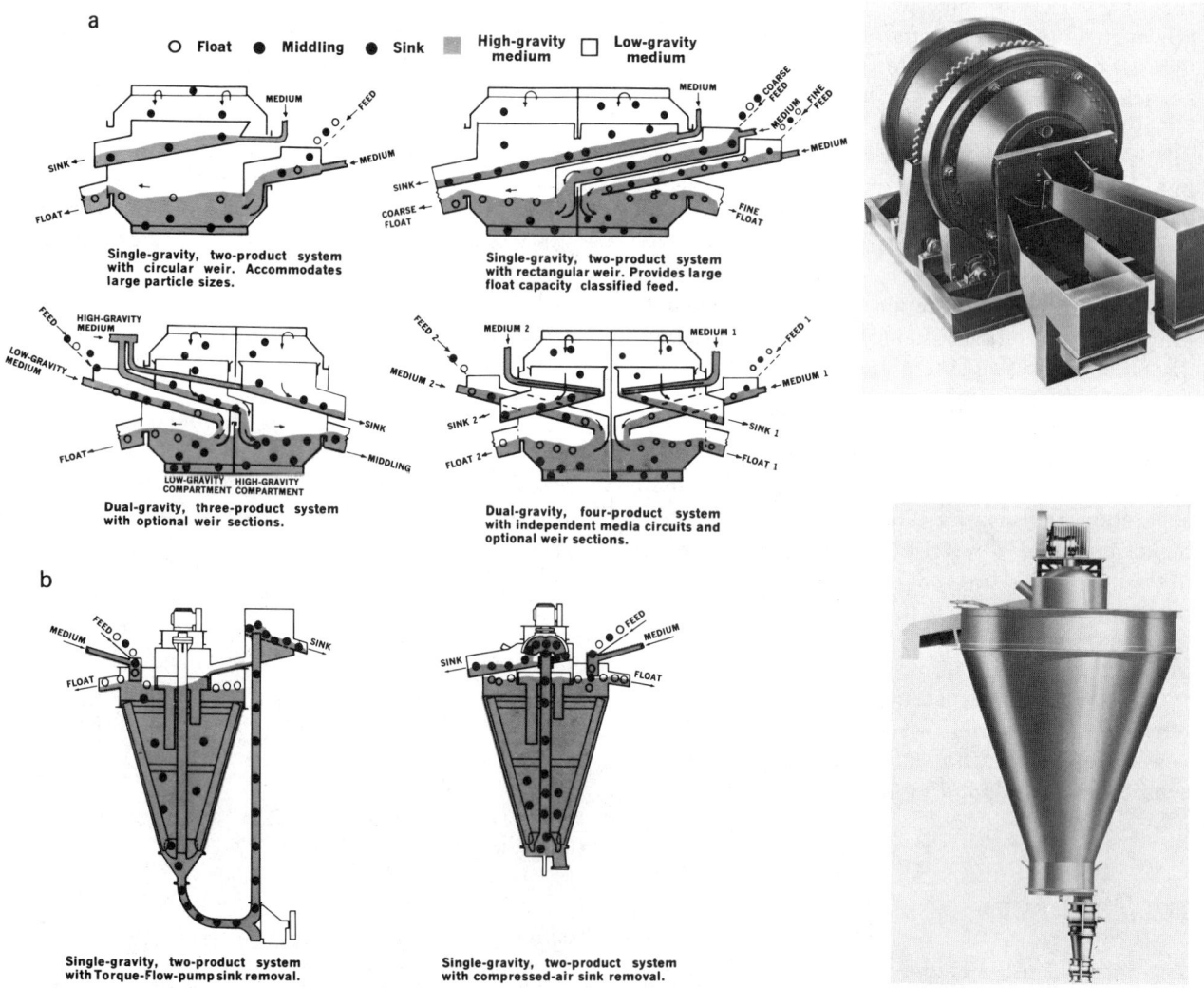

Figure 12.1. Gravity dense medium separators. (*a*) Drum separator. (*b*) Cone separator. (Courtesy Wemco Division, Envirotech Corp.)

Equipment and Applications

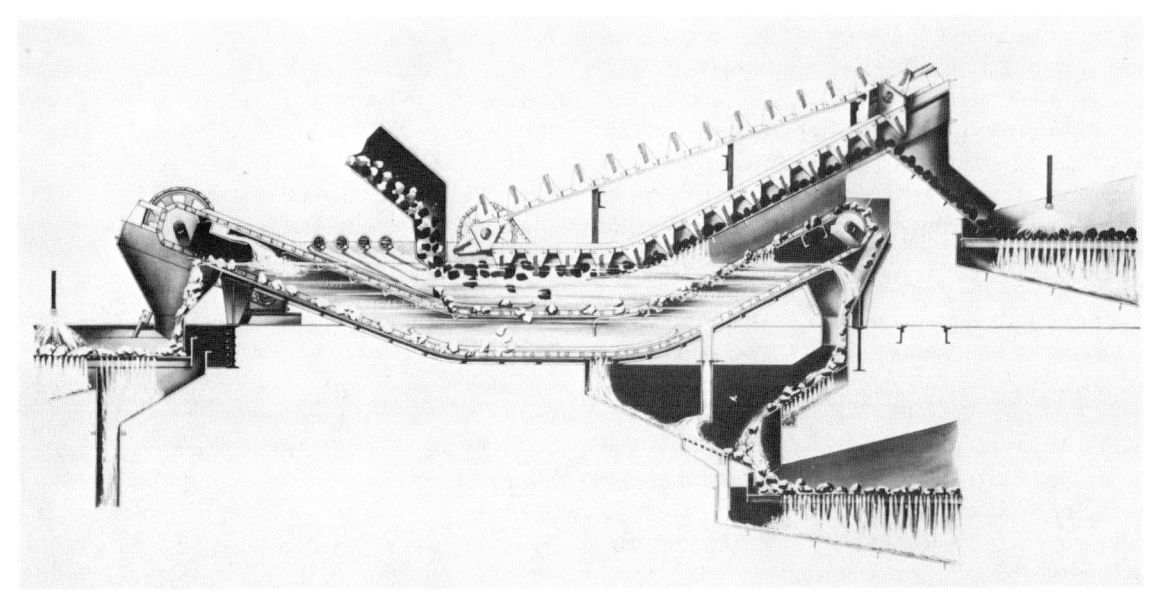

Figure 12.2. Three-product trough-type dense medium separator. (Courtesy McNally Pittsburg Manufacturing Corp.)

that can be separated in a cone separator is limited to 10 cm by the sink product discharge method.

Trough-type dense medium separators (Fig. 12.2), differ from the two just described in that both sink and float products (and middlings where separated) are removed by mechanical scrapers. Efficient separations have been reported at particle sizes as fine as 3 cm.

Gravity dense medium separators of the types above are widely used for cleaning coarse coal, but are also used with other minerals such as fluorite, lead-zinc ores, and iron ores.[4] Other gravity dense medium separators also in use in industry, particularly for coarse coal cleaning, are discussed elsewhere.[1]

12.1.2. Centrifugal Dense Medium Separators

For particle sizes finer than those treated in gravity dense medium separators, a greater acceleration must be applied to produce sufficient force to achieve a separation. Thus centrifugal separators are used, where an acceleration up to about 20 times that of gravity can be obtained.

Hydrocyclones, as used for classification (Chapter 10), are applied as centrifugal dense medium separators (Fig. 12.3). Design modifications have been introduced when using the hydrocyclone for dense medium separation, particularly to the inlet,[1] but they operate in the same manner as for classification (Section 10.4). They have been used since the 1950s for cleaning coal finer than 0.5 mm[1,5–9] and are now finding application with other minerals.[4,10–12]

Other centrifugal dense medium separators are the Vorsyl separator[13] and the Dynawhirlpool separator.[14] Both are substantially different in construction from the hydrocyclone, although as with

Figure 12.3. Hydrocyclones in use as centrifugal dense medium separators. Media separation screen also shown. (Courtesy Heyl & Patterson, Inc.)

the hydrocyclone the light (float) fraction is entrained in a rotating vortex and discharged. The heavy (sink) fraction, however, is discharged through a tangential outlet. Separations can be made with feed sizes in the range 0.5–30 mm, with capacities up to approximately 75 t/hr for the larger units. Although primarily used for coal, the Dynawhirlpool is used with a wide range of other minerals.

Another centrifugal separator, which may be considered to be an *autogenous dense medium separator*, is the "water-only" or autogenous dense medium hydrocyclone.[1,15-18] The design of an autogenous dense medium hydrocyclone (Fig. 12.4) differs from that of a conventional dense medium hydrocyclone or a classification hydrocyclone in that the cone is in stages and the final cone angle is much larger (up to 120°); moreover, the vortex finder is much longer, in some instances extending into the conical region. It is thought that fine particles of high and intermediate density collect and recirculate in this conical section, forming an autogenous dense medium barrier through which only particles of greater density can pass (within a limited range of particle sizes). The use of such hydrocyclones is becoming widespread in cleaning of coal in the 600–150 μm size range. A clear advantage is the lower capital and operating costs resulting from the elimination of medium separation and recovery costs.

12.1.3. Other Dense Medium Separators

Dry dense medium separation has been extensively tested but has not received wide attention by industry.[19] In this case, the dense medium is more commonly known as a pneumatically fluidized bed; the particles are suspended by continuous introduction of air across the bottom of the bed. Separations are possible, however, if the effective density of the bed is between that of the float and sink components of the feed.

An area of promise for future development in dense medium separation is in the use of magnetically stabilized dense medium beds.[20] By use of ferrofluids or suspensions of fine paramagnetic materials, the density can be very closely controlled within the magnetic field, and thus a separation effectively based on small density differences can be achieved.

12.2. MEDIUM CONTROL AND RECOVERY

For optimum separation to be achieved in the operation of a dense medium plant, a number of variables must be balanced. The two variables that are generally the most frequently checked and controlled are density and consistency of the medium. The former[1] can be controlled to within ±5 kg/m³ (i.e., ±0.005 specific gravity units), which is required to permit separations where a small difference in mineral density exists. The latter[21] is a function of the settling rate of the particles making up the medium. The medium is generally made up of finely ground ferrosilicon or magnetite, clay contaminants, and water. The clay frequently present in the medium acts as a stabilizing agent, since the coarser dense medium solids would quickly settle if they were mixed only in water. Clay concentrations in the medium vary between 3 and 7% by mass. Although clay is essential for reasonable medium stability, clay concentrations greater than 7–8% impart too high an apparent viscosity to the medium, and the feed can-

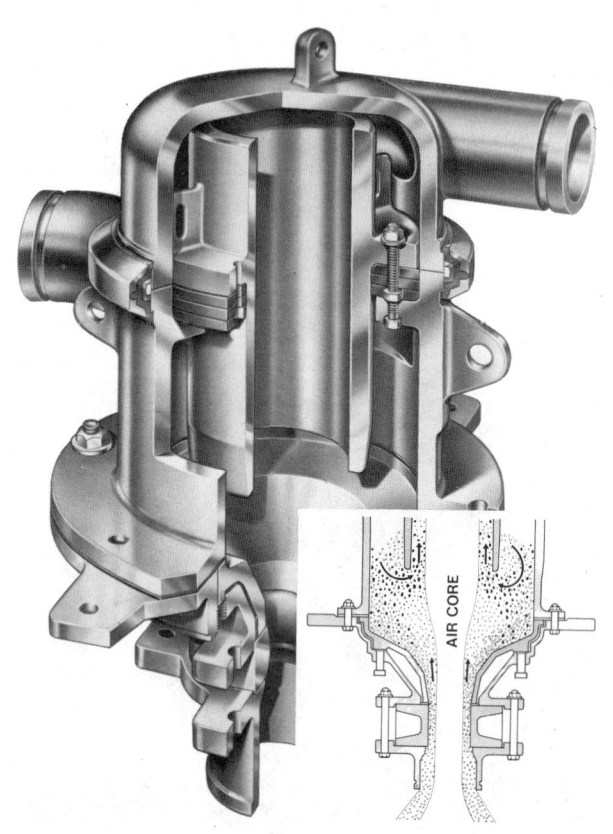

Figure 12.4. Autogenous dense medium or "water-only" hydrocyclone. Insert indicates separation occurring in autogeneous medium. (Courtesy McNally Pittsburg Manufacturing Corp.)

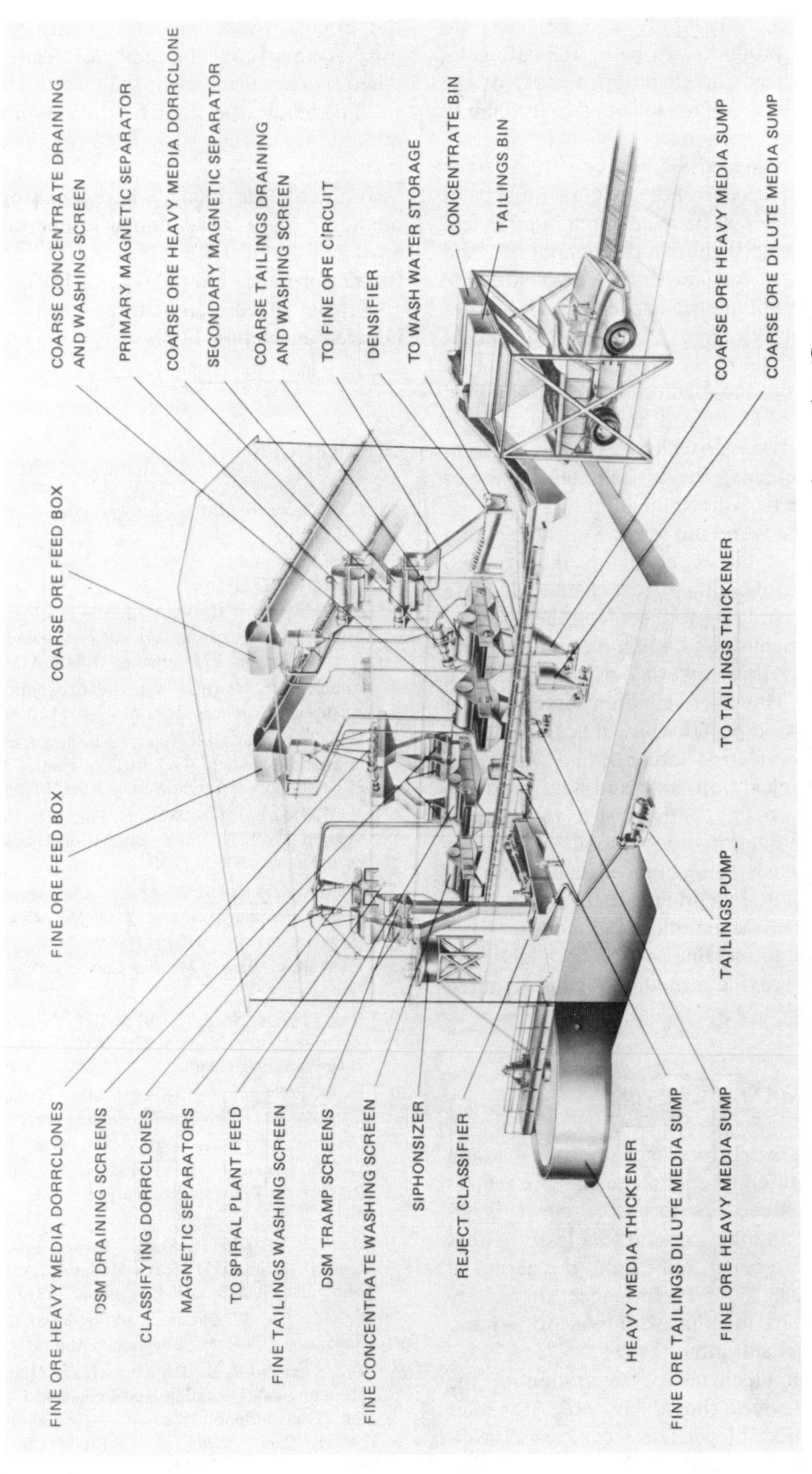

Figure 12.5. A typical dense medium plant showing medium recovery by magnetic separation. (Courtesy Dorr-Oliver Inc.)

not quickly and precisely stratify and then separate into the required products. To be on the safe side, plant operators clean the medium of most of the clays, and prefer to work at a lower stability than to have a very viscous medium.

While clearly hindered settling conditions exist (Section 5.1), the particles of the medium must be of such a size that they can be maintained in an essentially homogeneous distribution throughout the fluid (generally all < 200 μm, 50% < 74 μm). Medium suspensions, with volumetric concentrations greater than 25%, do not behave as Newtonian fluids but exhibit Bingham plastic behavior[22] (Section 4.2). Improved control of the medium has been obtained by use of spherical (atomized) ferrosilicon, and by the addition of polymers[21] which lower the apparent viscosity while retaining the desired consistency or suspension properties of the medium.

The efficient recovery and recycle of the medium is a major factor in the economics of the process. Losses will occur, of course, because of inadequate washing of the separated products, and the need for periodic replacement due to buildup of fine mineral particles resulting in a lowering of the attainable medium density. However, medium losses can be minimized. Common practice, as indicated in Fig. 12.5, is as follows: feed is screened and washed to remove fine particles; both sink and float products are screened in two stages, the first to recover the medium without dilution, the second with water sprays added to wash the medium particles from the products. This diluted medium is then concentrated by wet, low intensity magnetic separation (Chapter 14) to remove water (and fine particles originating in the feed) and to provide a medium of the required density for recycle.

12.3. PERFORMANCE CURVES

Data available on the performance of dense medium separators are limited to coal cleaning. The results show that the reduced performance curve is affected by the separator type, and to a lesser extent by the particle size being treated and the nature of the medium in use.[9,23-25] Performance curves are shown in Fig. 13.18, together with those for a number of gravity concentration devices.

The very high efficiency of separation of the dense medium devices should be noted, together with the small effect of particle size (Fig. 13.18c). This is to be expected, particularly in the vessels of the gravity dense medium separator type, since misplacement of liberated material occurs only when there is insufficient time for the finer particles to sink below the float fraction overflow, or to a minor extent, when there is entrapment or localized turbulence.

In the hydrocyclone, where an autogenous dense medium region is developed in operation, the separation is not nearly as efficient,[17,18,23,26] as shown by the performance curve (Fig. 13.18a).

Further discussion of these performance curves is given in Section 13.3.

REFERENCES

1. J. W. Leonard (Ed.), *Coal Preparation*, 4th ed., AIME/SME (1979).
2. J. H. Perry (Ed.), *Chemical Engineers' Handbook*, 4th ed., McGraw-Hill (1963).
3. C. G. Wilson, "Heavy Media Separation Flowsheet Development," Ch. 26 in *Mineral Processing Plant Design*, A. L. Mular and R. B. Bhappu (Eds.), AIME/SME (1978).
4. Anon., "Mt. Newman Iron Ore Concentrator in Western Australia," *Skillings' Min. Rev.*, 8–11 (Aug. 4, 1979).
5. M. Sokaski and M. R. Geer, "Cleaning Unsized Fine Coal in a Dense-Medium Cyclone Pilot Plant," U.S. Bureau of Mines Report of Investigations RI6274 (1963).
6. A. W. Deurbrouck, "Washing Fine Size Coal in a Dense-Medium Cyclone," U.S. Bureau of Mines Report of Investigations RI7982 (1974).
7. J. Mengelers and J. H. Absil, "Cleaning Coal to Zero in Heavy Medium Cyclones," *Coal Min. Process.*, **13**, 62–64 (May 1976).
8. E. Skolnik, "Heavy Medium Cleaning of —28 Mesh Coal," *Min. Eng.*, **32**, 1235–1237 (August 1980).
9. B. S. Taylor, W. L. Chen, and D. G. Chedgy, "Feed to Zero in Heavy Medium Cyclones," *Trans. AIME/SME*, **266**, 1985–1991 (1980).
10. H. L. D'Rozario, "The Application of the Heavy Media Cyclone as a Modern Processing Practice in Mineral Dressing," *Proc. Aust. IMM Conf.* (May 1973).
11. I. R. M. Chaston, "Heavy Media Cyclone Plant Design and Practice for Diamond Recovery in Africa," *Proc. 10th Int. Miner. Process. Congr. London, 1973*, IMM (1974).
12. H. Dreissen and J. H. Absil, "Theory, Practice and Development of the DSM Heavy Medium Cyclone Process for Minerals," AIME/SME Preprint 80-11 (1980).
13. J. Abbot, K. W. Bateman, and S. R. Shaw, "The Vorsyl Separator," *Proc. 9th Commonw. Min. Met. Congr.* (1969).
14. T. J. Lien and R. B. Bhappu, "Heavy Media Separation—Metallurgical, Operating, and Economic Characteristics of the Dynawhirlpool Processor," Ch. 25 in *Mineral Processing Plant Design*, A. L. Mular and R. B. Bhappu (Eds.), AIME/SME (1978).

References

15. J. Visman, "Bulk Processing of Fine Materials by Compound Water Cyclones," *CIM Bull.*, **59**, 333–346 (1966).
16. J. Visman, "Integrated Water Cyclone Plants for Coal Preparation," *CIM Bull.*, **61**, 365–370 (1968).
17. E. J. O'Brien and K. J. Sharpeta, "Water-Only Cyclones: Their Functions and Performance," *Coal Age*, **81**, 110–114 (January 1976).
18. E. A. Draeger and J. W. Collins, "Efficient Use of Water Only Cyclones," *Min. Eng.*, **32**, 1215–1217 (August 1980).
19. Anon., "The Dry Fluid Bed Separator," *Filtr. Sep.*, 535 (September 1972).
20. Y. Zimmels, I. J. Lin, and I. Yaniv, "Advances in Application of Magnetic and Electric Techniques for Separation of Fine Particles," Ch. 59 in *Fine Particles Processing*, P. Somasundaran (Ed.), AIME/SME (1980).
21. L. Valentyik, "The Rheological Properties of Heavy Media Suspensions Stabilized by Polymers," AIME/SME Preprint 71-B-9 (1971).
22. H. Kirchberg, E. Töpfer, and W. Scheibe, "The Effect of Suspension Properties on Separating Efficiency of Mechanical Classifiers," *Proc. 11th Int. Miner. Process. Congr., Cagliari, 1975*, pp. 219–244, Università di Cagliari (1975).
23. (a) B. S. Gottfried and P. S. Jacobsen, "Generalized Distribution Curves for Characterizing the Performance of Coal-Cleaning Equipment," U.S. Bureau of Mines Report of Investigations RI8238 (1977).
    (b) B. S. Gottfried, "A Generalization of Distribution Data for Characterizing the Performance of Float-Sink Coal-Cleaning Devices," *Int. J. Miner. Process.* **5**, 1–20 (1978).
24. A. W. Deurbrouck and J. Hudy, "Performance Characteristics of Coal-Washing Equipment: Dense-Medium Cyclones," U.S. Bureau of Mines Report of Investigations RI7673 (1972).
25. J. Hudy, "Performance Characteristics of Coal-Washing Equipment: Dense-Medium Coarse-Coal Vessels," U.S. Bureau of Mines Report of Investigations RI7154 (1968).
26. A. W. Deurbrouck, "Performance Characteristics of Coal-Washing Equipment: Hydrocyclones," U.S. Bureau of Mines Report of Investigations RI7891 (1974).

Chapter Thirteen

Gravity Concentration

Finagle's Rules for Scientific Research:
"Always verify your witchcraft."

Heinlein's Law: *"One man's magic is another man's engineering."*

Gravity concentration, the mainstay of concentration up until the 1920s, apparently dropped in significance with the advent of froth flotation. However, although gravity concentrators may not give the sharpness of separation possible with flotation, they generally have lower costs. Hence, with the increasing need for coal cleaning and the development of low cost, high capacity concentrators (some of which are effective on materials previously considered too fine to be treatable), gravity concentration has made a marked comeback within the last few decades. To help put gravity concentration in perspective, Table 13.1 gives some figures from the United States.[1]

This chapter considers the major concentrating methods that exploit differences in densities to bring about a separation. As with classification separations, the separation is carried out in a fluid, so that fluid dynamics are again an important aspect. As a consequence, the separations can be influenced by a density difference term ($\rho_s - \rho_f$) and a particle size term, and since the significance of the particle density can be best emphasized by high ρ_f, water is used in preference to air. Air can however be used when water is in very short supply, or when some other special benefit results from its use.[2] Because of the interrelation between density difference and particle size, gravity concentrators are best used with a restricted range of feed size, just as classifiers separate better with a narrow range of densities. In this chapter, we restrict our consideration to devices used primarily for concentration, but it should be remembered that under some conditions gravity concentrators can be used as classifiers, and vice versa.

It is a characteristic of all gravity concentrating devices that particles are held slightly apart so that they are able to move relative to each other and thus to separate, ideally into layers of dense and light minerals. How this interparticle spacing is achieved provides a convenient means of classifying gravity concentrators, as shown in Fig. 13.1 and Table 13.2. The first group, represented by jigs, applies an essentially vertical oscillatory motion to the solids-fluid stream. Shaking concentrators form the second group. These apply a horizontal shear to the solids-fluid stream by vibrating the surface under the stream. Included in this group are the shaking table, the vanner, the Bartles-Mozley concentrator, and the very famous miner's pan. Gravity flow concentrators such as sluices and troughs form the third group. Here the void space is basically maintained by the action of a layer of slurry flowing under the influence of gravity down an inclined surface. In this group are the oldest known concentrators (including those occurring in nature) and the comparatively recent Reichert cone, which has had an adoption rate probably unsurpassed by any device except the flotation cell.

13.1. EQUIPMENT AND APPLICATIONS

13.1.1. Jigs

The principle of the jig can most easily be illustrated by taking a laboratory sieve with about 1 cm of heterogeneous ground mineral in it, immersing the sieve in a bucket of water, and oscillating it up and down under water. This will result in the denser and

Equipment and Applications

TABLE 13.1: TONNAGES TREATED BY GRAVITY SEPARATION IN THE U.S.A IN 1973 (EXCLUDING DMS)

Ore Type	Ore Processed (millions of tonnes)	Approximate Value (millions $)
Coal	228	2900
Iron ore	50	900
Mineral sands	23	72
Gold	9	2
Tungsten	3	37
Total	313	3911
(c.f. Flotation)	(400)	(1500)

larger particles forming the lower layers, with the finer lighter, particles on top. Commercial jigs perform this operation on a larger scale. Jigs that move the screen surface have been used, but most modern jigs employ a stationary screen and pulse the water flow through it. The significant variations between the various types of modern jig therefore become the method used to cause the pulsation and the method of withdrawing the dense mineral product (Figs. 13.2–13.6; Table 13.2).

The essential features of a jig are illustrated in Fig. 13.2, where it can be seen that the tank or *hutch*

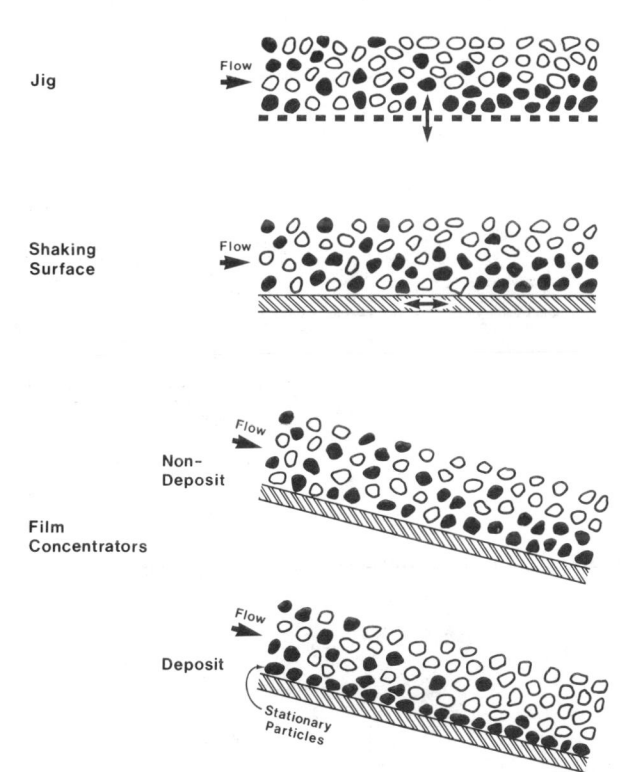

Figure 13.1. Classification characteristics of gravity separators.

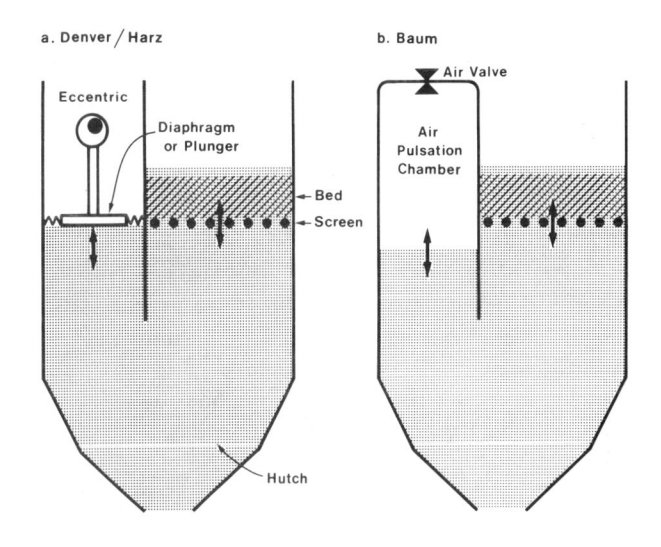

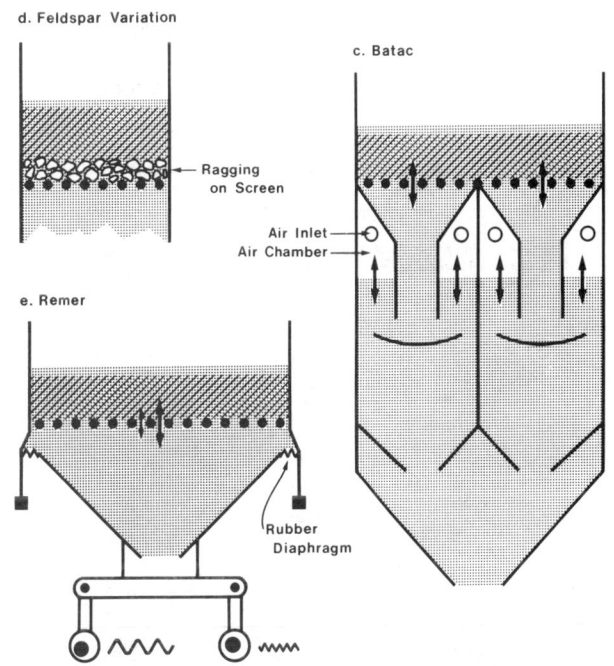

Figure 13.2. Schematic diagrams of the basic types of jig.

is divided by a wall into two main sections; one containing the support screen with the process bed on top of it and the other a section where the fluid pulse is generated. Earlier jigs[3] such as the Harz or Denver used reciprocating plungers or diaphrams to generate the water pulsations (Fig. 13.2a). About the turn of the century the Baum jig was developed using air pressure to generate the water pulsations (Figs. 13.2b and 13.3). Because more suitable pulsation cycles can be achieved using air instead of a

TABLE 13.2 GRAVITY SEPARATORS

	Machine	Size (m) (width x length)	Capacity* (average)	Speeds*	Feed Conc. (Vol. %)	Capital Cost** $/t/hr	Applications
J I G G I N G	Diaphragm or Plunger Mineral Jig	to 1.2 x 1.1	4.0 t/hr/m² (200 μm cassiterite)	300 r.p.m. (5 mm stroke)	10 including wash water	5	Roughing, cleaning, scavenging of relatively coarse cassiterite, gold, scheelite.
	Baum (Fig. 13.3)	to 17.6 m (2 x 6 cells in parallel)	20 t/hr/m² (-150 mm coal)	20 - 30 r.p.m.			Mainly for coal washing.
	Batac (Fig. 13.4)	30 m² (six 5 x 1 cells)	24 t/hr/m² (-150 mm coal) 12 t/hr/m² (-12 mm coal)	55 r.p.m.			Mainly for coal washing. Batac jig better able to treat finer coals than Baum jig.
	Wemco-Remer	1.5 x 4.9	7 t/hr/m² (-25 mm aggregate)	160 r.p.m. (400 r.p.m. secondary) (12 mm stroke)	~10 including wash water		Primarily used for aggregate production.
	Circular (Also known as I.H.C. Cleaveland Jig) (Fig. 13.5)	7.5 (diameter) (41.7 m²)	10 t/hr²/m² (200 μm cassiterite)				Extensively used on tin dredges.
	Pneumatic Jig (Fig. 13.6)	1.8 x 3.8	2-3 t/hr/m² width	600 r.p.m. (6 mm)	N.A.		Used for washing suitable coals, when dry product is an advantage.
S H A K I N G	Shaking Table (Fig. 13.9)	2.0 x 4.6	0.05 - 0.25 t/hr/m² (heavy minerals)	265 r.p.m. (20 mm stroke)	15	14	Treatment of coal, cassiterite, scheelite, and other heavy minerals.
	Slimes Table (Also known as Holman Slimes Table)	2.0 x 4.6	0.01 - 0.06 t/hr/m²	300 r.p.m. (10 mm stroke)	15	55	Used for particles too fine for conventional table. Suitable for cleaning concentrate from Bartles-Mozley concentrator.
	Bartles-Mozley Concentrator (Fig. 13.10)	1.2 x 1.5	2.5 t/hr	200 - 300 r.p.m.	1 - 4	10	Rougher concentrator for very fine heavy minerals.
	Bartles Crossbelt Vanner	2.75 x 2.4	0.5 t/hr		10	40	Similar applications to slimes table.
F L O W I N G	Spiral (Initially known as Humphreys Spiral) (Fig. 13.14)	0.6 (diameter) 2.9 (high)	1-5 t/hr per start	N.A.	6 - 20 (plus 30 - 60 l/min wash water)	0.5 to 1.4	Used for beach sands, iron ores, miscellaneous heavy minerals. Increasing tendency to use for cleaning cone concentrates. High capacity units suitable for roughing during grinding.
	Pinched Sluice (Fig. 13.12)	0.9 x 0.25 to 1.8 x 0.4	2-4 t/hr	N.A.	30 - 45		Concentration of beach sands, phosphate ore. Usable where throughput is insufficient for cone, or where more flexible circuit arrangement is necessary.
	Cone (Also known as Reichert cone) (Fig. 13.13)	2 (diameter)	65 - 90 t/hr	N.A.	35-40	1.0	Originally developed for beach sands, now used for coal, roughing and cleaning of iron ores, and roughing concentrator tailings to recover trace heavy minerals.

* Depends markedly on particle size of material being treated.
** Costs relative to cone.

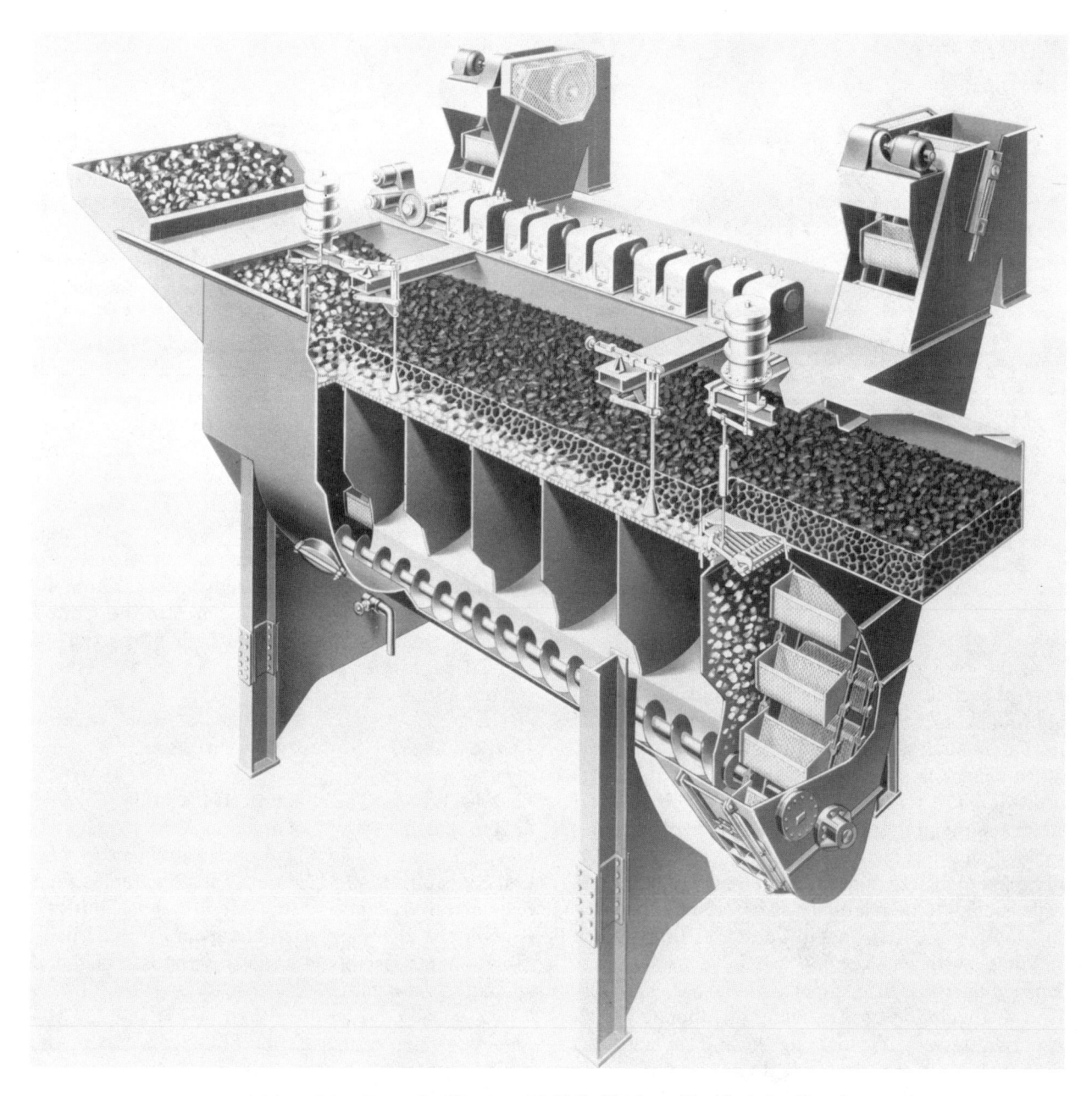

Figure 13.3. Baum jig. (Courtesy McNally Pittsburg Manufacturing Corp.)

mechanical device, the former is now the most widely used.[4]

As jig size increases the U-shaped hutch no longer gives an even flow across the whole bed. The Batac jig (Figs. 13.2c and 13.4) is a modification of the Baum jig and uses multiple air chambers under the screen to overcome this problem. It is now virtually the industry standard for coal cleaning.[5]

To supplement the pulsing action and maintain the bed in an open state for a longer time, additional water is generally fed to the hutch during the period when the bed is settling (the downward or *suction* stroke).

Two basic methods of dense mineral removal can be employed.[4] Over-the-screen jigging utilizes a screen aperture smaller than the mineral particle size, and the dense mineral discharges under a suitable weir while the lighter mineral overflows a different weir. A later modification, frequently called the English method, is applicable to finer materials.

Figure 13.4. Batac jig. The air valve mechanisms are on the right. (Courtesy KHD Humboldt Wedag AG.)

Here the screen aperture is larger than the dense mineral particles, which can then fall into the hutch where they are withdrawn by a suitable mechanical discharge such as a spiral or bucket conveyor. When using this type of discharge, a layer of large grains (*ragging*) must be kept on top of the screen if the dense mineral does not have sufficient particles larger than the screen aperture (Fig. 13.2d). Generally these ragging grains should be evenly sized and large enough to be relatively unaffected by the flowing water; also, they should be denser than coarse middling particles. Ragging can be denser or less dense than the dense mineral. Feldspar is widely used as ragging in coal cleaning jigs (hence the alternative term "feldspar jig"), and hematite is commonly used for separating denser minerals such as cassiterite and scheelite.

As with most concentrating devices, jigs seldom give sufficient separation in one stage; thus where throughput is sufficient they are generally built with a number of hutches in series. This also allows middling products to be produced with or without further grinding before another concentration stage. The circular jig (Fig. 13.5) is a radically different shape,[6] using a raking mechanism to ensure an even bed depth over its large area. Even so this jig is claimed to be mechanically simpler for a given capacity, since it achieves a fast compression/slow suction stroke with virtually no hutch water. The radial movement of the bed results in a decreasing horizontal velocity component and makes this jig particularly suitable for the recovery of fine minerals such as alluvial cassiterite.

The use of air instead of water for jigging also necessitates a significantly different mechanical arrangement. Pneumatic jigs (Fig. 13.6) basically consist of a sloping porous surface that is vibrated to assist flow.[4] Pulsed air is blown through the screen surface to produce segregation. A number of pockets occur along the surface for the removal of dense material. The use of this jig is generally restricted because air has such a low density. With suitable coals (those with few middlings), this method can have lower costs than a wet jig, particularly where water is short or a dry product is required. Pollution problems also are reduced.

In general jigs are more suitable for treating relatively large particle sizes, in the range 0.5–200 mm. Today, few minerals are liberated at these sizes, so that the *range* of applications is somewhat limited. Even so, about half the coal cleaned in the United States is treated in jigs, and a high proportion of alluvial cassiterite recovered in Southeast Asia is treated initially by jigging.[7]

13.1.2. Shaking Surface Concentrators

Shaking Table. The shaking table is a relatively old device that has slowly evolved into the modern tables, which have a small but important place in mineral concentration.[8,9] Generally, shaking tables treat finer materials than jigs are able to handle, but this is achieved at the expense of capacity; single deck tables have relatively low capacity for their cost and space requirements.

A typical table is illustrated in Fig. 13.7. Feed enters through a distribution box along part of the upper edge and spreads out over the table as a result of the shaking action and the wash water. Product discharge occurs along the opposite edge and end. The essentially rectangular table has an adjustable slope of about 0°–6° from the feed edge down to the discharge edge, with a much lower rise from the feed end to the discharge end. The surface is a suitably smooth material (e.g., rubber or fiberglass) and has an appropriate arrangement of riffles on it, which decrease in height along their length toward the discharge end. Various riffle arrangements can be used to emphasize grade or recovery, and their density can be increased by using a rhomboidal table (as in Fig. 13.7).

ORE BEARING MIXTURE

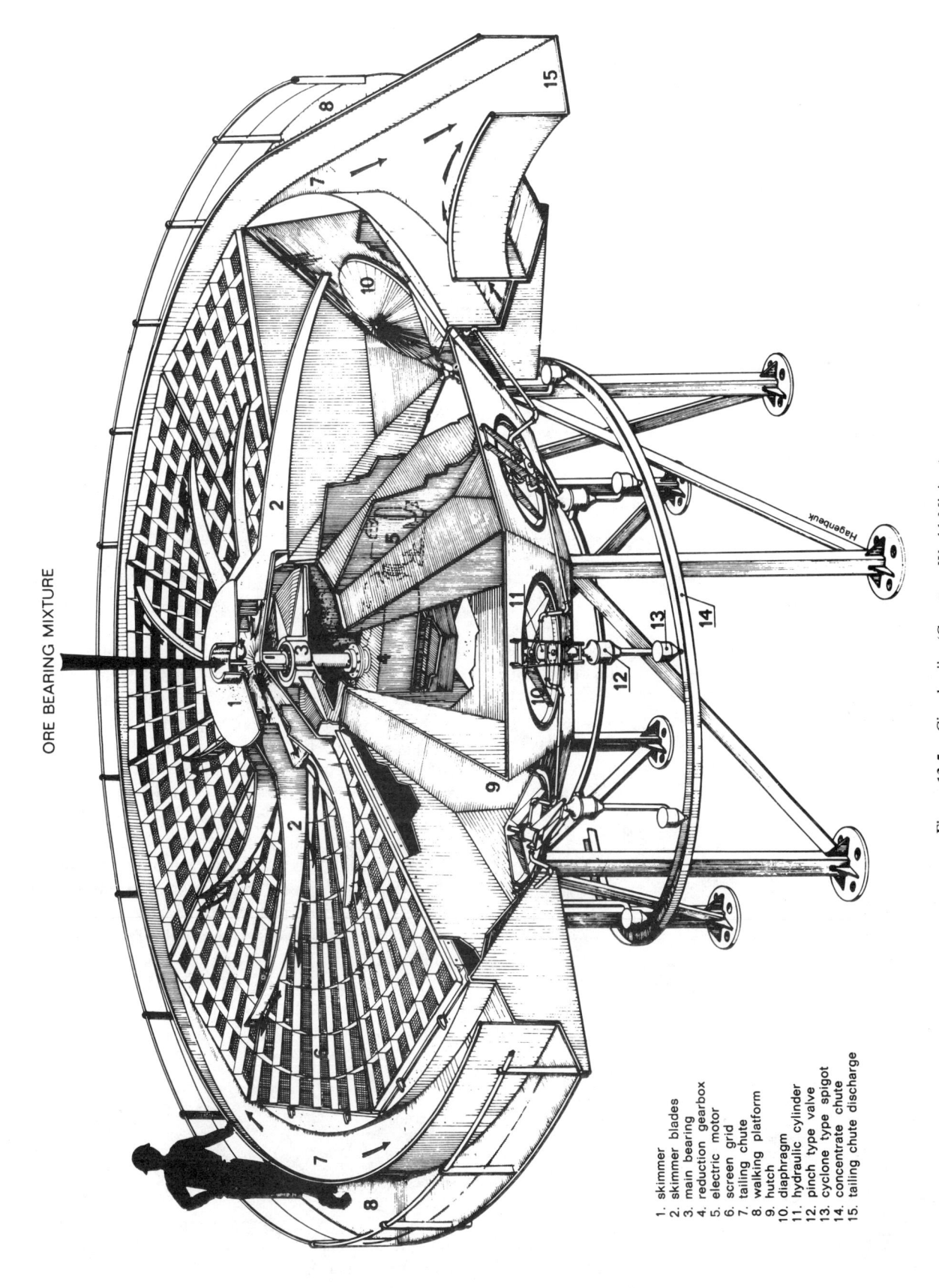

1. skimmer
2. skimmer blades
3. main bearing
4. reduction gearbox
5. electric motor
6. screen grid
7. tailing chute
8. walking platform
9. hutch
10. diaphragm
11. hydraulic cylinder
12. pinch type valve
13. cyclone type spigot
14. concentrate chute
15. tailing chute discharge

Figure 13.5. Circular jig. (Courtesy World Mining.)

255

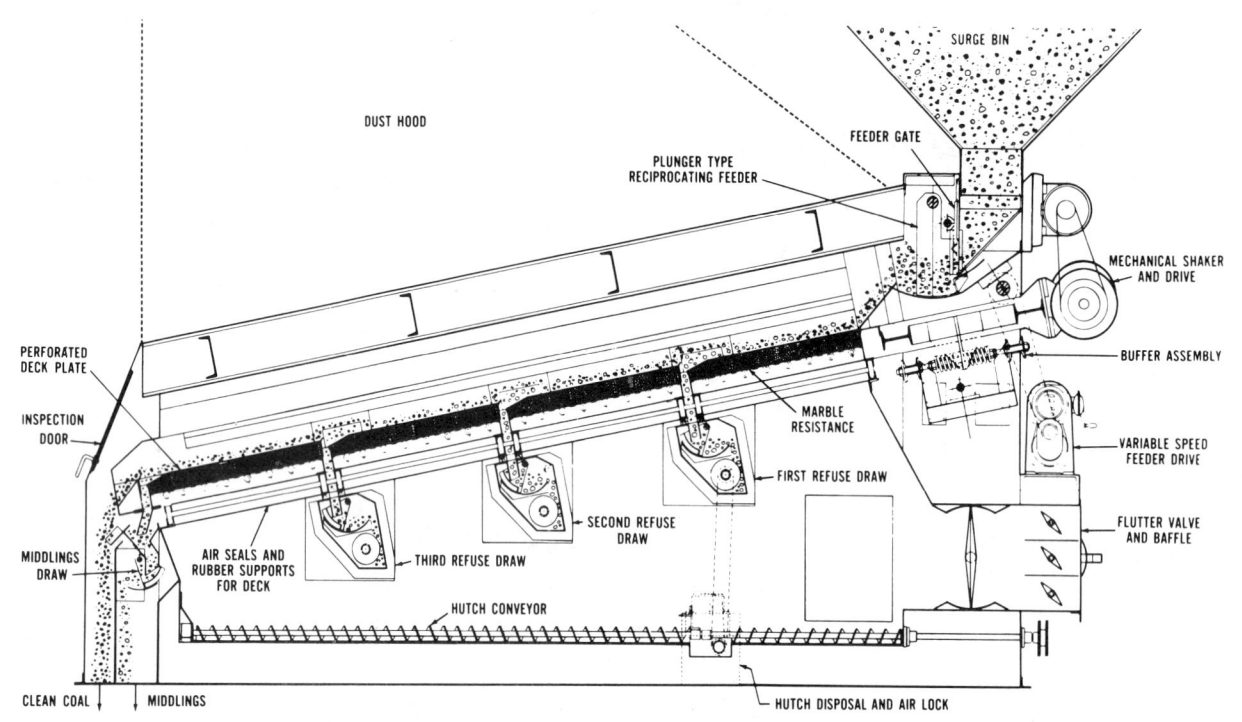

Figure 13.6. Pneumatic jig. (Courtesy Roberts & Schaefer Co.)

The differential shaking action (Fig. 13.8) is applied to the table along its horizontal axis. This action not only opens the bed to allow dense particles to sink, but by its asymmetry provides particle transport along the table. The Wilfley table was the first to use this differential shaking motion; its predecessor the bumping table[3] is no longer used because of its mechanical limitations. Further developments have since been made to the basic Wilfley table, and probably the most significant are those incorporated in the table shown in Fig. 13.9. Improvements include suspension of the table from the roof, which eliminates the need for heavy foundations to sustain the table motion; the use of rotating eccentric weights to provide table motion (instead of eccentric linkages); the stacking of tables (up to 6 in height) to save floor space and drive power; and the pool system of riffles, which has some of the riffles higher than others to make the flow down the surface calmer.[8] The shaking table has a considerable number of operating variables; these are listed in Table 13.3.

Tables are still widely used for coal cleaning,[4]

Figure 13.7. Schematic of a shaking table, showing the distribution of products.

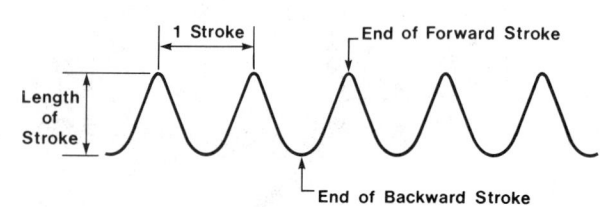

Figure 13.8. The differential shaking action of a shaking table. (After Sokaski et al.[46])

Equipment and Applications

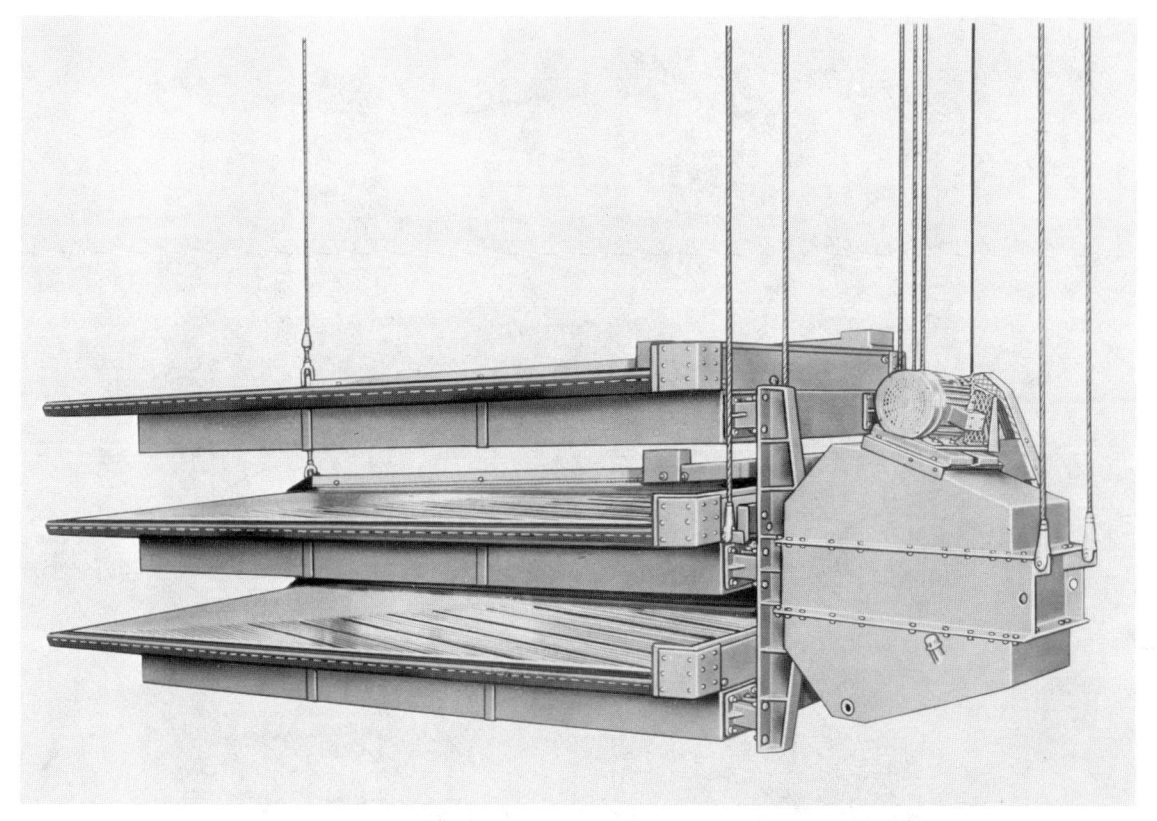

Figure 13.9. A three-deck shaking table. (Courtesy Deister Concentrator Co., Inc.)

typically treating materials nominally 0–6 mm. They are also used for concentrating heavy non-sulfide minerals such as cassiterite, scheelite, and gold. With these minerals, recovery normally falls off rapidly below 75 μm (and at larger sizes with less dense minerals). The Holman slimes table[10] has a particularly smooth riffle pattern to enable it to operate on finer material (15–75 μm), but because capacity is inversely proportional to the particle size, the capacity of these tables is very low.

An occasional use of the shaking table is as a visual control device for analyzing a tailings stream,[9] of for example, small flotation plants unable to justify more expensive instrumentation. In this method the spread-out mineral bed on the table enables the different minerals to be recognized, and the appearance of the valuable mineral in a band on the table indicates the need for corrective action.

Bartles-Mozley Concentrator. This shaking surface concentrator was developed to recover fine cassiterite that was irrecoverable by other means.[11] It consists of a suspended assembly of 40 fiberglass decks arranged in two sandwiches of 20 (Fig. 13.10). Each deck is 1.2 m wide by 1.5 m long and separated by a 13 mm space that also defines the pulp channel. Operation is semicontinuous, in that feed is distributed to all 40 decks for a period up to 35 min, after which the flow is interrupted briefly while the table

TABLE 13.3: VARIABLES OF SHAKING TABLE

Design Variables:
  Table shape
  Table surface material
  Shape of riffle
  Pattern of riffles
  Acceleration and deceleration
  Feed presentation

Running Speed:
  Motor speed
  Pulley size

Stroke:
  Toggle
  or vibrator setting

Operating Controls:
  Table tilt
  Pulp density of feed
  Wash water
  Position of product splitters

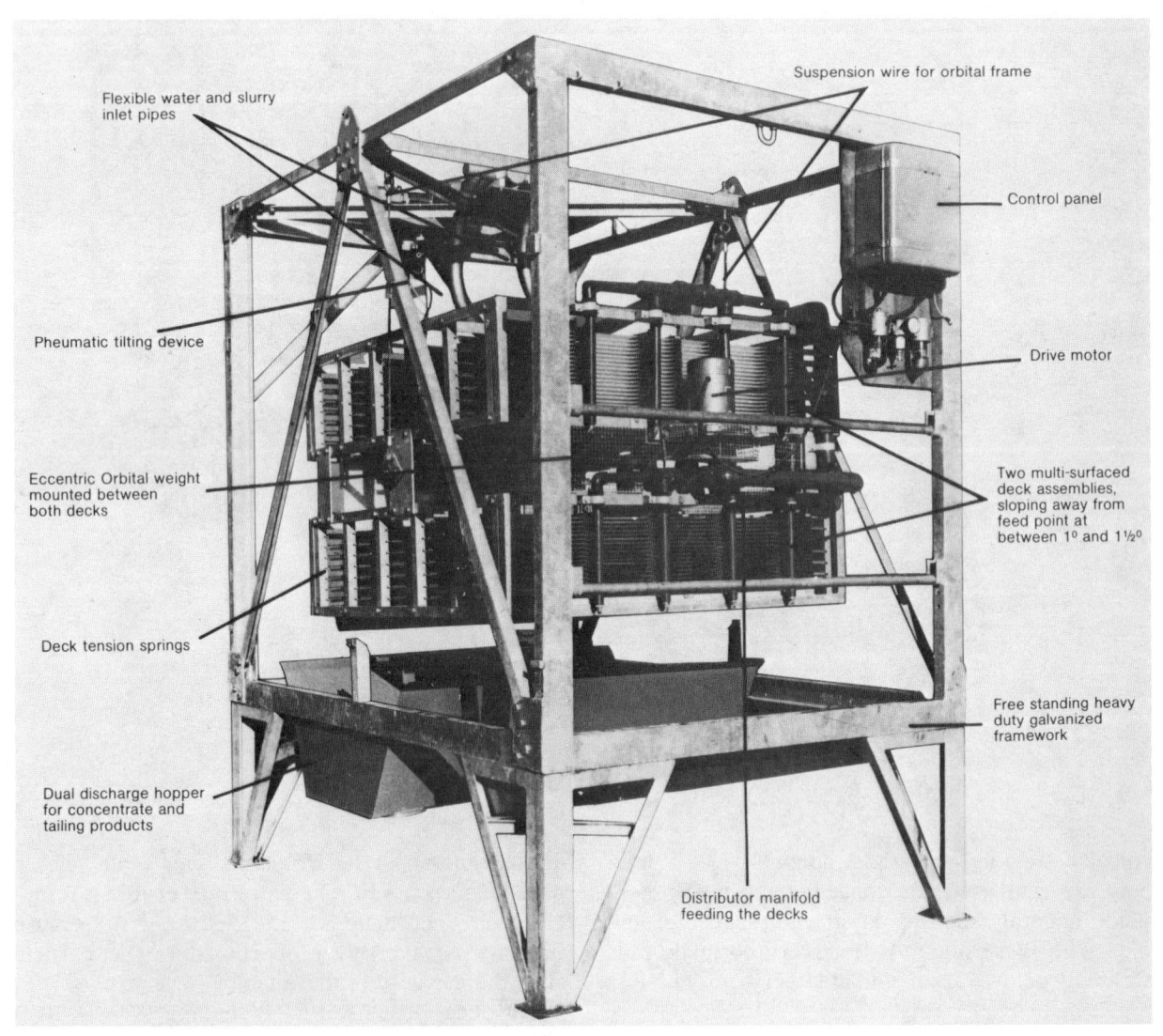

Figure 13.10. Bartles-Mozley orbital gravity concentrator. [Courtesy Bartles (Carn Brea) Ltd.]

is tilted so that the concentrate can be washed off. The cycle then repeats. During the loading cycle, the state of shear in the water is primarily maintained by means of an eccentric weight rotating in the space between the two sandwiches. This concentrator is capable of recovering over 60% of the 10 μm particles in low grade slurries of cassiterite or scheelite, at throughputs significantly higher than competitive devices such as the Holman table, although the latter device is suitable for upgrading the concentrate from the Bartles-Mozley concentrator.

Other Shaking Surface Concentrators. A number of other shaking devices have been either used[3,12] or proposed.[12–17] The vanner is a gently sloping continuous rubber belt that has a slurry spread over its vibrating surface.[3] Its virtual demise can be attributed to the success of flotation and to the low capacity of the vanner relative to shaking tables. It is in effect a continuous form of the old gold miner's pan (Fig. 13.11), which still survives today among old timers and in live museums. The Bartles crossbelt concentrator[13] is a modern form of the vanner, incorporating a vibrating mechanism similar to that used in the Bartles-Mozley concentrator. Of the new concentrators proposed, few have achieved commercial success even when they are technically successful, a feature that must in part be

Figure 13.11. The miner's pan.

attributed to the relatively small market for such equipment.

13.1.3. Film Concentrators

Concentration in a layer of liquid flowing under gravity down an inclined surface characterizes the film-type concentrators. Such concentration occurs in nature and is responsible for the so-called placer mineral deposits, which account for a large proportion of the world's tin, gold, and beach sand minerals (e.g., rutile, ilmenite, and zircon). Concentrators based on this principle predate recorded history, and yet some, such as the planilla, are still used today in the original form in some developing countries.[7,10] Although gravity concentration underwent a significant decline with the discovery of froth flotation, modern film concentrators have such low costs that they have found a range of applications.[1]

Two types of particle transport can be recognized in slurry flowing down an inclined surface: nondeposit transport (where the particles essentially remain in suspension) and deposit transport (where the particles may reside on the surface for appreciable lengths of time). On the basis of these two modes, three types of concentrator can be distinguished: those that feature nondeposit transport and include the pinched sluice group and the spiral; those such as troughs that feature deposit transport; and those that exhibit both deposit and nondeposit transport, such as the Rheolaveur. Of these, only the first type is widely used today.

Nondeposit Transport Concentrators. The basic pinched sluice (Fig. 13.12a) consists of a wedge-shaped sloping surface, with some form of splitter at the end for separating the heavy mineral which has concentrated in the bottom of the stream.[10,18] A curved splitter is now the preferred form, because it gives improved separation and is less prone to blockage.[19] The cone concentrator (Fig. 13.13) is an extension of this basic concept and achieves very high capacities by having the pinching effect occurring as the slurry flows down to the center of a conical surface.[20] A feature of this device is the upright cone above the concentrating cone which acts as a distributing surface. The cone unit actually consists of a number of cones in a stack to provide roughing, cleaning, and scavenging operations without intermediate pumping.

The spiral[10] is illustrated in Fig. 13.14. Here a slurry flows down a spiral surface and concentrates are split off and discharged at suitable points. Unlike the pinched sluices, this device uses a supply of secondary wash water to aid the separation process. Originally made of cast iron (and even old car tires), spirals are now made of fiberglass and can have two separate spirals intertwined to save floor space.

High capacity spirals are a more recent development. They operate at high pulp densities using no wash water, have a flatter profile, and have no splitters. Instead, the products are generally split as they discharge at the bottom of the spiral, although a limited number of slots may be distributed down the spiral when the proportion of dense mineral is high.

The modern sluices and spirals were originally developed for concentrating beach sands. The very high capacity of the cone also makes it particularly suitable as a low cost rougher concentrator, either for initial plant feed (to reject liberated gangue), or for recovering trace minerals from the tailings streams of flotation plants.[21] Spirals have been successfully used for concentrating iron ores, but in larger plants they have been replaced by cones.[22] However, the new high capacity spirals are cheaper than cones, and this trend could reverse. The

Lamex launder is a deep bed pinched sluice that has had some success in coal cleaning.[18]

Deposit Transport Concentrators. Trough washers or sluices are simply very long, parallel-sided channels. They are operated essentially on a semibatch basis, with the heavy materials collecting on the bottom and being removed during a period when the feed is stopped. Generally some form of texture is given to the bottom of the trough to better hold the dense mineral. Historically used for gold and cassiterite recoveries in particular, they have little commercial significance today.[3,7] The Buckman tilting concentrator is very similar in principle to the

Figure 13.12. (*a*) Pinched sluice rig (flow direction up the page). (*b*) Schematic circuit arrangement. (Courtesy Readings of Lismore Pty Ltd.)

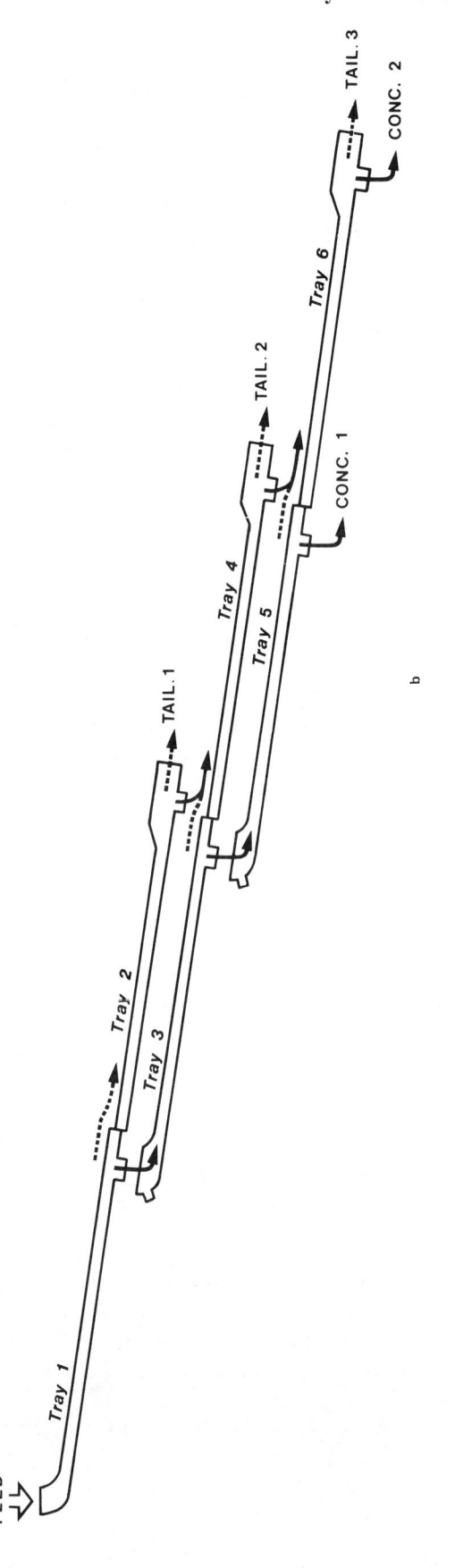

Theories of Gravity Concentrating Devices

Bartles-Mozley concentrator, except that the surface is not shaken.[10] Indications are that the latter is a significantly more efficient unit.[11]

Mixed Transport Concentrators. The Rheolaveur is a modification of a conventional trough, having dense mineral discharge ports at various positions along the bottom of the trough.[4] Hydraulic water may be added to the box surrounding the discharge port, to provide additional concentration. Although once used for coal cleaning, these devices are now being superseded.

Efforts to develop more modern equipment in this class also appear to have had limited commercial success.[23,24]

13.2. THEORIES OF GRAVITY CONCENTRATING DEVICES

A considerable range of theories exist for gravity concentrators,[3,10,15,25-28] and it is doubtful whether there is a universally accepted theory for any particular device. The isolated processes and mechanisms described in Chapter 5 are those that have been considered to have some relevance to one or more devices. The range of theories makes it clear that no one concept is adequate to explain the separation occurring in a given device. Rather it suggests that a number of processes occur, and that different processes may occur at different stages of the cycle, in different parts of the device, over different size ranges and under differing operating conditions.

13.2.1. Jigging

Jigging is probably the most complex gravity separation because of its continuously varying hydrodynamics. The mineral bed is repeatedly moved up by the water, expands, and then resettles, the resettlement occurring with the water flowing down at a lower rate (because of the addition of hutch water) than that which occurred on the upstroke. It follows that the wave form itself must be a significant parameter; the manner in which the bed expands is important, too, because it has a marked effect on the particle dynamics. A number of experimental studies have been reported in the literature,[29-36] but too often they fail to contribute to the understanding of jigging. This conflicting information appears to arise because many studies were

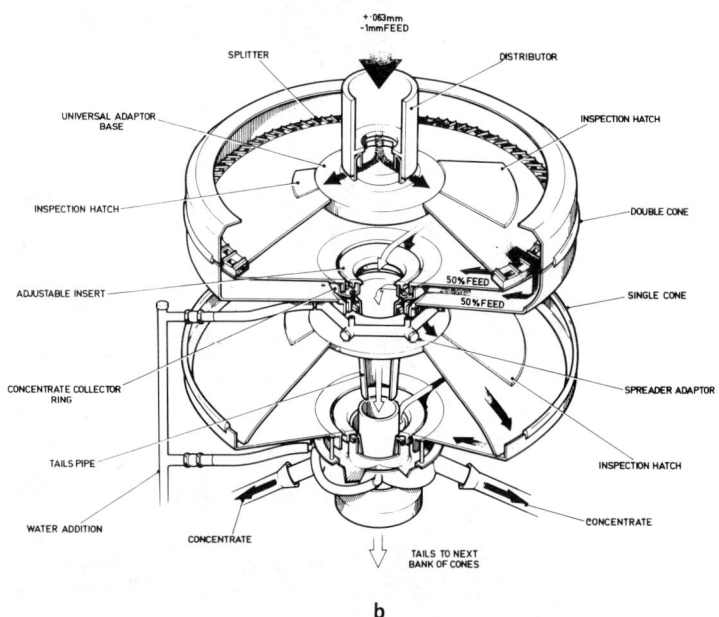

Figure 13.13. (a) Cone concentrator assembly in a floating plant (see, e.g., Fig. 1.5). (b) A DSV (double cone/single cone/variable insert) cone concentrator stage. (Courtesy Mineral Deposits Ltd.)

Figure 13.14. (*a*) A spiral installation. (Courtesy Sala International AB.) (*b*) Close-up of an operating spiral, showing wash water addition and heavy mineral splitting device. (Courtesy Mineral Deposits Ltd.)

Theories of Gravity Concentrating Devices

carried out on a narrow set of ideal conditions that resulted in behavior quite unlike that associated with practical jigs.[29-32]

Two approaches to jigging theory exist.[3,10,25,28,33] The classical concept considers the motion of individual particles,[3,10,25,33] and the center-of-gravity lowering concept forms the basis of the second.[26,37,38] Typical of the former is that described by Gaudin, who suggested that three mechanisms are involved[3]: hindered settling (Section 5.1), differential acceleration (Section 5.2), and consolidation trickling (Section 5.3). Most of the stratification supposedly occurs during the period when the bed is open, and results from differential hindered settling accentuated by differential acceleration. These processes put the coarse/dense grains on the bottom and the fine/light grains on the top. Consolidation trickling on the other hand occurs when the bed is compacted and places the fine/dense material on the bottom and the coarse/light material on top. Since the two effects arrange the particles in diametrically opposite ways, suitable adjustment of the cycle should supposedly balance the effects and result in an almost perfect stratification according to mineral density.

The center-of-gravity theory of jigging proposed by Mayer (Section 5.7) is probably now more widely accepted.[26,37-39] The notable feature of this theory is that the water pulsation is purely to open the bed, so that denser particles are able to move down through it; or put another way, the bed is opened up to permit the release of the potential energy of the mixed bed by the bed stratifying. Even though relatively little experimental evidence has been presented to substantiate this theory, it is able to explain a number of phenomena that occur with unusual particle shapes or size distributions.[37]

Although the center-of-gravity theory is generally presented as an alternative to the classical hydrodynamic approaches, it is debatable whether this is the case: rather the two theories should be considered to be complementary. This is because the center-of-gravity theory does not indicate how the denser particles move through the bed, and therefore cannot completely describe the rate of separation. Mayer considered that the rate would be simple first order, given by:

$$H_t = \Delta H \exp(-kt) \quad (5.31)$$

The rate constant k should be a function of several factors, including the jiggability ΔH; the form, amplitude, and frequency of the jig stroke; and the displacement and frictional resistance of particles. Logically, particle shape and hydrodynamics must be significant in the latter factor and thus some of the concepts developed in the classical theories must be relevant.

Vinogradov et al.[38] have presented a more rigorous analysis based on a total energy balance rather than just the potential energy as considered by Mayer. The shift in the center of gravity is given by the second-order equation:

$$\frac{d^2 H}{dt^2} + K_1 \frac{dH}{dt} + K_2 \, \text{fn}(H) + K_3 = 0 \quad (3.6)$$

where the first term accounts for the inertial properties of the system, K_1 the fluid resistance; K_2 the specific rate of separation (equivalent to k in Eq. 3.1, and primarily a function of quantities and densities of the two minerals), and K_3 the effect of gravity. Equation 3.6 can describe a process such as jigging that tends to an equilibrium position with a damped oscillation (Fig. 13.15).

There is apparently still some debate about how the bed expands during the jig cycle.[32,40] It is likely that in most jigs the bed lifts up *en masse* then, near the end of the pulse (lift) stroke, as the particles fall from the bottom of the bed, it expands by means of a loosening wave spreading upward through the bed. If the acceleration of the upstroke is too high, the water can break through the bed and cause turbulence. In addition, the finer particles are carried higher than the larger particles, which means that the dense fines have less chance of reaching the dense layer of the bed because they have an appreciably longer distance to travel—a distance large light particles could travel in a shorter time. It can be concluded therefore that provided the bed opens

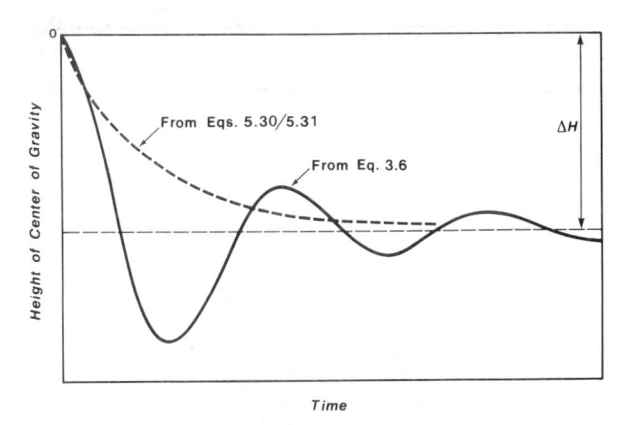

Figure 13.15. Comparison of two analyses of the jigging process.

from the bottom, separation occurs basically during the downstroke, that is, as the particles are resettling through the fluid. It is in fact during this period that hutch water is added to reduce the rate of the downstroke, and aid stratification. Clearly hydrodynamics must have an effect during this period.

Taggart has pointed out that the equisettling ratio[25] (Eq. 5.1), a significant factor in Gaudin's hindered settling concept, cannot explain the large range of particles that can be treated in the jig. Even the use of ρ_{sl} instead of ρ_l (Eq. 5.2) fails to give realistic ranges. However, as was discussed in Chapters 3 and 5 the use of ρ_{sl} instead of ρ_l is questionable where particle sizes are of the same order and Eqs. 5.4 and 5.5 are more realistic for considering relative hindered settling. However, these equations show that under certain conditions fine dense particles can have negative settling velocities, making the benefits of relative settling rates even more questionable.

On the other hand, differential acceleration provides a more tenable argument.[33] Consider, for example, a small heavy particle beside a large light particle in a bed, as the expanding wave passes them. According to Eq. 5.6, the denser particle will accelerate faster, thus moving ahead of the less dense particle (Fig. 5.1). Provided the fall distance and time are suitably restricted by the stroke length and hutch water addition, respectively, the large light particles can be kept behind the small dense particles. This is in line with normal jigging practice, where the amplitude used is proportional to the bed's particle size.[41] Furthermore, studies have shown that separations are primarily determined by relative densities[30] and that only a comparatively slow decrease in bed penetration rate occurs with a decrease in the size of the dense mineral particle[35] (Fig. 13.16). Although some studies on jigging have shown that the rate of penetration of the bed depends markedly on particle size, quantity, shape, and bed packing, most of these results were obtained on beds of uniformly sized particles and this must be the cause of this apparently abnormal behavior.[30,32] In some real beds, dense particles in the middle size ranges have been found to have severely retarded penetration rates, but this can be attributed to inadequate bed loosening.[35]

On the basis of the discussion above, it can now be argued that it is the settling velocities that *hinder* stratification rather than enhance it, in that the finest particles approach their terminal velocity relatively quickly, but are unable to *stay ahead* of the more massive particles, which eventually attain a higher terminal velocity. This also suggests, as is borne out in practice, *that screening of jig feeds is desirable and is preferable to classification.*

It follows that correct dilation or bed expansion is vital to good separation,[34-36,39,42] inadequate dilation being considered as the case of incomplete bed expansion rather than the magnitude of the average value being too low. Factors that can be adjusted in a given mineral system to achieve this are bed depth, amplitude and frequency of stroke, and water addition. Dilation is primarily achieved by adjustment of the amplitude and frequency of the stroke, but water addition provides supplementary dilation, mainly by giving the expansion wave more time to spread through the bed. On the other hand, its volume must be limited, since an excess prevents the descent of the finer dense minerals. The intensity of the stroke is the product of amplitude and frequency, and this must be increased as the bed depth increases to compensate for the higher mass to be lifted. While a range of combinations of amplitude and frequency may produce a given dilation measured at the surface, there will generally be a best combination for a given bed, which not only provides an *even* expansion wave up through the bed, but allows it to *just* reach the upper surface: these being the vital components of good bed dilation. In general, finer particles need more time for the expansion wave to pass through the bed and therefore require lower stroke frequencies. As mentioned above, the amplitude is also small. There is evidence to suggest that dilation can be aided by taking advantage of resonant frequencies of the pulses.[32,42,43]

As with most separations, the requirements of grade and recovery conflict in jigging. It has been suggested that for grade higher frequencies and

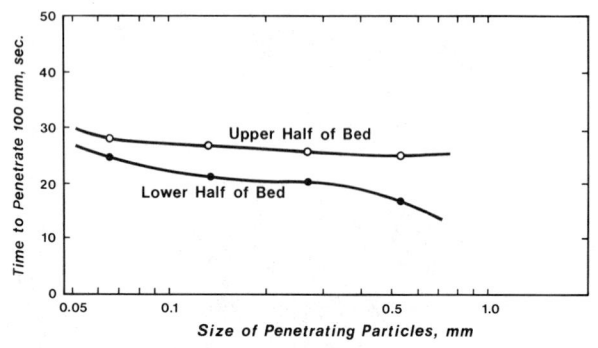

Figure 13.16. Penetration of high density particles under normal bed conditions in a jig. (After Michell and Suvarnapradip.[35]).

lower amplitudes are preferable, since small rapid movements provide the best absolute separation. Conversely, higher amplitude and lower frequencies give a more open bed and allow more rapid particle movement and thus enhance recovery.[34] A relatively deep layer of light mineral also enhances recovery of the dense mineral, while thickening of the dense mineral layer aids its grade.[29]

Ragging is used in jigs to allow finer particle sizes to be treated; as in fine coal washing or cassiterite concentration.[4,42] By using this method, of the fines, only the dense mineral is able to penetrate completely through the ragging interstitials during the downstroke; the lighter minerals penetrate only part way and are ejected by the upstroke. Some problems may arise when treating wide size ranges in this jig, since any coarse dense mineral left above the ragging is devoid of dense fines in the voids, and the system may lower its center of gravity by filling these voids with fine light particles.[37]

13.2.2. Shaking Concentrators

As in jigging, there is still no unified theory of shaking concentrators.[3,10,15,25,28,44–46] The generally accepted basis of separation is that material is spread out on a surface, where it becomes stratified by surface motion and gravitational flow. This stratification is primarily by density, with the denser material on the bottom; but secondary stratification, where larger particles rest on smaller particles, can occur in constant density layers.

Although Gaudin's analysis of thin film concentration (Section 5.4) explains some separations actually achieved on shaking tables, its validity and significance remain unproved. First, most concentration on modern tables occurs in the riffled area of the table; only in a comparatively small region outside the riffles can thin film conditions ($\delta \rightarrow d$) occur. (Although the older smooth top tables were capable of separations, their comparatively low capacity further indicates that concentration under such conditions is relatively inefficient). In one of the few experimental studies published, Carty[47] measured the terminal velocity of spherical particles rolling down a smooth plane boundary and found that for all materials

$$f_d = \frac{-215}{\text{Re}_p^{0.975}} \quad (13.1)$$

which leads to the same conclusions as Eq. 5.11; that is, the downtable velocity increases with increasing ρ_s. However, shaking tables place the denser mineral higher up the slope, which implies that if thin film concentration occurs, friction must have been a contributing factor (Eqs. 5.9 and 5.10). The most serious criticism of this mechanism relates to its being derived for an isolated particle in a laminar film. In practice the particles are essentially in contact, and though this could cause an apparent friction (actually a blockage to downtable motion), the limited experimental evidence[48] indicates that the drag conditions will be significantly different from those used in deriving Eq. 5.9.

Taggart's approach was a rollerskate analogy, with the fine particles being subjected to a more tortuous path during a lifting stage of the cycle, so that they tended to finish up below larger particles of the same density. Again, the center-of-gravity concept of Mayer[26] gives the most valid explanation of the density stratification and while it does not explain size stratification, it can explain why fine light particles may exist amongst large dense particles when there is a deficiency of small dense particles.

To explain the mechanisms of density stratification and reverse size stratification, it is necessary to consider other processes, particularly the shear analysis by Bagnold[49,50] (Section 5.5). One factor that has long been recognized in tabling is, like jigging, the importance of bed dilation[45]: until sufficient dilation has taken place, stratification, if it even occurs, is very slow. Bagnold's analysis shows that loosening occurs in a particulate bed and in all cases a certain minimum shear rate is required. According to Eqs. 5.20 and 5.21 this minimum shear will increase with mineral density, as is found in practice. (Note that the equations also show that particle size is a parameter.) In that the shear conditions in most shaking concentrators are in the viscous region, it seems likely that segregation therefore occurs by a trickling mechanism while the bed is in an expanded state generated by the oscillating shear. In this expanded state the denser particles can be expected to descend, to lower the bed center of gravity, and some degree of size segregation can be expected because the finer particles can penetrate the void spaces between larger particles of their own kind. Although Kirchberg and Berger's results[45] must be interpreted with caution because they were obtained in uniformly sized material, finer particles have been found to descend through beds faster than larger particles.

When relatively large particles are present, they

can be expected to rise by two mechanisms. First there is the wedging action (Section 5.6), although this is likely to be relatively unimportant because of the open state of the bed and lubrication. The second is suggested by Eq. 5.20 for inertia shear. This shows that larger particles will *rise* to the region of lower shear, and although this depends on a number of factors, it can be expected to have an effect on quartz particles of the order of 500 μm. It is worth noting that by Eq. 5.20

$$d \propto \frac{\rho_s - \rho_f}{\rho_s} \qquad (13.2)$$

which shows that if the particles become too large, segregation is determined primarily by size rather than by density. This explains not only the upper limit on particle size, but together with the trickling concept, why shaking tables operate better with *classified feeds* (where the average size of dense mineral is less than the average size of the light mineral).

Particle transport on the Bartles-Mozley concentrator is distinctly different from that on the shaking table.[11] On the former the shaking action is rotary on the plane of the surface, with the result that it causes bed expansion without horizontal transport. Dense particles therefore fall to the liquid/solid boundary and ideally remain there until the batch discharge cycle. Lighter particles on top are swept away by the net slurry flow down the surface. The minimum particle size captured is the particle that can penetrate the bed during the residence time and also remain stationary when it arrives at the tray surface. Whether a particle can be retained on a sloping surface has been considered in Eqs. 5.9–5.11 and 5.24. Equation 5.9 implies that only relatively flat particles are likely to remain stationary, whereas rounded particles are swept away at a velocity that increases with particle density (Eq. 5.11). Although the derivation of Eqs. 5.9–5.11 is questionable, the semiempirical analysis by Bagnold (Eq. 5.24) shows that denser particles can be retained on the surface without particle/surface friction. In fact, Eqs. 5.9 and 5.24 are similar in form, except that the latter has C_V terms to generate the "friction." At first sight, Bagnold's equation indicates relatively high critical angles, but it is possible to suggest bed conditions that give lower, more realistic values.[50] Furthermore even though Eq. 5.24 appears independent of d, this is not strictly true, since C_V and $\tan \theta_B$ are affected by d. Since a rigorous analysis is not yet possible, it is sufficient to note at this stage that there should be a critical inclination whereby the denser mineral can form a stationary bed against the solid surface, while the less dense mineral forms a flowing layer on top.

With the shaking table, the shaking action has the important secondary purpose of providing transport along the table. Again a net liquid flow down the table, resulting from the wash water, sweeps the light material away. Because of the general size stratification and the taper along the riffles, successively smaller light particles are exposed and are swept off the top of the bed as they become exposed above the riffle. It is perhaps worth considering at this point that it is not necessary for size stratification to have occurred in the bed for this action to occur. Rather, larger particles are exposed to the wash water first by virtue of their size and are subjected to high drag forces because they extend higher into the velocity profile of the flowing wash film. Even so, the net effect of the stratification processes is the characteristic distribution of products illustrated in Fig. 13.7.

As with other gravity concentrators only a limited size range of particles can effectively be separated by shaking concentrators. This arises not only because of the limits in the separating mechanism (whether or not they are considered to be determined by hydrodynamics) but also because the energy that must be applied to move the larger particles is also sufficient to bring about a degree of remixing, particularly of the finer particles. Because this type of concentrator is some times applied to the treatment of very fine particles (< 50 μm), its performance can be affected by additional particle-particle or particle-surface competing forces arising from pH-dependent electrical forces[51,52] (Chapter 6). In the case of the shaking table another factor limiting the separation of fines may be the eddying that occurs around the riffles.[45]

13.2.3. Pulsed-Shaking Concentrators

Although both jigs and shaking concentrators produce the same density segregation (dense material under the less dense material), they give (for the most part) opposite size segregations: the jig places the smallest particles on top of the bed, and the shaking concentrators place the smallest particles on the bottom of the bed. This implies that a concentrator that employs vertical liquid pulsation with a horizontal shaking action should be relatively efficient in treating materials with a range of particle

Theories of Gravity Concentrating Devices

sizes, because it should be possible to adjust the actions so that the size segregation effects cancel. Such a device has been tested on a laboratory scale, and although not used commercially, it clearly showed the expected superior performance.[16,53]

13.2.4. Film Concentrators

In terms of their concentrating mechanisms, the film concentrators are probably very similar to the shaking concentrators, in that stratification is determined by the center-of-gravity lowering and interparticle trickling. The significant difference is the manner in which the major fluid shear is generated to provide bed dilation: in the film concentrators this is provided by the velocity profile of the flowing film, rather than by a mechanical process. Because the slurry flow determines the shear, it is especially important with these devices to maintain constant feed conditions, particularly feed rate and slurry concentration. Slurry concentrations are relatively high, although they tend to decrease with decreasing particle size, which is in accordance with Eq. 5.21. The need for high concentrations can be attributed to the desirability of maintaining a low, relatively uniform void space over the full film depth (although in reality this ideal situation may not be possible because the shear rate is a function of depth).

Pinched sluices and cones have relatively low upgrading ratios, and consequently the products must be retreated in cleaner and scavenger circuits (Fig. 13.12b). This leads to another need for constant feed conditions; slight variations become magnified in the circulating loads, and the discharge slots are unable to successfully deal with this.[54]

In general the cone gives sharper separations than a pinched sluice because it has no wall effect to generate turbulence. However, in some cases the flow rates may not be high enough for the cone, and a pinched sluice must often be used to handle the relatively small concentrate stream.

Spirals also operate better with a stable feed.[55] The large number of discharge points makes the circuit more flexible, but the sheer number makes it impractical to change them too often.

The spiral has another concentrating parameter in common with the shaking table, the addition of secondary wash water at the center of the spiral. This aids the already present "river bend" action (Section 4.9.2) resulting from the circular path of the slurry.[15,46] Together these cause a component of the water flow to occur at right angles outward across the top of the slurry stream (Fig. 13.17). This results in the larger lighter particles reporting to the periphery of the spiral, with the denser material near the center, where it is removed through the discharge ports.

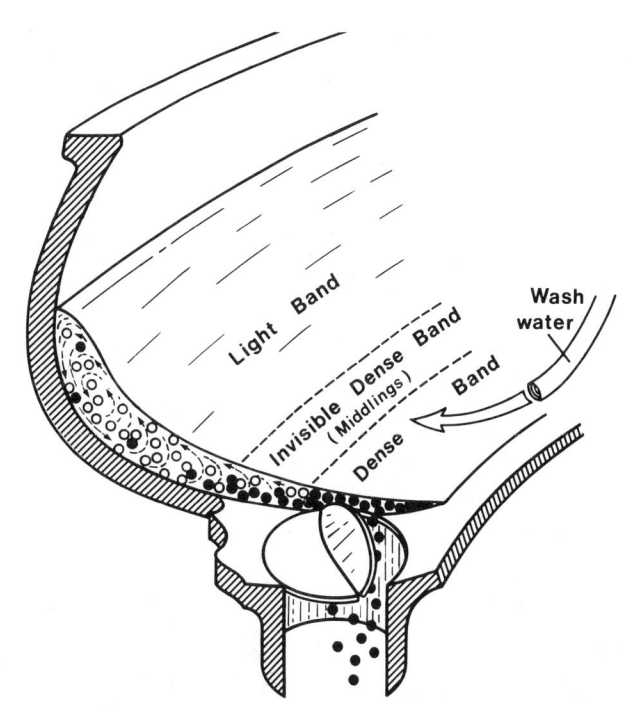

Figure 13.17. How the outward water flow effects segregation in a spiral.

Recently, some results on empirical studies of film concentrators have been published.[28e,54–56] Though providing little theoretical enlightenment, these references do give a good indication of the major factors that affect grade and recovery.

13.2.5. Hydrocyclone Concentrators

Few conclusive statements can be made about the mechanisms occurring in concentrating hydrocyclones. Analysis is difficult because these hydrocyclones operate with very high internal solids concentrations, induced by very large cone angles. Visman[57] has presented a qualitative description of the separation process, which is to some extent substantiated by the more recent studies by Renner and Cohen[58] (Section 10.4.1).

For the most part, the analytical approaches described in Chapter 10 become untenable at the high solids concentrations encountered. It can be argued that high solids concentrations enhance density effects because the $(\rho_s - \rho_f)$ terms can be replaced by

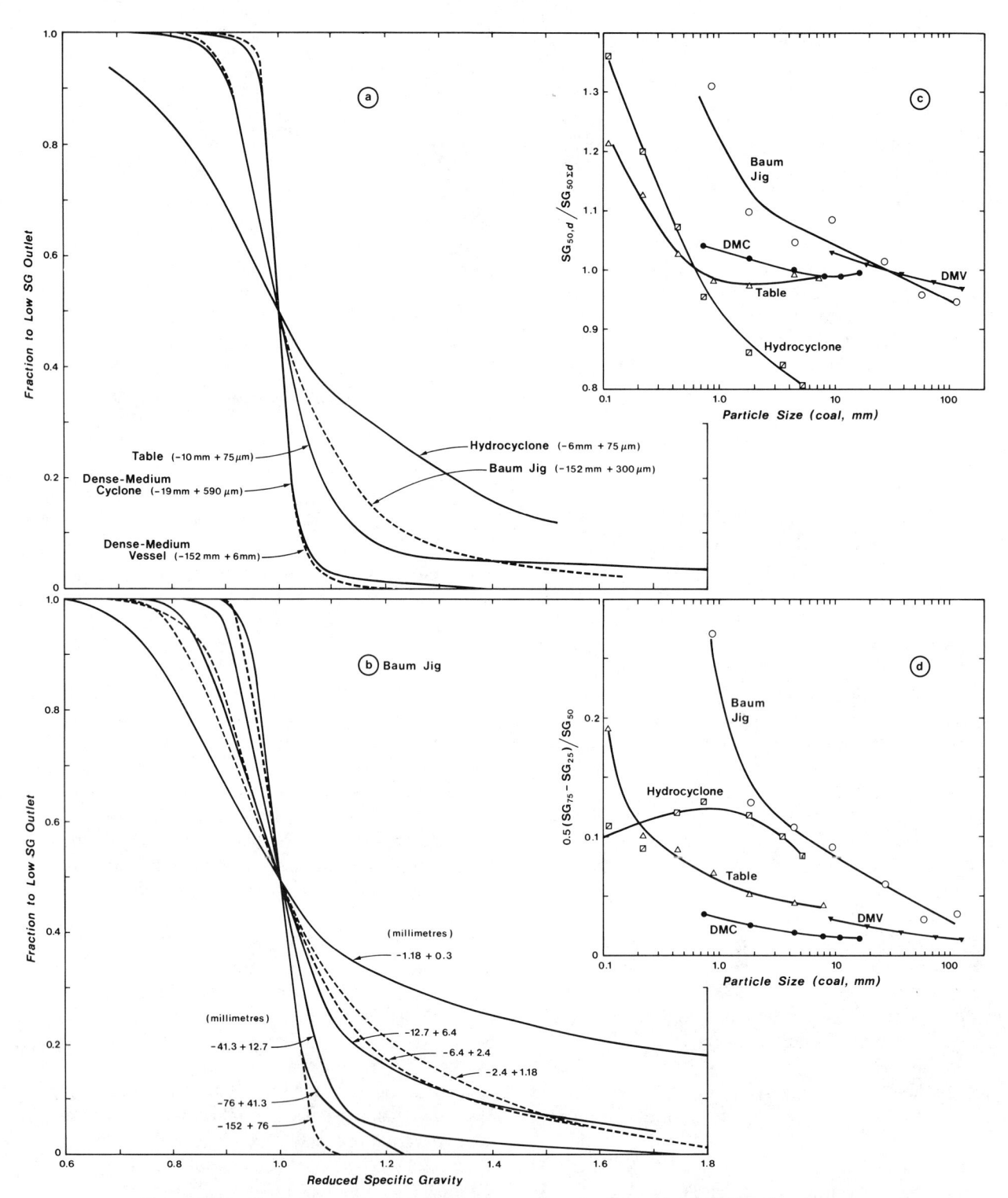

Figure 13.18. Performance data for gravity concentrators treating coal. (*a*) Reduced performance curves on wide size distribution feed. (*b*) The effect of particle size on the reduced performance curve of a Baum jig. (*c*) The effect of particle size on SG_{50}. (*d*) The effect of particle size on the probable error. (Data from Gottfried and Jacobsen.[59a])

($\rho_s - \rho_{sl}$) and that this may even occur to such an extent that ρ_{sl} lies between the density of the heavy and light material; that is, a dense medium separation occurs.

It is also possible to consider the mechanisms to be similar to those described for film concentrators, that is, one of shear forces maintaining the slurry in a dilated state so that the potential energy of the system can be lowered. Obviously the situation is considerably more complex in the case of the hydrocyclone because of the more involved shape of the separation chamber.

13.3. PERFORMANCE CURVES

Virtually all the information available on the performance curves of gravity concentrators has been obtained for coal washing separations, because of the difficulty in carrying out measurements on high density minerals. The coal results show conclusively that the reduced performance curve is primarily a characteristic of the device, and secondarily a characteristic of the particle size being treated.[59] Some of these basic curves are shown in Fig. 13.18a and b; correction factors for SG_{50} for decreasing particle size are given in Fig. 13.18c. Where the performance curve is applied to a range of particle sizes, separate curves must be combined in the appropriate proportions. In the case of jigs, it has been shown that the spread in the performance curve represents the density distribution in the jig bed, even though the bed's size distribution remains essentially that of the feed.[36,60] That the shape of the reduced performance curve is a function of the device and the size being separated emphasizes the need for caution in using probable error and area error (Fig. 3.8a) as measures of the quality of separation, because these are measured on the performance curve and are therefore functions of the separation density, a point all too often not appreciated. However, it is sometimes found that as the performance curve of a given system becomes steeper (because of a lowering of the separation density of the separation), the higher density end of the curve deviates from the ideal curve. Since this may be due to machine performance or mineral limitations, such a deviation is meaningful, and although it may show up in the area error, it is really the difference in shape between the curves that conveys the most information. Figure 13.18d gives a measure of the sharpness of separation produced by the different types of gravity concentrator used for coal separations. The failure of some devices to give sharp separations does not mean that such less efficient equipment is not useful; rather, these devices should be used only for separations where the difference in density of the two minerals is high and the quantity of middlings is low. Under these conditions an efficient separator may not be justified. On the other hand, close separations do require efficient separators.

It has yet to be established whether the data presented in Fig. 13.18 can be extended to high density separations, although until evidence to the contrary is available this would be a satisfactory approximation.

Although the basic reduced performance curves are now well established, there are as yet no correlations available for calculating the separation density SG_{50}, as there are for the d_{50} of classifiers (Chapter 10). Some of the significant factors likely to be involved in such a correlation are feed rate, concentrator size (area), and particle size. That particle size affects SG_{50} has already been shown in Fig. 13.18c. Limited data have been published[36] on the effect of feed rate: Fig. 13.19 suggests that in jigging at least $SG_{50} \propto$ capacity, but such a conclusion must be considered to be tentative in view of the limited data and the fact that the data does not apply to the same reduced performance curve. Since the scale-up of jigs and shaking concentrators

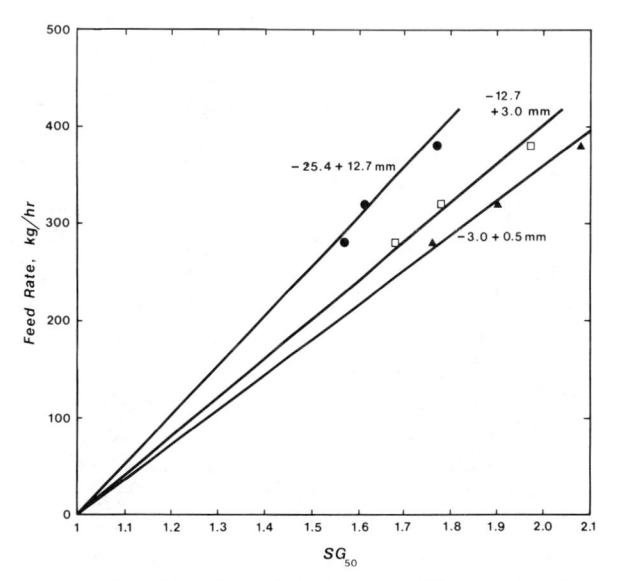

Figure 13.19. The effect of feed rate on SG_{50} in a jig. (Data from Whitmore.[36])

is based on the proportionality of capacity to area,[1] it can be tentatively suggested that $SG_{50} \propto$ area.

Little conclusive information is available on the importance of viscosity. In the pinched sluice it has been reported that a slight increase in viscosity (which can often be caused by clay particles in the water) appears to be beneficial.[61] Large amounts of fines are supposedly important in providing a heavy media effect (Chapter 12). Countering this, an increase in viscosity eventually slows down the rate at which particles can move, and as a consequence both the rate and the extent of separation decrease.[35]

Another factor that requires further elucidation is the effect of the ore's amenability to separation. Qualitatively, this has been described as dressability, washability, or in the case of jigging, jiggability, and it has been shown that the capacity is determined by the proportion of the feed that has a density close to the SG_{50}.[34] In quantitative terms it is the information contained in the separability curves of the minerals being separated. Dell has shown that the Mayer curve (effectively a cumulative grade/cumulative recovery form of the separability curves used by the coal washing industry) relates to the residence times,[62] but this concept has not yet been put in a form that can be used for predicting capacity.

Some indication of the effect of variables can be obtained through Eq. 3.13, which for gravity separations becomes

$$\mathcal{R} = \tfrac{1}{2} - \tfrac{1}{2}\,\mathrm{erf}\left\{\left[\frac{g\,\delta(m_{\rho I}/\Delta\rho)}{\mathcal{D}\mu^*}\right]^{1/2}(\rho - \rho_{50})\right\} \quad (13.3)$$

($m_{\rho I}$ = mass fraction of feed with density between ρ and $\rho + \Delta\rho$; ρ_{50} = separation density). The product $\mathcal{D}\mu^*$ can be interpreted as the degree of chaotic motion in the separating zone, and clearly one criterion for the sharpest separation is for this product to be a minimum. With a low agitation intensity the product will be high because the viscosity parameter predominates, whereas a high agitation intensity causes $\mathcal{D}$ to predominate. An intermediate intensity will minimize the product and, as described in Section 13.2.1, this implies in the case of jigging, for example, an optimum combination of amplitude and frequency (and hutch water addition). Indirect measures of $\mathcal{D}$ and μ^* can be obtained by carrying out suitable determinations in the separating zone; the vibration amplitude of a detector ball is proportional to $\mathcal{D}$, and an appropriate viscometer gives a measure of μ^*.[63]

In summary it can be said that reduced performance curves should be capable of predicting the separation that can be achieved on a given material in a given device.[59,64] As yet, there are no correlations for predicting the actual capacity of the device; thus empirical methods[28,54,55,63] or manufacturers' recommendations are necessary.

Example 13.1. What would be the recovery of coal, and its ash content, if the coal described in Example 3.1 were cleaned in a Baum jig at $SG_{50} = 1.5$? (Assume that the coal size range is -150 mm $+300$ μm).

Solution. To determine the quantities of coal and ash recovered, a performance curve must be applied to the separability data. After dividing the mean SG values by SG_{50} (1.5), the Baum jig reduced performance curve can be obtained from Fig. 13.18a (Table E13.1.1, columns 1 to 3). Columns 4 and 6 come from Table E3.1.1, and multiplication of these values by those in column 3 gives the recovery of each component in each SG fraction. The total recovery of coal [Σ column 7] is therefore 75.65%. Since the feed is 28.65% ash and 71.35% coal (Table E3.1.1), the ash content of the cleaned coal is:

$$\left[\frac{15.55 \times 28.65}{(15.55 \times 28.65) + (75.65 \times 71.35)}\right] \times 100 = 7.62\%$$

Thus, because the separator is incapable of a perfect separation, the ash content is raised by more than 50% (cf. 4.73% in Example 3.1).

If the ash had to be below 5%, a number of alternatives could be considered. First a more efficient separator, such as a dense medium vessel or cyclone, could be chosen. Second, one could lower the SG_{50}, which would lower the ash, but would also lower the coal recovery.

In reality, the coal described here is difficult to treat, and to determine the best solution more detailed information is essential. It is likely that particle size is an important parameter. Consider for example the -1.18 mm $+0.3$ mm size fraction. Figure 13.18c shows that for this size fraction,

$$\frac{SG_{50d}}{SG_{50\Sigma d}} = 1.3$$

TABLE E13.1.1: COAL CLEANING EXAMPLE

SG Fractions	Mean SG (1)	Mean SG Sep. SG (2)	Fract. to Outlet** (3)	% Ash in Feed*** (4)	Ash in Low SG Outlet (5)=(3)x(4)	% Coal in Feed*** (6)	Coal in Low SG Outlet (7)=(3)x(6)
Float to 1.30	1.29*	0.860	0.99	3.00	2.97	56.26	55.70
1.30 to 1.35	1.325	0.883	0.96	1.22	1.17	9.32	8.95
1.35 to 1.40	1.375	0.905	0.89	1.01	0.90	2.82	2.51
1.40 to 1.45	1.425	0.917	0.73	1.47	1.07	2.49	1.82
1.45 to 1.50	1.475	0.983	0.62	2.44	1.51	3.08	1.91
1.50 to 1.60	1.55	1.033	0.38	6.32	2.40	5.73	2.18
1.60 to 1.70	1.65	1.100	0.26	8.03	0.21	5.19	1.35
1.70 to 1.80	1.75	1.167	0.16	7.26	1.16	3.25	0.52
1.80 to Sink	2.0*	1.333	0.06	69.26	4.16	11.86	0.71
					15.55		75.65

* Estimated using assays and SG of ash in adjacent fractions
** From Fig. 13.18a
*** From Table E3.1.1

that is

$$SG_{50d} = 1.3 \times 1.5$$
$$\sim 1.95$$

Furthermore, the performance curve for this size fraction is more dispersed (Fig. 13.18b), with the result that most of the ash in this size fraction is being recovered. The ash content is unlikely to be constant with particle size, but it is obvious that either complete elimination or separate treatment of this size fraction would lower the ash content of the cleaned coal.

REFERENCES

1. C. Mills, "Process Design, Scale-Up and Plant Design for Gravity Concentration," Ch. 18 in *Mineral Processing Plant Design*, pp. 404–426, AIME (1978).
2. A. S. Joy et al., "Development of Equipment for the Dry Concentration of Minerals," *Filtr. Sep.*, **9**, 532–544 (September–October 1972).
3. A. M. Gaudin, *Principles of Mineral Dressing*, McGraw-Hill (1939).
4. J. W. Leonard and D. R. Mitchell (Eds.), *Coal Preparation*, 3rd ed., AIME (1968).
   (a) E. R. Palowitch et al., Ch. 9, "Wet Concentration of Coarse Coal."
   (b) M. Sokaski et al., Ch. 10, "Wet Concentration of Fine Coal."
5. R. E. Zimmerman, "Batac Jig," *Min. Congr. J.*, **60**, 43–49 (May 1975).
6. (a) Anon., "I.H.C.–Cleaveland Mineral Dressing Jigs," *Ports Dredging*, **69**, 13–15 (1971).
   (b) Anon., "New High-Capacity Circular Jig Recovers Fine Minerals," *World Min.*, **21**, 56–58 (October 1968).
7. A. G. Moncrieff and P. J. Lewis, "Treatment of Tin Ores" and "Discussion," *Trans. IMM (A)*, **86**, A56–A83 (April 1977).
8. R. L. Terry, "Minerals Concentration by Wet Tabling," *Miner. Proc.*, **15**, 14–19 (July–August 1974).
9. Anon., *Mineral Processing Flowsheets*, 2nd ed., Denver Equipment Co. (1965).
10. E. J. Pryor, *Mineral Processing*, 3rd ed., Elsevier (1965).
11. R. O. Burt and D. J. Ottley, "Fine Gravity Concentration Using the Bartles-Mozley Concentrator," *Int. J. Miner. Process.*, **1**, 347–366 (1974).
12. M. D. Bath, A. J. Duncan, and R. E. Rudolph, "Some Factors Influencing Gold Recovery by Gravity Concentration," *J. S. Afr. IMM*, **73**, 363–384 (1973).
13. R. O. Burt, "Development of the Bartles Crossbelt Concentrator for the Gravity Concentration of Fines," *Int. J. Miner. Process.*, **2**, 219–234 (1975).
14. M. Carta, C. Del Fa, and G. F. Ferrara, "A New Apparatus for Pneumatic Separation," in *Mineral Processing, Proc. 6th Int. Congr., Cannes, 1963*, A. Roberts, Ed., pp. 245–260, Pergamon (1965).
15. E. Douglas and D. L. R. Bailey, "Performance of a Shaken Helicoid as a Gravity Concentrator," *Trans. IMM*, **70**, 637–657 (August 1961).
16. L. D. Muller and J. H. Pownall, "Investigation and Devel-

opment of Some Laboratory Wet Gravity Mineral Concentrators," *Trans. IMM,* **71,** 379–392 (April 1962).

17. A. Thunaes and H. R. Spedden, "An Improved Method of Gravity Concentration in the Fine-Size Range," *Trans. AIME/SME,* **187,** 879–882 (1950).

18. L. Dupret, "Laundering Fine Coal," *Coal Age,* **59,** 92–96 (May 1954).

19. J. T. Woodcock, "Mineral Processing Progress in Australasia 1973," *Aust. Min.,* **66,** 1–20 (August 1974).

20. (a) R. A. Graves, "The Reichert Cone Concentrator; An Australian Innovation," *Min. Congr. J.,* **59,** 24–28 (June 1973).
    (b) T. J. Ferree, "An Expanded Role in Minerals Processing Seen for Reichert Cone," *Min. Eng.,* **25,** 29–31 (March 1973).

21. I. J. Terrill and J. B. Villar, "Elements of High-Capacity Gravity Separation," *CIM Bull.,* **68,** 94–101 (May 1975).

22. Sooi P. Chong, "Gravity Concentration Successfully Treats Iron Ore Fines at Carol Lake," *Min. Eng.,* **30,** 1639–1643 (1978).

23. R. C. Fisher, "New Equipment for Mineral Processing. Part 1," *Aust. Min.,* **59,** 39–43 (Apr. 15, 1967).

24. E. Condolios, G. Hoffnung, and C. Moreau, "Two Hydraulic Machines for Gravity Concentration of Ore," in *Mineral Processing, Proc. 6th Int. Congr., Cannes, 1963,* A. Roberts (Ed.), pp. 261–281, Pergamon (1965).

25. A. F. Taggart, *Elements of Ore Dressing,* Wiley (1951).

26. F. W. Mayer, "Fundamentals of a Potential Theory of the Jigging Process," *Proc. 7th Int. Miner. Process. Congr., New York, 1964,* Vol. I, pp. 75–97, Gordon and Breach (1965).

27. V. I. Revnivtsev et al., "Hydrodynamic Research into Gravity Concentration Processes and Methods for Their Improvement," *Proc. 10th Int. Miner. Process. Congr., London, 1973,* pp. 293–310, IMM (1974).

28. "Gravity Separation Technology Short Course," sponsored by the Mackay School of Mines, University of Nevada, Reno, October 1978.
    (a) C. Mills, "The Theory of Gravity Separation."
    (b) T. H. Nio, "Mineral Jigs."
    (c) C. Tiernon, "Wet Shaking Tables."
    (d) R. C. Hansen, "Spiral Concentration."
    (e) T. J. Ferree and I. J. Terrill, "The Reichert Cone Concentrator—An Update."
    (f) C. Mills and R. Burt, "Thin Film Concentrating Devices and the Bartles-Mozley Concentrator in Particular."

29. D. G. Armstrong, "Variations in Stroke Waveform in a Laboratory Jig," *Trans. IMM,* **73,** 643–662 (June 1964).

30. R. M. Horsley and P. F. Whelan, "The Mechanism of Coal Cleaning in Jigs—A Photographic Study," *J. Inst. Fuel,* **28,** 586–593 (December 1955).

31. D. J. Batzer, "Investigation into Jig Performance," *Trans. IMM,* **72,** 61–68 (1962–1963).

32. G. D. Lill and H. G. Smith, "A Study of the Motion of Particles in a Jig Bed," *Proc. 5th Int. Miner. Process. Congr., London, 1960,* pp. 515–535, IMM (1960).

33. G. A. Vissac, "Coal Preparation with the Modern Feldspar Jig," *Min. Eng. (Trans. AIME/SME),* **7,** 649–655 (July 1955).

34. G. F. Everson and S. C. Singhal, "The Stratification of Two Closely Sized Coal Fractions in a Baum-Type Jig," *J. Inst. Fuel,* **32,** 210–218 (May 1959).

35. F. B. Michell and P. Suvarnapradip, "A Study of Jigging as Applied to the Concentration of Alluvial Material . . . ," in *Mineral Processing: Proc. 6th Int. Congr., Cannes, 1963,* A. Roberts (Ed.), pp. 283–301, Pergamon (1965).

36. R. L. Whitmore, "Principles of Jig Washing: An Experimental Approach," *J. Inst. Fuel,* **31,** 3–11 (January 1958).

37. C. W. J. Van Koppen, "A Contribution to the Fundamentals of the Jigging Process," Paper B3, *5th Int. Coal Prep. Congr., Pittsburgh, 1966,* pp. 85–97 (1966).

38. N. N. Vinogradov et al., "Research on the Separation Kinetics of Gravity Processing in Mineral Suspensions," *Proc. 11th Int. Miner. Process. Congr., Cagliari, April 1975,* pp. 319–336, Università di Cagliari (1975).

39. H. Kirchberg and W. Hentzschel, "A Study of the Behaviour of Particles in Jigging," *Int. Miner. Dress. Congr., Stockholm, 1957,* pp. 193–215, Almqvist & Wiksell (1958).

40. E. Hoffman, Discussion in Ref. 32, p. 623 (1960).

41. R. C. M. van der Spuy, Discussion on Ref. 31, *Trans IMM,* **72,** p. 371 (1962–1963).

42. K. Ijuin, S. Takamura, and S. Omori, "Washing Fine Coal by Air-Pulsating Jig," *3rd Int. Coal Prep. Congr., Brussels, 1958,* pp. 443–449, Ann. Mines Belg. (1958).

43. G. F. Eveson and S. C. Singhal, "A Simplified Analysis of the Energies Involved in Jig Washing," *J. Inst. Fuel,* **32,** 398–408 (September 1959).

44. B. M. Bird and H. S. Davis, "The Role of Stratification in the Separation of Coal and Refuse on a Coal-Washing Table," U.S. Bureau of Mines Report of Investigations, RI2950 (1929).

45. H. Kirchberg and W. Berger, "Study of the Operation of Shaking Concentration Tables," Paper 25, *Int. Miner. Process. Congr., London, 1960,* pp. 537–551, IMM (1960).

46. C. R. Burch, Written Contributions to Discussion on Ref. 15, *Trans. IMM,* **71,** 406–415 (1961–1962).

47. J. J. Carty, Jr., "Resistance Coefficients for Spheres on a Plane Boundary," B.S. thesis, Massachusetts Institute of Technology (1957).

48. (a) H. Schlichting, *Boundary Layer Theory,* Ch. 21 (translated by J. Kestin), McGraw-Hill (1960).
    (b) H. Schlichting, "Experimentelles Untersuchungen zum Rauhigheitsproblem," *Ing.-Arch.,* **7,** 1–34 (1936); NACA TM 823 (1937).

49. R. A. Bagnold, "Experiments on a Gravity-Free Dispersion of Large Solid Spheres in a Newtonian Fluid Under Shear," *Proc. R. Soc., Sec. A,* **225,** 49–63 (1954).

50. R. A. Bagnold, "The Flow of Cohesionless Grains in Fluids," *Phil. Trans R. Soc., Ser. A,* **249,** 235–297 (1956).

51. R. O. Burt, "A Study of the Effect of Deck Surface and Pulp pH on the Performance of a Fine Gravity Concentrator," *Int. J. Miner. Process.,* **5,** 39–44 (1978).

52. S. R. Rao and L. L. Sirois, "Study of Surface Chemical Characteristics in Gravity Separation," *CIM Bull.,* **67,** 78–83 (June 1974).

References

53. L. D. Muller, C. P. Sayles, and R. H. Mozley, "A Pulsed Deck Gravity Concentrator and Comparative Performance Analysis," *Proc. 7th Int. Miner. Process. Congr., New York, 1964,* Vol. 1, pp. 63–74, Gordon and Breach (1965).
54. A. B. Holland-Batt, "Design of Gravity Concentration Circuits by Use of Empirical Mathematical Models," *Proc. 11th Commonw. Min. Met. Congr., Hong Kong, May 1978,* pp. 133–143, IMM (1979).
55. R. Dallaire, A. Laplante, and J. Elbrond, "Humphrey's Spiral Tolerance to Feed Variations," *CIM Bull.*, **71**, 128–134 (August 1978).
56. S. Abdinegoro and A. C. Partridge, "Flow Characteristics of a Pinched Sluice," *Proc. Australasian IMM Conf., Western Australia, August 1979,* pp. 79–83, Australasian IMM (1979).
57. J. Visman, "Bulk Processing of Fine Materials by Compound Water Cyclones," *CIM Bull.*, **59**, 333–346 (1966).
58. V. G. Renner and H. E. Cohen, "Measurement and Interpretation of Size Distribution of Particles Within a Hydrocyclone," *Trans. IMM (C)*, **87**, C139–C145 (June 1978).
59. (a) B. S. Gottfried and P. S. Jacobsen, "Generalized Distribution Curve for Characterizing the Performance of Coal-Cleaning Equipment," U.S. Bureau of Mines Report of Investigations, RI8238 (1977).
    (b) B. S. Gottfried, "A Generalisation of Distribution Data for Characterising the Performance of Float-Sink Coal Cleaning Devices," *Int. J. Miner. Process.*, **5**, 1–20 (1978).
60. F. Armstrong and W. M. Wallace, "Diagrammatic Representation of Jig Washing," *2nd Symp. Coal Prep., University of Leeds, 1957,* pp. 417–434, University of Leeds (1957).
61. M. Johnston, "The Performance Characteristics of the Reichert Cone Concentrator," Unpublished B. E. thesis, University of Queensland (1977).
62. C. C. Dell, "An Application of Mayer Curve Methods to the Study of Stratification," *3rd Int. Coal Prep. Congr., Brussels, 1958,* Paper F3, pp. 599–606, Ann. Mines Belg. (1958).
63. O. N. Tikhonov and V. V. Dembovsky, *Automation Process Control in Ore Treatment and Metallurgy. Part IV. Industrial Measuring and Control Systems,* Ch. 11, El-Tabbin Metallurgical Institute for Higher Studies (1973).
64. M. R. Geer, "How to Predict Results of Washing a New Coal," *Coal Age,* **57**, 96–97 (June 1952).

Chapter Fourteen

Magnetic Separation

Murphy's Law: "If anything can go wrong it will."

O'Toole's Commentary on Murphy's Law:
"Murphy was an optimist."

Magnetic separation has been used for almost 200 years in the concentration of iron ores, and is still used extensively today. A wide variety of devices have been used.[1] Since the start of this century there has been a steady expansion in both the equipment available and the range of ores for which magnetic separation is applicable.[2] The removal of small amounts of iron and iron-bearing minerals from industrial minerals has become an important application, as has the concentration of various ferrous and nonferrous minerals.

The property of a material that determines its response to a magnetic field is the *magnetic susceptibility*. Based on magnetic susceptibility, materials may be divided into two groups: *paramagnetic* materials, or those attracted by a magnetic field, and *diamagnetic* materials, those repelled by a magnetic field. A common practice (followed here) is to place very strongly paramagnetic materials in a separate category called *ferromagnetic*. Thus iron and magnetite are ferromagnetic materials; hematite, ilmenite, pyrrhotite, and many other minerals are paramagnetic; quartz and feldspar are among the diamagnetic minerals. The minerals that are separated using their magnetic properties are listed in Appendix F.

Before describing the principles underlying the operation of magnetic separators, the various types of equipment are discussed, together with their operating characteristics and applications.

14.1. EQUIPMENT AND APPLICATIONS

Magnetic separation equipment falls readily into two categories: *low intensity* and *high intensity* magnetic separators, the former being used primarily for ferromagnetic minerals but also for paramagnetic minerals of high magnetic susceptibility, and the latter for paramagnetic minerals of lower magnetic susceptibility. Both low and high intensity magnetic separations may be carried out either wet or dry. Wet processing predominates in low intensity magnetic operations, although large tonnage dry plants do exist. High intensity magnetic separators have traditionally been dry and of low capacity. Following advances in magnet design, very large wet high intensity magnetic separators have been introduced, and successfully applied in the concentration of paramagnetic minerals.

Table 14.1 lists the equipment available for magnetic separation and gives a brief explanation of its use. As well as low and high intensity separators, this table also includes as a third category, the magnetic separators used for tramp iron removal and coarse cobbing. Further details of magnetic separators are available in the literature[3,4] and from manufacturers' catalogues.

14.1.1. Tramp Removal and Cobbing

Tramp iron removal is used to protect mineral processing equipment such as crushers, screens, and conveyors. The equipment selected for removal of this tramp iron depends on many factors, including the size and amount of tramp iron occurring and the point at which protection is required. The basic types of equipment are shown in Fig. 14.1.

Coarse, strongly magnetic materials can be recovered using equipment similar to that used for tramp removal. Thus the dry *cobbing* (coarse rough-

TABLE 14.1: MAGNETIC SEPARATORS

	Separator Type	Description	Feed Rate (m³/min/m)	Field Strength (T at 5 cm)	-Size, mm -Speed	Applications
TRAMP REMOVAL & COBBING	Suspended Magnets (Fig. 14.1)	Fixed position electromagnet over conveyor belt or head pulley. Continuous removal possible.				Protection of crushers and other process equipment by removal of tramp iron. Also used to remove iron from foundry sands, recover iron from slags.
	Plate and Grate Magnets	Magnet below chute (plate); series of parallel magnetized bars (grate). Periodic cleaning of magnet necessary.				Tramp iron removal. Must be free flowing material.
	Magnetic Pulley	Magnets located within head pulley of conveyor belt.			150 m/min max belt speed	As for suspended magnets.
	Cobbing Drum (Fig. 14.1c)	Wide range of design, basically as shown in Fig. 14.1c. Operated dry.			300-916 (dia)	Coarse magnetic cobbing, tramp iron removal, iron recovery from slags, cleaning of scrap.
WET LOW INTENSITY	Concurrent (Fig. 14.3)	Feed slurry passes through trough in same direction as rotation of drum. Magnetic particles attracted to rotating surface by fixed magnets in drum. Non-magnetics sink to lowest point in trough and are drawn off. Magnetic concentrate carried by rotating drum through gap where compressed and dewatered before passing over weir. Agitation during long path through tank gives very good washing action and hence very clean concentrate.	50-350	0.06 to 0.07	760-1200 (dia) 1525-3050 (width)	High grade magnetite concentrate from relatively coarse ore; used on product from rod mill in taconite circuit. Widely used in heavy media recovery systems. Tailing often retreated in counter-rotation unit because of high remaining magnetics. Double or triple tank units (series) commonly used.
	Counter-Rotation (Fig. 14.3)	Feed slurry passed through trough in opposite direction to drum rotation. Magnetic particles picked up by drum and discharged almost immediately. Produces high recovery but low grade concentrate since little chance for trapped non-magnetics to be washed out. Recovery high as particles can be picked up further downstream by drum and returned as concentrate.	50-250	0.5-0.6 m/s	760-1200 (dia) 1525-3050 (width)	Good where surges in feed must be handled, where magnetic loses must be minimized, and where high grade is not required. Often used for retreating tailings from a concurrent drum such as in heavy media recovery where one must get maximum recovery of magnetite or ferro-silicon. With taconites, used on ball mill product. Double tanks often used.
	Counter-Current (Fig. 14.3)	Feed slurry introduced to about mid-point of magnetic section, so includes features of both concurrent and counter-rotation. Wash water increases agitation of magnetics to remove entrained particles and give clean concentrate. Tailings pass by magnet opposite to rotation to increase recovery.	20-250	0.05 to 0.06	760-1200 (dia) 1525-3050 (width)	Often used as finishing separator. Gives good recovery and extremely clean concentrate.
DRY LOW INTENSITY	High Speed Drum	High-speed separator with large number of permanent magnets (up to 44) placed in stationary 210° yoke within rotating drum (poles alternating). Non-magnetic thrown off by centrifugal force. Magnetics attracted to drum until out of field region. Agitation due to large number of poles gives clean product.	0.05-0.45	0.04 to 0.05	400-916 (dia) 300-3000 (width) 1-9 m/s	Used to concentrate magnetite when water is not readily available. Incorporated in circuits with dry grinding. May be used with recycle of middlings.
	Ball-Norton Type	Uses higher intensity magnetic field than high speed type but has only 8-10 alternating poles.	0.13-0.5		to 760 (dia) to 1525 (width)	Concentration of coarse magnetite in dry flowsheet. Gives high grade concentrate.
WET HIGH INTENSITY	Carousel Type (Fig. 14.6)	Carousel consists of annular box containing a ferro-magnetic matrix. Matrix can be steel spheres, grooved plates, expanded steel, or steel wool. Carousel rotates through high intensity magnetic field which induces high gradient magnetic field in matrix. Feed enters carousel in magnetic field where magnetics are retained; rotation carries magnetics out of field where they are flushed into a launder.	0.01-1.0 (t/min)	<2.0		Used for paramagnetic minerals. Concentration of hematite and chromite. Used in removing small amounts of iron bearing minerals from china clay and concentrates of cassiterite, scheelite, and ilmenite. Potential use in removal of pyrite from coal.
	Canister Type (Fig. 14.7)	Fixed steel wool or expanded steel matrix within magnetic field. Feed pulp followed by wash water with magnet on; flush magnetics with magnet off.	0.07-0.3 (t/min)	2.0	2130 (dia)	Primarily used in removal of fine iron minerals from china clay; similar uses to carousel type.
DRY HIGH INTENSITY	Induced Roll (Fig. 14.8)	Consists of a series of revolving laminated rolls formed of alternate magnetic and non-magnetic discs. Rolls are magnetized by induction from high intensity stationary electromagnet. Non-magnetics in natural trajectory from centrifugal action. Magnetics deflected by field. General practice is to have three successive rolls of increasing magnetic field strength, achieved by reducing gap. Sized feed desirable.	0.01-0.10	<2.1	250-1000 (dia)	Concentration of dry paramagnetic materials. Widely used in beach sands industry (see Appendix). Also used with wolframite, monazite, cassiterite, and in cleaning silica sand and feldspar.
	Cross Belt	Electromagnet with poles separated by belts. Magnet lifts magnetics from thin layer on feed belt to cross belt against upper pole. Upper pole comes to sharp edge, lower pole is flat; this gives high field gradient in gap.	0.02	2.0	450 (width)	Similar use to induced roll, but limited to high value minerals such as cassiterite, columbite, monazite, and tungsten minerals. Can be used to simultaneously separate a number of minerals with a range of susceptibilities. Used where extremely selective separation required.

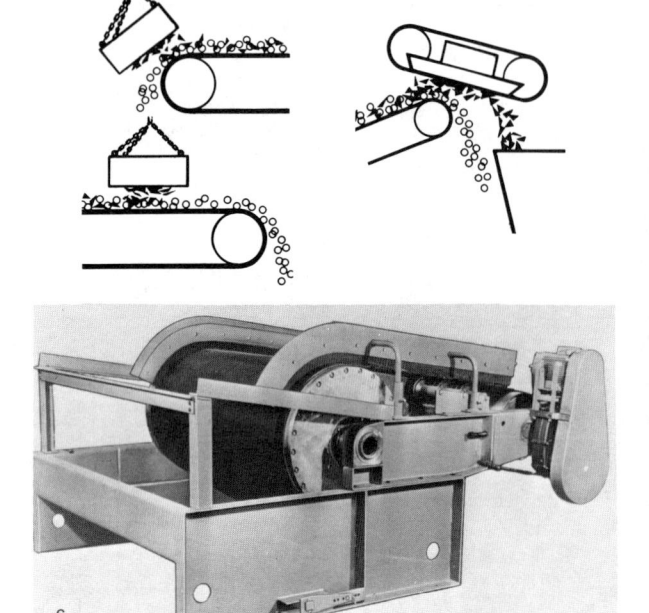

Figure 14.1. Equipment for removal of tramp iron and for magnetic cobbing. (*a*) Suspended magnets. (*b*) Suspended magnet, continuous removal. (*c*) Cobbing drum. (Courtesy Stearns Magnetics Inc.)

Figure 14.2. A wet drum low intensity magnetic separator. (Courtesy Eriez Magnetics).

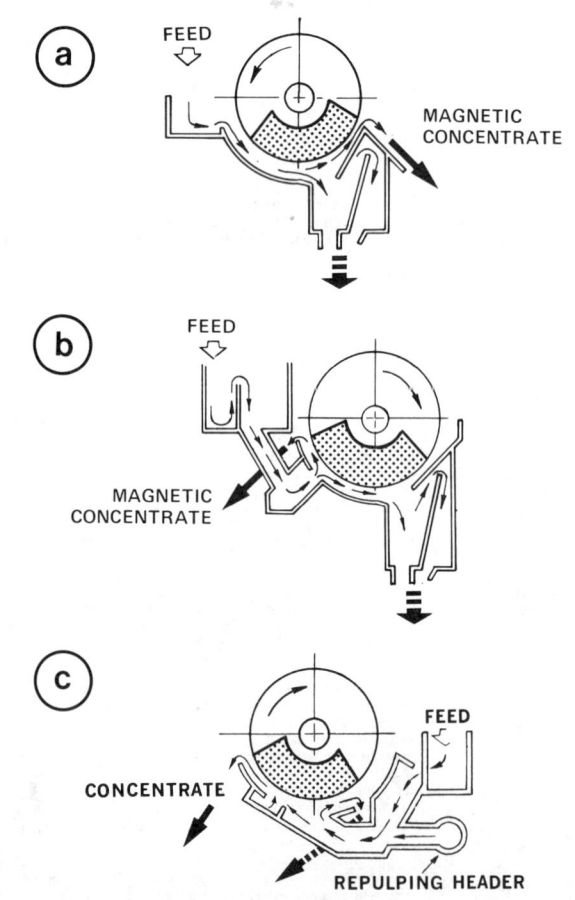

Figure 14.3. Wet drum low intensity magnetic separator tank designs. (*a*) Concurrent. (*b*) Counterrotation. (*c*) Countercurrent. (Courtesy Eriez Magnetics.)

ing) of magnetite, the recovery of iron values from blast furnace and steel making slags, and the cleaning of nonmagnetics from steel scrap all use similar equipment. In the cobbing of magnetite from iron ores, the aim is to *eliminate some of the gangue* mineral at a large particle size. This requires that the gangue mineral be liberated at a coarse size; the magnetic particles clearly contain unliberated gangue, hence vary widely in magnetic properties.

14.1.2. Low Intensity Magnetic Separators

Some low intensity magnetic separators are described in Table 14.1. The technical literature includes details of both wet[5-10] and dry[11,12] low intensity separators and their applications. Wet separators of the type shown in Figs. 14.2–14.4 are the most common. They employ either electromagnets or permanent magnets; electromagnets are primarily used where relatively high magnetic field strengths are required, but in general, permanent magnets of the Alnico or ceramic barium–strontium ferrite type are most widely used. Most drums have five poles symmetrical about the centerline of the magnetic yoke, with the poles alternating, since this

Equipment and Applications

Figure 14.4. Banks of wet drum low intensity magnetic separators consisting of three drums in series. (Courtesy Sala International AB.)

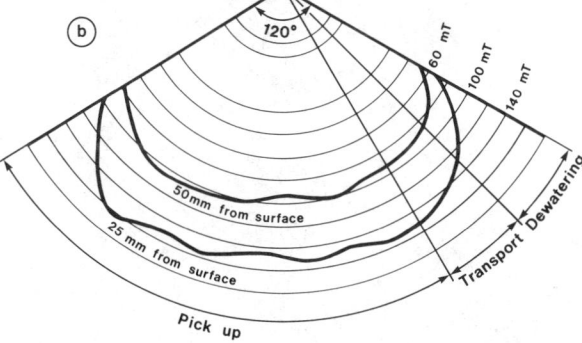

Figure 14.5. Magnetic field strength patterns as designed for wet drum separators. (a) Concurrent and countercurrent. (b) Counterrotation. (Courtesy Sala International AB.)

improves the agitation at the drum surface and so reduces the entrapment of nonmagnetics.

Traditionally the method of specification of a wet drum separator has been the measure of the magnetic field strength 5 cm from the drum surface, but the trend is to the more useful description of the magnetic field strength and gradient as shown in Fig. 14.5.[6,7,13,14]

14.1.3. High Intensity Magnetic Separators

Wet Separators. The operating components of a wet high intensity separator of the carousel type and

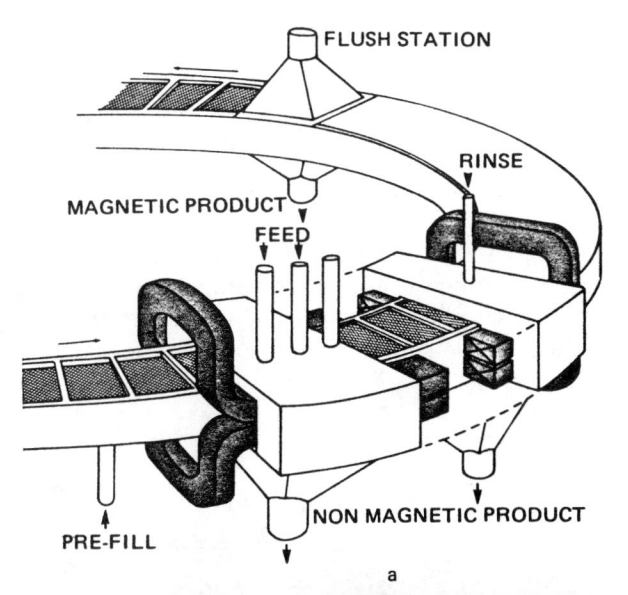

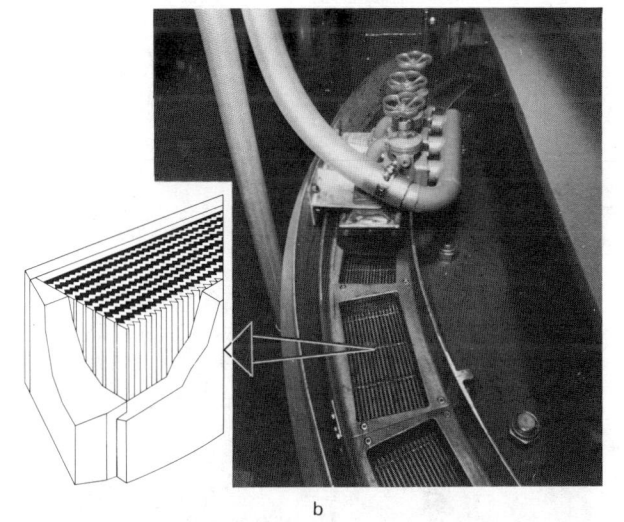

Figure 14.6. Carousel-type wet high intensity magnetic separator. (a) Operating components. (b) Photograph showing grooved plate ferromagnetic matrix. (Courtesy KHD Humboldt Wedag AG.)

the basic concept of the canister-type wet high intensity separator are shown in Figs. 14.6 and 14.7, respectively. The ferromagnetic matrix on which paramagnetic particles are collected may be in the form of grooved plates (Fig. 14.6), steel balls, steel wool, or sheets of expanded metal. Further general literature on wet high intensity magnetic separators is available,[15–17] as are applications to hematite,[18–20] industrial minerals,[21,22] and the removal of pyritic sulfur from coal.[22,23] It should be noted that the term "high gradient magnetic separation" is also widely used, as well as the acronyms WHIMS and HGMS.

Dry Separators. Figure 14.8 gives a schematic of an induced-roll dry magnetic separator such as the one shown in Fig. 14.9. More detailed descriptions of dry magnetic separators are available in the literature.[3,4,24]

14.1.4. Laboratory Test Devices

The Davis tube[25] (Fig. 14.10) is a laboratory separator commonly used to determine the fraction of strongly magnetic material (usually magnetite) present. Although normally operated wet, it can be operated dry.

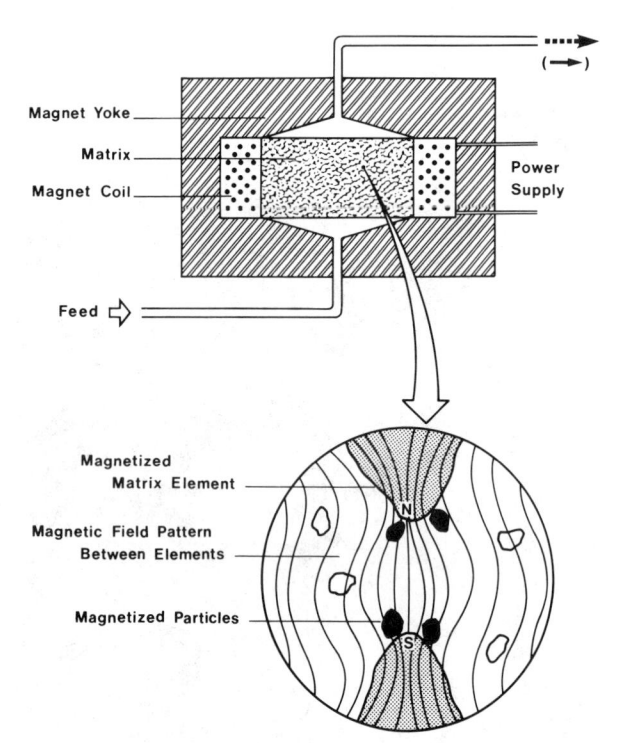

Figure 14.7. Schematic illustration of a canister-type high intensity magnetic separator.

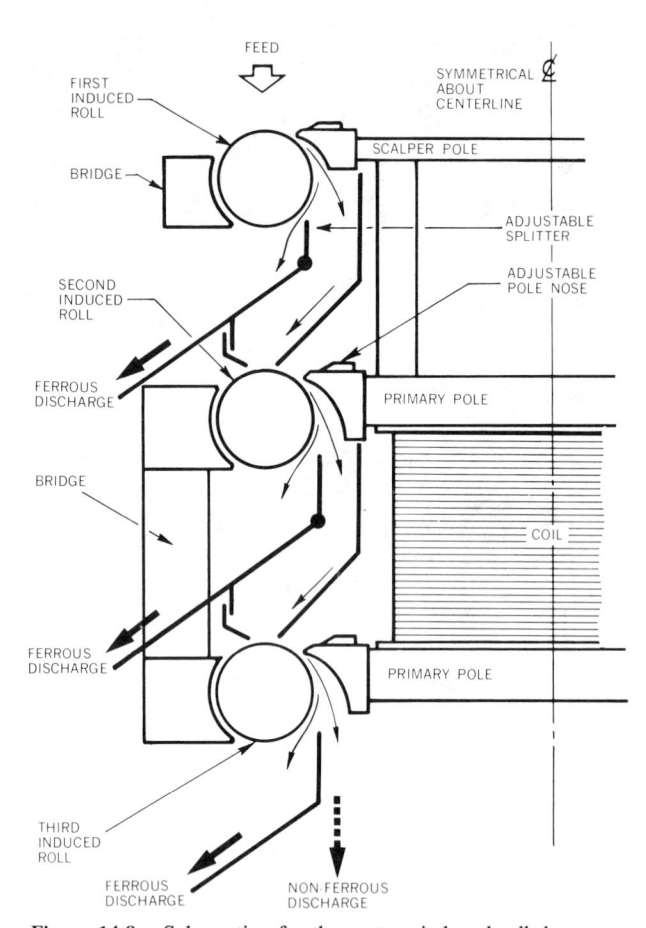

Figure 14.8. Schematic of a three-stage induced-roll dry magnetic separator. (Courtesy Eriez Magnetics.)

Figure 14.9. A three-stage induced-roll dry magnetic separator. (Courtesy Eriez Magnetics.)

Equipment and Applications

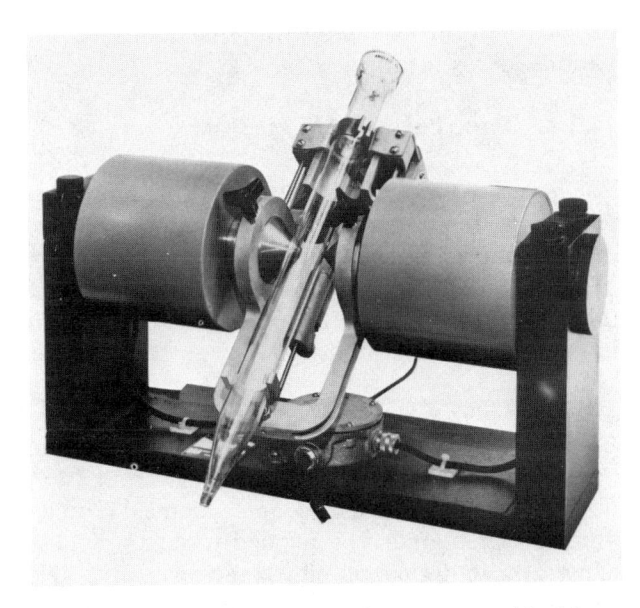

Figure 14.10. A Davis tube magnetic separator used for laboratory testing. (Courtesy Eriez Magnetics.)

The Frantz Isodynamic separator[26] (Fig. 14.11) is a device for fractionating weakly magnetic minerals.

14.1.5. Magnetic Flocculation and Demagnetizing

Magnetic flocculation[27,28] of fine ferromagnetic particles can be brought about by passing a slurry between two magnetic poles (normally permanent magnets), called magnetizing "blocks." Magnetic flocculation is used in the concentration of magnetite and in the cleaning of steel plant waste water (by increasing both the thickener settling rate and the filtration rate).

As ferromagnetic particles pass through a magnetic separator, there is a tendency for some magnetic flocculation to occur. This uncontrolled flocculation can reduce the efficiency of subsequent separations. Demagnetizing coils (Fig. 14.12) are used for the depolarization of ferromagnetic slurries. Such coils utilize ac current to generate an oscillating magnetic field with a diminishing strength as shown in Fig. 14.13.[27,28]

14.1.6. Potential Magnetic Separation Methods

Major advances in magnetic separation are anticipated when extremely high intensity magnets utilizing cryogenic superconductors are introduced.[29,30]

Extensive test work has been reported[31] on the application of magnetohydrodynamic and magnetohydrostatic methods to mineral separation. These methods functionally combine simultaneous separation by density, magnetic susceptibility, and electric conductivity.

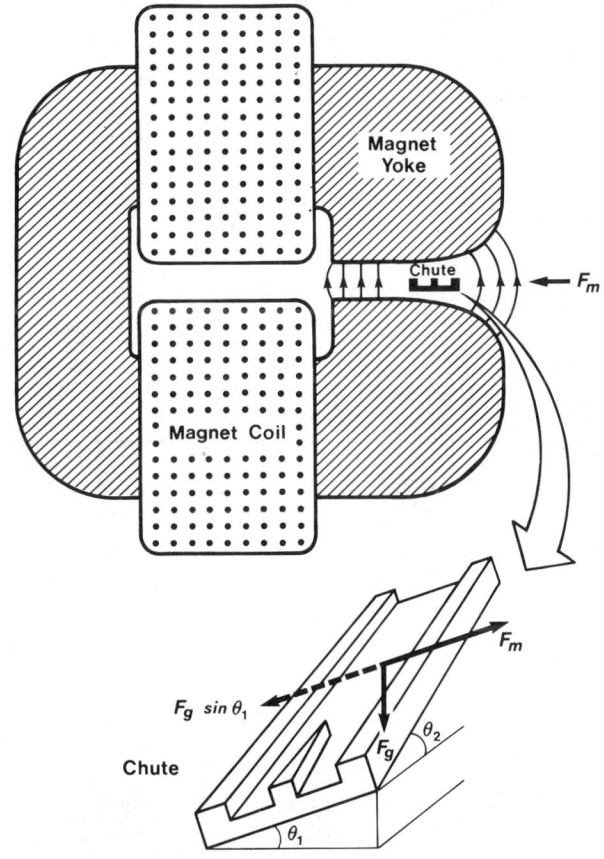

Figure 14.11. Schematic of a Frantz isodynamic separator for laboratory testing of paramagnetic materials.

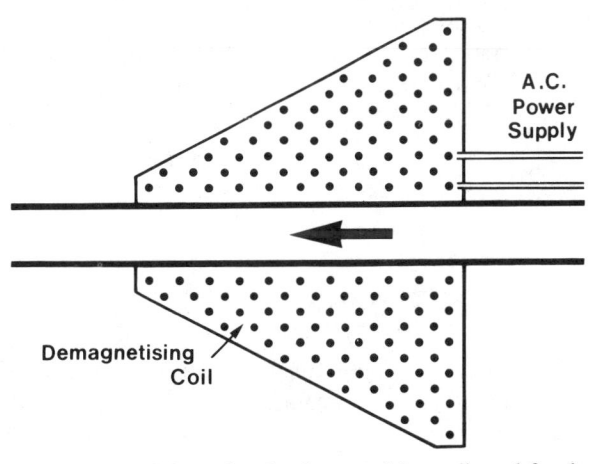

Figure 14.12. Schematic of a demagnetizing coil used for deflocculating magnetic slurries.

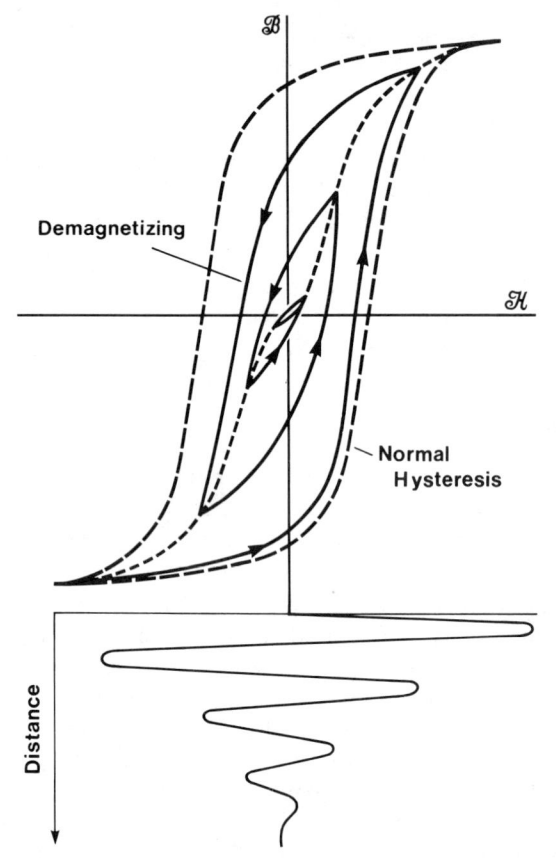

Figure 14.13. Magnetic field reduction occurring with distance through demagnetizing coil. (After Lantto.[28])

14.2. PRINCIPLES AND MECHANISMS OF MAGNETIC SEPARATION

The principles of magnetic separation, which form the basis of magnetic separator design and operation, are described in this section.

Magnetic separation is a physical separation of discrete particles based on a three-way competition between:

1. Magnetic forces.
2. Gravitational, centrifugal, frictional, or inertial forces.
3. Attractive or repulsive interparticle forces.

These magnetic, competing, and interparticle forces determine separator performance; they are dependent on both the nature of the feed and the character of the separator. The nature of the feed includes its size distribution, magnetic susceptibility, and other physical and chemical properties that may affect the various forces involved.

14.2.1. Basic Physics of Magnetism

The basics of magnetism are discussed in numerous physics textbooks and references works. A brief discussion only is given here.

When an object is placed in a magnetic field (*applied magnetic field* $\mathcal{H}$), the *magnetic induction* $\mathcal{B}$ in the object is given by

$$\mathcal{B} = \mathcal{H} + \mathcal{M} \quad (14.1)$$

($\mathcal{M}$ = *magnetization* or intensity of magnetization of the material of the object). The magnetization can be interpreted as a magnetic dipole moment per unit volume of material. The applied magnetic field is expressed in terms of either the magnetic field strength with units of amperes per meter (A/m), or the magnetic flux density in teslas (T). (The relationship[32] is 1 A/m = $4\pi \times 10^{-7}$ T.) The magnetization is normally expressed in amperes per meter, while the magnetic induction is normally expressed in teslas. Obviously, from Eq. 14.1, $\mathcal{H}$, $\mathcal{M}$, and $\mathcal{B}$ can all be expressed either in amperes per meter or in teslas, but for the equation to be valid, all must be the same.

Another quantity used in magnetism is the *magnetic flux*, expressed in webers (Wb). The weber is the magnetic flux which, linking a circuit of one turn, produces in it an electromotive force of one volt as it is reduced to zero at a uniform rate in one second. The tesla, then, is the magnetic flux density given by a magnetic flux of one weber per square meter.[32]

The relationship between the magnetization and the applied magnetic field is shown in Fig. 14.14[15] for typical ferromagnetic, paramagnetic, and diamagnetic minerals. A simple linear relationship exists for paramagnetic and diamagnetic materials. Such a relationship does not suffice for ferromagnetic materials; their behavior is much more complicated (Fig. 14.14a).

The slope of the magnetization curve is the *magnetic susceptibility* κ of the material, that is

$$\kappa = \frac{\mathcal{M}}{\mathcal{H}} \quad (14.2)$$

The magnetic susceptibility is a small positive constant for paramagnetic materials, a very small negative constant for diamagnetic materials, and for ferromagnetic materials it is variable and depends on

Principles and Mechanisms of Magnetic Separation

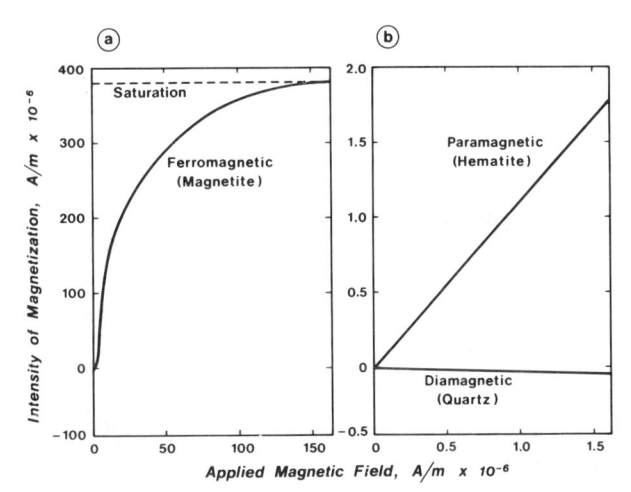

Figure 14.14. Typical magnetization curves. (*a*) Ferromagnetic minerals (magnetite). (*b*) Paramagnetic (hematite) and diamagnetic (quartz) minerals. (After Lawver and Hopstock.[15])

the applied magnetic field and the previous history of the sample (Fig. 14.14*a*).

In an applied magnetic field, the magnetization of ferromagnetic materials changes until the *saturation magnetization* is reached. If the applied field is reduced, the magnetization decreases but does not come back to its original value. Such an irreversible process of magnetization is called *hysteresis*. The residual magnetization of the material is called its magnetic *remanence* (Fig. 14.13 shows a hysteresis curve, and the way in which the remanence of ferromagnetic particles is reduced and ultimately eliminated).

The important variables that determine the magnetic force acting on a particle can be illustrated by reviewing an elementary experiment in magnetism.[15,33] The behavior of common mineral particles in a high intensity magnetic field is examined. To perform the experiment two devices are used: one to produce a magnetic field and one to measure the magnetic force on each particle. For the purpose of this illustration, a magnetic field is produced by the laboratory solenoid shown in Fig. 14.15. Also shown is a plot of the magnetic field along the axis of the solenoid when it is supplied with the maximum allowable steady state electric power. Consider the case where the magnetic field is 3 T at the center of the coil and 1.8 T just inside the ends of the coil. If the current through the solenoid is reduced to one-tenth of the maximum, the magnetic field at all points is proportionately reduced by one-tenth, and the power consumption is reduced to one-hundredth

of the maximum. Individual samples of quartz, hematite, and magnetite are now suspended from an analytical balance at various positions along the axis of the solenoid.

When the current is switched on, a change occurs in the apparent masses of the sample and sample holder, respectively. The change in the apparent mass in grams is equal to the magnetic force in gram-force units. To obtain the net magnetic force on the sample, the force on the empty holder must be measured and subtracted. It is found that the magnetic force is strongest not in the center of the coil where the field is strongest, but near the ends of the coil. In fact, at the center of the solenoid, the force is very close to zero. From examination of Fig. 14.15, it can be seen that the force is largest where the gradient (i.e., the rate of change of the magnetic field) is largest, and zero where the gradient is zero (as at the center of the solenoid). At maximum power, the gradient near the end of the solenoid is about 0.17 T/cm. When the sample is in the upper half of the solenoid, the apparent masses of hematite and magnetite increase when the magnetic field is turned on, while the apparent mass of quartz decreases. If the sample is in the lower half of the solenoid, the apparent masses of hematite and magnetite decrease, while that of quartz increases. This means that the magnetic force on

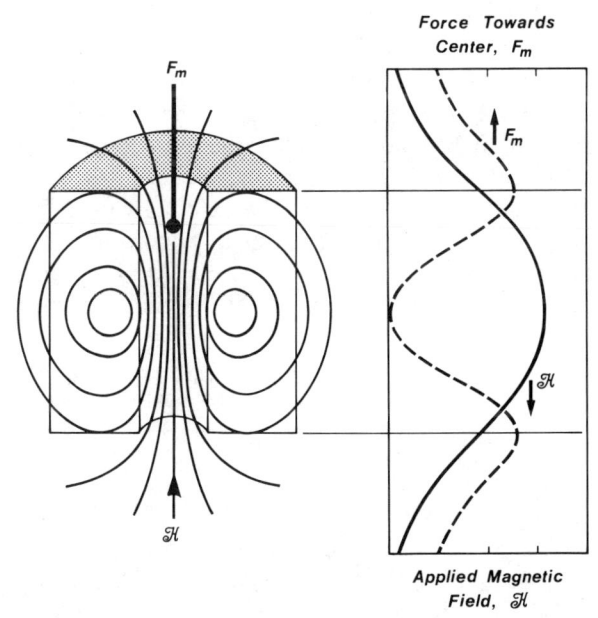

Figure 14.15. Experimental arrangement for measuring force in a magnetic field. Magnetic field strength and resulting force are shown. (After Lawver and Hopstock.[15])

hematite and magnetite is directed toward the center of the solenoid (the region of most intense magnetic field), while the force on the quartz is directed toward the region of weaker magnetic field. For all three minerals tested, the magnetic force is directly proportional to the mass or volume of the sample (provided the dimensions of the samples are small compared to the dimensions of the solenoid).

If the magnetic force on a 1 g sample placed just inside the upper edge of the coil is now measured, the results for the three minerals are given in Table 14.2. (A force is taken to be positive if it acts toward the region of more intense field; the forces are expressed in grams to provide a comparison with the weight of the sample.) Note that the magnetic force on hematite and quartz increases as the square of the magnetic field or field gradient, while the force on the magnetite increases only in proportion to the field strength. For quartz (a typical diamagnetic material), even with a field strength of 1.8 T, the magnetic force is a small fraction of the gravitational force. Hematite, on the other hand, is a paramagnetic substance. At a field strength of 0.18 T, the magnetic force is negligible compared to the sample weight, but when the field is increased to 1.8 T, the magnetic force approaches the order of magnitude of the gravitational force. For magnetite, the magnetic force is many times the sample weight even with the low magnetic field.

14.2.2. The Magnetic Force

It is difficult to determine the magnetic force on a particle. However, some simplifying assumptions can usually be made that allow the force to be calculated with sufficient accuracy in specific cases. If the particle is small enough for the externally produced magnetic field to be approximately uniform within the particle, the particle can be considered to be a point magnetic dipole located at the center of mass of the particle. The magnetic force on a particle in an applied magnetic field is then given by the vector equation

$$\vec{F}_m = (V\vec{\mathcal{M}} \cdot \nabla)\vec{\mathcal{B}} \qquad (14.3)$$

($V\vec{\mathcal{M}}$ = magnetic moment of the particle of volume V; $\vec{\mathcal{B}}$ = externally produced magnetic field or the magnetic induction at the center of mass of the particle).

From Eq. 14.2, the magnetization will be equal to $(\kappa_s - \kappa_m)\mathcal{H}$, with κ_s and κ_m being the magnetic susceptibilities of the particle and the carrying medium, respectively. Thus the vector equation (Eq. 14.3) can be rewritten in Cartesian coordinates as

$$F_{m,x} = V(\kappa_s - \kappa_m)\left(\mathcal{H}_x \frac{\partial \mathcal{B}_x}{\partial x} + \mathcal{H}_y \frac{\partial \mathcal{B}_y}{\partial x} + \mathcal{H}_z \frac{\partial \mathcal{B}_z}{\partial x}\right) \qquad (14.4)$$

for the magnetic force in the x-direction, with similar equations for the y- and z-directions. It is generally more useful to write Eq. 14.3 in terms of the radial coordinates. Thus

$$F_{m,r} = V(\kappa_s - \kappa_m)\left(\mathcal{H}_r \frac{\partial \mathcal{B}_r}{\partial r} + \mathcal{H}_\theta \frac{\partial \mathcal{B}_\theta}{\partial r}\right) \qquad (14.5)$$

$$F_{m,\theta} = \frac{V}{r}(\kappa_s - \kappa_m)\left(\mathcal{H}_r \frac{\partial \mathcal{B}_r}{\partial \theta} + \mathcal{H}_\theta \frac{\partial \mathcal{B}_\theta}{\partial \theta}\right) \qquad (14.6)$$

(r, θ, as defined in Fig. 14.16).

From Fig. 14.14 it can be seen that for paramagnetic and diamagnetic materials the magnetization is small compared to the applied magnetic field. This is not the case with ferromagnetic materials. Thus, $\mathcal{B}$ may be considered equal to $\mathcal{H}$ except when ferromagnetic particles are present. It is common in the literature to find the equations

TABLE 14.2: MAGNETIC FORCE ON 1 g SAMPLE SUSPENDED ALONG AXIS OF SOLENOID

	Test 1	Test 2
Applied Magnetic Field, $\mathcal{H}$ (T)	0.18	1.8
Magnetic Field Gradient, $\frac{\partial \mathcal{H}}{\partial x}$ (T/cm)	-0.017	-0.17
Magnetic Force, F_m (N)		
Quartz	0.015×10⁻⁴	1.51×10⁻⁴
Hematite	0.631×10⁻⁴	63.1×10⁻⁴
Magnetite	0.155	1.55

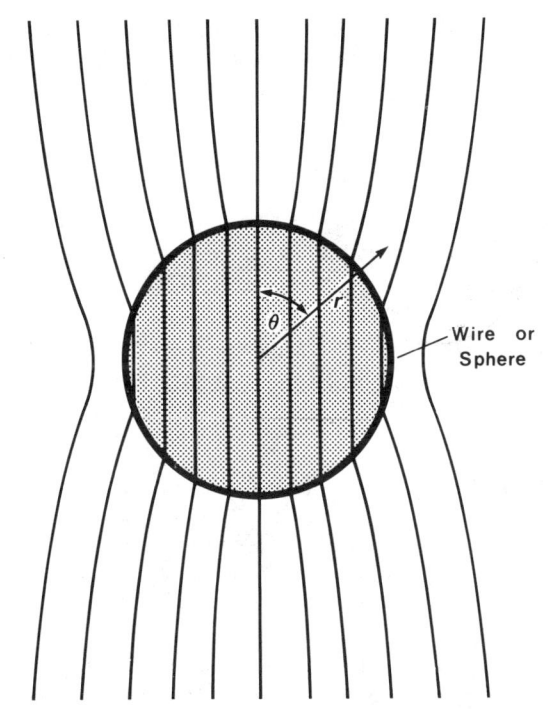

Principles and Mechanisms of Magnetic Separation

Figure 14.16. Definition of radial coordinates r and θ in uniform magnetic field.

above presented in greatly simplified form as

$$F_m = V(\kappa_s - \kappa_m)\mathcal{H}_r \frac{d\mathcal{H}_r}{dr} \quad (14.7)$$

From these equations, it can be seen that the magnetic force on a particle depends on both the applied magnetic field and the gradient of the induced magnetic field. The applied magnetic field and the field gradient that act on the particles in all magnetic separation devices may be produced in a variety of ways, and they result in widely varying field geometries and strengths. In high intensity magnetic separation, it is normal practice to provide a ferromagnetic matrix (Figs. 14.6 and 14.7) in which separation can take place. This matrix can take up a variety of forms, as discussed in Section 14.1.3. Two common forms are analyzed here: a matrix of steel balls[15] (spheres) and a matrix of fine wires[14] (cylinders) such as exist with steel wool.

Forces in a Spherical Matrix. To simplify the calculations involved, an isolated ferromagnetic sphere of diameter d_0 in a uniform applied magnetic field $\mathcal{H}$ is considered. The sphere develops a uniform magnetization in this applied magnetic field. The particles attracted to this sphere are paramagnetic, so the magnetization is small and the magnetic induction is approximately equal to the applied magnetic field. From elementary magnetic theory it can be shown that the components of the magnetic field outside the sphere are

$$\mathcal{H}_r = \left(\mathcal{H} + \frac{\pi}{3}\mathcal{M}\frac{d_0^3}{r^3}\right)\cos\theta \quad (14.8)$$

$$\mathcal{H}_\theta = \left(-\mathcal{H} + \frac{\pi}{6}\mathcal{M}\frac{d_0^3}{r^3}\right)\sin\theta \quad (14.9)$$

(r, θ, defined in Fig. 14.16). The corresponding radial magnetic field gradients are then

$$\frac{\partial \mathcal{H}_r}{\partial r} = -\pi\mathcal{M}\frac{d_0^3}{r^4}\cos\theta \quad (14.10)$$

$$\frac{\partial \mathcal{H}_\theta}{\partial r} = -\frac{\pi}{2}\mathcal{M}\frac{d_0^3}{r^4}\sin\theta \quad (14.11)$$

It can be shown that the magnetic force $F_{m,\theta}$ (as calculated from Eq. 14.6) is a force tending to move a particle to either the $\theta = 0$ or the $\theta = \pi$ position on the sphere, whichever is closer. The force $F_{m,r}$ is the force holding the particle onto the sphere, and can be calculated from Eq. 14.5. Since the force $F_{m,\theta}$ tends to move particles to the $\theta = 0$ or π positions, the equations for $F_{m,r}$ are determined only for these points of maximum force. The magnetic force $F_{m,r}$ evaluated at these positions is directed toward the center of the sphere with a magnitude

$$F_{m,r} = 16\pi V(\kappa_s - \kappa_m)\frac{\mathcal{M}}{d_0}\left(\mathcal{H} + \frac{8\pi}{3}\mathcal{M}\right) \quad (14.12)$$

The volume V can be written in terms of the particle diameter d_v, so Eq. 14.12 becomes

$$F_{m,r} = \frac{8\pi^2}{3}d_v^3(\kappa_s - \kappa_m)\frac{\mathcal{M}}{d_0}\left(\mathcal{H} + \frac{8\pi}{3}\mathcal{M}\right) \quad (14.13)$$

Forces in a Cylindrical Matrix. The case of a ferromagnetic wire in a uniform applied magnetic field can be analyzed in a manner similar to that of a sphere. For a constant wire diameter, the force on a particle is proportional to the square of the diameter of the particle.[34] This should be compared with the spherical matrix where the force on a particle is proportional to the cube of the diameter of the particle (Eq. 14.13).[15]

In Fig. 14.17 the magnetic force calculated for a particle at the surface of a cylindrical wire is plotted against the ratio of the wire diameter, D_w, to the particle diameter d_v. It is clear that there is a point at which the force is a maximum; Oberteuffer[34] shows

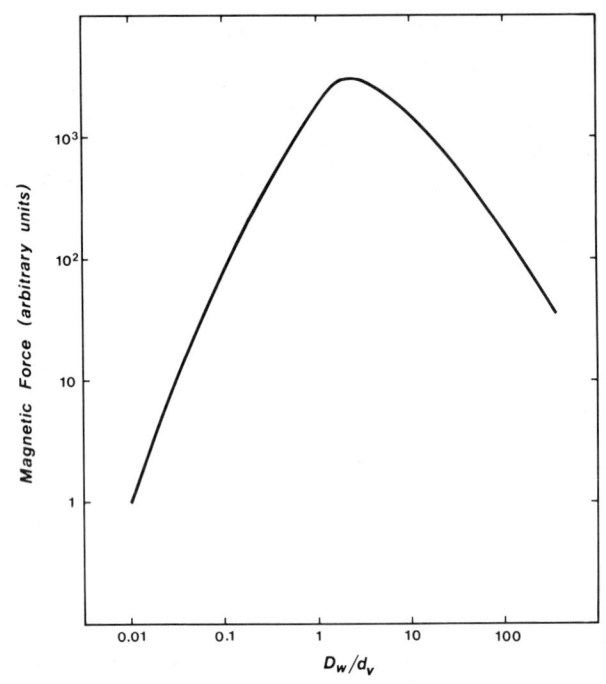

Figure 14.17. Relative magnetic force as a function of the ratio of wire diameter to particle diameter. (After Oberteuffer.[34])

that this maximum occurs at a particle diameter of $0.37 D_w$. If the particle is away from the wire surface, the size of particle on which the maximum force is exerted is inversely proportional to the distance from the center of the wire for any given wire diameter. Figure 14.18 shows a plot of the magnetic

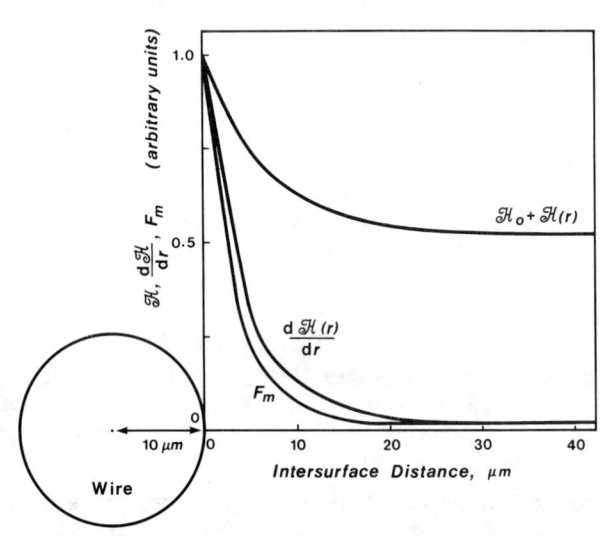

Figure 14.18. Dependence of magnetic field, magnetic field gradient, and resulting force on distance from ferromagnetic wire (for particle one-third wire diameter). (After Oberteuffer.[34])

field, magnetic field gradient, and resulting magnetic force on a particle one-third the diameter of a uniformly magnetized wire. The three functions are plotted against the distance from the surface of the wire. Note that the maximum change in gradient occurs adjacent to the wire surface and over a distance about equal to the diameter of the particle. Over this distance, the magnetic force drops to approximately 10% of its value at the surface of the wire.

For systems in which the particle diameter is approximately one-third that of a ferromagnetic wire (i.e., a matched system), the magnetic force is a maximum given by[34]:

$$F_{m,\max} = 0.46\,(\kappa_s - \kappa_m)\,\mathcal{H}^2 d_v^2 \qquad (14.14)$$

($F_{m,\max}$ = maximum magnetic force resulting when $D_w/d_v \simeq 3$). The magnetic force varies as the square of the particle diameter under these conditions.

Forces at the Surface of a Drum. When Eqs. 14.5 and 14.6 are solved for the surface of a drum, it is found[35] that $F_{m,\theta} = 0$ and the magnetic force in the radial direction is given by

$$F_{m,r} = -32\pi^2 \frac{\mathcal{H}_d}{\theta_d R} \exp\left(-\frac{2\pi(r-R)}{\theta_d}\right) \qquad (14.15)$$

($\mathcal{H}_d$ = magnetic field strength at the drum surface; R = radius of drum; r = radial distance from center of drum; θ_d = angular spacing between magnetic poles). At the drum surface, Eq. 14.15 reduces to

$$F_{m,r} = -32\pi^2 \frac{\mathcal{H}_d}{\theta_d R} \qquad (14.16)$$

Note that the magnetic force is independent of the particle size.

14.2.3. Competing Forces

The forces in magnetic separators that compete with magnetic forces and act on *all* particles in the separator are gravity, hydrodynamic drag, friction, and inertia. If the separation is occurring at the surface of a rotating drum, the centrifugal force may also be a factor. The relative importance of each force varies with separator design. *Gravity* and *hydrodynamic drag* forces are of major importance and are considered further here.

For a spherical particle with density ρ_s, the net gravitational force is given by

$$F_g = \frac{\pi}{6} d_0^3 (\rho_s - \rho_f) g \qquad (14.17)$$

Principles and Mechanisms of Magnetic Separation

(g = acceleration due to gravity; ρ_f = density of the fluid medium in use). In laminar flow, the hydrodynamic drag force is given by Stokes' law (Eq. 4.34):

$$F_d = 3\pi d_0 v \mu \qquad (4.34)$$

(v = velocity of particle relative to fluid; μ = viscosity of the fluid medium).

Thus the gravitational force is dependent on the third power of particle diameter, and is significant for large particles. The hydrodynamic drag force depends on the first power of the particle diameter when laminar flow conditions exist, so is more significant with small particles. Hence for a dry magnetic separator treating large particles, the magnetic forces must be sufficient to hold the magnetic particles against the competing force of gravity. In a wet magnetic separator for small particles, the magnetic forces must be larger than the hydrodynamic drag exerted by the flowing slurry.

In a dry low intensity drum magnetic separator,[35] the most important force acting to detach particles from the drum is generally the centrifugal force, given by

$$F_c = \rho_s V \omega^2 R \qquad (14.18)$$

(ω = angular velocity of a drum of radius R). Although the effect of gravity varies with the position of the particle on the drum, little error is introduced by neglecting gravity if the drum speed exceeds about 80 rpm. A particle will be detached from the drum if the magnitude of the *detachment* force (in this case, centrifugal) exceeds that of the *entrapment* force (magnetic), that is, if the *entrapment ratio* (i.e., entrapment force/detachment force) is less than one. The angular velocity ω_e at which entrapment just occurs [i.e., when the centrifugal force (Eq. 14.18) is balanced by the magnetic force (Eq. 14.16)] is

$$\omega_e = \frac{4\pi}{R}\left(\frac{2\mathcal{H}_d}{\rho_s V \theta_d}\right)^{1/2} \qquad (14.19)$$

Since particle size is not a factor, dry magnetic separators can successfully process feeds with a wide range of particle sizes.

The magnetic force on a paramagnetic particle in a cylindrical matrix was discussed in the preceding section. Let us now consider the magnetic and competing forces on a paramagnetic particle approximately one-third the size of the ferromagnetic wires making up a cylindrical matrix in a 1 T magnetic field (i.e., a matched system). Figure 14.19 shows the magnetic, gravitational, and hydro-

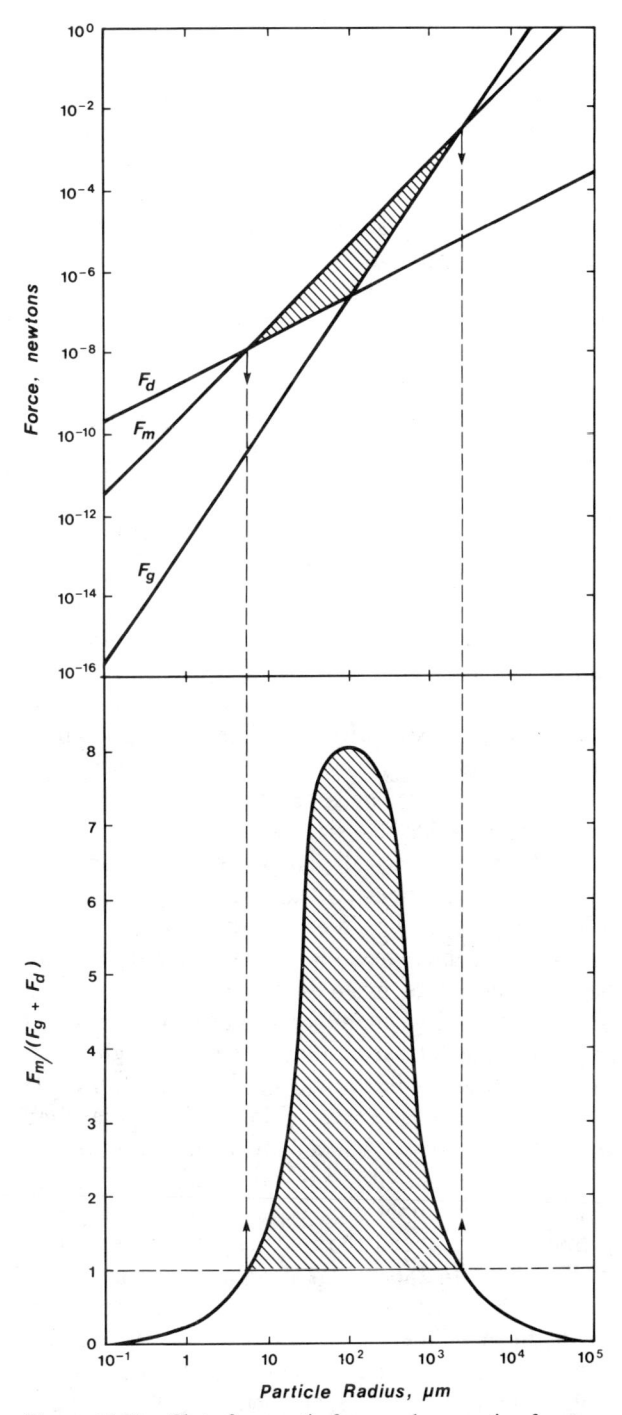

Figure 14.19. Plot of magnetic force and competing forces as function of particle size (ferromagnetic wire three times particle diameter; forces calculated for CuO particle in 8×10^5 A/m field with slurry velocity of 5 cm/sec). (After Oberteuffer.[34])

dynamic drag forces plotted against the particle size in such a situation. Also plotted is the entrapment ratio, $F_m/(F_g + F_d)$. In this figure, separation can occur in the cross-hatched region, that is, where the entrapment ratio is greater than one. It is clear that separation can occur only over a limited size range: with large particles gravitational forces become dominant, and with smaller particles drag forces become dominant. Note that if the system is not matched, the range of sizes that can be treated becomes even narrower, because the line for F_m in Fig. 14.19 moves down. Figure 14.19 also indicates that there is a size at which the entrapment ratio is a maximum; this will be the most effective particle size at which to achieve a separation, and this diameter is given by

$$d_0 = 3 \left[\frac{2v\mu}{(\rho_s - \rho_f)g} \right]^{1/2} \quad (14.20)$$

Since most mineral densities are similar, the optimum particle size is approximately the same for all minerals. Thus, separation of paramagnetic minerals is possible only within a restricted size range, and this is unlike the situation with ferromagnetic particles, which can be separated over a wide range of sizes. This limited size range in high intensity separation is highly dependent on the mineral's magnetic susceptibility. This can be seen in Fig. 14.20. However, the situation is not really this simple, because mechanical factors are also significant.

The concept of magnetic separability is further illustrated in Fig. 14.20 for a theoretical matched wet high intensity separator, a real wet high intensity separator, and a typical wet low intensity drum separator.[34] In each case, the lower limit is reached when the drag forces exceed the magnetic forces for fine particles. With the theoretical matched separator, the upper limit occurs when the gravitational forces exceed the magnetic forces for large particles. With the two real separators, the upper limit is the mechanical size limit of the separator. With the low intensity drum separator, the lower limit of magnetization is reached when the gravitational forces exceed the magnetic forces.

In summary, it can be seen that the use of the wet drum separator is limited to ferromagnetic materials and is not effective with particles finer than about 74 μm. The wet high intensity separator can treat paramagnetic materials with low magnetization over a narrow range of particle sizes, and only at smaller particle size. Recovery is thus dependent on magnetic susceptibility and particle size, and a device that treats a size distribution cannot be very selective between materials whose magnetic properties vary only slightly. Generally, however, minerals to be separated differ widely in their magnetic properties.

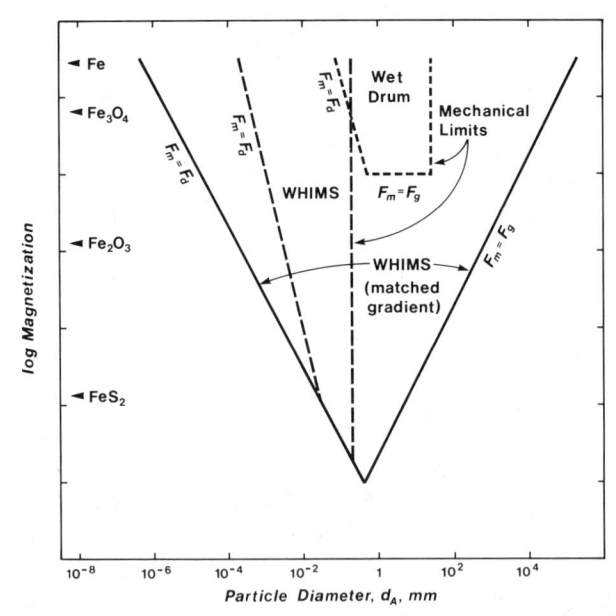

Figure 14.20. Plot showing range of mineral magnetization and particle size over which magnetic separation is possible. (After Oberteuffer.[34])

14.3. MAGNETIC SEPARATOR PERFORMANCE

14.3.1. Low Intensity Separators

General performance data for low intensity magnetic separators are limited. Some data are available in the literature on dry separator performance.[11,12,35] The factors affecting wet separators have been discussed by Lantto[36] and in manufacturers' literature,[14] but only limited quantitative information is provided.

The design of the magnetic separator tank is a major factor. Figure 14.21 shows that the countercurrent tank design is clearly superior to the concurrent tank design for the particular separation considered. However Fig. 14.22 indicates that although at fine sizes the countercurrent tank produces the better recovery, at coarser sizes the concurrent tank is superior. The effect of magnetic field strength and drum diameter on separator performance is also shown in these figures.

Magnetic Separator Performance

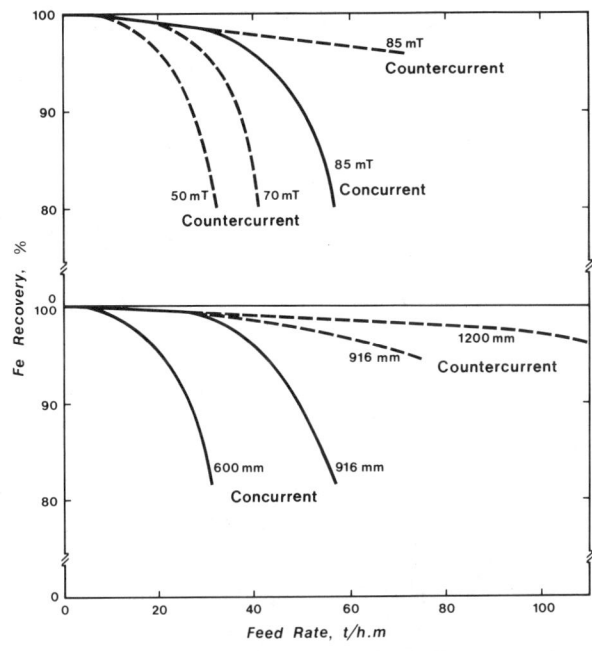

Figure 14.21. Effect of field strength, drum diameter, and tank design on iron recovery in a wet drum separator: magnetite ore, 75% −44 μm, 30% solids in slurry. (*Top*) Drum diameter 916 mm. (*Bottom*) Field strength, 85 mT. (Courtesy Sala International AB.)

Changes in feed conditions have a comparatively small effect on concentrate grade (Fig. 14.23).

When wet magnetic separators are used for iron ore concentration, normal practice is to use separators in series, with magnetics undergoing repeated cleaning and nonmagnetics being rejected as tailings at each stage. No data are available, but it may be expected that Eq. 3.3 could be applied, with a very high probability of a liberated magnetite particle being in the positive response (magnetics) stream, and a very low probability of a nonmagnetic gangue particle being in this stream. Middlings particles containing any significant amount of magnetite would also have a comparatively high probability of being in the positive response stream.

It has also been suggested[37] that Eq. 3.13 is applicable to magnetic separations. For magnetic separations, Eq. 3.13 becomes

$$\mathcal{R} = \tfrac{1}{2} - \tfrac{1}{2}\,\mathrm{erf}\left\{\left[\frac{\delta \mathcal{M} \nabla \mathcal{M}(m_{\kappa I}/\Delta\kappa)}{\mathcal{D}\mu^*}\right]^{1/2}(\kappa - \kappa_{50})\right\} \quad (14.21)$$

($m_{\kappa I}$ = mass fraction of feed with magnetic susceptibility between κ and $\kappa + \Delta\kappa$; κ_{50} = separation magnetic susceptibility).

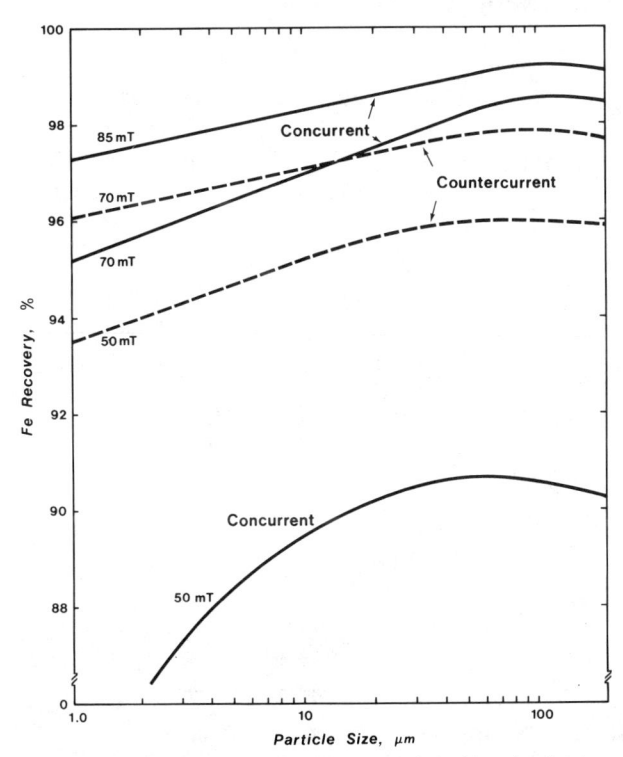

Figure 14.22. Effect of particle size on tank design selection in a wet drum separator. (Courtesy Sala International AB.)

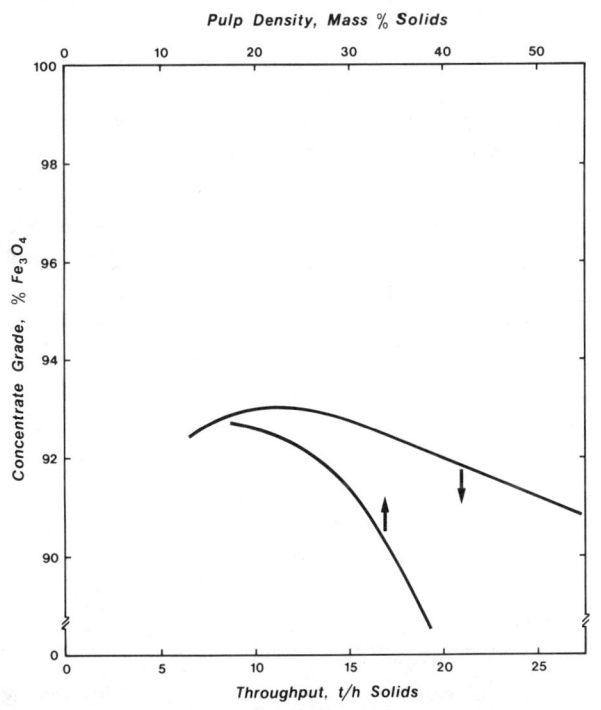

Figure 14.23. Effect of throughput and pulp density on concentrate grade in a wet drum separator. (After Lantto.[36])

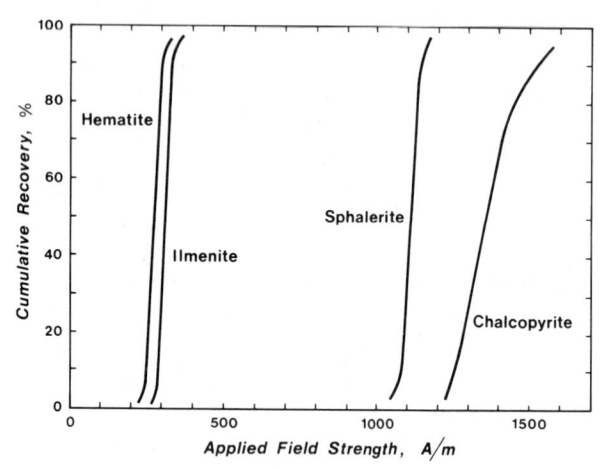

Figure 14.24. Separability curves for paramagnetic minerals. (After Dobby and Finch.[38])

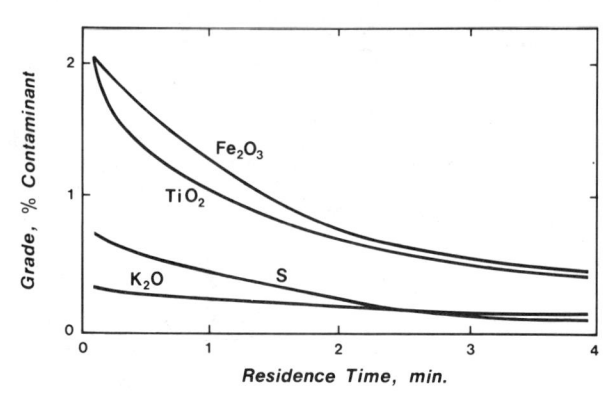

Figure 14.26. Effect of flow rate on removal of paramagnetic contaminants from kaolin in a canister-type wet high intensity separator. (After Murray.[41])

14.3.2. High Intensity Separators

Operating data are available for a number of high intensity magnetic separators. Separability curves (Fig. 14.24) have been developed for a number of paramagnetic minerals using a Frantz Isodynamic separator.[38,39] Using the same hematite and ilmenite samples as shown in Fig. 14.24, tests have been reported[38] that show the effects of particle size, slurry velocity, and loading time on the magnetic recovery in a canister-type high intensity magnetic separator (Fig. 14.25). Similar results have been reported for other mineral systems.[4,40,41] Figure 14.25a shows that a higher applied magnetic field is required to recover finer particles because the effect of the competing drag force is greater. Figure 14.25b shows that if a higher capacity is to be attained by increasing the pulp velocity through the canister, a higher magnetic field must be applied. This effect is also shown in Fig. 14.26 where a canister-type separator was used to remove magnetic contaminants from clay.[41] Figure 14.25c shows the effect of canister loading time, that is, the time during which new feed is passed through the canister before the flushing cycle is started. In another approach at describing the effect of loading time,[42] magnetics performance curves can be drawn, analagous to the breakthrough curves widely used in the analysis of ion-exchange operations.

In the only work thus far to give performance curves for magnetic separations,[43] a crossbelt separator was analyzed. Feed particle size distribution and separator setting were shown to have little effect on the fraction of material reporting to the

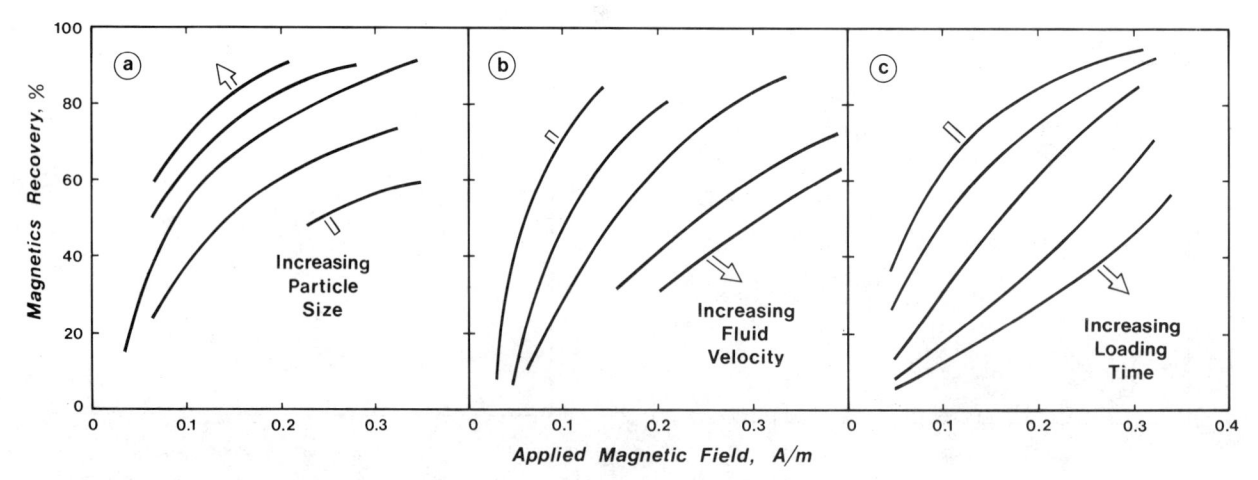

Figure 14.25. Effects of (a) particle size, (b) slurry velocity, and (c) loading time on recovery in a cannister-type wet high intensity separator. (After Dobby and Finch.[38])

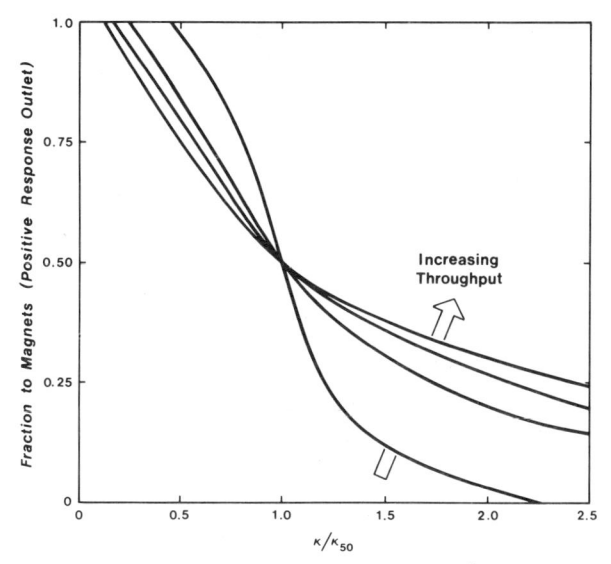

Figure 14.27. Performance curves for a crossbelt magnetic separator. (After Panou.[43])

magnetics product. The throughput, however, had a significant effect (Fig. 14.27).

REFERENCES

1. C. G. Gunther, *Electro-Magnetic Ore Separation*, Hill Publishing Co. (1909).
2. L. A. Roe, "Advances in Magnetic Separation of Ores," *Trans. AIME/SME*, **211**, 1261–1265 (1958).
3. W. J. Bronkala, "Magnetic Separation," in *Mineral Processing Plant Design*, A. L. Mular and R. B. Bhappu (Eds.), pp. 467–478, AIME (1978).
4. W. J. Bronkala, "Magnetic Separators and Their Application in the Mineral Industry," AIME/SME Preprint 69-B-364 (1969).
5. J. Suleski, "New Magnets and Tank Designs for Wet Magnetic Drum Separators," *World Min.*, **24**, 44–61 (April 1972).
6. A. Broadhurst, "New Magnetics Assist Wet Drum Separation," *Aust. Min.*, **60**, 26–27 (July 1968).
7. E. S. Twitchwell and J. A. Bartnik, "Eriez Unveils Permanent Wet Drum Separator," *Eng. Min. J.*, **168**, 94–96 (October 1967).
8. T. Maki and E. F. Furness, "Types of Commercial Wet Magnetic Separators and Their Applications in Mineral Dressing." Presented at AIME Meeting, 1955.
9. J. E. Forciea, L. G. Hendrickson, and O. E. Palasvirta, "Magnetic Separation for Mesabi Magnetite Taconite," *Trans. AIME/SME*, **211**, 1269–1276 (1958).
10. J. E. Forciea and R. W. Salmi, "Primary Magnetic Separator Specifications," *Trans. AIME/SME*, **232**, 339–345 (1965).
11. P. G. Kihlstedt and B. E. Skold, "Concentration of Magnetite Ores with Dry Magnetic Separators of the Mortsell-Sala Type," *5th Int. Miner. Process. Congr., London, 1960*, pp. 691–704, IMM (1960).
12. E. Lindgren, "Dry Magnetic Separation Practice in Europe on the Sala-Mortsell Separator," *Proc. 26th Mining Symp., Minnesota*, p. 111 (1965).
13. W. T. Barrett, J. E. Lawver, and J. L. Wright, "A Rapid Method of Evaluating Magnetic Separator Force Patterns," AIME/SME Preprint 70-B-98 (1970).
14. "Magnetic Wet Separators," Sala International Catalog 23 30-01-7742 GB (1977).
15. J. E. Lawver and D. M. Hopstock, "Wet Magnetic Separation of Weakly Magnetic Minerals," *Miner. Sci. Eng.*, **6**, 154–172 (1974).
16. G. H. Jones, "Wet Magnetic Separator for Feebly Magnetic Minerals: Description and Theory," *Proc. 5th Int. Miner. Process. Congr., London, 1960*, pp. 717–732, IMM (1960).
17. J. Iannicelli, "New Developments in Magnetic Separation," *IEEE Trans. Magn.* **MAG-12**, 436–443 (1976).
18. J. A. Bartnik, W. H. Zabel, and D. M. Hopstock, "On the Production of Iron Ore Superconcentrates by High-Intensity Wet Magnetic Separation," *Int. J. Miner. Process.* **2**, 117–126 (1975).
19. W. Jacobs, K. Brennecke, and H. D. Wasmuth, "Comparative Tests for Using Wet High Intensity Magnetic Separation," *Proc. 11th Int. Miner. Process. Congr., Cagliari, 1975*, pp. 337–362, Universita di Cagliari (1975).
20. D. M. Hopstock, "Wet High Intensity Magnetic Beneficiation of Oxidized Taconites," U.S. Bureau of Mines Report of Investigations RI 8363 (1979).
21. J. Iannicelli, "High Intensity High Gradient Magnetic Separation as a New Process Tool," Aquafine Corp. (1978).
22. H. H. Murray, "Beneficiation of Selected Industrial Minerals and Coal by High Intensity Magnetic Separation," *IEEE Trans. Magn.*, **MAG-12**, 498–502 (1976).
23. W. L. Freyberger, J. W. Keck, and D. J. Spottiswood, "Cleaning of Eastern Bituminous Coals by Find Grinding, Flotation, and High Gradient Magnetic Separation," *Proc. Symp. on Coal Cleaning*, Department of Energy/Environmental Protection Agency (1978).
24. E. S. Twitchwell, "High Intensity Magnetic Separation," *Miner. Process.*, 19–28 (April 1968).
25. N. F. Schultz, "Determination of Magnetic Separation Characteristics with the Davis Magnetic Tube," *Trans. AIME/SME*, **229**, 211–216 (1964).
26. J. McAndrew, "Calibration of a Frantz Isodynamic Separator and Its Application to Mineral Separation," *Proc. Aust. IMM*, **181**, 59–73 (1957).
27. W. H. Bensen, J. A. Bartnik, and G. D. Rose, "Demagnetizing Coils and Magnetic Flocculators Used in the Magnetite Beneficiation Industry," *Proc. 24th Mining Symp., Minnesota*, pp. 139–144 (1968).
28. H. Lantto, "The Effect of Magnetic Flocculation on the Beneficiation of Magnetic Materials," *Acta Polytech. Scand. Chem. Incl. Met. Ser.* No. 133 (1977).
29. J. H. P. Watson, N. O. Clark, and H. Windle, "A Superconducting Magnetic Separator and Its Application in Im-

proving Ceramic Raw Materials," *Proc. 11th Int. Miner. Process. Congr., Cagliari,* 1975, pp. 795–812, Università di Cagliari (1975).

30. H. E. Cohen and J. A. Good, "Principles, Design, and Performance of a Superconducting Magnet System for Mineral Separation in Magnetic Fields of High Intensity," *11th Int. Miner. Process. Congr., Cagliari,* 1975, pp. 777–793, Università di Cagliari, (1975).

31. V. Andres, *Magnetohydrodynamic and Magnetohydrostatic Methods of Mineral Separation,* Keter Publishing House (1976).

32. "Metric Practice Guide," ASTM Designation: E-380-70 (1970).

33. E. M. Purcell, *Electricity and Magnetism,* Vol. 2, p. 353, McGraw-Hill (1965).

34. J. A. Oberteuffer, "Magnetic Separation: A Review of Principles, Devices, and Applications," *IEEE Trans. Magn.,* **MAG-10,** 223–238 (1974).

35. D. M. Hopstock, "Fundamental Aspects of Design and Performance of Low-Intensity Dry Magnetic Separators," *Trans. AIME/SME,* **258,** 222–227 (1975).

36. H. Lantto, "Factors Affecting Low Intensity Magnetic Separation," *Acta Polytech. Scand., Chem. Incl. Met. Ser.* No. 135 (1977).

37. O. N. Tikhonov and V. V. Dembovsky, *Automatic Process Control in Ore Treatment and Metallurgy,* El-Tabbin Metallurgical Institute (1973).

38. G. Dobby and J. A. Finch, "Capture of Mineral Particles in a High Gradient Magnetic Field," *Powder Technol.* **17,** 73–82 (1977).

39. Anon., "Processing of Metalliferous Ores," *Mater. Tech.,* **3,** 113 (1978).

40. M. E. Arellano and G. Z. Zambrana, "High Gradient Magnetic Separation Applied to Tin Minerals," *IEEE Trans. Magn.,* **MAG-14,** 488–490 (1978).

41. H. H. Murray, "High Intensity Magnetic Beneficiation of Industrial Minerals—A Survey," AIME/SME Preprint 76-4-93 (1976).

42. I. Y. Aktao, "Mathematical Modeling of High-Gradient Magnetic Separation Devices," *IEEE Trans. Magn.,* **MAG-13,** 1486–1489 (1977).

43. G. Panou, "Use of Washability Curves in Magnetic Separation of Ores," *Proc. 7th Int. Miner. Process. Congr., New York, 1964,* pp. 375–389, Gordon & Breach (1965).

Chapter Fifteen

Electrostatic Separation

> *Hein's Law: "Problems worthy of attack prove their worth by hitting back."*

Electrostatic separators were first used in the late 1800s, primarily for the separation of high conductivity gold and metallic sulfides from low conductivity siliceous gangue. Early this century, the real breakthrough occurred with the development of the first corona discharge-type separator. This separator was based on the same principles as the high tension electrostatic separator in use today.

These early machines were also used with sulfides such as in the separation of sphalerite from galena. However, the advent of flotation put an end to most of the early electrostatic operations.[1] It was not until the 1940s, when the demand for rutile increased rapidly, that interest in electrostatic separation was renewed. The rutile deposits generally contained minerals of similar specific gravity, and with surface properties so similar to rutile that selective flotation was not possible. However, rutile is considerably more conductive, and so the development of new separation machines was stimulated.[2] Carpenter[3-5] and his associates developed the focusing or beam-type electrode that is the basis for all high tension separators in use today. Advances followed quickly in both techniques and equipment, and the use of electrostatic separation spread to other minerals.[2,6]

Electrostatic separation is applied as a concentration process with only a small number of minerals; where it is applied, however, it has proved highly successful. It is frequently combined with gravity and magnetic separation devices for separating the nonsulfide minerals from each other. Gravity separation is used to remove silica and produce a bulk concentrate, then a combination of electrostatic and magnetic separation used to separate the valuable minerals remaining. Appendix F shows for a number of minerals the combination of specific gravity with conductivity and magnetic properties that can be used to obtain a separation.

In this chapter the various types of electrostatic separators are discussed, together with their applications and the principles on which they are based. Detailed discussions of electrostatic separator operations may be found in the literature generally. The principles of operation have been best described by Fraas[2] and Moore.[7]

15.1. EQUIPMENT AND APPLICATIONS

The major application of electrostatic separation has been in the processing of beach sands and alluvial deposits containing titanium minerals. There are few beach sand plants in the world today that do not use the electrostatic separation process to separate rutile and ilmenite from zircon and monazite. The majority of the world's rutile and zircon is produced in Australia, and every producing plant there uses electrostatic separation. A number of plants in Florida, some capable of producing 1000 t/day of heavy mineral concentrate, are using a combination of electrostatic separation and high intensity magnetic separation. In the sample flow sheet for a plant processing beach sands shown in Appendix B-8, the initial concentration by gravity separation removes silica, and following drying, electrostatic separation is used to separate rutile and zircon.

Electrostatic separation is used with a number of other minerals. Almost every major tin mining operation in Malaysia, Thailand, and Nigeria uses the electrostatic separation process to separate cassiterite, columbite, and ilmenite from gangue minerals.[8]

Wabush Mines in Canada uses electrostatic separation to produce 1000 t/hr of high grade specular hematite. LKAB in Sweden also uses electrostatic separation to produce a high grade, low phosphorus specular hematite concentrate.[8] Electrostatic separation has been shown capable of producing iron ore concentrates containing as little as 0.1% silica.[9] Other minerals also are treated by electrostatic separation, but to a much lesser extent. Also, electrostatic separators are being used in industrial waste recovery.

There are two basic types of electrostatic separator. Various names are used by the manufacturers, but basically the devices can be described as *electro-dynamic* and *electro-static* separators. Although most early separators were of the electrostatic type, most of those in use today are electro-dynamic.

15.1.1. "Electro-Dynamic" Electrostatic Separators

Electrostatic separators of the electro-dynamic type are all based on the original Carpenter design.[5] They are commonly called *high tension* separators. Figure 15.1 shows the major features of this separator. The feed is carried by the grounded rotor into the field of a charged ionizing electrode. The feed particles accept a charge by ion bombardment. The conductor particles lose their charge to the grounded rotor and are thrown from the rotor surface by centrifugal force; they then come under the influence of the electrostatic field of the nonionizing electrode and are further attracted from the rotor surface. The nonconductor particles are not able to dissipate their charge rapidly to the rotor, and so are held to the rotor surface by their own image forces (Section 15.3.1). As the rotor carries the nonconductor particles on its surface, their charge is slowly lost and they drop from the rotor, middlings particles losing their charge faster and dropping first. The residual nonconductors are removed from the rotor surface by a brush. In some high tension separators the removal of nonconductor particles from the rotor is assisted by high voltage "wiping"; this is carried out by means of an additional electrode placed on the brush side of the rotor.

The size of a particle influences its action in the separator, since the charge on the surface of a coarse particle is lower in relation to its mass than that on a fine particle. Thus, a coarse particle is more readily thrown from the rotor surface. The separators are operated with multiple layers of particles on the rotor surface, so fine conductor particles will tend to be trapped by nonconductors because of their low mass. Thus the fine particles tend to report preferentially with the nonconductor fraction. As a result it is normal operating practice to use multiple stages of cleaning; as can be seen in Figs. 15.2 and 15.3, the separators are manufactured in banks. Figure 15.2 shows that a bank of separators can be arranged for cleaning of the conductor fraction, the nonconductor fraction, or both. Often, to be able to produce clean conductor and nonconductor fractions simultaneously, it is necessary to take a middlings fraction, which contains fine conductor particles and coarse nonconductors. If unliberated particles are present, they will tend to report to this middlings fraction.

Plant throughput is limited by the circulating load, which operating experience suggests largely depends on feed composition. There are two common causes for increases in circulating load: one is an increase in the number of particles that are inherently middlings, that is, those of intermediate conductivity between that of the "conductor" and the "nonconductor"; the other is a broadening of the particle size distribution. High tension separators operate over a wide range of particle sizes. With beach sands, a size range of 50–1000 μm is possible, although a smaller size range is both desirable and more common. With hematite ores, all sizes less than 1 mm have been accepted. However, particles less than 75 μm cause problems.

Separators are available in a number of sizes. Typically, diameters are 150–240 mm,[10] and lengths

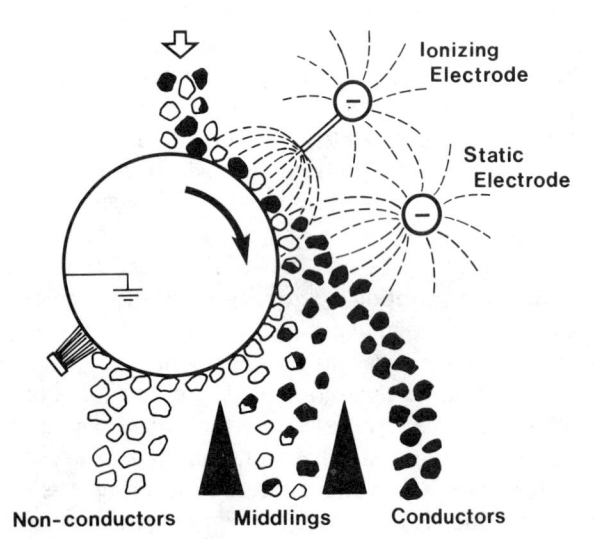

Figure 15.1. An electro-dynamic or high tension separator.

Equipment and Applications

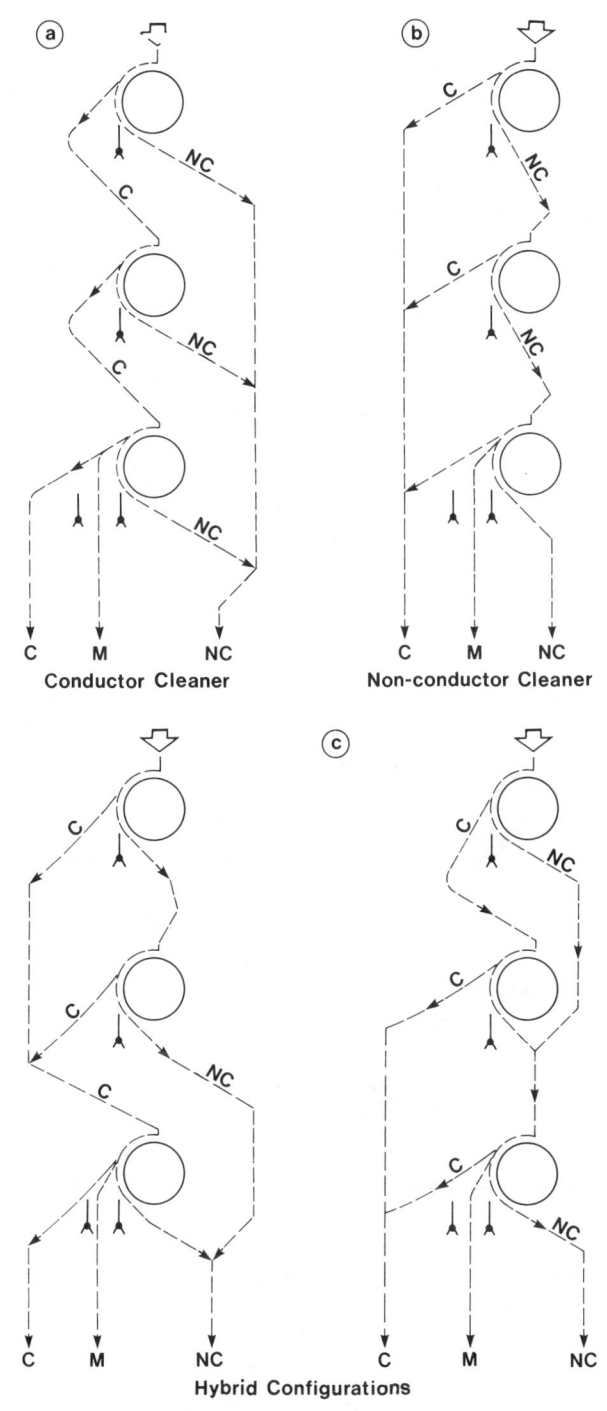

Figure 15.2. Schematic of possible configurations of electrostatic separators arranged in banks. (Courtesy Mineral Deposits Ltd.)

for example, handle up to 54 t/hr of specular hematite ore in a one pass operation.[8]

A number of operating variables are available to optimize the separator performance. Feed rate, rotor speed, electrode position and voltage, and product splitter position can be varied. Operating are up to 3 m. The capacity of a high tension separator depends on many factors; capacities up to 2500 kg/hr · m of rotor length are obtained with iron ores and approximately 1000 kg/hr · m with beach sand minerals. A separator with six 3 m rotors can,

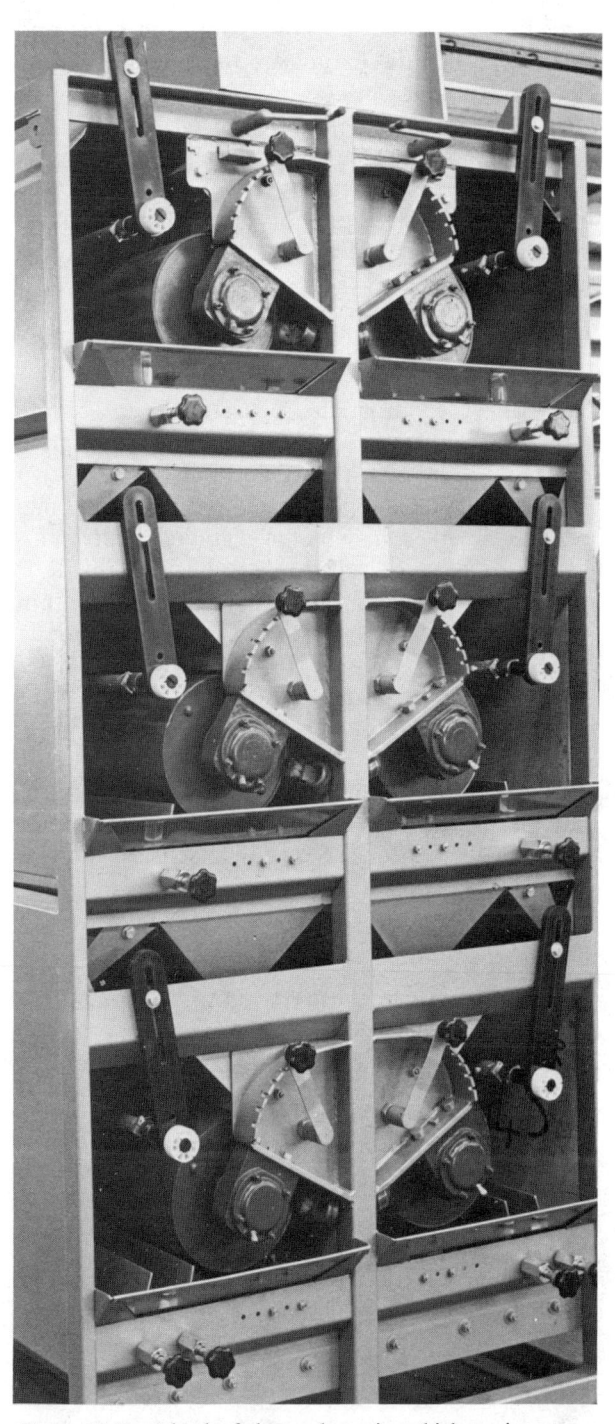

Figure 15.3. A bank of electro-dynamic or high tension separators. (Courtesy Mineral Deposits Ltd.)

strategy is frequently to increase feed rate until the product quality reaches the minimum acceptable level. The other operating settings are in general determined by testing, either on a pilot scale unit or on the production unit. Electrode voltage can be varied only within a limited range; the voltage range of stable corona discharge is quite narrow for any particular electrode position. Both positive and negative corona discharges are used, and each has special advantages in certain separations. A negative electrode, however, is preferred, since it can produce a more intense corona in air before arcing. Separators are supplied with 40 kV dc power supplies and operate up to this level. Current flow is low, and varies from 5 to 15 mA per meter of rotor length.

15.1.2. "Electro-Static" Electrostatic Separators

The earliest electrostatic separators were of the plate electro-static type.[1,2] Free-fall plate separators,[11] in which particles fall between two near-vertical plates (one charged positively and one negatively with a high voltage gradient between), have been used to separate sylvite from halite,[12] feldspar from quartz,[13] and phosphate pebble from quartz.[14] However, these devices are no longer in use.

There are two types of electro-static separator manufactured, a rotor type and a plate type.[10,15]

Rotor-Type Separator. The rotor-type electro-static separator (Fig. 15.4) is similar in appearance to the high tension separator discussed above. However, there is no ionizing electrode in the electro-static separator. Instead, there is a large single electrode producing an electric field. When a particle is placed on the grounded rotor in the presence of the electric field, the particle rapidly develops a surface charge by induction. Whether the particle is a conductor or a nonconductor, it can be considered to have become polarized. However, as indicated in Fig. 15.8 a conductor particle rapidly becomes an equipotential surface and has the same potential as the grounded rotor; therefore it is attracted toward the electrode. Thus, the conductor particle is drawn from the surface by attraction to the electrode while the nonconductor particle continues to adhere to the rotor surface until gravity causes it to fall. Thus, a separation is achieved, but with a different particle charging mechanism from that of the high tension separator.

Plate-Type Separator. There are two separators of the plate type manufactured: the plate electro-static separator (Fig. 15.5a) and the screen electro-static

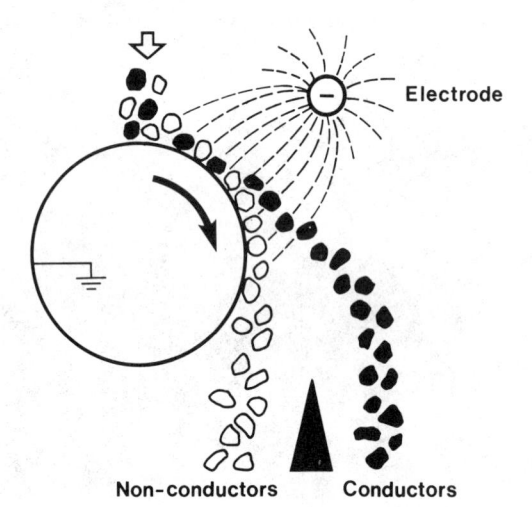

Figure 15.4. A rotor-type electro-static separator.

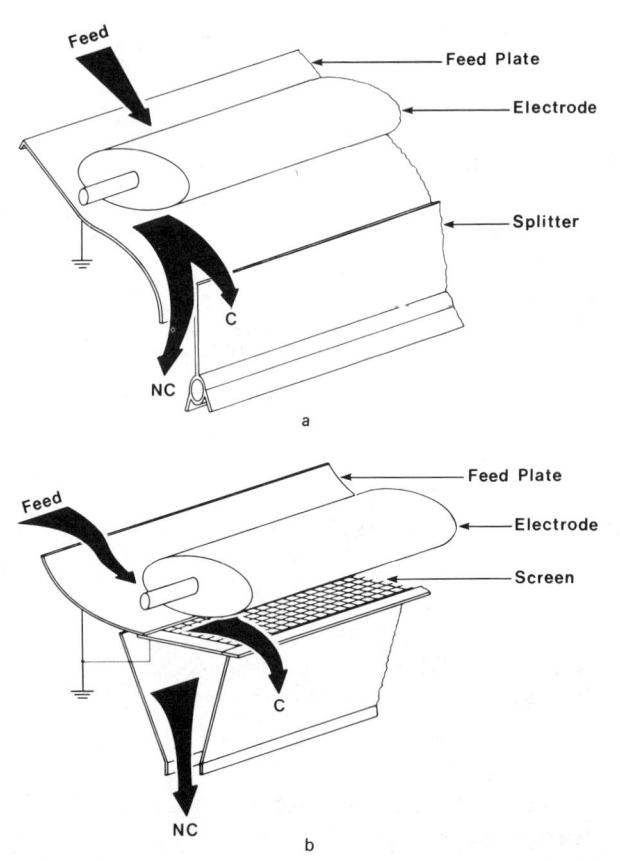

Figure 15.5. Electro-static separators. (a) Plate (b) Screen. (Courtesy Mineral Deposits Ltd.)

Operating Environment

separator (Fig. 15.5b). Their operating principles are similar. The feed particles slide down a grounded plate into the divergent electric field induced by the large curved electrode. The particles are charged by induction, the conductor particles acquiring a charge opposite to the electrode. Thus the conductor particles are attracted towards the electrode. The nonconductor particles continue down the plate or through the screen.

In electro-static separators, it is the fine particles that are most affected by the weak forces involved, and so the conductor product preferentially contains fine conductors. At the same time, there is no tendency for the coarse nonconductors to enter the conductor stream. Because these separators act primarily on the conductor particles, they are mainly used for cleaning a small quantity of conductor particles from a large quantity of nonconductors. In particular, their main use is in cleaning small amounts of rutile and ilmenite out of zircon concentrates. They are generally used in banks, as shown in Fig. 15.6, with the nonconductors being recleaned through successive separators. Plate and screen electro-static separators can be seen in Fig. 15.6a and 15.6b, respectively.

The discussion above has dealt in terms of the separation of conductor particles from nonconductor particles. However, with electrostatic separation, it is possible to separate nonconductor particles with differing electrical properties.[15]

15.1.5. Other Electrostatic Separators

Electrostatic separation in a medium other than air has been considered for many years.[1,16-18] The process has been called *dielectric* separation[16] or *electrophoretic* separation.[19] Basically the separation is made in a "dielectric liquid," a liquid with a dielectric constant that lies between those of the minerals to be separated. Various designs of equipment have been proposed, including one successfully tested in the laboratory for the separation of chromite ores, which used a rotating drum as one electrode.[19]

Another device that has not reached commercial application is the vibrated bed electrostatic separator.[2,20] More efficient separation and mechanical simplicity are claimed for this device.

15.2. OPERATING ENVIRONMENT

The size of particles fed to electrostatic separators is not readily controlled. With beach sand minerals, the feed is accepted as is because the material is naturally classified. The effect of particle size has been discussed in reference to each type of separator and is not treated further here.

15.2.1. Particle Coatings

For a separation to be achieved, all particle surfaces must be free of moisture. Beach sand particles are often coated with organic materials, and if not removed these cause difficulties in separation. Various cleaning techniques are used, ranging from simple washing, through attrition scrubbing, to caustic scrubbing. Also, when slimes are present, they must be removed before drying ahead of electrostatic separation. For example,[9] before desliming, a hematite had a resistivity of 5×10^{-8} ohm-m, whereas after desliming the resistivity was 2×10^{-2} ohm-m.

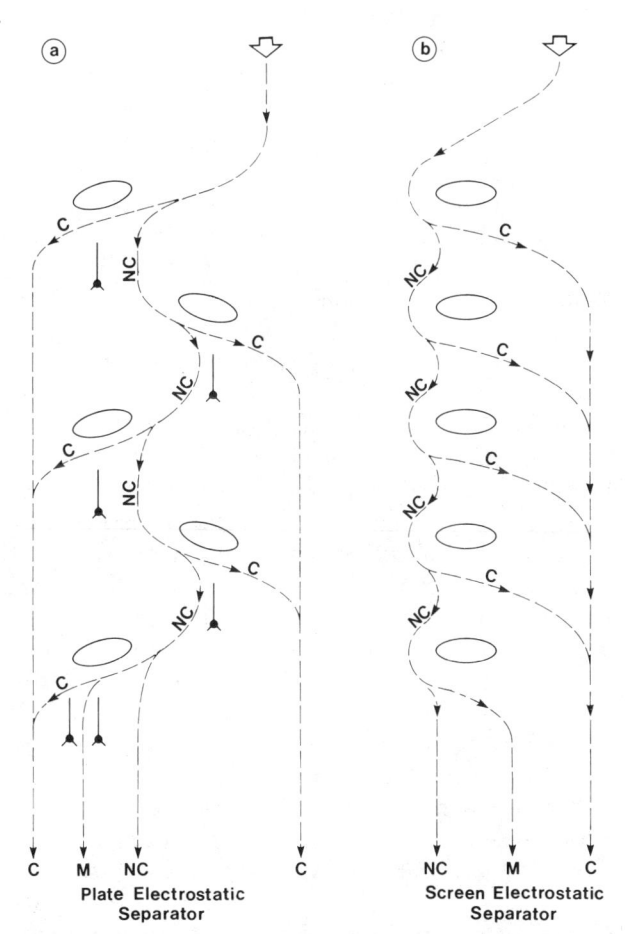

Figure 15.6. Schematics of bank arrangement for plate (a) and screen (b) electro-static separators. (Courtesy Mineral Deposits Ltd.)

Dedusting may also be necessary when very fine particles are present.

A detailed report on surface treatment is provided by Fraas.[2]

15.2.2. Feed Temperature

Mineral temperature has a marked effect on mineral conductivity. Fraas[2] gives separator performance data for a wide range of minerals with increase in temperature. His data, for example, indicate that although the conductivity of zircon and monazite (nonconductors) remains essentially unchanged by temperature increases, that of rutile increases considerably even in the 50–300°C range. This means that the separation would be enhanced by operating at an elevated temperature. Normal operating practice in the separation of rutile is to operate at a particle temperature of 90°C or above.[8]

Electrostatic separators are manufactured with heating coils in the feed hopper. Heating is also provided in the electric field zone, generally by heating lamps.

15.2.3. Separation Environment

It has already been stated that there must be no surface moisture on the particles if a separation is to be obtained. Avoiding such moisture is not a serious problem, since surface drying is straightforward. The difficulty is in maintaining moisture-free surfaces when the humidity is high.

The effect of increased humidity is to reduce the selectivity in any given separation. Fraas[2] has discussed the effect of humidity on particular applications in great detail. In general, the effect of humidity is less in high tension separation than in electro-static separation, primarily because of the higher charge levels employed.

15.3. MECHANISMS OF ELECTROSTATIC SEPARATION

There are three distinct stages of electrostatic separation processes that must be considered: the charging of the particles, the separation that occurs at a grounded surface, and the separation caused by the trajectory of the particles. The mechanisms involved in each stage are briefly discussed here.

15.3.1. Particle Charging Mechanisms

In the electrostatic separation of minerals, particles can be charged by:

1. Contacting dissimilar particles.
2. Ion bombardment.
3. Induction.

Each of these mechanisms gives rise to a surface charge. Each is a distinct mechanism, but in every separation process two or more occur. These mechanisms have been identified in the literature by a variety of names: widely used terms for the three mechanisms listed are, respectively, "contact electrification," "electrification by ion bombardment," and "electrification by conductive induction."[8,21]

Charging by Contacting Dissimilar Particles. When the surfaces of two dissimilar particles are brought into contact and then separated, one generally becomes positive and the other negative. The area of contact between particles is quite small, so to build up any appreciable surface charge it is necessary to cause repeated contacts. This occurs whenever there is bulk movement of particles. If the particles are composed of poorly conducting minerals, the surface charge density of the particles can become high enough for this mechanism to be used as the basis for mineral concentration.

The theory of charging by contacting of dissimilar materials is complex and is not discussed here. It is summarized by Lawver,[21] where further references are given. There is no clear statement in the literature about the mechanism by which the charge transfer is occurring. It is thought that the charge transfer is due to electron transfer, although there is evidence in some systems that the charge is transferred by ions.[22] As a general rule, if two dielectric materials are contacted and separated, the material with the higher dielectric constant becomes positively charged. However, this is not particularly useful with minerals because the electrical properties of a particular mineral can vary widely because of the presence of trace impurities.

This mechanism was made use of in free-fall electrostatic separators. It is not a major mechanism in any of the electrostatic devices in current use.

Charging by Ion Bombardment. The "bombardment" of ions or electrons through air is nothing more than the conduction of electricity through air.

Mechanisms of Electrostatic Separation

Gases differ from liquids and solids in the manner by which they conduct electricity. Metals, both liquid and solid, conduct by movement of electrons; in insulating solids, such as most oxides and silicates, and in aqueous solutions, the charge is carried by ions. But in gases under ordinary conditions there are neither ions nor electrons, and the gas molecules are far apart; thus gases act as good insulators. But if the potential between two electrodes is raised high enough, there is an electric break down of the gas and the gas "discharges."[22] By appropriate shaping of the electrodes, a "corona" discharge can be obtained. It has been established[23] that for discharge to a flat surface (or large diameter cylinder surface), a fine wire parallel to the cylinder gives the optimum corona discharge.

If the mineral particles are caused to pass within the corona, they will be struck by electrons or anions (if a negative corona), and the solids will emerge charged. This effect will be limited by leakage of the charge if the particles enter the corona field in contact with a conducting surface such as a metal drum. The rate of this leakage will depend on the conductivity of the particles, or at least their surface conductivity. Particles that are good conductors lose their charge to the grounded surface. Particles that are poor conductors or nonconductors retain their charge, and are held to the surface by their own "image" force, F_i.[21] (The theory of image forces is discussed in more detail elsewhere.[24]) This force represents the attraction between the charged nonconductor particle and the grounded surface, which is equivalent to a similar charge of opposite sign in the mirror-image position (Fig. 15.7). The image force is given by

$$F_i = \frac{1}{K_\epsilon} \frac{e^+ e^-}{y^2} \quad (15.1)$$

This equation is of course Coulomb's law, in which e^- is the total negative electric charge on the particle and e^+ is the corresponding total positive image charge, y is the distance from the charge to the grounded surface, and K_ϵ is the dielectric constant. This equation can be written[25] as

$$F_i = 9 \times 10^9 \frac{e^+ e^-}{y^2} \quad (15.2)$$

for electrostatic separation in air.

If these equations are to be used, it is necessary to be able to measure the surface charge or charge density of mineral particles. Techniques have been developed to measure surface charge density and also electrical conductivity in particles.[26,27]

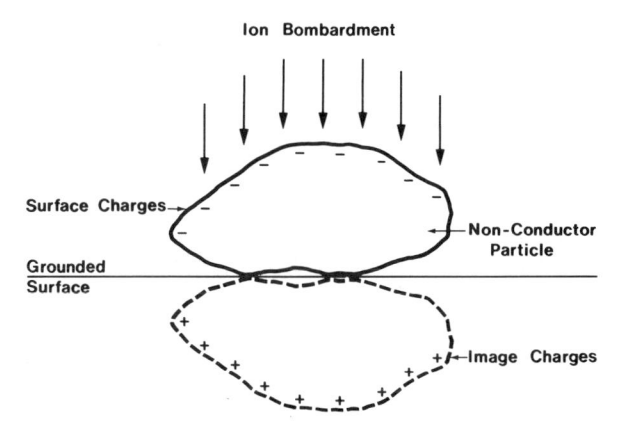

Figure 15.7. The concept of the "image" force.

Charging by Induction. If a particle is placed on a grounded conductor in the presence of an electric field, the particle will rapidly develop a surface charge by induction. Both conductor and nonconductor particles will become polarized, but a conductor particle will have an equipotential surface through its contact with the grounded conductor. The nonconductor particle will remain polarized. This is shown in Fig. 15.8.

There is a finite time required for a particle to be charged. This is given by[2,21]

$$e^\pm = K_{Cp} \mathscr{V} \left[1 - \exp\left(\frac{-t}{K_r K_{Cp}}\right)\right] \quad (15.3)$$

($e^\pm$ = electric charge; K_{Cp} = capacitance of particle; K_r = resistance of particle; $\mathscr{V}$ = applied potential). For a conductor particle in which conductivity is high, the particle quickly reaches its total charge $K_{Cp}\mathscr{V}$, hence experiences an electrical force F_e, given by

$$F_e = e^\pm \mathscr{E} \quad (15.4)$$

($\mathscr{E}$ = electric field strength or intensity).

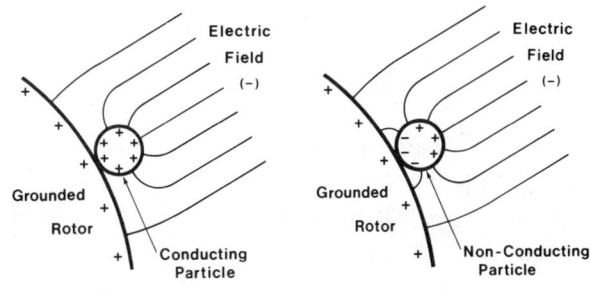

Figure 15.8. Charging particles by induction.

15.3.2. Separation at a Grounded Surface

The separation that occurs at the grounded surface in electrostatic separators results from the combination of electrical, centrifugal, and gravity forces.[2,21,28,29] The forces acting on a particle on the surface of a grounded rotor are shown in Fig. 15.9. If the particles are assumed to be spherical and to have an even distribution of the surface charge, then the triboadhesive or friction force F_f is small and can be ignored. The remaining forces acting on a particle will now be considered.

The *electrical force* F_e acts in the direction of the electrical field. It is the force of attraction between the charged particle and the electrode. It is given by Eq. 15.4 when the particle is in an electric field of constant strength, as may be approximated near the drum surface.

The concept of the *image force* was introduced in Section 15.3.1. The image force F_i is given by Eq. 15.1, with y becoming the particle diameter d_v. It is the force of attraction between a charged particle and a grounded surface such as the rotor.

The *centrifugal force* is due to the rotation of the grounded surface. For a particle of density ρ_s, it is given by

$$F_c = \tfrac{1}{6}\pi d_v^3 \rho_s \omega^2 R \qquad (15.5)$$

(ω = rotor angular velocity; R = radius).

The *gravitational force* is given by

$$F_g = \tfrac{1}{6}\pi d_v^3 \rho_s g \qquad (15.6)$$

Balance of Forces. In high tension separators where the major charging mechanism is ion bombardment, the important forces are the image and the centrifugal. A balance between these two forces yields a *pinning factor*[28]:

$$\frac{F_i}{F_c} = \frac{6}{\pi \rho_s K_\epsilon} \cdot \frac{(e^\pm)^2}{d_v^5 \omega^2 R} \qquad (15.7)$$

The pinning factor is a measure of the tendency of a particle to be pinned to the rotor surface. If the pinning factor is greater than unity, the particle will adhere to the rotor surface; that is, the particle is classified as a nonconductor. If the pinning factor is less than unity, the particle will be thrown from the rotor surface. It should be noted that in this analysis the effect of multilayers of particles, as found in industrial separation, has not been taken into account. However, the approach has been found to provide an acceptable prediction of the rotor radius and angular velocity that permit a separation between two minerals.[28]

In electro-static separators of the rotor type, the significant forces are the electrical and the centrifugal. A *lifting factor* can be defined similar to the pinning factor discussed earlier as

$$\frac{F_e}{F_c} = \frac{6\mathscr{E}}{\pi \rho_s} \cdot \frac{e^\pm}{d_v^3 \omega^2 R} \qquad (15.8)$$

In this case, the lifting factor is a measure of the tendency of a particle to be lifted from the rotor surface and attracted towards the electrode.

The operation of electro-static separators of the plate type can also be characterized by a lifting factor. In this case the significant forces are electrical and gravity. The lifting factor here is given by

$$\frac{F_e}{F_g} = \frac{6\mathscr{E}}{\pi \rho_s g} \frac{e^\pm}{d_v^3} \qquad (15.9)$$

Particle Discharge Characteristics. In high tension separation, the rate at which a particle discharges should determine where it leaves the rotor. Previous discussion has been in terms of conducting and nonconducting particles. In fact there is a distribution of particle conductivities that results in a range of charge decay curves as indicated in Fig. 15.10. The charge decay on a particle is found to be exponential with time.

15.3.3. Separation by Trajectory

The charge on a particle is assumed to remain constant after it escapes from the grounded conductor in electrostatic separation. Thus the forces acting on the particles are electrical, gravity, and drag, as

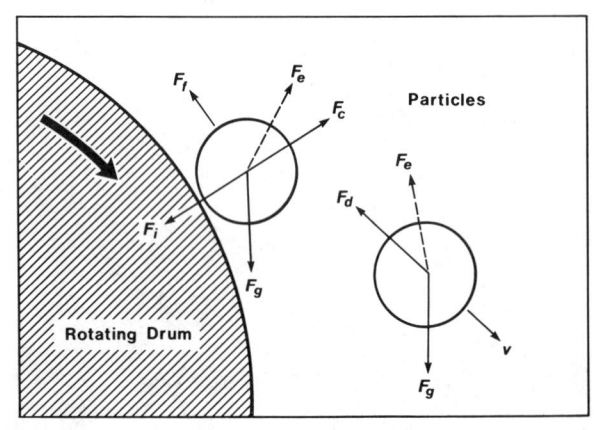

Figure 15.9. Forces acting on particle in electrostatic separators.

Mechanisms of Electrostatic Separation

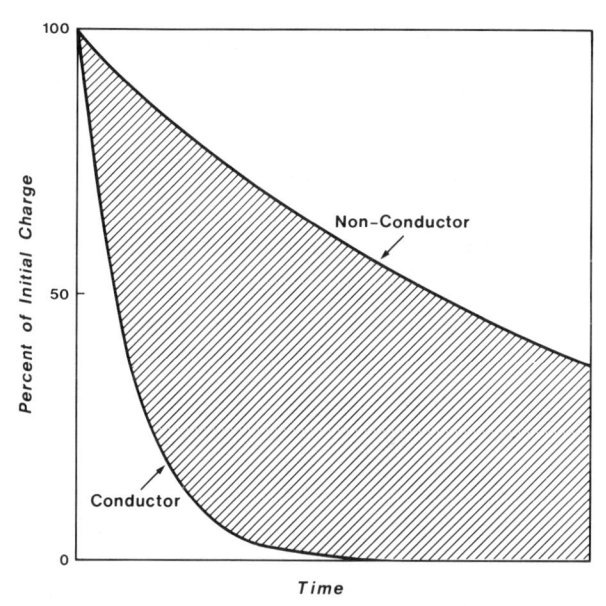

Figure 15.10. Particle discharge curves indicating distribution of conductivities. (After Morrison.[28])

shown in Fig. 15.9. Fraas[2] and Hopstock[30] calculate the trajectory for a particle moving at v m/sec initially from the following equations written for the x- and y-directions:

$$F_x = -F_d \frac{v_x}{v} + e^{\pm}\mathcal{E}_x \quad (15.10)$$

$$F_y = -F_d \frac{v_y}{v} + e^{\pm}\mathcal{E}_y - Mg \quad (15.11)$$

The solution of these equations requires numerical integration.

15.3.4. Multistage Separations

As shown in Figs. 15.2 and 15.3, electrostatic separators are normally stacked in banks for cleaning either the conductor or the nonconductor fractions. When material is passed through a single separator, a complete separation is seldom obtained.

Fraas[2] reports the results of a stagewise cleaning of garnet from chromite. The feed material was passed through a rotor-type electro-static separator. The highly conductive chromite was removed from the less conductive garnet, and the nonconductor residue repassed through the separator an additional seven times. The data obtained are shown in Fig. 15.11, along with the results obtained when the chromite concentrate is repassed through the separator.

The data for the chromite alone can be represented by

$$\log\left(\frac{M}{M_I}\right) = n \log(1 - p) \quad (3.4)$$

(M_I, M = initial mass of feed and mass remaining in the nonconductor product after n passes, respectively; p = probability of a particle being deflected into the conductor product). The probability of a particle being deflected into the conductor product is a single constant value (p = 0.49 in this case). This means that every time a chromite particle passes through the separator, it has the same probability of being extracted as a conductor particle.

The interpretation of the data for the chromite and garnet mixture is more complex. In this case, the data deviate from a straight line for more than four passes. This is because the data represent the deflection of two minerals; the curve is a composite of two straight lines, one representing the chromite and the other the garnet. The garnet particles will have the same probability of being deflected to the conductor product on every pass. The probability of deflection will be much lower than that for the conductive chromite.

This analysis clearly shows why multiple stages of separation are required in electrostatic separation. It would be desirable to be able to relate the

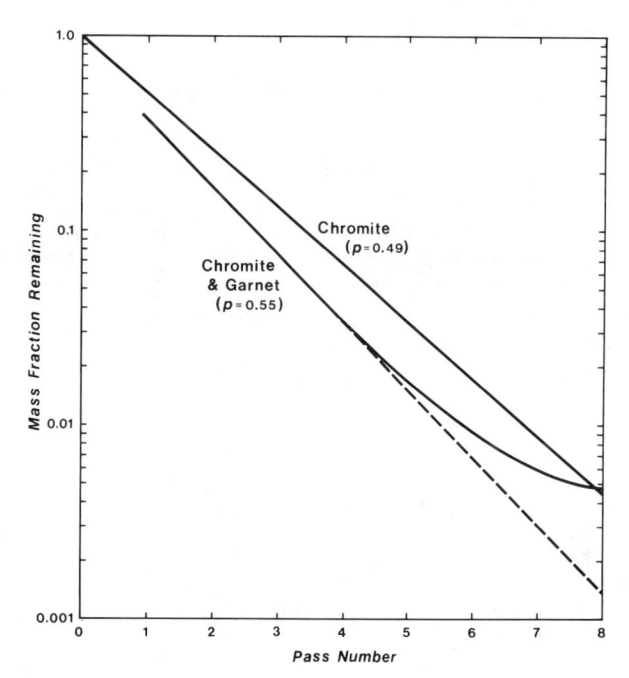

Figure 15.11. Electrostatic separation data. Note constant probability of deflection with each pass. (After Fraas.[2])

value of the probability of deflection for each mineral to the operating variables of the separation. Obviously, a high probability of deflection is required for one mineral and a low probability value for the other. However, unless the probability of one mineral is close to 1 and the other close to 0, multiple stages will give enhanced separation.

Because both electrostatic and flotation separations can be described by Eq. 3.4, Examples 16.1 and 23.1 effectively illustrate the analysis of electrostatic separations.

REFERENCES

1. A. M. Gaudin, *Principles of Mineral Dressing,* McGraw-Hill (1939).
2. F. Fraas, "Electrostatic Separation of Granular Minerals," U.S. Bureau of Mines Bulletin 603 (1962).
3. J. H. Carpenter, "Technical Aspects of High Tension Separation." Presented at AIME Meeting, Florida, 1949.
4. J. H. Carpenter, "Electrical Methods for the Separation of Minerals," *Miner. Sci. Eng.,* **2,** 23–30 (1970).
5. J. H. Carpenter, "Electrostatic Separator," U.S. Patent 2,548,771 (1951).
6. J. H. Carpenter and W. P. Dyrenforth, "Advances in Technique and Equipment for Iron Ore Separation," *Annual Mining Symposium,* University of Minnesota (1965).
7. A. D. Moore, *Electrostatics and Its Applications,* Ch. 10, Wiley (1973).
8. W. P. Dyrenforth, "Electrostatic Separation," in *Mineral Processing Plant Design,* A. L. Mular and R. B. Bhappu (Eds.), pp. 479–489, AIME (1978).
9. R. M. Funk and J. E. Lawver, "Production of Iron-Ore Superconcentrates by High-Tension Electrostatic Separation," *Trans. AIME/SME,* **247,** 23–26 (1970).
10. "Electrostatic Separation System," Mineral Deposits Ltd. Catalogue No. ESS.1 (1976).
11. O. C. Ralston, *Electrostatic Separation of Mixed Granular Solids,* Elsevier (1961).
12. I. M. LeBaron and W. D. Kropf, "Application of Electrostatics to Potash Beneficiation," *Trans. AIME/SME,* **211,** 1081–1083 (1958).
13. E. Northcott and I. M. LeBaron, "Applications of Electrostatics to Feldspar Beneficiation," *Trans. AIME/SME,* **211,** 1087–1093 (1958).
14. E. Northcott and F. N. Oberg, "Application of Electrostatics to Concentration of Coarse Pebble Phosphate," *Trans. AIME/SME,* **211,** 1084–1086 (1958).
15. "High Tension Separation," Carpco HT Gen. Bulletin 75.
16. H. S. Hatfield, "Dielectric Separation: A New Method for the Treatment of Ores," *Bull. IMM,* **233,** 335–342 (1924).
17. V. T. Andres, "Separation of Solid Particles in Liquid Dielectrics," *Powder Technol.,* **23,** 85–97 (1979).
18. I. J. Lin, I. Yaniv, and Y. Zimmels, "On the Separation of Minerals in a High Gradient Electric Field." Presented at 13th Int. Miner. Process. Congr., Warsaw, 1979.
19. C. E. Jordan and C. P. Weaver, "Development of a Continuous Dielectrophoretic Separator," SME Preprint 77-H-323 (1977).
20. R. Elsdon, "Vibrated Bed Electrostatic Separation," *Powder Technol.,* **23,** 121–129 (1979).
21. J. E. Lawver, "Fundamentals of Electrical Concentration of Minerals," *Mines Mag.,* 20–23 (January 1960).
22. A. M. Gaudin, "The Principles of Electrical Processing with Particular Application to Electrostatic Separation," *Miner. Sci. Eng.,* **3,** 46–57 (April 1971).
23. L. B. Loeb, "Recent Development in Analysis of Positive and Negative Coronas in Air," *J. Appl. Phys.,* **19,** 882–897 (1948).
24. A. Sommerfeld, *Electrodynamics,* Academic Press (1952).
25. J. E. Lawver and J. L. Wright, "Basic Studies of Electrostatic and Electrodynamic Separation of Iron Ore," University of Minnesota Experiment Station, Progress Report No. 4 (1968).
26. J. E. Lawver and J. L. Wright, "A Cell for Measuring the Electrical Conductivities for Granular Materials," *Trans. AIME/SME,* **244,** 78–82 (1968).
27. J. E. Lawver and J. L. Wright, "The Design and Calibration of a Faraday Pail for Measuring Charge Density of Mineral Grains," *Trans. AIME/SME,* **244,** 445–449 (1968).
28. R. D. Morrison, "Review of High Tension Roll Separation," JKMRC Report, Australia (1974).
29. J. F. Delon, "Théorie de la Separation Electrostatic avec l'Aide de l'Effet Corona," *Ann. Mines,* No. 3 (1966).
30. D. M. Hopstock, "Analysis of the Performance of a Rotating Drum-Type Electrostatic Separator," AIME Annual Meeting (1971).

Chapter Sixteen

Flotation and Other Surface Separations

Finagle's Rules for Scientific Research:
"Do not believe in miracles—rely on them."

The introduction of the flotation process early in this century revolutionized the minerals industry, and must be considered to be a major step in man's technological advance. In the 25 years following its first commercial application at Broken Hill in Australia, great strides were made both in the chemical aspects of flotation and in equipment development. Today flotation is clearly the dominant mineral concentration method; it is used for almost all sulfide minerals and is widely used for nonsulfide metallic minerals, industrial minerals, and coal. Historical reports of the development of this process are available.[1-4]

Flotation can be applied to low grade ores and to ores that require fine grinding to achieve liberation. Since it is a relatively selective process, an important application is in the separation and concentration of the minerals in complex ores, for example, sulfide ores containing copper, lead, and zinc.

Although flotation is the primary concentration process based on the interfacial chemistry of mineral particles in solution, and is emphasized in this chapter, a number of other processes are in commercial use, and there is the potential for further commercial processes to be developed. Such processes as selective flocculation, selective agglomeration, and various modifications of the flotation process are also treated in this chapter.

16.1. FLOTATION EQUIPMENT

Since flotation was first developed as a concentration process, many designs of flotation machine have been introduced. They may all be considered in either of two categories: *mechanical* flotation machines, which from the time of the early Minerals Separation flotation machine to today have clearly been the most widely used, and *pneumatic* flotation machines, which had early success in the form of the Callow pneumatic flotation cell but are today limited in use to special applications. Within each category there are two types, those that are operated as a single tank, and those that are operated as a bank of tanks (or cells) in series. Most flotation is carried out in *banks of flotation cells*, for reasons discussed in Section 16.3 and in Chapter 23.

Concentration by flotation may be considered in terms of two groups of variables. First, chemical conditions: the interaction of chemical reagents with the mineral particles to make one selectively hydrophobic. Second, physical-mechanical conditions, which are determined by the flotation machine characteristics.

Although there are many different designs of flotation machine (Figs. 16.1–16.3 and Table 16.1), they all have the primary function of making the particles that have been rendered hydrophobic contact and adhere to air bubbles, then allowing those particles to rise to the surface and form a froth, which can be removed. To achieve this function, a flotation machine must:

1. Maintain all particles in suspension. This requires that upward pulp velocities exceed the settling velocity of all particles present (including largest/heaviest).
2. Ensure that all particles entering the machine

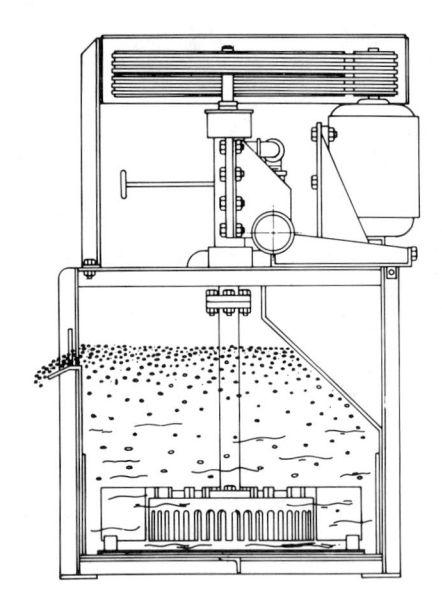

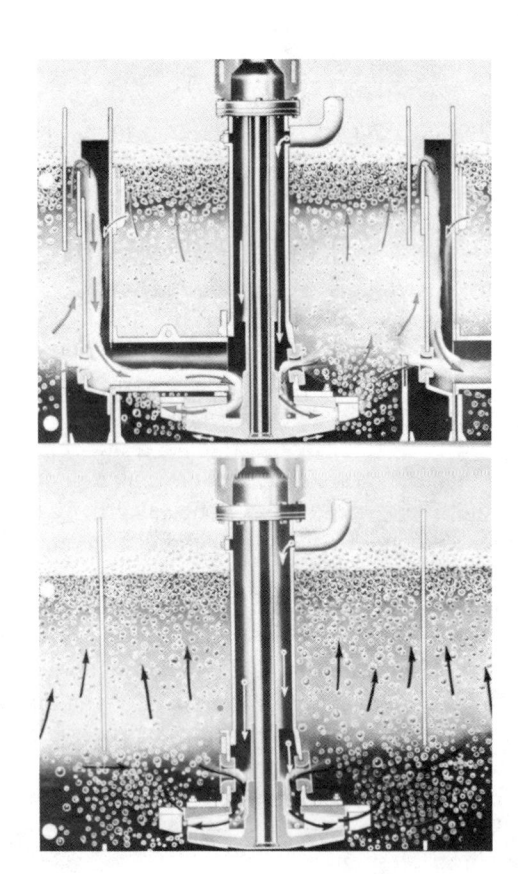

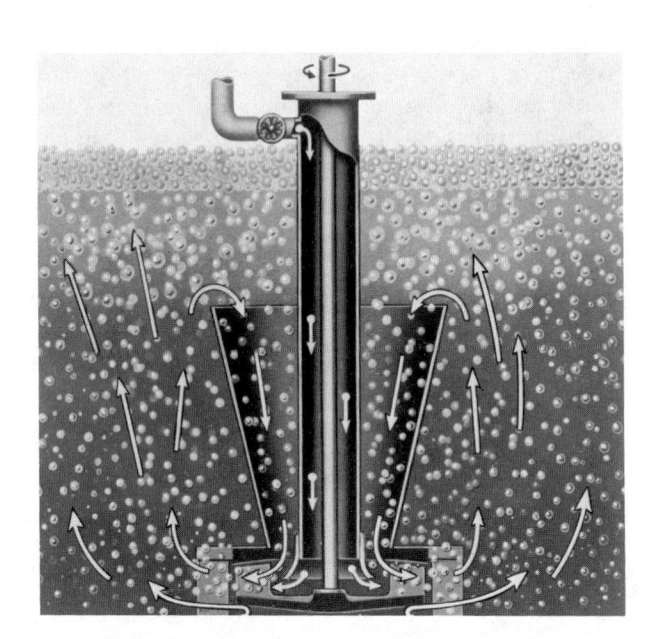

Figure 16.1. A selection of industrial flotation machines. (*a*) Agitair flotation cells. (Courtesy The Galigher Co.) (*b*) Denver flotation cell mechanisms. (Courtesy Joy Manufacturing Co.) (*c*) Outokumpu flotation cell agitator. (Courtesy Outokumpu Oy.) (*d*) Wemco-Fagregren flotation cell schematic. (Courtesy Wemco Division of Envirotech Corp.) (*e*) Cyclo-cell. (Courtesy Heyl & Patterson Inc.) (*f*) Dissolved air flotation machine. (Courtesy The Permutit Co., Inc.)

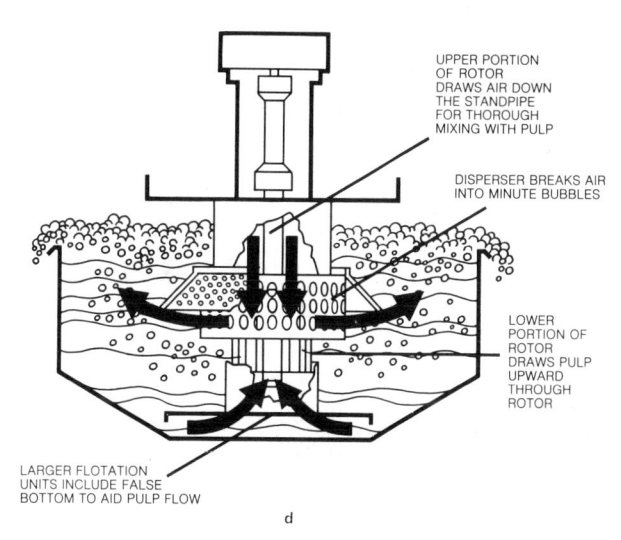

UPPER PORTION
OF ROTOR
DRAWS AIR DOWN
THE STANDPIPE
FOR THOROUGH
MIXING WITH PULP

DISPERSER BREAKS AIR
INTO MINUTE BUBBLES

LOWER
PORTION OF
ROTOR
DRAWS PULP
UPWARD
THROUGH
ROTOR

LARGER FLOTATION
UNITS INCLUDE FALSE
BOTTOM TO AID PULP FLOW

c

d

e

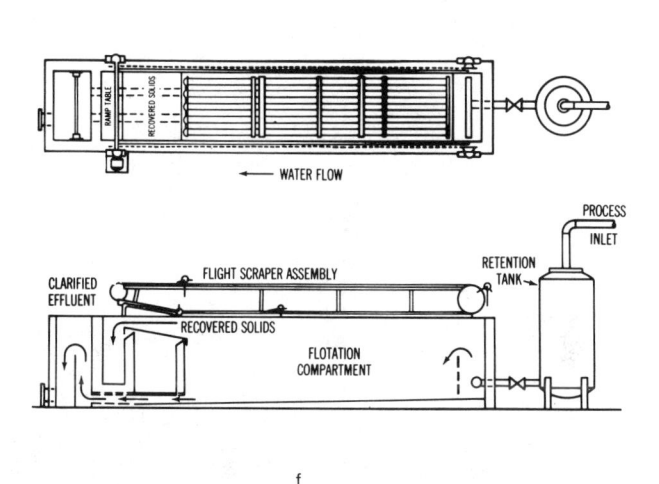

RAMP TABLE   RECOVERED SOLIDS

← WATER FLOW

PROCESS
INLET

FLIGHT SCRAPER ASSEMBLY   RETENTION
                            TANK
CLARIFIED
EFFLUENT
           RECOVERED SOLIDS
                            FLOTATION
                            COMPARTMENT

f

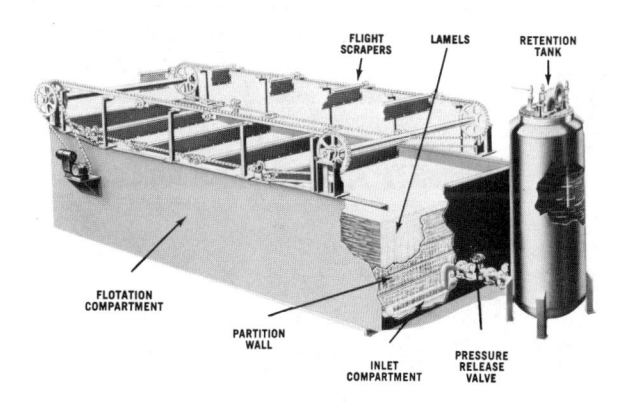

FLIGHT   LAMELS   RETENTION
SCRAPERS   TANK

FLOTATION
COMPARTMENT

PARTITION
WALL
        INLET   PRESSURE
        COMPARTMENT   RELEASE
                      VALVE

303

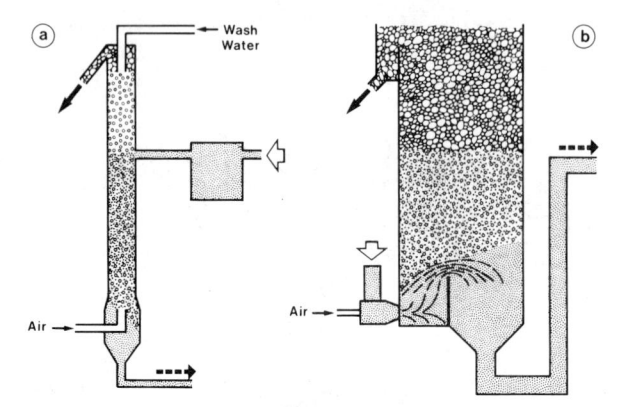

Figure 16.2. Schematic illustrations of (*a*) flotation column and (*b*) Davcra flotation machine.

have the opportunity to be floated. Bypassing or short-circuiting through the machine should be minimized. Likewise, dead space is undesirable, because it reduces the effective volume of the machine.

3. Disperse fine air bubbles throughout the pulp. The extent of aeration required depends on the particular mineral system and mass fraction to be floated.
4. Promote particle-bubble collision so that hydrophobic particles may attach to bubbles and rise to the froth. This can be done by vigorous agitation, countercurrent flow, or dissolved air (gas) precipitation.
5. Provide a quiescent pulp region immediately below the froth to minimize pulp entrainment in the froth and turbulent disruption of the froth layer.
6. Provide sufficient depth of froth to permit drainage of entrained particles to occur.

The characteristics of a number of the available flotation machines are listed in Table 16.1. More extensive details are presented in the literature.[5,6] Selection of a particular type of flotation machine for an application is clearly not a straightforward task. The main factors to be considered in rating machine performance are:[5,7]

1. Metallurgical performance, as represented by grade and recovery.
2. Capacity, as tonnes per hour of feed per unit volume.
3. Operating costs per tonne of feed, including power consumption, maintenance, and direct labor.
4. Ease of operation (which may well be subjective, based on past experience).

16.1.1. Mechanical Flotation Machines

Mechanical flotation machines of the cell type (see Figs. 16.1*a*–*d* and Table 16.1 for selected examples) are the most commonly used in the flotation of metallic minerals. Each machine (an individual flotation cell) has an impeller that rotates within baffles. Air is introduced through the impeller to provide good dispersion and sufficient mixing to cause the particle-bubble collisions that are the essential prerequisite to particle-bubble attachment. The air can be drawn through the impeller shaft or through a standpipe surrounding the shaft by the suction created at the impeller, in which case the aeration rate is limited by the impeller design and speed, or it can be introduced under pressure. The latter approach has been used for many years in Agitair flotation machines (Fig. 16.1*a*) for flotation separations in which stronger frothing is required. It

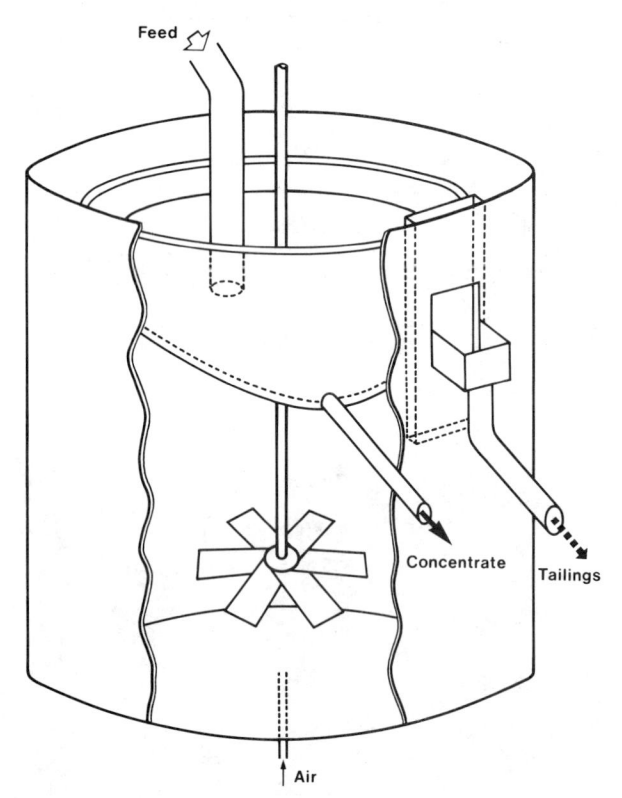

Figure 16.3. Schematic illustration of a tank-type Maxwell flotation machine.

TABLE 16.1: THE MAJOR TYPES OF FLOTATION MACHINES

	MACHINE	SIZE (m^3)	IMPELLER -Diameter (cm) -Peripheral Speed (m/min)	CAPACITY* (m^3/hr)	POWER* (kw)	AERATION -Rate (m^3/min) Fl_g (x 10^3)	CHARACTERISTICS AND APPLICATIONS
MECHANICAL — CELL	Agitair	0.008 to 18.4 (28.3)	11.4 to 101.6 / 259 to 518	1.3 to 3800	0.11 to 32.8	0.057 to 15.6 / 7 to 26	All cells approximately constant depth. High froth area to cell volume ratio. Impeller disc with posts. Extensively baffled. Largest cells have four impellers. Air introduced at low pressure through impeller shaft. Widely used with ores and coal (Fig. 16.1a).
	Denver	0.007 to 14.2 (36.1)	11.4 to 83.8 / 381 to 503		0.19 to 23.1	0.14 to 11.3 / 9.5 to 12.6	Cell-to-cell or open trough design available. Impeller flat-turbine type. Air introduced at low pressure through impeller shaft (cell-to-cell may be self-aerating). Increased recirculation by use of collar improves suspension of coarse particles. Wide use with ores and coal (Fig. 16.1b).
	Krupp	0.50 to 3.5 (5.0)	38.1 to 71.9 / 509 to 543	3.2 to 21.6	1.6 to 10.5	0.13 to 0.88 / 1.2 to 2.4	Self-aerating cell. Impeller of flat-turbine type. Used with ores and coal.
	Outokumpu	2.7 to 16.0	37.8 to 75.9 / 314 to 381	36 to 1540	5.0 to 22.0	1.0 to 20.0 / 4.4 to 21.1	Impeller (flat-turbine) curved to give even dispersion of air and to maintain uniform coarse solids suspension throughout cell. Air introduced at low pressure through impeller shaft. Used with ores (Fig. 16.1c).
	Sala	2.7 to 10.7	77.1 / 445 to 473	12.7 to 38.1	4.7 to 24.6	2.3 to 8.0 / 5 to 5.8	Shallow, flat-turbine impeller. Air introduced at low pressure through stand pipe and aerator ring to periphery of impeller blades. Used with ores.
	Wemco-Fagregren	0.026 to 12.0 (28.3)	8.9 to 66.0 / 274 to 396	0.19 to 106	0.30 to 25.4	0.057 to 9.9 / 5 to 11.4	Impeller at shallow depth to allow for self-aeration through stand pipe. Impeller of turbine-rotor type with surrounding disperser. With very large cells, false bottom to cell provides improved recirculation and suspension of coarse particles. Widely used with ores and coal (Fig. 16.1d).
MECHANICAL — TANK	Maxwell	4.2 to 56.6	45.7 to 107 / 390 to 402	32 to 422	3.4 to 23.1	1.4 to 8.5 / 4.1 to 5.8	Cylindrical tank with froth launder internal (annulus) or external. Impeller can be turbine (flat or pitched) or propeller (downwards). Aeration by air-sparger below impeller or by air distribution ring. Low aeration and low power intensity not in wide use; used with ores and coal (Fig. 16.3d).
	Nagahm	0.045 to 8.1	10.2 to 83.2 / 351 to 823	0.013 to 3.4	0.33 to 7.5	0.16 to 4.2 / 4.6 to 15.1	Cylindrical tank with froth plow to both internal and external launders. Impeller is curved-blade turbine. High aeration rates, minimum pulp turbulence. Used for fine ores and coal; also used for ultra-fine particles and in ion and precipitate flotation.
PNEUMATIC — COLUMN	Column	2.55 to 40.1	None	14 to 230	3.7 to 22.4	0.28 to 4.5 / —	Square cross-section column 13.7m in height. Large units divided internally into (effectively) four smaller units. May have wash water addition to top of froth column. Gentle agitation by air flow countercurrent to pulp flow. Potential use with ores and coal (Fig. 16.2).
PNEUMATIC — TANK or CELL	Cyclo-cell	6.8 to 22.6	None				Pulp pumped tangentially through submerged vortex chamber to provide agitation. Low pressure air introduced to vortex to give air dispersion on discharge. No baffles in cell. Used extensively for coal (Fig. 16.1e).
	Davcra	5.4 to 34	None				Pulp pumped tangentially into external nozzle, then discharged into cell to provide agitation. Air injected into vortex within nozzle for intimate mixing. Baffle in cell to prevent short-circuiting to tailings. Used with ores and coal (Fig. 16.2).
	Flotaire	12 to 42	None			Aspirated	Cylindrical tank with external launder. Water manifold around tank connects to aspirators, the water and air introduced through distribution plate at bottom of tank. Feed introduced below froth at top of tank. Flow countercurrent in 4.6m high tank. Good for coarse particles. Introduced for use with coal and ores.

*Based on ore having a density of 2700 kg/m^3 in a 25% solids slurry, normal impeller speed and aeration, medium floating ore.

has also found favor more recently because it allows aeration rate to be used as a variable in the automatic control of flotation operations.

The design of impeller and dispersion baffles varies considerably from one machine to another. In each machine, a highly turbulent region is required to provide particle-bubble contact, but also a quiescent zone adjacent to the froth layer in which the mineral-laden bubbles can rise without turbulent disruption of particles adhering to the bubble surface. Differences in design used to create these regions can be seen in Fig. 16.1a–d; details of the designs and the particular advantages of each are discussed in the literature.[6,8–10]

In recent years, there have been dramatic increases in the size of individual flotation cells.[11–13] These large cells have been designed using hydrodynamic scale-up criteria (Sections 4.8.1 and 16.3.1), and have had wide industry acceptance.

Froth product discharge in cell-type mechanical flotation machines may be either by direct overflow, or with mechanical assistance in the form of paddles. Pulp flow is from one cell to the next, with an overflow weir between cells, a partial baffle, or no baffle at all between cells, as in the "hog trough." In the first arrangement, mixing is close to perfect, whereas in the latter two, bypassing and even back-mixing can occur between cells.[14]

Single tank mechanical flotation machines such as the Maxwell[6,15,16] and the Nagham,[17] are not widely used but have been successful in particular applications (Fig. 16.3 and Table 16.1).

16.1.2. Pneumatic Flotation Machines

In pneumatic flotation machines of both cell and tank types,[6,18,19] mixing of air and pulp occurs in injection nozzles (Figs. 16.1 and 16.2, Table 16.1). In the flotation column (Fig. 16.2 and Table 16.1) countercurrent flow is established in the lower section of the column. If water sprays are used at the top of the froth column, enhanced drainage of entrained particles from the froth can be expected. Although extensively tested, flotation columns have not yet found industrial acceptance.[6,19–21]

Other methods are also used for introducing the air into the pulp, although they are in general restricted to applications outside the traditional mineral processing area, such as the treatment of industrial effluents. *Dissolved-air flotation* involves the dissolution of air (or other gas) into the liquid while under pressure, followed by the precipitation of bubbles on fine particles on return to atmospheric pressure (Fig. 16.1f). In *vacuum flotation*, the liquid is saturated with air at atmospheric pressure, then a vacuum applied to achieve the same result as above. A third method is *electroflotation*, in which the gas bubbles are generated by electrolysis.

16.1.3. Laboratory Flotation Machines

Laboratory flotation testing is required in the initial process design, for process improvement during the life of a plant, and to continually optimize the pro-

Figure 16.4. Laboratory test flotation machine.

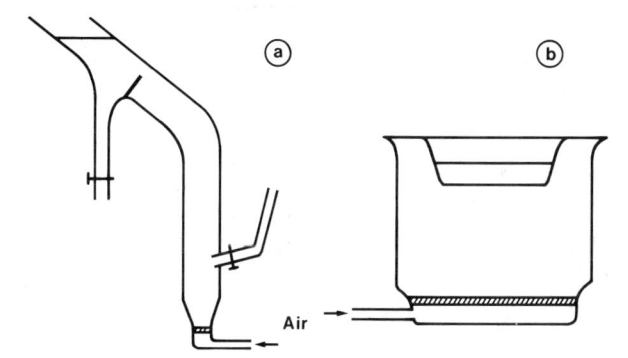

Figure 16.5. Schematic illustration of "micro" flotation test equipment. (a) Hallimond tube. (b) Microflotation cell.

cess as changes in ore grade or mineralogy occur. Most testing is carried out in batch flotation cells (Fig. 16.4), which accept between 500 and 2000 g of ore. Test procedures using such cells are described elsewhere.[1,22] Data from such laboratory cells can be used to specify reagent requirements and retention times in full-scale operations.

For more basic research purposes, a number of very small flotation cells have been designed. Most widely used are the Hallimond tube[23,24] and the microflotation cell[25] (Fig. 16.5), although other cells have been reported.[26,27]

16.2. CHEMISTRY OF FLOTATION

The principles of interfacial chemistry and the basis for their utilization in the selective flotation of minerals were discussed in Chapter 6, in books on the subject by Sutherland and Wark,[2] Gaudin,[3] Klassen and Mokrousov,[28] and others,[4,29,30] and in numerous papers to be found in the mineral processing literature. Although the chemistry of flotation can be discussed in general terms, it must be kept in mind that each application is a special case, in which a unique combination of mineral chemistry and water chemistry is involved. Thus it is impossible to select the flotation reagents, reagent levels, and conditions required at a mine site solely from the separation of the same minerals at another, although this may be a good starting point.

16.2.1. Flotation Reagents

Most minerals are naturally hydrophilic. To achieve a separation by flotation, the surfaces of one mineral must be selectively rendered hydrophobic (Section 6.1). This may be achieved by the *regulation of the solution chemistry*, followed by the addition of a *collector* that selectively adsorbs and provides the required hydrophobic surface. A small number of minerals are naturally hydrophobic, such as coal[31] and molybdenite.[32-34] It is possible to float these minerals without any collector, although it is common practice to add a "supplemental collector" as discussed later. Other minerals also can be rendered hydrophobic without the use of collector. Additions of large concentrations of sodium sulfide to sulfide minerals can result in hydrophobic surfaces and subsequent "collectorless" flotation;[35,36] in this case it is believed that the sulfide ion has replaced oxidation and hydration products at the mineral surface.

The regulation of the solution chemistry can include the addition of reagents; either *activators*, that increase selectivity by enhancing collector adsorption, or *depressants*, that retard or prevent collector adsorption. It can also include the addition of a *dispersant* to ensure that mineral surfaces are free of fine particles,[37] or the use of activated carbon to remove unwanted ions or molecules from the solution.[38] Finally, regulation of solution chemistry can include control of pH to ensure that the collector (often a weak acid) will exist in solution in the desired form.

Another group of reagents that must be considered in flotation is the *frothers*. Frothers have two functions: to improve the dispersion of fine bubbles in the pulp, and to control the characteristics of the froth.

Collectors. The collector is the most critical of the flotation reagents. Collectors are organic molecules or ions that selectively adsorb on mineral surfaces. They must render the mineral surface hydrophobic, so that at equilibrium (as indicated by the contact angle) there is bubble-particle attachment, and they must also reduce the induction time, to ensure that attachment occurs on bubble-particle collision (Section 6.1).

Most collectors are weak acids, bases, or their salts. They are *heteropolar* and may be considered to have two functional "ends," one ionic, which can be adsorbed at the mineral surface either by chemical reaction with the ions of the mineral surface (chemisorption) or by electrostatic attraction to the mineral surface (physical adsorption), and the other an organic chain or group, which provides the hydrophobic surface to the mineral. Some collectors, however, are non-ionizing compounds; once adsorbed, they render the mineral hydrophobic in the same manner as heteropolar collectors.

Figure 16.6 shows most of the widely used collectors, both anionic and cationic. A detailed discussion of each is beyond the scope of this book.[3,30,39]

Anionic collectors of thiol type, in which the polar group contains bivalent sulfur, are used for sulfide mineral flotation. Xanthates[40] and dithiophosphates are the most widely used; dialkyl thionocarbamates have been well accepted in recent years (e.g., Dow reagent Z-200); thiocarbanilide and mercaptobenzothiozole (e.g., Cyanamid reagent R-404) are used as supplementary collectors; dithiocarbamates and alkyl mercaptans[41] have found limited use; and xanthogen formates (e.g., Minerec) have not been widely used, but are expected to be more significant sulfide collectors in the future. Dixanthogens,[40] and in fact the dithiolates that result from the oxidation of other thiol collectors as well, act as collectors but are in general formed in solution or at the mineral surface rather than added.[42]

Hydrocarbon chain length in thiol collectors is quite short. With most of these collectors, decreasing solubility as chain length increases limits chain length to about 6 carbon atoms. It should be noted, however, that in general longer chain length results in stronger adsorption of the collector but reduced selectivity between sulfides. For maximum selectivity it is normal practice to use a short chain collector.

Oxyhydryl anionic collectors are used for the flotation of nonsulfide minerals. Included in this group are the carboxylates, organic sulfates, and sulfonates. The carboxylates [fatty acids and the corresponding soaps (Na^+ salts)] are the most widely used, normally as a combination that has been extracted as a by-product from a natural plant or animal unsaturated fat source. The most frequently used fatty acid collector is marketed as "tall oil," a by-product of the wood pulp industry. Tall oil is approximately 50% oleic acid, with lesser amounts of linoleic, linolenic, and rosin (abietic) acids (Fig. 16.7). Sodium salts are often used as collectors, since they are more soluble than the associated acids. It is largely because of solubility that fatty acids used as collectors are limited to the 18 carbon chain of oleic acid (or shorter), and to unsaturated hydrocarbon chains.

The organic sulfates and sulfonates are not used as frequently as the fatty acids. Although they act much like the fatty acids, they tend to adsorb less

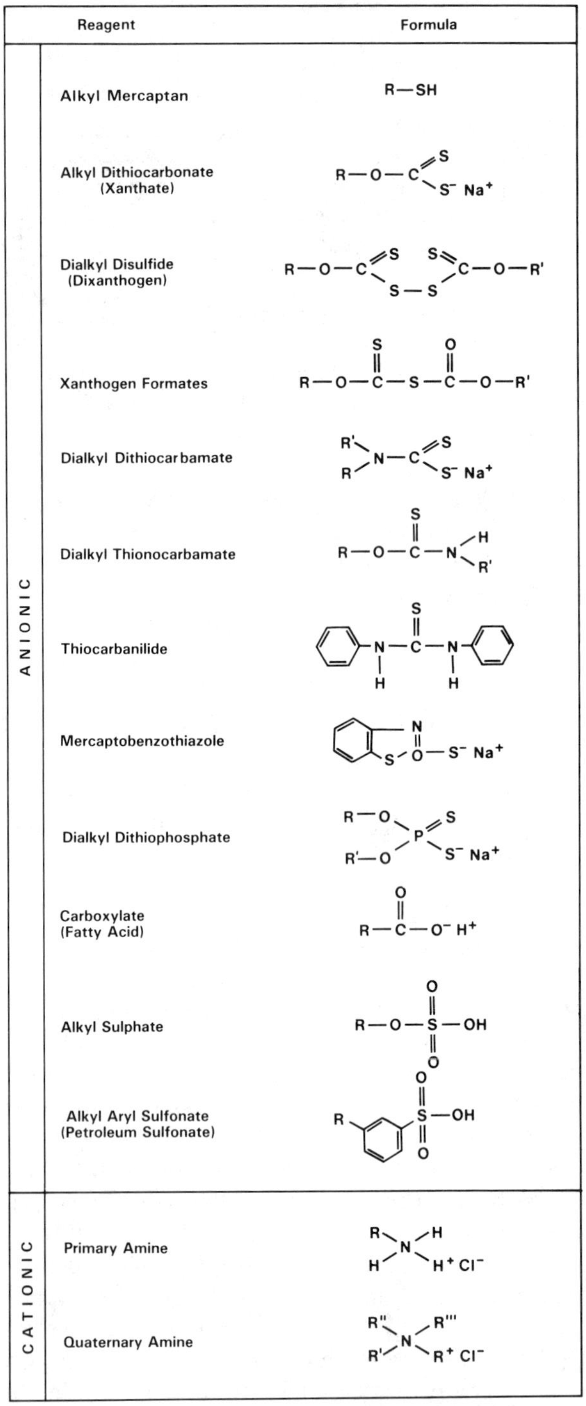

Figure 16.6. Common flotation collectors.

strongly and so have application where greater selectivity is required. Other collectors, such as the hydroxamates, are also in this group; they have not yet, however, found successful commerical application.[43,44]

Chemistry of Flotation

Oleic	$CH_3(CH_2)_7CH=CH(CH_2)_7COOH$
Linoleic	$CH_3(CH_2)_4CH=CHCH_2CH=CH(CH_2)_7COOH$
Linolenic	$CH_3CH_2CH=CHCH_2CH=CHCH_2CH=CH(CH_2)_7COOH$
Abietic (rosin)	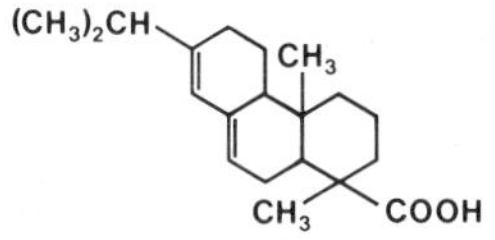 (structure of abietic acid)

Figure 16.7. Carboxylic acids found in tall oil.

The concentration in solution of longer chain collectors (e.g., the oxyhydryl collectors in common use) is limited by the association of ions and/or molecules that can occur. This association or *micellization* occurs at a particular concentration for each chemical species, known as the critical micelle concentration (CMC). If the concentration of collector added to a solution exceeds the CMC, micellization occurs and the concentration of collector available for mineral adsorption is greatly reduced. CMC values for most collectors are available.[44-46]

Cationic collectors have a positively charged polar group associated with the hydrophobic hydrocarbon chain or group. They are usually amines (Fig. 16.6), although ether amines are also in use. Amines from primary to quaternary have been used, but it is the primary and secondary that are in common use. These collectors are often derived from natural fats and are often marketed under the name of the particular fat source (e.g., tallow amine acetate). Both alkyl and aryl hydrocarbon groups are used, the group chain or length being limited by the amine solubility. To assist in solubility, the amine collectors are normally available as chlorides or acetates.

Non-ionic supplemental collectors are also used. However, these are not collectors in the sense intended here: in fact, they adsorb onto another collector, which is adsorbed on the mineral surface, rather than onto the mineral surface itself (Fig. 6.10d). Thus they are used to increase the hydrophobicity of the particles that have a collector already adsorbed on them; they are not in any way mineral selective. Fuel oil is commonly used in this way.

Additional collectors are continually being developed and tested,[47] and further developments are to be expected. Of particular interest are the possibilities of "designing" collectors for specific minerals.[48]

Frothers. Frothers are water-soluble organic reagents that absorb at the air-water interface. They are heteropolar molecules (Fig. 16.8), with a polar group to provide water solubility, and a nonpolar hydrocarbon group.

A frother is required to provide a froth above the pulp that is stable enough to prevent undue froth breakage and subsequent return of particles to the pulp before the froth is removed. It is important, however, that the froth break down rapidly once removed; otherwise problems occur in slurry pumping and in subsequent processing steps.

Another important requirement in a frother is that it should not adsorb on mineral particles: if a frother were to act as a collector, the selectivity of the collector in use would be reduced. Some collectors, such as the fatty acids, do exhibit frothing properties. However, for good plant control the interaction of frothing agent and collector should be minimized.

Alcohols and related compounds such as the glycol ethers are most widely used as frothers, largely because of their inability to adsorb on mineral particles, hence to act as collectors. Aromatic alcohols from natural sources, such as pine oil or cresylic acid, have been extensively used. Synthetic frothers are now widely used; they have the advantage of closely controlled composition, which assists in maintaining plant stability. Methyl isobutyl carbinol, and the polypropylene glycol ethers, are in this category of frothers.

Reagent	Formula
Polypropylene Glycol Ether	$CH_3-(O-C_3H_6)_n-OH$
Methyl Isobutyl Carbinol (MIBC)	$CH_3-CH(CH_3)-CH_2-CH(OH)-CH_3$
Terpineol (Pine Oil)	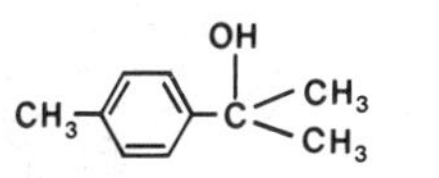 (aromatic structure with OH and C(CH_3)_2 group)
Xylenol (Cresylic Acid)	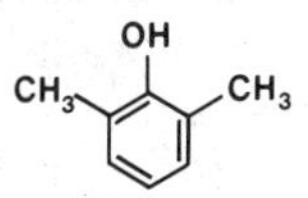 (xylenol structure: OH on benzene with two CH_3 groups)

Figure 16.8. Common flotation frothers.

Any compound that absorbs at the air-water interface will exhibit frothing properties. In fact, solutions of high ionic strength exhibit frothing properties as a result of the depletion of ions that occurs at the interface (Fig. 6.3).

16.2.2. Sulfide Flotation

The flotation of sulfides is of major economic importance and has been carried out for many years; thus it has been extensively analyzed and reported in the literature.[2-4,29] In the brief treatment given here, the chemical aspects of sulfide flotation are considered in two parts: the adsorption of collector on sulfide minerals, and the use of activators and depressants to improve selectivity between sulfides.

Collector Adsorption. After long and often controversial debate, it is now widely accepted that there are two separate mechanisms by which collectors adsorb on sulfide minerals.[29,49-52] One mechanism is chemical, and results in the presence of chemisorbed metal xanthate (or other thiol collector ion) at the mineral surface. The other mechanism is electrochemical and gives an oxidation product (dixanthogen if collector added is xanthate) that is the hydrophobic species adsorbed at the mineral surface. It is thought that with galena, chalcocite, or sphalerite, the chemisorption mechanism occurs. With pyrite, arsenopyrite, or pyrrhotite, the mechanism is electrochemical oxidation. With chalcopyrite, both mechanisms occur. It should be noted, however, that electrochemical oxidation has been reported to occur to some extent when chemisorption is the major mechanism. The two mechanisms are as follows:

Chemisorption Mechanism. Galena is used as an example. In collector adsorption by this mechanism, the following steps occur:

1. Oxidation of surface sulfide to sulfate (or thiosulfate) by dissolved oxygen in the pulp, according to the reaction:

$$PbS_{(s)} + 2O_{2(g)} \rightleftharpoons PbSO_{4(s)} \quad (16.1)$$

The equilibrium constant for this reaction is 10^{126}. Thus at equilibrium, an oxygen partial pressure of 10^{-63} atm would be required; this is clearly exceeded by the partial pressure of oxygen in an aerated pulp (0.2 atm).

2. Ion-exchange replacement of surface sulfate ion by carbonate ion.

$$PbSO_{4(s)} + CO_3^{2-} \rightleftharpoons PbCO_{3(s)} + SO_4^{2-} \quad (16.2)$$

In aerated pulps, carbonate ions will be present. Solubility products of the lead sulfate and carbonate indicate this reaction will occur. Experimental evidence indicates the presence of both sulfate and carbonate ions.[53,54]

3. Ion-exchange replacement of surface carbonate and sulfate ions (and hydroxyl ion, depending on pH) by xanthate ion (indicated by X^-).

$$PbSO_{4(s)} + 2X^- \rightleftharpoons PbX_{2(s)} + SO_4^{2-} \quad (16.3a)$$

$$PbCO_{3(s)} + 2X^- \rightleftharpoons PbX_{2(s)} + CO_3^{2-} \quad (16.3b)$$

$$Pb(OH)_{2(s)} + 2X^- \rightleftharpoons PbX_{2(s)} + 2OH^- \quad (16.3c)$$

Since lead xanthates are more stable than the carbonate or sulfate, lead xanthates form at the galena surface. This has been shown experimentally.[55]

4. Bulk precipitation of lead xanthate at the mineral surface as indicated by greater than monolayer coverage.[55,56]

Electrochemical Oxidation Mechanism. When a conductive mineral, such as a sulfide, is in an aqueous solution, it develops a potential called the "rest potential." Consider pyrite as an example. At the surface of the pyrite, two independent electrochemical reactions occur.[57-60] Dixanthogen (or dimer of other collector ion) forms by anodic oxidation of the xanthate ion,

$$2X^- \rightleftharpoons X_2 + 2e^- \quad (16.4)$$

while the cathodic reaction is the reduction of oxygen adsorbed at the surface.

$$\tfrac{1}{2}O_{2(ads)} + H_2O + 2e^- \rightleftharpoons 2OH^- \quad (16.5)$$

Electron transfer is through the sulfide mineral. The overall reaction is:

$$2X^- + \tfrac{1}{2}O_{2(ads)} + H_2O \rightleftharpoons X_2 + 2OH^- \quad (16.6)$$

for which the reversible potential can be determined. If the rest potential for the mineral-solution system is greater than the reversible potential for xanthate oxidation, as it can be with pyrite, dixanthogen will form. When the rest potential is less than the reversible potential for xanthate oxidation, dixanthogen cannot form and the metal xanthate is the only possible adsorbed species.

Chemistry of Flotation

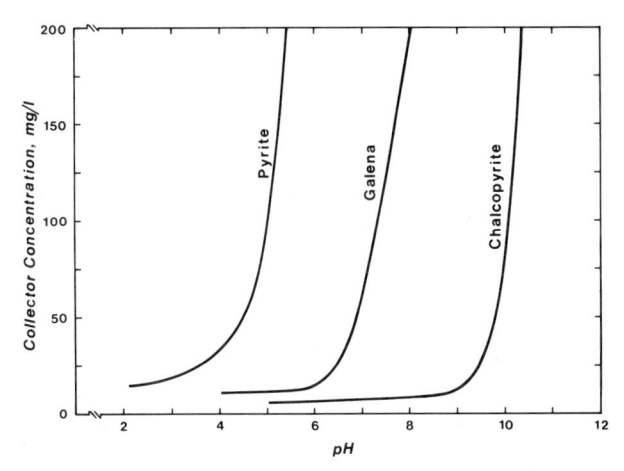

Figure 16.9. Critical pH curves for flotation with sodium diethyl dithiophosphate as collector. Flotation possible to left of curves. (After Wark and Cox.[61])

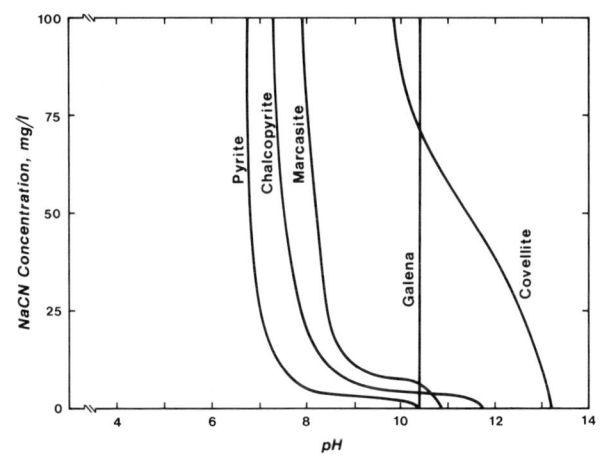

Figure 16.10. Critical pH curves for flotation with potassium ethyl xanthate (25 mg/l) in the presence of sodium cyanide. Flotation possible to left of curves. (After Sutherland and Wark.[2])

Sulfide Selectivity. Selectivity between sulfide minerals is possible when one can adsorb collector by one or both of the mechanisms above, and the others cannot. Figure 16.9 shows contact angle data for three sulfide minerals.[2,61] In this figure, flotation is possible to the left of each curve, but not to the right. Clearly, as the pH is increased, depression occurs. For pyrite, when the pH is raised above about 6, the rest potential no longer exceeds the reversible potential for the oxidation of the collector ion; and also, the Fe-collector compound is not as stable as competing Fe-hydroxy species. With galena, the Pb-collector compound prevails until the pH reaches about 8; above this value, it is Pb-hydroxy species that predominate. From inspection of Fig. 16.9, it is clear that selective flotation is possible based on the depressant action of the hydroxyl ion alone. However, although these results give an indication of possible flotation operating conditions, it must be kept in mind that they were obtained by contact angle tests on clean specimens under ideal conditions. In many sulfide systems, pH regulation alone is insufficient for acceptable selectivity.

The principle of using competing ions, however, can be used to improve selectivity. In Figs. 16.10 and 16.11 the depressant effect of cyanide ions and sulfide ions, respectively, are shown by contact angle curves, flotation being possible to the left of each curve.[2] The mechanisms of depression are more complex than was thought in the past.[2,4] Extensive examination of cyanide depression indicates it to be electrochemical in nature (Fig. 16.12); further analysis, however, is beyond the scope of this text.[49,62,63]

Other ions have also been successfully applied as depressants.[2,3] For example, sulfite, chromate, and dichromate have been used as depressants for galena;[64] ferrous ion has a depressant effect on pyrite flotation, presumably because it reduces the potential of the solution.[57]

Selectivity in sulfide minerals can in some cases be improved by activation or control of activation. Of particular importance in this regard is sphalerite flotation; without activation, sphalerite recovery is very poor with thiol collectors.[65] Activation can

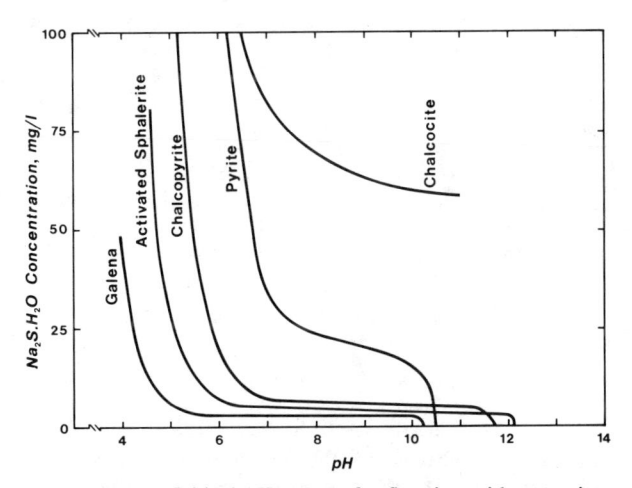

Figure 16.11. Critical pH curves for flotation with potassium ethyl xanthate (25 mg/l) in the presence of sodium sulfide. Flotation possible to left of curves. (After Wark and Cox.[61])

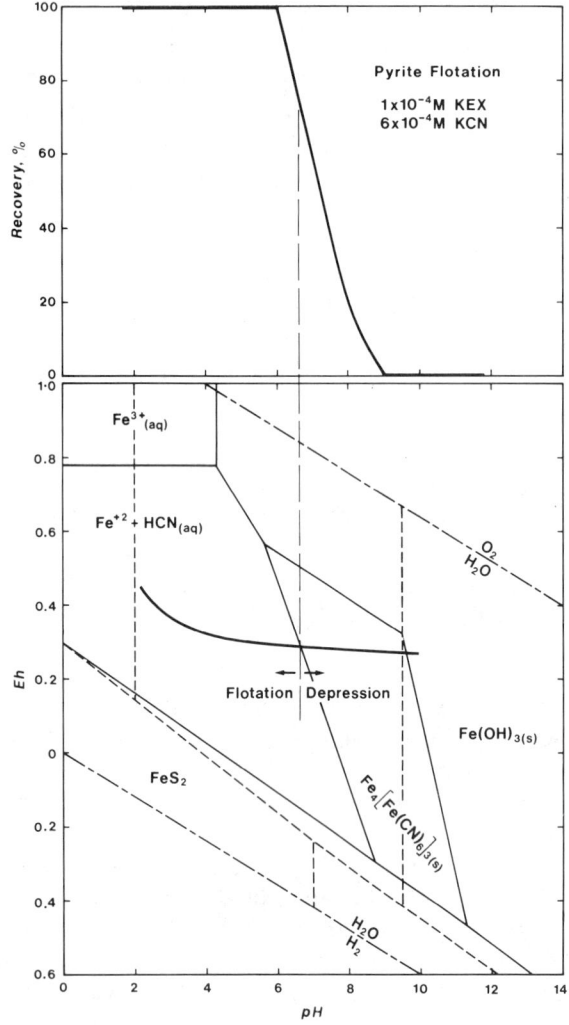

Figure 16.12. Pyrite flotation results and corresponding stability diagram for FeS_2, $Fe_4[Fe(CN)_6]_3$, and $Fe(OH)_3$ at 3×10^{-4} M total dissolved sulfur, 5×10^{-5} M total dissolved iron, and 6×10^{-4} M cyanide addition. Note that flotation occurs in $Fe^{2+}_{(aq)} + HCN_{(aq)}$ region, not in $Fe_4[Fe(CN)_6]_3$ region. (After Elgillani and Fuerstenau.[62])

occur by replacement of zinc ions in the mineral surface by metal ions of more stable sulfides, such as copper or lead, according to the reaction:

$$ZnS_{(s)} + Cu^{2+}_{(aq)} \rightleftharpoons CuS_{(s)} + Zn^{2+}_{(aq)} \quad (16.7a)$$

$$ZnS_{(s)} + Pb^{2+}_{(aq)} \rightleftharpoons PbS_{(s)} + Zn^{2+}_{(aq)} \quad (16.7b)$$

Thus the sphalerite particle (Eq. 16.7a) has a surface of covellite, and the particle behaves like a covellite particle. To activate sphalerite, common practice is to add copper sulfate. This reaction can however, occur, whenever activating ions are present in solution. Thus it is necessary, with some ores, to deactivate sphalerite, that is, to prevent natural activation from occurring. With lead-zinc ores, in which activation is due to lead ions, the addition of zinc sulfate is adequate for the reversal of Eq. 16.7b, hence deactivation. If copper minerals are present, it may be necessary to add cyanide ions to complex the copper ions present and prevent activation. That this reaction is not simple, however, is indicated by reports of sphalerite activation occurring with cyanide additions.

Another sulfide mineral that does not float strongly with thiol collectors is pyrrhotite. It can also be activated by copper ion substitution at the mineral surface.[66]

16.2.3. Oxide and Silicate Flotation

A wide range of oxide and silicate minerals are concentrated by flotation, using both anionic and cationic collectors. Detailed reports[67,68] and reviews[44,46] of flotation methods are available.

Collector Adsorption. Both anionic and cationic collectors are physically or electrostatically adsorbed on oxides and silicates.[6] With some minerals, chemisorption of anionic collectors can also occur.

As discussed in Section 6.1 and illustrated in Fig. 6.8, the potential of the surface (as represented by the zeta potential) of oxide and silicate minerals is dependent on the solution pH. For collector adsorption to occur, the potential of the mineral surface must be negative for cationic collectors and positive for anionic collectors.

For effective flotation, collectors with 10 or more carbon atoms in the hydrocarbon chain are required. It is important for sufficient adsorption and hemi-micelle formation to occur (Section 6.1.4, Fig. 6.10) at collector concentrations below that of the CMC.[44–46]

For chemisorption to occur, the formation of hydroxy complexes of the metal ions comprising the mineral is necessary. This has been demonstrated with a large number of oxides;[44,69] it has also been noted that for anionic flotation of quartz the presence of an activating polyvalent metal ion hydroxy complex is required.[70,71]

Selectivity Enhancement. Where collector adsorption is electrostatic, competition will exist between collector ions and inorganic ions in the solution. Thus, for example, where a cationic collector is

Chemistry of Flotation

used, inorganic cations can depress flotation and inorganic anions can act as activators.

Fluoride ion is widely used in silicate activation. Despite some uncertainty regarding the mechanism of activation, it is widely accepted that the fluosilicate complex (SiF_6^{2-}) and its adsorption at metal ion sites on the mineral is significant. It should be noted that with quartz, where no metal ion site exists, fluoride ion acts as a depressant rather than as an activator.

Hydrophilic organic colloidal materials (e.g., starches, dextrin, tannin, quebracho) are also used as depressants. In the cationic flotation of silica gangue from iron ores, for example, starches are used to selectively depress the iron minerals.[46,71]

With very insoluble oxides and silicates, such as hematite, sufficient conditioning time must be allowed for surface conditions to stabilize both before and after collector addition. Temperature can also be an important factor, involving collector solubility; raising pulp temperature can give greatly increased recoveries. And recently, careful control of dispersion with sodium silicate has been shown to give improved separation in fine iron ores.[37]

16.2.4. Salt-type Mineral Flotation

Minerals with solubilities greater than those of most oxides and silicates, but lower than those of "soluble salt" minerals, are categorized for convenience as salt-type minerals. Included in this category are apatite, scheelite, barite, fluorite, magnesite, calcite, and gypsum. These minerals can be readily separated from oxides and silicates, but they are extremely difficult to separate from each other because of the great similarity in their surface chemical and physical properties.[68] A comprehensive review of the flotation of these minerals is available elsewhere,[72] and only a brief summary is presented here.

Both anionic and cationic collectors as used for oxide and silicate flotation are used with the salt-type minerals. The mechanism of adsorption is complex and not well understood; however, it has been shown that chemisorption as well as physical adsorption can occur.

Improved selectivity can be obtained by the controlled application of both inorganic and organic modifiers. Polyvalent cations and inorganic anions affect the physical adsorption of collector on salt-type minerals in the same way as oxides and silicates. The selectivity, however, is not significantly changed. Sodium silicate is commonly used as a depressant for calcite; it has the added advantage that it also acts as a depressant for silica if present.[28,72] However, sodium silicate will depress all calcium minerals to some extent. The addition of aluminum ions improves selectivity by reducing the depressant action of sodium silicate on calcium salts other than calcite (e.g., fluorite or scheelite). Other anions are also used to improve selectivity, but to a lesser extent.

Organic colloids have been used for many years to improve selectivity between the salt-type minerals,[72,73] but understanding of the mechanisms involved is limited. Starch, tannin, and quebracho are all used for calcite depression, but different mechanisms are believed to be involved.

16.2.5. Soluble Salt Flotation

The flotation separation of sylvite (KCl) from halite (NaCl) in potash ores is carried out in saturated brine solutions. Either the sylvite or the halite can be floated, although it is normal practice to use a long-chain amine for the flotation of the positively charged sylvite from the negatively charged halite. A number of mechanisms have been proposed for this collector adsorption, and it is thought that more than one must be operative.[74] However, it is believed that associated aqueous species are acting as collectors as shown in Fig. 16.13.

16.2.6. Fine and Coarse Particle Flotation

Recovery of minerals by flotation is most successful in the 10–100 μm size range.[75] However, a number of flotation-related techniques are available below

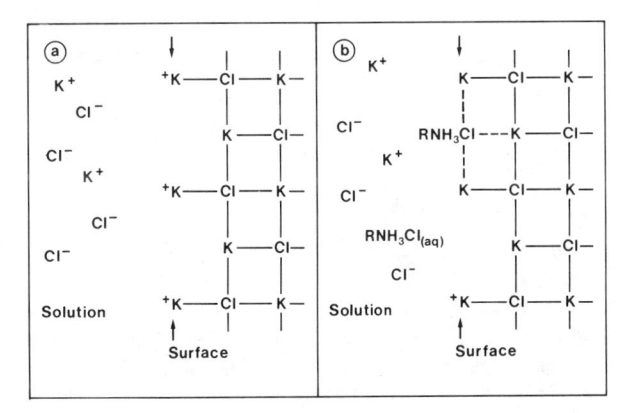

Figure 16.13. Potash (KCl) surface in (a) absence and (b) presence of a primary amine cationic collector. (After Roman et al.[74])

this size range. These have been discussed in detail in the literature and reviewed elsewhere.[76] Some of these techniques are illustrated in Fig. 16.14, including carrier or ultraflotation (Fig. 16.14b), oil or emulsion flotation (Fig. 16.14c), agglomerate flotation or floto-flocculation (Fig. 16.14d), and liquid-liquid or two-liquid extraction (Fig. 16.14e).

Other processes related to the above are ion and precipitate flotation.[77]

For especially large particles, special flotation techniques are also available. Agglomerate or skin flotation is used, particularly for coarse phosphate particles.[78] In this flotation, the particles float on the pulp surface rather than attach to bubbles as in conventional flotation. The separation is actually carried out in equipment normally used for gravity separations, such as spirals or tables (Chapter 13).

16.3. MECHANICS OF FLOTATION

Although the chemistry of flotation is the major aspect of the process, it cannot be considered in isolation. In Section 16.2 flotation was treated as an equilibrium process; however, the kinetics of flotation can play a significant role in attaining a separation.

16.3.1. Machine Design

Differences in flotation cell design were discussed in Section 16.1. The principles of design, and hydrodynamic scale-up criteria (Section 4.8.1), have been extensively reported by Harris[5,6,79-81] and others.[82-84] Scale-up to the large flotation cells introduced in recent years has been based on two dimensionless numbers, the power number **Po**, given by

$$\mathbf{Po} = \frac{\mathscr{P}}{\rho_f \mathscr{N}^3 D_a^5} \quad (4.86)$$

and the air flow number

$$\mathbf{Fl}_g = \frac{I_{V,g}}{\mathscr{N} D_a^3} \quad (4.88)$$

The air flow number when applied to flotation cells is often represented in modified form to take into account the surface area of the cell through which air may flow, A_f. In this case, Eq. 4.88 is multiplied by D_a^2/A_f, to give

$$\mathbf{Fl}_g = \frac{I_{V,g}}{A_f \mathscr{N} D_a} \quad (16.8)$$

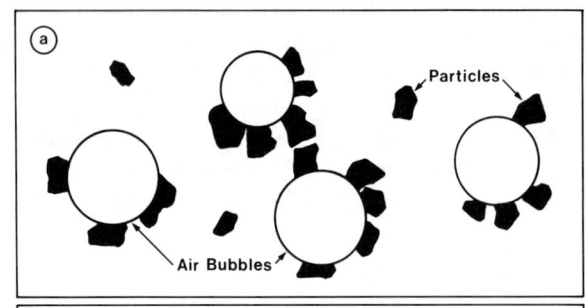

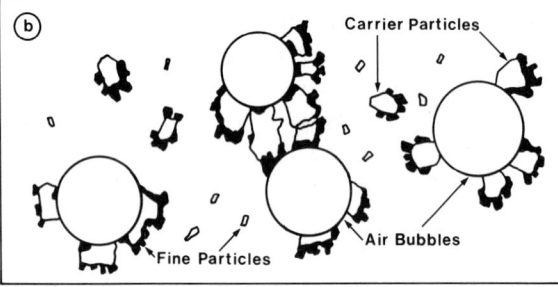

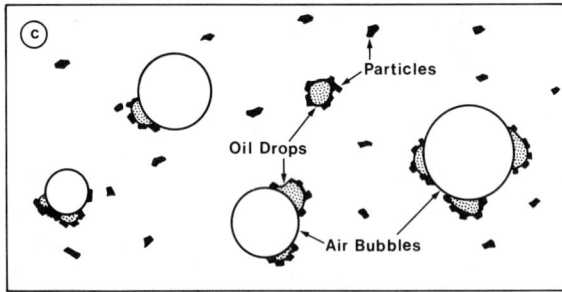

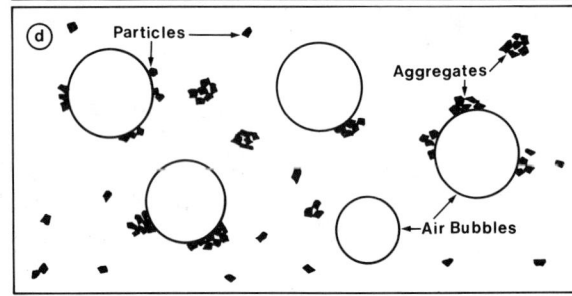

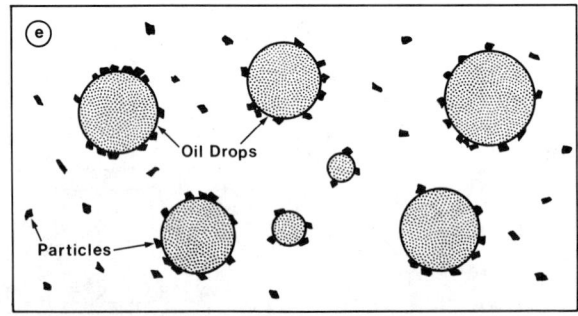

Figure 16.14. Schematic representation of flotation processes. (a) Froth flotation. (b) Carrier or ultraflotation. (c) Oil or emulsion flotation. (d) Agglomerate or floc flotation (floto-flocculation). (e) Liquid-liquid or two-liquid extraction. (After Fuerstenau.[76])

Mechanics of Flotation

16.3.2. Froths and Entrainment

Particles can enter a flotation froth by either of two mechanisms. The hydrophobic particles attach to bubbles and enter the froth with those bubbles. Other particles, however, also enter the froth; they do so by entrainment in the water carried along by the rising bubbles. In this way gangue and other unwanted particles enter the froth, particularly at the finer particle sizes (Fig. 16.15). Particles entrained in this manner may be washed out of the froth as the froth drains (Fig. 16.16). However, as the froth walls become thinner, these particles, although not attached to bubbles, become physically trapped in the froth.

Studies have been carried out to determine the significance of the water entrainment mechanism to flotation results.[85-87] It has been shown that the rate at which nonhydrophobic minerals enter the froth product is approximately proportional to the rate of water flow in that product (Fig. 16.17). Also, the greater the degree of drainage from the froth, which occurs at greater froth heights, the greater the removal of entrained particles (Fig. 16.18). This indicates the potential value of water sprays above a froth, as has been reported for conventional flotation cells[28] and as is possible with a flotation column.

As froth drainage takes place, some froth breakdown occurs. However, the presence of particles on

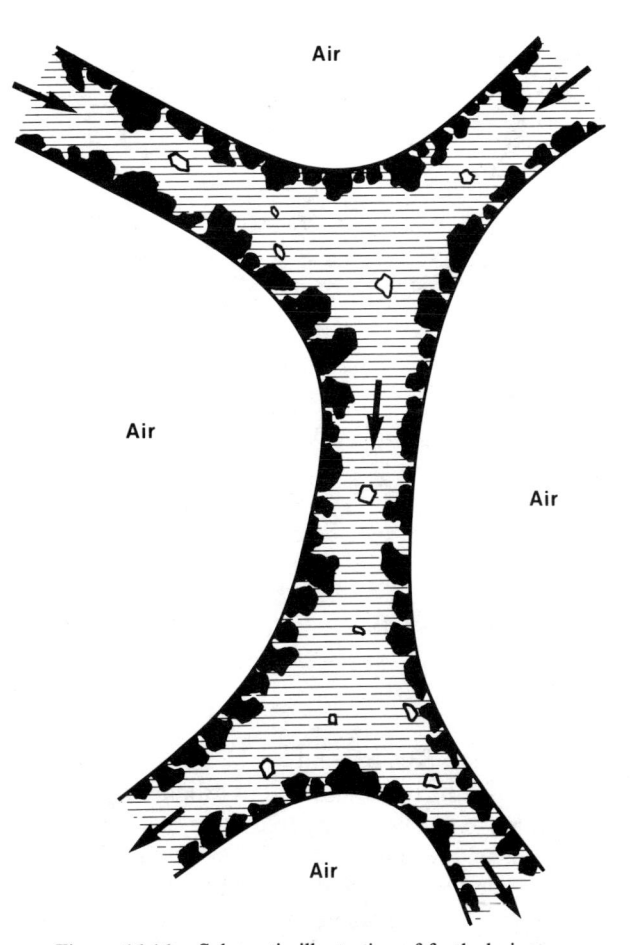

Figure 16.16. Schematic illustration of froth drainage.

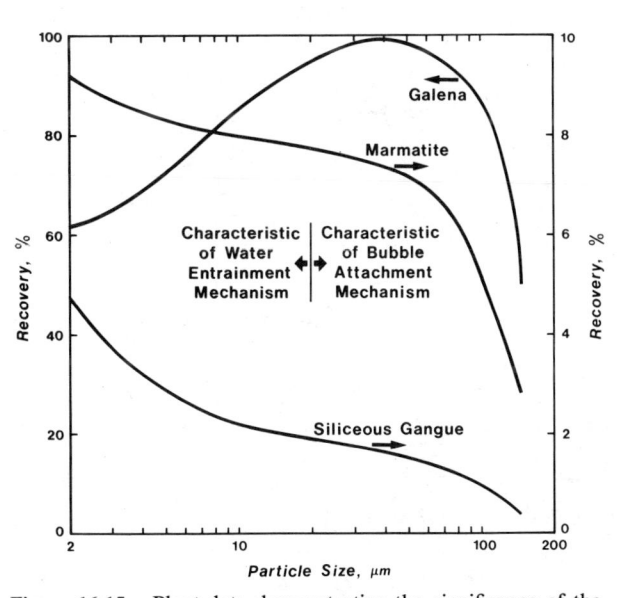

Figure 16-15. Plant data demonstrating the significance of the water entrainment mechanism in flotation. (After Thorne et al.[86])

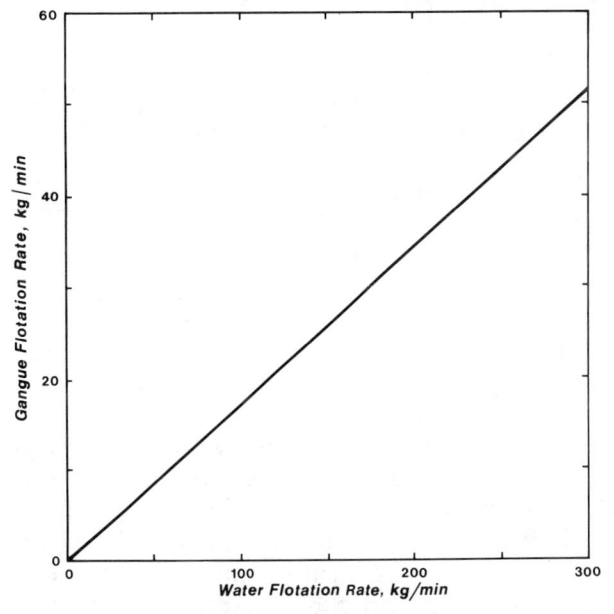

Figure 16.17. Water entrainment determines the flotation rate of nonhydrophobic particles. (After Engelbrecht and Woodburn.[87])

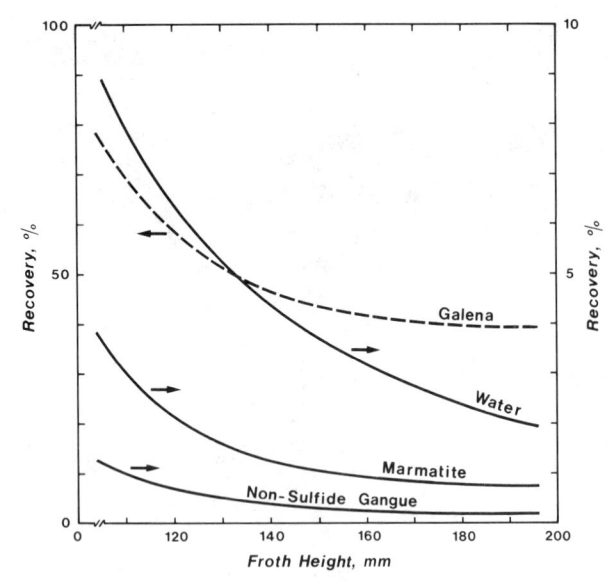

Figure 16.18. Importance of froth drainage on concentration during the flotation process. (After Thorne et al.[86])

the bubbles tends to stabilize the froth.[28,88] When a bubble rises through a pulp of hydrophobic solids, it becomes extensively covered with particles. The draining of the films between bubbles in the froth thus formed is restricted by the thin layer of solids in the film. The volume of this particle-stabilized froth is therefore roughly proportional to the amount of hydrophobic solids present. If this froth is handled mechanically, it will break down. If, however, there is appreciable flocculation in the pulp, fairly thick walls of interlocked particles are formed between the bubbles, and the froth will persist beyond the flotation cell launder.

16.3.3. Conditioning: Particle Size Effects

The effect of particle size on flotation recovery (Fig. 16.15) was attributed in the preceding section to particle entrainment into the froth. But other factors are also involved. The presence of a large fines or slime fraction in a flotation feed is in general detrimental to the flotation of the coarser particles, because of the excess reagent take-up by the high surface-area-to-mass fines and possible adsorption of these fines on the larger ore particles. It has been known for many years that by separate conditioning of fine and coarse fractions, improved recoveries and selectivity are possible in the coarse fraction.[3,89] More recently, this finding has been confirmed in a comprehensive series of tests (Fig. 16.19).[75,90]

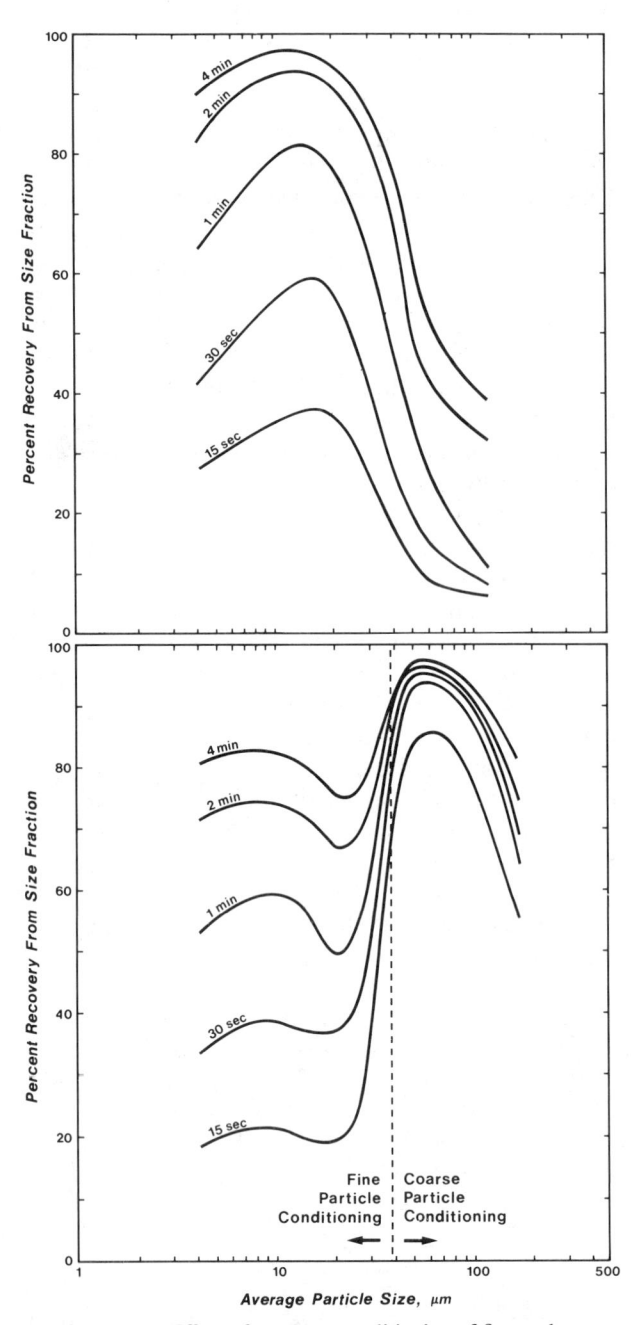

Figure 16.19. Effect of separate conditioning of fine and coarse fractions on flotation results. Sphalerite flotation with potassium ethyl xanthate as collector: (a) activated with 0.13 kg/t $CuSO_4$ and (b) +37 μm conditioned with 0.10 kg/t $CuSO_4$, −37 μm at 0.03 kg/t $CuSO_4$. (After Trahar.[90])

16.3.4. Flotation Kinetics

Although flotation has been recognized as a rate process for many years, the deliberate control and use of differences in flotation rate for improved

Flotation Operations

selectivity has been limited. Recently, however, it has been noted that the control of flotation kinetics by reagent addition may be as important as the control of "equilibrium" or "maximum recovery."[91,92]

As indicated in Section 3.2, two similar approaches are used to mathematically describe the rate of separation in mineral processes. There was disagreement in the past, but it is now widely accepted[86,91-94] that flotation can be represented by a first-order rate equation that gives

$$m = m_I \exp(-kt) \qquad (3.2)$$

(m_I, m = mass of particles present in pulp initially and at time t, respectively; k = rate constant) or by

$$m = m_I(1-p)^n \qquad (3.3)$$

(m_I, m = mass of particles present in pulp initially and after n stages of flotation; p = probability of flotation in any stage).

Both k and p in the equations above are complex. Each particle size fraction of each mineral

TABLE 16.2 : FLOTATION PROBABILITIES

Size (μm)		% Slow Floating	p(Slow)	p(Fast)
	+147	50	0.04	0.72
-147	+104	19	0.04	0.73
-104	+74	8	0.04	0.73
-74	+53	4	0.04	0.74
-53	+43	3	0.04	0.73
-43	+19	3	0.04	0.74
-19	+14.3	3	0.04	0.68
-14.3	+10.1	3	0.04	0.63
-10.1	+6.8	3	0.04	0.59
-6.8	+4.7	3	0.04	0.50
-4.7		13	0.04	0.38

type may have a different value of the rate constant at any given set of chemical and physical operating conditions. Also, it has been found that Eqs. 3.2 and 3.3 do not represent most experimental data, but must be modified by considering each mineral (in total, or in individual size fractions) to consist of two components, one "fast floating" and the other "slow floating." Equations 3.2 and 3.3 can then be written as

$$m = m_I[\mathfrak{F}_S \exp(-k_S t) + (1-\mathfrak{F}_S)\exp(-k_F t)] \quad (16.9)$$

($\mathfrak{F}_S$ = fraction slow floating; k_F, k_S = rate constants for fast and slow floating fractions, respectively) and

$$m = m_I[\mathfrak{F}_S(1-p_S)^n + (1-\mathfrak{F}_S)(1-p_F)^n] \quad (16.10)$$

(p_F, p_S = probability of flotation of fast and slow floating fractions, respectively).

If experimental data are plotted as in Fig. 16.20, $\mathfrak{F}_S$ and either k_S and k_F or p_S and p_F can be determined as indicated. Representative data for individual size fractions are given in Table 16.2.[94]

16.4. FLOTATION OPERATIONS

The traditional method for the analysis of flotation circuits has been to provide a metallurgical balance that shows the performance of each section. A constant difficulty has always been the determination of the value (or otherwise) of recycle streams within the circuit, for example, the common practice of recycling cleaner tailings to rougher feed.[95] Using the equations representing the flotation process introduced in Section 16.3, where the equation constants are experimentally determined, the performance of a flotation section can be predicted (Example 16.1). As further discussed in Chapter 23, the performance of more complex circuits can then

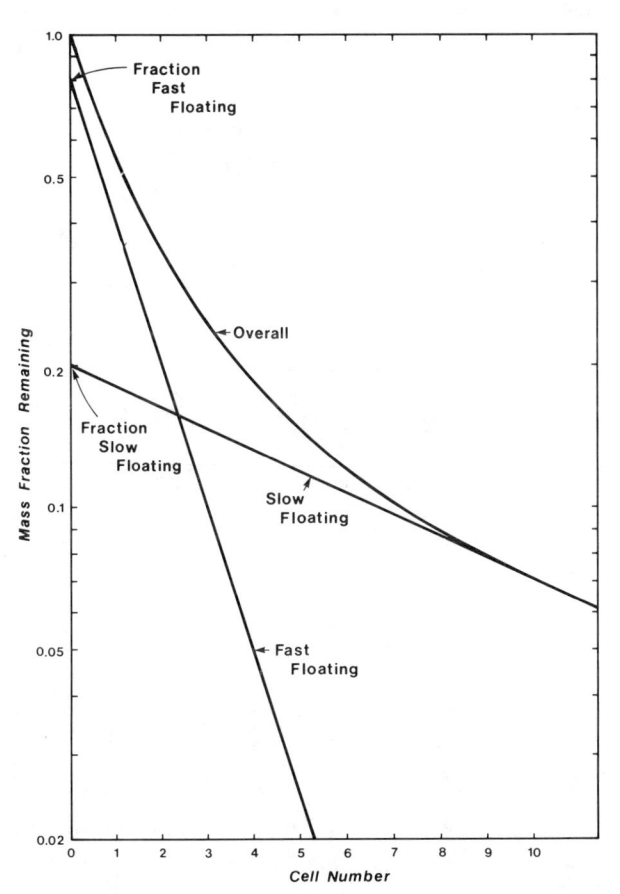

Figure 16.20. Typical flotation plant data showing first-order kinetics of both fast floating and slow floating fractions.

be analyzed. This approach has been successfully applied in a number of flotation plants throughout the world.[86,94,96,97]

Example 16.1. The PbS ore from Example 2.4 is treated in a bank of flotation cells. What are the cumulative grade/cumulative recovery characteristics of the concentrate along the bank? Assume that the following flotation probabilities p apply

$$\text{liberated PbS} = 0.5$$

$$\text{locked PbS} = 0.1$$

$$-38 \ \mu\text{m liberated gangue} = 0.03$$

(A limited range of probabilities is given for simplicity. More rigorous analysis would require different values for each size, and for variations in middlings grade. The single value for liberated gangue is to account for fine particles being "floated" by water carryover.)

Solution

Basis. 100 t of feed (i.e., 10 t of PbS, 90 t gangue)
The quantity floated at each stage can be calculated by Eq. 3.3. Thus for the first cell:

mass of liberated PbS floated
$$= [1 - (1 - p)^1]M$$
$$= [1 - (1 - 0.5)] \times 6.414$$
$$= 3.21 \text{ t}$$

mass of locked PbS floated
$$= [1 - (1 - 0.1)] \times 3.586$$
$$= 0.36 \text{ t}$$

$$\therefore \quad \text{recovery} = \frac{3.21 + 0.36}{10} \times 100$$
$$= 35.7\%$$

Since the middlings particles average 64.7% PbS, the mass of locked gangue floated

$$= 0.36 \times \frac{1 - 0.647}{0.647}$$
$$= 0.196 \text{ t}$$

Mass of liberated gangue floated:

$$(-38 \ \mu\text{m}) = [1 - (1 - 0.03)] \times 29.05$$
$$= 0.872 \text{ t}$$

$$\therefore \quad \text{total gangue floated} = 0.196 + 0.872$$
$$= 1.07 \text{ t}$$

$$\therefore \quad \text{grade of material floated} = \frac{3.57}{3.57 + 1.07} \times 100$$
$$= 77.0\% \text{ PbS}$$

For two cells:

mass of liberated PbS floated
$$= [1 - (1 - 0.5)^2] \times 6.4$$
$$= 4.81 \text{ t}$$

mass of locked PbS floated
$$= [1 - (1 - 0.1)^2] \times 3.6$$
$$= 0.681 \text{ t}$$

$$\therefore \quad \text{recovery} = \frac{4.81 + 0.681}{100} \times 100$$
$$= 54.9\%$$

mass of locked gangue $= 0.681 \times \dfrac{1 - 0.647}{0.647}$
$$= 0.372 \text{ t}$$

mass of liberated gangue floated
$$= [1 - (1 - 0.03)^2] \times 29.07$$
$$= 1.717 \text{ t}$$

$$\therefore \quad \text{total gangue floated} = 1.717 + 0.372$$
$$= 2.089 \text{ t}$$

TABLE E16.1.1: FLOTATION PROBABILITY EXAMPLE

Number of Cells	1	2	4	8	12	16	20	30
Mass liberated PbS	3.207	4.811	6.013	6.389	6.412	6.414	6.414	6.414
Mass locked PbS	0.359	0.681	1.233	2.042	2.573	2.932	3.150	3.434
Mass locked gangue	0.196	0.372	0.673	1.114	1.404	1.594	1.719	1.874
Mass liberated gangue	0.872	1.717	3.332	6.282	6.894	11.206	13.253	17.401
Recovery	35.7	54.9	72.5	84.3	89.9	93.4	95.6	98.5
Grade	77.0	72.4	64.4	53.3	46.6	42.2	39.0	33.8

∴ grade of material floated = $\frac{5.49}{5.49 + 2.089} \times 100$
= 72.4% PbS

For the nth cell:

mass of liberated PbS floated
$$= [1 - (1 - 0.5)^n] \times 6.414$$

mass of locked PbS floated
$$= [1 - (1 - 0.1)^n] \times 3.586$$

mass of liberated gangue
$$= [1 - (1 - 0.03)^n] \times 29.05$$

Results are tabulated in Table E16.1.1.

16.5. OTHER SURFACE SEPARATIONS

Although flotation (plus the closely related processes listed in Section 16.2.6) is the major separation method based on the surface chemistry of mineral particles, it is not the only method. Selective flocculation and agglomeration methods are both used commercially to a limited extent, the former for hematite[98,99] and the latter for coal[100] and fine metallic oxide minerals.[101]

Both processes utilize the same principles as discussed for flotation to achieve selectivity. In selective flocculation, polymeric flocculants (Fig. 16.21) as used in thickening (Chapter 17) are applied; in this case the flocculants selectively adsorb on the hematite particles, resulting in hematite flocs that settle quickly and so can be separated from the siliceous gangue.

In selective agglomeration, one mineral is rendered hydrophobic (or is naturally hydrophobic, as with coal) so that it can adsorb a hydrocarbon (often fuel oil). By using high shear conditions, the hydrocarbon-coated particles can come in contact and agglomerate, the agglomerate size being increased until a size separation is possible.

Other surface separations are in use, but not in wide use in the mineral industry.

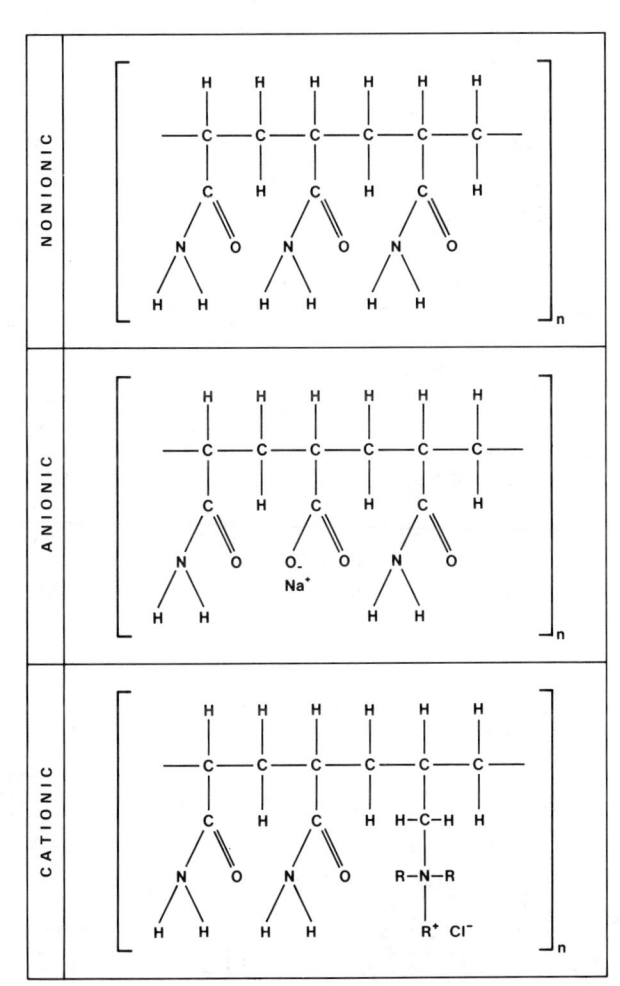

Figure 16.21. Structure of selected polyelectrolyte flocculants.

REFERENCES

1. A. F. Taggart, *Handbook of Mineral Dressing*, Wiley (1945).
2. K. L. Sutherland and I. W. Wark, *Principles of Flotation*, Australasian IMM (1955).
3. A. M. Gaudin, *Flotation*, McGraw-Hill (1957).
4. D. W. Fuerstenau (Ed.), *Froth Flotation—50th Anniversary Volume*, AIME/SME (1962).
5. N. Arbiter and C. C. Harris, "Flotation Machines," Ch. 14 in *Froth Flotation—50th Anniversary Volume*, D. W. Fuerstenau (Ed.), AIME/SME (1962).
6. C. C. Harris, "Flotation Machines," Ch. 27 in *Flotation—A. M. Gaudin Memorial Volume*, M. C. Fuerstenau (Ed.), AIME/SME (1976).
7. A. C. Dorenfeld, "Flotation Circuit Design," Ch. 15 in *Froth Flotation—50th Anniversary Volume*, D. W. Fuerstenau (Ed.), AIME/SME (1962).
8. K. Fallenius, "Outokumpu Flotation Machines," Ch. 29 in *Flotation—A. M. Gaudin Memorial Volume*, M. C. Fuerstenau (Ed.), AIME/SME (1976).
9. V. R. Degner and H. B. Treweek, "Large Flotation Cell Design and Development," Ch. 28 in *Flotation—A. M. Gaudin Memorial Volume*, M. C. Fuerstenau (Ed.), AIME/SME (1976).
10. T. M. Plouf, "Large Volume Flotation Machine Development Criteria and Plant Testing; Denver Equipment Division's Single Mechanism 1275 ft³ (36 m³) DR Flotation Machine," *Proc. 13th Int. Miner. Process. Congr. Warsaw, 1979*, Elsevier (1981).

11. M. A. Anderson, C. I. Wilmot, and C. E. Jackson, "A Concentrator Improvement and Modernization Program Utilizing Large-Volume Flotation Machines," AIME/SME Preprint 75-B-77 (1975).
12. N. Arbiter and C. C. Harris, "Design and Operating Characteristics of Large Flotation Cells," AIME/SME Preprint 79-370 (1979).
13. J. A. McAllister and J. C. Coburn, "Flotation," *Eng. Min. J.*, **181**, 114–117 (June 1980).
14. W. R. Bull and D. J. Spottiswood, "A Study of Mixing Patterns in a Bank of Flotation Cells," *Colorado School Mines Q.*, **69**, 1–26 (1974).
15. J. R. Maxwell, "Large Flotation Cells in Opemiska Concentrator," *Trans. AIME/SME*, **252**, 95–98 (1972).
16. J. D. Dinsdale and Y. Berube, "A Characterization of Hydrodynamics in a 700 ft Maxwell Flotation Cell," *Can. Met. Q.*, **11**, 507–513 (1972).
17. T. Nagahama, "Treatment of Effluent from the Kamioka Concentrator by Flotation Techniques, Including the Development of the Nagham Machine," *CIM Bull.*, **67**, 79–89 (April 1974).
18. Anon., "Innovative Design Revamps Flotation Cells," *Eng. Min. J.*, **172**, 78–79 (July 1971).
19. L. R. Flint, "Factors Influencing the Design of Flotation Equipment," *Miner. Sci. Eng.*, **5**, 232–241 (July 1973).
20. D. A. Wheeler, "Big Flotation Column Mill Tested," *Eng. Min. J.*, **167**, 98–103 (November 1966).
21. P. Boutin and D. A. Wheeler, "Column Flotation," *World Min.*, **120**, 47–50 (1967).
22. R. D. MacDonald and R. J. Brison, "Applied Research in Flotation," Ch. 12 in *Froth Flotation—50th Anniversary Volume*, D. W. Fuerstenau (Ed.), AIME/SME (1962).
23. A. F. Hallimond, "Laboratory Apparatus for Flotation Tests," *Min. Mag.*, **70**, 87–90 (1944).
24. D. W. Fuerstenau, P. H. Metzger, and G. D. Seele, "How to Use the Modified Hallimond Tube for Flotation Testing," *Eng. Min. J.*, **158**, 93 (March 1957).
25. M. C. Fuerstenau, "An Improved Micro-Flotation Technique," *Eng. Min. J.*, **165**, 108–109 (November 1964).
26. C. C. Dell and M. J. Bunyard, "Development of an Automatic Flotation Cell for the Laboratory," *Trans. IMM*, **C81**, C246–248 (1972).
27. A. C. Partridge and G. W. Smith, "Flotation and Adsorption Characteristics of the Hematite-Dodecylamine-Starch System," *Can. Met. Q.*, **10**, 229–234 (1971).
28. V. I. Klassen and V. A. Mokrousov, *An Introduction to the Theory of Flotation*, Butterworths (1963).
29. M. C. Fuerstenau (Ed.), *Flotation—A. M. Gaudin Memorial Volume*, AIME/SME (1976).
30. V. A. Glembotskii, V. I. Klassen, and I. N. Plaskin, *Flotation*, Primary Sources (1963).
31. J. W. Leonard (Ed.), *Coal Preparation*, 4th ed., AIME/SME (1979).
32. S. Chander and D. W. Fuerstenau, "On the Natural Flotability of Molybdenite," *Trans. AIME/SME*, **252**, 62–69 (1972).
33. R. D. Cuthbertson, "New Facets in Flotation at Climax," *Colorado School Mines Q.*, **56**, No. 3, 197–214 (1961).
34. R. M. Hoover and D. Malhotra, "Emulsion Flotation of Molybdenite," Ch. 16 in *Flotation—A. M. Gaudin Memorial Volume*, M. C. Fuerstenau (Ed.), AIME/SME (1976).
35. G. W. Heyes and W. J. Trahar, "The Natural Flotability of Chalcopyrite," *Int. J. Miner. Process.*, **4**, 317–344 (1977).
36. R. H. Yoon and A. I. Stemerowicz, "Collectorless Flotation of Chalcopyrite and Sphalerite," *Proc. 11th Can. Miner. Process. Meeting*, pp. 186–204 (1979).
37. D. C. Yang, "Flotation in Systems with Controlled Dispersion-Carrier Flotation, etc.," Ch. 26 in *Beneficiation of Mineral Fines*, P. Somasundaran (Ed.), AIME/SME (1979).
38. J. A. Meech and J. G. Paterson, "Improvements in Copper/Lead Separation with Activated Carbon," *Trans. AIME/SME*, **264**, 1758–1767 (1979).
39. G. W. Poling, "Reactions Between Thiol Reagents and Sulfide Minerals," Ch. 11 in *Flotation—A. M. Gaudin Memorial Volume*, M. C. Fuerstenau (Ed.), AIME/SME (1976).
40. S. R. Rao, *Xanthates and Related Compounds*, Dekker (1971).
41. D. R. Shaw, "Dodecyl Mercaptan: A Superior Collector for Sulfide Ores," AIME/SME Preprint 79-338 (1979).
42. N. P. Finkelstein and G. W. Poling, "The Role of Dithiolates in the Flotation of Sulfide Minerals," *Miner. Sci. Eng.*, **9**, No. 4, 177–197 (1977).
43. M. C. Fuerstenau, R. W. Harper, and J. D. Miller, "Hydroxamate vs. Fatty Acid Flotation of Iron Oxide," *Trans. AIME/SME*, **24**, 69–73 (1970).
44. M. C. Fuerstenau and B. R. Palmer, "Anionic Flotation of Oxides and Silicates," Ch. 7 in *Flotation—A. M. Gaudin Memorial Volume*, M. C. Fuerstenau (Ed.), AIME/SME (1976).
45. P. Mukerjee and K. J. Mysels, "Critical Micelle Concentrations of Aqueous Surfactant Systems," U.S. National Bureau of Standards Report NSRDS-NBS 36 (February 1971).
46. R. W. Smith and S. Akhtar, "Cationic Flotation of Oxides and Silicates," Ch. 5 in *Flotation—A. M. Gaudin Memorial Volume*, M. C. Fuerstenau (Ed.), AIME/SME (1976).
47. D. R. Nagaraj and P. Somasundaran, "Commercial Chelating Extractants as Collectors: Flotation of Copper Minerals Using "LIX" Reagents," *Trans. AIME/SME*, **266**, 1892–1897 (1979).
48. B. Yarar, personal communication.
49. R. Woods, "The Oxidation of Ethyl Xanthate on Platinum, Gold, Copper, and Galena Electrodes: Relation to the Mechanism of Mineral Flotation," *J. Phys. Chem.*, **75**, 354–362 (1971).
50. A. Granville, N. P. Finkelstein, and S. A. Allison, "Review of Reactions in the Flotation System Galena-Xanthate-Oxygen," *Trans. IMM (C)*, **81**, C1–C30 (1972).
51. J. Leja, "Some Electrochemical and Chemical Studies Related to Froth Flotation with Xanthates," *Miner. Sci. Eng.*, **5** (4), 278–286 (1973).
52. E. E. Maust, P. E. Richardson, and G. R. Hyde, "A Conceptual Model for the Role of Oxygen in Xanthate Adsorp-

References

tion on Galena,'' U.S. Bureau of Mines Report of Investigations RI 8143 (1976).

53. M. C. Fuerstenau, "Thiol Adsorption Collector Processes," in "The Physical Chemistry of Mineral-Reagent Interactions in Sulfide Flotation," U.S. Bureau of Mines Information Circular IC 8818, pp. 7–24 (1980).

54. O. Mellgren, "Heat of Adsorption and Surface Reactions of Potassium Ethyl Xanthate on Galena," *Trans. AIME/SME,* **235,** 46–59 (1966).

55. J. Leja, L. H. Little, and G. W. Poling, "Xanthate Adsorption Studies Using Infrared Spectroscopy," *Trans. IMM (C),* **72,** C407–C423 (1963).

56. N. P. Finkelstein et al., "Natural and Induced Hydrophobicity in Sulfide Mineral Systems," in Advances in Interfacial Phenomena of Particulate/Solution/Gas Systems: Application to Flotation Research, P. Somasundaran and R. B. Grieves (Eds.), AIChE Symposium Series, No. 150, Vol. 71, pp. 165–175 (1975).

57. M. C. Fuerstenau, M. C. Kuhn, and D. A. Elgillani, "The Role of Dixanthogen in the Xanthate Flotation of Pyrite," *Trans. AIME/SME,* **241,** 148–156 (1968).

58. H. Majima and M. Takeda, "Electrochemical Studies of the Xanthate-Dixanthogen System on Pyrite," *Trans. AIME/SME,* **241,** 431–436 (1968).

59. R. Woods, "Electrochemistry of Sulfide Flotation," Ch. 10 in *Flotation—A. M. Gaudin Memorial Volume,* M. C. Fuerstenau (Ed.), AIME/SME (1976).

60. H. H. Haung and J. D. Miller, "Kinetics and Thermochemistry of Amyl Xanthate Adsorption by Pyrite and Marcasite," *Int. J. Miner. Process.,* **5,** 214–266 (1978).

61. I. W. Wark and A. B. Cox, "Principles of Flotation. III. An Experimental Study of Influence of Cyanide, Alkalis, and Copper Sulfate on Effect of Sulfur-Bearing Collectors at Mineral Surfaces," *Trans. AIME/SME,* **112,** 267–302 (1934).

62. D. A. Elgillani and M. C. Fuerstenau, "Mechanisms Involved in the Cyanide Depression of Pyrite," *Trans. AIME/SME,* **241,** 437–445 (1968).

63. B. Ball and R. S. Rickard, "The Chemistry of Pyrite Flotation and Depression," Ch. 15 in *Flotation—A. M. Gaudin Memorial Volume,* M. C. Fuerstenau (Ed.), AIME/SME (1976).

64. J. Shimoiizaka et al., "Depression of Galena Flotation by Sulfite or Chromate Ion," Ch. 13 in *Flotation—A. M. Gaudin Memorial Volume,* M. C. Fuerstenau (Ed.), AIME/SME (1976).

65. N. P. Finkelstein and S. A. Allison, "The Chemistry of Activation, Deactivation, and Depression in the Flotation of Zinc Sulfide: A Review," Ch. 14 in *Flotation—A. M. Gaudin Memorial Volume,* M. C. Fuerstenau (Ed.), AIME/SME (1976).

66. C. S. Chang, S. R. B. Cooke, and I. Iwasaki, "Flotation Characteristics of Pyrrhotite with Xanthates," *Trans. AIME,* **199,** 209–217 (1954).

67. R. M. Manser, *Handbook of Silicate Flotation,* Warren Spring Laboratory (1975).

68a. R. A. Wyman, "Solving Industrial Mineral Flotation Problems at the Mines Branch, Ottawa, Canada," *Trans. AIME/SME,* **250,** 231–236 (1971).

68b. R. A. Wyman, "The Flotability of Twenty-one Non-Metallic Minerals," Canadian Mines Branch, Technical Bulletin TB 108 (1969).

68c. R. A. Wyman and J. H. Colborne, "The Flotability of Ten Non-Metallic Minerals (Supplement to TB 108)," Canadian Mines Branch, Technical Bulletin TB 186 (1974).

69. R. B. Palmer et al., "Mechanisms Involved in the Flotation of Oxides and Silicates with Anionic Collectors," *Trans. AIME/SME,* **258,** 257–263 (1975).

70. M. C. Fuerstenau and W. F. Cummins, "The Role of Basic Aqueous Complexes in Anionic Flotation of Quartz," *Trans. AIME/SME,* **238,** 196–200 (1967).

71. S. R. Balajee and I. Iwasaki, "Adsorption Mechanism of Starches in Flotation and Flocculation of Iron Ores," *Trans. AIME/SME,* **244,** 401–406 (1969).

72. M. S. Hanna and P. Somasundaran, "Flotation of Salt-Type Minerals," Ch. 8 in *Flotation—A. M. Gaudin Memorial Volume,* M. C. Fuerstenau (Ed.), AIME/SME (1976).

73. H. Baldauf and H. Schubert, "Correlations Between Structure and Adsorption for Organic Depressants in Flotation," Ch. 39 in *Fine Particles Processing,* P. Somasundaran (Ed.), AIME/SME (1980).

74. R. J. Roman, M. C. Fuerstenau, and D. C. Seidel, "Mechanisms of Soluble Salt Flotation. Part I," *Trans. AIME/SME,* **241,** 56–64 (1968).

75. W. J. Trahar and L. J. Warren, "The Flotability of Very Fine Particles—A Review," *Int. J. Miner. Process.,* **3,** 103–131 (1976).

76. D. W. Fuerstenau, "Fine Particle Flotation," Ch. 35 in *Fine Particles Processing,* P. Somasundaran (Ed.), AIME/SME (1980).

77. J. W. Perez and F. F. Aplan, "Ion and Precipitate Flotation of Metal Ions from Solution," AIChE Symposium Series No. 150, Vol. 71, pp. 34–39 (1975).

78. B. M. Moudgil and D. H. Barnett, "Agglomeration-Skin Flotation of Coarse Phosphate Rock," *Trans. AIME/SME,* **266,** 283–289 (1979).

79. N. Arbiter, C. C. Harris, and R. Yap, "Hydrodynamics of Flotation Cells," *Trans. AIME/SME,* **244,** 134–148 (1969).

80. C. C. Harris and A. Raja, "Flotation Machine Impeller Speed as Scale-Up Criteria," *Trans. IMM,* **C79,** C295–C297 (1970).

81. C. C. Harris, "Impeller Speed, Air, and Power Requirements in Flotation Machine Scale-Up," *Int. J. Miner. Process.,* **1,** 51–64 (1974).

82. K. Fallenius, "A New Set of Equations for the Scale-Up of Flotation Cells," *Proc. 13th Int. Miner. Process. Congr., Warsaw, 1979,* Polish Scientific Pub. (1979).

83. H. Schubert and C. Bischofberger, "On the Hydrodynamics of Flotation Machines," *Int. J. Miner. Process.,* **5,** 131–142 (1978).

84. H. Schubert and C. Bischofberger, "On the Optimization of Hydrodynamics in Flotation Process," *Proc. 13th Int. Miner. Process. Congr., Warsaw 1979,* Elsevier (1981).

85. A. Jowett, "Gangue Mineral Contamination of Froth," *Br. Chem. Eng.,* **11,** 330–333 (1966).

86. G. C. Thorne et al., "Modeling of Industrial Sulfide Flota-

tion Circuits," Ch. 26 in *Flotation—A. M. Gaudin Memorial Volume,* M. C. Fuerstenau (Ed.), AIME/SME (1976).

87. J. A. Engelbrecht and E. T. Woodburn, "The Effects of Froth Height, Aeration Rate, and Gas Precipitation on Flotation," *J. S. Afr. IMM,* **76,** 125–132 (October 1975).

88. A. Dippenaar, "The Effect of Particles on the Stability of Flotation Froths," S. Afr. NIM Report No. 1988 (1978).

89. J. A. Rogers, D. Simpson, and P. F. Whelan, *J. Inst. Fuel,* **29,** 545 (1956).

90. W. J. Trahar, "The Selective Flotation of Galena from Sphalerite with Special Reference to the Effects of Particle Size," *Int. J. Miner. Process.,* **3,** 151–166 (1976).

91. R. R. Klimpel, "Selection of Chemical Reagents for Flotation," AIME/SME Preprint No. 80-34 (1980).

92. F. F. Aplan, "Use of Flotation Process for Desulfurization of Coal," ACS Symposium Series No. 64, pp. 70–82 (1977).

93. N. Arbiter and C. C. Harris, "Flotation Kinetics," Ch. 8 in *Froth Flotation—50th Anniversary Volume,* D. W. Fuerstenau (Ed.), AIME/SME (1962).

94. A. W. Camerson et al., "A Detailed Assessment of Concentrator Performance at Broken Hill South Limited," *Proc. Aust. IMM,* **240,** 53–67 (December 1971).

95. G. E. Agar and W. B. Kipke, "Predicting Locked Cycle Flotation Test Restults from Batch Data," *CIM Bull.,* **71,** 119–125 (November 1978).

96. P. S. B. Stewart, D. N. Sutherland, and K. R. Weller, "An Optimization Study of Stage Flotation at the New Broken Hill Consolidated Lead-Zinc Concentrator." Presented at 12th Int. Miner. Process. Congr., Brazil, 1977.

97. R. P. King, "The Use of Simulation in the Design and Modification of Flotation Plants," Ch. 32 in *Flotation—A. M. Gaudin Memorial Volume,* M. C. Fuerstenau (Ed.), AIME/SME (1976).

98. A. D. Read, "Selective Flocculation Separations Involving Hematite," *Trans. IMM (C),* **80,** C24–C31 (1971).

99. A. F. Colombo and H. D. Jacobs, "Beneficiation of Nonmagnetic Taconites by Selective Flocculation-Cationic Flotation," U.S. Bureau of Mines Report of Investigations RI 8180 (1976).

100. K. V. S. Sastry (Ed.), *Agglomeration 77,* AIME/SME Ch. 54–56 (1977).

101. G. Zambrana Z., "Recovery of Minus Ten Micron Cassiterite by Liquid-Liquid Extraction," *Int. Miner. Process. J.,* **1,** 335–345 (1974).

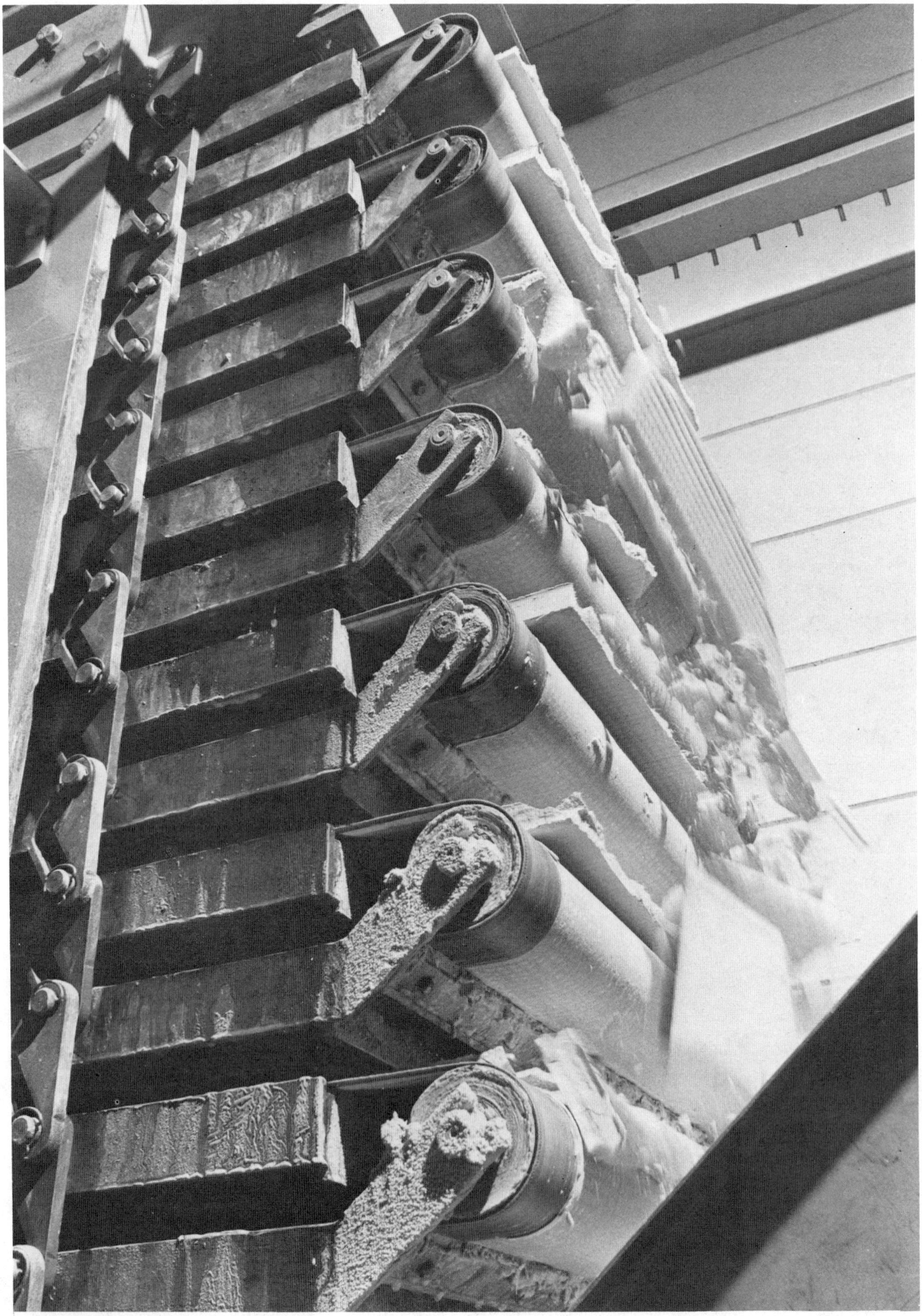

Part V

Dewatering

Earlier chapters have shown the importance of water in mineral processing operations. At some stage it is necessary to separate the water from the solids: exactly when it is removed depends on the process. For example, a concentrate is normally dewatered before transportation to a smelter. Tailings, and in an increasing number of situations concentrates, are transported as slurries, in which case dewatering may occur both before and after transportation.

Figure V.1 shows a typical sequence of process operations, with possible water contents at each stage and the limits of variation that could probably be tolerated. A process stream contains about 80–92% water by volume and the *degree* of dewatering also depends on the process.

In common with other separations, most dewatering processes involve at least two stages. Thickening (Chapter 17) and filtration (Chapter 18) are normally used for concentrate dewatering. This combination is capable of giving a solid that is suf- ficiently low in moisture for delivery to a smelter, as well as a filtrate that is essentially solids free, or at least low enough in solids for recycle to the plant.

Thickeners are generally of the continuous cylindrical type, but the advent of lamella thickeners could mean a changing emphasis as their potential for space and cost savings are realized. Sludge blanket fed thickeners, which use the sediment as a "filter," allow clearer overflows from dilute pulps than do conventional clarifiers. Disc filters, and to a lesser extent, drum filters, are the mainstay for most final dewatering because of their ability to remove most fine particles from a process stream. Fine particles, notoriously difficult to treat with vacuum filters, can now be handled with automatic filter presses.

Tailings are normally dewatered by a sedimentation device in the form of a tailings pond, this being the most economical method of handling the vast quantities of low value material that may be involved. In some cases the pond stage may be preceded by a conventional thickener.

Alternative dewatering systems are possible in

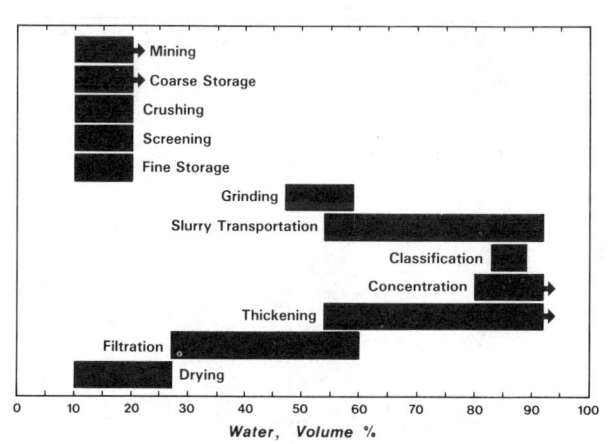

Figure V.1. Variation in water content during various stages of processing. (After Holland-Batt and Apostolides.[1])

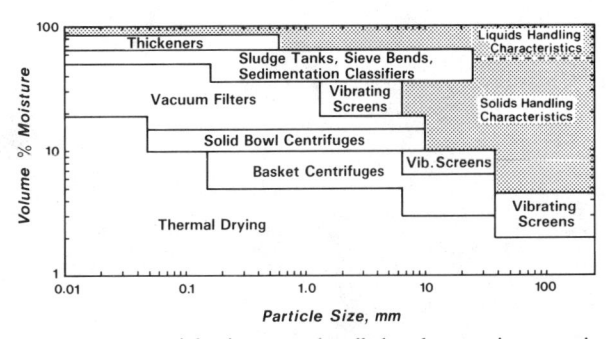

Figure V.2. Particle size range handled and approximate moistures produced by various forms of dewatering equipment. (After Sandy.[2])

certain circumstances, particularly when fines recovery is not important (Fig. V.2). Under these conditions, screens (Chapter 9), hydrocyclones (Chapter 10), centrifuges (Chapter 19), or a combination thereof (Chapter 19) can be used more cheaply than filtration.

REFERENCES

1. A. B. Holland-Batt and G. A. Apostolides, "Particulate Material Processing," *Chem. Process. Eng.,* **49,** 63–66 (1969).
2. E. J. Sandy, "Mechanical Dewatering," Ch. 12 in *Coal Preparation,* 3rd ed., AIME (1968).

Chapter Seventeen

Sedimentation

> *Rule of Accuracy:* "When working toward the solution of a problem, it always helps if you know the answer."
>
> *Corollary:* "Provided, of course, you know there is a problem."

Sedimentation is the removal of suspended solid particles from a liquid by gravitational settling. Such operations may be divided into thickening and clarification. Although governed by similar principles, these processes differ in that the primary purpose of thickening is to increase the solids concentration, whereas clarification serves to remove solids from a relatively dilute stream. Table 17.1 describes the classification of slurries for sedimentation. Slurries of class 1 and part of class 2 are treated in clarifiers; the remainder are treated in thickeners.

Clarification separations are characterized by sedimentation without a clearly defined interface between clear liquid and sediment, and as a consequence capacity is limited by the amount of solid that can be accepted in the overflow. Performance therefore is characteristic of a wet classifier and can be best analyzed as such. *Thickening* operations on the other hand are characterized by a clear liquid/sediment interface and capacity is limited by the underflow conditions. From the discussion presented in Section 4.7, it follows that equisettling particles or a minimum concentration are necessary for thickening.

17.1. SEDIMENTATION EQUIPMENT

17.1.1. Cylindrical Continuous Thickeners

The most common type of sedimentation unit is the cylindrical continuous thickener with mechanical sludge-raking arms,[1] as illustrated in Fig. 17.1. Feed enters the thickener through a central feed well and clarified liquor overflows into a launder around the periphery. Thickened sludge (*the sludge blanket*) collects in the conical base and is raked by the slowly revolving mechanism to a central discharge point.

Tanks. Tanks are normally cylindrical, smaller units being constructed in steel or wood, and larger ones (> 30 m) of concrete. The tank base is a shallow cone to facilitate sludge removal at its apex and is generally constructed of the same materials as the walls, although large concrete thickeners may have an earthen base.[2] The base slope is typically 80–140 mm/m, but higher slopes (up to 45°) may be used near the center of very large thickeners or for "settling" pulps in very small thickeners where height is not much of a problem.

Feed Wells. In conventional thickeners the cylindrical feed well (Fig. 17.2a and b) carries approximately 1 m^3/min · m^2. One of its primary functions is to act as a baffle to absorb the energy of the feed stream, and to this end it is often more than a simple cylinder (Fig. 17.2c).

The other function of the feed well is to allow, where necessary, flocculation. Although it is sometimes possible to do this by providing the well with sufficient holding time, a more effective approach is to provide some form of back-mixing. This can be achieved with a perforated well wall[3] (Fig. 17.2d), so that density gradients in the well provide a stirring action; or by providing an agitating device in the well. Sludge blanket fed thickeners deliver the

TABLE 17.1: CLASSIFICATION OF SEDIMENTATION

Pulp Description	Description of Sedimentation	Examples	Test Methods
Dilute (Clarification)	Particles or flocs initially settle independently, with no sharp interface. Settling depends largely on particle or floc size, but also on concentration	Turbid water, silt, trade waste.	Long tube and variations thereof.
Intermediate (Clarification)	Upper zone of independent particle subsidence. Lower zone of collective subsidence. Line of demarcation not sharp.	Chemical and metallurgical slurries, raw sewage, flue dust.	Long tube (and variations thereof). Kynch flux Curves.
Concentrated (Thickening)	Slurry settles with sharp interface. Ideally, settling rate is function of concentration only. In practice, initial settling rate may increase as flocs form, or decrease as faster settling particles move ahead of interface.	Chemical and metallurgical slurries, activated sludge.	Kynch flux curves.
Compressible (Thickening)	Settles initially with a sharp interface. Settling in sludge blanket non-ideal; depends also on time and blanket depth.	Particularly flocculated slurries.	Extensive Kynch flux curves.

feed (with flocculant) below the sludge level. This is done either with a deep feed well[4] (Fig. 17.2e), or by feeding from the bottom[5,6] (Fig. 17.2f, not to be confused with Fig. 17.2b). Such methods tend to fluidize the sludge in this zone and, because of the high solids concentration, to enhance flocculation and provide a filtering effect. The technique is particularly beneficial in what is normally a clarifying operation, since the sedimentation process then becomes one of thickening, and area requirements are reduced by factors of 2–10.

Rakes. The primary purpose of the rakes is to deliver the sludge to the central discharge. This movement generally has an important secondary effect in that flocs are broken down, resulting in a higher pulp density. Although rakes can be arranged to rake 1–4 times per revolution, the most common method is to use two large rakes and two shorter ones, to provide 4 rakes per revolution in the central heavier loaded area. Typical tip speeds are 5–8 m/min.

Three basic types of thickener can be distinguished by the method of supporting and driving the rake mechanism.[1] In the first type, the rake mechanism is supported by a superstructure across the tank, which also provides a carriage for the feed delivery system, and a walkway. Such a support is normally restricted to tanks smaller than 20 m in diameter. The second method, used for tanks 20–150 m in diameter, has a central column for support and for the drive mechanism. Traction thickeners form the third type, their drive being provided by a motorized carriage riding on the tank wall.

Various features are used to minimize or even out the load on the drive mechanism. The normal drag load can be minimized by keeping the structural members of the arm above the mud zone, and having the rake blades hanging on posts. Even so, some mechanism is also desirable to allow for excessive loads due to sludge or "iceberg" buildup. The most common method is to provide a mechanism (preferably automatic) that raises the entire rake mechanism as the load increases. An alternative approach is to hinge the arms at the center so that they ride

Figure 17.1. Cutaway view of thickener. (Courtesy Environmental Equipment Division of FMC Corp.)

Sedimentation Equipment

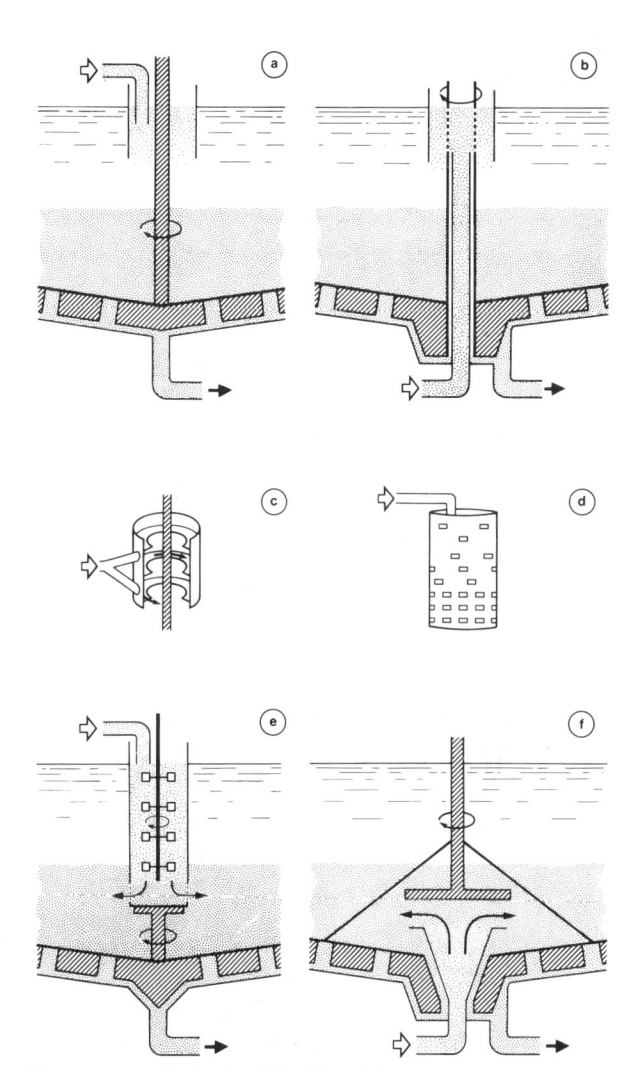

Figure 17.2. Methods of feeding thickeners. (*a*) Conventional feed well. (*b*) Bottom-fed feed well. (*c*) Counterflow feed well. (*d*) Perforated feed well. (*e*) Sludge blanket fed by deep feed well; (*f*) Sludge blanket fed by bottom feed.

over obstructions. Besides overload protection, these safety mechanisms allow the thickener to store extra slurry, providing circuit surge capacity.

Overflow Launders. Efficient thickener operation also requires control of the liquid velocity at the overflow weir. Clarifiers should be restricted to 0.2 m³/min · m; metallurgical thickeners commonly operate at 0.1 m³/min · m. If flow rates are likely to be higher than this, additional weirs can be placed inside the tank area. When flows are very low, notches can be used. Deep notches have an advantage in that constant flows can be maintained even with high surface winds.

17.1.2. Lamella Thickeners

Lamella thickeners have been designed to reduce the space requirement of conventional thickeners; a 5.5 m × 3.7 m × 5.2 m high lamella module can contain up to 230 m² of settling area equivalent to a 17 m conventional thickener.[7,8] They are characterized by packs of sloping parallel plates in the settling area (Fig. 17.3). This results in very short settling distances (less than the vertical distance between the plates), and the effective settling area is the vertically projected surface of all the plates.

The slope of the plates is a significant parameter, in that it must ensure an even flow of the settled solid down the plates to the discharge. In some cases vibrators may be used to ensure such a flow.

17.1.3. The Deep Cone Thickener

The deep cone thickener[9] (Fig. 17.4) uses comparatively high levels of flocculant (200 g/tonne). This, together with compression toward the bottom of the cone allows a plastic-like underflow discharge, which can be handled by conveyor. The 4 m diameter vessel has a capacity of about 80–110 m³/hr, and

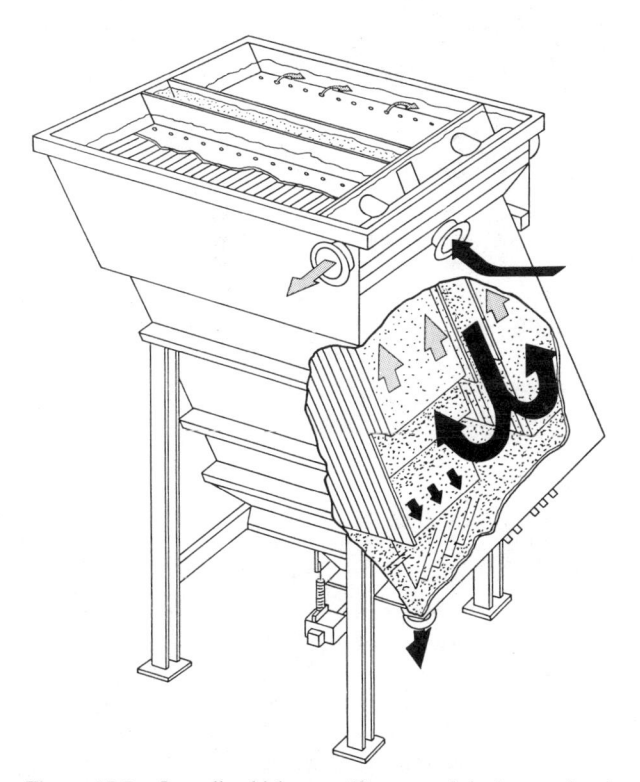

Figure 17.3. Lamella thickener. (Courtesy Sala International AB.)

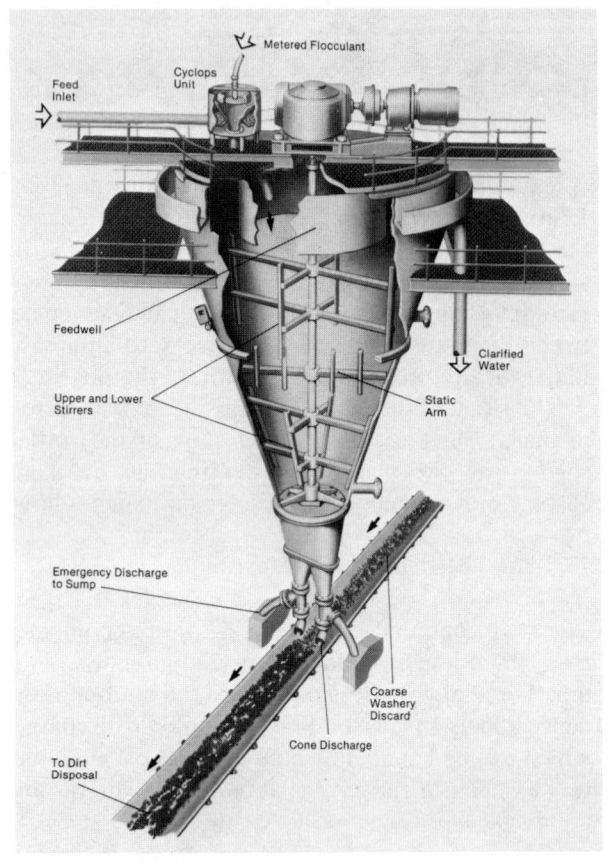

Figure 17.4. Deep cone thickener. (Courtesy Denver Equipment Division, Joy Process Equipment Ltd.)

produces a product with 60–70% solids (by volume).

17.1.4. Other Thickeners and Clarifiers

The principle of settling between inclined surfaces is employed in a variety of clarifier designs.[10] In fact, one of the sludge blanket fed thickeners also has inclined surfaces within the tank to aid sedimentation.[4]

Another form of space-saving thickener is the tray thickener.[1] It is essentially a stack of conventional thickeners and may be operated in series, parallel, or countercurrent. The latter arrangement is useful in leaching or washing operations.

In some clarifying operations, such as those in underground mines where space is at a premium, tanks may be rectangular (Fig. 17.5), 2–20 m long with widths 20–30% of the length. Sludge is raked to one end of the tank by scrapers fixed to a continuous chain.

Batch thickeners are occasionally used, but they are of little practical importance in mineral processing operations.

17.2. SEDIMENTATION BEHAVIOR

Two criteria are used to specify a sedimentation device: surface area and depth (although it is debatable whether the latter can or need be rigorously evaluated). The surface area however, must be determined and must be large enough to ensure that the upward velocity of liquid leaving through the overflow is not greater than the settling velocity of the slowest settling particle to be recovered. Although current theory as presented in Chapter 4 is of value, reliable information generally must be obtained from empirical batch test methods, even though most sedimentation is carried out in continuous devices. The area of a clarifier is determined by holding samples of a slurry in a graduated cylinder for various times, and withdrawing the uppermost liquid for suspended solids analysis.[1,11]

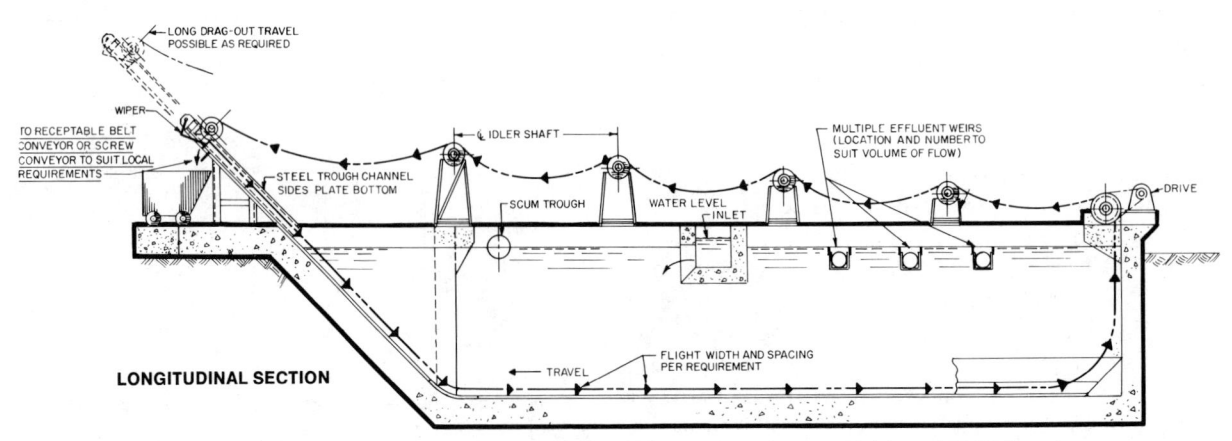

Figure 17.5. Rectangular thickener. (Courtesy Environmental Equipment Division of FMC Corp.)

Sedimentation Behavior

In practice, this method does not give a true indication of the performance of a clarifier, since it fails to account for the *size distribution* of the particles in the overflow, and thus their ability to remain suspended in the discharge stream. Realistically the clarifier is a classifier and should be analyzed as such, that is, by describing its performance in terms of a performance curve with a d_{50} cut point (Chapter 10). To do this, the samples collected in the conventional test must be subjected to a size analysis.

Thickener areas are also evaluated by testing slurry in a graduated cylinder: in this case by analyzing the rate at which the interface between clear liquid and slurry falls.[1,12] This chapter concentrates on thickener analysis, since it requires the use of concepts we have not considered before.

17.2.1. Sedimentation Flux Curves

A good general description of the various ways in which sedimentation can occur was presented in Section 4.7, along with relationships for free and hindered settling. Unfortunately, these relationships are normally inadequate for rigorous design calculation because the solids in most practical slurries are too heterogeneous. However, the relationships are useful in that they indicate the important variables, and their relative significance.

It is apparent that if the slurry is dilute enough, free settling may occur; but most settling occurs under hindered conditions, where upward flowing liquid, displaced by settled solids, slows the effective settling rate. In the latter stages of settling there is significant particle-to-particle contact. Further complexity is caused by the existence of flocs. These loose agglomerates of particles of varying sizes are effectively large particles with low density arising from entrapped water. (In some cases the average density is so low that the floc actually settles more slowly than the individual particles.) Because of the weak, variable strength of their bonding, flocs are capable of considerable deformation.

Figure 17.6a illustrates the variations in settling velocity v_h that occur with slurry concentration. Although true free settling can be considered to occur at very low concentrations ($< 1\%$), the curve can be approximated by Eq. 4.70:

$$v_h = v_\infty \varepsilon^n \quad (4.70)$$

In the case of thickening, instead of considering the settling rate, it is more useful to consider the flow rate of particles per unit area, the solids flux ψ,

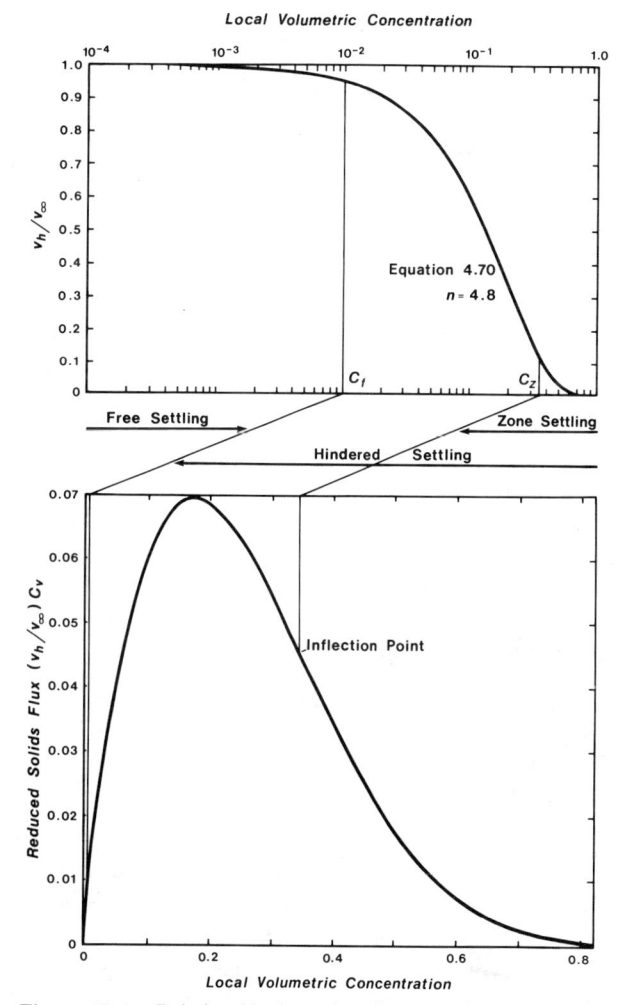

Figure 17.6. Relationship between hindered settling and the sedimentation flux curve.

which is given by

$$\psi = Cv \quad (4.78)$$

(ψ = mass flux of particles, kg/m² · sec; C = concentration, kg/m³). Since $1 - C/\rho_s = \varepsilon$ and v can be approximated by Eq. 4.70, it follows that

$$\psi = v_\infty \varepsilon^n (1 - \varepsilon) \rho_s \quad (17.1)$$

which predicts the characteristic form of the sedimentation flux curve (Fig. 17.6b). This shows that a low solids flux occurs at low concentrations (because few particles exist) and at high concentrations (because the settling velocity is severely reduced). In practice, because n is difficult to predict, Eq. 17.1 is seldom used to generate flux curves; instead they are determined from batch experimental tests.

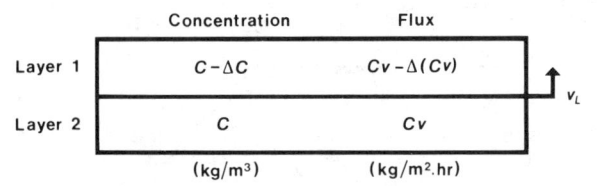

Figure 17.7. Sedimentation between two elemental layers of slurry.

Consider now a slurry that has an infinitely thin layer having a concentration C, with an infinitely thin layer immediately above it of concentration $C - \Delta C$, as illustrated in Fig. 17.7. Consider further that layer 2 is moving upward with a velocity v_L. At any time the amount of solid leaving layer 2 is, as a result of sedimentation, Cv, and, as a result of upward movement of the interface, Cv_L. For the layer above, the corresponding values are $(Cv - \Delta(Cv))$ and $(C - \Delta C)v_L$. If the concentration C remains constant, a material balance gives

$$Cv + Cv_L = (Cv - \Delta(Cv)) + (C - \Delta C)v_L$$

that is,

$$v_L = -\frac{\Delta(Cv)}{\Delta C} \qquad (17.2)$$

Consider now the batch sedimentation test illustrated in Fig. 4.13a, with an even initial concentration C_I of uniform particles. The upper interface will settle at velocity $v_{L,I}$ and zone B will settle with a flux $\psi_I = C_I v_I$ throughout. If Fig. 17.8 represents the flux curve of the slurry, it follows from Eq. 17.2 that $v_{L,I}$ is the slope of the line from the origin to the point $[C_I, C_I v_I]$. (Since the slope is positive, the velocity will be negative, i.e., downward.) However, at the bottom of the vessel after an infinitely short time layers with all concentrations from C_I to C_{max} exist. Each of these concentration layers will move up from the bottom at a velocity also given by Eq. 17.2.

For example, in Fig. 17.8 the layer of concentration C_A will move upward at a velocity $v_{L,A}$ equal to $\Delta(Cv)_A/\Delta C_A$ (since the slope IA is negative, the velocity in Eq. 17.2 becomes positive, i.e., upward). It follows that there is one concentration layer C_K that moves up faster than any other layer. This layer will move up at $v_{L,K}$ and will overtake all layers having concentrations between C_I and C_K, resulting in a discontinuity or sharp interface between the two concentration layers C_I and C_K (zones B and D in Fig. 4.13a) (provided C_I and C_K remain unchanged.)

Figure 17.9 plots the heights of the ascending and descending interfaces as a function of time. Until the two interfaces meet, the concentration of the slurry between the upper and lower interfaces remains at C_I. Below the lower interface, the concentration changes from C_K at the top to C_{max} at the bottom. After time t_Z there is only one interface,

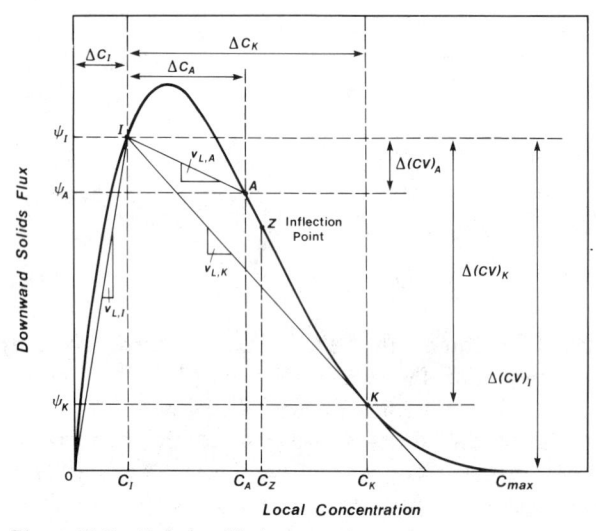

Figure 17.8. Relationship between the propagation velocity of a layer and the batch flux curve.

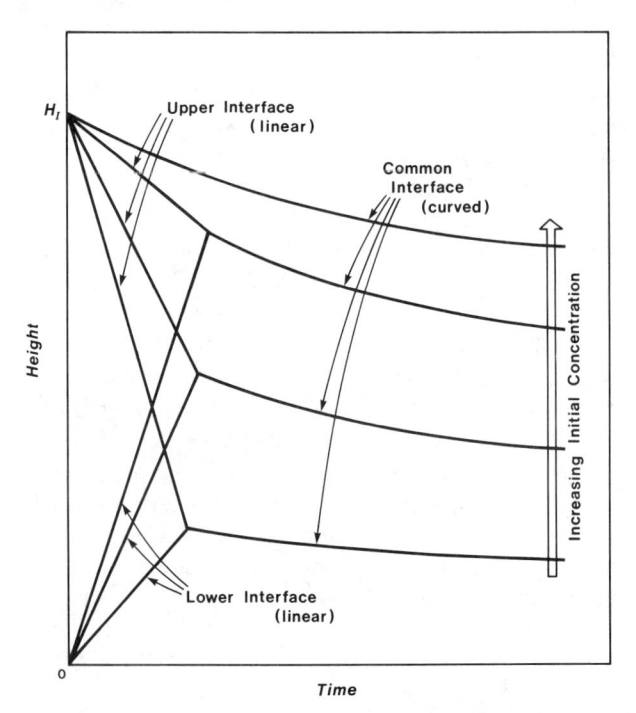

Figure 17.9. The effect of initial concentration on the batch settling curve.

above which there are no particles and which from then on, falls at the settling velocity of the slurry just below the interface. The settling velocity of the interface will now slow down as successively slower moving concentration layers with $C > C_K$ gradually reach the interface from the bottom. These layers will have risen with their velocity given by $d(Cv)/dC$, that is, the negative slope of the tangent to the flux curve.

The maximum possible value of $v_{L,K}$ is $v_{L,Z}$, and clearly this occurs when the initial concentration is at the inflection point of the flux curve, that is, $C_I = C_Z$ as shown in Fig. 17.8. It follows that if C_I exceeds C_Z, there will not be an upper and a lower interface because all concentration layers will be rising from the bottom more slowly than $v_{L,K}$ and under these conditions the settling curve will be a continuous function of time (top curve in Fig. 17.9). The characteristic hindered settling that occurs when C exceeds C_Z is described as *zone settling* (Fig. 17.6). Under these conditions, even if the solid is heterogeneous, the material settles with a sharp interface between clear liquid and slurry,[13,14] and the settling velocity is assumed to be a function of concentration only. The last assumption is the basis of most thickener design, and its limitation must be appreciated.[15] In practice some slurries, particularly flocculated ones, exhibit compression resulting from the weight of solids above,[16-20] and under these conditions the concentration layers do not propagate from the bottom at constant velocity. Such nonideal behavior can generally be detected by producing experimental flux curves using different initial concentrations and slurry heights.

With initial concentrations between C_f and C_Z hindered settling does occur (Fig. 17.6), but if the particles are heterogeneous, they may settle at differing rates, giving a diffuse interface that in a continuous thickener may cause fine material to report to the supposedly clear overflow.

17.2.2. Jernqvist Construction of Flux Curves

The Jernqvist construction provides a convenient method of producing the flux curve from a batch settling curve.[12] It is based on the analysis of Kynch,[15] which assumes zone settling [$v = \text{fn}(C)$ only] and that the concentration layers propagate upward from the vessel bottom at constant velocity (i.e., no compression).

The method is illustrated in Fig. 17.10, which consists of two curves: the experimental batch settling curve of H versus t, and the constructed flux curve of ψ ($= Cv$) versus C.

The construction procedure (steps ii–viii are so identified on Fig. 17.10):

i. Choose convenient scales for the time, height, and concentration axes.

ii. From the chosen scales, calculate the ψ scale, for example:

   Time scale: 1 cm of graph paper represents 5 min
   Height scale: 1 cm of graph paper represents 0.05 m (5 cm)
   Concentration scale: 1 cm of graph paper represents 25 kg/m³

   $$\therefore \text{scale} = \frac{25 \text{ kg}}{\text{m}^3} \left| \frac{0.05 \text{ m}}{1 \text{ cm}} \right| \frac{1 \text{ cm}}{1 \text{ cm}} \left| \frac{1 \text{ cm}}{5 \text{ min}} \right.$$
   $$= \frac{0.25 \text{ kg}}{\text{m}^2 \text{ min}}$$
   represented by 1 cm of graph paper

iii. Draw in a vertical line through the initial uniform constant concentration C_I in the batch test.

iv. Draw a horizontal line through H_I to produce axis AB (this can be made to correspond to the concentration axis, and eliminates a common construction error).

   To obtain each point of the solids flux curve:

v. Draw a tangent to the H versus t curve and draw a parallel line through the origin of the ψ versus C graph.

vi. From the point D where the tangent cuts the H axis, draw a horizontal line to cut the C_I line at E.

vii. Produce a line from the origin F of the H versus t curve through E to cut AB at G.

viii. Draw a vertical line through G to cut the line on the ψ versus C graph that was drawn parallel to the tangent. This intersection H defines a point on the flux curve.

ix. Repeat steps v–viii until enough points have been obtained to define the flux curve.

Notes

1. The curved portion of the H versus t curve gives points on the flux curve only for $C > C_Z$.

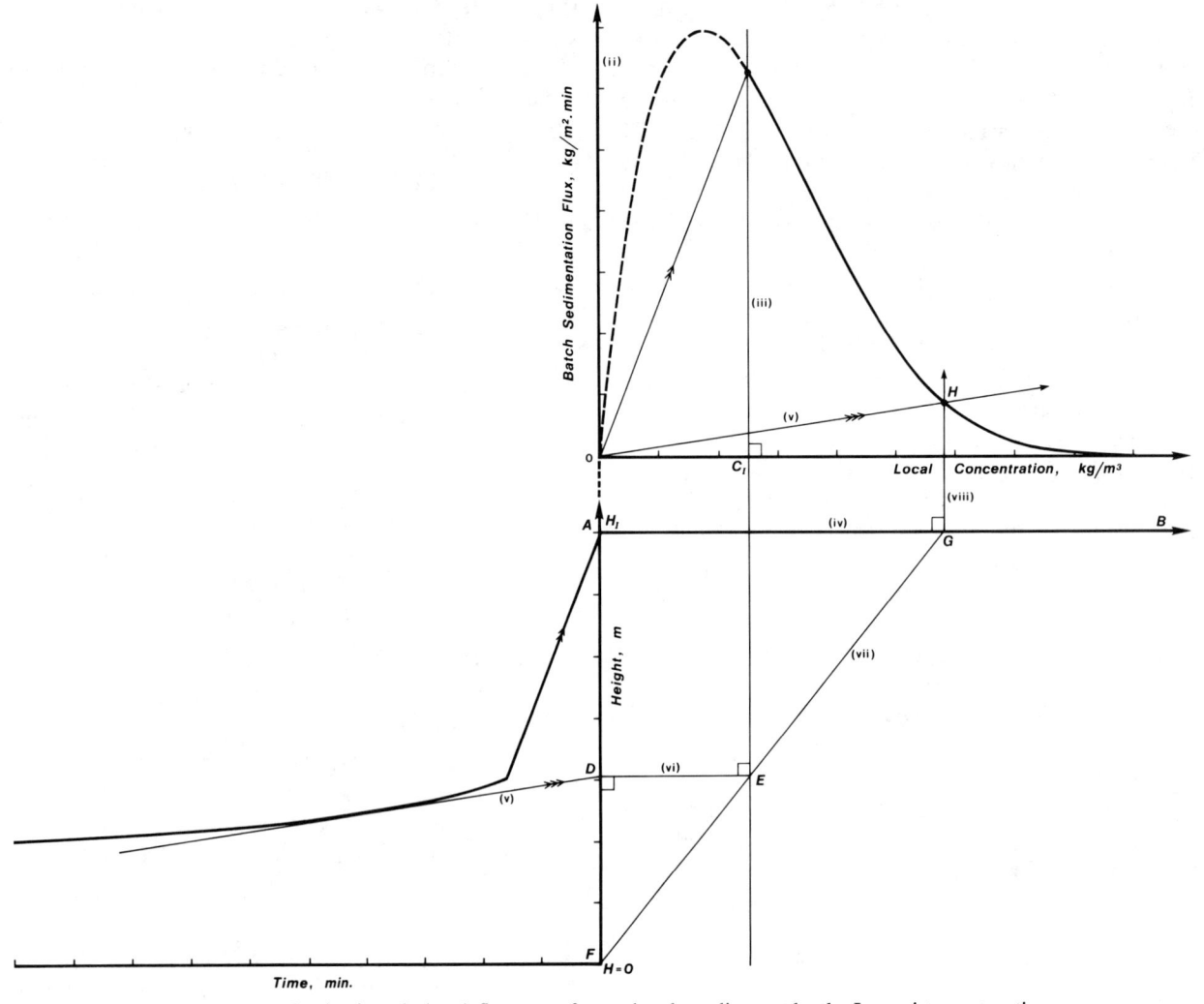

Figure 17.10. Derivation of a batch flux curve from a batch settling test by the Jernqvist construction. (After Jernqvist.[12])

2. When the H versus t curve has an initial linear portion, this gives only *one* point on the flux curve, and it will occur at $C < C_Z$.
3. Since the same concentrations are transcended, the same solids flux curve should be produced from settling curves having different initial concentrations of the same slurry.
4. If the settling curve has an initial linear section, the discontinuity in the curve should be sharp. No attempt should be made to draw tangents at this point.
5. The construction assumes $v = \text{fn}(C)$ only, and takes no account of compression where $v = \text{fn}(C, t)$. If different initial concentrations result in different flux curves at $C > C_Z$, this is an indication of compression. Consequently, although it is possible to design a thickener from a single batch test, it is always desirable to study a range of concentrations and initial heights to build up the flux curve at $C < C_Z$ to see whether nonideal behavior is occurring.
6. Settling tests should be carried out over sufficient time to establish H at t_∞ so that $C_{\max}$ can be obtained.
7. Test conditions should be as close as possible to those likely to be encountered in practice. The aging of slurries is a particularly common problem and source of error.

17.3. STEADY STATE OPERATION OF A CONTINUOUS THICKENER

17.3.1. Maximum Capacity

Figure 17.11 is a suitable representation of a continuous thickener and defines the symbols used here. Sedimentation behavior can be assumed to be given by the characteristic flux curve obtained from a batch test; but to this must be added the bulk flows within the thickener. Although there may not be a sharp horizontal cutoff, for convenience the bulk flow flux can be considered to split at the feed inlet into underflow (positive response) and overflow (negative response).

Flux Above the Feed Inlet. Since the cross-sectional area of the inlet is relatively small, the bulk flow above the inlet can be taken as $O_{(-)}/A_T$. Consequently, for an arbitrary concentration layer above the feed inlet, the overflow withdrawal causes an upward solids flux $\psi_{(-)}$ given by

$$\psi_{(-)} = \frac{O_{(-)}}{A_T} C_{(-)} \quad (17.3)$$

as shown in Fig. 17.12. Opposing this is the gravitation sedimentation flux ψ_s ($= Cv$) so that the net upward flux ψ_u is

$$\psi_u = \psi_{(-)} - \psi_s \quad (17.4)$$

Thus above the feed inlet the net flux is downward for all concentrations below C_2 (Fig. 17.12) and a

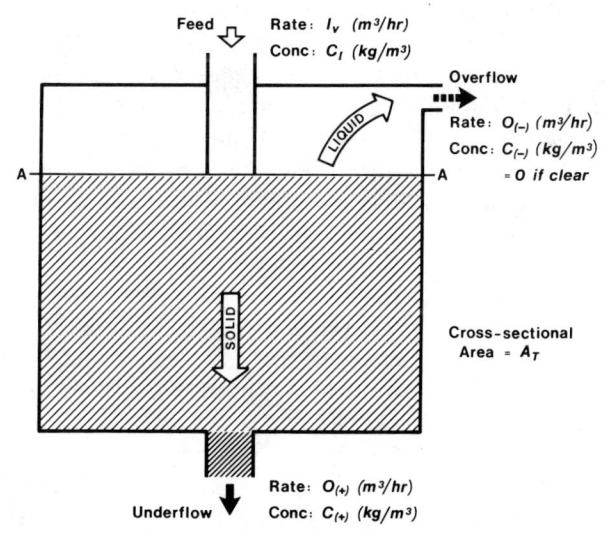

Figure 17.11. Schematic representation of a continuous thickener, and definition of nomenclature.

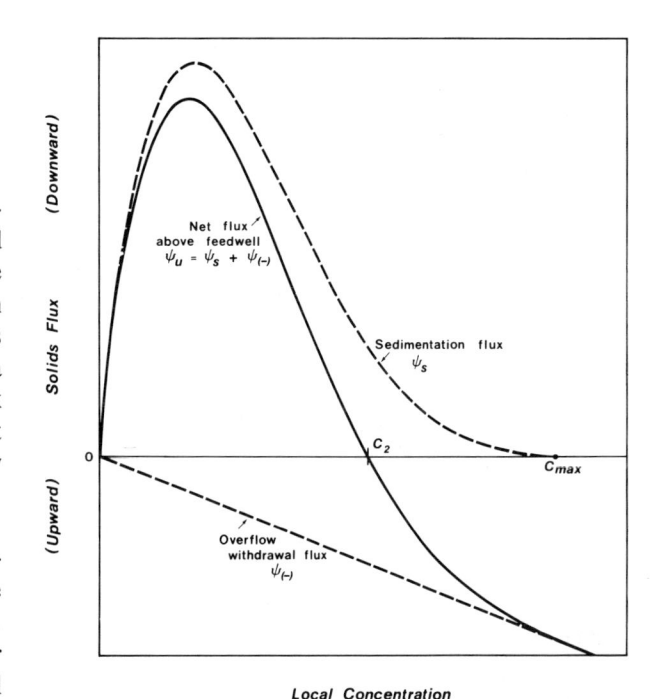

Figure 17.12. Flux in thickener above feed well.

design limitation[21] is $C_I > C_2$. However in normal practice concentrations above the feed inlet are very low, so that $\psi_{(-)}$ is very low, in turn making the critical value C_2 high, with the result that this is not a realistic design criterion in most cases.

Flux Below the Feed Inlet. For any arbitrary concentration layer below the feed inlet, the underflow withdrawal produces a solids flux given by

$$\psi_{(+)} = \frac{O_{(+)}}{A_T} C_{(+)} \quad (17.5)$$

When added to the sedimentation flux, the net downward flux ψ_d is

$$\psi_d = \psi_{(+)} + \psi_s = \frac{O_{(+)}}{A_T} C_{(+)} + Cv \quad (17.6)$$

This situation is represented in Fig. 17.13a, and it can be seen that there is a critical concentration in the thickener ($C_I \leq C_{\text{crit}} \leq C_{\text{max}}$) through which the flux of solid particles is a minimum ψ_{crit}. Since all solids must pass through this layer, the value of ψ_{crit} must be the maximum the thickener can accommodate. This means that any concentration layer below the feed inlet whose concentration is less than C_{crit} recovers more particles than it loses until the concentration reaches C_{crit}. Similarly all layers with

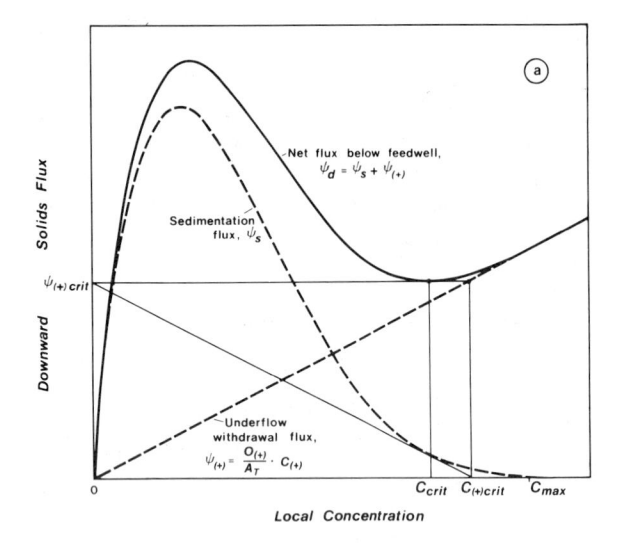

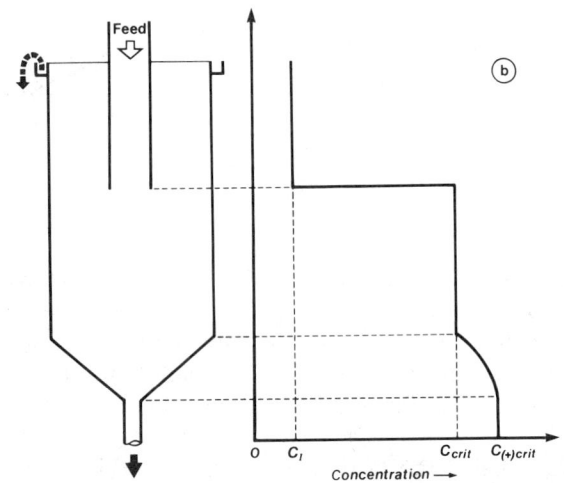

Figure 17.13. Equilibrium conditions in a thickener below the feed well at maximum loading. (*a*) Flux curves. (*b*) Concentrations.

concentrations exceeding C_{crit} receive fewer particles than they lose until their concentration reaches C_{crit}. It follows that for equilibrium ψ_{crit} must equal the input and underflow fluxes (ψ_I and $\psi_{(+)}$, respectively) so that a thickener can be designed from

$$\psi_{crit} = \frac{I_V}{A_T} C_I = \frac{O_{(+)}}{A_T} C_{(+)crit} \quad (17.7)$$

(I_V = volumetric feed rate, m³/hr).

For ψ_d to have a minimum value, $d\psi_d = 0$, which from Eq. 17.6

$$\frac{O_{(+)}}{A_T} \cdot dC_{(+)} + d(Cv) = 0$$

$$\frac{d(Cv)}{dC_{(+)}} = -\left(\frac{O_{(+)}}{A_T}\right) \quad (17.8)$$

Thus the minimum in the ψ_d curve occurs when the ψ_s curve has a slope of $-(O_{(+)}/A_T)$. Hence, as illustrated in Fig. 17.13*a*, the required area of a thickener, from the point of view of underflow conditions, can also be determined by drawing the tangent to the batch test solids flux curve through the point $C_{(+)crit}$ on the concentration axis. This tangent cuts the ψ-axis at the point $(I_V/A_T)C_I = (O_{(+)}/A_T)C_{(+)crit}$, and is frequently referred to as the *operating line*. If I_V and C_I are known, the area can be determined.

The discussion above also implies that a thickener operating at maximum capacity would have C_{crit} existing at all levels below the feed inlet, with a sudden jump to $C_{(+)crit}$ at the feed outlet, as illustrated in Fig. 17.13*b*. That this situation generally does not occur is explained further in the next section.

The main advantage of this design approach is its flexibility. As is considered below, *if* the flux curve truly represents the slurry (as it ideally should), it shows the complete behavior of the thickener, and over- or undercapacity operation can be analyzed.[12,22,23] Alternatively, although one batch test is sufficient to design a thickener, nonideal behavior can generally be detected if testing under different conditions produces variations in the flux curve, thus indicating potential problems.

It has been claimed that this method is not always reliable.[24] When the method has been found to be inadequate, however, there has generally been failure to obtain reliable flux curves. For example, in full-scale thickeners, the flow patterns may cause large particles to settle out rapidly in the central zone, leaving a slurry of finer particles that consequently require a larger sedimentation area. In such a situation the problem is then really one of obtaining a representative sample of slurry. Furthermore, even though flocculated pulps are clearly nonideal, the work of Scott has shown that this design philosophy gives very satisfactory, although slightly conservative, results for such materials.[16-18]

To bypass the difficulties of obtaining reliable flux curves, Robins has recommended the use of a semicontinuous thickener to obtain design data.[23] It has not been shown that the extra effort achieves any better designs.

Example 17.1. Laboratory batch tests were carried out on a slurry, and, using the Jernqvist construction, gave the batch flux curve shown in Fig. E17.1.1. The slurry is to be thickened in a 20 m

Steady State Operation of a Continuous Thickener

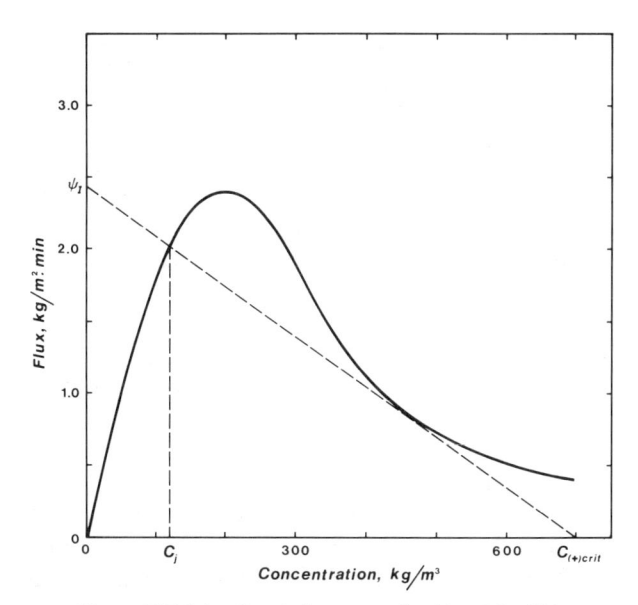

Figure E17.1.1. Batch flux curve for Example 17.1.

(diameter) circular thickener, at a feed rate of 3.0 m³/min. Because the underflow is to be filtered, it is considered that its concentration should be at least 700 kg/m³. What is the maximum feed concentration the thickener can accept and still produce a clear overflow?

Solution. Draw the operating line (tangent to the batch flux curve) from $C_{(+)} = 700$ kg to the vertical flux axis, to obtain the maximum feed flux ψ_I.
Hence

$$\psi_I = 2.45 \text{ kg/m}^2 \cdot \text{min}$$
$$= \frac{I \cdot C_I}{A}$$

Solving:

$$C_I = \frac{2.45 \text{ kg}}{\text{m}^2 \cdot \text{min}} \left| \frac{\pi \, 20^2 \text{ m}^2}{4} \right| \frac{\text{min}}{3.0 \text{ m}^3}$$
$$= 257 \text{ kg/m}^3 \text{ (maximum)}$$

17.3.2. Normal Operation with Slight Underload

Normal operation of a thickener will feature clear overflow and slight underloading. At steady state in such a situation, $\psi_I = \psi_{(+)} < \psi_{\text{crit}}$, and the situation can initially be considered as that represented by curve 1 in Fig. 17.14a, where the thickener contains two conjugate concentrations, $C_{(+)}$ in the underflow and C_j in the tank. The concentrations existing in a thickener are therefore those shown in Fig. 17.14b

rather than those in Fig. 17.13b. Comparison of Figs. 17.13 and 17.14 also shows that $C_{(+)}$ is slightly less than $C_{(+)\text{crit}}$ when the thickener is operating in this condition.

The apparently sudden increase in concentration at the outlet actually occurs in the conical section of the thickener and can be explained as follows.[25] At a cross-sectional area $A(A < A_T)$ the withdrawal flux through A increases to $\psi_{(+),A} = (A_T/A) \cdot (O_{(+)}/A_T) \cdot C_{(+)}$ so that new curves apply as shown in Fig. 17.14a (e.g., curve 2). The new flux through A is $\psi_A = \psi_I(A_T/A)$, and since $\psi_{(+),A}$ and ψ_A are moved

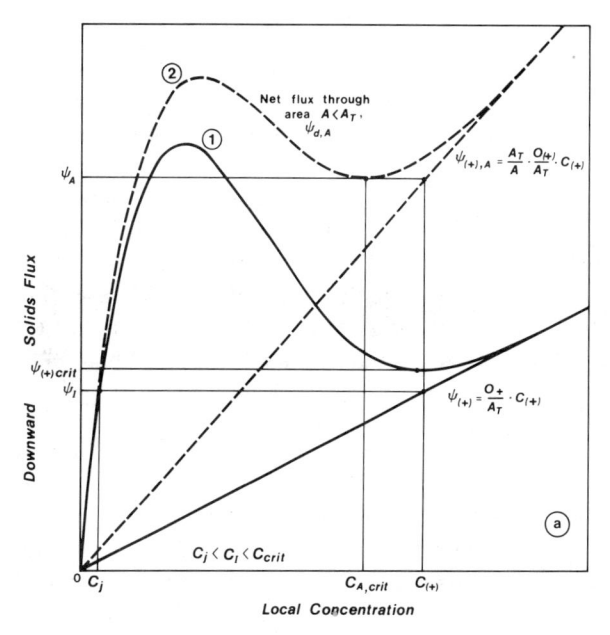

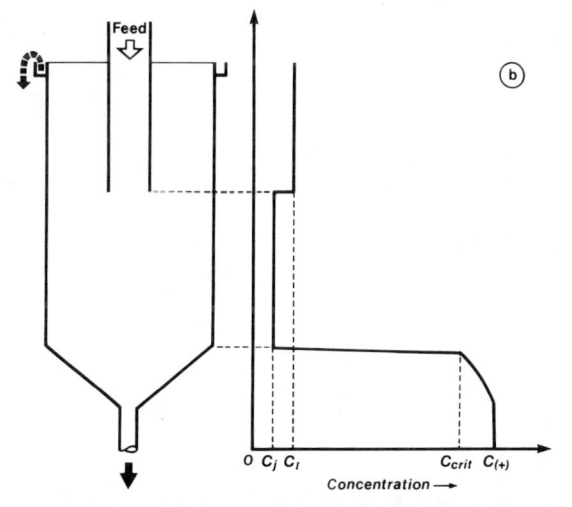

Figure 17.14. Equilibrium conditions in a thickener below the feed well at slight underloading. (*a*) Flux curves (*b*) Concentrations.

up by the same ratio (A_T/A), they will always intersect at $C_{(+)}$. However, because the ψ_s curve is unchanged as A decreases, the horizontal ψ_A will eventually contact the minimum on the new $\psi_{d,A}$ curve, since this is not displaced upward as rapidly. At the particular value of A where they touch, the lower conjugate concentration will no longer exist but rather will be replaced by $C_{A,\text{crit}}$. This concentration will exist at the top of the sludge zone, and from there down will increase to $C_{(+)}$ at the discharge.

Thus in normal operation, the sludge line would lie somewhere in the conical section of the thickener and the situation would be that shown in Fig. 17.14b. As maximum loading is approached, the sludge line rises to the top of the conical section, $C_{A,\text{crit}}$ tends to $C_{(+)\text{crit}}$, and the discharge concentration rises to $C_{(+)\text{crit}}$. On the basis of this argument, the thickener in reality needs negligible height, and the main purpose of the vertical height above the conical section is to absorb transitory overloading.[22] In practice, flocculated pulps may require compression to give high pulp densities, and in this case, the height in the sludge zone determines the amount of compression so that thickener height becomes important.[16]

17.3.3. Overload

If ψ_I exceeds ψ_{crit}, the thickener is in an overloaded condition and will eventually reach the steady state situation illustrated in Fig. 17.15.[22] The excess solid will be rejected to the overflow according to

$$\psi_I - \psi_{\text{crit}} = \frac{O_{(-)}}{A} C_{(-)} \quad (17.9)$$

The concentration $C_{(-)}$ of the overflow bears no relation to that inside the thickener above the feed inlet. Rather, a step dilution occurs, analogous to the concentration increase at the bottom discharge.

In practice sudden overload does not occur. Rather, the interface between C_I and C_{crit} moves up slowly according to Eq. 17.2; in this case the slope of the tangent from the point $[\psi_I, C_I]$ to the flux versus concentration curve for discharge (Fig. 17.15). The result is that since this is such a low velocity, the thickener may be able to tolerate an overload for many hours before solids appear in the overflow. If ψ_I is greater than the maximum value of ψ_d, the rate of rise of an interface above the bottom of the feed well becomes significant. Analysis of this situation is best considered on a flux diagram such as Fig. 17.12 after calculating $C_{(-)}$ from Eq. 17.9.

Example 17.2. While operating with a marginal underload, the thickener in Example 17.1 is subjected to a 20% increase in the feed concentration.

1. If the thickener has a shallow feed well, the bottom of which is 1.0 m above the edge of the conical section, how long can the overload be tolerated?

2. If the overload persists, what would be the eventual concentration of solids in the overflow?

Solution

1. The maximum time is that required for the sludge interface to rise from the conical section to the bottom of the feed well. The rise velocity of this interface is given by the slope of the line, on the net flux curve, that joins the two concentrations forming the interface, that is, A and B on Fig. E17.2.1. The new feed flux $\psi_{O/L} = 1.2\,\psi_I$ (i.e., 2.94 kg/m^2 · min). The slope of the line A-B on Fig. E17.2.1 gives

$$v = \frac{(2.94 - 2.45)\text{ kg}}{\text{m}^2 \cdot \text{min}} \bigg| \frac{\text{m}^3}{(415 - 170)\text{ kg}}$$
$$= 2.0 \times 10^{-3} \text{ m/min}$$

$$\therefore \quad t \text{ for 1 m} = \frac{1 \text{ m}}{} \bigg| \frac{\text{min}}{2.0 \times 10^{-3} \text{ m}}$$
$$= 500 \text{ min}$$

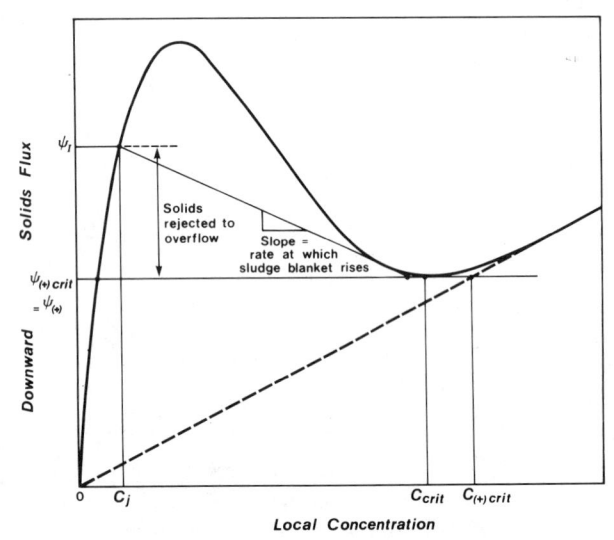

Figure 17.15. Overloading of a thickener.

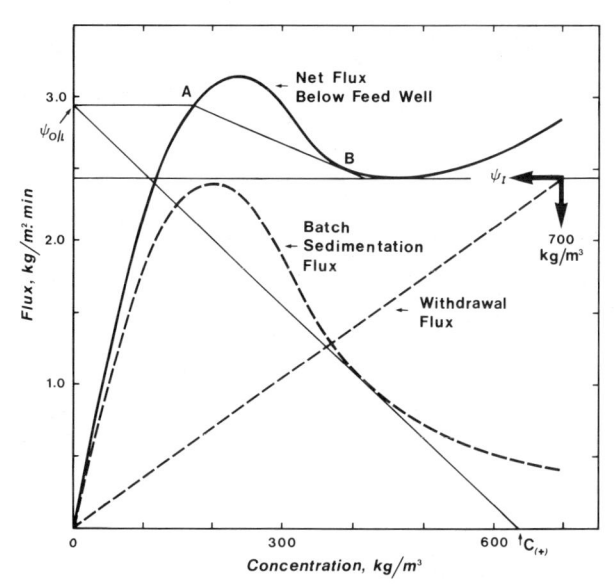

Steady State Operation of a Continuous Thickener

Figure E17.2.1. Net flux curve for Example 17.2, with overload conditions shown.

2. To solve $\psi_{(+)} = O_{(-)}C_{(-)}/A$, $O_{(-)}$ must first be determined by a materials balance. Because the underflow conditions are unchanged, $\psi_{(+)} = \psi_I$

$$\therefore O_{(+)} = \frac{\psi_I A}{C_{(+)}}$$
$$= \frac{2.45 \text{ kg}}{\text{m}^2 \cdot \text{min}} \left| \frac{\pi\, 20^2 \text{ m}^2}{4} \right| \frac{\text{m}^3}{700 \text{ kg}}$$
$$= 1.10 \text{ m}^3/\text{min}$$

Materials balance:

$$\text{in} = \text{out}$$
$$I = O_{(+)} + O_{(-)}$$
$$O_{(-)} = 3.0 - 1.10$$
$$= 1.90 \text{ m}^3/\text{min}$$

Since $\psi_{(-)} = 0.2\psi_I$, the excess flux is

$$C_{(-)} = \frac{0.2 \times 2.45 \text{ kg}}{\text{m}^2 \cdot \text{min}} \left| \frac{\text{min}}{1.90 \text{ m}^3} \right| \frac{\pi\, 20^2 \text{ m}^2}{4}$$
$$= 81.0 \text{ kg/m}^3$$

17.3.4. Overdilute Feed

If the feed concentration is less than the lower conjugate concentration, some solids will be rejected in the overflow.[25] This situation, which also effectively represents the operation of a clarifier, is illustrated in Fig. 17.16. It can be seen that although $\psi_I < \psi_{\text{crit}}$, $\psi_I > \psi_{(+)}$, and the difference $\psi_I - \psi_{(+)}$ will again be the flux to the overflow $\psi_{(-)}$. With dilute slurries of heterogeneous material it may be difficult to obtain a reliable flux curve, in which case the design may be based on the long tube method, or on the simpler suspension method.[1] A better approach however is to use some technique such as flocculation or recycling, or a device such as the sludge blanket fed thickener, to change the sedimentation mechanism from clarification to thickening.

Example 17.3. What is the minimum feed concentration for a clear overflow from the thickener in Example 17.1?

Solution. The intersection of the operating line with the flux curve gives the conjugate concentration C_j, which represents the minimum concentration. From Fig. E17.1.1 this is 120 kg/m³.

17.3.5. Flocculated Pulps

Most mineral slurries have a wide range of particle sizes, and as a consequence frequently do not settle with a sharp interface (Fig. 4.13c). Today, flocculants are extensively used as a sedimentation aid, ostensibly to increase the settling rate. Although the rates are in fact increased, the significant achievement is the grouping together of the particles into fairly uniformly sized flocs, *so that the slurry settles with a sharp interface.* Thus, the sedimentation mechanism is changed from clarification to thicken-

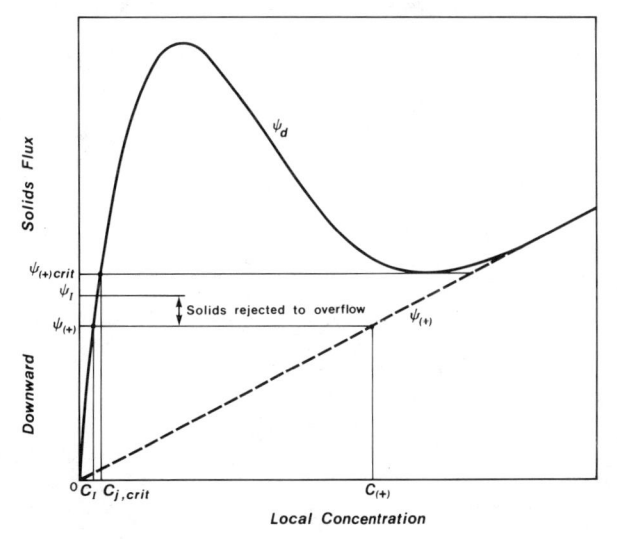

Figure 17.16. Overdilute feed to a thickener.

ing, and in principle can be analyzed by the methods previously described. In practice, the settled slurry undergoes compression and the Kynch analysis is strictly speaking no longer valid.

Scott[16-18] used Eq. 6.45 to estimate the effective floc size and to derive a "reduced" flux curve. He also found that the flux curve could be divided into three concentration ranges. At low concentrations independent spherical flocs occur with apparent diameters many times the particle size. At intermediate concentrations the flocs deform slightly, and channels form between them, allowing relatively rapid liquid upflow. Gentle raking in batch tests and operating thickeners will break up these channels and significantly reduce the settling rate. (It is this situation that makes design by the Coe and Clevenger[26] method unreliable; but when the Jernqvist flux curve analysis is used, the nonideal conditions appear to cancel each other.) At high concentrations the channels have closed up and compression of the flocs occurs changing the flux curve. The amount of compression depends on the height of slurry, and consequently under these conditions thickener depth is a design parameter.[17,20] The existence of these different ranges can be determined by a simple graphic technique.[27]

At this stage it must be emphasized that the mechanisms involved in the thickening of flocculated slurries are not fully understood,[20] and the foregoing must be considered to be simply an introduction to the topic. However, it does appear likely that such slurries can in fact be represented by a flux envelope[28] (Fig. 17.17) rather than a single flux curve. This suggests that a reasonably reliable estimate of the envelope can be obtained for an operating thickener by carrying out batch tests in tall cylinders, using a stirrer (to break down the flocs and simulate the raking action) and different initial concentrations (to give different final heights, and thus a measure of the degree of compression that can occur).

Some other aspects of flocculation are worth mentioning. In many instances, flocculation may not be perfect, and a small proportion of the finest particles remain in suspension to leave through the overflow. Provided this proportion is acceptably low, the situation can be treated as thickening. Second, flocculation does require an induction time. Some batch tests, for example, may show a settling rate that initially increases with time, an effect that can be attributed to flocculation occurring after the start of the test. Such increased flocculation largely results from the faster settling particles overtaking the slower settling ones, thereby raising the local concentration and bringing about flocculation.

When the particles are very fine, flocculation may actually result in an even slower settling rate, because the increased average "particle" (i.e., floc) size is more than nullified by a decrease in the floc's average density. One method of overcoming this difficulty is to add a small proportion of larger particles to the slurry; a method that has been suggested for the notoriously difficult-to-settle Florida phosphate slimes.[29]

17.3.6. Lamella Thickeners

Theoretically, a lamella thickener can be sized using the previously described methods, assuming that the effective area A_e is the total vertically projected area of the lamella, that is,

$$A_e = NA \cos \theta \qquad (17.10)$$

(A = area of lamella; θ = inclination of lamella to the horizontal; N = number of lamellae). In practice, a correction factor K_{La} must be included thus

$$A_e = NK_{La}A \cos \theta \qquad (17.11)$$

This correction arises from two main effects; a factor of < 1.0 to allow for turbulence effects resulting from the closeness of the lamellae;[8] and a factor of > 1.0 to account for the significantly higher settling rates that occur between sloping surfaces because of channeling in the slurry.[30] No reliable information is available on the separate effects, but experience suggests that $1 < K_{La} < 1.7$, with $K_{La} = 1.4$ being a suitable average value.[31]

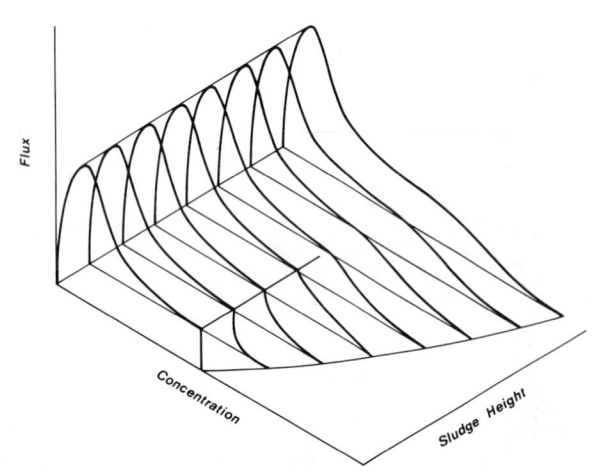

Figure 17.17. Flux envelope. (After Tory.[28])

17.3.7. Flow Patterns

Actual flow patterns[24,32] in sedimentation devices are not as simple as those proposed earlier (Fig. 17.10). It appears that the true flows are of more significance in clarification than in thickening, since published information shows that the flux curve method (based on the simple flows of Fig. 17.10) gives remarkably good agreement between operating thickeners and correctly interpreted batch tests.

In some cases, the true flows may produce beneficial effects,[3,24] but in other instances, they significantly reduce performance.[32] Constructional details will obviously influence the flows. For example, feed wells can be designed to enhance flocculation.[3,4] In another instance, a modified overflow design, claimed to be theoretically justified because it effectively enlarged the tank, may in fact have simply improved the flow patterns.[33]

17.3.8. Thickener Operation

A thickener is generally designed on the basis of a certain feed flux and underflow concentration. In operation, however, the feed flux varies, and although a thickener can often tolerate significant surges, some corrective measures will eventually be necessary if the feed flux is persistently high or low. Control can be carried out by altering the underflow pump rate and is best achieved with a variable rate, positive displacement pump, such as a diaphragm pump.[34] Adjusting the pump discharge rate changes the $\psi_{(+)}$ flux component of the net downward flux ψ_d (Eq. 17.6). Thus in cases of persistent overload, the thickener discharge will have a lower concentration (Fig. 17.15), whereas persistent underloading will result in a higher discharge concentration (Fig. 17.14). If no pump adjustment were made, these nondesign loadings would eventually result in solid in the overflow or low discharge concentrations, respectively.

Example 17.4. If the overload in Example 17.2 persists, how should the operation be changed to produce a clear overflow?

Solution. To maintain a clear overflow, the withdrawal flux must be increased to equal the overload feed flux. Such conditions can be obtained from Fig. E17.2.1 by drawing a new operating line from $\psi_{O/L}$. This gives an intersection on the concentration axis of

$$C_{(+)} = 635 \text{ kg/m}^3$$

which can be obtained by increasing the underflow pumping rate. Thus

$$\psi_{O/L} = \psi_{(+)} = \frac{C_{(+)} O_{(+)}}{A}$$

$$\therefore O_{(+)} = \frac{2.94 \text{ kg}}{\text{m}^2 \cdot \text{min}} \left| \frac{\text{m}^3}{635 \text{ kg}} \right| \frac{\pi \, 20^2 \text{ m}^2}{4}$$

$$= 1.45 \text{ m}^3/\text{min}$$

That is, an increase of $(1.45 - 1.10)/1.10 \times 100 = 32\%$.

Care should be taken in pumping excess inventory from overloaded thickeners: excessive pumping rates may result in very thin underflows even before the sludge inventory is depleted. This phenomenon arises because of "rat-holing," where the pumping rate is so fast in the vicinity of the outlet that the viscous outer sludge cannot flow into the discharge zone.[34] Careful control therefore necessitates knowledge of underflow pulp density and slurry levels within the thickener.[35]

REFERENCES

1. R. H. Perry and C. H. Chilton (Eds.), *Chemical Engineers' Handbook*, Section 19, 5th ed., McGraw-Hill (1973).
2. E. S. Hsia and F. W. Reinmiller, "How to Design and Construct Earth Bottom Thickeners," *Min. Eng.*, **29**, 36–39 (August 1977).
3. H. E. Cross, "A New Approach to the Design and Operation of Thickeners," *J. S. Afr. IMM*, **63**, 271–298 (February 1963).
4. R. C. Emmett and R. P. Klepper, "The Technology and Performance of the Hi-Capacity Thickener." Presented at AIME Annual Meeting, Denver, 1978.
5. A. A. Terchick, D. T. King, and J. C. Anderson, "Application and Utilisation of the Enviro-Clear Thickener in a U.S. Steel Coal Preparation Plant," *Trans. AIME/SME*, **258**, 148–151 (June 1975).
6. N. P. Chironis, "New Clarifier/Thickener Boosts Output of Older Coal Preparation Plant," *Coal Age*, **81**, 140–145 (January 1976).
7. R. L. Cook and J. J. Childress, "Performance of Lamella Thickeners in Coal Preparation," AIME/SME Preprint 77-F-135 (1977).
8. W. E. Schlitter and W. Markl, "Cross-Flow Lamella Thickeners," *Min. Mag.*, **134**, 291–297 (April 1976).
9. J. Abbott et al., "Coal Preparation Plant Effluent Disposal by Means of Deep Cone Thickeners," *Mine Quarry*, **2**, 37–50 (October 1973).
10. K. M. Yao, "Design of High-Rate Settlers," *J. Environ. Eng. Div., Proc. ASCE*, **99**, 621–637 (October 1973).

11. E. B. Fitch and D. G. Stevenson, "Gravity Separation Equipment," in *Solid/Liquid Separation Equipment Scale-Up,* pp. 81–153, Upland Press (1977).
12. A Jernqvist, "Experimental and Theoretical Studies of Thickeners. Parts I–IV," *Svensk Papperstidn.,* **68,** 506–511 (Aug. 15, 1965); **68,** 545–548 (Aug. 31, 1965); **68,** 578–582 (Sept. 15, 1965); **69,** 395–398 (June 15, 1966).
13. R. Davies and B. H. Kaye, "Experimental Investigation into the Settling Behaviour of Suspensions," *Proc. Powtech '71, Int. Powder Technol. and Bulk Solids Conf.,* pp. 73–79, Powder Advisory Centre, London (1971).
14. M. J. Lockett and H. M. Al-Habbooby, "Relative Particle Velocities in Two Species Settling," *Powder Technol.,* **10,** 67–71 (1974).
15. G. J. Kynch, "Theory of Sedimentation," *Trans. Faraday Soc.,* **48,** 166–176 (1952).
16. K. J. Scott, "Continuous Thickening of Flocculated Suspensions," *Ind. Eng. Chem. Fundam.,* **9,** 422–427 (1970).
17. K. J. Scott, "Experimental Study of Continuous Thickening of a Flocculated Silica Slurry," *Ind. Eng. Chem. Fundam.,* **7,** 582–595 (November 1968).
18. K. J. Scott, "Theory of Thickening," *Trans. IMM (C),* **77,** C85–C97 (June 1968).
19. L. A. Adorjàn, "A Theory of Sediment Compression," *Proc. 11th Int. Miner. Process. Congr., Cagliari, 1975,* pp. 297–318, Università di Cagliari (1975).
20. M. J. Pearse, "Gravity Thickening Theories: A Review," Warren Springs Laboratory Report LR 261 (MP) (1977).
21. J. M. Coulson and J. F. Richardson, *Chemical Engineering,* Vol. 2, 2nd ed., Pergamon (1968).
22. N. J. Hassett, "Thickening in Theory and Practice," *Miner. Sci. Eng.,* **1,** 24–40 (January 1969).
23. W. H. M. Robins, "The Theory of the Design and Operation of Settling Tanks," *Trans. Inst. Chem. Eng.,* **42,** T158–T163 (1964).
24. J. P. S. Turner and D. Glasser, "Continuous Thickening in a Pilot Plant," *Ind. Eng. Chem. Fundam.,* **15,** 23–30 (1976).
25. N. J. Hassett, "Concentrations in a Continuous Thickener," *Ind. Chem.,* **42,** 29–33 (January 1964).
26. H. S. Coe and G. H. Clevenger, "Methods for Determining the Capacities of Slime Settling Tanks," *Trans. AIME/TMS,* **55,** 356–384 (1916).
27. E. Barnea, "New Plot Enhances Value of Batch-Thickening Tests," *Chem. Eng.,* **84,** 75–78 (Aug. 29, 1977).
28. E. M. Tory, private communication quoted by K. J. Scott in ref. 17.
29. P. Somasundaran, E. L. Smith, and C. C. Harris, "Dewatering of Phosphate Slimes Using Coarse Additives," *Proc. 11th Int. Miner. Process., Congr., Cagliari, 1975,* pp. 1301–1322, Università di Cagliari (1975).
30. K. W. Pearce, "Settling in the Presence of Downward-Facing Surfaces," *Symp. on Interaction Between Fluids and Particles,* pp. 30–39, Institution of Chemical Engineers (1962).
31. Chemie und Metall GmbH., personal communication (1979).
32. J. W. de Villiers, "An Investigation into the Design of Underground Settlers," *J. S. Afr. IMM,* **61,** 501–521 (June 1961).
33. C. A. Lee, "New Launder Design Provides a Way of Increasing Settling Tank Efficiency," *Plant Eng.,* **27,** 126–127 (Apr. 19, 1973).
34. F. R. Weber, "How to Select the Right Thickener," *Coal Min. Process.,* **14,** 98–104, 116 (May 1977).
35. W. Johnson, "Atlas Minerals Saves Money with Automatic Thickener Controls," *Eng. Min. J.,* **167,** 92–93 (October 1966).

Chapter Eighteen

Filtration

Allen's Axiom: "When all else fails, read the instructions."

Boyle's Law: "Clearly stated instructions will consistently produce multiple interpretations."

Filtration is the removal of solid particles from a fluid, by passing the fluid through a filtering medium on which the solids build up. Industrial filtration is analogous to laboratory filtration and differs basically in the bulk of material handled and the necessity of treating it at low cost. Many factors can be important in selecting a filtration process, but since mineral processing operations are concerned primarily with recovering the solids at large throughputs, the selection of equipment is considerably narrowed.

Filtration equipment size is specified by the surface area necessary to produce the required product. As with sedimentation, particle properties cannot be adequately measured, and small-scale filtration tests must be carried out to obtain basic data. Although simple tests such as that illustrated in Fig. 18.1 can be used, data become more reliable as the test equipment approaches the operation and size of that to be used for the full-size equipment.[1]

Filters can be operated in two basic modes. *Constant pressure filtration* maintains a constant pressure, so that the flow rate falls slowly from a maximum at the start of the cycle. Most continuous filters can be considered to operate on this principle, using a vacuum to provide the pressure difference. (Some additional variable pressure may come from the hydraulic head of the system.) *Constant rate filtration* requires gradually increasing pressure as the cake builds up and increases the resistance to flow. A common approach is to use a constant flow rate until the pressure builds up to a certain level, and use constant pressure filtration for the remainder of the time. This cycle can conveniently be achieved using centrifugal pumping and has the advantage of forming a rather loosely knit initial cake, which minimizes the quantity of solids forced into and through the medium: typically it is used for pressure filtration, where pressures greater than 1 atm are possible.

18.1 FILTRATION EQUIPMENT

Continuous vacuum filters[2,3] are the most widely used for treating concentrates. Although restricted to a filtration pressure of less than 1 atm, they generally provide the most economical continuous operation. They can be divided into three classes: drums, discs, and horizontal filters such as the belt type. Although substantially different in design, they are all characterized by a filtration surface that moves by mechanical or pneumatic means from a point of slurry deposition under vacuum to a point of filter cake removal. Their continuous nature is somewhat deceptive because in reality these filters operate with an endless series of batch events that only approximate a continuous pattern.

In the past, pressure filters have seldom been economical in mineral processing operations, because of the large throughputs occurring. Now that continuous filter presses are available, this situation could change.

Figure 18.2 shows the operating capabilities of the commonly used filters and serves as a starting point for selection.

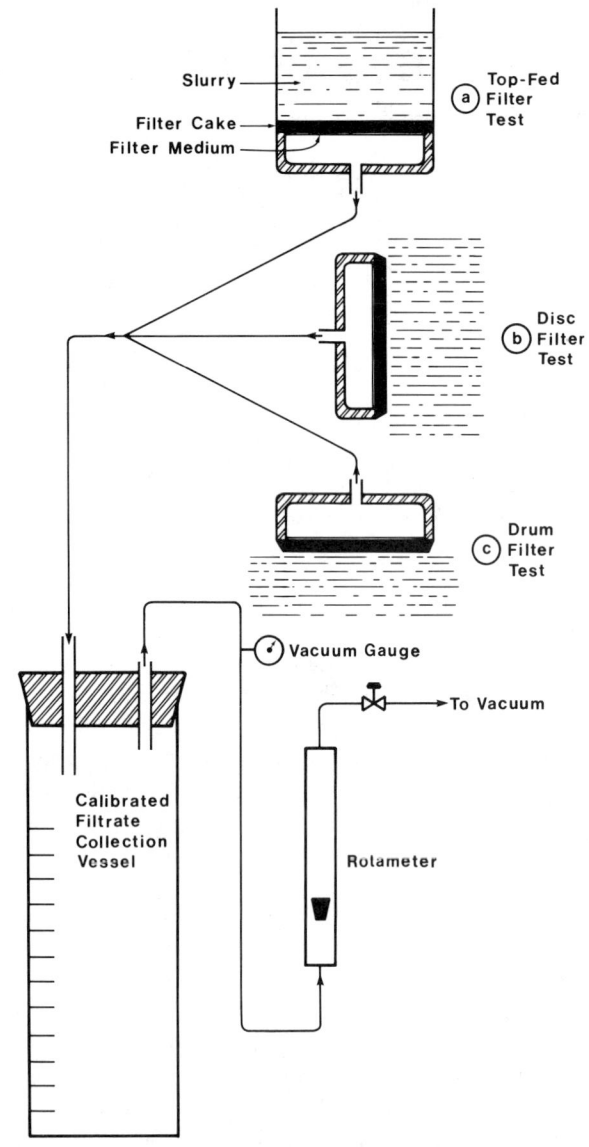

Figure 18.1. Typical filter test arrangement. The filter head selected should be appropriate for the type of filter being considered. For top-fed filters, the head must have walls high enough to contain sufficient slurry. Heads (*b*) and (*c*) are used immersed in the slurry, which should have realistic agitation.

(a)

Slurry characteristics		Filtration characteristics				
Slurry dry solids cont %		20	10 20	1 10	1	0.1
Cake-forming speed		~20 mm/s	~20 mm/min	~10 mm/min	<1 mm/min	no cake
D S -forming capacity kg/m²h		~3000	~3000	~500	<50	<5
Filtration speed l/m²h		~15000	~1000	~100	50 5000	50 5000
Recommended filtration equipment						
Gravity filter						
Top feed filter		■■■■				
Drum filter		■■■■	■■■■	■■■■	■■■■	
Belt filter			■■■■	■■■■	■■■■	
Disc filter			■■■■	■■■■		
Precoat filter					■■■■	■■■■
Automatic pressure filter			■■■■	■■■■	■■■■	

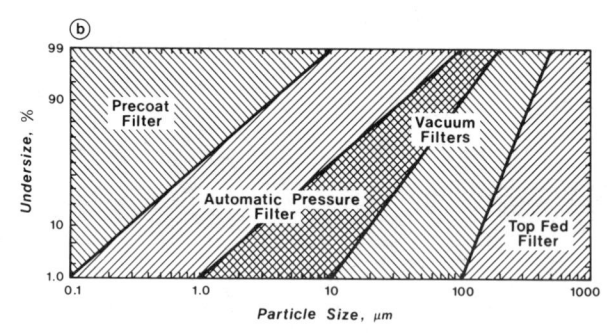

Figure 18.2. Filter type selection. (Courtesy Larox Oy.)

18.1.1. Drum Filters

A typical drum filter is illustrated in Fig. 18.3. Essentially it consists of a horizontal cylindrical drum that rotates while partially immersed in an open tank, the bottom of which is curved to match the drum. In this zone most filters have some means of agitating the slurry. Drum diameters vary from about 1 to 4.5 m, with filtration areas of 1–80 m².

The drum shell itself consists of a series of shallow compartments about 20 mm deep covered with a drainage grid and a filter medium. The interior of each compartment is in turn connected by a conduit to a valve mechanism on the central drum shaft, which allows vacuum or pressure to be applied to the compartment at various stages of the cycle.

By the action of the automatic valve on the drum shaft, vacuum is applied to the immersed sections and results in cake buildup on the filter medium surface. As the drum revolves, the cake is raised above the liquid level and wash water, if required, is sprayed on the surface. The vacuum is maintained at this stage and beyond, when cake dewatering occurs. Before the cake can reenter the slurry on the opposite side of the drum, some form of discharge is used. This can be achieved with a comparatively simple doctor blade close to the filter cloth. To remove thin filter cake, more sophisticated discharge methods are necessary. One approach uses continuous strings (spaced about 10–20 mm apart) around the drum. These pass out over a roller and so pull the cake off the drum. A similar approach passes the filter medium over a set of rollers out from the drum. This also allows the filter medium to be back-washed, but it requires more complex construction to keep the medium aligned and wrinkle-free. Both these methods increase the dead area of the drum.

Filtration Equipment

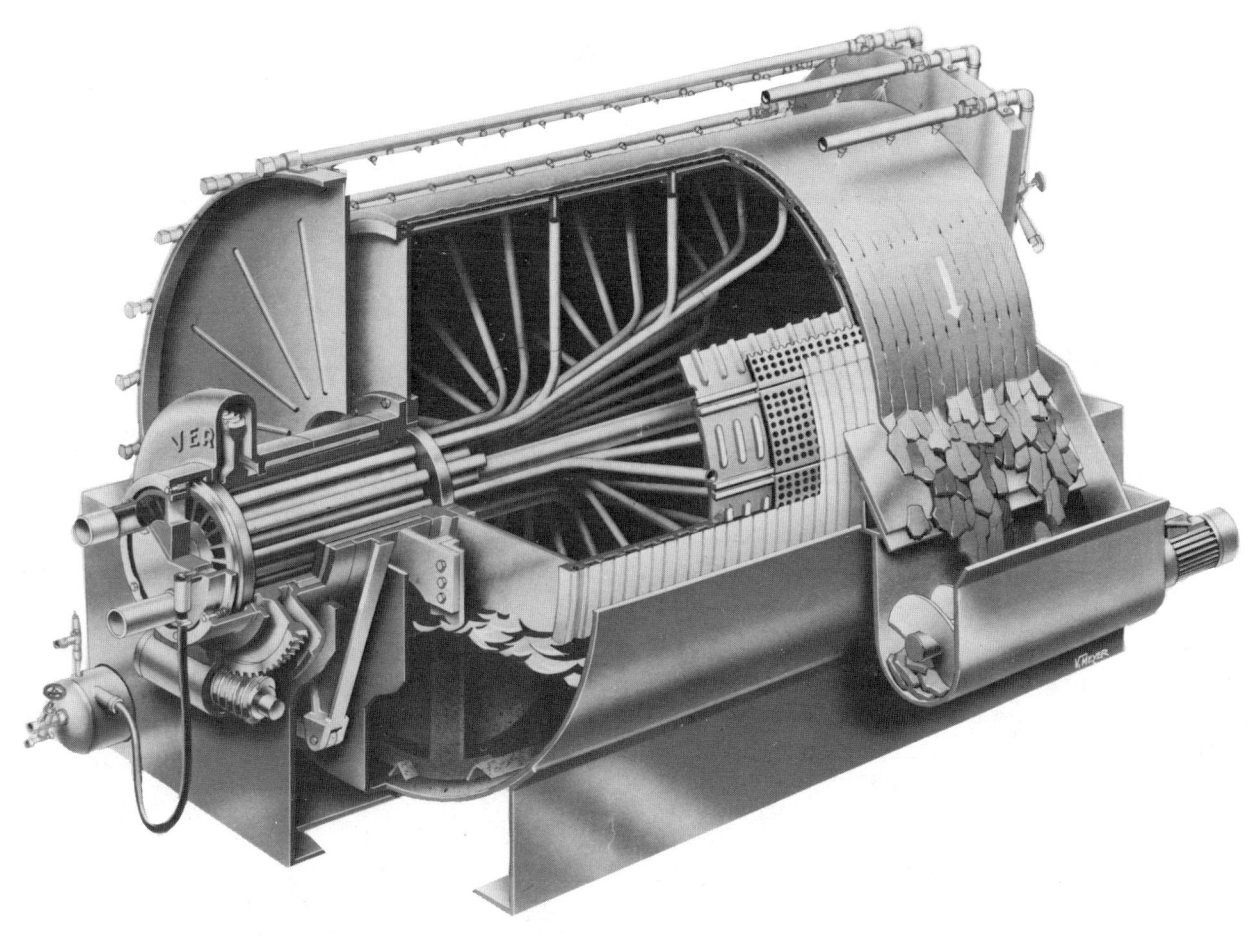

Figure 18.3. Cutaway of drum filter with scraper discharge. (Courtesy Filters Vernay.)

If the solids are fast settling and might not form a satisfactory cake, a drum filter can be top fed onto the ascending face of the drum. Since the cake has to be discharged at the bottom of the drum, only half the drum is effectively used.

When it is essential to obtain a clear filtrate from a suspension of very fine particles, it may be necessary to precoat the filter medium with a layer of suitable cake (although such clarity is seldom required in mineral operations).

Other variations of the drum filter are the internally fed filter and the single compartment filter. The former has the filter medium on the inside of the drum, and is well suited to heterogeneously sized materials in that the coarse particles can be caused to form cake first. The latter filter has the entire inside of the drum subjected to vacuum, except for a small discharge zone. Although more expensive and inflexible in operation, it is suitable for very slow filtering slurries.

Construction materials of filters vary widely: because of the comparatively low pressures used, construction is relatively light, and even expensive corrosion-resistant materials can economically be justified in critical areas. Size was originally limited by difficulties in producing metal castings; now factory or on-site fabrication is possible for larger filters. Even so, a number of smaller filters in parallel may be preferable to a single filter to provide adequate availability.

18.1.2. Disc Filters

A typical disc filter is illustrated in Fig. 18.4. Essentially it consists of a number of discs partly immersed in a slurry, and mounted at regular intervals along a hollow shaft. Each disc is divided into segments and is ribbed on both sides to provide support for the filter medium. Again, the central shaft is connected by a set of valves to a vacuum and

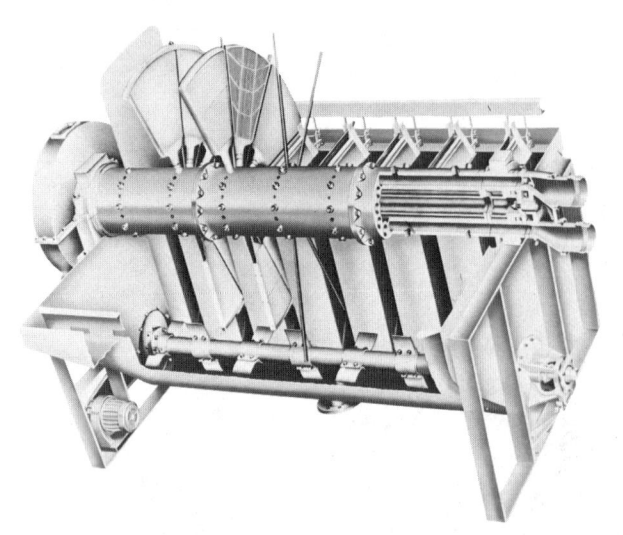

Figure 18.4. Cutaway of disc filter. (Courtesy of Eimco Process Machinery Division, Envirotech Corp.)

pressure system to allow cake formation and discharge respectively. As the disc sections submerge during rotation, the vacuum is applied to form a cake on both sides of the disc. As the segment emerges from the slurry, vacuum is maintained to provide cake dewatering, but a wash stage may be applied in between if necessary. Before the cake-carrying segment reaches the slurry again, a light blast of air is applied, which causes the cloth to inflate slightly and discharge the cake. If necessary the discharge may be assisted by the use of a scraper.

Disc filters may have 1–12 discs, which can be up to 5 m in diameter, thus resulting in about 30 m² of filter surface per disc. Discs may have 12–30 segments; the larger number provides considerably better performance. Although the trough is common to all discs, with an agitator to maintain the suspension, better cake formation can be achieved by having a separate trough for each disc and allowing about 10% of the slurry to overflow and be recycled. These measures ensure a continuous, even flow of slurry close to the filter media.[4]

Disc filters are the cheapest and most compact of the continuous filters. Their major disadvantage is considered to be their inability to effectively wash, but this is relatively unimportant in concentrate filtration.

18.1.3. Horizontal Continuous Vacuum Filters

The horizontal continuous vacuum filters[2,3] are characterized by a horizontal filtering surface in the

Figure 18.5. Plane filters. (Courtesy KHD Humboldt Wedag AG.)

form of a belt, table, or series of pans in a circular (Fig. 18.5) or linear arrangement (Fig. 18.6). In spite of their varied forms, they have common advantages and disadvantages. The advantages are independent choice of cake formation and wash and drying times, effective filtration of heavy dense solids, effective filtration of sludges, the ability to flood wash, and an adaptability to countercurrent washing or leaching. However, horizontal continuous vacuum filters are more expensive to install and operate than drum filters, and they require relatively large floor areas for a given filter area. In general, this restricts these filters to hydrometallurgical operations where their washing ability justifies their higher costs. The belt filter (Fig. 18.6) has the additional disadvantage that only half the filtering surface is effectively used, although this may be offset by the opportunity to back-wash the idle filter medium.

18.1.4. Pressure Filters

Batch pressure filters such as filter presses are widely used in the chemical industry[2] but have seldom been adopted for mineral processing because

Filtration Equipment

Figure 18.6. Belt filter. (Courtesy Straight Line Filters, Inc.)

the batch operation is too labor intensive.[5] On the other hand, filter presses offer the advantages of considerably higher operating pressures (i.e., higher throughputs), low capital cost, low floor space requirements, and greater flexibility in operation.

Recently, continuous filter presses have been developed (Fig. 18.7), providing a viable method of obtaining filtration pressures up to 16 atm. Strictly speaking, these machines are not continuous; rather, they automatically cycle through a series of operations (Fig. 18.8).

18.1.5. Other Filters

A number of other filter types are available and find applications in special circumstances.[2] The Moore filter is a simple batch vacuum filter that is sometimes used for slow settling suspensions. The Artisan continuous filter is a recent development particularly suited to fine particles such as clay, pigments, and precipitates.[6,7] Filter cake is maintained as a thin layer by rotating wiper blades and after the slurry has passed a number of filtering surfaces, it discharges from the pressure vessel as a thick, plastic-like cake.

In the past, the only practical method of continuous pressure filtration was to enclose a drum or disc filter in a pressure vessel. Such an approach was expensive and mechanically complex, and was limited to pressures of about 2–3 atm. Centrifugal filters also find some application and are considered briefly in the next chapter.

18.1.6. Filter Media

Correct selection of the filter medium is an essential part of efficient filter operation. Although the primary purpose of the medium is to retain solids, other factors are significant. A good filter medium should:

Have the ability to bridge solids across the pores.

Have a low resistance to filtrate flow.

Avoid wedging in its pores particles, which would significantly increase the resistance to flow.

Resist chemical attack.

Have sufficient strength to withstand the filtration pressure and mechanical wear.

Allow efficient discharge of the cake.

Figure 18.7. Automatic pressure filter. (Courtesy Larox Oy.)

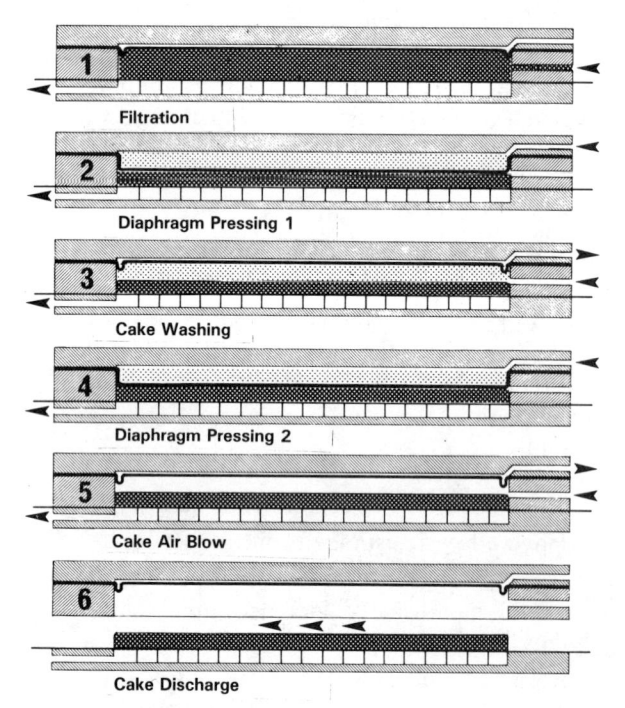

Figure 18.8. Operating cycle of automatic pressure filter. (Courtesy Larox Oy.)

1. Filtration
2. Diaphragm Pressing 1
3. Cake Washing
4. Diaphragm Pressing 2
5. Cake Air Blow
6. Cake Discharge

Filter media can be made from a variety of materials.[2,8] Cotton fabrics, which are widely used, are cheap, have good mechanical strength and resistance to wear, and discharge cakes readily. On the other hand, tight weaves have high flow resistance, plug quickly, and have poor chemical resistance. Felt overcomes some of these disadvantages and has proved to be suitable for disc filters.[3] Synthetic materials are more suitable for chemical resistance applications, but tend to have poor mechanical properties. Metal fabrics from a variety of common metals are available and can be used under severe conditions, where their long life justifies the higher initial cost.

18.2. FILTRATION CALCULATIONS

18.2.1. Filtration Resistances

By means of a pressure difference applied between the filter inlet and outlet, filtrate is forced through the equipment where the solids are retained and

Filtration Calculations

form a filter cake. The filtrate passes through three kinds of resistance in series, which together produce the total pressure drop:

1. Resistance in conduits conveying the slurry to the upstream face of the filter cake, and filtrate away from the filter medium.
2. Resistance of the filter cake.
3. Resistance associated with the filter medium.

In a well-designed system, the pressure drop in the inlet and outlet is negligible in comparison with the other pressure drops and may be neglected. The resistance of the filter medium to filtrate is actually higher than that offered by a clean filter medium to clear filtrate, because solid particles rapidly become embedded in the meshes of the medium and cause a significant increase in the resistance to flow. The entire resistance in the medium, including the embedded particles, represents the *medium resistance*, and requires a pressure ΔP_m to overcome it. The resistance offered by all the solids not associated with the filter medium is the *cake resistance*, and requires pressure ΔP_c. Obviously at the start of filtration the cake resistance is zero, but as filtration proceeds this increases and becomes the predominant resistance.

18.2.2. Pressure Drop Through Filter Cakes

Although the flow of filtrate through a filter cake can be treated as flow through a packed bed and therefore described by Eq. 4.62, in most practical cases flow is laminar and consequently the Carman-Kozeny equation (4.63) can be adapted.

$$\frac{\Delta P_c}{\delta_c} = \frac{150(1 - \varepsilon)^2}{\varepsilon^3 d_{vs32}^2} \mu v_s \qquad (18.1)$$

(ΔP_c = pressure drop across cake; δ_c = cake thickness). This equation therefore relates the cake pressure drop ΔP_c to cake porosity and thickness, and to the particle diameter. However, to allow the introduction of the measurable parameters of filtration, the equation is modified.[9] First, the particle size is expressed in terms of the specific surface area, given by

$$d_{vs} = \frac{6}{S_p/V_p} = \frac{6}{S_0} \qquad (18.2)$$

Provisionally, the cake thickness δ_c can be expressed in terms of the mass of cake M_c,

$$M_c = \rho_s(1 - \varepsilon)A_c\delta_c \qquad (18.3)$$

(A_c = area of filter cake). Combining Eqs. 18.1–18.3:

$$\Delta P_c = \frac{4.17\mu v_s M_c S_0^2(1 - \varepsilon)}{A_c \rho_s \varepsilon^3}$$
$$= \mu v_s \cdot \frac{\alpha M_c}{A_c} \qquad (18.4)$$

where α is the *specific cake resistance*, defined as

$$\alpha = \frac{4.17(1 - \varepsilon)S_0^2}{\rho_s \varepsilon^3} \qquad (18.4a)$$

The grouping of terms describing the filter cake properties into specific cake resistance does not imply that α will be constant for a given slurry. When α is insensitive to ΔP_c, the filter cake is described as *incompressible*, although probably no cake is actually this ideal. Even for a constant pressure drop ΔP_c, α may change because of variations in the void fraction ε and the specific surface S_0.[10] For instance, ε usually varies with the compacting stress applied across the bed. Since this stress is proportional to $\Delta P_c/\delta_c$, ε may also vary as δ_c increases during filtration. Both ε and S_0 may also be sensitive to the degree of flocculation, which may itself vary during the process. However, in most constant pressure filtrations, α is constant after the initial period of high flow rate, which occurs before the form of the filter cake is fixed.

As mentioned above, the filtration pressure drop ΔP_f also must overcome the resistance of the filter medium, which is in series with the cake resistance. Since Eq. 18.4 is in the familiar form of a driving force proportional to a resistance times a rate, the medium resistance Ω_m can be included in the equation to give

$$\Delta P_f = \Delta P_c + \Delta P_m = \mu v_s \left(\frac{\alpha M_c}{A_c} + \Omega_m\right) \qquad (18.5)$$

Strictly, the cake resistance is a function of ΔP_c rather than ΔP_f, but since ΔP_c greatly exceeds ΔP_m over most of the filtration cycle, this approximation is adequate.

In filtration it is generally more convenient to have the superficial velocity v_s and the total mass of cake M_c expressed in terms of V_f, the total volume of filtrate collected. Thus,

$$v_s = \frac{dV_f}{dt} \frac{1}{A_c} \qquad (18.6)$$

A mass balance may be used to correlate M_c and V_f. If the feed slurry is dilute

$$M_c = V_f C_c \qquad (18.7)$$

(C_c = the mass of solids deposited as cake per unit volume of filtrate collected). However, in reality, some filtrate is retained in the pores of the filter cake so that the volume of filtrate collected is less than the liquid in the feed slurry. This results in C_c being greater than the concentration of solids per unit volume of liquid in the feed slurry C_I. A rigorous mass balance therefore results in the correction

$$C_c = \frac{C_I}{1 - (M_w/M_c - 1)(C_I/\rho_l)} \quad (18.8)$$

(M_w = mass of wet cake, including pore liquid; ρ_l = density of filtrate).

Combining Eqs. 18.5–18.7 gives:

$$\Delta P_f = \frac{dV_f}{dt} \frac{\mu}{A_c} \left(\frac{\alpha C_c V_f}{A_c} + \Omega_m \right) \quad (18.9)$$

or

$$\Delta P_f = \frac{dV_f}{dt} \frac{\alpha C_c \mu}{A_c^2} (V_f + V_e) \quad (18.10)$$

(V_e = volume of filtrate necessary to build up a fictitious amount of filter cake, the resistance of which is equal to the filter medium and the piping between the pressure tapping points). This filter medium resistance will of course be much higher than that of filtrate flow through clean filter medium, because of the presence of solids in the medium pores.

18.2.3. Interpretation of Filtration Data

Equation 18.10 provides the basis for interpreting filtration: either constant pressure, constant flow, or continuous operations. However, use of the equation requires knowledge of V_e and α. In principle the latter may be estimated from the properties of the solids, but in practice it is more convenient to evaluate both α and V_e together from experimental data,[2,11] particularly since α is likely to vary during the filtration cycle.

Rearrangement of Eq. 18.10 gives

$$\frac{dt}{dV_f} = \frac{\alpha C_c \mu}{A_c^2 \Delta P_f} (V_f + V_e) \quad (18.11)$$

so that if the slurry in question is subjected to a constant pressure filtration test, a plot of reciprocal filtration rate versus filtrate volume will have a slope $\alpha C_c \mu / A_c^2 \Delta P_f$, and an intercept of

$$\left[\frac{V_e \alpha C_c \mu}{(A_c^2 \Delta P_f)} \right] = \text{slope} \times V_e \quad (18.12)$$

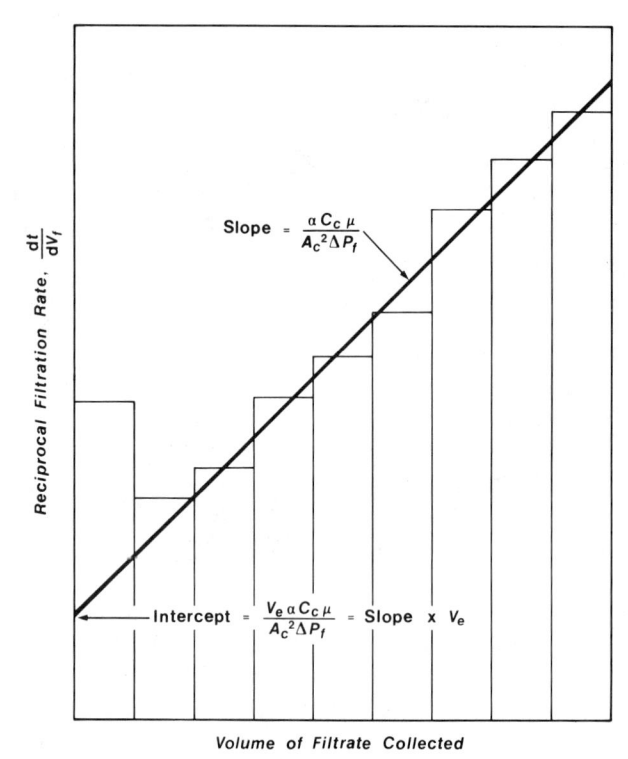

Figure 18.9. Typical constant pressure filtration data, and the evaluation of filtration coefficients α and V_e.

thus allowing α and V_e to be evaluated. A typical set of data is illustrated in Fig. 18.9. The histogram presentation is used because the values of reciprocal filter rate are normally obtained as $\Delta V_f/\Delta t$. Typically, the initial values are irregular because of incomplete cake formation.

Most filter cakes are to some extent compressible. By conducting constant pressure experiments at various pressure drops, the variation of α with ΔP_f may be found. A number of empirical equations have been presented, but the most widely used is

$$\alpha = \alpha_0 (\Delta P_f)^n \quad (18.13)$$

where α_0 and n are empirical constants: the latter being called the *compressibility coefficient*. It is zero for incompressible cakes but normally falls between 0.2 and 0.8. Evaluation of the constants from typical data is illustrated in Fig. 18.10.

Compressibility data can also be obtained with compression-permeability cells, but such a method should be used with care because pressure transmission through the solids is poor in such a device.[12] Fortunately, continuous filtration uses such rela-

Filtration Calculations

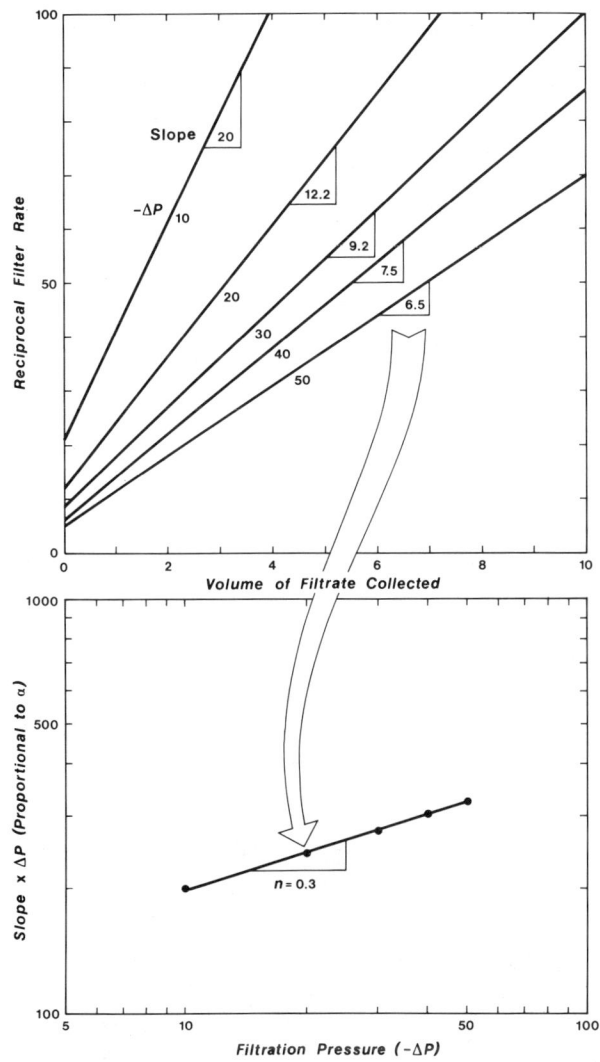

Figure 18.10. Determination of compressibility coefficient.

tively small pressure differences that adequate data can be collected by either method.

18.2.4. Continuous Filtration

Although the feed slurry, filtrate, and cake move at a steady constant rate in a continuous filter, the conditions at any particular element on the filter surface are transient because the process goes through cake formation, washing, drying, and scraping. However the pressure drop across the filter during cake formation can be treated as constant so that Eq. 18.11 can be applied.[11] This equation is in the form

$$\frac{dt}{dV_f} = K_1 V_f + K_2 \qquad (18.14)$$

where

$$K_1 = \frac{\alpha C_c \mu}{A_c^2 \Delta P_f} \qquad (18.14a)$$

$$K_2 = \frac{\alpha C_c \mu V_e}{A_c^2 \Delta P_f} \qquad (18.14b)$$

Integrating Eq. 18.14 between the limits gives

$$\int_0^t dt = \int_0^{V_f} (K_1 V_f + K_2) \, dV_f$$

which becomes

$$t = K_1 \frac{V_f^2}{2} + K_2 V_f \qquad (18.15)$$

Solving this quadratic equation for V_f and dividing by A_c and t gives

$$\frac{V_f}{A_c t} = -\frac{K_2}{A_c t K_1} + \left(\frac{K_2^2}{A_c^2 t^2 K_1^2} + \frac{2}{A_c^2 t K_1}\right)^{1/2} \qquad (18.16)$$

In a continuous filter, t is always less than the total cycle time t_{cy}, so that

$$t = \mathfrak{F}_{cy} t_{cy} \qquad (18.17)$$

($\mathfrak{F}_{cy}$ = fraction of cycle available for cake formation). With rotary drum and disc filters $\mathfrak{F}_{cy}$ turns out to be slightly less than the fractional submergence in the slurry of the drum or disc. If the specific cake resistance varies with ΔP_f according to Eq. 18.13, this may be combined with Eq. 18.16.

When the medium resistance is negligible, Eq. 18.16 reduces to

$$\frac{V_f}{A_c t} = \frac{1}{A_c}\left(\frac{2}{K_1 t}\right)^{1/2} = \left(\frac{2 \Delta P_f}{\alpha C_c \mu t}\right)^{1/2} \qquad (18.18)$$

and the filtrate flow rate is inversely proportional to the square root of the specific cake resistance, the fluid viscosity, and the cycle time. Thus, it follows that the filtration rate increases as the drum speed increases and the cycle time t_{cy} decreases because of thinner cake formation. A limit is reached, however, when the cake becomes too wet or thin to allow satisfactory discharge, and typical minimum thicknesses are shown in Table 18.1.

Although this analysis is not strictly valid for disc filters where ΔP_f varies across the filtration area, addition of the maximum hydrostatic head in ΔP_f produces results similar to those obtained by more rigorous analysis.[13]

TABLE 18.1: MINIMUM THICKNESS FOR FILTERCAKE DISCHARGE

Filter Type	Minimum Thickness (mm)
Drum	
Belt	3 - 5
Roll discharge	1
Knife scraper	6
Coil	3 - 5
String discharge	6
Horizontal Belt	3 - 5
Horizontal Table	19
Tilting Pan	19 - 25
Disk	9 - 13

Example 18.1. Laboratory filter tests on a slurry have produced the following data:

$$\alpha = 4.7 \times 10^9 \, (\Delta P)^{0.3}$$

$$\frac{M_w}{M_c} = 1.67$$

$$\rho_s = 4000 \text{ kg/m}^3$$

$$\rho_f = 1000 \text{ kg/m}^3$$

$$\mu = 10^{-3} \text{ kg/m} \cdot \text{sec}$$

Media resistance negligible

What drum filters would be required to treat 0.5 m³/min if the feed slurry concentration is 350 kg/m³?

Solution. Since the media resistance is negligible, Eq. 18.18 can be used. However, because A_c and t are unknown, a second equation in A_c and t must be developed from the volume of cake discharged. Assuming a blade discharge, the minimum discharge thickness is 6 mm (Table 18.1). Thus

Volume of cake formed during filtration time t
$$= A_c \times 6 \times 10^{-3} \text{ m}^3$$

but
volume of slurry treated in time t

$$= \frac{0.5 \text{ m}^3}{\text{min}} \left| \frac{\text{min}}{60 \text{ sec}} \right| t \text{ sec}$$

$$= (8.33 \times 10^{-3} \, t) \text{ m}^3$$

∴ volume of solid deposited in time t

$$= \frac{(8.33 \times 10^{-3} \, t) \text{ m}^3 \, | \, 350 \text{ kg} \, | \, \text{m}^3}{\text{m}^3 \, | \, 4000 \text{ kg}}$$

$$= (7.29 \times 10^{-4} \, t) \text{ m}^3$$

Since $M_w/M_c = 1.67$

$$C_{V,c} = \frac{0.6/4000}{0.6/4000 + 0.4/1000}$$

$$= 0.273$$

∴ volume of cake formed in time t

$$= \frac{7.29 \times 10^{-4} \, t}{0.273}$$

$$= (2.67 \times 10^{-3} \, t) \text{ m}^3$$

$$= A_c \times 6 \times 10^{-3} \text{ m}^3$$

∴ $\quad t = 2.25 \, A_c \text{ sec} \quad$ (E18.1.1)

For Eq. 18.18: assume filter can produce 0.75 atm vacuum; that is,

$$\Delta P_f = 0.75 \times 1.01 \times 10^5 \text{ N/m}^2$$

∴ $\quad \alpha = 4.7 \times 10^9 \, (0.75 \times 1.01 \times 10^5)^{0.3}$

$$= 1.37 \times 10^{11} \text{ m/kg}$$

also, from Eq. 18.8

$$C_c = \frac{350}{1 - (1.67 - 1)(350/1000)}$$

$$= 457 \text{ kg/m}^3$$

while V_f/t = feed rate $(1 - C_V)(C_I/C_c)$

$$= \frac{0.5 \text{ m}^3}{\text{min}} \left| \frac{\text{min}}{60 \text{ sec}} \right| 1 - (350/4000) \left| \frac{350}{457} \right.$$

$$= 0.00582 \text{ m}^3/\text{sec}$$

Substituting in Eq. 18.18:

$$\frac{1}{A_c} \frac{V_f}{t} = \left(\frac{2 \, \Delta P_f}{\alpha \, C_c \, \mu \, t} \right)^{1/2}$$

$$\frac{0.00582}{A_c} = \left(\frac{2 \times 0.75 \times 1.01 \times 10^5}{1.37 \times 10^{11} \times 457 \times 10^{-3} \times t} \right)^{1/2}$$

∴ $\quad A_c = 3.74 \sqrt{t} \quad$ (E18.1.2)

Combining Eqs. E18.1.1 and E18.1.2,

$$A_c = 3.74 \sqrt{2.25 \, A_c}$$

∴ $\quad A_c = 31.5 \text{ m}^2$

A number of filter combinations could provide this area. For example, since no washing is required, 40% submergence should be possible. Thus

$$A_D = \frac{31.5}{0.4}$$

$$= 79 \text{ m}^2$$

$$= N \times \pi \, D_D \times L_D$$

say three 3 m × 3 m filters, which will give a safety margin of about 7%. Tests should also be carried out

Filtration Calculations

on the filter cake to see whether sufficient dewatering occurs and what vacuum pumping rates would be required.

18.2.5. Batch and Centrifugal Filtration Calculations

Batch filtration may be considered in terms of constant flow rate or constant pressure operations. However, many batch filters have a centrifugal pump providing the pressure for filtration, and this type of operation often approximates constant rate followed by constant pressure. Equation 18.10 can be applied to such operations,[2,9,11] but because they are of little importance in mineral processing they are not considered here.

Centrifugal filtration calculations are of limited use. A rate equation can be derived but it is restricted to flows through the cake after formation, and has a number of simplifying assumptions.

18.2.6. Washing

Washing after filtration is seldom practiced on concentrates, but it may be necessary in other operations where pore liquor either is likely to contaminate the cake or is too valuable not to be partly recovered. When wash liquid enters the filter cake, it displaces some pore liquor, without dilution, by plug flow. This displacement stage seldom removes more than 50% of the original pore liquor. Further washing occurs by a breakthrough phase where a substantial amount of liquor is still being removed, but with increasing dilution. This stage ends when the wash volume is about twice the void volume and the overall removal efficiency rises to 70–95%. Further washing produces a very inefficient diffusional stage. Eventually a stage may be reached where it is more economical to repulp and refilter; but any wash operation must be discontinued when the cost of extra washing exceeds the gain in value in the products.

Representative experimental curves of fraction of liquor remaining versus wash ratio (the volumetric ratio of wash liquid to cake voids) are shown in Fig. 18.11.[14] The central band represents a typical cake; high efficiency is more typical of low porosity, high resistance cake, which consequently has a low wash rate.

Cake wash time is difficult to correlate, but satisfactory results can be obtained in many cases by plotting wash time versus $M_c V_w/A^2$ (V_w = volume of wash liquid), although the linearity often falls off after large wash volumes[14] (Fig. 18.12).

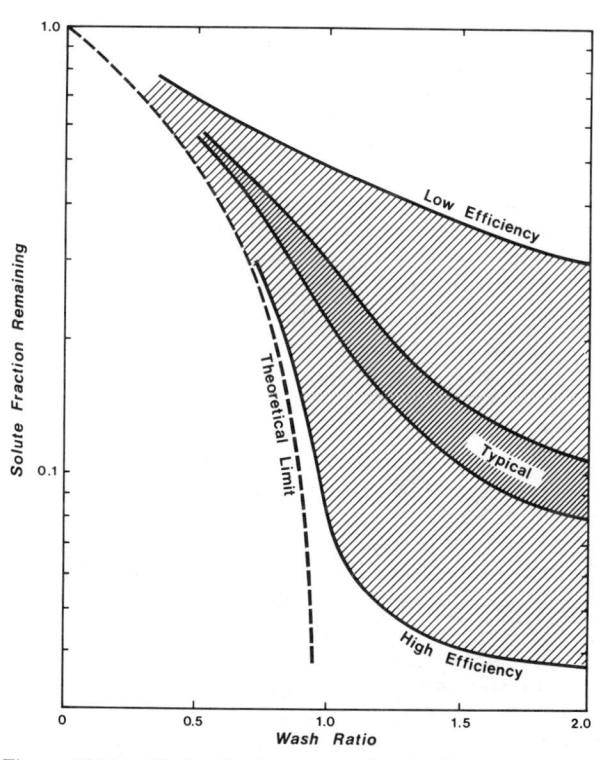

Figure 18.11. Wash effectiveness. (After Dahlstrom and Silverblatt.[14])

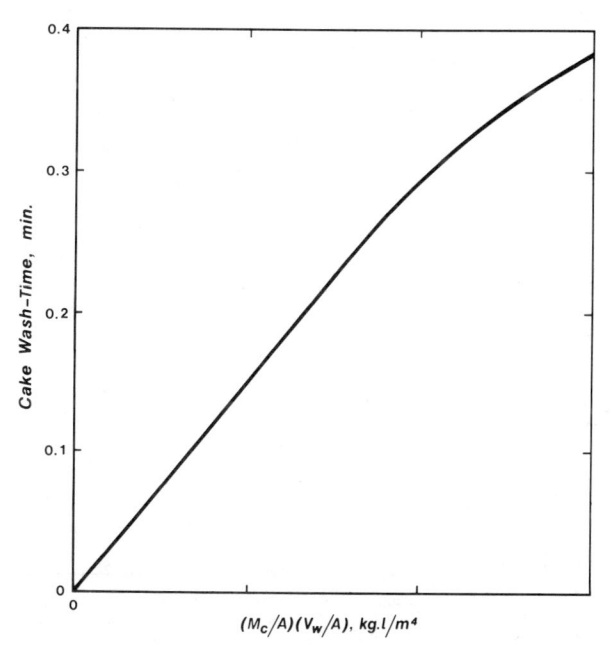

Figure 18.12. Typical cake wash-time data. (After Dahlstrom and Silverblatt.[14])

18.2.7. Cake Dewatering

After the filter cake leaves the slurry (or the wash stream if used), some dewatering of the cake is possible by maintaining the vacuum. During this operation, the moisture content falls as a function of the air rate through the cake, $\Delta P_f/\mu$, A_c/M_c and the drying time t_d. Normally ΔP_f and μ are constant, whereas the air rate is dependent on the particle properties, so that design data must be collected only in the form shown in Fig. 18.13. This requires relatively simple test work, although the data can be expected to show significant scatter.[15]

By conducting vacuum filtration tests with a rotameter in the air line, measurements of air flow with time are obtained to allow vacuum pump capacities to be determined.

Satisfactory cake dewatering can also be achieved by cake compression, but this is practical only in a limited number of situations, such as with the automatic filter press (Fig. 18.8).

Even lower cake moisture levels can be achieved by using steam instead of air drying[16,17] (Fig. 18.13). Under these conditions the steam lowers the viscosity of the filtrate and thus uses less energy to remove water than does thermal drying (Chapter 19).

18.2.8. Flocculation

Although flocs deform under the pressures used for filtration, flocculants can be beneficial by producing a more even pore distribution and higher cake porosity, both of which result in higher filtration rates.[18] When carrying out test work, it should always be remembered that even if flocculant is not added as a deliberate filter aid, it may be present as a carryover from sedimentation where it is commonly added.

REFERENCES

1. D. G. Osborne, "Scale-Up of Rotary Vacuum Filter Capacities," *Trans. IMM (C)*, **84**, C158–C166 (1975).
2. R. H. Perry and C. H. Chilton (Eds.), *Chemical Engineers' Handbook,* 5th ed., Section 19, McGraw-Hill (1973).
3. R. Bosley, "Vacuum Filtration Equipment Innovations," *Filtr. Sep.*, **11**, 138–149 (March–April 1974).
4. B. Wetzel, "Disc Filter Performance Improved by Equipment Redesign," *Filtr. Sep.*, **11**, 270–274 (May–June 1974).
5. D. W. Hutchinson and E. M. Duralia, "Pilot Plant Pressure Filtration of Coal-Preparation-Plant Refuse Slurry." Presented at AIME Annual Meeting, Georgia, 1977.
6. A. Bagdasarian and F. M. Tiller, "Operational Features of Staged, High-Pressure, Thin-Cake Filters," Filtration Soc. Conf. on Filtration, Productivity and Profits at Filtech/77, September 1977, Olympia, London, pp. 22–26 (1977).
7. A. Bagdasarian, F. M. Tiller, and J. Donovan, "High-Pressure, Thin-Cake, Staged Filtration," *Filtr. Sep.*, **14**, 455–460 (September–October 1977).
8. D. B. Purchas, "Filter Media, A Survey," *Filtr. Sep.*, **2**, 465–474 (1965).
9. W. L. McCabe and J. C. Smith, Ch. 30 in *Unit Operations of Chemical Engineering,* 3rd ed, McGraw-Hill (1976).
10. F. M. Tiller and H. Cooper, "The Role of Porosity in Filtration. Part 5. Porosity Variation in Filter Cakes," *AIChE J.*, **8**, 445–449 (1962).
11. C. E. Silverblatt, H. Risbud, and F. M. Tiller, "Batch, Continuous Processes for Cake Filtration," *Chem. Eng.*, **81**, 127–136 (Apr. 29, 1974).
12. F. M. Tiller, S. Haynes, and W. M. Lu, "The Role of Porosity in Filtration. Part 7. Effect of Side-Wall Friction in Compression-Permeability Cells," *AIChE J.* **18**, 13–19 (1972).
13. A. Rushton, "Design Throughputs in Rotary Disc Vacuum Filtration with Incompressible Cakes," *Powder Technol.*, **21**, 161–169 (1978).
14. D. A. Dahlstrom and C. E. Silverblatt, "Continuous Vacuum and Pressure Filtration," Ch. 11 in *Solid/Liquid Sep-*

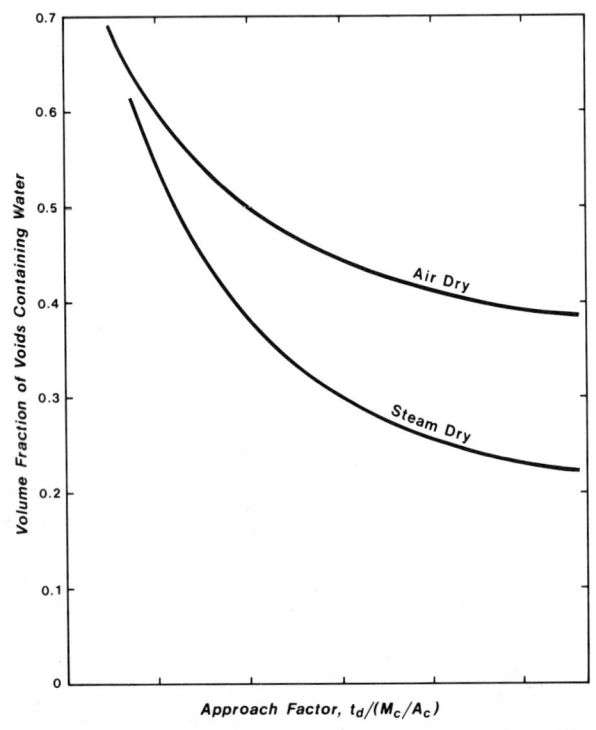

Figure 18.13. Typical cake residual moisture correlation. (After Emmett and Dahlstrom.[16])

References

aration Equipment Scale-up. (D. Purchas, Ed.), Uplands Press (1977).

15. P. A. Nelson and D. A. Dahlstrom, "Moisture-Content Correlation of Rotary Vacuum Filter Cakes," *Chem. Eng. Prog.*, **53**, 320–327 (July 1957).

16. R. C. Emmett and D. A. Dahlstrom, "Steam Drying of Filter Cake," *Chem. Eng. Prog.*, **68**, 51–55 (January 1972).

17. K. W. Daykin et al., "Steam-Assisted Vacuum Filtration," *Mine Quarry*, **7**, 59–65 (March 1978).

18. M. Clement and J. Bonjer, "Investigation on Mineral Surfaces for Improving the Dewatering of Slimes with Polymer Flocculants," *Proc. 11th Int. Miner. Process. Congr., Cagliari, 1975*, pp. 271–295, Università di Cagliari (1975).

Chapter Nineteen

Dewatering: Systems and Miscellaneous Methods

Epstein's Law: "If you think the problem is bad now, just wait until we've solved it."

The two preceding chapters discussed the two most widely used dewatering methods. Under specific circumstances, other forms of dewatering may be adequate or required, and these are introduced here. As previously mentioned, adequate dewatering is seldom possible in a single stage. Therefore this chapter also briefly considers overall dewatering systems, that is, combinations of units commonly used.

19.1. CENTRIFUGAL SEPARATIONS

Centrifuges are expensive but versatile units that can be used as classifiers, clarifiers, thickeners, and filters.[1-3] In mineral processing they are used mainly for dewatering when gravitational settling rates would otherwise be too slow, or when comparatively low residual water levels are required—for example, with coal or before thermal drying (Section 19.2). The extent of the increased settling force can be expressed in multiples of gravity N_g, given by

$$N_g = 2.8 \times 10^{-4} \mathcal{N}^2 R_b \qquad (19.1)$$

($\mathcal{N}$ = speed of rotation, rpm; R_b = radius of centrifuge bowl, m).

19.1.1. Equipment

Two basic types of centrifuge can be recognized: solid bowl and perforated basket. Figure 19.1 illustrates the solid bowl type. In essence it consists of a high speed (1000–6000 rpm) rotating bowl that has a slurry centrifuged against its inner surface. The heavier solids settle through the liquid to the inner surface of the bowl. In the machine illustrated, a helical scroll inside the bowl rotates at a slightly slower speed, and conveys the solids out of the liquid pond up to the discharge opening. Once above liquid level, further drainage of the solids occurs and washing, if required, can be carried out. Drained solids and liquids leave the machine at opposite ends. Solid bowl centrifuges are well suited to treatment of dilute feeds and finer particle sizes.

The vibrating form of perforated basket centrifuges (Fig. 19.2) is the one most commonly seen in the mineral industry, generally for the dewatering of coal. Material transport through this centrifuge is achieved by vertical vibrations in the basket. These vibrations also loosen the bed of particles, aiding drainage, so allowing lower speeds (550–750 rpm) than are necessary with solid bowl centrifuges. Because of the perforations in the basket, these centrifuges are not suitable for feeds having a significant proportion of fines.

19.1.2. Principles of Centrifugal Sedimentation

Selection of a centrifuge is seldom possible from laboratory data and is therefore best left to a manufacturer, who will have suitable test facilities and experience.[1-3] Consequently, this section considers the centrifuge only briefly.

Like tank sedimentation equipment, a continuous centrifuge can behave as a clarifier or as a thickener.

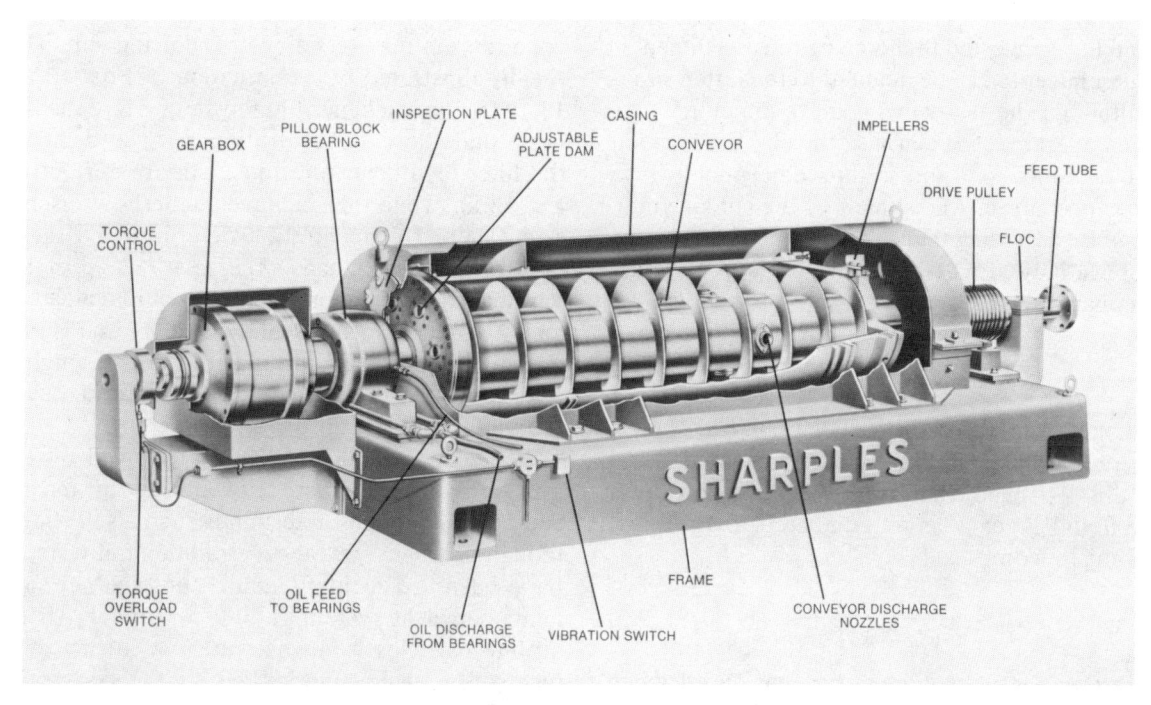

Figure 19.1. Solid bowl centrifuge. (Courtesy Pennwalt.)

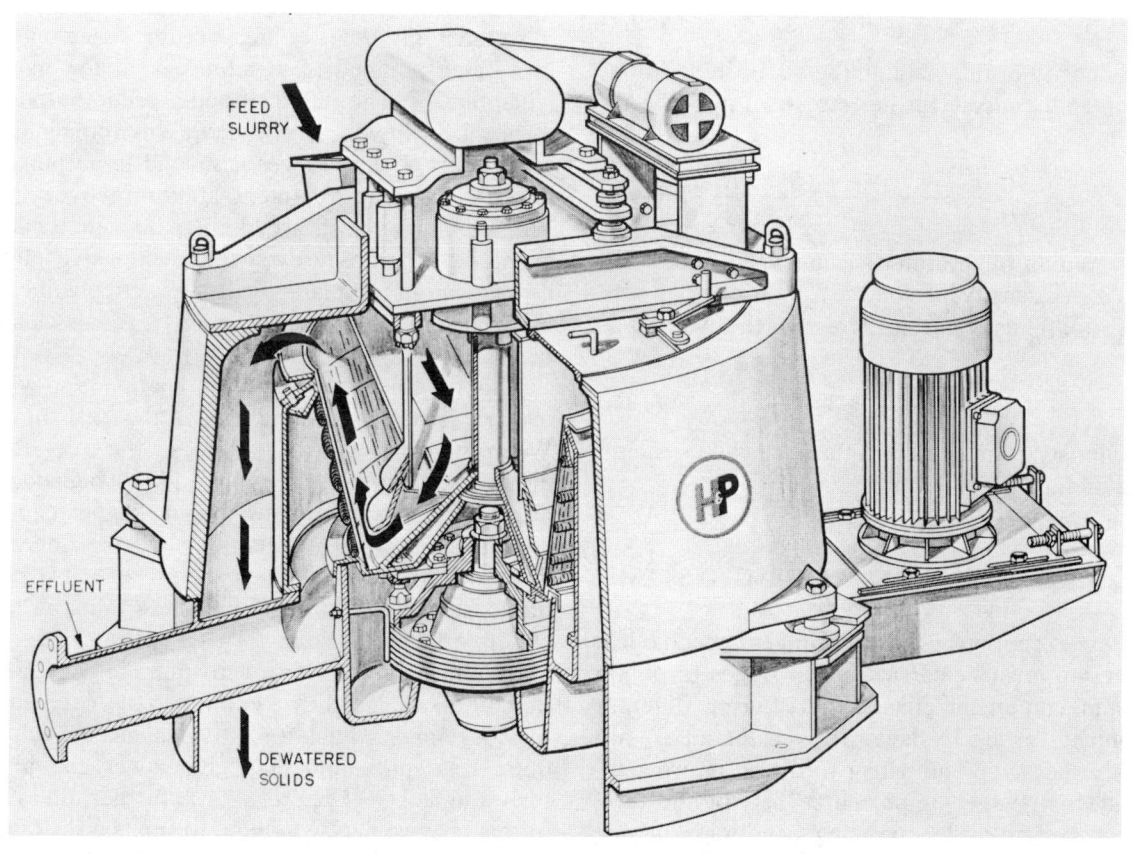

Figure 19.2. Vibrating basket centrifuge. (Courtesy Heyl & Patterson, Inc.)

357

Vesilind has suggested that conventional sedimentation flux concepts can be adapted to the latter situation, although the test work is more difficult.[4]

When operated as a clarifier, the process is again one of classification, and is thus determined by a performance curve. An estimate of the cut size may be made by assuming that the particle is at all times moving radially at its terminal velocity, and leads, for laminar flow, to

$$d_{50}^2 = \frac{9 I_V \mu \ln[2 R_b^2/(R_l^2 + R_b^2)]}{\pi \omega^2 L (\rho_s - \rho_l)(R_b^2 - R_l^2)} \quad (19.2)$$

(I_V = volumetric feed rate; L = length of bowl; R_b = inside radius of bowl; R_l = inside radius of liquid surface; ω = speed, rads/sec). When the thickness of the liquid layer is small compared to the bowl radius this becomes

$$d_{50}^2 = \frac{4.5 I_V \mu}{\pi \omega^2 L R_b^2 (\rho_s - \rho_l)} \quad (19.3)$$

A further factor that must be considered when operating a centrifuge is the stress in the bowl. Rotation of the bowl generates a stress in the bowl wall known as the self stress $\mathscr{S}_s$, given by

$$\mathscr{S}_s = \rho_m R_b^2 \omega^2 \quad (19.4)$$

(ρ_m = density of the metal in the wall). In addition, the contents cause a further stress $\mathscr{S}_c$ in the wall, given by

$$\mathscr{S}_c = \omega^2 R_b \frac{\rho_l(R_{s/l}^2 - R_l^2) + \rho_s(R_b^2 - R_{s/l}^2)}{2\delta} \quad (19.5)$$

($R_{s/l}$ = radius of the interface between liquid and solid; ρ_s = density of solid *layer*; δ = bowl wall thickness). Thus, the total stress in the wall $\mathscr{S}_b$ is given by

$$\mathscr{S}_b = \mathscr{S}_s + \mathscr{S}_c \quad (19.6)$$

Obviously $\mathscr{S}_b$ must be less than the yield strength of the metal.

19.2. MECHANICAL DEWATERING SYSTEMS

With few exceptions (e.g., centrifuges), mechanical dewatering must be carried out in stages to obtain optimum equipment efficiency, requiring different equipment for each stage. It is essential to appreciate that such equipment forms an interdependent system (which is of course part of the total process system): the operating conditions in one component cannot be considered independent of the conditions in the preceding or following unit. This is readily illustrated by examination of Eqs. 17.7 and 18.18 for the thickener/filter system. Here the thickener underflow concentration $C_{(+),\text{crit}}$ determines the filter feed concentration C_c (more correctly C_l, Eq. 18.8), while the thickener underflow discharge rate $O_{(+)}$ and concentration determine V_f, the volume of filtrate. Because the relationship of these variables to each other and to the equipment size (A_T and A_c) is not linear, selection of A_T and A_c is an optimization problem. In general terms though, the thickener provides the cheaper dewatering, so it is the better unit to run the *hardest*.

For concentrates, thickening followed by filtration provides adequate dewatering in most instances. Economics determines such a selection (although in many instances tradition may actually have been the deciding factor), but another significant point is the system's ability to recovery virtually all fine particles from the water. A further advantage, often easily overlooked, is the conventional thickener's ability to act as a surge tank for solids. A more thorough discussion of the practical aspects of operating a thickener/filter system has been given by Silverblatt and Emmett.[5]

In many instances tailings are thickened only before being pumped to a tailings pond for disposal (Chapter 22). The tailings pond's prime purpose is disposal; however, with today's environmental requirements, it also serves a second important purpose as a water treatment and/or recovery unit. Where pumping costs are likely to be high, it may be economic to employ even more extensive tailings dewatering in the plant. Such a practice is increasingly becoming characteristic of coal cleaning plants.[6,7]

That the thickener/filter combination is so widely used does not mean it is always the best method. When the slurry is low in fines, or the fines do not have to be removed, the comparatively high costs of filtration can be avoided by using higher capacity equipment such as screens or hydrocyclones.[8-10] Even where excessive fines are present, it is possible to devise circuits without filters that can be as effective at solids removal as thickener/filter systems. One example is that shown in Fig. 19.3, where the filter is replaced by a combination of screen and hydrocyclones, which for a given solids capacity require less space and may give lower capital and operating costs[8] (Chapter 25). A further alternative to this arrangement would be to replace the lamella thickener with a cone thickener[7] (Fig. 17.4), and so

Thermal Dewatering (Drying)

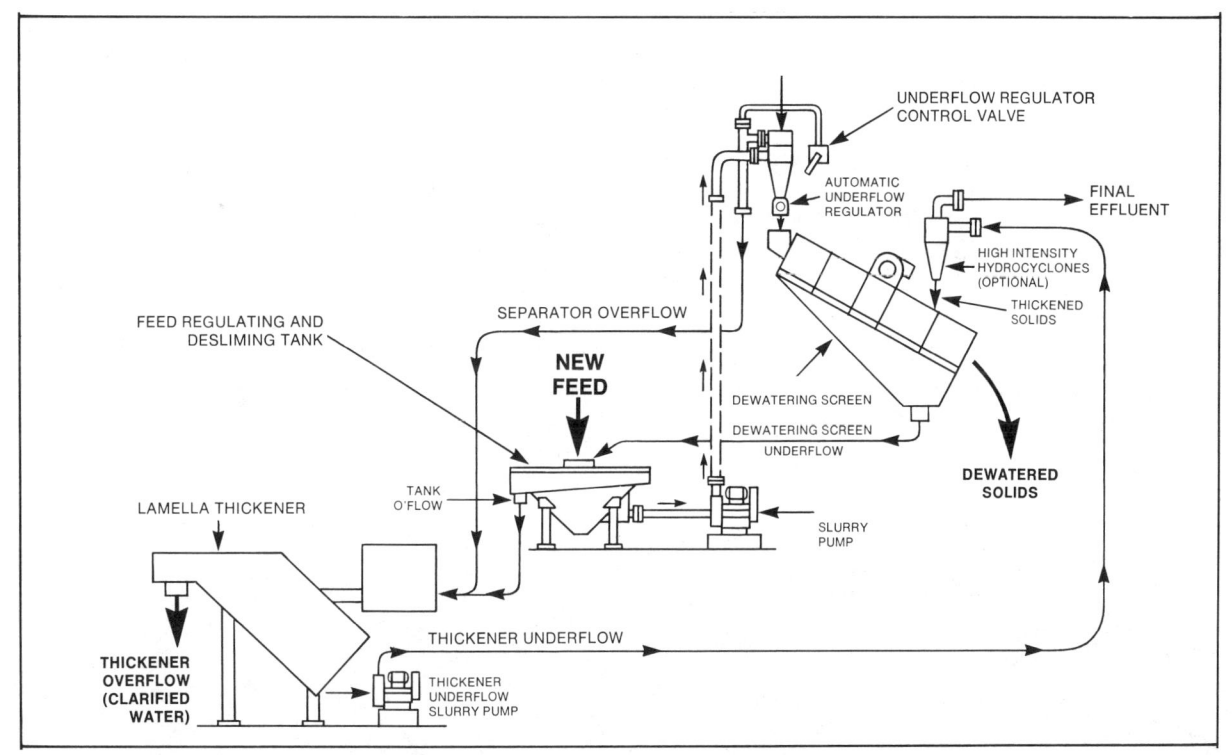

Figure 19.3. Dewatering system with no filter. (Courtesy Lawjack Equipment Co. and *CIM Bulletin*.)

produce two solids products that could be mechanically conveyed (as distinct from slurry pumping): one coarse (suitable for, say, underground mine backfill) and one fine (suitable for landfill).

19.3. THERMAL DEWATERING (DRYING)

The methods of dewatering discussed so far are relatively cheap mechanical separations that seldom reduce the water below 25% by volume. When lower levels are required, it is necessary to dry the material by thermal methods. These are comparatively expensive, since the solid must not only be heated, but the water must be evaporated so that it can be carried away in a gas stream.

The term "drying" is a relative one, and simply means that there is a further reduction in the moisture content from some initial level provided by mechanical dewatering to some acceptable lower level. For example, a moisture content of 10–20% by volume would normally allow particles to flow freely, yet suppress dust formation.

The necessity for drying may be to make the product suitable for sale (e.g., paint pigments), or for subsequent processing (e.g., in pyrometallurgical operations).

Some pyrometallurgical operations have large amounts of heat available in the off gas that can be used for drying, in some cases in the same furnace. However, there are many such processes today that have very tight heat balances, so that an auxiliary drying unit is necessary.

19.3.1. Equipment

Again, because of the large throughputs, most dryers of interest are continuous. Hearth and tray driers have a series of hearths or trays stacked one on top of the other. The solids are raked from level to level down the stack, while hot gas passes up countercurrently.

Rotary driers are revolving (horizontal) cylindrical shells with a slight inclination toward the discharge end to facilitate solids transport. Hot gas is normally used for drying and passes through the shell cocurrent or countercurrent to the solids (Fig. 19.4). Since the solids spend much of their time suspended in the gas, the gas velocity is limited by the carryover of fines. Alternatively, rotary driers may

Figure 19.4. Indirect-direct dryer. Hot gases pass initially through inner cylinder, then countercurrently to solids in outer cylinder. (Courtesy Combustion Engineering, Inc.)

have the heating supplied to an external jacket on the shell.

Flash driers expose the solids to a hot gas stream for only a few (3–4) seconds. Drying is so rapid in this situation that even heat sensitive materials such as coal can be dried in very hot gases. Spray driers are similar in principle, except that the material is fed as a slurry and dispersed into hot gas as a spray. Although rapid, these driers can be relatively inefficient in heat utilization. They do offer the advantage that other unit operations such as classification or size reduction can be incorporated into them.

Fluidized beds are suitable for drying fine material where the particles can be suspended in an upward flow of hot gas (Fig. 19.5). A feature of the fluidized bed is its even temperature and good mixing, and because each particle is entirely surrounded by hot gas, the drying is far more rapid (30–60 sec) than hearth or tray driers. The process does require considerable dust collection equipment because of the elutriation of fines.

With fine, heat sensitive material, indirect heat exchanger driers can be used. These have a hot liquid flowing in hollow discs (Fig. 19.6) or spirals that are in contact with the moist solids. With suitable oils as the heating liquid, temperatures up to 300°C may be attained.

19.3.2. Drying Behavior

Typical drying behavior of a solid dried under constant drying conditions is illustrated in Fig. 19.7a.

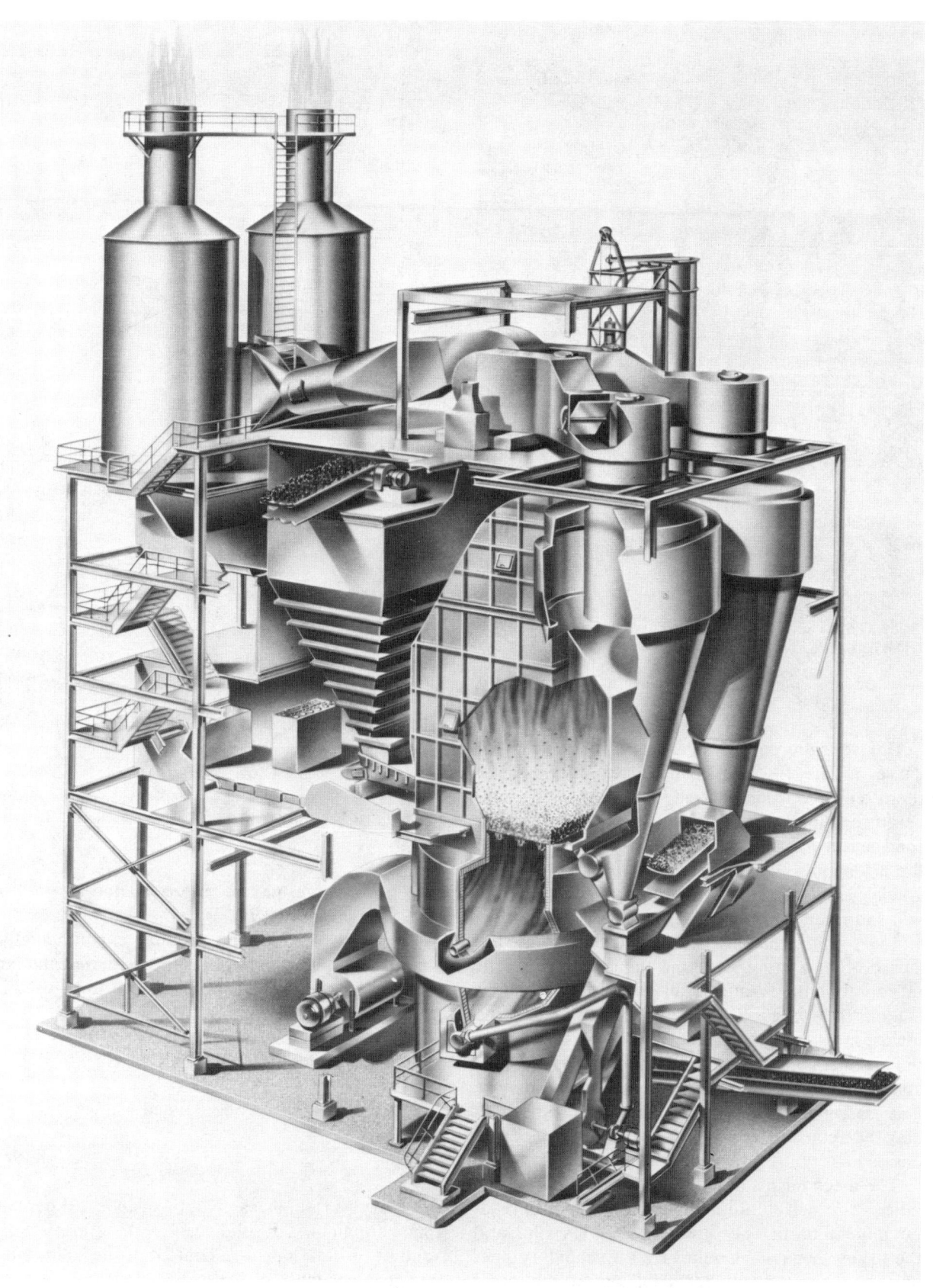

Figure 19.5. Fluidized bed dryer. (Courtesy McNally Pittsburg Manufacturing Corp.)

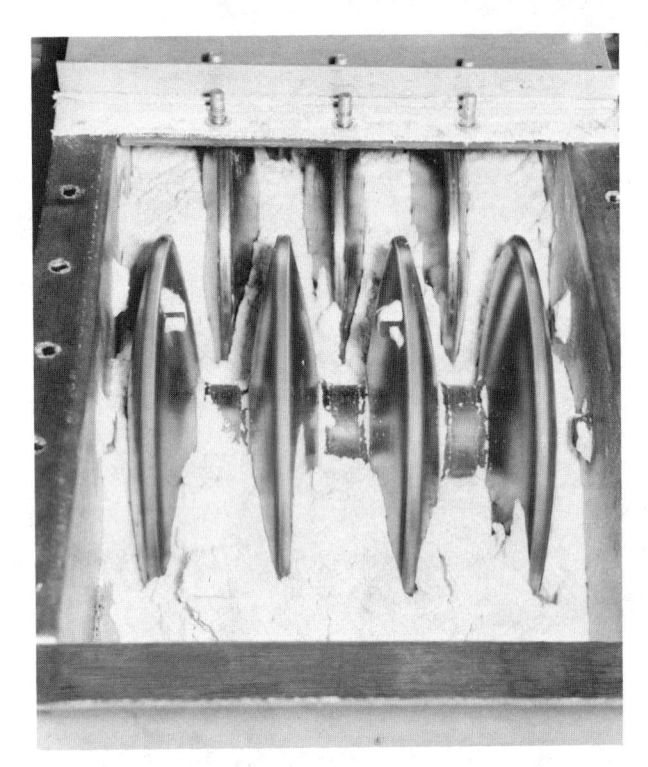

Figure 19.6. Indirect heat exchanger dryer. (Courtesy BSP Division of Envirotech Corp.).

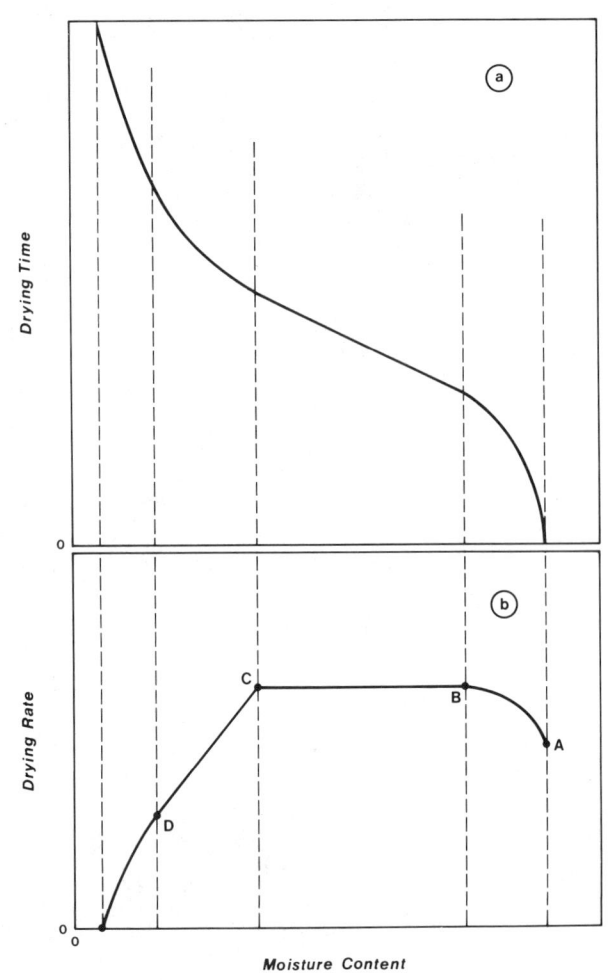

Figure 19.7. (*a*) Typical drying curve for constant drying conditions. (*b*) Drying rate curve derived from (*a*).

Differentiation of the data allows the drying rate curve in Fig. 19.7*b* to be produced. Region A-B represents a warming period where the solids temperature adjusts to a steady state at which the wet-solid surface reaches the wet bulb temperature of the drying gas. The region B-C is the *constant rate drying period* where the entire exposed surface is still saturated with moisture. It ends at the *critical moisture content*, point C. Beyond this point, the surface temperature rises and the drying rate falls off rapidly until eventually an equilibrium moisture content is reached. This equilibrium content is the lowest moisture obtainable with the material under the given conditions, and in the case of most minerals it is close to zero. Even though less moisture may be removed during the falling rate period, it may take considerably longer than the constant rate period.

The mechanisms of moisture removal during the falling rate period are complex. Layers of spread-out mineral particles behave like a porous solid, and the drying behavior is influenced by capillary phenomena. In this situation, the falling rate section of the drying curve may be approximated by one, or two, straight lines.

Drying curves must be obtained experimentally, and it is desirable to have the experimental arrangement as close as possible to that to be used in practice.

19.3.3. Calculation of Drying Times

The drying rate $\mathfrak{D}$ can be defined as:

$$\mathfrak{D} = -\frac{M_s}{A}\frac{dX}{dt} \qquad (19.7)$$

($\mathfrak{D}$ = drying rate, mass of liquid evaporated per unit time per unit area of solid surface; M_s = mass of dry solid; X = bulk moisture content of the solid, mass of liquid per mass of dry solid). This can be rearranged and integrated to give the drying time:

$$\int_0^t dt = -\frac{M_s}{A}\int_{X_I}^{X_t} \frac{dX}{\mathcal{D}} \quad (19.8)$$

(X_I = initial bulk moisture content; X_t = bulk moisture content at time t). Since the drying mechanism varies, different relationships between X and $\mathcal{D}$ must be used for each section of the drying rate curve.

For the constant rate period, $\mathcal{D}$ is constant at $\mathcal{D}_c$ so that Eq. 19.8 integrates to

$$t_c = -\frac{M_s}{A\mathcal{D}_c}(X_c - X_I) \quad (19.9)$$

(t_c = time to dry to the critical point; X_c = bulk moisture content at the critical point). $\mathcal{D}_c$ depends on the driving force available for heat and mass (moisture) transfer from the drying surface to the drying medium and can be described by

$$\mathcal{D}_c = K_M(mw)_g(\mathfrak{H}_i - \mathfrak{H}_g) = \frac{K_M}{l_{\text{vap}}}(T_g - T_i) \quad (19.10)$$

(K_M = mass transfer coefficient; $(mw)_g$ = molecular weight of drying gas; $\mathfrak{H}_i$ = humidity of gas at interface; $\mathfrak{H}_g$ = humidity of drying gas; K_H = heat transfer coefficient; l_{vap} = latent heat of vaporization of water; T_g = temperature of drying gas; T_i = temperature of liquid/gas interface). The so-called heat and mass transfer coefficients are generally described by empirical correlations.[1] For example,

$$K_M = K I_g^n \quad (19.11)$$

(I_g = mass flow rate of gas, per unit area kg/m² · hr; K = dimensional coefficient) when drying with air flowing parallel to the solid surface.

If the falling rate period can be approximated by a single straight line to the origin, then

$$\frac{\mathcal{D}}{X} = \frac{\mathcal{D}_c}{X_c} \quad (19.12)$$

and Eq. 19.8 can be integrated to give

$$t - t_c = -\frac{M_s}{A}\frac{X_c}{\mathcal{D}_c}\ln\left(\frac{X}{X_c}\right) \quad (19.13)$$

REFERENCES

1. R. H. Perry and C. H. Chilton, *Chemical Engineers' Handbook*, 5th ed., McGraw-Hill (1973).
2. M. E. O'K. Trowbridge, "Problems in the Scale-Up of Centrifugal Separation Equipment," *The Chem. Engr.*, No. 162, A73–A87 (August 1962).
3. C. M. Ambler, "How to Select the Optimum Centrifuge," *Chem. Eng.*, **76**, 96–103 (Oct. 20, 1969).
4. P. A. Vesilind, "Estimating Centrifuge Capacities," *Chem. Eng.*, **81**, 54–57 (Apr. 1, 1974).
5. C. E. Silverblatt and R. C. Emmett, "Industrial Thickeners and Filters—Putting It All Together." Presented at Eng. Foundation Conf.; Theory, Practice and Process Principles for Physical Separations, Asilomar, October–November 1977).
6. D. A. Dahlstrom, "Closing Coal Preparation Plant Water Circuits with Classifiers, Thickeners and Continuous Vacuum Filters," *2nd Symp. on Coal Preparation*, pp. 151–179, University of Leeds (1957).
7. J. Abbott et al., "Coal Preparation Plant Effluent Disposal by Means of Deep Cone Thickeners," *Mine and Quarry*, **2**, 37–50 (October 1973).
8. L. D. A. Jackson and T. A. Kirk, "The Linatex Solids Recovery System," *CIM Bull.*, **69**, 79–85 (August 1976).
9. E. J. Sandy, "Mechanical Dewatering," Ch. 12 in *Coal Preparation*, 3rd ed., (J. W. Leonard and D. R. Mitchell, Eds.), AIME (1968).
10. D. Cooper, "New Vortex Sieve Works for Quarto Mining," *Coal Min. Process.*, **11**, 48–50 (September 1974).

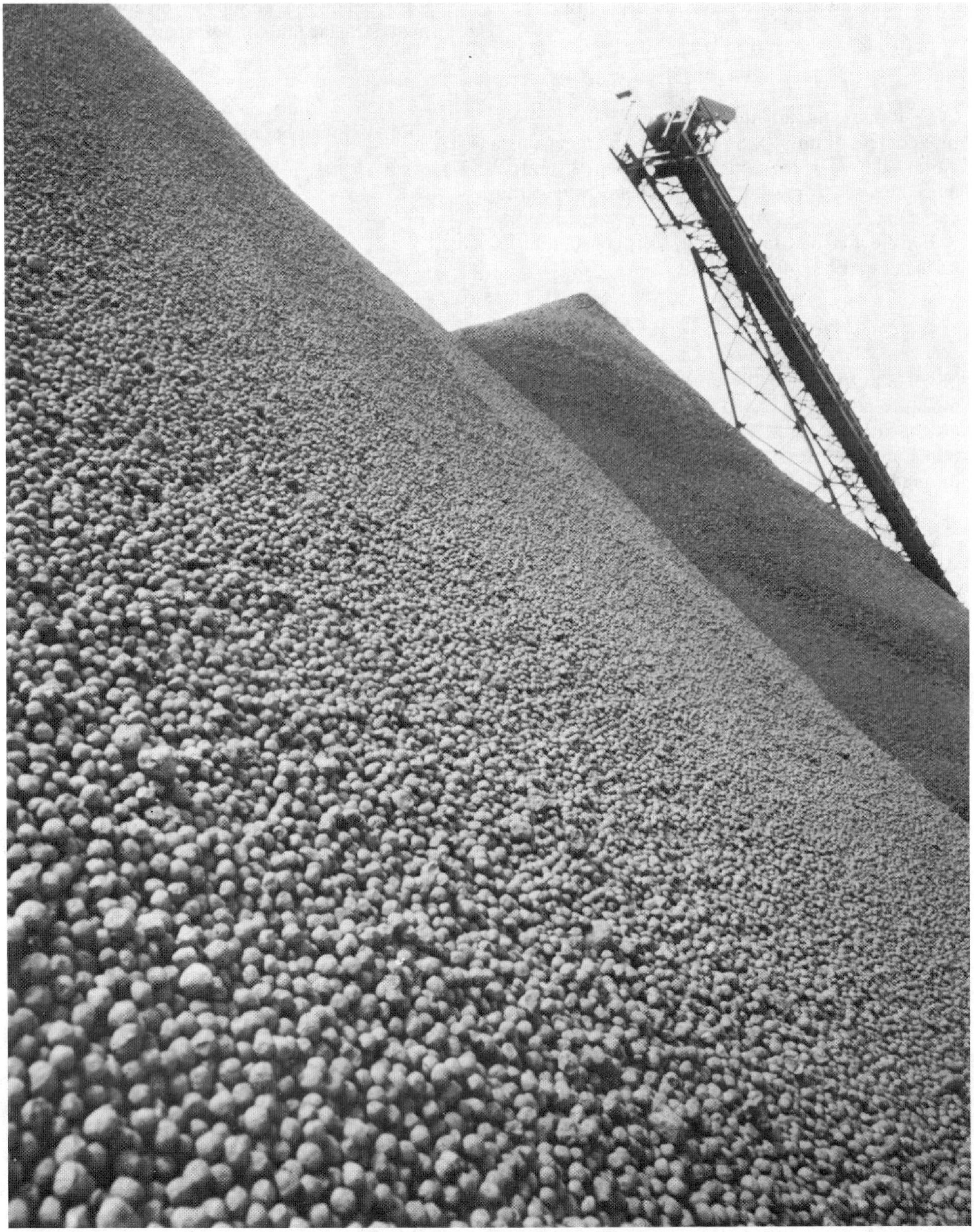

Part VI

Materials Handling

The transportation and storage of materials constitute a major function in all mineral processing plants. Materials transport is required between each processing step in a plant, and often as part of a processing step. Surge capacity or storage is also required, both ahead of the plant and within it. The importance of this materials handling in a plant is often not adequately realized, and as a result it is often some aspect of materials handling that limits the capacity of a plant.

The handling of dry solids includes outdoor or covered stockpiles with their reclaiming systems, bins for crushed products and the feeders to recover the material from them, and conveying methods, from the large belt conveyor for coarse material to pneumatic conveying of fine particles.

Most concentrating operations are carried out in water, and so the handling of slurries is involved. Slurries must be moved over numerous short distances by gravity flow or by pumping; and in some cases, such as is often the case with tailings disposal, long-distance slurry pipelines are also involved. The holding of slurries in tanks is also common in mineral processing plants, and it is essential that this be done in the most economic manner.

Associated with materials handling is another vital part of any operation, the obtaining of a representative sample. This is a requirement at large and small particle sizes, and in both dry and wet handling systems.

The final topic included in Part VI is tailings disposal, since this is largely a materials handling operation.

Chapter Twenty

Dry Solids Handling

Booker's Law (original): "*A gram (ounce) of application is worth a tonne (ton) of abstraction.*"

All mineral processing operations require at some stage the handling of bulk dry solids. Examples include ore as mined, crushed ore, and dried concentrates. In this chapter, the two distinct phases of bulk solid handling are discussed: the holding of materials and the transportation of materials.

For large capacities and very coarse ore, *stockpiles* are used: *bins* are used for smaller capacities and finer ore. Although it is often implied that both stockpiles and bins serve for storage, "storage" is actually used to provide *surge capacity* between various phases of an operation (e.g., between mine and mill, or between crushing and grinding) or to allow *blending*.[1]

Bulk solids transportation is required initially to move material from the mine face using ore passes, trucks, rail cars, or conveyors. Extensive transport facilities may also be required in the mill, although with finer particles, slurry transportation is more common.

20.1. STOCKPILES

Stockpiles are generally used for ore that has passed through the primary crushing stage, and sometimes also for coarse products such as crushed-and-sized coal or iron ore. Stockpiles are treated only briefly here; details are available elsewhere.[1-4]

The surge capacity of a coarse ore stockpile has a number of advantages for a mill: it helps provide a uniform flow of ore to the mill, it allows a means of blending to provide a consistent feed grade to the mill, and it allows the operation of mine and mill to be independent so that the mill can operate continuously, even when the mine is operating fewer shifts. It has been estimated[4] that the efficiency of a mill is increased by 10–25% when a feed stockpile is included.

Stockpiles are formed on a concrete or earthen pad and occasionally they are covered by a roof. A number of methods are used to form the stockpile; included are fixed stackers, tripper conveyors, reversing shuttle conveyors, traveling winged stackers, and radial stackers.[1,4] The resulting stockpiles are conical, elongated, or radial. All stockpiles have sloping sides; the angle of this slope is a property of the bulk solids known as the *angle of repose*.

Pronounced segregation often occurs in stockpiles (Fig. 20.1), and although it cannot be eliminated, it can be minimized by suitable selection of a reclaiming system.

Reclaiming from stockpiles can be carried out using bottom tunnels, bucket-wheel reclaimers, scraper trucks, or front-end loaders. The latter two are used with very large and very small stockpiles, respectively, and are not further discussed here. Bucket-wheel reclaimers are restricted to large operations; they are discussed in detail by Wohlbier.[1] The layout of a bottom-tunnel reclaiming system is shown in Fig. 20.1. The use of multiple feeders perpendicular to a common collecting conveyor promotes the remixing of fine and coarse particles segregated when the stockpile was formed. The mass-flow hopper design shown (Fig. 20.1) is used to increase the live capacity of the stockpile (Section 20.2.2). For a flat-bottomed stockpile, the *live capacity* is limited to 20–25% of the total with conical stockpiles, and 30–35% with elongated stockpiles.[4]

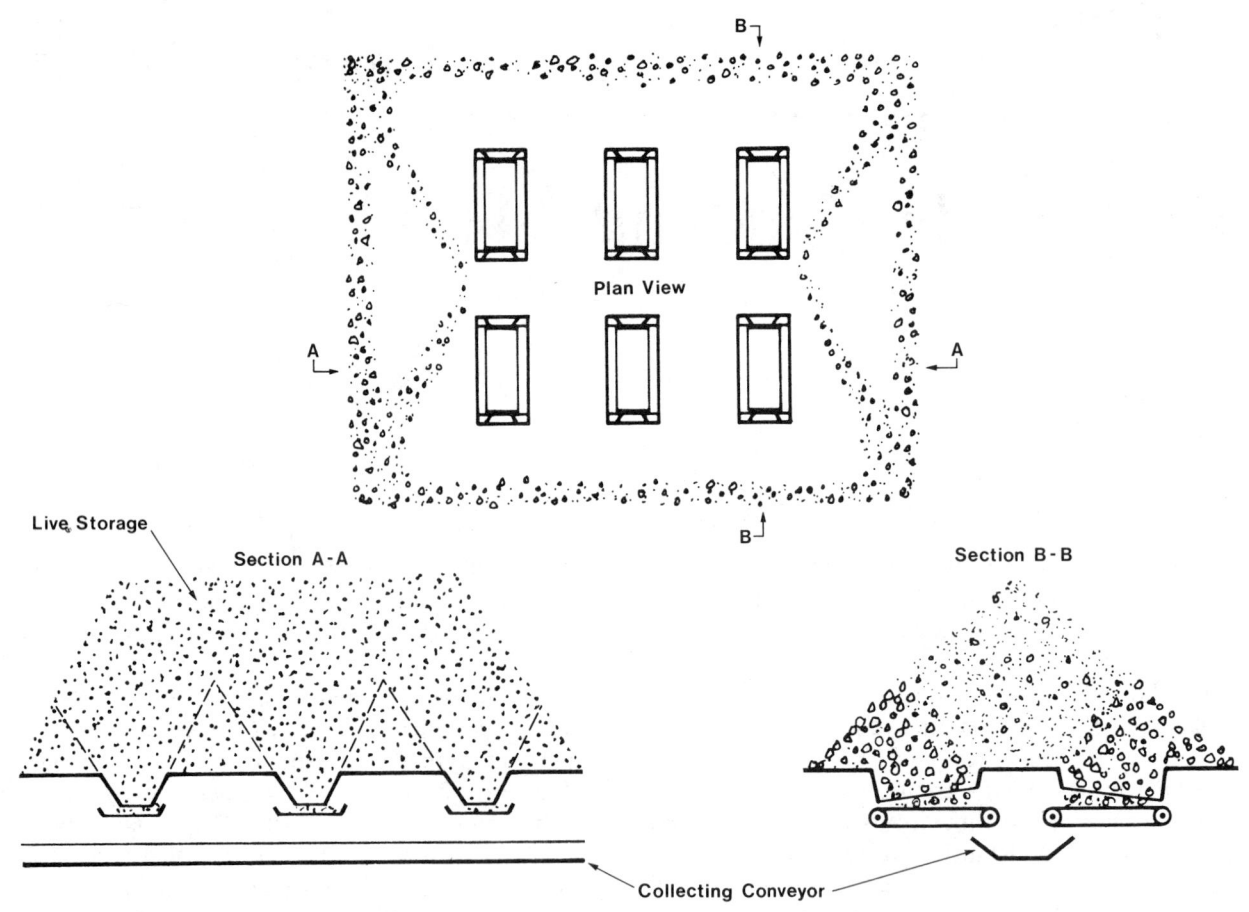

Figure 20.1. Coarse ore stockpile with bottom tunnel reclaim system.[2] (Courtesy *Chemical Engineering*.)

20.2. BINS AND HOPPERS

All mineral processing operations require and use bins. These bins may be very large, such as the "coarse ore" or "fine ore" bins that are frequently found ahead of a crushing or grinding circuit; or they may be quite small, such as those used for dry reagent "storage" or for receiving dry concentrates.

The function of the bin is to provide a large surge capacity and to allow for the feeding of solids to the subsequent processing stage at a *controlled, specified feed rate*. Any interruption in flow from the bin prevents efficient operation of subsequent stages. And yet, until recent years, bins and hoppers have been treated as simple storage devices with little understanding of the manner of their operation. The problems encountered in their operation are numerous, as indicated by the battered sides that have been pounded with sledge hammers, jetted with air, or even blasted, to encourage flow. Although major advances have been made in bin design and operation, flow-aid devices (such as vibrators) are still frequently required.

A "bin" consists of two parts: a converging section at the bottom known as a *hopper* (which may be either conical or wedge shaped), and a vertical section above the hopper called the *bin*, which provides most of the storage volume.

20.2.1. Flow Patterns in Bins

The major problems associated with bins and hoppers are:

Arching: a no-flow condition caused by bridging of the bulk materials over the hopper opening (Fig. 20.2a).[5,6]

Piping: A restricted-flow condition (sometimes called rat-holing) in which flow is limited to a

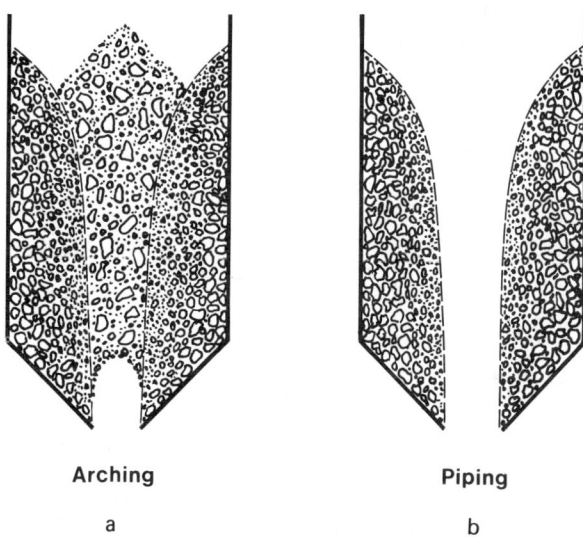

Figure 20.2. Flow limitations in bins. (*a*) Arching. (*b*) Piping.

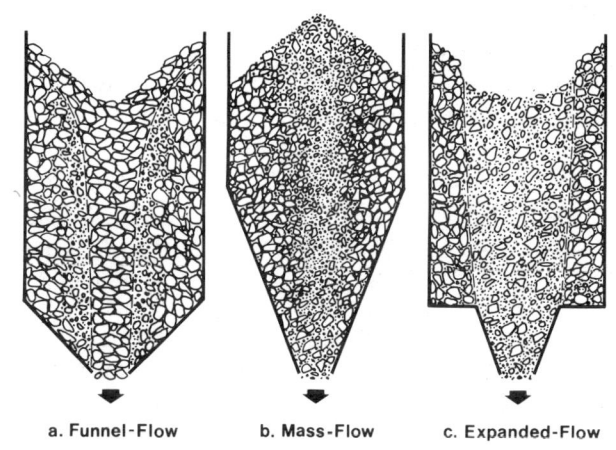

Figure 20.4. Flow patterns in funnel-flow and mass-flow bins (After Johanson.[8])

a. Funnel-Flow b. Mass-Flow c. Expanded-Flow

vertical channel above the discharge opening, and only the material in this channel is removed (Fig. 20.2*b*).[5,6] Piping substantially reduces the effective capacity of the bin.

Particle segregation: When a bin is charged, the coarser particles tend to move to the outside of the bin, resulting in wide variations in the bin discharge (Fig. 20.3).[7]

Each of these problems can be largely eliminated by appropriate bin design. Bins are of two basic designs; they are the *funnel-flow* bin (Fig. 20.4*a*) and the *mass-flow* bin (Fig. 20.4*b*). A third design (Fig. 20.4*c*), sometimes called an expanded-flow bin, is a combination, consisting of a mass-flow hopper below a funnel-flow bin.

Most existing bins are of the funnel-flow type.[8] As indicated in Fig. 20.3, the size distribution of the discharge varies widely, depending on whether the level in the bin is rising or falling. Only if the bin level is constant is the discharge uniform.

All bins segregate material at the point of charging. In funnel-flow bins there is no remixing of this segregated material. Thus funnel-flow bins are acceptable for use only with ores that do not segregate, are free-flowing, and are chemically stable.

With the mass-flow bin, all the material in the bin is in motion whenever any is discharged (Fig. 20.4*b*). In this case, the first material in is the first out, and on discharge the material is remixed, thus eliminating the detrimental effects of the segregation that occurred on charging.[8]

20.2.2. Bin and Hopper Design

A method for the systematic design of bins and hoppers has been developed by Jenike, Johanson, and Colijn,[5-12] based on both theoretical and experimental analyses. In this method, a bin is designed so that there is no arching and no piping, and mass-flow conditions are achieved.

Flow Properties of Bulk Solids. When bulk solids flow in a bin or hopper, continuous shear deformation occurs. The pressure exerted on the bulk solids changes as the solids move down the flow channel, and as the pressure changes, the strength and den-

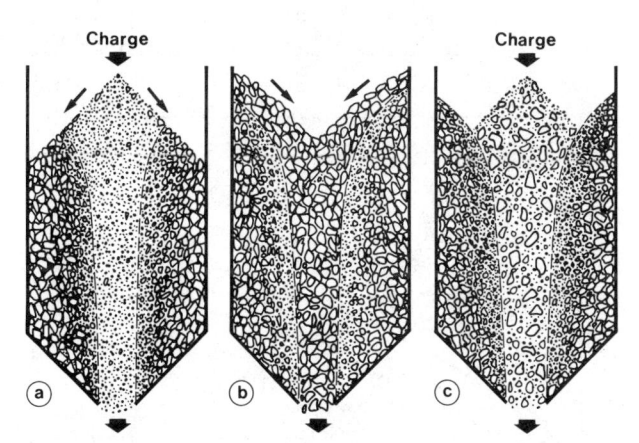

Figure 20.3. Flow patterns and segregation in a funnel-flow bin. (*a*) Flow in greater than out: draw mostly fines. (*b*) Flow out greater than flow in: draw mostly coarse. (*c*) Flow in and out equal: draw same as charge. (After Jenike.[7])

sity of the bulk solid changes. When flow stops, the pressures exerted during flow remain.

If the pressure acting on the bulk solids flowing in a hopper causes consolidation to the point at which its strength can support an arch, flow will stop. Thus it is necessary to know the strength developed by the bulk solids under the continuous deformation and pressure existing in the flowing mass. Experimental data[5,10] indicate that continuous deformation during steady flow occurs only for certain stress conditions in the bulk solid; these conditions are represented by the *effective angle of friction* of the bulk solid, which is given by (Fig. 20.5):

$$\sin \theta_{fe} = \frac{\mathfrak{S}_1 - \mathfrak{S}_2}{\mathfrak{S}_1 + \mathfrak{S}_2} \qquad (20.1)$$

(θ_{fe} = effective angle of friction; $\mathfrak{S}_1, \mathfrak{S}_2$ = major and minor principal compressive stresses, respectively). This angle is essentially constant for a given material at a specific temperature and moisture content; it is measured using a shear tester.[10]

Associated with each state of consolidation is a yield locus, specified by the *angle of internal friction* θ_{fi} (Fig. 20.5). The *unconfined yield strength* of the bulk solid at this particular consolidation is indicated by $\mathfrak{S}_y$ (Fig. 20.5).

When the bulk solids move against the wall of the bin or hopper, the friction condition is represented by the wall yield locus (Fig. 20.5) and is specified by the *wall angle of friction* θ_{fw}. The effective angle of friction, the unconfined yield strength, and the bulk density of the bulk solids for which the bin is to be used must be determined over the range of consolidating pressures expected to exist in the bin (Fig. 20.6). It should be noted that the moisture content of the bulk solids affects its flow properties, as does the time allowed for consolidation in the bin (Fig. 20.7).

Once the properties of the bulk material are known, critical conditions can be established for mass flow, arching, and piping.

Mass Flow. A detailed analysis of the flow conditions and the radial stresses in the hopper has

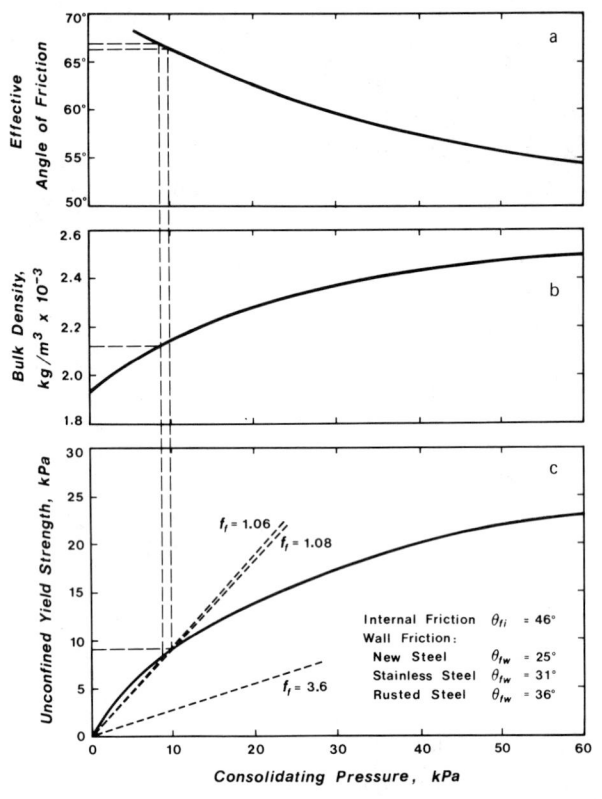

Figure 20.6. Flow properties for an iron ore containing 10% moisture. (After Johanson.[6])

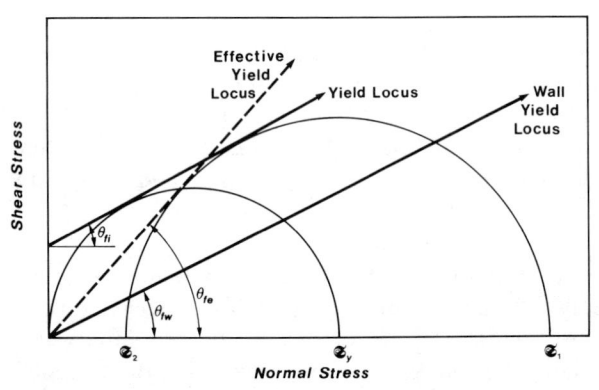

Figure 20.5. Specification of flow properties for a bulk solid.

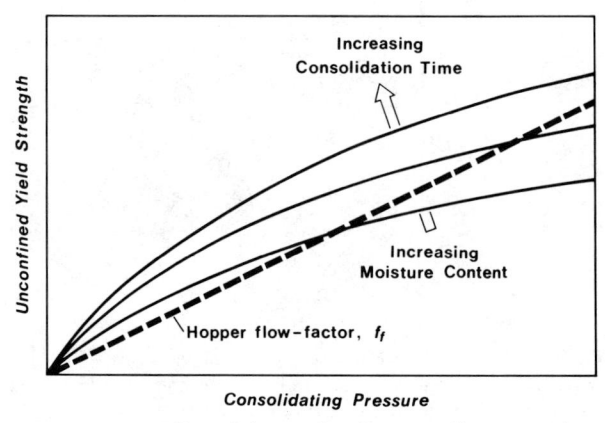

Figure 20.7. Effect of time and moisture on flow properties.

Bins and Hoppers

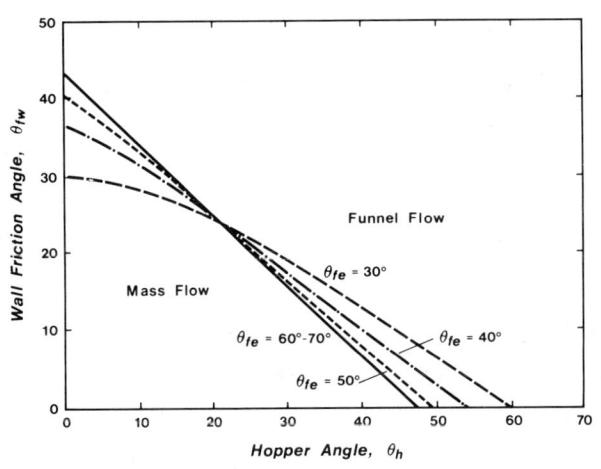

Figure 20.8. Critical angles for mass flow in conical hoppers. (Johanson and Colijn.[5])

shown[5,13] that there are limits on the compatibility of hopper angles, bulk solid wall friction angle, and effective angle of friction (Fig. 20.8).

Arching. For a slot opening, the minimum dimension to prevent arching is:[5]

$$B_a \geq \frac{\mathfrak{S}_y}{\rho_b g} \quad (20.2a)$$

(B_a = slot width [length $>2.5 \times$ width]; $\mathfrak{S}_y$ = unconfined yield strength; ρ_b = bulk density) and for a circular opening the minimum diameter is[5]:

$$B_a \geq \frac{2\mathfrak{S}_y}{\rho_b g} \quad (20.2b)$$

The bulk solids strength could be determined from Fig. 20.6 if the consolidating pressure for steady flow were known. The consolidating pressure in the region of the opening can in fact be calculated,[5] and this has been done for a range of material properties and hopper angles. The critical condition for arching is expressed in terms of a *flow factor* (Figs. 20.9a and 20.10),

$$f_f = \left[\frac{\mathfrak{S}_1}{\mathfrak{S}_y}\right]_{\text{crit}} \quad (20.3)$$

(f_f = flow factor; $\mathfrak{S}_1$ = consolidating pressure [major principal compressive stress]). These flow factors are represented on the unconfined yield strength versus consolidating pressure graph (Fig. 20.7) as straight lines of slope f_f. If a bulk solid with certain strength properties is placed in a hopper with a certain flow factor, the critical values of $\mathfrak{S}_y$ and $\mathfrak{S}_1$ for arching are given by the intersection of the two curves.

Piping. The opening dimension to prevent piping is[5]:

$$D_p \geq 4\frac{\mathfrak{S}_y}{\rho_b g} K_p \quad (20.4)$$

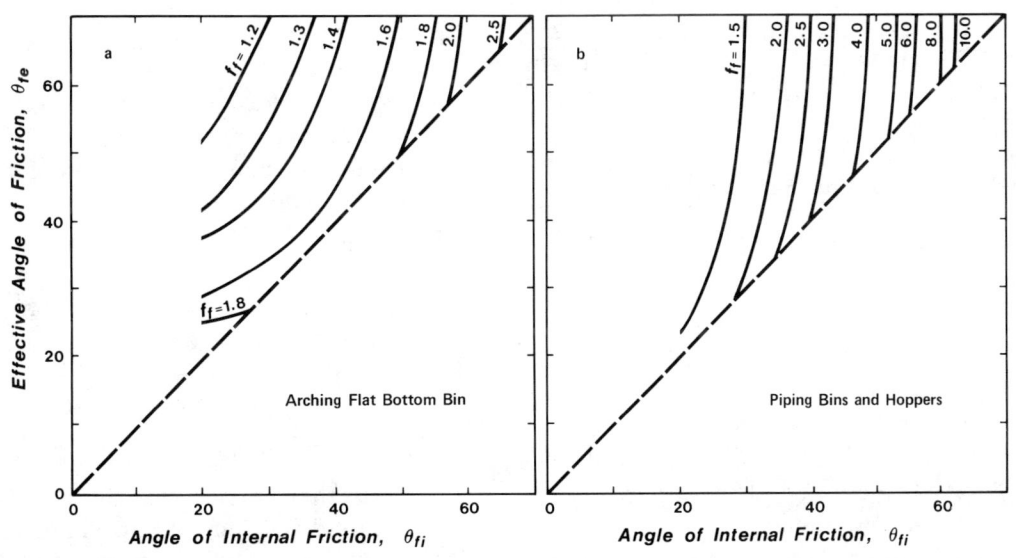

Figure 20.9. Critical flow factors for (*a*) arching in a flat bottom bin and (*b*) piping in all bins and hoppers. (After Johanson and Colijn.[5])

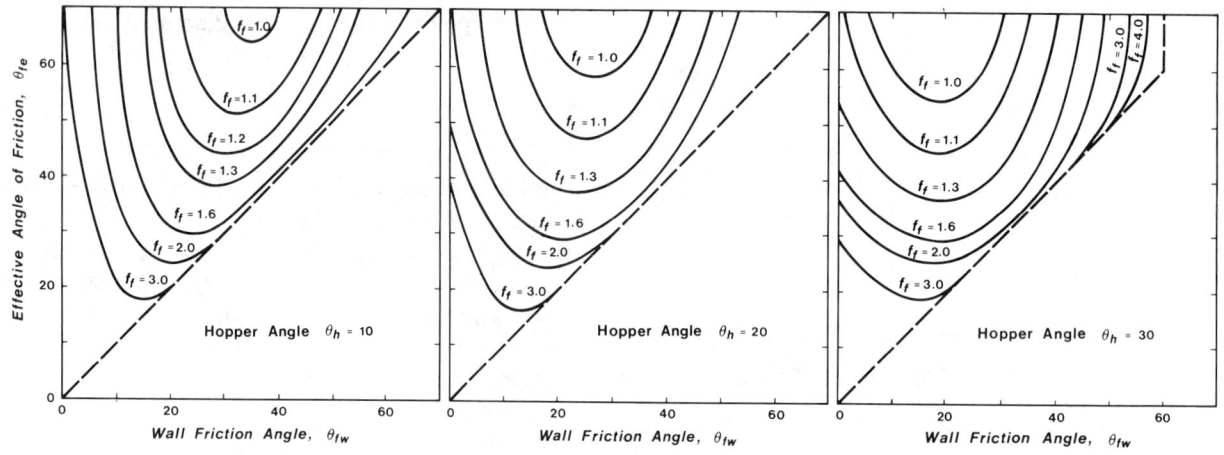

Figure 20.10. Critical flow factors for arching in hoppers. (After Johanson and Colijn.[5])

(D_p = diagonal or diameter of opening [piping]; K_p = piping factor [Fig. 20.11]). A flow factor for piping (Fig. 20.9b) is defined and used in the same manner as for arching.

It should be noted that piping is a limited-flow condition; if mass flow exists, piping cannot occur.

Example 20.1. Flow properties for a crushed iron ore containing 10% moisture are shown in Fig. 20.6. What is the minimum discharge opening that could be used for this ore in a steel-lined bin with a wedge-shaped hopper?

Solution

i. To minimize overall bin height for a required volume, the largest possible hopper angle should be used. For mass flow, however, Fig. 20.8 indicates that no hopper angle greater than 20° may be used for this material, and a 20° angle only for new steel. Select 10° hopper angle.

ii. Data: $\theta_h = 10°$
$\theta_{fi} = 46°$
$\theta_{fw} = 25°$ (assuming new steel hopper)

iii. To design hopper must initially assume value of θ_{fe}. Since the range in Fig. 20.6a is 55–70°, use midrange value of 60°.

iv. From Fig. 20.10a, for $\theta_{fw} = 25°$ and $\theta_{fe} = 60°$, $f_f = 1.08$ ($= \mathfrak{S}_1/\mathfrak{S}_y$).

v. Draw line representing $f_f = 1.08$ on Fig. 20.6c. This line intersects properties curve at $\mathfrak{S}_1 = 10$ kPa, $\mathfrak{S}_y = 9$ kPa. Corresponding to $\mathfrak{S}_1 = 10$ kPa, $\theta_{fe} = 66°$, thus original assumption of $\theta_{fe} = 60°$ not acceptable.

vi. Repeat steps iv and v with $\theta_{fe} = 66°$. This results in $f_f = 1.06$ from Fig. 20.10a; the new line for f_f on Fig. 20.6c intersects properties curve at $\mathfrak{S}_1 = 9$ kPa; the corresponding $\theta_{fe} = 67°$, thus the assumed value of $\theta_{fe} = 66°$ is acceptable. Thus $\mathfrak{S}_y = 8.5$ kPa and $\rho_b = 2120$ kg/m³.

vii. Check arching limiting condition. The minimum slot width to prevent arching is (Eq. 20.2a)

$$B_a \geq \frac{8500}{2120 \times 9.8}$$
$$\geq 0.41 \text{ m}$$

viii. Check piping limiting condition. Figure 20.9b shows that for $\theta_{fi} = 46°$ and $\theta_{fe} = 66°$, the flow factor for piping $f_f = 3.6$. Plotting line representing $f_f = 3.6$ on Fig. 20.6c indicates by extrapolation that $\mathfrak{S}_y = 25$ kPa and $\rho_b = 2500$ kg/m³.

ix. Figure 20.11 indicates that for material with $\theta_{fi} = 46°$, the piping factor $K_p = 1.1$.

x. The minimum slot diagonal to prevent piping is (Eq. 20.4)

$$D_p \geq 4 \times \frac{2500}{2500 \times 9.8} \times 1.1$$
$$\geq 0.45 \text{ m}$$

Thus the minimum width of slot that could be used is 41 cm. Piping will be prevented in any slot with length greater than width.

Bins and Hoppers

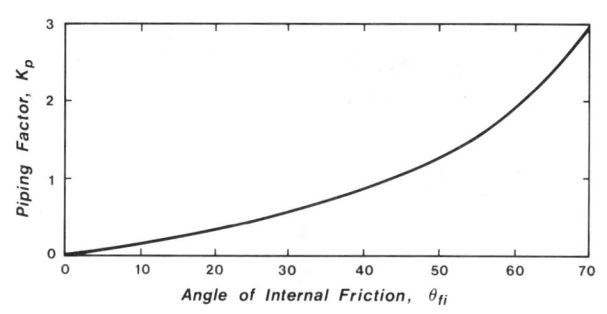

Figure 20.11. The piping factor. (After Johanson and Colijn.[5])

20.2.3. Hopper Discharge Rate

Even though the critical opening size to prevent flow obstruction (arching or piping) may be exceeded, the flow rate may not be sufficient for the process requirements. With bulk solids composed predominantly of particles greater than 250 μm, this flow rate is a function of bulk properties and hopper geometry only.[8]

Johanson[6] has shown that the steady state discharge rate from a slotted-type hopper is

$$O_V = BL \left(\frac{Bg}{2 \tan \theta_{ch}}\right)^{1/2} \left(1 - \frac{f_f}{f_{f,a}}\right)^{1/2} \quad (20.5a)$$

(B = slot width; L = slot length; f_f = flow factor [critical for arching]; $f_{f,a}$ = actual flow factor at hopper outlet; θ_{ch} = flow channel [half] angle; O_V = volumetric [bulk] discharge rate). For a conical-type hopper:

$$O_V = \frac{\pi B^2}{4} \left(\frac{Bg}{4 \tan \theta_{ch}}\right)^{1/2} \left(1 - \frac{f_f}{f_{f,a}}\right)^{1/2} \quad (20.5b)$$

(B = slot width, in this case slot diameter). The actual flow factor at the hopper outlet is given by

$$f_{f,a} = \frac{\mathfrak{S}_1}{\mathfrak{S}_y} \quad (20.6)$$

with the consolidating pressure at this point being calculated by

$$\mathfrak{S}_1 = \rho_b B g f_f \quad (20.7)$$

The use of these equations is demonstrated in the following example.

Example 20.2. What is the slot length required in the hopper of Example 20.1 for a maximum discharge rate of 50 m³/min (slot width is 50 cm)?

Solution. From Example 20.1, $f_f = 1.06$, $\mathfrak{S}_y = 8.5$ kPa, $\mathfrak{S}_1 = 9$ kPa, $\rho_b = 2120$ kg/m³. The consolidating pressure at the bin discharge is given by Eq. 20.7,

$$\mathfrak{S}_1 = 2120 \times 0.5 \times 9.8 \times 1.06$$
$$= 11.0 \times 10^3 \text{ Pa}$$
$$= 11.0 \text{ kPa}$$

At a consolidating pressure of 11.0 kPa, the bulk density is infact 2180 kg/m³ (Fig. 20.6b); the corrected consolidating pressure at the discharge is then

$$\mathfrak{S}_1 = 2180 \times 0.5 \times 9.8 \times 1.06$$
$$= 11.3 \text{ kPa}$$

The corresponding unconfined yield strength $\mathfrak{S}_y = 9.5$ kPa. The actual flow factor at the discharge is (Eq. 20.6)

$$f_{f,a} = \frac{11.3}{9.5} = 1.2$$

From Eq. 20.5a, the required length to give a discharge rate of 50 m³/min is

$$\frac{50}{60} = 0.5 \times L \times \left[\frac{0.5 \times 9.8}{2 \times \tan 10}\right]^{1/2} \left[1 - \frac{1.06}{1.2}\right]^{1/2}$$

that is, length required = 1.31 m.

With a mass-flow bin, the *flow channel angle* is the same as the hopper angle. The effect of hopper angle on discharge rate can be seen from Eq. 20.5 and Fig. 20.12. This figure also indicates the effect of moisture content on the discharge rate.

With very fine powders, the discharge rate is highly dependent on both particle size and bin and hopper geometry. The discharge rate of fine powders can be very much lower than that of coarse solids in an identical bin and hopper. To increase discharge rates, and to provide a steady discharge rate, it is sometimes necessary to inject air into the bin at certain levels to eliminate the vacuum that would otherwise develop.[8,14–16]

20.2.4. Multiple Outlet Bins

Multiple outlet bins[17] are commonly used. Hopper slopes and outlet dimensions must satisfy the minimum requirements to prevent arching and piping as discussed in Section 20.2.2.

Segregation patterns associated with multiple outlets are as shown in Fig. 20.1. If a discharge is directly under a central feed point, it receives fine material, whereas outer discharge openings receive

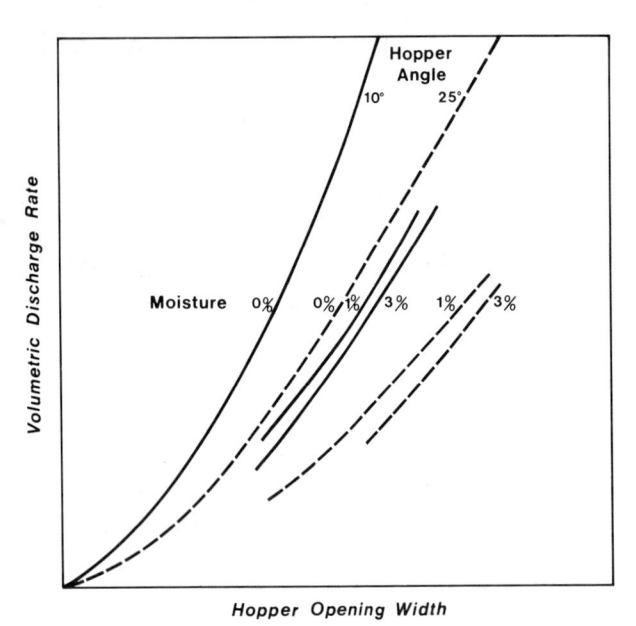

Figure 20.12. Discharge rate of an iron ore from a slotted hopper. (After Johanson.[6])

coarser particles. This is true for both funnel flow and mass flow. It can be eliminated by removing the central discharge opening.

20.3. FEEDERS

Bin, hopper, and feeder must be treated as an integral unit. Once a bin and a hopper have been selected, the feeder must be designed to produce a flow across the whole hopper opening. The feeder size must exceed the critical dimensions determined in the bin and hopper design or it might cause limited flow in the bin.

There are many feeder designs in use.[18-20] The major ones are indicated in Table 20.1. Band-type feeders (belt and apron) are most common (about 50% of all feeders), but vibratory feeders (Fig. 20.13) are also very widely used (about 25% of all).[3] Sometimes feeders have a sizing function added to them (Figs. 20.14 and 9.5b). A notable feature of all

Figure 20.13. A vibratory feeder. (Courtesy FMC Corp.)

Transport of Bulk Solids

TABLE 20.1: FEEDERS FOR BULK SOLIDS

Type	Capacity m³/hr	Description and Applications
BELT OR APRON	20-700	Belt feeder consists of flat-belt driven by head pulley. Widely used for feeding from slotted hoppers (length not limited, Fig. 20.15). Limited to top particle size of approximately 15 cm. Particles must not impact on belt. Apron feeders have a "belt" of overlapping steel plates for heavy duty applications: large rocks, direct impact, hot solids. Both widely used.
SCREW	5-100	Variable pitch helix or screw, pitch increasing in flow direction. Pitch variation between 0.5 and 1.5 diameters. Particles must be less than smallest pitch. Not suitable for abrasive solids. Can be totally enclosed, so used where spillage and dusting a problem.
ROTARY TABLE	5-300	Rotating circular table. Skirt raised above table in helical pattern. Solids squeeze out from hopper, removed by fixed plow. Limited to small particles. Spillage low.
VIBRATORY	15-600	Source of vibration can be mechanical, electromechanical, or electromagnetic (Fig. 20.13). Used with round, square, or short slotted openings. Not for use with cohesive solids. Precise feeding control possible. Can be combined with screen (Fig. 20.14).
STAR		Provides uniform withdrawal along slot opening. Can be used with long slotted opening.
ROTARY PLOW	200-1900	Draws from one point at a time, but can be used with long slotted openings. Used with large volume, coarse ore stockpiles.

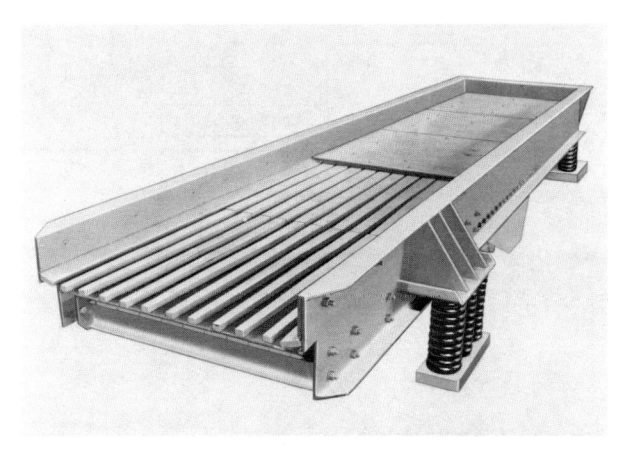

Figure 20.14. A vibratory feeder with grizzly. (Courtesy Allis-Chalmers Corp.)

feeders is that carrying capacity must increase in the flow direction, to draw material from the entire hopper opening (Fig. 20.15).

The feeding of fine particles ($-74\ \mu$m) is more difficult than the feeding of coarse. A feeder might be required to provide a positive shutoff to prevent the uncontrolled release of fines from the hopper (surging and flooding). Solutions to particular fine particle feeding problems have been reported in the literature.[15]

20.4. TRANSPORT OF BULK SOLIDS

The transport of bulk solids from a bin or stockpile is a critical operation in a processing plant. Feeders (Section 20.3) are used to control and regulate the rate of removal of the bulk solids from storage. Following the feeder, a means of transporting the bulk solids to the next processing step is normally required.

A variety of methods are used. The choice depends on a number of factors, including the nature

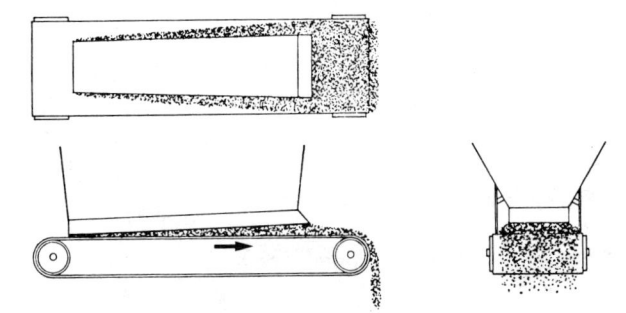

Figure 20.15. Belt or apron feeder showing increased carrying capacity in flow direction. (After Johanson.[18])

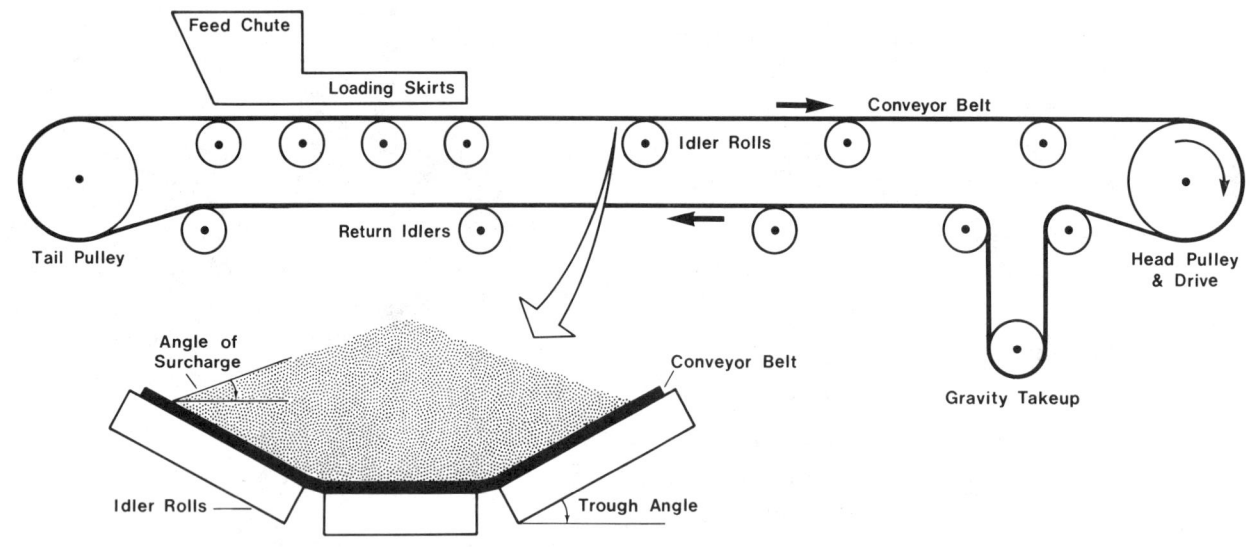

Figure 20.16. Representation of a belt conveyor.

and size of the solids, the distance the solids are to be moved, and the change in elevation required. Detailed descriptions of the equipment available are provided by the manufacturers, and in-depth surveys have been presented in the literature.[21]

The classification of the equipment for transporting bulk solids is somewhat arbitrary; however, it is convenient to classify it as chutes, mechanical conveyors (horizontal and inclined), pneumatic conveyors, and elevators.

20.4.1. Chutes

Chutes are flat plates used to transfer bulk solids for short distances by gravity flow. It is normal to use a chute at transfer points, such as between feeders and conveyors, or from one conveyor to another.

A chute must have adequate capacity for its location; also, the solids must not remain on the chute. Bulk solids will slide on a chute (or any flat plate) as long as the slope of the chute exceeds the wall angle of friction (Section 20.2.2).

Chute surfaces are commonly lined to prolong their life. Wear is particularly high because of the angle of impingement on the chute surface. Liners, often of rubber, are either flat or serrated to reduce this impingement angle.[3]

20.4.2. Mechanical Conveyors

This discussion of mechanical conveyors is restricted to those operated either horizontally or on an incline; vertical conveyors are discussed in Section 20.4.4.

Belt Conveyors. The most widely used conveying method consists of an endless belt moving over a series of rolls (Fig. 20.16). Conveyor belts are available in a wide range of sizes and materials. They can be designed to operate horizontally or inclined, either ascending or descending. Techniques for conveyor belt design are available in the literature.[4,21,22] The major features of all designs are shown in Fig. 20.16, and an example of manufacturers' data is provided in Appendix G; its use is demonstrated in Example 20.3.

For the design of a conveyor belt to meet a particular requirement, the properties of the material to be conveyed must be determined. These properties are size and size distribution, bulk density, moisture content, temperature, abrasive or corrosive nature, and *angle of repose*. The angle of repose, or more correctly the *dynamic angle of repose* (also called the angle of surcharge when related to conveyor belts), is the angle formed naturally by the bulk material when it is loaded onto the moving conveyor belt (Fig. 20.16). It is a property of the particular bulk solid.

Apron conveyors are also manufactured, for heavy duty use (see apron feeders, Table 20.1).

Example 20.3. Select a conveyor belt to transport 4 m³/min of iron ore from a bin to a secondary crusher plant, a horizontal distance of 56 m and a vertical lift of 18 m. The bulk density of the ore is 2200 kg/m³, the ore is 100% − 15 cm.

Solution

i. Bulk material characteristics (Table G-1): Iron ore, lumpy, + 1 cm (max. size 15 cm); average flowing, angle of repose 30–45°; abrasive; angle of repose, 35°.
ii. Surcharge angle is 15° less than angle of repose, that is, surcharge angle = 20° (Table G-2). Could select belt with 20, 35, or 45° trough angle. Select 20° trough angle because it is recommended with 20° or higher surcharge angle and it increases capacity of belt.
iii.
   a. Minimum belt width recommended for this material is 61 cm (Table G-2).
   b. For 20° surcharge angle, 20° trough angle, capacity of belt is 0.91 m³/min on a belt moving at 30 m/min.
   c. Required capacity is 4 m³/min; thus a belt speed of (4/0.91) × 30 = 132 m/min is required with a 61 cm wide belt.
   d. Recommended maximum belt speed with this material is 106 m/min (Table G-3). Proposed belt speed thus is too high, so must use wider belt.
   e. Use 76 cm belt. This belt has a capacity of 1.49 m³/min at a speed of 30 m/min. Thus a belt speed of (4/1.49) × 30 = 81 m/min will have the required capacity. Recommended maximum belt speed (Table G-3) is 122 m/min; thus 76 cm belt is acceptable.
iv. Power required (at shaft of head pulley) to drive belt conveyor must be sufficient to overcome all friction and drive empty belt, to move load horizontally, and to lift load vertically.
   a. Distance between end pulleys of belt is 59 m. Power requirement for empty conveyor (Table G-4) is 0.69 kW at 30 m/min, thus power requirement at 81 m/min is 0.69 × 81/30 = 1.9 kW.
   b. Power requirement to move 4 m³/min of material horizontally 56 m is 4.2 kW (Table G-5).
   c. Power requirement to lift 4 m³/min of material 18 m is 27.1 kW (Table G-6).

   total power requirement = 1.9 + 4.2 + 27.1
   = 33.2 kW

v. *Summary.* A 76 cm wide belt (with a 20° trough angle) moving at 81 m/min is recommended. Power required is 33.2 kW. For mechanical aspects of conveyor belt design, manufacturer's literature should be referred to.

Screw Conveyors. The screw conveyor[21,23] is similar to the screw feeder (Table 20.1), except that it has a constant pitch helix.

Screw conveyor design is empirical; design methods are described by manufacturers and in the literature.[21,23] They can be operated either horizontally or inclined, but their capacity is greatly reduced when inclined.[21]

Chain Conveyors. Chain conveyors are somewhat similar to belt conveyors, but with continuous chains replacing the belts. The solids may be carried directly by the chains or by attachments to the chains; flight conveyors, drag conveyors, and some apron conveyors are types of chain conveyor.

Chain conveyors suffer from high maintenance requirements; however, they are used with very abrasive materials such as some ores, slags, and coke.[21] The design is again empirical.

Vibratory Conveyors. Vibratory or oscillating conveyors have limited use in the mineral industry, although they are more widely used in the chemical industry. Their design, as with other mechanical conveyors, is empirical.[21]

20.4.3. Pneumatic Conveyors

Pneumatic conveyors have not been widely employed in the mineral industry, but their use is increasing, particularly with low density minerals. They are of particular interest when fine dry particles are to be conveyed such as in a dry grinding circuit.

Pneumatic conveying is essentially an extension of the flow of fluids through a packed bed (Fig. 4.12) introduced in Section 4.6. A more detailed description is beyond the scope of this book, and the reader is referred elsewhere for further information.[24,25]

20.4.4. Elevators

Elevators are mechanical conveyors that move bulk solids vertically or on inclines close to vertical.

Bucket Elevators. Bucket elevators[21] are manufactured in a range of designs and sizes. They are used with particles up to about 10 cm. Sizing is based on empirical relationships provided by manufacturers.

Skip Hoists. Skip hoists are widely used to haul ore from underground mines.[24] They generally discharge into coarse ore bins on the surface.

Skip hoists are also used to elevate coarse bulk materials over a limited distance, such as when feeding a blast furnace or handling coal in a power plant. They provide an intermittent flow of material and thus are unsuitable where continuous flow is required.

20.5. SAMPLING

For a process to be operated efficiently, it is essential to know the composition, size distribution, and feed rate for that process. Thus, it is necessary to extract a suitable sample of the feed material. Samples are also frequently required for accounting or mine control purposes.

The sample of a bulk solid must be *representative*. That this is a difficult requirement can be seen by inspecting Fig. 20.1 or Fig. 20.3. Segregation is normal in stockpiles, bins, and other containers, such as rail cars used in the bulk delivery of ore, or drums used to deliver samples of ore for testing.

Sampling methods from stockpiles and bulk carriers are set by government regulation.[26,27] Sampling for in-plant or laboratory investigation is normally carried out using the methods described by Taggart.[28] The minimum size recommended for such samples can be obtained from Fig. 20.17.

Sampling from a flowing stream is achieved by cutting a sample from that stream. A sampler must be designed so that every particle, regardless of size or composition, has an equal chance of becoming part of the sample. Thus, to avoid sampling bias, the following rules must be observed:

1. The cutter must make a complete traverse of the flowing stream with each cut.
2. The cutter must have parallel edges and move at right angles to the flowing stream.
3. The cutter opening must have a width at least three times that of the largest particle size being sampled.
4. The cutter speed must be constant.

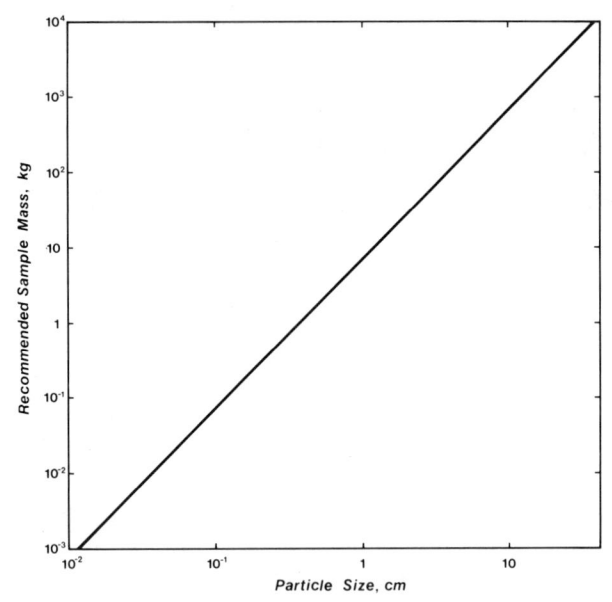

Figure 20.17. Recommended minimum sample mass based on top particle size. (After Taggart.[27])

5. The flowing stream to be sampled must be in free fall.
6. The sample must be passed quickly through the cutter to prevent buildup or blockage.

It is not unusual for the primary sample to be further reduced in quantity by secondary and possibly tertiary samplers.

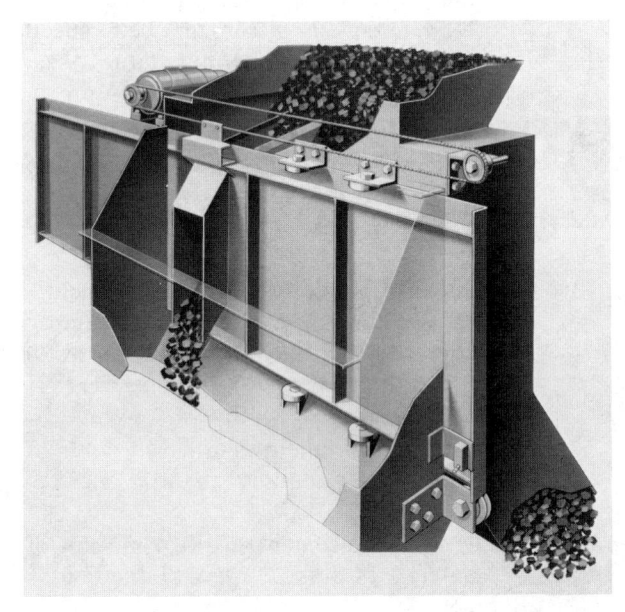

Figure 20.18. A cross-chute sampler for dry bulk solids. (Courtesy McNally Pittsburg Manufacturing Corp.)

Many designs of sampler are marketed. The most common for dry bulk solids are the cross-chute sampler (Fig. 20.18) and the rotary sampler. In all designs, the rules above must be considered.

REFERENCES

1. R. H. Wohlbier (Ed.), *Stacking, Blending, Reclaiming of Bulk Materials*, Trans. Tech Publications (1977).
2. "Materials Handling," *Chem. Eng., Deskbook Issue*, **85** (Oct. 30, 1978).
3. Anon., "Dry Solids Handling in Plants," *Eng. Min. J.*, **180**, 113–141 (1979).
4. F. D. Dietiker, "Belt Conveyor Selection and Stockpiling and Reclaiming Applications," in *Mineral Processing Plant Design*, A. L. Mular and R. B. Bhappu (Eds.), pp. 618–685, AIME (1978).
5. J. R. Johanson and H. Colijn, "New Design Criteria for Hoppers and Bins," *Iron Steel Eng.*, **41**, 85–104 (October 1964).
6. J. R. Johanson, "Method of Calculating Rate of Discharge from Hoppers and Bins," *Trans. AIME/SME*, **232**, 69–80 (1965).
7. A. W. Jenike, "Why Bins Don't Flow," *Mech. Eng.*, **86**, 40–43 (May 1964).
8. J. R. Johanson, "Why Bulk Powders Flow—Or Don't," *Chem. Tech.*, **5**, 572–576 (1975).
9. A. W. Jenike, "Storage and Flow of Solids," Bull. 123, Utah Engineering Experiment Station, University of Utah (1964).
10. A. W. Jenike, P. J. Elsey, and R. H. Woolley, "Flow Properties of Bulk Solids," *Proc. ASTM*, **60**, 1168–1181 (1960).
11. A. W. Jenike, "Quantitative Design of Mass Flow Bins," *Powder Technol.*, **1**, 237–244 (1967).
12. A. W. Jenike and J. R. Johanson, "Quantitative Design of Bins for Reliable Flow," *Miner. Sci. Eng.*, **4**, 3–13 (April 1978).
13. J. R. Johanson and A. W. Jenike, "Stress and Velocity Fields in Gravity Flow of Bulk Solids," Bull. 116, Utah Engineering Experiment Station, University of Utah (1962).
14. J. R. Johanson and A. W. Jenike, "The Effect of the Gaseous Phase on Pressures in a Cylindrical Silo," *Powder Technol.*, **5**, 133–145 (1971–1972).
15. G. B. Reed and J. R. Johanson, "Feeding Calcine Dust with a Belt Feeder at Falconbridge," *Trans. ASME, J. Eng. Ind.*, **95**, 72–74 (February 1973).
16. W. Bruff and A. W. Jenike, "A Silo for Ground Anthracite," *Powder Technol.*, **1**, 252–256 (1967–1968).
17. J. R. Johanson, "Design for Flexibility in Storage and Reclaim," *Chem. Eng., Deskbook Issue*, **85**, 19–26 (Oct. 30, 1978).
18. J. R. Johanson, "Feeding," *Chem. Eng., Deskbook Issue*, **76**, 75–83 (Oct. 13, 1969).
19. Z. F. Oszter, "Comments on Specific Feeder Applications," *CIM Bull.*, **59**, 363–382 (1966).
20. F. M. Thompson, "Smoothing the Flow of Materials Through the Plant: Feeders," *Chem. Eng., Deskbook Issue*, **85**, 77–87 (Oct. 30, 1978).
21. H. Colijn, "Mechanical Conveyors and Elevators—CPI Workhorses," *Chem. Eng., Deskbook Issue*, **85**, 43–58 (Oct. 30, 1978).
22. "Belt Conveyors for Bulk Materials," *CEMA Book*, Conveyor Equipment Manufacturers Association (1966).
23. "Screw Conveyors," *CEMA Book 350*, Conveyor Equipment Manufacturers Association (1971).
24. M. N. Kraus, "Guide to Pneumatic Conveying," *Chem. Eng., Deskbook Issue*, **85**, 63–73 (Oct. 30, 1978).
25. A. S. Foust et al., *Principles of Unit Operations*, 2nd ed., Wiley (1980).
26. (a) "Chemical Analysis of Metals and Metal Bearing Ores," *Annual Book of ASTM Standards*, Part 12 (1980).
    (b) "Coal and Coke," *Annual Book of ASTM Standards*, Part 26 (1980).
27. A. F. Taggart, *Handbook of Mineral Dressing*, Wiley (1945).
28. F. Jordison, "Tips on Automatic Sampling," *Chem. Eng., Deskbook Issue*, **85**, 103–107 (Oct. 30, 1978).
29. P. M. Gy, *Sampling of Particulate Materials: Theory and Practice*, Elsevier (1979).

Chapter Twenty-One

Slurry Handling

Murphy's Law of Thermodynamics:
"Things get worse under pressure."

Most mineral processing plants operate wet. Not only is water the preferred medium for carrying out most concentration processes (Part IV), it is also the most convenient medium for many grinding and classification operations. This means that the material within a plant can be transported from one operation to another in slurry form.

Hydraulic mining is used for some ores, such as pebble phosphates, beach sands, or alluvial deposits of gold, cassiterite, and other minerals. In these cases the ore is transported to the mill in slurry form. In a number of operations, concentrates are transported from mills over long distances in slurry pipelines, frequently across rugged terrain where other forms of transport have not been viable. And all tailings are discharged from wet plants through a pipeline to the tailings pond.

Slurry handling includes not only the transport of slurries from one point to another, but also the suspension and agitation of slurries in tanks during processing. The principles of the operations discussed in this chapter were introduced in Chapter 4.

21.1. PUMPS AND PUMPING

In a mineral processing plant, pumps are used for the movement of liquids and, more important, for the movement of slurries. Slurries must be pumped through numerous short distances within a plant. Also, tailings must be pumped to a tailings pond, or underground as backfill. And in some instances, concentrates must also be pumped.

These slurries are often highly abrasive and corrosive, and they may contain coarse particles at high pulp densities. The design of the pumps, together with the associated valves and pipelines, constitutes a major task in the design of a plant.

21.1.1. Equipment

Pumps. Pumps (Fig. 21.1) are normally considered in two categories, *positive-displacement pumps* and *centrifugal pumps*. Both types have their specific fields of application. Positive-displacement pumps deliver a constant volume of liquid or slurry for each stroke or revolution, independent of the discharge or suction head, whereas centrifugal pumps deliver a volume dependent on the discharge head. In general, positive-displacement pumps are used where very high heads are required, and centrifugal pumps where low heads but high flow rates are required.

There are a number of types of positive-displacement pump available,[1-3] but they may be considered in two groups. *Rotary pumps* are limited to use with liquids (including highly viscous liquids) and cannot be used with abrasive solids. Several varieties are available, including the gear pump and the screw pump. A major advantage of rotary pumps is their ability to discharge at constant flow rates.

Reciprocating pumps are also in the positive-displacement category. They are widely used in slurry pumping where a high head is required, such as with long-distance slurry pipelines, or pumping water from mines. They may be considered in three groups: a direct action plunger type, in which the plunger is in contact with the slurry; an indirect

Pumps and Pumping

action type, in which the plunger acts on an intermediate liquid that transfers the pressure to the slurry through a diaphragm (to eliminate slurry wear on plunger and glands); and a diaphragm pump, in which a diaphragm is activated either mechanically or pneumatically to create a positive displacement.

In all reciprocating pumps, flow is intermittent. All require nonreturn valves operating in the slurry, and these can require considerable maintenance.

Centrifugal pumps are very widely used in the mineral industries.[1-4] They are available in a wide range of sizes, from very small to in excess of 1000 m³/sec. Centrifugal pumps in general have a lower efficiency than positive-displacement pumps. However, their operation is simple, no valves are necessary, and capital and maintenance costs are low.

The centrifugal pump consists of an impeller and a casing. A number of impeller designs are available.[3,4] Larger diameter impellers are used with slurries to reduce the speed required, hence to minimize wear. Flow passages through the impeller, and between the impeller and the volute casing, must be large enough to pass the largest particle and also to prevent excessive internal velocities (which cause rapid wear). For slurry use, it is common practice to line both impeller and casing with rubber; the lining is normally replaceable.

Piping and Valves. The general approach to the design of pipelines has been discussed in Section 4.4 and is described elsewhere.[1,3] With pipelines for slurry application,[5-7] abrasion becomes a major problem. Pipelines tend to exhibit wear along the lower pipe surface, at any change of direction, and following small interruptions to flow (such as welds). Thus bends or elbows should be of large radius, and lateral (45° angle) rather than tee fittings should be used.

Valves also have been described elsewhere.[1,3] When used with slurries, valves must be designed for abrasive service. Those commonly used with liquids, such as globe valves, are not suitable for slurries because of their restriction to flow, which would cause wear in the valve and downstream of the valve. The valve should provide a full line opening. Although a gate valve does this, it relies on machined metal surfaces for closure and so it also is not suitable. The most successful valve type (Fig. 21.2) consists of a rubber sleeve that can be compressed. For high pressure applications where such a system cannot be used, a modified ball or plug valve is a suitable alternative.

21.1.2. Pump Sizing

Positive-displacement pumps are best selected using manufacturers' literature. It is a question of selecting a pump with the required capacity, capable of delivering the required head, which is determined by the methods described in Sections 4.4 and 21.3.

Centrifugal pump selection is more complex.[8] Typical pump operating characteristics are shown in Fig. 21.3. For slurry use, the pump performance curves change as indicated in Fig. 21.4. Stable operation of a centrifugal pump is possible only if there is a single value of flow rate (capacity or discharge) for each value of head, as is the case in Figs. 21.3 and 21.4.

If a centrifugal pump is operated at a flow rate approaching its maximum capacity, the pressure at the impeller entrance or at the vane tips can fall below the vapor pressure of the liquid being pumped. Bubbles of vapor may form, then move to a region of high pressure and collapse. This phenomenon[7] known as *cavitation*, can occur at such a high speed that pitting of the impeller results. Cavitation can of course be eliminated by reducing the pumping rate; it is, however, better prevented at the design stage.

For an incompressible fluid, such as a liquid or slurry, the general energy balance can be written for a unit mass as[3]:

$$\Delta(\tfrac{1}{2}\bar{v}^2) + g\Delta H + \frac{1}{\rho_f}\int_{P_1}^{P_2} dP + \frac{1}{\rho_f}\Sigma P = -W_p \quad (21.1)$$

(ΔH = difference in height; ΣP = friction loss in pipe; W_p = work done by system per unit mass; ρ_f = fluid density [liquid or slurry]). If a simple pumping system (Fig. 21.5) is now considered, the energy balance is

$$\tfrac{1}{2}\bar{v}_1^2 + gH_1 + \frac{1}{\rho_f}P_1 = \tfrac{1}{2}\bar{v}_2^2 + gH_2 + \frac{1}{\rho_f}P_2 + \Sigma P \quad (21.2)$$

($\bar{v}_1$, H_1, P_1 = properties at point 1; $\bar{v}_2$, H_2, P_2 = properties at point 2). If the datum plane is considered at point 2, then $H_2 = 0$, and if $\bar{v}_1$ is negligible compared to $\bar{v}_2$, the total head at point 2 is

$$\tfrac{1}{2}\bar{v}_2^2 + \frac{1}{\rho_f}P_2 = gH_1 + \frac{1}{\rho_f}P_1 - \Sigma P \quad (21.3)$$

Thus the pressure at the suction inlet is

$$P_2 = \rho_f g H_1 + P_1 - \Sigma P - \tfrac{1}{2}\rho_f \bar{v}_2^2 \qquad (21.4)$$

The *net positive suction head* is the difference between the head at the suction inlet and the vapor pressure of the liquid at the impeller entrance. Thus from Eq. 21.3

$$\begin{aligned}P_{nps} &= (\tfrac{1}{2}\rho_f \bar{v}_2^2 + P_2) - P_v \\ &= (\rho_f g H_1 + P_1 - \rho_f \Sigma P) - P_v\end{aligned} \qquad (21.5)$$

(P_{nps} = net positive suction head; P_v = vapor pressure of liquid). The pressure at the entrance to the impeller will be less than that at the suction inlet, and the pressure difference can be assumed to be related to the velocity at the inlet (point 3 in Fig. 21.5) by[3]

$$P_2 - P_3 = K_1(\tfrac{1}{2}\rho_f \bar{v}_3^2) \qquad (21.6)$$

(K_1 = constant, characteristic of the pump design).

Cavitation is likely to occur if the total head at the impeller entrance is less than or equal to the vapor pressure, that is, $P_3 \leq P_v$. To prevent cavitation, the net positive suction head must be kept greater than a value that depends on the velocity at the impeller entrance and the pump design (K_1). This value is provided by pump manufacturers, as can be seen in Fig. 21.6.

Manufacturers' data on pumps generally are provided for the pumping of liquids, and so must be adapted for use with slurries. The procedure for adaption is empirical,[2] utilizing a slurry pump factor (Fig. 21.7) as follows:

$$\mathfrak{R}_{sp} = 1 - K_{sp}\mathfrak{F}_V \qquad (21.7)$$

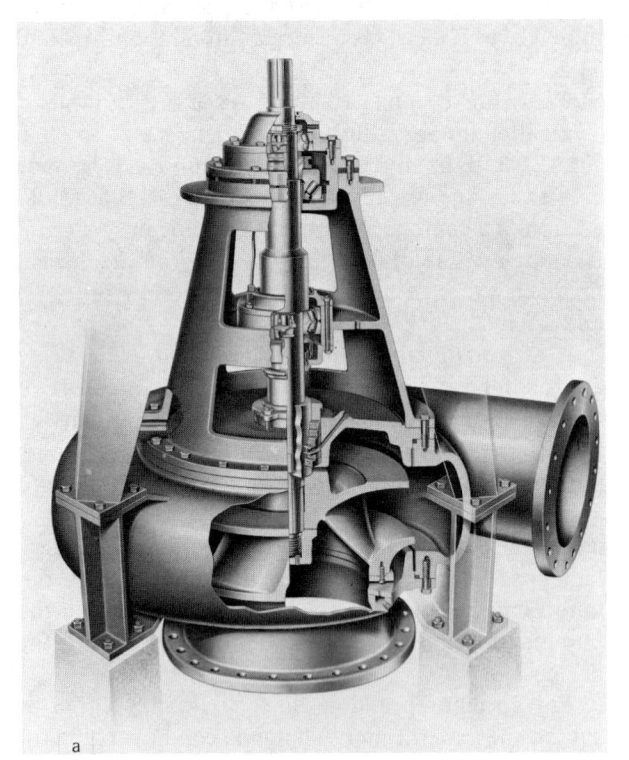

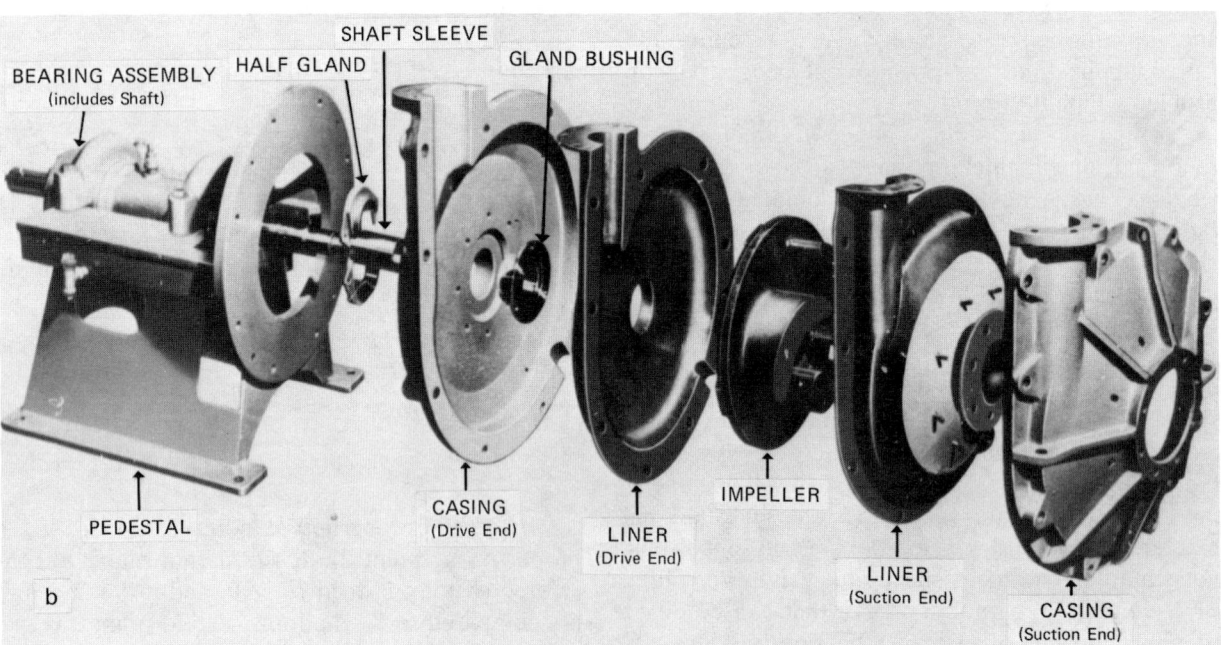

Figure 21.1. (*a*) Cutaway of a vertical centrifugal pump. (Courtesy Morris Pumps, Inc.) (*b*) Expanded view of a horizontal centrifugal pump. (Courtesy The Galigher Co.) (*c*) A diaphragm positive-displacement pump. (Courtesy Dorr-Oliver Inc.)

[$\tilde{\phi}_v$ = volume fraction of solids in slurry; K_{sp} = slurry pump factor; $\Re_{sp}$ = head (slurry)/head (water) and also = pump efficiency (slurry)/pump efficiency (water)]. This approach is used in the following example.

Example 21.1. Select a pump to move 200 m³/hr of tailings slurry at a pulp density of 15% by volume (solids density = 2700 kg/m³) against a total head of 22 m of slurry (average particle size = 150 μm).

Solution

$$\text{flow rate} = 200 \text{ m}^3/\text{hr}$$

From Fig. 21.7:

$$K_{sp} = 0.35$$

From Eq. 21.7:

$$\Re_{sp} = 1 - 0.35 \times 0.15$$
$$= 1 - 0.0525$$
$$= 0.95$$

$$\text{head of slurry} = 22 \text{ m slurry}$$

$$\text{equivalent head of water} = \frac{22}{0.95}$$
$$= 23.2 \text{ m}$$

From a pump selection chart (e.g., Appendix H), a pump is selected (type 6/4 DAM). The performance curves for this pump (Fig. 21.6) are now used for determining pump efficiency and speed.

$$\text{efficiency (water)} = 64\%$$
$$\text{pump speed} = 1025 \text{ rpm}$$
$$\text{efficiency (slurry)} = 64 \times \Re_{sp}$$
$$= 64 \times 0.95$$
$$= 61\%$$

power

$$= \frac{200 \text{ m}^3}{\text{min}} \left| \frac{1 \text{ min}}{60 \text{ sec}} \right| \frac{23.2 \text{ m water}}{1 \text{ m water}} \left| \frac{977.1 \text{ N/m}^2}{} \right| \frac{1 \text{ kW}}{1000 \text{ W}}$$
$$= 75.6 \text{ kW}$$

$$\therefore \quad \text{pump power} = \frac{75.6}{0.61}$$
$$= 124 \text{ kW}$$

21.2. AGITATED TANKS

Tanks are used in a variety of locations in a mineral processing operation. However, although surge capacity within a plant would often be desirable, tanks are rarely used for this purpose because of the energy required to maintain particles in suspension. Except for sumps used for the collection of

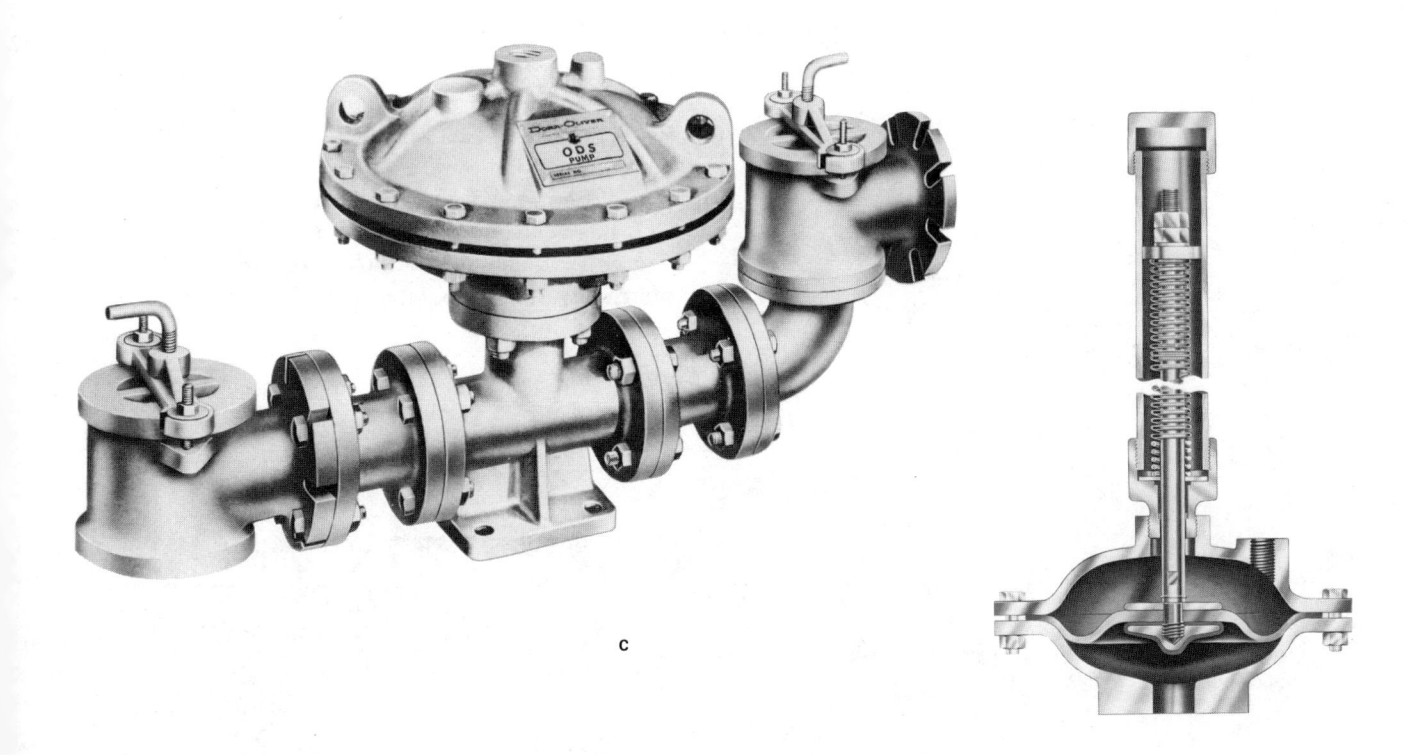

c

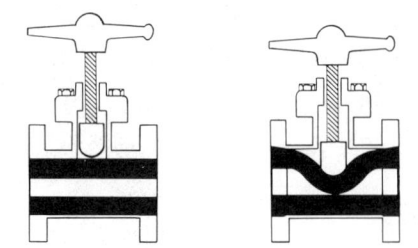

Figure 21.2. Valve for use with mineral slurries.

process streams ahead of pumps, most tanks in a plant require agitation.

21.2.1. Equipment

Agitated tanks are available in a wide range of sizes and designs. Figure 21.8 shows agitated tanks designed for specific requirements: a conditioning tank for flotation (Fig. 21.8a) and a scrubbing tank where vigorous agitation is required (Fig. 21.8b).

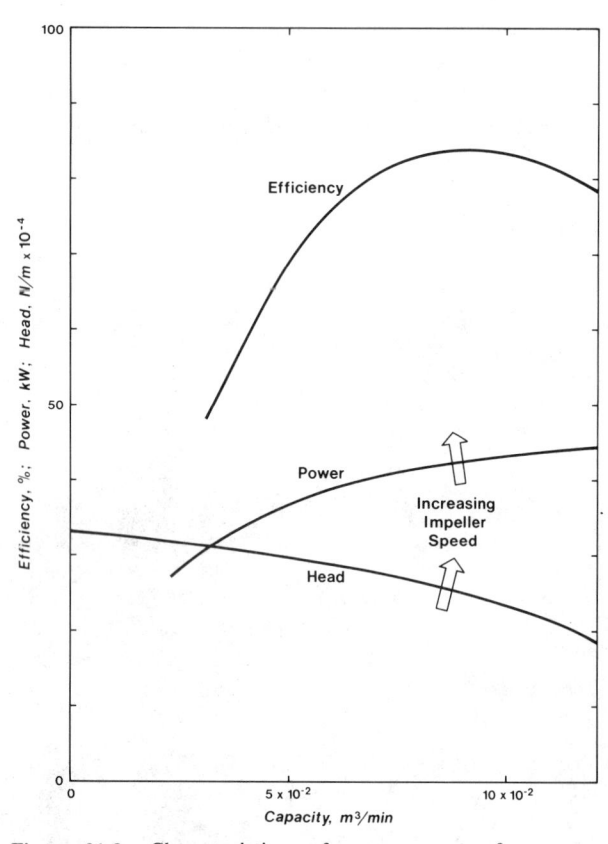

Figure 21.3. Characteristic performance curves for a centrifugal pump (water only).

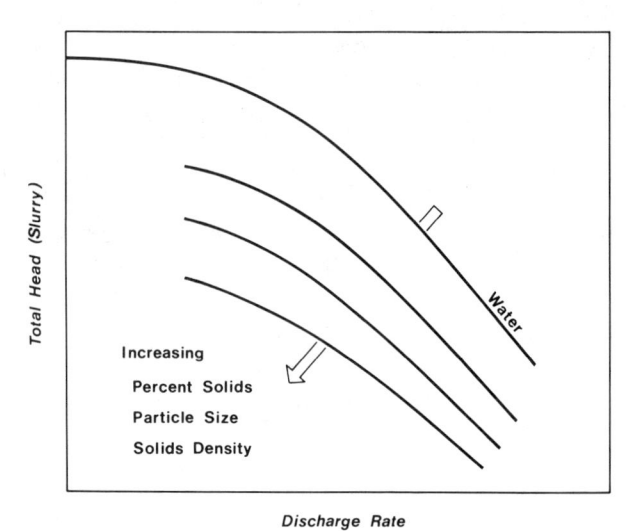

Figure 21.4. Effect of slurry on pump performance curves. (After McElvain and Cave.[19])

21.2.2. Agitated Tank Design

The principles of agitated tank design[9,10] are discussed in Section 4.8; however, as mentioned in that section, the design approach discussed is aimed at complete dispersion. The agitator power in this case is calculated by[9]:

$$\frac{\mathcal{P}}{V_T} = 0.092 \; g \; v_\infty \left(\frac{D_T}{D_a}\right) \left(\frac{1-\varepsilon}{\varepsilon}\right)^{1/2} \exp\left(\frac{5.3 \; H_a}{D_T}\right) (\rho_s - \rho_l) \quad (4.91)$$

Most agitation tanks in mineral processing operations are used to maintain particles in suspension while chemical reactions take place; therefore complete dispersion is rarely required, and an alternative design approach is preferable.

It has been shown experimentally[11] that the same degree of dispersion of solids occurs in two geometrically similar agitated tanks if the power input

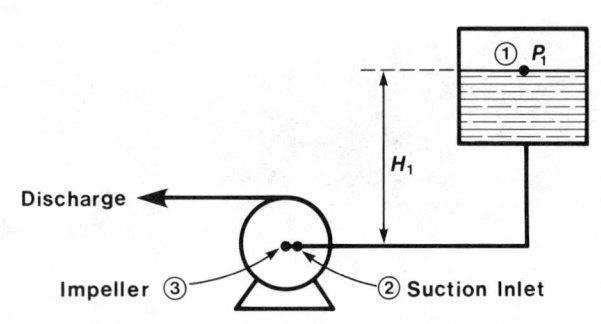

Figure 21.5. A simple pumping system.

Agitated Tanks

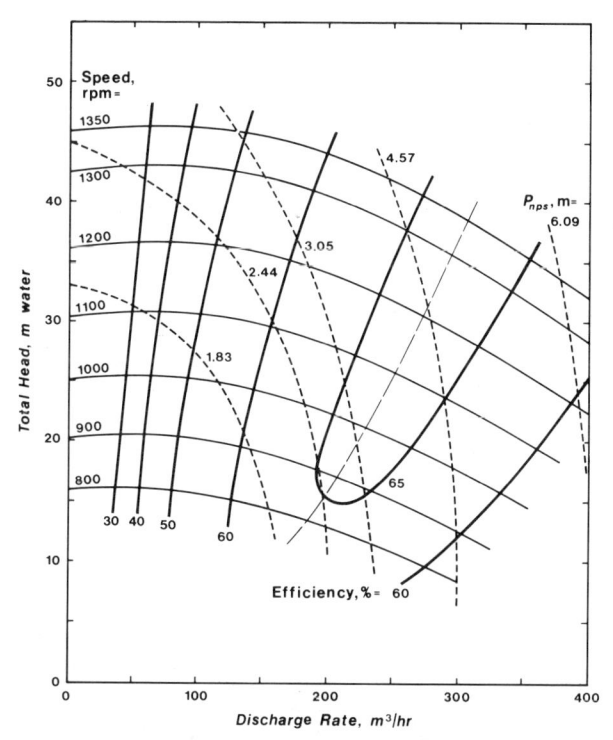

Figure 21.6. Manufacturer's data for pump performance. (Impeller diameter, 36.5 cm; maximum power, 60 kW; maximum particle size, 33 mm.)

per unit volume is maintained at the same level. Skelland[12] shows that this results in

$$\frac{\mathcal{N}_1}{\mathcal{N}_2} = \left(\frac{D_{a2}}{D_{a1}}\right)^{2/3} \qquad (21.8)$$

(D_{a1}, D_{a2} = diameter of agitator in tanks 1 and 2, respectively; $\mathcal{N}_1$, $\mathcal{N}_2$ = impeller rotational speed in

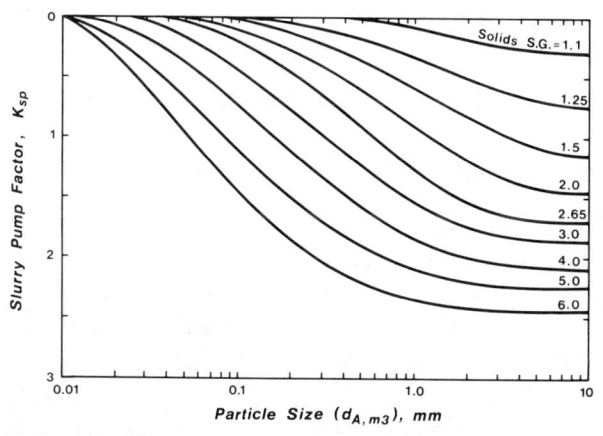

Figure 21.7. Slurry pump factor chart (for adaptation of manufacturer's head and efficiency data to slurry application). (After McElvain and Cave.[19])

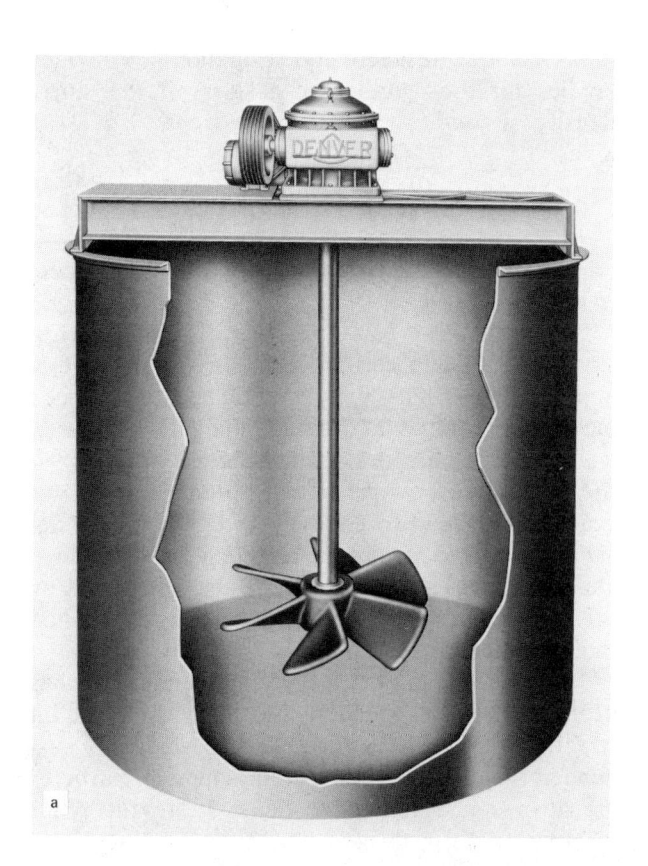

Figure 21.8. Agitated tanks for mineral applications. (a) A typical flotation conditioning tank. (b) A two-stage scrubbing tank. (Courtesy Joy Manufacturing Co.)

tanks 1 and 2, respectively). Equation 4.86, which applies for the agitation of a slurry if the slurry density is used,[12] when combined with Eq. 21.8, gives:

$$\frac{\mathscr{P}_1}{\mathscr{P}_2} = \left(\frac{D_{a1}}{D_{a2}}\right)^5 \left(\frac{\mathscr{N}_1}{\mathscr{N}_2}\right)^3 = \left(\frac{D_{a1}}{D_{a2}}\right)^3 \quad (21.9)$$

($\mathscr{P}_1, \mathscr{P}_2$ = power required in tanks 1 and 2, respectively).

Although these equations (Eqs. 21.8 and 21.9) may be of use in certain situations, their general applicability is limited by the high agitator speed inevitably required in any small diameter test tank.

As was stated with regard to the use of Eq. 4.91, the use of the power per unit volume criteria for scale-up can lead to excessively conservative designs. Energy is required to maintain the solid particles in suspension, but energy is also consumed by viscous dissipation, and it is this viscous dissipation that is dependent on tank size.[6]

In another approach, a "scale of agitation" has been proposed,[13] which ranges from the point at which all particles are moving and periodically suspended (level 1), through a point where all particles are always just suspended (level 3), to the point where a uniform dispersion is reached (level 10). The midpoint in the scale (level 5) corresponds approximately to a point at which all particles are completely suspended and slurry uniformity has been achieved in the lower half of the agitated volume. In this approach, an empirical chart (Fig. 21.9) has been developed[13] to relate the values of this scale of agitation (obtained by visual observa-

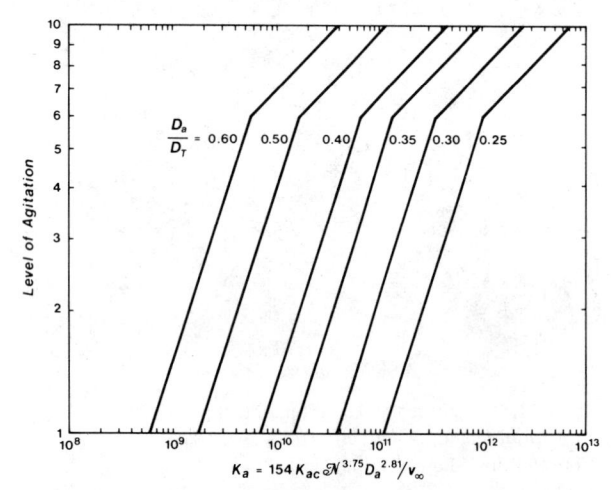

Figure 21.9. Chart relating observed "level of agitation" to slurry properties and tank design parameters. (After Gates et al.[13])

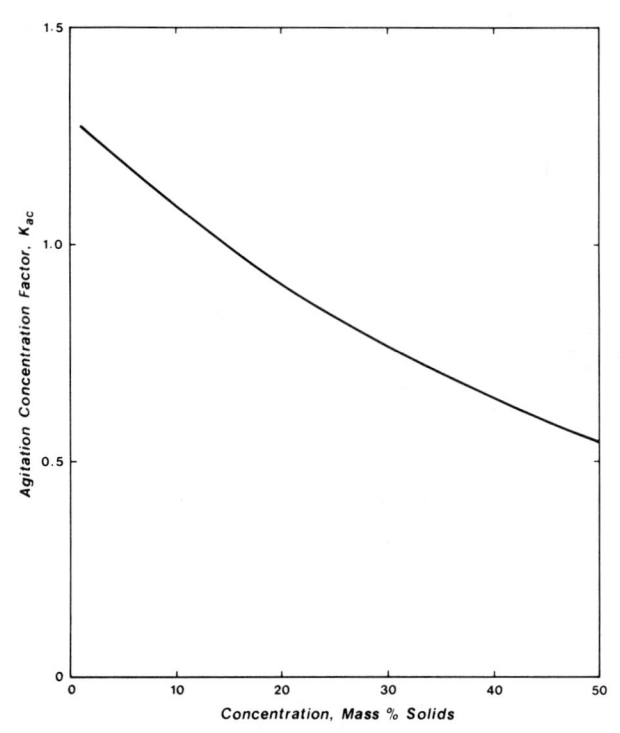

Figure 21.10. Agitation concentration factor for use in Eq. 21.10. (After Gates et al.[13])

tions) to the design parameters of the tank and the properties of the slurry, represented by:

$$K_a = \frac{154 K_{ac} \mathscr{N}^{3.75} D_a^{2.81}}{v_\infty} \quad (21.10)$$

(K_a = agitation factor defined by Eq. 21.10; K_{ac} = agitation concentration factor given by Fig. 21.10; $\mathscr{N}$ = rotational speed [rpm]; v_∞ = [turbulent] terminal settling velocity of particles). The agitation concentration factor (Fig. 21.10) indicates the increasing power requirement with higher solids concentrations.[13]

Example 21.2. The hydrocyclone overflow from Example 1.5 is to be conditioned in a tank before flotation (Example 16.1). The conditioning tank is baffled, 3.0 m in diameter, and the slurry depth is 3 m. The impeller is pitched blade, 90 cm in diameter, installed 45 cm above the bottom of the tank. What agitator power is required? (From Example 1.5, ρ_s = 3145 kg/m³, C_V = 15.3%. From Example 10.1, assume $d \simeq 150$ μm, and from Example 4.2, v_∞ = 0.0193 m/sec).

Solution. The power can be determined from Eq. 4.91. D_T = 3.0 m; D_a = 0.9 m; H_a = 0.45 m; C_V = 0.153; ε = 0.847; ρ_s = 3145 kg/m³.

From Eq. 4.33, calculate v_∞ (turbulent):

$$v_\infty = \left[\frac{4}{3} \times \frac{9.81 \times 150 \times 10^{-6}}{0.44}\left(\frac{3145 - 1000}{1000}\right)\right]^{1/2}$$
$$= 0.098 \text{ m/sec}$$

From Eq. 4.91,

$$\frac{\mathscr{P}}{V} = 0.092 \times 9.81 \times 0.098 \times \frac{3}{0.9} \times \left(\frac{0.153}{0.845}\right)^{1/2}$$
$$\exp\left(\frac{5.3 \times 0.45}{3}\right)(3145 - 1000)$$
$$= 595$$

$$\therefore \mathscr{P} = 595 \times 3 \times \pi \times (1.5)^2 \text{ W}$$
$$= 12.6 \text{ kW}$$

Example 21.3. Repeat Example 21.2 using the "scale of agitation" method.

Solution. Assume a level of agitation scale value of 5 gives acceptable agitation.

i. For $D_a/D_T = 0.9/3 = 0.3$, from Fig. 21.9, $K_a = 2.8 \times 10^{11}$

ii. A volume concentration of 15.3% is equivalent to a mass concentration of 36.2%. Thus from Fig. 21.10, $K_{ac} = 0.69$

iii. From Eq. 21.20,

$$2.8 \times 10^{11} = \frac{154 \times 0.69 \times \mathscr{N}^{3.75}(0.9)^{2.81}}{0.0193}$$

$$\therefore \mathscr{N}^{3.75} = 6.84 \times 10^7$$

$$\therefore N = 123 \text{ rpm}$$

iv. From Eq. 4.85,

$$\mathbf{Re}_a = \frac{(0.9)^2 \times (123/60) \times 1000}{10^{-3}}$$
$$= 1.7 \times 10^6$$

v. From Fig. 4.18, Po = 1.1

vi. From Eq. 4.86,

$$1.1 = \frac{\mathscr{P}}{1000 \times (123/60)^3 \times (0.9)^5}$$

$$\therefore \mathscr{P} = 5596 \text{ W}$$
$$= 5.6 \text{ kW}$$

Comment. It can be seen that the lower degree of agitation appreciably reduces the power requirement. A scale value of 10 gives a power requirement of approximately 30 kW.

Impellers. A wide range of impeller designs are used for agitation. In most common use are turbine impellers, either flat blade or pitched blade (Fig. 4.18). With a flat-blade turbine impeller, a radial flow pattern is produced; that is, the primary direction of flow of fluid from the impeller is in the radial direction. The pitched-blade turbine, however, produces an axial flow pattern, which is clearly of considerable assistance in keeping solids suspended. An indication of the difference in performance of the two turbines can be seen in Fig. 4.18. It must be kept in mind, however, that the application of this figure is restricted to the turbine impellers with the specifications listed. Similar data for other turbine designs are available elsewhere.[12,14]

For solids suspension, impellers are located at a height of one-fourth the slurry depth above the tank bottom.

Baffles. The effect of baffles in agitated tanks is also seen in Fig. 4.18.

For solids suspension, frequent practice is to install four evenly spaced vertical baffles, with a width equal to $D_T/12$.

21.3. SLURRY TRANSPORTATION

The subject of slurry flow in pipes has been introduced in Section 4.8.2, and is treated in detail elsewhere.[6,7] Slurry flow takes place in numerous pipelines in any wet mineral processing plant, and in tailings disposal. Also, longer distance pipelines are being used for concentrate transportation, in particular, for iron and copper concentrates, and for coal.[6,7,15–17]

Flow in horizontal pipes can be homogeneous, that is, the particles essentially dispersed uniformly across the pipe, or heterogeneous, with a significant particle concentration gradient across the pipe. Homogeneous suspensions occur only with very fine particles (such as clays or fine limestone). Although the flow of these suspensions may be laminar, turbulence is normally required to maintain the particles in suspension. Homogeneous flow, then, is the flow of homogeneous suspensions, and also that of normally heterogeneous suspensions at high velocities (sometimes referred to as pseudohomogeneous flow).

Most mineral slurries, however, consist of heterogeneous suspensions. Of necessity, the flow of all heterogeneous suspensions must be turbulent. Het-

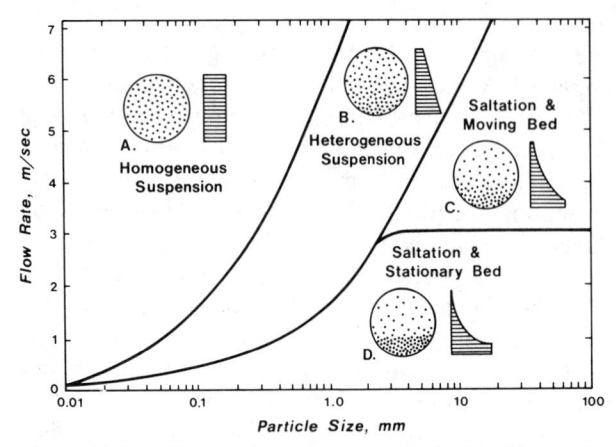

Figure 21.11. Flow regimes in a slurry pipeline (drawn for specific solids density, concentration, and pipe diameter). (After Newitt et al.[20])

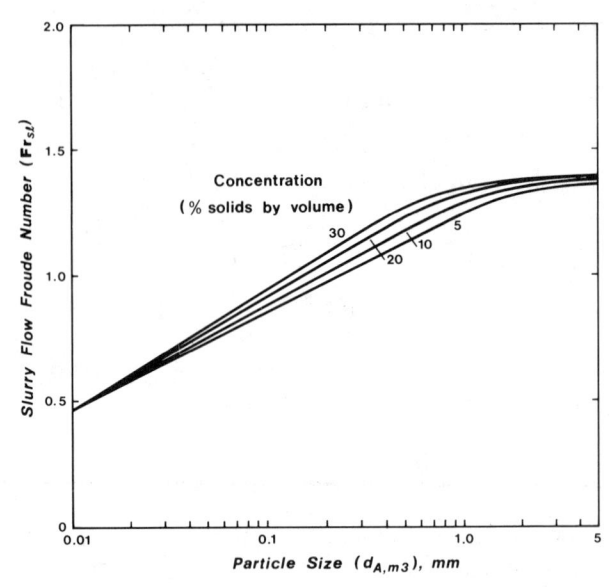

Figure 21.12. Correlation of Froude number for slurry flow with particle size. (After McElvain and Cave.[19])

erogeneous flow can either be full pipe flow of a heterogeneous suspension, or partial pipe flow above a moving or stationary bed of the coarser particles from the slurry. In the latter two cases, the larger particles effectively "bounce" along the pipe, alternately settling and being lifted back into the flowing stream (saltation). These flow regimes are shown in Fig. 21.11.

To determine the minimum required transportation velocity, Eq. 4.93 could be used (Sections 4.4 and 4.8.2). However, for heterogeneous flow this equation becomes similar to the equation of Durand (Eq. 4.92), which can be written as

$$v_d = \mathbf{Fr}_{sl} \left[2gD \left(\frac{\rho_s - \rho_l}{\rho_l} \right) \right]^{1/2} \quad (21.11)$$

($\mathbf{Fr}_{sl}$ = Froude number for slurry; v_d = deposition velocity, the minimum velocity to prevent deposition; ρ_s, ρ_f = density of solid and liquid, respectively).

With heterogeneous flow, the friction factor is increased and can be obtained by

$$\frac{\mathbf{f}_{sl} - \mathbf{f}_p}{\mathbf{f}_p(1 - \varepsilon)} = 82 \left[\frac{\bar{v}^2}{gD} \cdot \left(\frac{\rho_l}{\rho_s - \rho_l} \right) \sqrt{\mathbf{f}_d} \right]^{-3/2} \quad (21.12)$$

($\mathbf{f}_{sl}$ = friction factor for slurry). This shows that the increase in friction is due to $\mathbf{Fr}$, $\mathbf{f}_d$, the concentration, and the density difference. Durand[18] correlated the slurry Froude number with particle size and concentration of solids by volume, and this correlation is widely used.[6] However, this correlation was developed for particles of uniform size, and although applicable for closely sized material, it has been found to result in excessively high values of the calculated deposition velocity. Other correlations[19] have subsequently been developed for a distribution of sizes (Fig. 21.12).

To determine the pressure loss due to friction in the pipe, the method presented in Section 4.4 is used. **Re** is calculated (for heterogeneous suspensions, viscosity and density of the liquid are used), and the friction factor is determined from Fig. 4.7.

TABLE 21.1a: EQUIVALENT PIPE LENGTHS OF VALVES AND FITTINGS

Item	Equivalent Lengths in Pipe Diameters (L/D)
Gate Valves (Conventional)	
Fully open	13
¾ open	35
½ open	160
¼ open	900
Globe Valves (Conventional)	
Fully open	340-450
Elbows	
90° standard	30
90° large radius	20
45° standard	16
Tees	
Flow straight through	20
Flow through branch	60

TABLE 21.1b: PRESSURE LOSS AT PIPE ENTRANCE AND EXIT

Item	Equivalent Velocity Head ($2\rho \bar{v}^2$)
Pipe Exit	1.0
Pipe Entrance	
Well rounded	0.04
Slightly rounded	0.23
Sharp edged	0.50

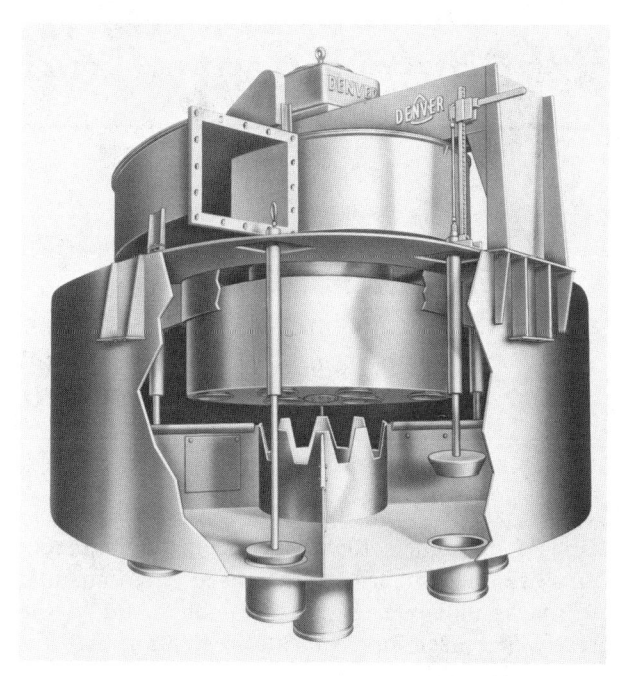

Figure 21.13. Cutaway of a slurry distributor. (Courtesy Joy Manufacturing Co.)

The total effective length of pipe is determined by

$$\Sigma L = L + L_e \quad (21.13)$$

(L = pipe length; L_e = equivalent pipe length of values and fittings [Table 21.1a]; ΣL = effective pipe length). The pressure loss in the pipe due to friction is then calculated from (Eq. 4.18):

$$\Delta P = \mathbf{f}_{sl} \left(\frac{\Sigma L}{D}\right)\left(2\rho_l \bar{v}^2\right) \quad (21.14)$$

(ΔP = pressure loss due to pipe friction). If the pipe enters or exits from any tanks in the circuit, the additional pressure loss should be included (Table 21.1b).

21.3.1. Slurry Distribution

There is often a need to split a slurry stream into a number of equal streams. This need is most frequently seen when several flotation rougher banks are fed from a single grinding line.

Normal practice is to use some form of slurry distributor (Fig. 21.13); even with such a device, obtaining equal flow rates remains a problem.

REFERENCES

1. R. H. Perry and C. H. Chilton (Eds.), *Chemical Engineers' Handbook*, 5th ed., McGraw-Hill (1973).
2. T. G. Hicks and T. W. Edwards, *Pump Application Engineering*, McGraw-Hill (1971).
3. A. S. Foust et al., *Principles of Unit Operations*, 2nd ed., Wiley (1980).
4. J. E. Fatzinger, "The Design and Application of a Centrifugal Slurry Pump," in *Mineral Processing Plant Design*, A. L. Mular and R. B. Bhappu (Eds.), pp. 665–678 AIME/SME (1978).
5. J. C. Loretto and E. T. Laker, "Process Piping and Slurry Transportation," in *Mineral Processing and Plant Design*, A. L. Mular and R. B. Bhappu (Eds.), pp. 679–702, AIME/SME (1978).
6. E. J. Wasp, J. P. Kenny, and R. L. Gandhi, *Solid-Liquid Flow—Slurry Pipeline Transportation*, Trans Tech Publications (1977).
7. *The Transportation of Solids in Steel Pipelines*, Colorado School of Mines Research Foundation (1963).
8. R. F. Neerken, "Selecting the Right Pump," *Chem. Eng.*, Deskbook Issue, **85**, 87–98 (Apr. 3, 1978).
9. J. Weisman and L. E. Efferding, "Suspension of Slurries by Mechanical Mixers," *AIChE J.*, **6**, 419–426 (1960).
10. T. N. Zwietering, "Suspending of Solid Particles in Liquid by Agitators," *Chem. Eng Sci.*, **8**, 244–253 (1958).
11. R. E. Johnstone and M. W. Thring, *Pilot Plants, Models, and Scale-up Methods in Chemical Engineering*, McGraw-Hill (1957).
12. A. H. Skelland, *Non-Newtonian Flow and Heat Transfer*, Wiley (1967).
13. L. E. Gates, J. R. Morton, and P. L. Fondy, "Selecting Agitator Systems to Suspend Solids in Liquids," *Chem. Eng.*, **83**, 144–150 (May 24, 1976).
14. S. Nagata, *Mixing: Principles and Applications*, Kodansha (1975).
15. J. M. Link, G. A. Ponska, and N. W. Kirshenbaum, "Mineral Slurry Transport—An Update," in *Fine Particle Processing*, P. Somasundaran (Ed.), AIME/SME (1980).
16. R. M. Turian and T. F. Yuan, "Flow of Slurries in Pipelines," *AIChE J.*, **23**, 232–243 (1977).
17. W. F. McDermott, "Savage River Mines—The World's First Long Distance Iron Ore Slurry Pipeline," *CIM Bull.*, **73**, 340–345 (December 1970).
18. R. Durand, "The Hydraulic Transportation of Coal and Other Minerals in Pipes," Colloq. of Nat. Coal Board, London (November 1952).
19. R. E. McElvain and I. Cave, "Transportation of Tailings," *World Mining Tailings Symposium*, (1972).
20. D. M. Newitt et al., "Hydraulic Conveying of Solids in Horizontal Pipes," Trans. IChemE, **33**, 93–102 (1955).

Chapter Twenty-Two

Tailings Disposal

Sturgeon's Law: "*Ninety percent of everything is crud.*"

The disposal of tailings from milling operations is a major task. The problems involved vary widely from one ore to another, and from one operation to another. In most cases the major fraction of the ore mined eventually becomes tailings. For example, a porphyry copper ore may have a copper grade of 0.5%; if a concentrate of about 30% Cu is obtained in the mill, over 98% of the mined ore ends up as tailings. This material has no present commercial value, and must be disposed of as economically as possible.

Such tailings do not include mine overburden, which may itself be an even larger volume.

In early mining operations, tailings were frequently dumped in nearby lakes or rivers. This practice has in some instances continued until recent times.[1-3] However, the two most common practices have been:

1. To size the tailings, and use the coarse fraction (sands) for construction of a *tailings dam*, while placing the fine fraction (slimes) in the resulting *tailings pond*.
2. To size the tailings, and use the coarse fraction for *backfill* underground (possibly mixed with cement), while sending the fine fraction to a surface tailings pond.

Increases in tonnage mined, due to greater demand and lower grades, have resulted in more tailings. The more extensive grinding required to liberate values in low grade, fine grained ores results in a tailings product that is less satisfactory for both backfill and tailings dam construction. These factors have resulted in changes in the techniques of tailings disposal. However, the major changes in the disposal of tailings that have occurred in recent years are due primarily to increased environmental awareness and interest by the public, and to the greatly increased governmental regulation that has resulted.

Two major aspects are of importance with tailings disposal: safety and economics. In turn, two areas of safety are of concern. First, the dam structure must be designed and constructed so that it will not fail either during construction and use, or in later years when it is no longer in use; this is vital, since failure would result in the release of the vast quantities of unconsolidated slimes from the tailings pond. The second area of safety concerns pollution: pollution of natural lakes or rivers by suspended particles, organics, or inorganic ions must be prevented.

The disposal of tailings must, however, be carried out at the lowest cost compatible with safety. This requires the selection of a tailings pond of sufficient capacity, and in close proximity to the mine.

It should also be pointed out that increasingly, reclamation and revegetation involve additional costs that must be included in tailings disposal.

22.1. TAILINGS PONDS AND DAMS

Although tailings ponds are basically similar, there are notable exceptions. With the tailings from uranium ores, and from ores such as gold or silver leached with cyanide solutions, special precautions are required to prevent groundwater contamination—for example, the lining of the complete tailings pond area with polythene sheeting and consolidated clay.

22.1.1. Tailings Ponds

A site selected for a tailings pond must meet certain criteria: it must be within an economical distance of the mill, it must provide for adequate capacity, the subsurface geology and hydrology must meet certain minimum specifications, and it must reflect compliance with all pertaining environmental regulations. The site selection process is long and complex, and although the mill engineer is vitally involved in the selection, he is but one of a large team.

Extensive reports on tailings pond site selection are available[4-6] and therefore are not discussed further here.

22.1.2. Tailings Dam Design

When tailings dams are constructed using the coarser fraction of tailings, normal practice is to have a series of hydrocyclones spaced along the top of the tailings dam, with the coarse fraction discharging directly onto the dam, and the fine fraction discharging into the pond behind the dam. Periodically the hydrocyclones are raised or moved to another part of the dam. The dam thus rises ahead of the pond level. A limited secondary sizing is also taking place; the fine fraction from the hydrocyclone enters the tailings pond, where the coarser particles of this fraction settle out while the finer particles remain suspended and flow further from the dam. Even so, the coarser fraction contains fines which reduce consolidation.

In an attempt to minimize the cost of tailings disposal, a once widely used method for constructing tailings dams was the *upstream* method (Fig. 22.1a). As the height of the tailings dam rose, each successive dyke moved further upstream, and so overlay an unstable bed of unconsolidated tailings. This resulted in a phreatic line (surface of water saturation) close to the dam outer surface. Any change that resulted in saturation of the lower dykes (e.g., a rise in the pond level, or torrential rain) could quickly lead to dam failure. This type of dam no longer meets the requirements for slope stability, seepage control (i.e., internal drainage), or resistance to seismic shocks.

Downstream methods of dam construction (Fig. 22.1b) are an obvious improvement. With this construction, underdrains can be constructed below the dam, and each successive layer of coarse particles from the tailings is deposited on a base of coarse, free-draining particles. The location of the phreatic line with the downstream method (Fig. 22.1b) indicates the improved stability of this design.

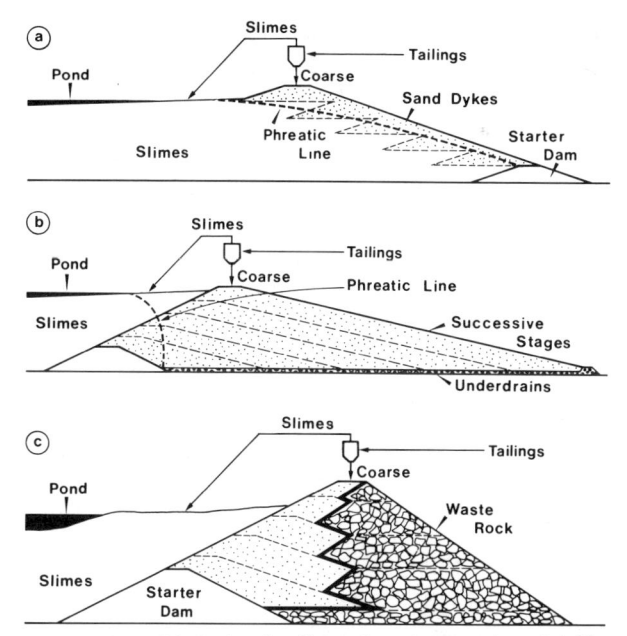

Figure 22.1. Methods of tailings dam construction. (*a*) Upstream method. (*b*) Downstream method. (*c*) Mine waste rock dam construction. (After Klohn.[7])

A major disadvantage of the downstream method is the large volume of coarse particles required. Where fine grinding is required in the mill, sufficient coarse material may not be available. In this situation, an acceptable alternative is to use mine waste rock for dam construction (Fig. 22.1c). Many methods are used, ranging from a dam utilizing only a small fraction of waste rock, to one constructed primarily of waste rock. Further consideration of such designs is beyond the scope of this book.[4-7]

22.1.3. Difficult to Settle Solids

In a number of situations (the Florida phosphate tailings being the most notable example[8]), problems can be encountered when particles are too fine to settle to an acceptable level of consolidation. Satisfactory solutions to these problems have yet to be found, although a number of approaches have been attempted.[9-11]

22.2. TAILINGS WATER

In mineral processing plants today, the trend is clearly toward maximum recycle of process water.

In dry areas, this has always been the practice for economic reasons, but today it is not restricted to these areas: government regulation of plant effluent streams effectively mandates the recycle of all possible process water.

The water balance about a typical mineral processing plant is shown in Fig. 22.2. Clearly if more water enters the system than is lost by evaporation, a tailings water discharge will be required. For the most part, water entering the system can be controlled. Water enters with the ore, and though difficult to regulate, it is a comparatively small amount. Water may also enter as groundwater flows into the tailings pond, but this can be almost eliminated by the use of diversion channels around the tailings area.[4,5]

22.2.1. Water Recycle

The composition of the water used in a mineral processing operation is an important factor. This applies primarily to processes such as flotation, flocculation, and agglomeration, but composition can also be significant in gravity and magnetic separations, as well as in grinding, classification, thickening, and filtration. Essentially, water composition can be relevant whenever the chemical nature of the mineral surface is important. This subject has been considered in earlier chapters, but some further points are worthy of consideration here.

The influence of water supply in flotation is a suitable example. External water entering a plant always contains a variety of ions, some of which may be directly involved in the solution chemistry of a flotation process. Water hardness (Ca^{2+} and Mg^{2+}) and pH must always be considered (Section 16.2); other ions can become important in particular flotation processes. External water, however, is generally free from organic compounds such as those that cause frothing or flocculation, or will adsorb on mineral surfaces.

Water recycled from tailings provides a different situation (Table 22.1). This water can be recycled from a tailings thickener (added to flow sheet to reduce the volume of tailings transported to the tailings pond), or it can be recycled from a decant point in the tailings pond itself. If the retention time in the tailings pond is sufficient (and at several weeks, it generally is), organic reagents have in general either decomposed or been adsorbed. Likewise a significant fraction of the heavy metal cations will have

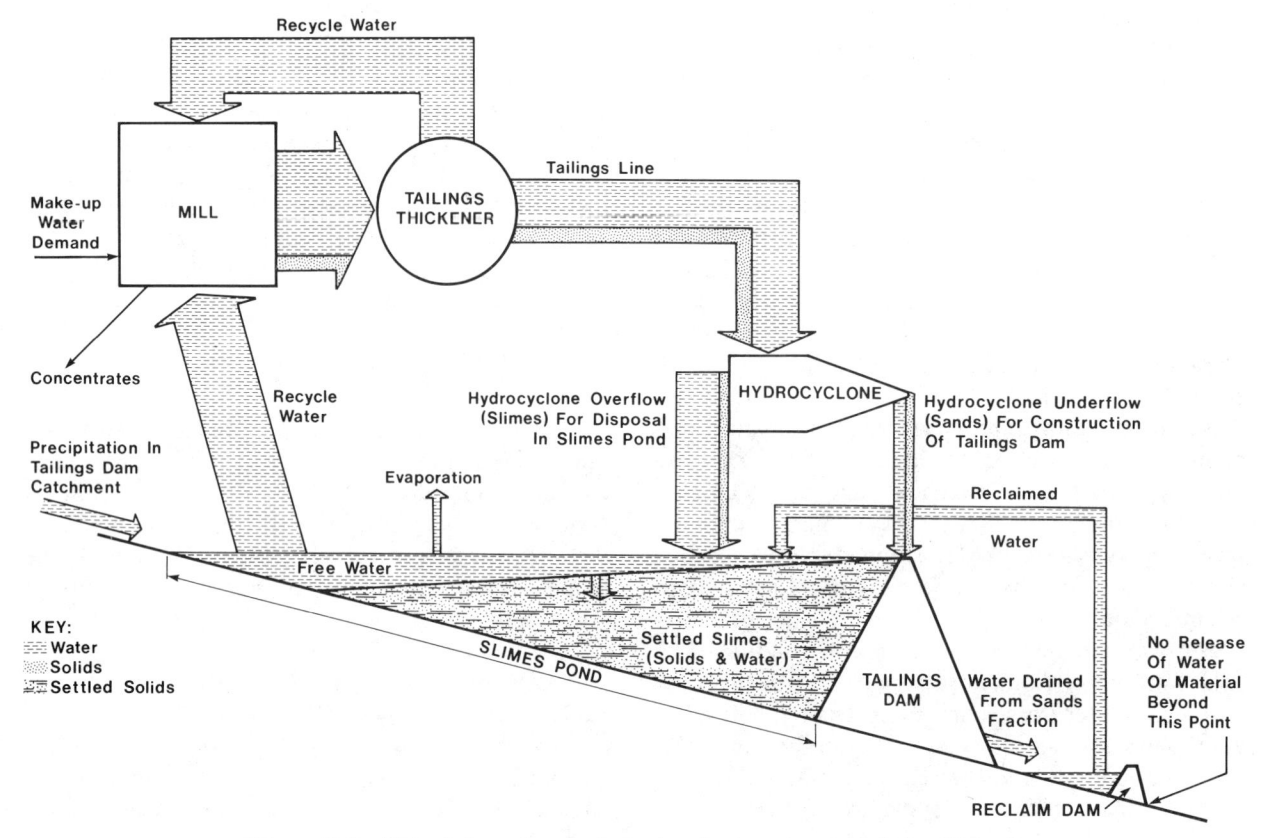

Figure 22.2. Water balance for a typical mineral processing plant. (After Klohn.[7])

TABLE 22.1: REAGENT LEVELS IN TAILINGS WATER

	Reagent	Normal Addition kg/t (mg/l)	Conc. in Tailings mg/l	Recommd. Acceptable Levels	
				Public Water Supply mg/l	Fresh Water Aquatic Life mg/l
FROTHERS	Alcohols	0.007-0.25			350-1700
	Pine Oil	0.025-0.75			10-100
	Cresylic Acid	0.025-0.90			1.5-10
	Surface Active Agents	0.02-0.2		<0.5	0.2-10
COLLECTORS	Amines	0.25-2.5 (15-150)			0.5 1000
	Fatty Acids	1.0-7.5 (60-450)			5
	Fuel Oil			none	0.5 x 96-hr LC50*
	Xanthates	0.1-0.5 (5-40)	0.1-2.0		10 75 (96-hr LC50)
	Dithio-phosphates	0.1-0.5 (5-40)			400-1000 (96-hr LC50)
MODIFIERS	Cu^{++} ($CuSO_4$)	0.5-12.5	0.06-50	1.0	0.1 x 96-hr LC50
	Zn^{++} ($ZnSO_4$)	0.25-7.5	0.01-9	5.0	0.005 x 96-hr LC50
	Cyanide (NaCN)	0.25-1.25	<0.01-0.03	0.2	0.005
	pH			5-9 (pH)	6-9 (pH)
	$NaSiO_3$	1.25-7.5		75	250
	Soluble Sulfides			none	0.002
	Total Dissolved Solids			5000	

* The 96-hour LC50 concentration is the lethal concentration that will kill 50% of the individuals in a population or organisms in 96 hours.

precipitated, although this depends significantly on the pH of the tailings water. However, continued recycle can result in the buildup of ions to equilibrium levels, and very stable organic ions have been found to cause problems in flotation.[12–14]

Although the quantity of water recycled from a concentrate thickener may be comparatively small, it can be particularly troublesome because it may have appreciably higher levels of chemicals.

A simple example will show the possible complications of water recycle. Consider a copper-zinc sulfide ore. The conventional flow sheet would require the flotation of the copper mineral first (with possibly CN^- added to prevent activation of the zinc mineral by Cu^{2+}), followed by flotation of the zinc mineral (with Cu^{2+} added as an activator). The final tailings thus would contain copper ions, and if this water were recycled, additional CN^- would have to be added to prevent these ions from activating the zinc mineral in the copper flotation circuit; however, this tailings water could be used to advantage if additional water were required in the zinc flotation stage.

In summary, the recycle of plant water can have the beneficial result of reducing reagent consumption if recycle is possible to an appropriate point in the plant circuit. On the other hand, it can have the effect of increasing reagent consumption if ions present in the water must be removed.

This brief discussion has not included the addition of water treatment steps in the recycle line, although such steps are clearly a technical possibility.

REFERENCES

1. E. J. Klohn and D. Dingeman, "Tailings Disposal System for Reserve Mining Company," in *Tailing Disposal Today*, Vol. 2, G. O. Argall (Ed.), pp. 178–209, Miller Freeman Publishing (1979).
2. Anon., "Reserve Mining: An Epic Battle Draws to a Close," *Environ. Sci. Technol.*, **11**, 948–950 (October 1977).
3. J. R. Oxberry, P. Doudoroff, and D. W. Anderson, "Potential Toxicity of Taconite Tailings to Aquatic Life in Lake Superior," *Water Pollut. Control Fed. J.*, **50**, 240–251 (1978).
4. *Tailings Disposal Today*, Vol. 1, Miller Freeman Publishing (1972).
5. G. O. Argall (Ed.), *Tailings Disposal Today*, Vol. 2, Miller Freeman Publishing (1979).
6. J. D. Jones, "Design and Construction of Tailings Ponds and Reclamation Facilities—Case Histories," in *Mineral Processing Plant Design*, A. L. Mular and R. B. Bhappu (Eds.), pp. 703–713 AIME/SME (1978).
7. E. J. Klohn, "Design and Construction of Tailings Dams," *CIM Bull.*, **65**, 28–44 (April 1972).
8. "The Florida Phosphate Slimes Problems—A Review and Bibliography," U.S. Bureau of Mines Information Circular IC 8668 (1975).
9. B. M. Moudgil, "Mined Land Reclamation by the Florida Phosphate Industry," *Trans. AIME/SME*, **260**, 187–191 (1976).
10. R. H. Sprute and D. J. Kelsh, "Limited Field Tests in Electrokinetic Densification of Mill Tailings," U.S. Bureau of Mines Report of Investigations, RI 8034 (1975).
11. P. Somasundaran, E. L. Smith, and C. C. Harris, "Dewatering of Phosphate Slimes Using Coarse Additives," *Proc. 11th Int. Miner. Process. Congr., Cagliari, 1975*, pp. 1301–1322, Università di Cagliari (1975).
12. A. D. Read and R. M. Manser, "Residual Flotation Reagents: Problems in Effluent Disposal and Water Recycle," *Proc. 11th Int. Miner. Process. Congr., Cagliari, 1975*, pp. 1323–1344, Università di Cagliari (1975).
13. R. D. Gott, "Development of Waste Water Treatment at the Climax Mine," *Am. Min. Congr. J.*, **64**, 28–34 (April 1978).
14. M. J. A. Vreugde and G. W. Poling, "The Effect of Flocculants on Reclaim Water Quality for Flotation," *CIM Bull.*, **68**, 54–59 (December 1975).

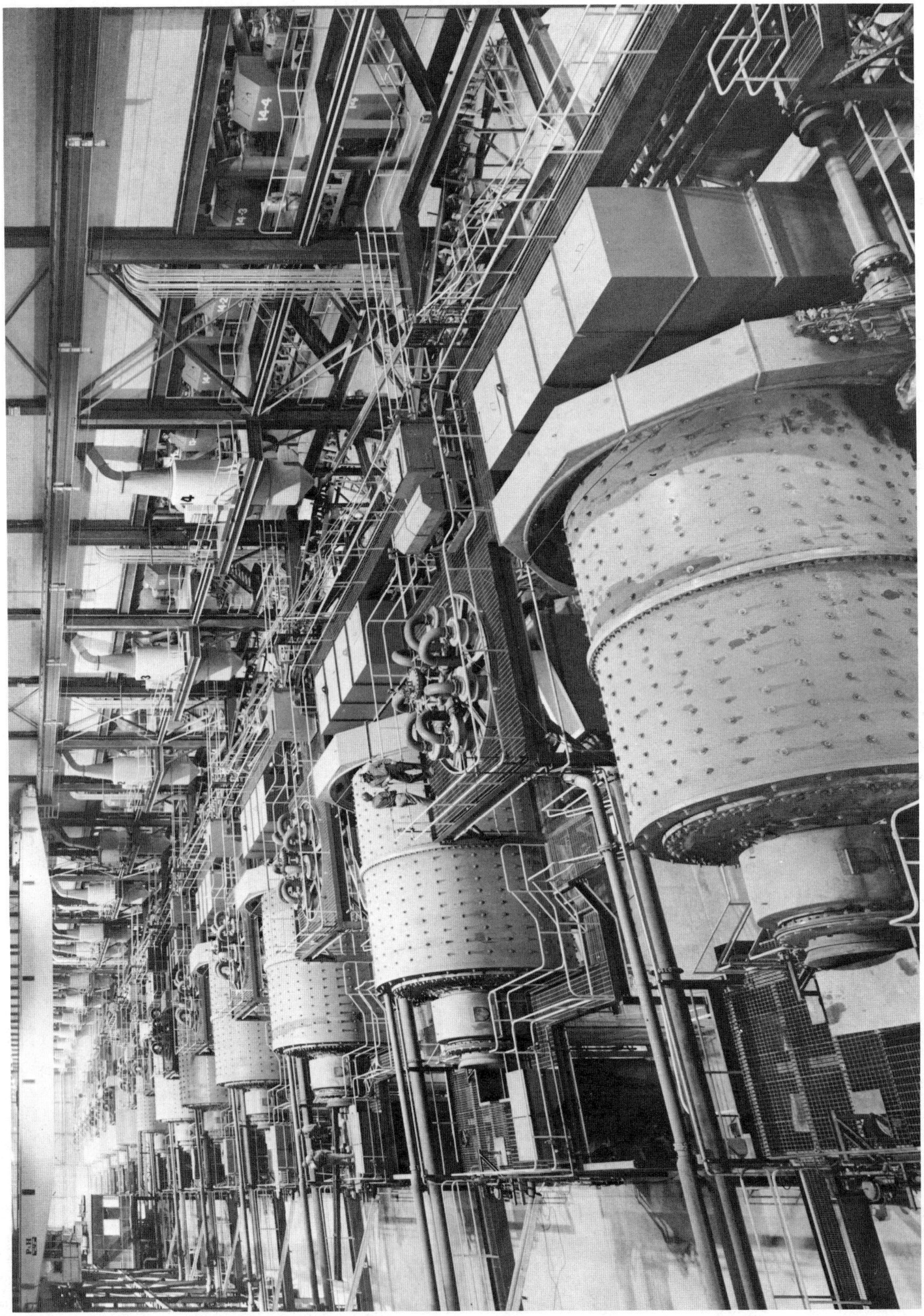

Part VII

Plant Practice

Previous chapters have considered primarily the various "unit operations" of mineral processing, for the most part in isolation from one another. This final part is more concerned with whole plants, in particular, with how the various unit operations combine and interact with each other. Chapter 23 emphasizes the important principle that any item of equipment is limited in what it can efficiently achieve. Thus, with even a comparatively simple ore, efficient treatment requires a range of equipment, and an appreciable amount of recycle.

Because ores (as fed to the mill) are not homogeneous, equipment settings must be regularly adjusted to obtain optimum performance. Development of suitable controls requires not only suitable sensors for detecting changes in the input and/or output streams, but also a certain understanding of the operating principles of the equipment. Chapter 24 explores some of the philosophies used in designing control systems.

The economic basis of mineral processing was introduced in Chapter 1. Chapter 25 takes up further aspects of the subject, including capital and operating costs, the viability of projects, and the concept of economic design.

Chapter Twenty-Three

Process Integration and Flowsheet Analysis

Cliff-Hanger Theorem: "Every problem solved introduces a new unsolved problem."

This chapter reconsiders some of the material previously introduced, with particular reference to its application in processing circuits. Because of the variety of equipment and mineral systems, this can be only an introduction, details are available elsewhere.[1-7]

When considering mineral processing circuits, two factors are particularly significant. First, each ore is unique, hence it requires a unique flowsheet to obtain optimum performance (which, as described in Chapter 3, and later in Section 25.5, may be hard to define). The second concerns the interrelations between the various process operations. Although each unit of equipment is present to serve a specific purpose, any change in the operation of one unit affects the behavior in subsequent operations, or, where recycle occurs, in preceding operations. Such interactions would not be important if conditions were stable and each unit of equipment were operated at its optimum level; but this seldom occurs, and it thus becomes necessary to be satisfied with an overall optimum. In fact, this argument should be carried further, since mineral processing is itself frequently one stage in a larger operation that may include mining, smelting, and metal forming.

The first section of this chapter considers a complex flowsheet and justifies its equipment and arrangement on the basis of the ore mineralogy. Subsequent sections cover some aspects of circuit optimization: that is, how equipment can be best utilized.

23.1. FLOWSHEET CASE STUDY

A more detailed description of the plant discussed here has been presented elsewhere.[8] The ore contains about 1.7% WO_3 (as scheelite, $CaWO_4$), 0.25% Cu (as chalcopyrite, $CuFeS_2$), and between 6 and 8% other sulfide minerals (mainly pyrrhotite, FeS). The scheelite occurs in four associations:

Disseminated in massive sulfides.
Disseminated in massive pyroxenite.
Disseminated in limestone.
Fine intergrowths with limestone, pyroxenite, chalcopyrite, and pyrrhotite.

The majority of the scheelite is sized between 400 and 500 μm, but ranges from 100 to 600 μm. Typical liberation with particle size is shown in Table 23.1, where it can be seen that about half the scheelite is liberated at 420 μm, although high liberation does not occur until below 150 μm.

Because of its high density and relatively poor floatability, gravity methods are preferred for concentrating scheelite. With this particular ore, separation is complicated by the presence of dense sulfides and the relatively fine liberation size. These factors contribute to the complexity of the flowsheet (Fig. 23.1).

The crushing plant has a conventional layout, using jaw crushers for primary and secondary crushing, and a cone crusher for tertiary crushing.

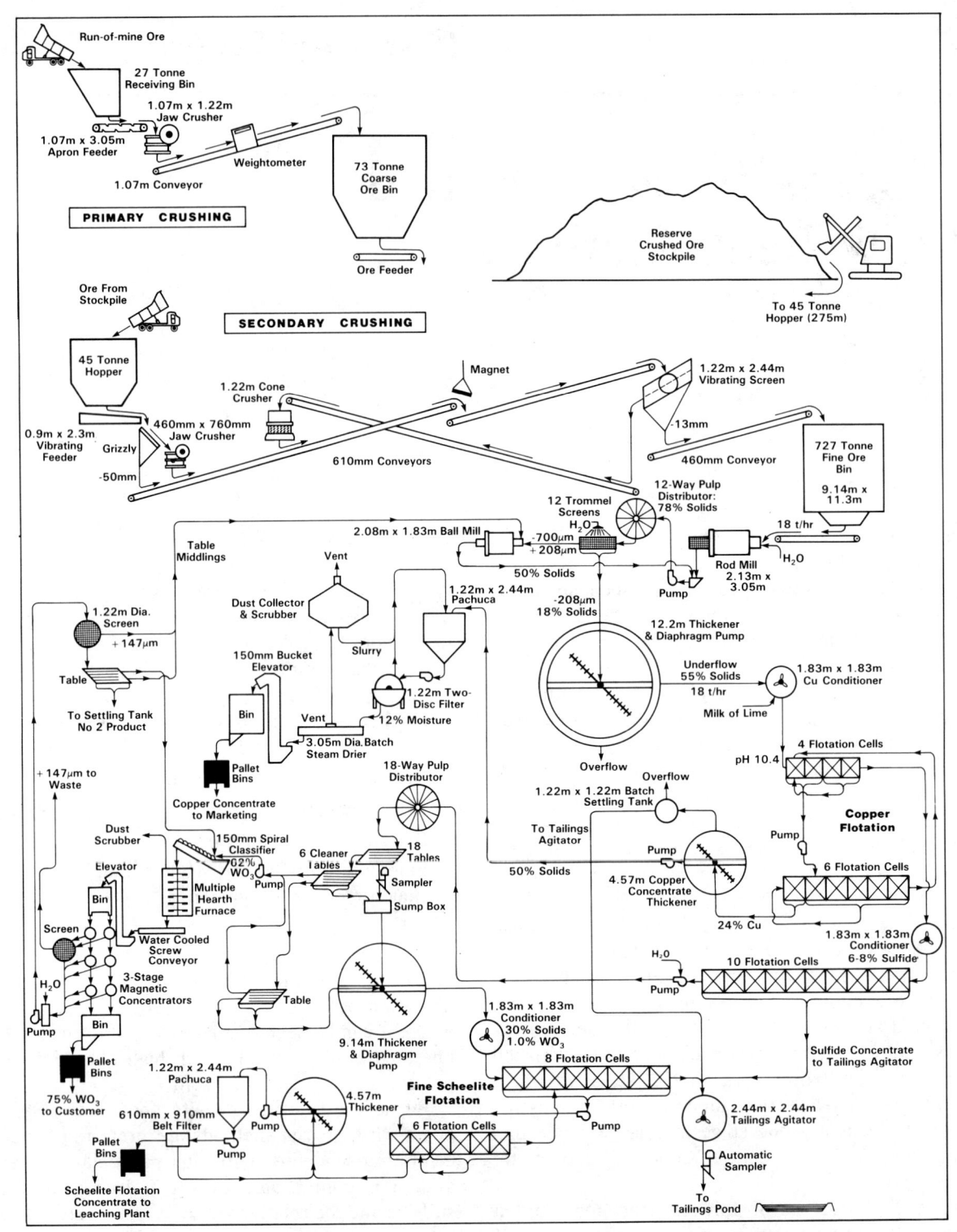

Figure 23.1. Flow sheet of Canada Tungsten Mining Corp. Ltd. (Courtesy Joy Manufacturing Co.)

TABLE 23.1: SCHEELITE LIBERATION

Size (μm)		% Distribution of Scheelite				Total
		Scheelite + Sulfides	Scheelite + Silicates	Scheelite + Calcite	Liberated Scheelite	
-600	+420	47.7	32.3		20.0	100
-420	+300	32.0	19.5		48.5	100
-300	+210	28.8	25.2	1.6	44.4	100
-210	+150	23.0	10.0		67.0	100
-150	+105	5.9	14.0		80.1	100
-105	+75	7.1	3.2		89.7	100
-75	+53	2.8	5.9		91.3	100

Jaw crushers have been used instead of gyratory crushers primarily because of the low throughput. The apparently high capacity of the larger primary crusher results because its operation is restricted to a 3.5 month mining season; hence the large intermediate stockpile. The secondary crusher is in open circuit, but with a grizzly to bypass material already undersize, while the tertiary crusher is in closed circuit. The total crushing operation reduces run-of-mine ore to -13 mm suitable for rod mill grinding.

The grinding circuit is particularly critical to the successful treatment of this ore because scheelite is a very friable mineral and because good liberation is achieved only near the lower size limit of the concentrating equipment. Thus a sizing separation is used immediately after the rod mill, since the -210 μm fraction, although far from fully liberated, contains sufficient liberated scheelite to justify concentration. Wet classification is not used, since it would result in the very dense scheelite building up in the oversize, and thereby being subjected to detrimental further grinding.

Since the sulfides would report with the scheelite during gravity concentration, about 80% of them are floated off first. By initially floating at a high pH (10.4) with a low collector level (0.01 kg/t), only the chalcopyrite absorbs sufficient collector to float, and a conventional rougher/scavenger/cleaner circuit produces a marketable copper concentrate. The tailings from the copper flotation then pass to a conditioning tank, where further collector additions render most of the remaining sulfides floatable (a decreased pH is not used because this would give corrosion problems). Although most of the sulfide mineral is pyrrhotite, some sphalerite is present, and this requires $CuSO_4$ activation. Scheelite losses in the sulfide concentrates are comparatively low because middling sulfide/scheelite particles are generally too large and heavy to float.

Tailings from the sulfide flotation pass to shaking tables where, after cleaning, a coarse scheelite concentrate is produced. Some of the particles in this coarse concentrate are middlings of scheelite and sulfide, so further treatment of the concentrate is necessary to reduce the sulfide content. This is achieved, after partial dewatering in a spiral classifier, by roasting the concentrate to make the pyrrhotite more magnetic so that it can be removed by three stages of magnetic separation. The most magnetic particles (i.e., from the first stage) are screened at 150 μm and the oversize is discarded, since it contains negligible scheelite. The remaining magnetic concentrate is also screened, and the +150 μm particles (assumed middlings because they were not strongly magnetic) are returned to the regrind ball mill, which also receives the +210 μm material discharged from the rod mill/trommel. The -150 μm magnetic concentrate is tabled; the concentrate is considered to be misplaced material and is therefore returned to the spiral classifier. Middlings concentrate is returned to the ball mill for regrinding to increase the liberation, before recycling through the circuit.

Because, on a table, recovery of particles less than 75 μm is very low, the tailings from the main table circuit still contain 50% of the scheelite, mainly as liberated fine particles. As such they must be treated by flotation, using roughers and cleaners. Even so, this flotation separation is not as good as that attainable with sulfide minerals; the fine flotation concentrate is only about 38% WO_3 (and > 50% calcite, $CaCO_3$), compared to the 75% WO_3 of the coarse table concentrate. This fine concentrate is eventually acid leached to remove the calcite and produce a 70% WO_3 concentrate and raise the overall recovery to 80%.

It can be seen that the overall treatment philosophy develops from the unique characteristics of this particular ore, although different equipment may offer some slight improvements. Many other descriptions of flowsheets are available in the literature, and a selection is given in Appendix B. The reader should study a variety of these to become familiar with the philosophies involved.[1-4]

23.2. SELECTION BETWEEN ALTERNATIVE TYPES OF SEPARATING EQUIPMENT

The essential prerequisite to deciding on separating equipment is a full knowledge of the ore's mineralogy, particularly, the minerals that are present and the size at which adequate liberation occurs. Once this information is available in the form of separability curves, choice of equipment is rapidly narrowed down (e.g., see Appendix F). For example, magnetic minerals such as magnetite (Fe_3O_4) are comparatively easy to separate from nonmagnetic minerals because of the vast difference in magnetic properties (Fig. 3.7). Gravity concentration is also attractive when density differences are high, but it may be impractical for selective separation between a number of heavy minerals, or when the liberation size is very fine: this is why flotation is so widely used for sulfide minerals. On the other hand, nonsulfide minerals are comparatively difficult to separate by flotation because of relatively small differences in floatability.

In most cases, the type of separator selected is the same for similar ores; the number of units, and their size, arrangement, and separator setting however depend on the particular ore. This is not to say that equipment selection should always follow previous examples. New equipment developed often has a superior performance; for example, the high capacity of the cone makes it an ideal substitute for large banks of spirals. All too often some new equipment does not receive the attention it deserves, while other equipment is placed on the market and described in glowing terms that turn out to be exaggerated.

One other aspect needs mentioning: the versatility of some types of equipment. We have repeatedly emphasized that any separator operates best over a comparatively narrow set of conditions; largely because separation seldom depends on a single force, but rather is affected by secondary (generally opposing) forces. In some cases these secondary effects can be utilized to such an extent that the equipment can have other important applications. Examples are the use of the hydrocyclone and the screen. Both are primarily used for size separations. However, because the separation in a hydrocyclone is also affected by density, it may be used as a gravity separator. Also, provided the quantity of fines is negligible, both screens and hydrocyclones can be used to separate solids from liquids (Fig. 19.3). Such dewatering separations may not be as complete as that obtainable with "conventional" thickening and filtration; nevertheless, it may be more than sufficient in certain circumstances (i.e., an acceptable product is obtained at lower cost).

23.3. ROUGHING, CLEANING, AND SCAVENGING

A consequence of the limited separation generally attainable in a single device is the need for roughing, cleaning, and scavenging components in a circuit. Generally speaking, roughing is the primary operation, which uses a moderate separating force to remove fully liberated valuable; scavenging uses a strong separating force to recover as much as possible of the remaining valuable (which is generally incompletely liberated); and the cleaner uses a low separating force to upgrade the rougher concentrate by removing misplaced waste material. The extent to which each phase is employed can be considered from two viewpoints (Chapter 3): the "thermodynamic" aspects, which represent the limitations imposed by the mineral properties (separability curves), and the kinetic aspects, largely restricted by the time the material is exposed to the separating forces (performance curves).

23.3.1. Thermodynamic Aspects

While the thermodynamic aspects of the separation are represented by the separability curves, a useful form of illustration is the release analysis type of plot,[9] since any product obtained from an ore can be analyzed in terms of vectors joining points on this graph. On such a graph, the horizontal component of a vector represents the product's mass, the vertical component, the recovery of the valuable in it, and the slope of the vector is the grade. Figure 23.2 reproduces the release analysis curve previously derived in Example 3.2. Thus, from this ore (assuming at this stage no machine limitations) it should be possible to produce an 80% PbS concentrate, represented by vector $\overrightarrow{OR}$ (recovery of PbS = 83%, mass = 10.4%), and a tailings represented by vector $\overrightarrow{RT}$ (loss of PbS = 100% − 83%, mass = 100% − 10.4%, grade = 1.9%). One further assumption is made at this point: the ore has been ground to give optimum liberation (the optimum liberation is essentially that at which the extra valuable released by further liberation is counterbalanced by the loss arising from overgrinding material already liberated, which becomes too fine to be separated).

The circuit for such a separation is illustrated in Fig. 23.3a. Because the tailings grade ($\overrightarrow{RT}$ in Fig. 23.2) is still relatively high (i.e., the recovery is too

Roughing, Cleaning, and Scavenging

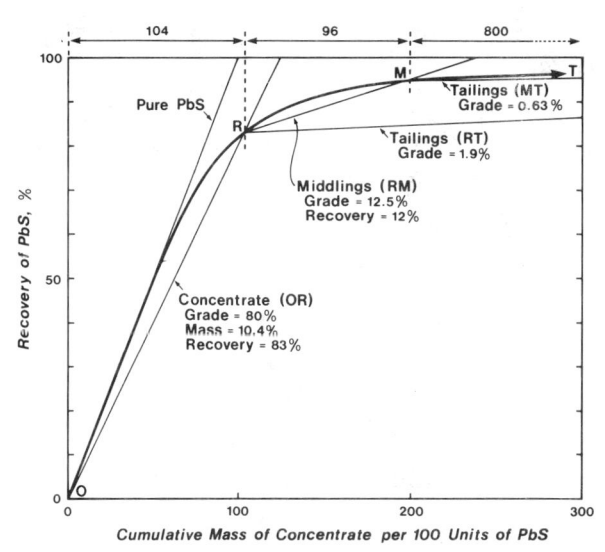

Figure 23.2. Release analysis curve, from Example 3.2.

low), an alternative approach would be to apply a scavenger operation to the tailings, dividing it into a middlings concentrate ($\overrightarrow{RM}$, Fig. 23.2) and a low grade tailings ($\overrightarrow{MT}$, Fig. 23.2, 0.63% PbS). The necessary flow sheet is now of the form shown in Fig. 23.3b. (The likely need for cleaning is neglected in Fig. 23.3a and 23.3b.) In general, the success or otherwise of a separation largely centers around this scavenger concentrate. It normally is unsalable in this initial low grade form and must be further processed. However, since it consists largely of middlings particles, simply recycling it to the rougher circuit (the logical point) will achieve one of three things: a continually increasing circulating load (clearly hazardous); eventual rejection of the middlings particles to the tailings (back to Fig. 23.3a), or rejection of the middlings to the concentrate, which will give a concentrate represented by vector $\overrightarrow{OM}$ (Fig. 23.2). However, further regrinding of the scavenger concentrate *by itself* will increase its liberation without overgrinding the previously liberated material (since this has already been recovered in the rougher concentrate). That is, an optimum liberation release analysis curve can also be obtained for the scavenger concentrate. This may be analyzed separately, or combined with the initial release curve to give a new curve that will indicate a better overall performance than the optimum obtained with the initial single stage grind. This stage separation with intermediate grinding can achieve superior separation, and such an approach is essen-

tial when liberation occurs at fine sizes. Until recently, normal practice in flotation was to return the regrind mill product to the rougher circuit (Fig. 23.3.c), but more recent practice is to use an additional set of separators (Fig. 23.3d), since this reduces problems arising from the large fluctuating circulating loads that can occur in the roughers.

Note this discussion treats only the maximum separation possible: no consideration is given to kinetic aspects, nor do the basic circuits in Fig. 23.3 include the one or more stages of cleaning that are normally necessary. In practice once all these stages are included a number of circuit arrangements become possible, and finding the best one may involve a large amount of test work,[10] although computer simulation may make it possible to narrow down the number of viable alternatives.[11]

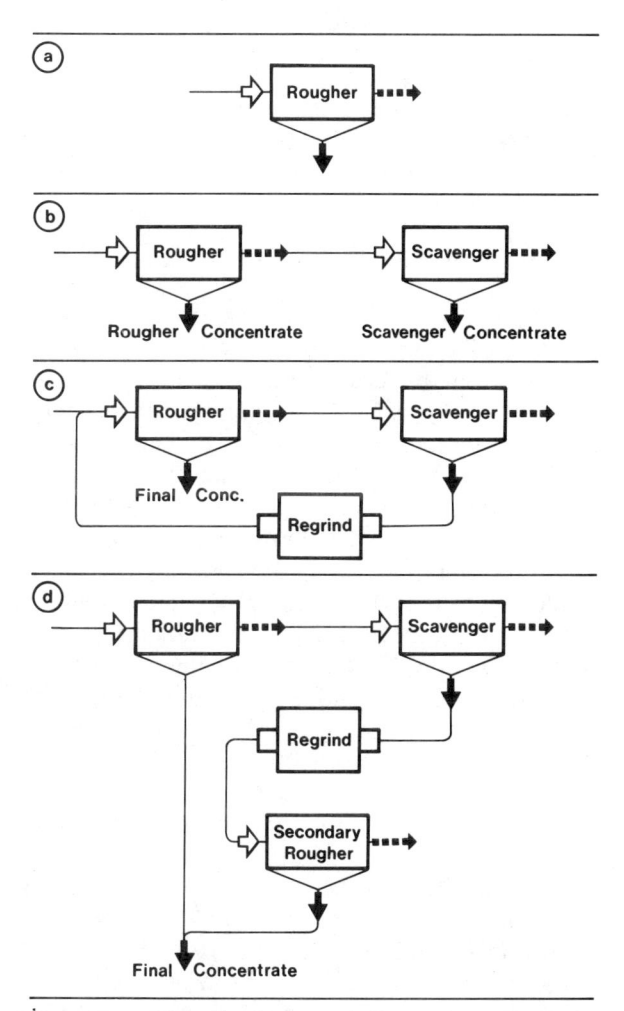

Figure 23.3. Basic circuits for producing products vectored on Fig. 23.2 (cleaning operations neglected).

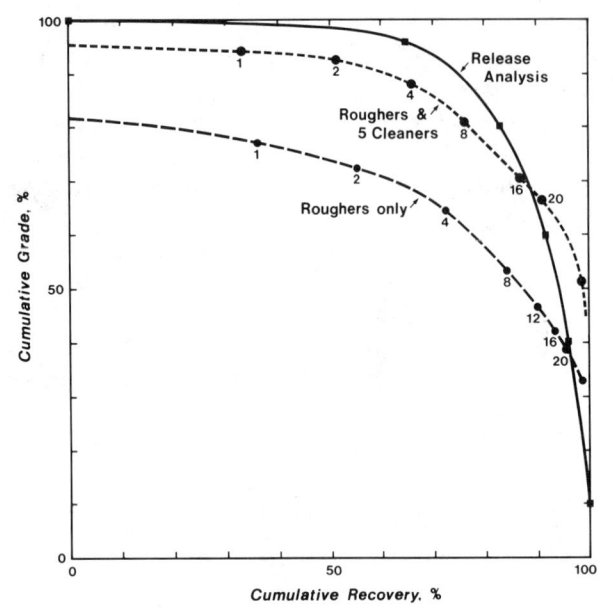

Figure 23.4. The cumulative grade versus cumulative recovery curves of PbS from Examples 3.2, 16.1, and 23.1 illustrate the significance of circuit arrangements. Numbers indicate the number of rougher cells.

23.3.2. Kinetic Aspects

The significance of kinetics can be illustrated by extending Examples 3.2 and 16.1. Figure 23.4 shows the data from the former example plotted as a grade recovery curve (solid line), representing the maximum separations possible using various separator settings (i.e., the "thermodynamic" limitations imposed by the mineral properties: the separability data). One significant feature of this curve is that initial recovery (i.e., $\mathscr{R} \to 0$) is at the pure mineral grade, but as recovery rises to 100% the cumulative grade of the concentrate falls to the feed grade. If this ore has the kinetic behavior described in Example 16.1, the cumulative grade/cumulative recovery along a bank of rougher cells is as shown in Fig. 23.4 (long-dashed line). Note that at all recoveries, the actual grade is well below the "ideal" represented by the laboratory determined grade/recovery curve.

How separation efficiency can be improved by incorporating a cleaner circuit is illustrated in the following example.

Example 23.1. How are the grade/recovery characteristics of Example 16.1 changed by having the concentrate cleaned in a bank of five cells? (For simplicity, assume that the flotation probabilities do not change.)

Solution. To solve the problem, it is first desirable to derive an expression for the mass of cleaner concentrate produced. Let:

$$\text{mass of fresh feed} = M_I$$

$$\text{mass of total rougher feed} = M_\Sigma$$

$$\text{recovery in rougher concentrate} = \mathscr{R}_{(+)R}$$

$$\text{recovery in cleaner concentrate} = \mathscr{R}_{(+)C}$$

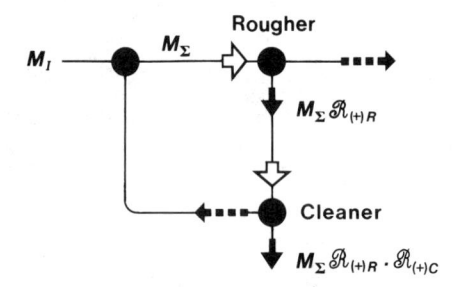

For the recycle stream:

$$\begin{aligned}\text{mass} &= M_\Sigma - M_I \\ &= \text{cleaner tailings} \\ &= M_\Sigma \mathscr{R}_{(+)R}(1 - \mathscr{R}_{(+)C})\end{aligned}$$

$$\therefore \quad M_\Sigma = \frac{M_I}{1 - \mathscr{R}_{(+)R}(1 - \mathscr{R}_{(+)C})}$$

$\therefore$ mass of cleaner concentrate

$$= \frac{M_I \mathscr{R}_{(+)R} \mathscr{R}_{(+)C}}{1 - \mathscr{R}_{(+)R}(1 - \mathscr{R}_{(+)C})} \quad (E23.1.1)$$

As with Example 16.1, the recovery of any component after n cells is given by

$$\mathscr{R}_{(+)R} = 1 - (1 - p)^n$$

while the recovery in the cleaner cells is given by

$$\mathscr{R}_{(+)C} = 1 - (1 - p)^5$$

Sample calculation: 10 rougher cells

Basis. 100 units of fresh feed
For liberated PbS:

$$\mathscr{R}_{(+)R} = 1 - (1 - 0.5)^{10} = 0.999$$

$$\mathscr{R}_{(+)C} = 1 - (1 - 0.5)^5 = 0.969$$

Substituting in Eq. E23.1.1:
$\therefore$ mass in cleaner concentrate

$$= \frac{6.414 \times 0.999 \times 0.969}{1 - 0.999(1 - 0.969)}$$

$$= 6.407 \text{ tonnes}$$

Roughing, Cleaning, and Scavenging

TABLE E23.1.1: FLOTATION CLEANING EXAMPLE

Number of Roughers	1	2	4	8	16	20	40
Mass liberated PbS	3.156	4.772	6.001	6.388	6.414	6.414	6.414
Mass locked PbS	0.156	0.314	0.634	1.260	2.305	2.680	3.459
Mass locked gangue	0.085	0.171	0.346	0.688	1.258	1.468	1.887
Mass liberated gangue	0.126	0.256	0.522	1.090	2.367	3.078	7.313
Recovery	33.1	50.9	66.4	76.5	87.2	90.9	98.8
Grade	94.0	92.3	88.4	81.1	70.6	66.7	51.8

For locked PbS:

$$\mathscr{R}_{(+)R} = 1 - (1 - 0.1)^{10} = 0.651$$

$$\mathscr{R}_{(+)C} = 1 - (1 - 0.1)^{5} = 0.410$$

Substituting in Eq. E23.1.1:

∴ mass in cleaner concentrate

$$= \frac{3.586 \times 0.651 \times 0.410}{1 - 0.651(1 - 0.410)}$$

$$= 1.554 \text{ tonnes}$$

$$\text{Recovery} = \frac{6.407 + 1.554}{10} \times 100\%$$

$$= 79.6\%$$

mass of locked gangue floated

$$= 1.554 \times \frac{1 - 0.647}{0.647}$$

$$= 0.848 \text{ tonne}$$

For liberated gangue:

$$\mathscr{R}_{(+)R} = 1 - (1 - 0.03)^{10}$$

$$= 0.263$$

$$\mathscr{R}_{(+)C} = 1 - (1 - 0.03)^{5}$$

$$= 0.141$$

∴ mass in concentrate

$$= \frac{29.05 \times 0.263 \times 0.141}{1 - 0.263(1 - 0.141)}$$

$$= 1.392 \text{ tonnes}$$

∴ total mass of gangue in cleaner concentrate

$$= 0.848 + 1.392$$

$$= 2.24 \text{ tonnes}$$

∴ grade $= \dfrac{7.96}{7.96 + 2.24} \times 100\%$

$$= 78.0\%$$

Results for various numbers of roughers cells (plus five cleaner cells) are given in Table E23.1.1.

It can be seen that the addition of cleaner cells (short-dashed line, Fig. 23.4) raises the separation efficiency toward the limitations imposed by the mineral characteristics (the release analysis curve, Fig. 23.4).

A number of other features in Fig. 23.4 are also significant. First, although efficiency is raised by including cleaner cells, it is achieved at the expense of some recovery. This can be seen by comparing the two dashed curves at a given number of rougher cells. However *overall* efficiency is still higher with the cleaner cells because the relatively large increase in grade has moved the grade/recovery curve toward the top right-hand corner of the diagram (see Section 3.5.4).

Second, it can be seen that the rougher/cleaner curve has three distinct sections, which can be explained as follows. The first three rougher cells are recovering predominantly fully liberated PbS. The second section of the curve results from the next 16 or so cells which, while they are recovering the last traces of fully liberated PbS, are recovering predominantly middling particles. Any additional cells result in a third section where the last of the middling particles are recovered, but only at the expense of proportionately large amounts of gangue, which rapidly lower the concentrate grade. (In this example it is not possible to determine precisely where the third region begins, because of the simplifications made.)

Example 23.1 omitted consideration of the production of a separate middlings concentrate; instead middlings particles were forced into the tailings (< 4 cells), or increasingly to the cleaner concentrate as further rougher cells were added. This suggests that cells 4 onward should be arranged to give a separate scavenger concentrate containing most of the middlings particles. Such a concentrate should then be reground to increase liberation and then refloated either in the original three rougher cells, or in a new bank of cleaner cells.

As mentioned earlier, current practice is to not treat the reground scavenger concentrate in the original rougher cells because it can introduce large fluctuations in the circulating load. One cause for this can be seen in the examples. If it is assumed that the rougher cells (cells 1–3, say) are designed for a certain circulating load, a decrease in the recycle will decrease the throughput and thus raise the flotation probabilities. This will result in more middling particles floating in the rougher cells instead of in the scavenger/regrind circuit, and will produce a lower grade of concentrate. Alternatively an increase in the circulating load will cause more liberated material to be recovered in the scavenger cells, where it will be subjected to overgrinding and possible loss to the tailings. In practice both effects are eventually self-correcting, and are not this extreme: but the net effect is still undesirable—slow fluctuation in the circulating load, with the perturbation peaking an appreciable time after the causal event. This makes control difficult because by the time the effect is detected the cause may have dissipated.

By reworking these examples with alternative parameters, it is possible to illustrate one other point: any separator is capable of only a limited amount of work. This point is often overlooked when a lower grade of ore is fed to a concentrator. If throughput is maintained, such a situation must produce a drop in either grade or recovery. Frequently, in an effort to maintain a constant amount of metal in the output, throughput is increased, an effect that will even further lower the grade or recovery. Eventually of course the system will stabilize, but with much lower separation efficiency.

A word of caution: Examples 3.2, 16.1, and 23.1 are intended to illustrate the overall characteristics of rougher/cleaner/scavenger arrangements. Because the data are fictitious, and because considerable simplifications have been made, specific conclusions cannot be drawn. For example, the overlap of the dashed and solid lines at high recoveries in Fig. 23.4 would not occur in practice. The selection of five cleaner cells is arbitrary, and five is not necessarily an optimum number; a different number of cleaner cells, and their combination with re-cleaner cells should also be considered. Such detailed considerations are beyond the scope of this book, although they have been described elsewhere.[11]

23.3.3. Circuit Analysis

The literature abounds with descriptions of operating plants, and in many cases copious plant statistics are given. Unfortunately, much of the information is of little use in assessing true efficiencies and does little to explain *why* inefficiencies occur; thus it also fails to indicate how improvements can be made. We have continually emphasized the importance of using performance curves rather than single efficiency coefficients to assess the quality of a separation, whether the separating property is size, density, magnetic susceptibility, or anything else. However, because of the effect of competing forces, bulk performance curves may not be sufficient. For example, in nonsizing separations (gravity, flotation, etc.) performance is also affected by particle size. Consequently it is often desirable to assess this effect by determining the performance curves for individual size fractions. Comparatively little data have been published along these lines, but the papers by Gottfried[12] on gravity separations of coal are a notable exception.

One of the best examples of the type of investigation necessary, and a demonstration of its value, is the study by Cameron et al. of a sulfide flotation concentrator.[13,14] This analysis was carried out on the grinding and flotation circuits. Particle size and flow rate analysis on the grinding circuit allowed breakage rates in the mills to be estimated, and by using the data in a computer simulation it was shown that the capacity could be raised by allowing the circulating load to increase. In the lead flotation circuit, flotation probabilities were determined for individual size fractions and the results showed that each size fraction contained a fast and a slow floating fraction. The fast floating galena was presumed to be fully liberated, and except for significant proportions of the -4.7 μm fraction, was readily recovered. Middling particles were assumed to make up the slow floating fractions, and it was these that contributed to losses in the larger size fractions. Part of the zinc was found to exist as a readily floatable mineral that consistently reported to the lead concentrate. Results from the zinc circuit suggested that the main contribution to the zinc loss in the tailings was the -7 μm particles.

Although one might perhaps guess that the losses could be expected to result from large middling and extremely fine particles, such a study pinpoints the *extent of the problems*. Particularly notable in this example was just how well most of the circuit was performing.

One other aspect is highlighted in this study: that relating to mathematical complexity. There is not much doubt that ores are extremely complex, and in many cases rigorous analysis of their processing systems will never be practical. The result is that in most situations various levels of complexity are possible, and an appropriate level should be used for a given problem: in general, that which provides adequate information is the best. For example, in the study above a relatively complex size reduction simulation, incorporating mill flow characteristics, was used, since it was considered that this was necessary to adequately represent the operation. On the other hand, the flotation representation used is one of the simplest available, since it was sufficient to extract the relevant information. A more detailed discussion on the subject of analytical complexity has been given by Kelsall and Stewart.[15]

In the case of size separations, density is frequently a factor, and analysis of density fractions (or composition fractions) becomes necessary. An example of this situation has been described by Carpenter.[16] A classifier was treating a scheelite-molybdenite ore, and in terms of overall performance, the classifier appeared to be operating satisfactorily. Analysis of the classifier performance in terms of the different minerals provided a different story. This showed that the scheelite, by virtue of its high density, was being unnecessarily recycled for further regrinding. The harm caused by this regrinding of liberated mineral was further compounded by the very friability of the scheelite, which resulted in relatively large size reductions with each passage that eventually contributed to excessive scheelite losses to the tailings because the material was too fine to recover (cf. Section 23.1).

23.4. PLANT DESIGN

By applying the principles of mineral processing, as has been done in this book, a start has been made on the development of plant design. Although a detailed examination of plant design is beyond the scope of this book, some aspects are worth mentioning here (further information can be obtained from many of the references in the various chapters, as well as other published material on the subject[5,17-26]).

Plant design is not just a matter of calculating the size of equipment needed for a given job. Rather, it is a long process that starts with a feasibility study that involves sampling of the ore body, evaluation and testing of ore processing methods, equipment selection and sizing, plant layout (Fig. 23.5), economic evaluation (Chapter 25), and finally, plant construction, start-up, and optimization.[5,17,18]

Reliable sampling of an ore body is necessary, since no ore body is homogeneous with respect to grade and mineralization. Variations with depth can be particularly important in sulfide ore bodies, since near the surface the ore may be more oxidized, making its flotation more difficult. Where variations are significant, it must be decided whether the ore from various sections is to be blended or treated separately.

Testing of different processing methods must initially be carried out on a laboratory scale, where it is (relatively) easy to narrow down potential processing procedures and to compare alternative processing methods. While pilot plants are occasionally necessary to decide between processing methods, they are generally essential to evaluate the better potential circuit arrangements.[19] Even so, in most cases the best circuit is found only by modifications to the full size plant.

At all stages of process development it is critical that the experimental work be carried out on representative material: over and beyond the necessity for reliable samples is the need for consideration of the material's *condition*. For example, ground sulfide ore may oxidize if left for any length of time before flotation, or the behavior of slurry for sedimentation tests may change simply because the mixture has been allowed to settle in a container. Factors such as these immediately mean that there will be a degree of uncertainty in the sizing of full-scale equipment, even in the few cases where reliable scale-up criterion are available. What is more, even pilot plant tests may not indicate the best operating conditions. This of course does not imply that equipment should be overdesigned. True, in some instances it is necessary to allow safety factors, but oversized equipment means excessive costs and can also be detrimental to processing (e.g., discussion following Example 23.1). In summary it may be stated that it is impossible to accurately design a full size plant, and a good design should have sufficient flexibility to allow the plant to be tuned once in operation. In fact, plant tuning is essentially a matter of isolating the bottleneck in the plant and correcting it so that a subsequent bottleneck can be attended to.

Current practice is to use the largest possible equipment consistent with satisfactory performance, since for most equipment the cost per unit

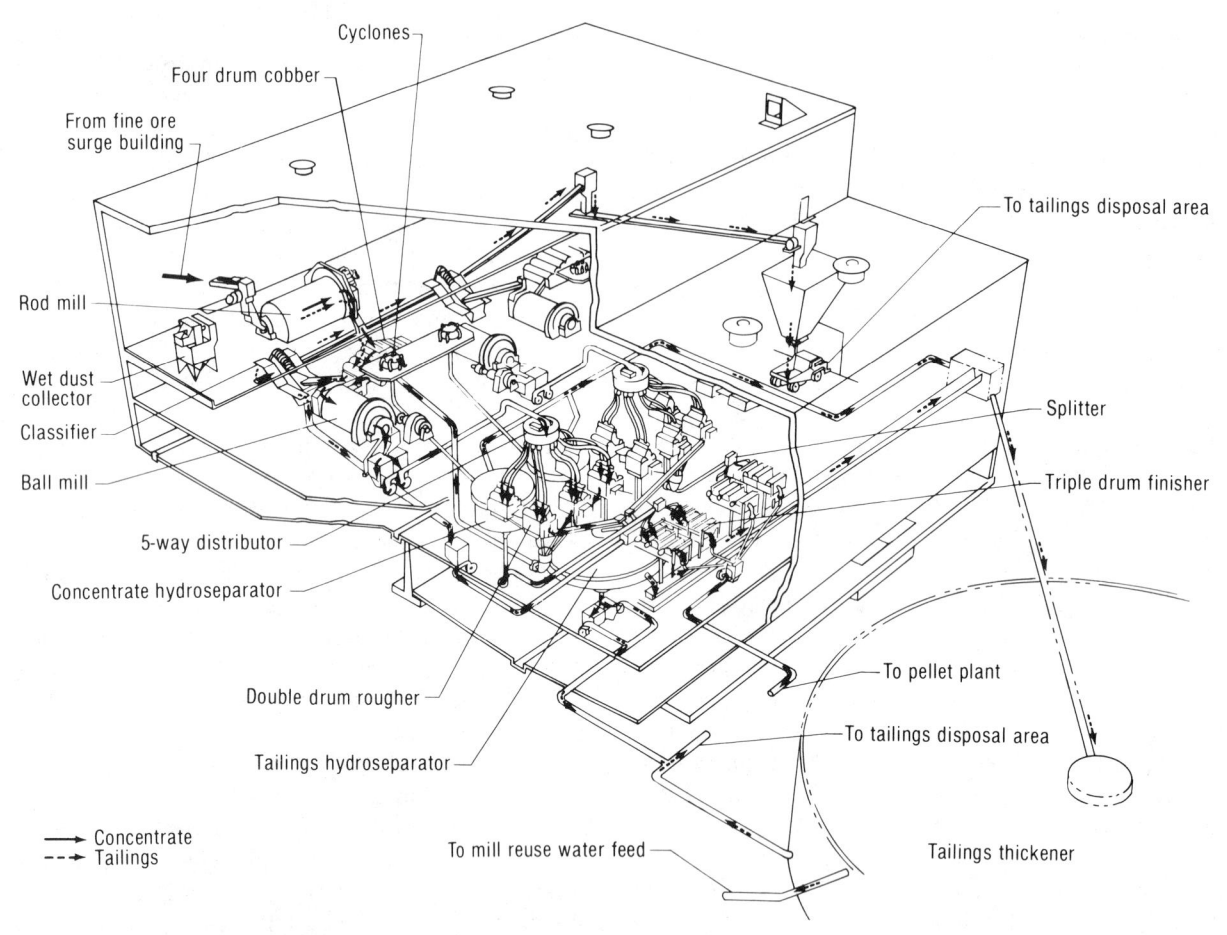

Figure 23.5. Plant layout of a concentrator treating a magnetite ore. (Courtesy Engineering and Mining Journal.)

of capacity decreases as the capacity increases (Chapter 25). Multiple process lines may suggest higher availability; but such is not the case.[20] In today's plants, availability is primarily a function of ease of maintenance. Modern equipment has a high degree of reliability, and sufficient maintenance can be carried out during routine shutdowns to the extent that overall utilization is generally higher than it would have been if multiple lines had been employed. Typically equipment reliability is of the order of 90%+, and the higher costs of multiple lines cannot justify even a possible few percent improvement in availability.

Where processing is done in stages with some stages having recycle (e.g., size reduction), equipment sizing becomes an even more complex optimization problem, since all items of equipment are interdependent. Under these conditions, a number of circuit arrangements are capable of producing a given product. By considering capital and operating costs as a function of capacity for each item of equipment, the lowest cost circuit can be determined by the application of dynamic programming.[21-23] This approach has been shown to be useful in the particularly difficult case of crushing plants, which is complicated because the cost per unit of capacity of a crusher actually rises with rising capacity.[21]

In-process surge capacity is an essential component of all processes. It is used between two major pieces of equipment so that they are not rigidly tied together. This may be necessary because two stages do not operate at the same rate (e.g., crushing and grinding plants), because transport between stages is intermittent (e.g., rail transport between mine and concentrator), or because natural surges in process equipment have resulted from fluctuating recycle. When solids storage is necessary, considerable savings may be made by using open conical storage instead of bins.[20] Thickeners are commonly used to hold slurries.[20]

Attention must also be paid to plant layout, to minimize the number of buildings, allow for easy maintenance, reduce costs of materials transportation, and allow for modifications and future expansions (which always occur).[20,24-26] Often these are conflicting requirements, and the lowest first cost may not necessarily be best in the long run.

REFERENCES

1. Anon., *Modern Mineral Processing Flowsheets*, 2nd ed., Denver Equipment Co. (1965).
2. R. Thomas (Ed.), *E/MJ Operating Handbook of Mineral Processing*, McGraw-Hill (1977).
3. J. T. Woodcock, "Mineral Processing Progress in Australasia," *Aust. Min.* (an annual review article, occurring in an issue about midyear).
4. Flowsheets in *Deco Trefoil* magazine.
5. A. L. Mular and R. B. Bhappu (Eds.), *Mineral Processing Plant Design*, AIME/SME (1978).
6. "Reference Manual and Buyers Guide," *Can. Min. J.* (published annually).
7. Anon., "A Portfolio of Flowsheets", *Eng. Min. J.*, **170**, 171-202 (June 1969).
8. R. Ellerman, "Operations of Canada Tungsten Mining Corp. Ltd," *Deco Trefoil*, 9-24 (Summer 1969). Also published as Denver Bull. M4-B 138, Denver Equipment Co. (1969).
9. C. C. Dell, "Release Analysis, A New Tool for Ore Dressing Research," in *Recent Developments in Mineral Dressing*, pp. 75-84, IMM (1953).
10. R. H. Lamb and L. R. Verney, "Investigations into RST Group Concentrator Practices," *Trans. IMM (C)*, **76**, C154-C167 (1967).
11. R. P. King, "The Use of Simulation in the Design and Modification of Flotation Plants," Ch. 32 in *Flotation: A. M. Gaudin Memorial Volume*," Vol. 2, M. C. Fuerstenau (Ed.), AIME/SME (1976).
12. (a) B. S. Gottfried, "A Generalization of Distribution Data for Characterizing the Performance of Float-Sink Coal Cleaning Devices," *Int. J. Miner. Process.*, **5**, 1-20 (1978).
    (b) B. S. Gottfried and P. S. Jacobsen, "Generalized Distribution Curve for Characterizing the Performance of Coal-Cleaning Equipment," U.S. Bureau of Mines Report of Investigations, RI 8238 (1977).
13. A. W. Cameron et al., "A Detailed Assessment of Concentrator Performance at Broken Hill South Ltd," *Proc Australasian IMM*, No. 240, 53-67 (December 1971).
14. D. F. Kelsall et al., "The Effects of a Change from Parallel to Series Grinding at Broken Hill South," *Australasian IMM Newcastle Conf.*, pp. 333-347 (1972).
15. D. F. Kelsall and P. S. B. Stewart, "A Critical Review of Applications of Models of Grinding and Flotation," *Symp. on Automatic Control Systems in Mineral Processing Plants, Brisbane*, pp. 213-232, Australasian IMM (May 1971).
16. R. D. Carpenter, "Preparation of Flotation Plant Feed. Part I. Analysis of the Grinding Circuit at the Pine Creek Mill," Ch. 19-1 in *Froth Flotation—50th Anniversary Volume*, D. W. Fuerstenau (Ed.), pp. 494-505, AIME/SME (1962).
17. G. Schultz, "Aggregate Plant Design: The Planned Approach. Part I," *Rock Prod.* **82**, 72-80 (September 1979); Part II, **82**, 80-84 (October 1979); Part III, **82**, 92-95 (November 1979).
18. (a) G. H. Arrowsmith, A. Duncan, and M. Manackerman, "Investigation Leading to the Design of a New Coal-Preparation Plant," *Mine and Quarry*, **5**, 47-52 (December 1976).
    (b) G. H. Arrowsmith, A. Duncan, and M. Manackerman, "Preparing for a New Coal Preparation Plant," *Mine and Quarry*, **6**, 39-42 (January 1977).
19. R. D. MacDonald and F. M. Stephens, "Designing Ore-Treatment Pilot Plants," *Min. Eng.*, **11**, 509-512 (1959).
20. R. S. Shoemaker and A. D. Taylor, "Mill Design for the Seventies," *Trans. AIME/SME*, **252**, 131-136 (June 1972).
21. E. M. Calanog and G. H. Geiger, "How to Optimize Crushing and Screening Through Computer-Aided Design," *Eng. Min. J.*, **174**, 82-87 (May 1973).
22. (a) A. M. Gerrard and C. J. Liddle, "The Optimal Selection of Multiple Hydrocyclone Systems," *The Chem. Engr.*, No. 297, 295-296 (May 1975).
    (b) A. M. Gerrard and C. J. Liddle, "How to Get the Most Out of Your Hydrocyclone Systems," *Process Eng.*, 105-107 (June 1976).
23. M. D. Flavel, "Scientific Methods to Design Crushing and Screening Plants," Presented at AIME/SME Annual Meeting, Atlanta, 1977; published in *Min. Eng.*, **29**, 65-70 (June 1977).
24. I. R. M. Chaston, "Radial Layout for Increased Treatment Plant Productivity," *Min. Eng. (Trans. AIME/SME)*, **30**, 1564-1569 (1978).
25. J. C. Loretto, "Design of Mills for Developing Countries," *Trans. IMM (C)*, **80**, C98-C111 (1971).
26. H. Ruff and J. F. Turner, "Coal Preparation and Mineral Processing," *Mine and Quarry*, **6**, 42-48 (April 1977).

Chapter Twenty-Four

Plant Control

Computer Maxim: "*To err is human but to really foul things up requires a computer.*"

The control of each individual process in a processing plant is a complex task; controlling the overall operation of that plant is extremely complex. In the chemical industry, automatic control has been in wide use for many years. However, in mineral processing plants, automatic control is a recent introduction, primarily because of:

1. The lack of instrumentation suitable for measuring the required process variables, or rugged enough to tolerate the severe conditions existing in such a plant.
2. The need for reliable representative sampling systems.

Mineral processing plant control has thus traditionally relied on the skill of the plant metallurgists and operators. Frequently, such vital data as head, concentrate, and tailings grades have been unavailable until the following shift or even the following day. As a result, plant control has been highly variable from plant to plant. It is the complexity of mineral processes, however, that makes the effective manual control of plants so difficult, and so some means of automatic control must be introduced if optimum plant performance is to be approached.

In the 1960s and 1970s, the introduction of appropriate instrumentation (including computers) provided operators with vastly increased information on plant operation. This in turn has led to the development and installation of automatic control systems, stimulating further demand, hence improvements in the instrumentation available.

In this chapter, the aims, methods, and achievements of automatic control are discussed.

24.1. CONTROL OBJECTIVES

The objectives of any automatic control system must be clearly specified if full value for the investment involved is to be achieved. These objectives can vary considerably.

In general terms, automatic control in a concentrator is aimed at one or more of the following:

1. Increased throughput.
2. Improved recovery of the valuable minerals.
3. Improved concentrate grade (resulting in cost savings in subsequent processing).
4. Reduced operating costs (e.g., by savings in reagents, more productive use of mill personnel).

Although it may be possible to achieve all the goals above, it is normal practice to set more specific control objectives, and these can vary from plant to plant, and within a plant from one time to another. For example, in a copper concentrator, when copper demand is high, the objectives may be to maximize tonnage, whereas if the copper demand is reduced, improved recovery and/or lower operating costs may well become the control objective. Thus it may be necessary to build sufficient flexibility into a control scheme to allow for periodic changes in the objective.

The control objectives for an individual process within a concentrator must also be clearly specified. For example, in a grinding circuit, possible objectives might be:

1. Product size distribution maintained constant at constant throughput.

2. Product size distribution maintained constant and throughput maximized.

These are clearly different objectives and can require significantly different control strategies.

However, the first and *primary objective of any control scheme* is to *make the plant operate in a stable manner*. Only after stable operation can be maintained can there be any attempt at optimization of a plant.

24.2. PRINCIPLES OF AUTOMATIC PROCESS CONTROL

In the other chapters of this book, steady state operations have been considered. Process control, however, is inherently concerned with process behavior under unsteady state or dynamic conditions, where the accumulation terms in mass (or heat) balances must be included.

In this section, the basic principles of automatic process control are introduced. For a more detailed treatment, the reader is referred to books on the subject.[1-5]

24.2.1. The Control System

The control of a simple process is used to introduce the concepts of automatic control.

If the flow of water through a pipe is to be maintained at some desired rate, one method would be to have an operator measure the rate, then adjust a valve to increase or decrease the rate until the desired value is obtained. With time, the operator would learn how much to move the valve to compensate for any deviation of the flow rate from the desired value. However, this simple task can be readily (in general, less expensively and more reliably) performed by the use of machines (Fig. 24.1a). A flow meter sends a signal (proportional to the flow rate) to a *controller*, where a comparison is made with the *set point* (the desired value); then a signal is sent from the controller to a control valve, which indicates either the required new position of the valve or how much the valve should open or close. This control configuration is known as *feedback control*.

A block diagram for the *control loop* of Fig. 24.1a is shown in Fig. 24.1b. This block diagram is in fact quite general, and applies for all simple feedback control loops. The process variable that is to be

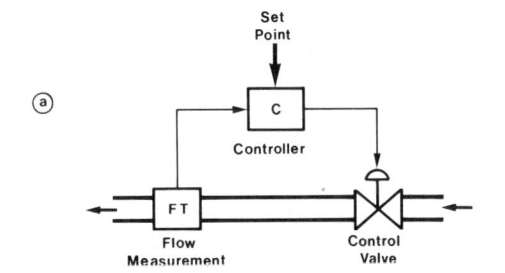

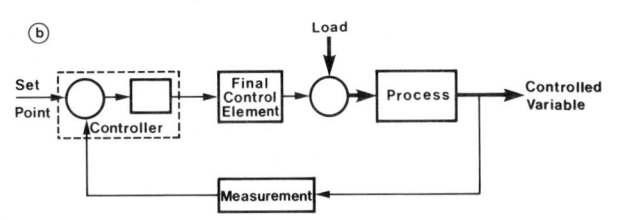

Figure 24.1. The basic feedback control system. (*a*) The control loop and its elements. (*b*) Block diagram to represent control loop.

maintained at some set point, that is, the *controlled variable* (flow rate), is measured (flow meter) and the measurement compared with the set point in the controller. The controller determines the value of the *manipulated variable* (valve setting) of the *final control element* (control valve) required to maintain the controlled variable at or near the set point value, even when there are changes in the *load* or *disturbance variable* (e.g., line pressure).

In the example used here, the *process* is a very simple one (flow of water through a pipe). However, the concepts and terminology remain the same for more complex "processes."

A feedback control system must be able to respond to two different types of change: change in set point (servomechanism control) and change in load (regulator control). Both types of change are significant in mineral processing operations, and control systems must be designed accordingly.

24.2.2. Analysis of Process Dynamics

For the design of a control scheme, the time dependency or dynamics of a process must be analyzed. This analysis of process dynamics can be:

1. Theoretical, that is, writing the unsteady state equations for the process and analyzing their characteristics.
2. Experimental, that is, testing the operating process.

Clearly, the latter is not possible if the control scheme is designed before the process unit has been constructed; also, limitations may be placed on the testing of an operating plant. The theoretical approach, though quite reliable for simple process loops, is often limited by the assumptions that must be made with more complex processes and processes in which there are recycle streams. The combination of the two is often the most successful: theoretical analysis to obtain the basic control system design, followed by experimental analysis for fine tuning the system.

A simple process is now analyzed as an example. A solution of a dissolved salt flows at a constant rate into a perfectly mixed (Section 3.3.2) tank of constant hold-up volume. The concentration of the salt entering the tank varies with time. If the concentration in the tank discharge is to be controlled at a particular level, the relationship between inlet and outlet concentrations must be known.

Since the hold-up volume is constant, the volume flow rate in must equal the volume flow rate out (assuming density is constant). Thus a transient mass balance may be written.

flow rate$_{salt\ in}$ − flow rate$_{salt\ out}$
= rate of accumulation of salt in the tank (24.1a)

that is,

$$I_V C_{V,I} - O_V C_{V,o} = \frac{d(VC_{V,o})}{dt} \quad (24.1b)$$

($C_{V,I}$, $C_{V,o}$ = volume concentration in inlet stream and outlet stream at time t, respectively; $I_V = O_V$ = volume flow rate in inlet and outlet streams). Clearly, at steady state, $d(VC_{V,o})/dt = 0$ and Eq. 24.1b becomes

$$I_V C_{V,I} - O_V C_{V,o} = 0 \quad (24.2)$$

In analyzing a process for control purposes, it is a convenient (and common) practice to consider the process variables in terms of their deviation from steady state values, thus Eq. 24.1b becomes:

$$I_V(\Delta C_{V,I}) - O_V(\Delta C_{V,o}) = V \frac{d(\Delta C_{V,o})}{dt} \quad (24.3)$$

($\Delta C_{V,I}$, $\Delta C_{V,o}$ = deviations of $C_{V,I}$ and $C_{V,o}$ from their steady state values at time t, respectively), or

$$\Delta C_{V,I} - \Delta C_{V,o} = \tau \frac{d(\Delta C_{V,o})}{dt} \quad (24.4)$$

(τ = residence time or *time constant* for the process). If the change occurring in the inlet concentration is known, this ordinary linear first-order differential equation can be solved. However, for control purposes, it is common practice to write Eq. 24.4 in terms of the Laplace transform of each variable as

$$\Delta C_{V,I}(s^*) - \Delta C_{V,o}(s^*) = \tau s^* \Delta C_{V,o}(s^*) \quad (24.5)$$

[The Laplace transform of a function fn(t) is defined to be fn(s^*) according to the equation

$$\text{fn}(s^*) = \int_0^\infty \text{fn}(t) \exp(-s^* t)\, dt$$

This transformation changes a function of the independent variable t to a function of the independent variable s^*. Rather than using this equation, it is frequently convenient to use tables of Laplace transforms as found in most texts on process control and in mathematical tables.[1,2,6]]

Equation 24.5 can then be rearranged to give

$$\frac{\Delta C_{V,o}(s^*)}{\Delta C_{V,I}(s^*)} = \frac{1}{\tau s^* + 1} \quad (24.6)$$

which is called the *transfer function* of the process. In fact, many simple processes are *first-order processes* and can be represented by the more general transfer function

$$\frac{K_{(s)}}{\tau s^* + 1}$$

with $K_{(s)}$ being the steady state *gain* of the process defined as

$$K_{(s)} = \frac{\text{change of output steady state}}{\text{change of input steady state}} \quad (24.7)$$

Any process that can be represented by the perfect mixing equation (Eq. 3.8), and there are many in a concentrator, can also be represented by the first-order transfer function. Also, other processes can be represented by a series of perfect mixing vessels, hence by the product of first-order transfer functions. Another simple process that can be represented by a first-order equation is the holding tank with outflow through a resistance that either is linear or can be linearized in a narrow operating range.[1,2]

More complex processes may follow a higher order differential equation, hence higher order transfer functions are required. Although the mathematics is more complicated, the general principles are the same. Another type of process is one with dynamics represented by a *transportation lag* or *dead time*. Flow through a pipe or transportation by conveyor are examples of such a process. Using the same approach as for the first-order system above,

Principles of Automatic Process Control

it can readily be shown that the transfer function for such a process is

$$\exp(-\tau s^*)$$

with the time constant τ now being the transportation time.

Most of the elements or "processes" in a plant can be represented either by a first-order transfer function, a transportation lag transfer function, or a combination. In general this is true, however, only when one controlled variable is considered at a time. If there is interaction between the controlled variables, a more sophisticated approach to control system analysis may be warranted (Section 24.2.6).

Experimental methods for the analysis of process dynamics require the introduction of a known change in the load variable and the measurement of the resulting response in the controlled variables. Normal practice is to introduce an impulse (such as a tracer) or a step change in load variable. Figure 24.2 shows the effect of introducing a step change in feed rate to a pilot plant ball mill in closed circuit with a hydrocyclone. In this example, the ball mill discharges to a sump from which the slurry is pumped to the hydrocyclone, the hydrocyclone underflow returning to the ball mill. A simple feedback control loop is included in the process to control the sump at a preset level by change of water addition to the sump. The response of the grinding circuit product size, and also other response variables, can be seen.

24.2.3. Controllers and Control Algorithms

A controller has two components, the first where the measured value of the controlled variable is compared with the set point value, to determine the error, ϵ, and the second where the error signal undergoes some control action. These two components exist whether the controller is pneumatic, electronic or digital.

A number of basic control actions are in wide use. The simplest, an *on-off controller* has limited use. The control logic in an on-off controller is

$$U(t) = \begin{cases} 1 & \text{for } \epsilon(t) \geq \epsilon_{DB} \\ 0 & \text{for } \epsilon(t) \leq -\epsilon_{DB} \end{cases} \quad (24.8)$$

(U = controller output signal, i.e., value of manipulated variable; ϵ = error signal; ϵ_{DB} = deadband of on-off controller). Since no control action is taken until $\epsilon \geq |\epsilon_{DB}|$, such a controller is generally acceptable only on control loops of lesser importance. In most controllers, however, control action commences as soon as an error is identified. In the *proportional controller*, the controller output is proportional to the error, that is,

$$U(t) = K_{\mathcal{P}}\epsilon(t) \quad (24.9)$$

($K_{\mathcal{P}}$ = proportional control constant, often expressed as the *proportional band* $[= 100/K_{\mathcal{P}}]$). The effect of such a controller is shown in Fig. 24.3a. If a step change were introduced to a first-order process, then without any control (i.e., $K_{\mathcal{P}} = 0$), the response would be as shown. With the introduction of proportional control (i.e., $K_{\mathcal{P}} > 0$ in Fig. 24.3a), the controlled variable would be closer to the set point. However, with proportional control action only, there is always *offset*; that is, the controlled variable will never reach the set point value.

The addition of *integral* or *reset control*, however, eliminates this offset (Fig. 24.3b). The control equation for proportional plus integral control (PI control) is

$$U(t) = K_{\mathcal{P}}\epsilon(t) + K_{\mathcal{R}}\int\epsilon(t)dt \quad (24.10)$$

($K_{\mathcal{R}}$ = integral control constant, often expressed in terms of the *reset time*, $\tau_{\mathcal{R}}$, where $K_{\mathcal{R}} = K_{\mathcal{P}}/\tau_{\mathcal{R}}$).

An additional mode of control is sometimes introduced to speed the response and improve the stability of a control system. In this case, *derivative* or *rate control*, control action is proportional to the

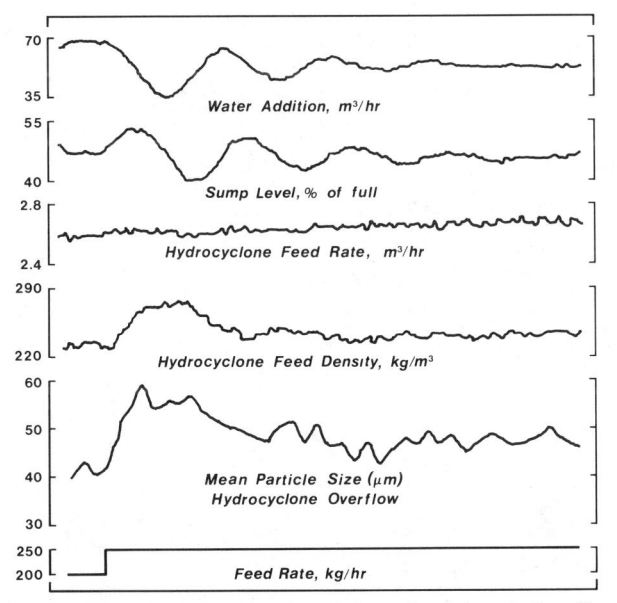

Figure 24.2. Response curves for a pilot plant ball mill–hydrocyclone circuit with step change introduced to feed. Elapsed time is one hour.

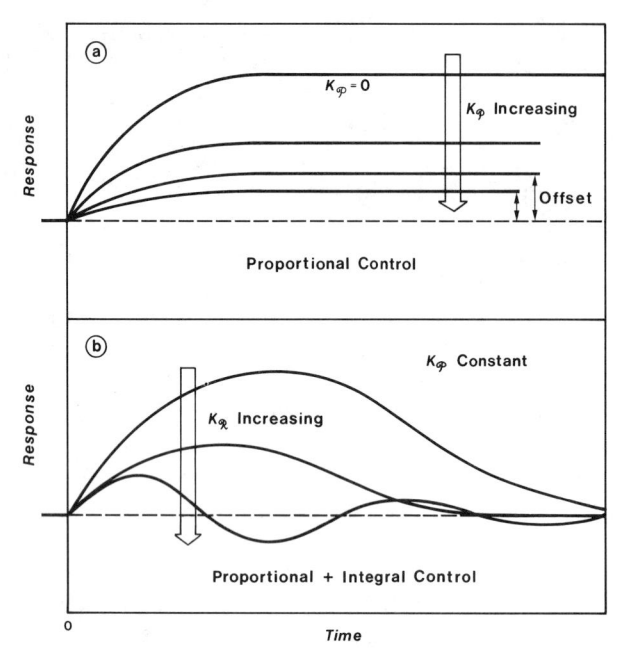

Figure 24.3. Response curves for process with (a) proportional control only and (b) proportional plus integral control modes.

rate of change of the error and so only becomes significant when the error is increasing (or decreasing) rapidly. The control equation for proportional plus integral plus derivative control (PID) is

$$U(t) = K_{\mathcal{P}}\epsilon(t) + K_{\mathcal{R}}\int \epsilon(t)\,dt + K_{\mathcal{D}}\frac{d\epsilon}{dt} \quad (24.11)$$

($K_{\mathcal{D}}$ = derivative control constant). It is also possible to have proportional plus derivative control (*PD*), although it is not used widely.

If a digital computer is used for control, Eq. 24.11 must be modified. Digital control algorithms must be in the form of difference equations, with a time base equal to the *sampling interval*. Equation 24.11 becomes

$$U_i = K_{\mathcal{P}}\epsilon_i + K_{\mathcal{R}}\sum_{j=0}^{i}\epsilon_j \Delta t + \frac{K_{\mathcal{D}}(\epsilon_i - \epsilon_{i-1})}{\Delta t} \quad (24.12)$$

(Δt = sampling interval [time]; subscript *i* designates present value of variable). In fact, with digital control, the actual value of the manipulated variable is not calculated, but only the change in this variable.

24.2.4. Stability of Control Systems

The selection and adjustment of the settings on process controllers (i.e., $K_{\mathcal{P}}$, $K_{\mathcal{R}}$, and $K_{\mathcal{D}}$) depends on the characteristics of the process. The control scheme has been introduced to "control" the process; however, it is possible for a control system to have the reverse effect. Figure 24.4 shows three types of response that are possible. An unstable response is clearly unacceptable; a stable response is required. In the selection of controller settings, it is necessary to determine the point of *critical stability* (Fig. 24.4), then apply an appropriate safety margin. A number of techniques are in use for this determination.

Perhaps the most widely used is that of *frequency response analysis*. In this method, the transfer functions of all elements in the control loop are required. Just as transfer functions were developed for processes in Section 24.2.2, so transfer functions can be written for controllers (transfer function for PID controller is $[K_{\mathcal{P}} + K_{\mathcal{R}}/s^* + K_{\mathcal{D}}s^*]$), measurement

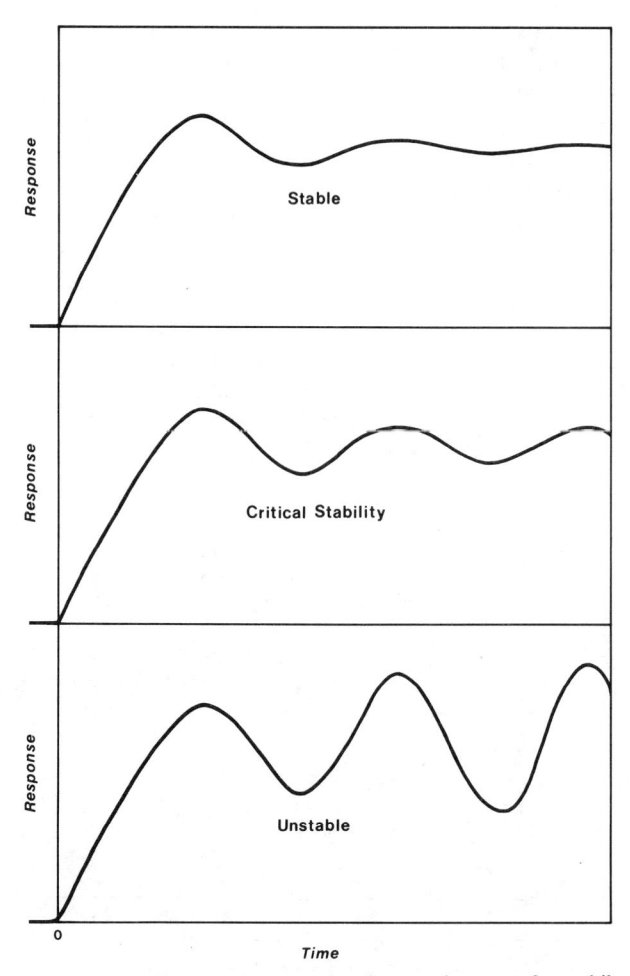

Figure 24.4. Response curves showing requirements for stability.

Principles of Automatic Process Control

devices, and final control elements. Using these transfer functions (and other techniques beyond the scope of this text), control settings can be determined to give stable control.[1-4] Furthermore, techniques have been developed, such as that of Ziegler and Nichols,[7] for selecting optimum controller settings.[1-5]

24.2.5. Complex Control Systems

In the development of a control system for a plant, it is often desirable to use the output of a controller as the set point for another controller. This is called *cascaded control*. For example, if the level in a sump is maintained by water addition (Fig. 24.12a) two control loops may be used. One is a level control loop, where the level is measured and the desired water flow rate determined. The second or inner loop is for controlling the flow rate to that set point.

A generalized block diagram for cascaded control is shown in Fig. 24.5. Such a system may be analyzed using the techniques discussed earlier.

24.2.6. Modern Control Concepts

An alternative approach to control has been developed in other industries, and recently its applicability to mineral processes has been demonstrated.[8-10] Known as the "state space" approach (or simply "modern control theory"), the single input to single manipulated variable of classical control is superseded by the more flexible *multivariable control*. In multivariable control,[3] one or more variables can be manipulated based on the measured values of a number of controlled variables.

In the state space approach, the entire set of variables required to define the state of a process at any time t is represented by a vector $\vec{J}$ called the state vector. The state vector is composed of state variables, which for a process in a concentrator may be flow rates, pulp densities, particle size distributions, and component grades at all points in the process circuit. The dynamic response of $\vec{J}$ to changes in a vector of manipulated variables $\vec{U}$ (e.g., feed rate, water addition rate, pumping rate within a process circuit, or reagent addition rate) is given by

$$\frac{d\vec{J}}{dt} = [K]_s \cdot \vec{J} + [K]_u \cdot \vec{U} \qquad (24.13)$$

($[K]_s, [K]_u$ = process matrices {where each element is a transfer function}, with $[K]_s \cdot \vec{J}$ and $[K]_u \cdot \vec{U}$ representing the influence of state variables and manipulated variables on process performance, respectively).

The design of a control system using the state space approach, then, requires the specification of a control strategy that results in the desired response to a disturbance. The state space control equation for a multivariable feedback control loop (cf. Eqs. 24.9–24.11) is then

$$\vec{U} = [K]_c \cdot (\vec{J}_M - \vec{J}_{M,SP}) \qquad (24.14)$$

($[K]_c$ = matrix of controller constants or operators; $\vec{J}_M$ = vector representing measured values of the state variables; $\vec{J}_{M,SP}$ = set point).

Almost all multiple-input/multiple-output processes exhibit some degree of interaction. A number of methods may be used to cope with these interactions. One method is that of *pairing* the controlled and manipulated variables and determining the relative gain of each pair.[3] Another method is by *decoupling* the interacting processes.[3,8] In this method, decoupling elements are inserted in the control configuration to eliminate the process interaction. For

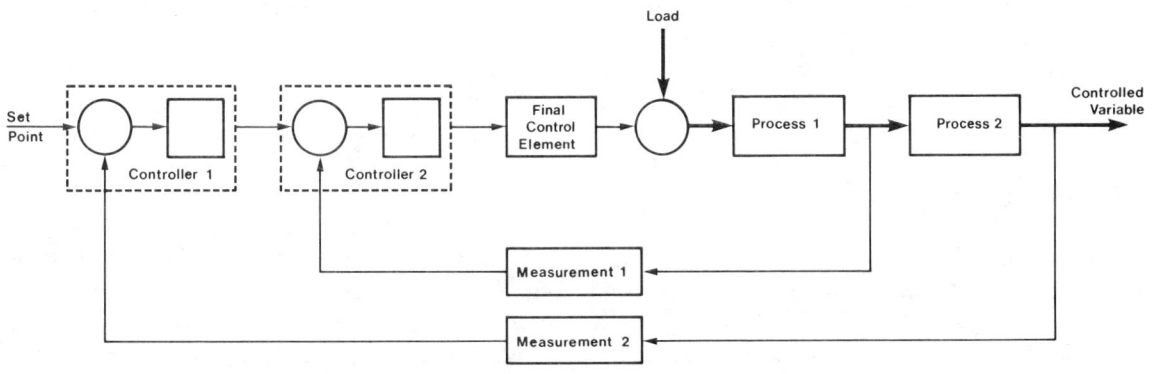

Figure 24.5. Block diagram for a cascaded control system.

example, the process matrix $[K]_s$ could be modified

$$[K]_{s,DC} = [K]_s \cdot [K]_{DC} \qquad (24.15)$$

($[K]_{DC}$ = matrix of decoupling elements; $[K]_{s,DC}$ = apparent process matrix resulting from decoupling [a diagonal matrix]). The result is a process in which there appears to be no interaction.

The state space approach also provides a basis for adaptive control, optimal control, and the filtering of noise from controlled variable measurements.[3,11-13] It can be expected to become more widely used as more sophisticated control strategies are introduced. A detailed explanation, however, is beyond the scope of this text.

24.3. INSTRUMENTATION

As with the chemical process industries, both pneumatic and electronic analogue instrumentation have been installed in mineral concentrators. However, it is only with the advent of the digital computer, with its capabilities for monitoring, data storage, and real-time control, that process control has become widespread. Thus most control systems installed today are based on a digital computer, either with the computer in a supervisory control capacity, with analogue controllers for individual control loops, or with direct digital control.

The range of instrumentation available (and in fact required) for any control system installation is very broad, and clearly beyond the scope of this text. Information is readily available in manufacturers' literature. Discussion here is limited to those instruments developed primarily for process measurements in concentrators.

24.3.1. Dry Solids Conveying

The use of *belt scales* to determine the mass transported by conveyor belt was one of the early on-line measurements in a mill. The measurement is needed for both accounting and control purposes. A wide range of belt scales are available.[14]

24.3.2. Slurry Flow Rate

The flow rate of slurries in pipes can be measured[15] using either *magnetic* or *ultrasonic flowmeters* (Fig. 24.6).

Magnetic flowmeters, the more widely used, require a conductive liquid (or slurry) to flow through

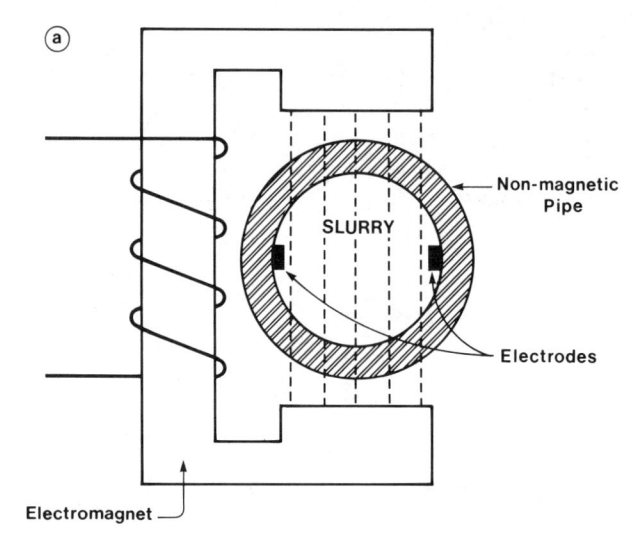

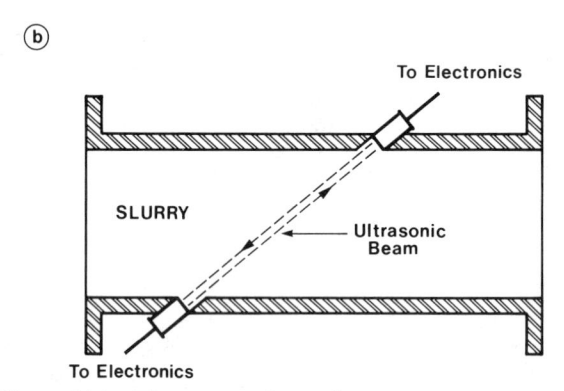

Figure 24.6. Flow meters for on-line measurement of slurry flow rates. (*a*) Magnetic flowmeter. (*b*) Ultrasonic flowmeter.

a magnetic field; a voltage is induced between the electrodes (Fig. 24.6*a*) that is proportional to the velocity of flow. Thus a volume flow rate can be determined; to determine the mass flow rate of contained solids, the slurry density must also be determined. Thus it is common practice to combine a magnetic flowmeter with a density gauge (Section 24.3.3). In ultrasonic flowmeters, pressure pulses are transmitted in two directions (Fig. 24.6*b*); the difference between the two pulse time-delay measurements is proportional to flow.

24.3.3. Pulp Density

The density, or percent solids, of a flowing slurry can be determined using a *nuclear density meter* or *gamma gauge* (Fig. 24.7). Gamma rays from a radioactive source pass through the slurry; the transmission of the gamma rays to the ionization

Instrumentation

chamber-type radiation detector is inversely proportional to the slurry density.

Other techniques, such as the absorption of ultrasonic energy, are possible but have not been widely used.

24.3.4. Particle Size

The measurement of the size distribution of coarse particles (3–300 mm) in transport on a conveyor belt has recently become possible.[16] In an optical instrument (Fig. 24.8a), the surface of the material on a moving conveyor belt is illuminated with light at low angles, thereby casting shadows along the line of observation. The light intensity pattern generated allows discrimination of the particles, and the chord length can be measured. This instrument is suitable for measuring size distributions of feed to a crushing or autogenous grinding circuit.

A number of methods are used for fine particle size distribution measurement. An inferred-size technique[17-19] requires the development of an empirical equation that relates the particle size distribution (as a single parameter) to the operating conditions of a hydrocyclone. Although this technique has been successfully applied, it has the limitation that any change in ore properties requires changes to be made in the empirical equation.

Direct measuring instruments are available for fine particle size distribution measurement. The most widely applied (Fig. 24.8b) utilizes the attenuation of ultrasonic energy that occurs on transmission through a slurry. This attenuation depends on both density and particle size distribution; however, it is possible to discriminate between the two.[20-22] Another instrument[23,24] utilizes the size dependence of the light scattering that results from passing a monochromatic light source (laser beam) through a dilute slurry (Fig. 24.8c).

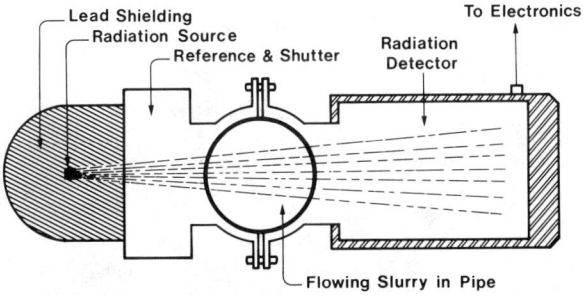

Figure 24.7. Nuclear density meter for on-line measurement of slurry density.

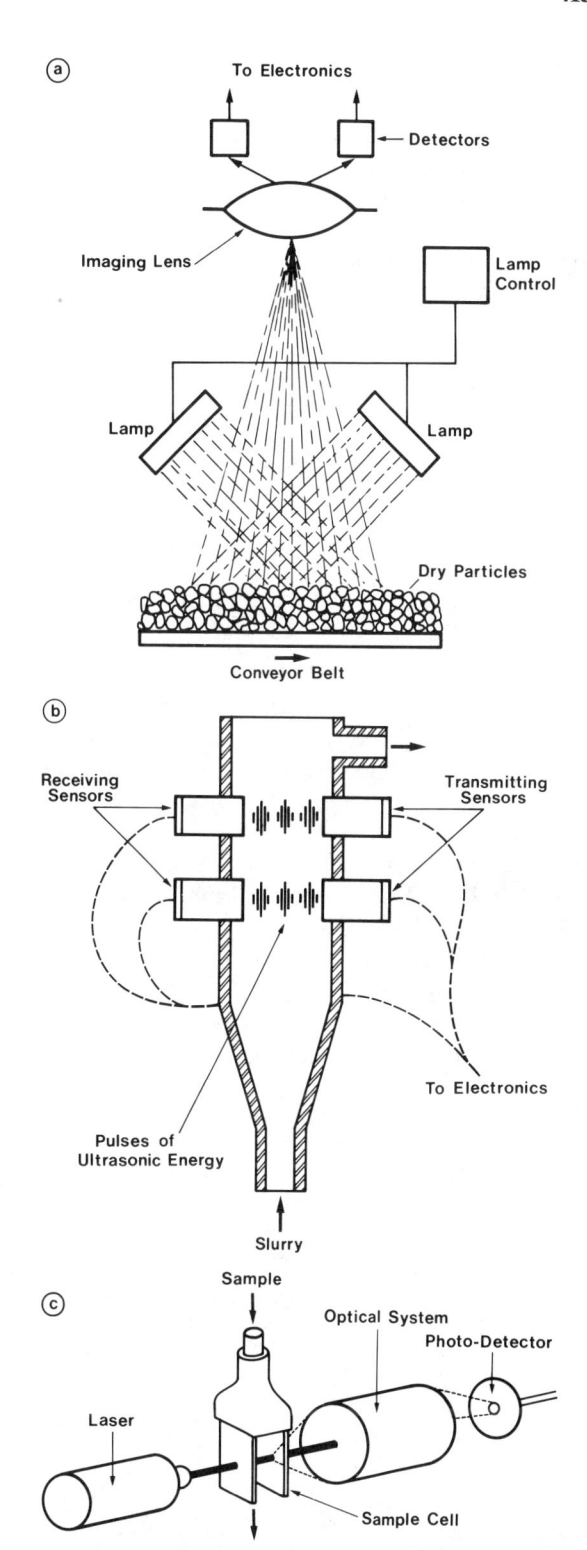

Figure 24.8. On-line measurement of particle size distributions. (a) Optical measurement of coarse particle sizes (3–300 mm); material on conveyor belt. (b) Ultrasonic measurement of size distribution of particles in slurry. (c) Laser light scattering measurement of particles in slurry.

24.3.5. On-Stream Assaying

Perhaps the most significant development in instrumentation for mineral process control was that of the on-stream X-ray fluorescence analyzer. This provided a method for obtaining assays of slurry streams quickly enough (about 10–15 min) for use in control. Such analyzers are in wide use and have been described in detail elsewhere.[25]

In X-ray fluorescence analysis, a source of radiation causes excitation in sample elements and subsequent emission of response radiation characteristic of each element. The emitted characteristic radiation is then analyzed quantitatively.

Two basic techniques of on-stream X-ray fluorescence analysis are in use; a centralized system and an in-stream probe system. The former is the most widely used. A single high energy excitation source (X-ray generator) is provided for the analysis of a number of slurry samples pumped to a central location where the complete analyzer is installed. The samples are analyzed sequentially. Such a system requires careful sampling for successful analysis. With the in-stream probe,[26,27] sample excitation is carried out with low energy sources, usually radioactive isotopes (Fig. 24.9). The detector is housed inside the probe, so only an electrical signal is sent to a centralized control room. Although in-stream probes are not as widely used as centralized systems, it can be anticipated that their use will increase as improved designs become available.

X-ray fluorescence analysis is restricted to heavier elements. For the lighter elements, such as aluminum and silicon in iron ores, another technique that has been successfully applied on-stream is neutron activation analysis.[28,29]

24.4. CONTROL APPLICATIONS

A number of applications of automatic control in concentrators are briefly discussed in this section. Detailed descriptions are available elsewhere.[18,30–32] Of course processes other than those discussed in this section are also controlled.

Automatic control of individual processes in a plant is only a first step toward overall plant control and optimization. Separate control of a grinding circuit and a flotation circuit is normally of value; however, for example, if changes in liberation characteristics of an ore occur, the result is seen in diminished flotation circuit performance. But it is in the grinding circuit that changes must be made, possibly by changing the set point of the product particle size distribution.

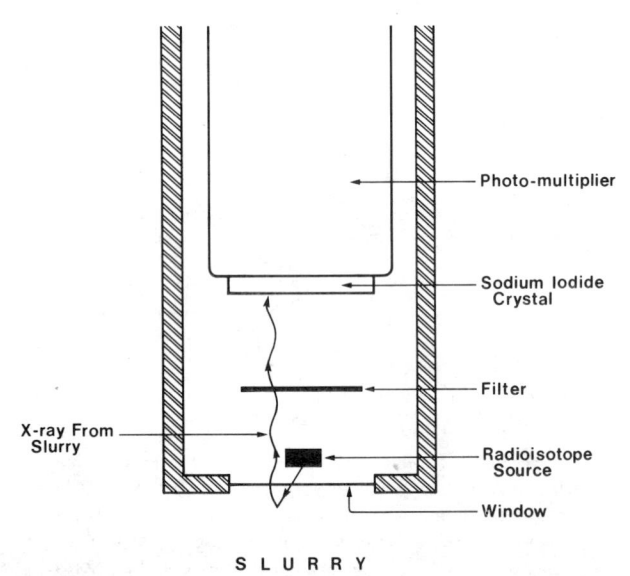

Figure 24.9. In-stream X-ray fluorescence probe for on-line assaying.

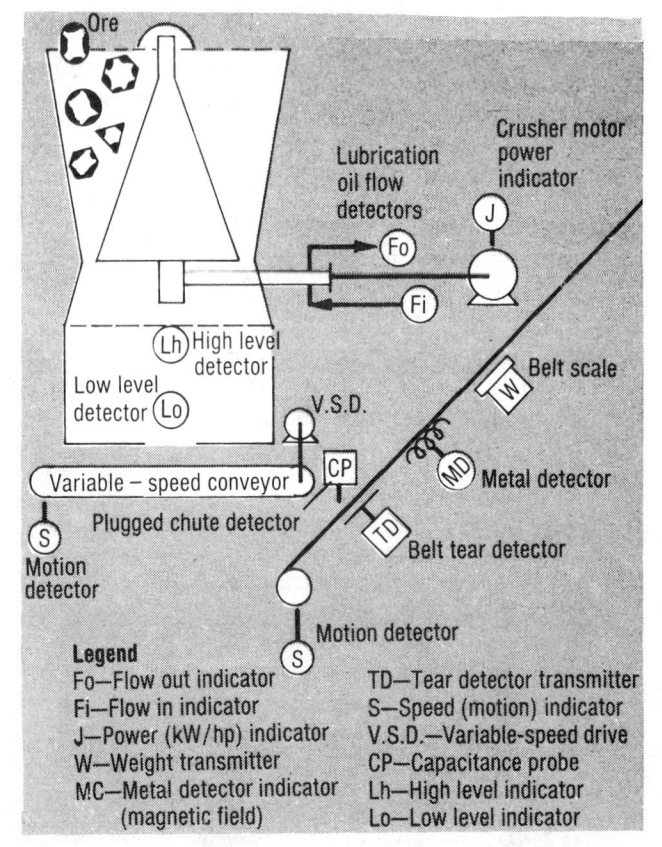

Figure 24.10. Instrumentation for primary crusher control. (Courtesy *Engineering and Mining Journal*.)

Control Applications

24.4.1. Crushing Control

Primary crusher controls are limited to protective alarms and shutdown mechanisms. Bearing pressure and temperature, crusher power, and ore level in the chamber below the crusher, are usually monitored, and alarm or automatic shutdown levels set (Fig. 24.10).[30]

With secondary and tertiary crushing, using cone crushers,[18,30-35] there are a number of possible objectives, the control objective depending on the function of the crushing plant. For example, a crushing plant producing a final sized product (e.g., a high grade iron ore) may have as a control objective the minimizing of undersize while maintaining a constant throughput. In the crushing plants of most concentrators, however, the objectives are one of the following, the choice largely being based on the throughput that can be accepted in the subsequent processing steps:

1. To maximize throughput while producing a product finer than some specified top size.
2. To produce as fine a product as possible at constant throughput.

In all cases, of course, the protection of the crusher from plugging or overload is a primary consideration.

There are three variables that can be manipulated in normal crusher operation: ore feed rate, feed size (possibly), and crusher opening. Maximum throughput can be achieved when the crusher is choke fed, so by detecting level in the crusher bowl, feed rate can be adjusted to maintain this condition.

The throughput of a crusher increases with increasing power draft. At higher throughputs, the power consumption per tonne also increases, resulting in a finer product.[18,33,34] Thus for a specified screen opening in closed circuit crushing, an effective method for maximizing throughput (objective 2) is to maintain the highest possible power draft.

An example of a crushing circuit control scheme is shown in Fig. 24.11.

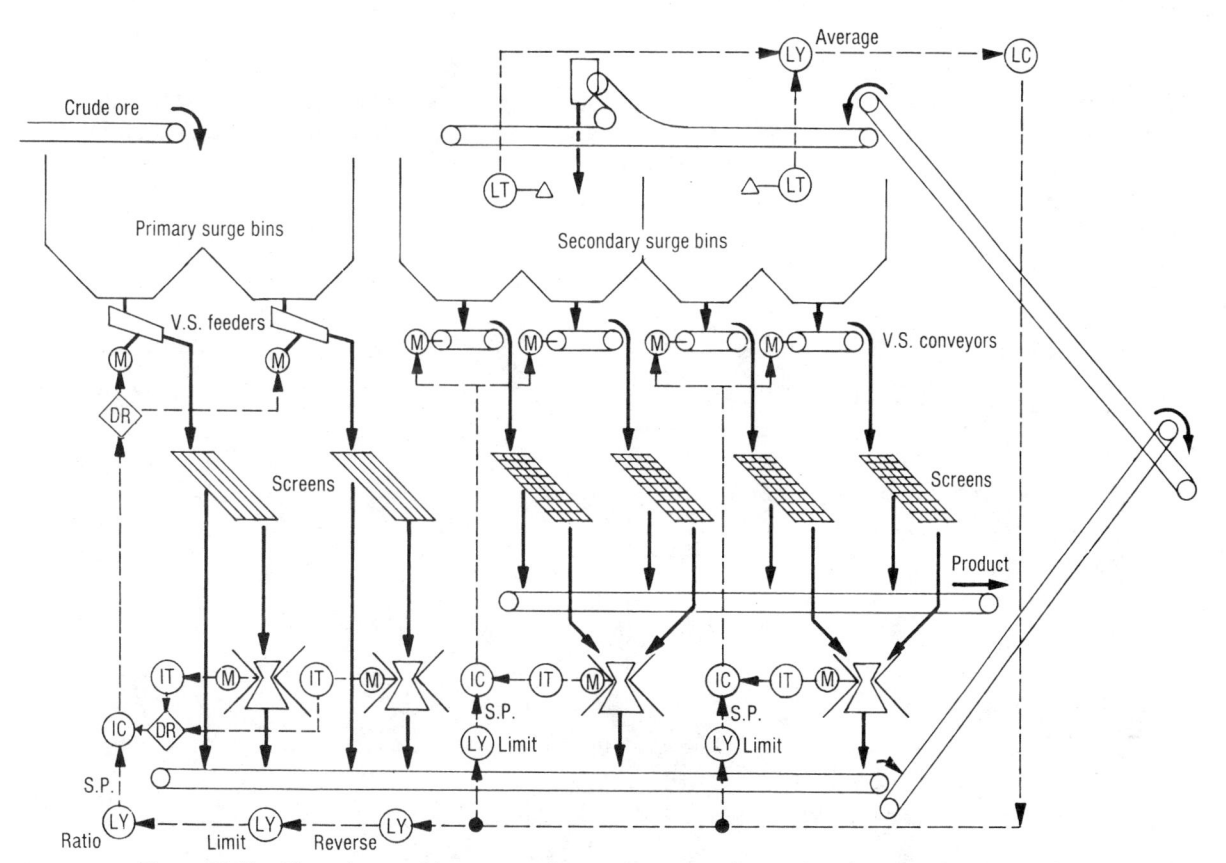

Figure 24.11. Flow sheet and instrument diagram for a secondary and tertiary crushing plant: LC. level controller; LT, level transmitter; IC, current controller; IT, current transmitter.[18,30] (Courtesy *Engineering and Mining Journal.*)

24.2.2. Grinding Control

In a grinding circuit, grinding and classification are interdependent, so for control purposes the whole circuit must be treated as one process.[18,30]

Possible control objectives for a grinding circuit were introduced in Section 24.1. To meet these objectives, the following process measurements can be made: ore feed rate, water addition rate, classifier feed rate and pulp density, circuit product particle size distribution, mill power draft, and mill load (by bearing pressure or sound level measurement). Only ore feed rate, water addition rate, and classifier feed rate (using a variable-speed pump) can be varied independently, that is, are possible manipulated variables.

Single-Stage Ball Mill Circuits. Three basic control schemes are used with a single-stage ball mill in closed circuit with a classifier.[18,19,22,32,36,37] They are based on measurement and control of classifier feed density (Fig. 24.12a), circulating load (i.e., classifier feed rate: Fig. 24.12b), and circuit product particle size distribution.

All three schemes have been successfully applied. The first, however, though responding well to long-term changes in ore characteristics, gives poor short-term response.[18] In this scheme, any change in ore feed characteristics results in a change in the circulating load, hence in the level in the sump. This sump level is measured and maintained by varying the water addition to the sump, which changes the classifier feed density. Classifier feed density is then controlled by appropriate changes in the ore feed rate.

Rod Mill Ball Mill Circuits. Two basic control schemes are used for rod mill–closed circuit ball mill grinding circuits,[18,20,21,38] one based on the measurement and control of product particle size distribution, the other on maintaining circulating load.

In the size-based control scheme, two primary control loops are used:

1. A hydrocyclone loop, which controls the hydrocyclone overflow size by varying the water addition to the classifier feed sump (water addition increased if size greater than set point value).
2. A circulating load loop, in which the classifier feed sump level is controlled by varying the ore feed rate to the rod mill (with ratio control of water addition to the rod mill).

In the alternative control scheme, the objectives of maximizing throughput at a preset grind is achieved by controlling both the rate of water addition to the rod mill and the classifier feed sump, and the classifier feed sump level, to maintain the circulating load at the maximum power draft.

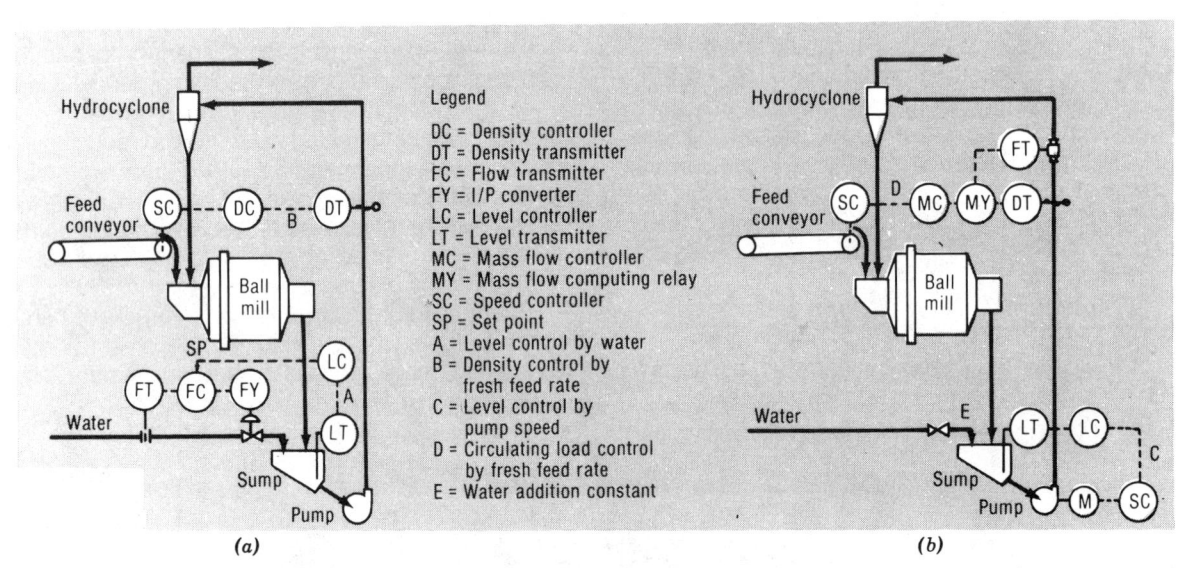

Figure 24.12. Ball mill grinding circuit control schemes. (a) Based on maintaining constant cyclone feed density and using a fixed speed pump. (b) Based on maintaining constant circulating load and using a variable speed pump.[18,30] (Courtesy *Equipment and Mining Journal*.)

References

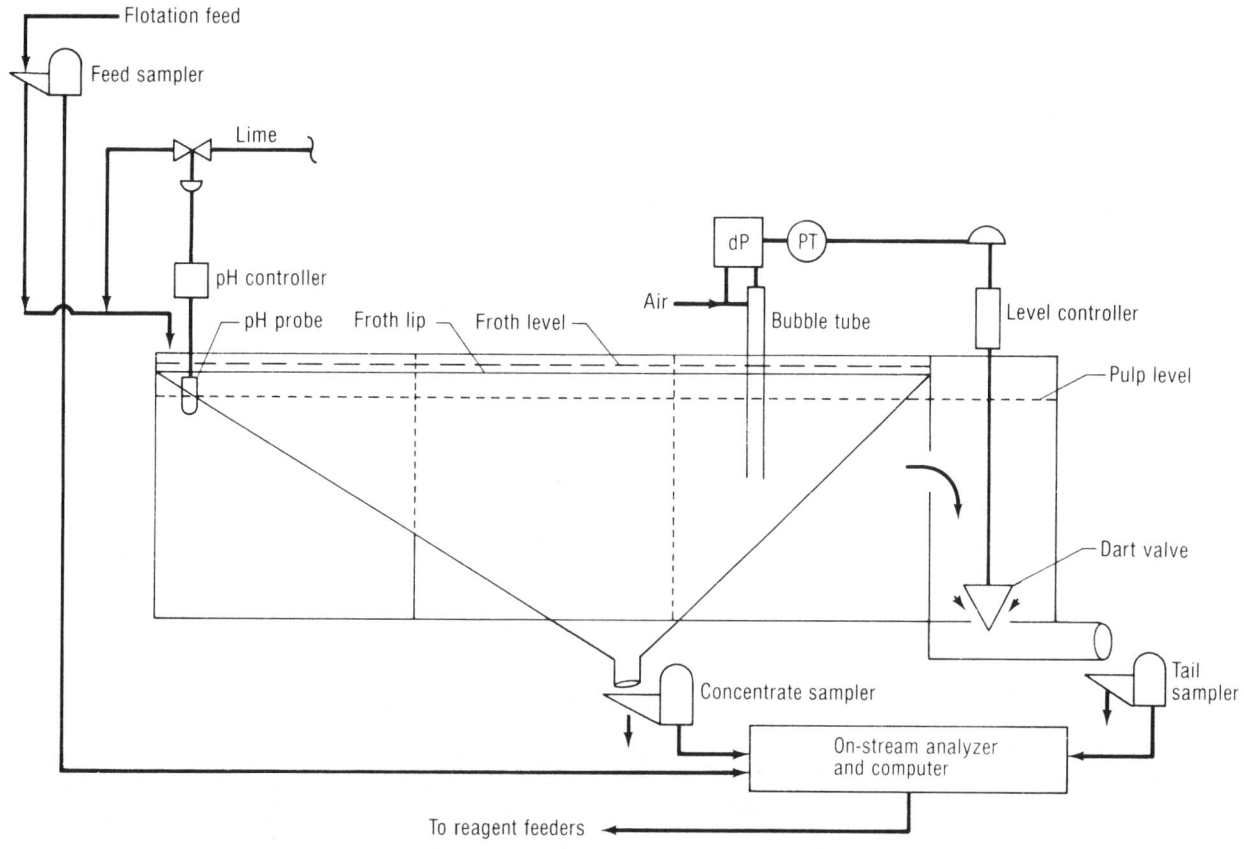

Figure 24.13. A flotation circuit control scheme: dP, differential pressure cell; PT, pressure transmitter.[30] (Courtesy *Engineering and Mining Journal*.)

Autogenous Grinding Circuits. With autogenous mills, grinding efficiency is a maximum at maximum power draft. A typical control loop used, then, is that of varying the ore feed rate to maintain a constant power draft.[18,31] However, the power draft increases with mill loading to a maximum and then decreases with further mill loading (cf. Fig. 8.11); thus it is important for the power draft set point to always be less than the maximum. To prevent overloading, such as may occur when a harder ore is encountered, mill load may be determined by measuring the mill bearing pressure.

Since autogenous mills, operated in closed circuit, may be followed by either pebble or ball mills also operated in closed circuit, the control scheme required becomes extremely complex.

Semi-autogenous mills have been controlled in a similar manner to autogenous mills.[18,39-41]

24.4.3. Flotation Control

Many flotation circuit control schemes have been reported in detail elsewhere.[30,42-44]

For an individual bank of flotation cells, aeration rate and pulp level (or froth depth) can be controlled (Fig. 24.13). Also, the reagent addition and pH of a flotation circuit can be controlled.

The response of a flotation circuit is normally measured by assay of the products; both grade and recovery are of interest. As discussed in Chapter 16, changing the chemical conditions clearly will affect the performance of a flotation circuit, and so control of reagent addition is widely used. Physical changes, such as aeration rate or pulp level, however, often give a faster response, and so are also used for control.

REFERENCES

1. P. Harriot, *Process Control*, McGraw-Hill (1964).
2. D. R. Coughanowr and L. B. Koppel, *Process Systems Analysis and Control*, McGraw-Hill (1965).
3. F. G. Shinskey, *Process Control Systems*, McGraw-Hill (1967).
4. Y. Takahashi et al., *Control and Dynamic Systems*, Addison-Wesley (1970).

5. M. Athans and P. L. Falb, *Optimal Control,* McGraw-Hill (1966).
6. R. H. Perry and C. H. Chilton (Eds.), *Chemical Engineers' Handbook,* 5th ed., McGraw-Hill (1973).
7. J. G. Ziegler and N. B. Nichols, "Optimum Settings for Automatic Controllers," *Trans. ASME,* **64,** 759–768 (1942).
8. H. L. Wade and J. W. Chang, "A Decoupling Technique with Application in Grinding Control," in *Instrumentation in the Mining and Metallurgy Industries,* Vol. 4, T. O. Arney and A. L. Mular (Eds.), pp. 1–7, Instrument Society of America (1976).
9. A. L. Hinde and R. P. King, "Minimal Variance Control Strategies for Wet Milling Circuits," AIME/SME Preprint No. 77-B-86 (1977).
10. J. A. Herbst and K. Rajamani, "Control of Grinding Circuits," in *Computer Methods for the 80's in the Mineral Industry,* A. Weiss (Ed.), pp. 770–786, AIME (1979).
11. W. D. T. Davies, *System Identification for Self-Adaptive Control,"* Wiley (1970).
12. D. M. Anslander et al., *Introducing Systems and Control,* McGraw-Hill (1974).
13. A. E. Brysen and Y. Ho, *Applied Optimal Control,* Blaisdell Publishing (1969).
14. H. Colijn, *Weighing and Proportioning of Bulk Solids,* Trans Tech Publications (1975).
15. P. Harrison, "Flow Measurement—A State of the Art Review," *Chem. Eng.,* **87,** 97–104 (Jan. 14, 1980).
16. S. J. Bruno and J. H. Vignos, "On-Line Instrument for Coarse Particle Size Distribution Measurement," AIME/SME Fall Meeting (1979).
17. A. J. Lynch, T. C. Rao, and W. J. Whiten, "Technical Note on On-Stream Sizing Analysis in Closed Circuits," *Proc. Aust. IMM,* **223,** 71–73 (1967).
18. A. J. Lynch, *Mineral Crushing and Grinding Circuits—Their Simulation, Optimization, Design, and Control,* Elsevier (1977).
19. K. L. Manning, "Computer Control of Grinding at Climax," *Am. Min. Congr. J.,* **64,** 47–50 (November 1978).
20. C. B. Webber and L. S. Diaz, "Automatic Particle Size and Rod Mill Control at Craigmont," *Can. Min. J.,* **94,** 36–37 (June 1973).
21. C. B. Webber and L. S. Diaz, "Automatic Particle Size and Rod Mill Tonnage Control at Craigmont Mines Ltd," *Proc. 5th Can. Miner. Process. Meeting,* pp. 31–60 (1973).
22. R. E. Hathaway, "A Proven On-Stream Particle Size Monitoring System for Automatic Grinding Circuit Control," *Mining and Metallurgy Symposium of ISA* (1972).
23. E. L. Weiss and H. N. Frock, "Rapid Analysis of Particle Size Distributions by Laser Light Scattering," *Powder Technol.,* **14,** 287–293 (1976).
24. T. Bay, H. J. Kortright, and E. C. Muly, "Continuous Particle Size Analysis and Grinding Control," in *Instrumentation in the Mining and Metallurgical Industries,* Vol. 3, Paper 668, ISA (1975).
25. H. R. Cooper, "On-Stream X-Ray Analysis," in *Flotation—A. M. Gaudin Memorial Volume,* M. C. Fuerstenau (Ed.), pp. 865–894, AIME/SME (1976).
26. N. W. Stump and A. N. Roberts, "On-Stream Analysis and Computer Control at the New Broken Hill Consolidated Ltd. Concentrator," *Trans. AIME/SME,* **255,** 143–148 (1974).
27. R. A. Fookes et al., "Plant Trials of Radioisotope Immersion Probes for On-Stream Analysis of Mineral Process Streams," *Trans. IMM (C),* **82,** C21–C25 (1973).
28. M. Borsaru and R. J. Holmes, "Determination of Silica in Bulk Iron Ore Samples by Neutron Activation Analysis," *Anal. Chem.,* **50,** 296–298 (1978).
29. R. J. Holmes, A. J. Messenger, and J. G. Miles, "Dynamic Trial of an On-Stream Analyzer for Iron Ore Fines," *Proc. Aust. IMM,* **274,** 17–22 (June 1980).
30. W. E. Horst and R. C. Enochs, "Instrumentation and Process Control," *Eng. Min. J.,* **181,** 70–95 (June 1980).
31. G. Bourque and D. Runnels, "The Mount Wright Story—Total Plant Control by a Process Computer at Quebec Cartier's Mount Wright Property," *CIM Bull.,* **70,** 89–96 (April 1977).
32. J. L. Bolles, A. J. Broderick, and H. R. Wampler, "Morenci Concentrator Process Control System," *Trans. AIME/SME,* **262,** 310–319 (1977).
33. J. H. Fewings and W. J. Whiten, "Crushing Control Systems Development at Mount Isa Mines Limited," *IFAC Symp. Automatic Control Min., Miner. Met. Process.,* pp. 119–124 (1974).
34. R. C. Kellner and K. J. Edmiston, "Investigation of Crushing Parameters at Duval Serrita Corporation," *Trans. AIME/SME,* **258,** 345–352 (1975).
35. M. D. Flavel, "Control of Crushing Circuits Will Reduce Capital and Operating Costs," *Min. Mag.,* **138,** 207–213 (1978).
36. W. D. Gould and T. O. Arney, "Design and Start-up of the Pinto Valley Digital Control System," *Trans. AIME/SME,* **262,** 76–80 (1977).
37. S. R. Holsinger, "Computer Controlled Grinding Circuit at Silver Bell," *Min. Eng.,* **30,** 1661–1664 (1978).
38. T. E. Perkins and L. Marnewecke, "Automatic Grind Control at AMAX Lead Co.," *Min. Eng.,* **30,** 166–170 (1978).
39. J. McManus, A. R. Paul, and F. Yu, "Process Control at the Lornex Grinding Circuit," *CIM Bull.,* **71,** 146–151 (1978).
40. J. E. Bailey and H. B. Carson, "Cyprus Pima Mining Company's Computer Based Control System," AIME/SME Preprint 75-B-22 (1975).
41. J. H. Bassarear, "Panel Discussion on Grinding Circuit Control (Cyprus Pima)," AIME/SME Preprint 78-B-110 (1978).
42. H. R. Cooper, "Trends in Computer Automation of Flotation Circuits," AIME/SME Preprint 79-101 (1979).
43. E. V. Manlapig and D. J. Spottiswood, "Present Practices in the Computer Control of Copper Flotation Plants," *Trans. AIME/SME,* **265** (1980).
44. H. W. Smith, "Computer Control in Flotation Plants," in *Flotation—A. M. Gaudin Memorial Volume,* M. C. Fuerstenau (Ed.), pp. 963–993, AIME/SME (1976).

Chapter Twenty-Five

Economics

Fiedler's Rule: "*Forecasting is very difficult, especially if it's about the future.*"

Cheops's Law: "*Nothing ever gets built on schedule or within budget.*"

This chapter is intended as an introduction to the economics of a mineral processing operation. More detailed discussions are readily available in a number of excellent texts,[1-4] although detailed data on actual mineral processing plants are comparatively limited.[5-11]

Before any plant can be built, an estimate of the capital and operating costs must be made so that it can be established that the operation should operate at a net profit (equal to the total income minus *all* expenses). By itself though, a net profit is not enough: the project must also be worthwhile in terms of the amount invested. Thus in this chapter we consider three aspects: capital investment, operating costs, and investment worth.

The major components of a cost estimate are set out in Fig. 25.1 and are shown in more detail in Tables 25.1[3] and 25.3.[3] In determining any of these costs, there are three likely sources of error: those due to the unreliability of the data, those due to inexperience, and those due to omitted costs. The latter factor can lead to disproportionately large errors, and for this reason checklists such as Tables 25.1 and 25.3 should always be used. Costing exercises should also be carried out in a clear, detailed, tabular form. This is because the data normally must be analyzable by other people, and tables make this easier (in fact, additional summary tables should be used to enhance readability). Also, a tabular format allows easy adjustment or updating of data, which is necessary as any project progresses and is difficult without a concise, systematic approach.

Cost estimates must be made at all stages of any proposed project, be it a potential ore body, or a new process being developed in a laboratory. Even comparatively rough costing methods give an indication of whether a project is, or ever can be, worthwhile. On the other hand, economic analysis of apparently nonviable projects can highlight the areas where cost savings are necessary, or possible, to make the project viable.

25.1. CAPITAL COSTS

The capital cost of a project is the total amount of money invested. It consists of two main components: the fixed capital, that used to build the plant, and the working capital, that tied up in day-to-day operating expenses. In turn, the fixed capital can be further subdivided as shown in Fig. 25.1 and Table 25.1.

Three basic approaches are used to estimate capital costs. The simplest extrapolates from the cost of an existing plant. This may be used when very little information about the flow sheet is available. Even assuming that the estimator's experience is sufficient, this method can be expected to have an accuracy of only ±50%. The most complex approach is a detailed costing of all items that must be purchased; here the accuracy can be as close as ±5-10%. This method is expensive and requires considerable experience; thus it is restricted to final planning in preparation for actual construction, and as such is beyond the scope of this introductory discussion. It

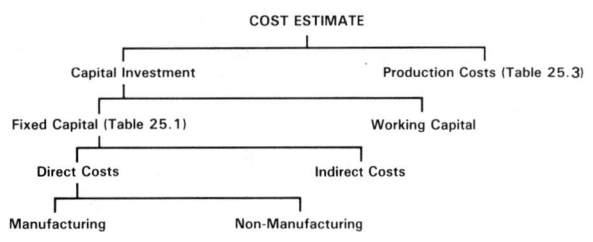

Figure 25.1. The major components of a cost estimate.

therefore is not considered further except to note that the principle of the method is self-explanatory.

In between these two extremes there is a variety of methods that use factors to convert the total equipment cost into a plant cost, and these methods are emphasized, since they are of most use to engineers. But before describing these methods, it is necessary to describe cost indexes. These are essential because most cost data are out of date, and indexes allow such data to be used for the *estimation* of current or future costs.

25.1.1. Cost Indexes

Cost indexes are used to update costs by the following formula:

$$\text{present cost} = \text{original cost} \times \frac{\text{present cost index}}{\text{cost index at time of original purchase}} \quad (25.1)$$

A number of cost indexes exist, because cost increases occur in a variety of areas, such as the cost of materials (steel, concrete, wood, etc.), energy, labor (including productivity). The various indexes differ in the way in which these components are proportioned. It is therefore important to select the index that is relevant to the given situation. It should always be remembered that because cost indexes represent *average* cost increases, they are particularly prone to error when applied to a single item of equipment only. However, when applied to a number of items of equipment individually the "highs" can be expected to cancel out the "lows." Even then, data updated by more than 10 years (or say $> 50\%$ increase in cost) are likely to be of questionable accuracy.

M. & S. Cost Indexes. This index gives trends for *installed equipment* costs and consequently is one of the most useful. It is available in a number of forms,

TABLE 25.1: FIXED-CAPITAL INVESTMENT

DIRECT COSTS
1. *Purchased equipment*
    All equipment listed on a complete flow sheet
    Spare parts and non-installed equipment spares
    Surplus equipment, supplies, and equipment allowance
    Inflation cost allowances
    Freight charges
    Taxes, insurance, duties
    Allowance for modification during startup

2. *Purchased equipment installation*
    Installation of all equipment listed on complete flow sheet
    Structural supports, insulation, paint

3. *Instrumentation and controls*
    Purchase, installation, calibration

4. *Piping*
    Process piping
    Pipe hangers, fittings, valves
    Insulation — piping, equipment

5. *Electrical equipment and materials*
    Electrical equipment — switches, motors, conduit, wire, fittings, instrument and control wiring, lighting, panels
    Electrical materials and labor

6. *Buildings (including services)*
    Process building — substructures, superstructures, platforms, supports, stairways, ladders, access ways, cranes, monorails, hoists
    Auxiliary buildings — administration and office, medical or dispensary, cafeteria, garage, product warehouse, parts warehouse, guard and safety, fire station, change house, personnel building, shipping office and platform, research laboratory, control laboratory
    Maintenance shops — electric, piping, machine, welding, carpentry, instrument
    Building services — plumbing, heating, ventilation, dust collection, air conditioning, building lighting, elevators, telephones, intercommunication systems, painting, sprinkler systems, fire alarm

7. *Yard improvements*
    Site development — site clearing, grading, roads, walkways, railroads, fences, parking areas, wharves and piers, recreational facilities, landscaping
    Town site and housing

8. *Service facilities*
    Utilities — steam, water, power, compressed air, fuel
    Tailings disposal — dam
    Facilities — boiler plant incinerator, river intake, water treatment cooling towers, water storage, electric substation, refrigeration plant, air plant, fuel storage, waste disposal plant, fire protection
    Nonprocess equipment — office furniture and equipment, cafeteria equipment, safety and medical equipment, shop equipment, automotive equipment, yard material-handling equipment, laboratory equipment, locker-room equipment, garage equipment, shelves, bins, pallets, hand trucks, housekeeping equipment, fire extinguishers, hoses, fire engines, loading stations
    Distribution and packaging — raw-material and product storage and handling equipment, product packaging equipment, blending facilities, loading stations

9. *Land*
    Surveys and fees
    Property cost

INDIRECT COSTS
1. *Engineering and supervision*
    Engineering costs — administrative, process, design and general engineering, drafting, cost engineering, procuring, expediting, reproduction, communications, scale models, consultant fees, travel

2. *Construction expenses*
    Construction, operation and maintenance of temporary facilities, offices, roads, parking lots, railroads, electrical, piping, communications, fencing
    Construction tools and equipment
    Construction supervision, accounting, timekeeping, purchasing, expediting
    Warehouse personnel and expense, guards
    Safety, medical, fringe benefits
    Permits, field tests, special licenses
    Taxes, insurance, interest

3. *Contractor's fee*

4. *Contingency*

Capital Costs

including:

All industry index, which includes industrial, commercial, and housing equipment.

An average index for processing industries.

A variety of indexes for individual processing industries including one for mining and milling.

These have a basis of 100 in 1926, and are published every 2 weeks in *Chemical Engineering*.

Chemical Engineering Plant Construction Cost Index. Although weighted heavily to equipment costs, this *Chemical Engineering* index also reflects *plant construction* costs, that is, it is also a material/equipment/labor index. Because plant costs have risen more slowly than have equipment costs, this index is recommended for updating total plant costs. Its basis is 100 in 1957–1959, and is published in *Chemical Engineering*.

Engineering News Record Construction Cost Index. This shows variations in labor rates and materials costs for industrial construction and includes weightings for steel, lumber, cement, and labor. It is published in *Engineering News Record*.

Labor and Materials Price Indexes. Where indexes are not available, they can be developed by combining suitable proportions of materials costs and wage rates (e.g., on a 50/50 basis), using appropriate government statistics.

Area Indexes. Cost indexes are based on average data from a whole country. In reality materials and labor costs vary from area to area, so that each area actually has its own index.[4] This problem is normally surmounted by applying a single area index to final plant cost.

25.1.2. Fixed Capital Costs Extrapolated From Existing Plant Costs

The simplest method uses the cost of a previously constructed plant as the basis for determining the cost of a new plant. The relationship is:

$$\mathscr{C}_n = \mathscr{C}_e \left(\frac{I_n}{I_e} \right)^n \left(\frac{CI_n}{CI_e} \right) K_{Lo} \qquad (25.2)$$

($\mathscr{C}_n$, $\mathscr{C}_e$ = costs of new and existing plants respectively; CI_n, CI_e = cost indexes at time of construction of new and existing plants, respectively; I_n, I_e = capacities of new and existing plants, respectively; n = capacity exponent; K_{Lo} = location factor). The accuracy of the method depends significantly on the comparability of the two plants. Ideally the plants should use not only the same process, but the same flow sheet, and if not, the reliability can be expected to fall off rapidly. For chemical plants the exponent n is in the range 0.4–0.9, with an average value about 0.7. Reliable values for mineral processing plants have never been published, presumably because of the difficulty in comparing plants. However, Eq. 25.2 is also valid for individual items of equipment, and in this situation n varies from 0.1 to > 2.0, with an all-equipment average of 0.6 (hence the "six-tenths factor"). Since individual items of mineral processing equipment tend to have above-average values of n, it is likely that the value of n for most mineral processing plants should therefore be greater than 0.7.

In reality, for any particular plant, n is only approximately constant over a specified range. It does in fact gradually increase with capacity. Clearly, when $n = 1.0$ there is much less advantage in building a larger plant. However, this apparent economic limit is in fact a technological one, and improvements in process design gradually raise the capacity at which n tends to unity.

The factor K_{Lo} should be included to allow for increasing costs arising from local conditions, such as isolation from a major production center or port, freight costs, local productivity, and overseas construction.

25.1.3. Equipment Costs

The more accurate costing methods involve determination of the cost of the major items of process equipment: to do this, three approaches are used; cost graphs, detailed estimates, and quotes.

Various sources of graphed equipment costs are available;[3–5,12] they are log-log plots of cost versus capacity, of the form illustrated in Fig. 25.2a. Often the graphs are approximately linear over a suitable range, so that they can be described by an equation of the form

$$\mathscr{C} = \mathscr{C}^* (I)^n \qquad (25.3)$$

($\mathscr{C}$ = cost; $\mathscr{C}^*$ = reference cost; I = a measure of equipment size, normally capacity; n = cost exponent). Obviously Eq. 25.3 is analogous to Eq. 25.2, and as mentioned above n averages 0.6. Typical

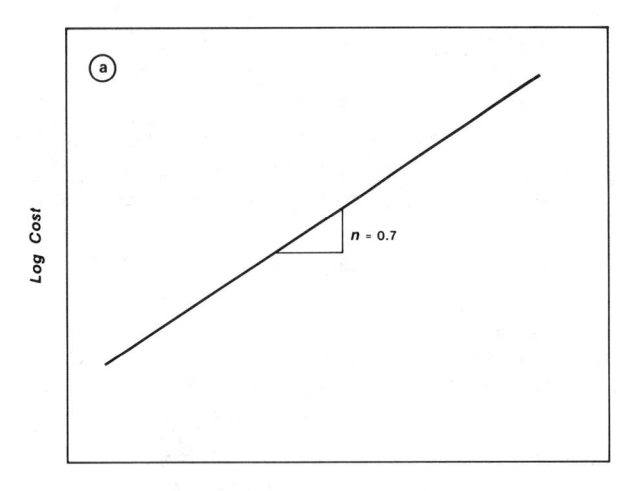

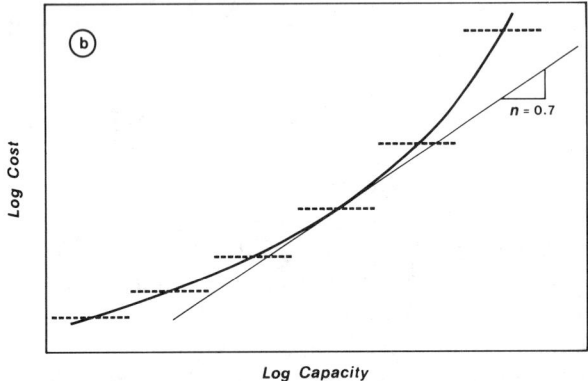

Figure 25.2. Schematic example of cost versus capacity data. (*a*) Commonly used form. (*b*) True nature of the data: a limited number of sizes having a range of capacities.

values of n (and $\mathscr{C}^*$) for some commonly used mineral processing equipment are shown in Table 25.2.[10]

In reality these cost graphs are of the form shown in Fig. 25.2*b*, because equipment is generally made in a range of discontinuous sizes that have a range of capacities. Note also that the slope may not be constant, but rather may increase with increasing capacity. Unlike the situation found with plants, the value of n for equipment may be well over 1.0. This is possible because under these conditions the total operating cost per unit of capacity may still be decreasing as the capacity increases. Furthermore, in the case of items such as jaw and gyratory crushers (which have $n > 1.0$), the physical size of the machine may be the overriding factor.

Other factors necessitate care when using these graphs. Some graphs include costs such as installation, foundations, or state tax, whereas others exclude items such as drive motors. Data are generally given f.o.b. (free on board), and do not include land or sea freight, insurance, duty, and so on.

The cost of fabricated equipment (tanks, bins, etc.) can alternatively be estimated by the "shell plus fittings method" or by a *detailed method* that directly costs materials, labor, overhead, and profit margins. In all cases, to obtain reliable information current manufacturers' quotes are essential.

25.1.4. Equipment Factored Methods

The equipment factored methods derive the total plant cost from the total purchase cost of the processing equipment.[13] The process equipment includes size reduction equipment, separators, conveyors, storage and mixing vessels, pumps, and the like. It does not include their installation, piping, electrical wiring, instrumentation, buildings, land development, and so on. Equipment costs should be on a "delivered to site" basis, based on current prices of new equipment. However, savings can be made by using standard units, idle equipment, second-hand equipment, and flexible design. Unexpected savings may also be possible because of company policies, such as reciprocal purchasing agreements. When any special savings are made (e.g., the use of idle or secondhand equipment), the full price of new equipment must be used in all calculations, any saving made being deducted from the total fixed capital.

The simplest factor method uses a single average factor to convert the total equipment cost into a fixed plant cost.[3,14–18] The more widely quoted method uses separate factors for each of the various

TABLE 25.2: REPRESENTATIVE VALUES OF n IN EQ. 25.3

Equipment	Measure of Size	n
Jaw crusher	Area of feed opening	1.28
Primary gyratory	Receiver opening x mantle diameter	1.41
Cone crusher	Diameter of discharge annulus	1.80
Hammer mill	Area of feed opening	0.67
Ball mill	Power	0.54
Rod mill	Power	0.57
Hydrocyclone	Capacity	0.91
SRL Pump	Capacity	0.64
Vibrating screen	Screen area	0.76
Spiral classifier	Spiral diameter	1.53
Table	Deck area	0.35
Flotation cell	Volume	0.74
Jig	Capacity	0.17
Disc filter	Filter area	0.44
Drum filter	Filter area	0.33
Thickener	Diameter	1.38
Vibrating feeder	Feeder area	0.99
1 m Belt conveyer	Length between centers	1.13
Motor	Power	1.11

"integrating" components, such as equipment installation, piping, electrical, and instrumentation. These factors are published as average values and the likely range of values.[3] (If the average value is used for each item, this method reduces to the single average factor method.) With either of these methods, slightly improved reliability can be obtained by using different factors for each of three basic types of plant; solids, solids/fluids, and fluids.[3] Typical values for mineral processing plants are 2.3–3.4.[11]

The factor concept was further improved by Miller,[17] who related factors to the magnitude of the cost of the project, since integration costs become a smaller proportion of the total cost as the project size increases. The most versatile approach is the *module* method described by Guthrie.[4,12] The basic module concept is illustrated in Fig. 25.3. Not only can a bare module be a single item of equipment, or an appropriate grouping; modules can also be designated for items such as site development, buildings, and indirect costs. In essence, the method is a combination of *all* costing methods, and therein lies one of its greatest advantages. Initially, relatively large modules can be used; but as more information becomes available, the modules can be subdivided into smaller and smaller modules, with each successive split allowing more accurate factors or cost information to be incorporated. This splitting can be carried out to the point where the costing becomes fully detailed, and no estimate factors are used at all. Unfortunately, at present, comparatively few data have been published for mineral plants, so it is necessary to rely on data from chemical plants.

It should be noted that all the factors used in these various methods change with time. Although this change may not be as significant as that in equipment costs, it does warrant the use of the most up-to-date data available.

Utilities. Utilities include items such as steam, water, power, compressed air, fuel, waste disposal, fire protection, first aid, and cafeteria. They may be included in the fixed capital, or as part of the operating cost (e.g., steam "purchased" at a certain cost).

Contingencies. This is a legitimate item of fixed capital. It is included to allow for unpredictable events, such as floods, storms, strikes, small design changes, and cost escalations. It does *not* cover incompetent costing.

25.2. WORKING CAPITAL

The working capital is necessary for the inventory of raw materials, work-in-progress, finished products, extended credit, and miscellaneous day-to-day cash requirements. It normally ranges from 10 to 20% of the total fixed capital, but can be much higher if the plant has large storage facilities. A better assessment is[18]

$$\begin{aligned} \text{working capital} &= \\ &= \mathscr{C}_{rm} + \mathscr{C}_{wip} + \mathscr{C}_{fp} + \mathscr{C}_{ec} + \mathscr{C}_c \\ &\sim \mathscr{C}_R + \tfrac{1}{4}\mathscr{C}_M + \mathscr{C}_M + \tfrac{1}{4}\mathscr{C}_M + \mathscr{C}_M \quad (25.4) \\ &\sim \mathscr{C}_R + 4\mathscr{C}_M \\ &\sim 3\mathscr{C}_s \end{aligned}$$

($\mathscr{C}_{rm}$ = value of raw material, normally taken as one month's requirements; $\mathscr{C}_{wip}$ = value of work in progress, taken as a week in normal cases; $\mathscr{C}_{fp}$ = value of finished product, normally taken as one month's output; $\mathscr{C}_{ec}$ = value of extended credit, taken as one month's production; $\mathscr{C}_c$ = cash for wages, utilities, etc., about one month's production; $\mathscr{C}_R$ = value of month's raw materials; $\mathscr{C}_M$ = value of monthly production cost; $\mathscr{C}_s$ = value of monthly sales).

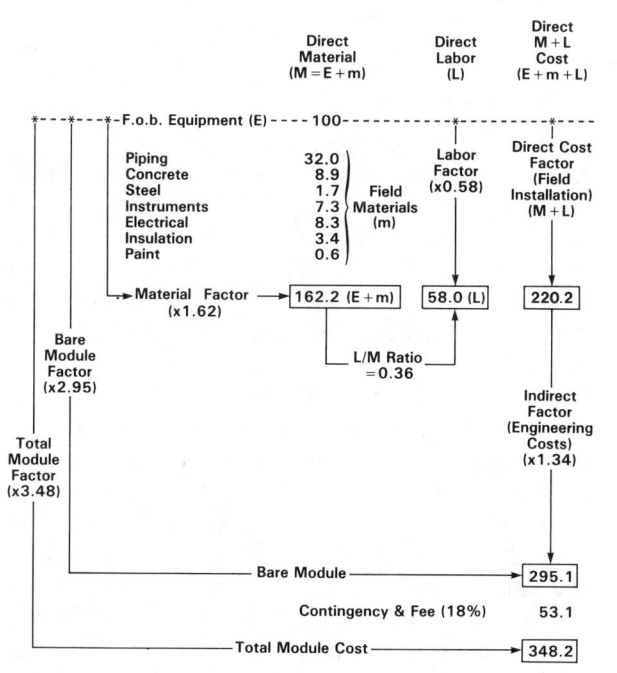

Figure 25.3. Costing module (symbols are specific to this figure only). (After Guthrie.[4,12])

25.3. OPERATING AND PRODUCTION COSTS

A detailed checklist of all the costs that make up the total production cost is given in Table 25.3, and it is always advisable to use such a list to ensure that all costs have been included. Production costs also can be divided into two classes: those dependent on production rate and those independent of production rate. In general, the former are themselves of two types, those that are directly proportional to throughput (e.g., raw materials) and those that increase at a faster rate (e.g., maintenance). If all production is sold, the relationship between these costs and income is of the form illustrated in Fig. 25.4.

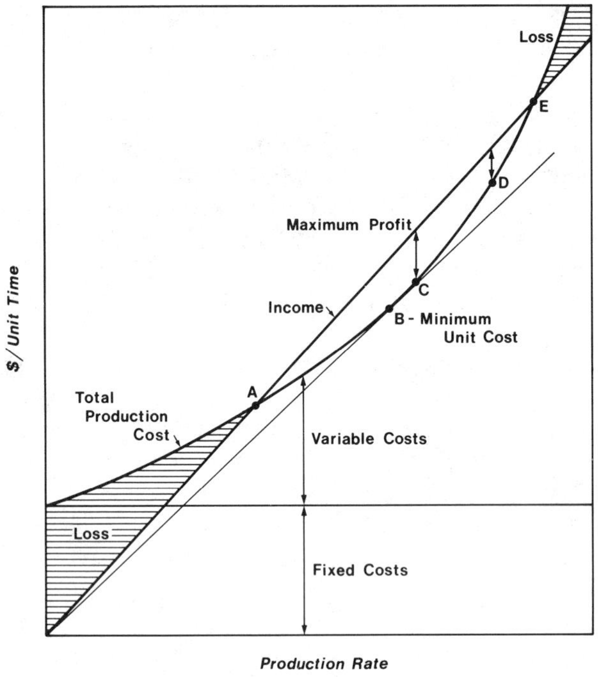

25.3.1. Labor Costs

Labor costs are a major item in mineral processing. Two components are involved: the number of people employed and the rate at which they are paid. Information on the latter can generally be obtained from government publications and technical journals. Estimation of the number of employees is more difficult. Charts (Wessel correlation[19]) have been published that relate the number of man-

Figure 25.4. The break-even curve for a manufacturing plant.

hours/day required for each processing step, but these are not very reliable. A better method is to base labor requirements on those used in a similar type of plant already operating.[20] The so-called one-fourth rule gives:

$$\frac{\text{operating labor in new plant}}{\text{operating labor in existing plant}} = \left(\frac{I_n}{I_e}\right)^{0.25} \quad (25.5)$$

Obviously this equation is similar to Eqs. 25.1 and 25.2, but since it is less sensitive to local conditions, it is more useful. Even so, the most reliable method is to use work studies to assess actual manpower requirements. Supervisory and clerical labor are normally estimated separately. If their numbers are low, the requirements can be enumerated. In larger operations, they can be assessed from the operating labor costs, typical values being 10–20%. Administration costs (top management, secretaries, accountants, etc.) can also be estimated this way: typically they are 40–60% of operating labor costs. On top of all labor costs, indirect payroll costs must be applied to cover workers' compensation, pensions, insurance, holiday pay, and other fringe benefits. These vary from country to country but in the United States they can be as high as 50% of the labor costs.

TABLE 25.3: PRODUCTION COSTS

Mining Cost			
Operating labor			
Operating supervision			
Steam	} Power and utilities	} Direct production costs	
Electricity			
Fuel			
Water			
Maintenance and repairs			
Operating supplies			
Laboratory charges			
Royalties (if not on lump-sum basis)			
Depreciation	} Fixed charges		} Manufacturing costs
Taxes (property)			
Insurance			
Medical		} Plant overhead costs	
Safety and protection			
General plant overhead			
Payroll overhead			
Packaging			
Restaurant			
Recreation			
Salvage			
Control laboratories			
Plant superintendence			
Storage facilities			
Executive salaries		} Administrative expenses	} General expenses
Clerical wages			
Engineering and legal costs			
Office maintenance			
Communications			
Shipping			
Research and development			

25.3.2. Energy and Other Utilities

Electricity is a significant cost, especially with low grade ores requiring fine grinding. The cost depends on the net price of a kilowatt-hour, and the power consumption per tonne, the latter depending mainly on the ore hardness and the product fineness. Significant amounts of power are used elsewhere (Fig. 25.5), although these costs can be minimised by good plant design and layout.[21]

Other utilities such as fuel, steam, and increasingly, water, must also be allowed for in a cost estimate.

In many cases, utilities can be costed in two ways. The easiest method is to consider the utility to be purchased at a specific rate from an outside utility company. In other instances, mainly because of isolation, it may be necessary for a company to provide its own utilities such as power generation or water supply. Under these conditions, the equipment and its operation can be costed as part of the plant. Either way, the cost should be included *once* only.

25.3.3. Supplies

The most commonly used supplies are chemical reagents, grinding media, and liners. For the most part they are proportional to throughput, and can be costed accordingly. Typical reagent dosages are published regularly[22] and typical wear rates are given in Table 8.3.

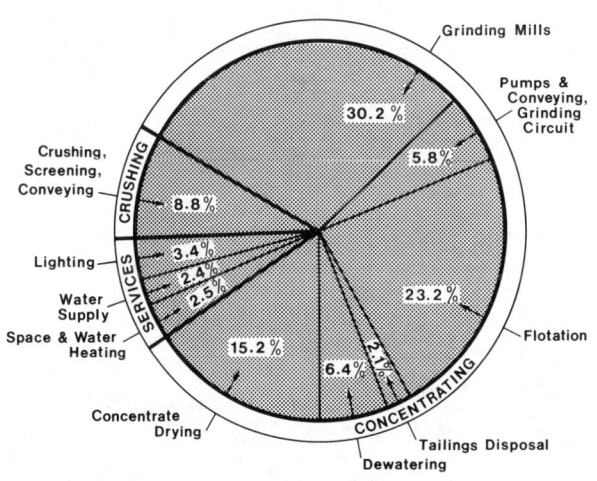

Figure 25.5. Typical distribution of power consumption in a flotation concentrator. (After Barrat et al.[21])

25.3.4. Maintenance

For the most part maintenance depends on the original total equipment cost and can be taken as a fixed percentage of this (2–10%). In a plant having crushing and grinding operations, the costs of replacement liners and wear surfaces are in addition to this maintenance cost.

25.3.5. Depreciation

Physical assets, such as plant, decrease in value with age (depreciate) because of factors such as physical deterioration or functional depreciation due to technological advances. In simple terms, depreciation is a cost item allowed to recover the cost of the plant during its lifetime. There are various methods of assessing depreciation, and although rigorous methods must be used for operating plants or detailed estimates, they are seldom justified in preliminary estimates. In this situation it is generally adequate to take the annual depreciation as a fixed percentage (10–20%) of the plant investment. In that some mineral operations (e.g., small concentrators) have a known projected life, the whole plant can be depreciated at the same rate given by 100% divided by the life in years. Some plants may justify higher than average values because they have a relatively short life, whereas buildings have below average values. Land does not depreciate and must be excluded from the fixed capital cost before depreciation is calculated.

25.3.6. Other Items

Most of the other costs listed in Table 25.3 are self-explanatory. Further information on them can be obtained from suitable references.[3,7,23,24]

25.4. INVESTMENT WORTH

Having estimated the cost of operating a prospective project, it becomes necessary to consider whether the project is worthwhile. The following discussion covers only the financial aspects. One should always remember that other intangible factors may be involved (e.g., such social factors as creating employment, or providing a reliable source of supply for another operation). Such factors are difficult to assess, and are not considered here.

In a simplistic way, investment worth (or profitability, or return on investment) is a measure of the amount of money made relative to the amount of money invested, and as such can be expressed as a percentage that is analogous to the interest paid on a bank deposit or loan. Thus the *simple interest rate of return* (*SIRR*) can be defined as

$$SIRR\ (\%) = \frac{\text{net annual profit}}{\text{total capital investment}} \times 100 \quad (25.6)$$

Today, it would be unwise to state an "acceptable" rate of return. Some sources quote figures as high as 100% for high risk projects, but examination of large companies show that many of them operate in the region of 0–20% and some times even at a loss (e.g., a subsidized operation). Put another way, the sheer fact that a large company operates is, today, a significant social benefit that can compensate for a low return on investment.

Another simple method of measurement is to consider the time required for the cumulative profits to recover the capital invested (not to be confused with depreciation), often called the *payback time*. Though widely criticized, these two simple approaches are normally all that can be justified in most preliminary cost estimates. Only when it has been decided whether a project should actually be started is it worthwhile to consider more sophisticated measures of investment worth that incorporate compound interest and the cash flow.

The need for consideration of the cash flow can be seen by examination of Fig. 25.6. Initially there is an outflow (negative flow) of capital as the plant is built; then as the plant starts up and begins to operate at a profit, the operation (ideally) recovers the capital and goes on to make a profit. The point at which the cash flow returns to zero represents the payback time. A line from the origin to the curve at a particular time has a slope that gives the average simple interest rate of return up to that time (as distinct from the slope of the cash flow line, which gives the simple interest rate of return defined in Eq. 25.6). The inclusion of compound interest allows for the fact that interest is itself capable of earning further interest, or that profits as they are made can be invested to make further profits.

Figure 25.7 shows the effect of including compound interest on the cumulative cash flow. The compound interest rate (often described as the *discount rate*) on this graph can be interpreted in a number of ways. Possibilities are to consider it as the rate of borrowing money for the project, the

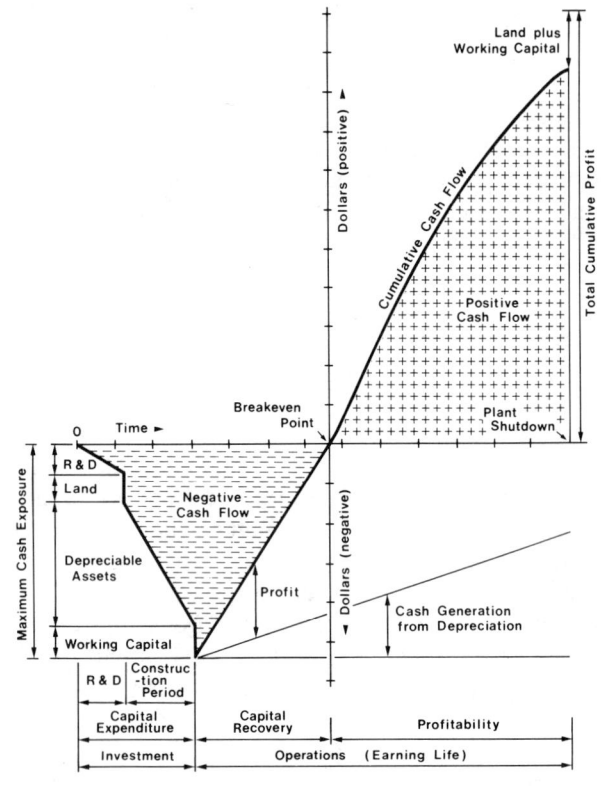

Figure 25.6. Elements of project cash flow. (After Guthrie.[4])

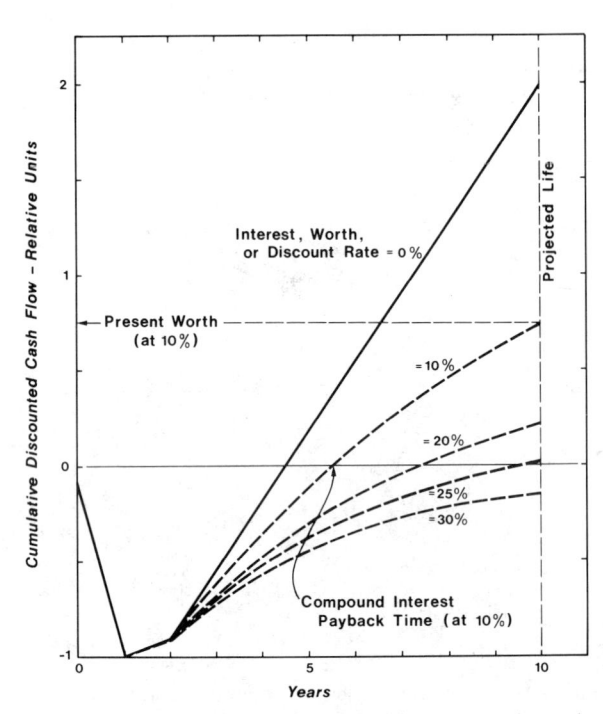

Figure 25.7. The effect of compound interest on the cash flow.

average rate the company is making on all its investments, or the rate of return the company considers to be acceptable for the given project. Clearly, as the interest rate or "worth" of money increases, any given project becomes less attractive.

Some of the compound interest measures of investment worth can also be illustrated by Fig. 25.7. The *compound interest payback time* is simply the number of years taken for the cash flow to return to zero. The *compound rate of return* is the interest rate that brings the cash flow to zero at the end of the project (~25% in this example). The latter can be given more meaning by comparing it to the rate for borrowing money for the project, to the current average return for the company, and to inflation. The *present worth* is the value of the cumulative cash flow at the end of the project.

These and other measures of investment worth have been extensively discussed in the literature.[2,3,18,25] In spite of their claimed superiority, the compound interest measures are themselves open to one serious criticism, in that they are using a single number to describe a nonlinear graph; a device that in fact requires at least three parameters. The inadequacies are easily demonstrated by Fig. 25.8, which shows two projects with equal lives, (compound) payback times, (compound) rates of return, and present worths. Yet the projects are obviously different. It can be convincingly argued that project A is the better proposal, for two reasons: first the outlay occurs later, and second, income builds up faster, so that if the project must be terminated early, the reduction in value will be less. On the other hand, if at the planned end of the project it could be kept going another year, indications are (by extrapolation) that project B would be the better project.

Thus, no single number by itself can describe the graph, and a critical analysis of the cash flow diagram is the best measure of investment worth. Even so, two points must be borne in mind: the shape of the graph depends significantly on the interest rate chosen, and any conclusions reached must also consider the reliability of the forecasted cash flow. Although the apparent limited lives of ore bodies may suggest reasonably reliable forecasting, rapidly changing inflation, metal prices, and government regulations tend to neutralize this. On the other hand, the life of most ore bodies can be altered by changing the cut-off grade, so that the cash flow can in turn be changed to maintain a profitable operation.[26]

While discussing investment worth, it is desirable to consider the importance of incremental investments.[3] In brief, this means that *all* money invested should give an acceptable return on investment and is best illustrated with an example, such as dewatering a concentrate. By increasing the amount of investment (e.g., larger equipment, or more stages of processing), more water can be removed, and the value of the concentrate increased. Ultimately, of course, the cost of additional water removal will cost far more than the increase in value, and the investment/income relationships will be of the form shown in Fig. 25.9. The peak in the curve represents

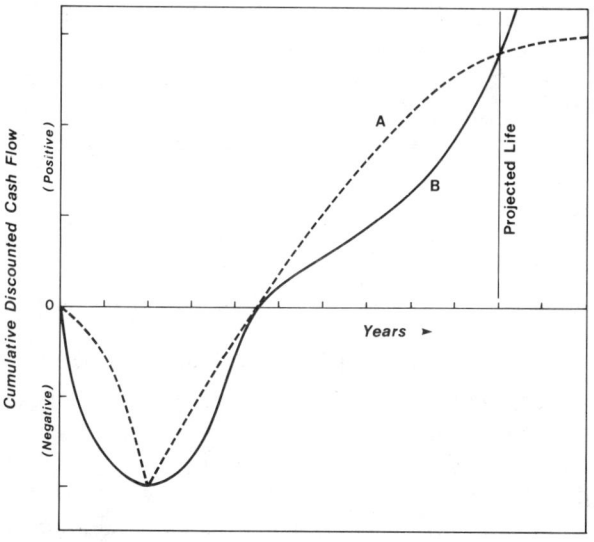

Figure 25.8. Comparison of two investment proposals.

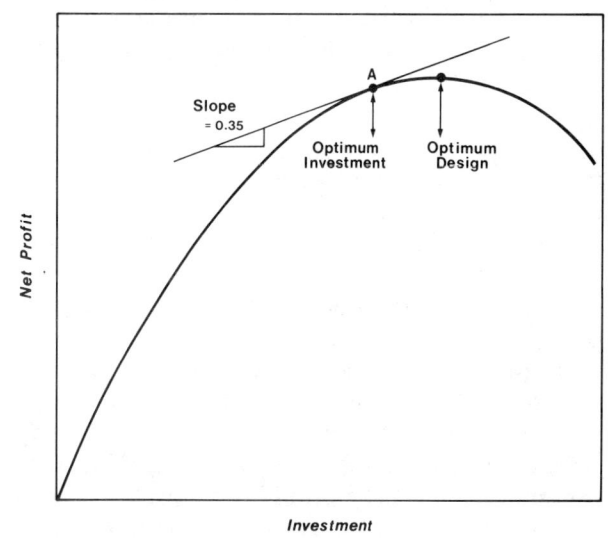

Figure 25.9. Optimum investment.

the optimum design point. It is not however the optimum economic point. This is given by the point of the curve that has a slope equal to the minimum acceptable rate of return: in this example a 35% minimum gives point A in Fig. 25.9. The reason for this is that all *increments* of investment above this point have rates of return below 35%; until at the peak, the last increment of investment has a (quite unacceptable) zero rate of return. (Beyond the peak, the returns would be negative.)

25.5. OPTIMUM OPERATING CAPACITY

There are four significant points on the graph in Fig. 25.4. The first is the break-even point (point A), where income just balances the total production cost. Production rates less than this run at a loss, although under certain circumstances it can be argued that an operation can still be considered "profitable" in this region.[27] At point B a tangent from the origin contacts the total production cost, representing the minimum cost (and thus maximum profit) *per unit of output*. The greatest vertical distance between income and production costs (point C) is the maximum *profit per unit of time* and therefore represents the maximum return on investment. Higher production rates may still be justified as long as the reduced total profit continues to give a rate of return greater than a certain minimum acceptable to the company (e.g., point D: strictly speaking it is the incremental return on the incremental cost increase that is significant[3]). Under certain circumstances, a plant may operate above point D, or even above point E, where production runs at a loss.

In practice, it is difficult to decide what is the optimum operating capacity (as distinct from its nominal capacity) of a mineral processing plant. Occasionally, where the output is sold on the open market, one may not be able to select the operating point, because it is limited by market conditions. All too often, the capacity of a captive concentrator may be set by smelter demand: the smelter calls for so much concentrate per day. This the concentrator produces with little regard for efficiency, which in turn results in the smelter using a comparatively low quality feed that lowers its efficiency (even though the throughput may seem an impressive figure). Such a practice is quite undesirable, and mine, concentrator, and smelter must be considered as a combined unit, with adjustments made to optimize the *overall* operation.

Even deciding *what* is the optimum is a complex issue. Conventional economic practice is to operate at the level that gives maximum profit (point C, Fig. 25.4); but although this is realistic for a manufacturing plant, the situation is not so simple for a processing plant such as a concentrator. One reason for this is that the selling price of a concentrate is affected by its quality (i.e., grade), so that the income line on Fig. 25.4 is not straight, but rather decreases in slope as the production rate increases.

One method of treating the problem was illustrated in Example 3.3, where a grade/recovery curve was combined with a smelter schedule to give an apparent optimum operating point; that is, the return per tonne of feed was maximized (i.e., point B, Fig. 25.4). In practice, this would be attained by suitable adjustment of the separator settings (which would only slightly affect total throughput). Such an approach is actually used by a number of concentrators for control purposes.[28,29] (This economic analysis can also be used to indicate avenues of research that offer the greatest potential reward.[30])

However, in reality the grade/recovery curve is not fixed. Rather it can be significantly altered by variations in ore properties such as hardness, grain size, and grade, and even with uniform ores it is affected by the feed rate. (These effects can be seen by making appropriate changes in the Examples.) Furthermore, unlike a manufacturing plant (where the total raw material cost is virtually proportional to throughput), the raw material cost of a concentrator is determined by the mining cost, which is a function of variables such as mine capacity, cut-off grade, and technology. Thus, for a concentrator, Fig. 25.4 at the best represents only a cross-section from a very complex multidimensional plot.

In an attempt to analyze optimum economic operation, Steane[31] suggested that significantly higher throughputs could be achieved by a marked decrease in liberation and that although concentrator recoveries would be much lower, total profits would be higher. However, his analysis applied to new rather than operating mills and failed to consider all relevant parameters, particularly the larger investment required.

More thorough analyses have been attempted.[32-37] These suggest that for a total mining, concentrating, smelting operation it is more appropriate to consider the incremental increase in present worth in a given time, rather than just maximization of present worth. Schaap has shown how this can be related to a number of parameters, which include ore hardness, grain size, cut-off grade, mine and concentrator recovery, throughput, concentrate value, and costs.[34-36] It is significant that large capital investment (typical of modern operations) favor lower overall (mine + concentrator) recoveries, and that toward the end of the mine life, recovery should gradually be increased. Schaap also considers the ways in which taxation could be used to encourage higher extraction from ore bodies.[35]

In summary, it can be said that determination of an optimum operating level is complicated by the many variables involved in the *total* mining, mineral processing, smelting operation. Thus, it is hardly surprising that most concentrator staff tend to aim for optimum separation efficiency, which may in fact be fairly close to the optimum economic level. Where these two optima do differ, an optimum separation efficiency will generally mean a higher recovery and in a world of increasingly scarce resources,[38] this is not entirely without merit.

The reader requiring more information is referred to more detailed discussions on the subject,[34-36] which also consider the even more complex issue of projected operations.[37]

25.6. CONCLUDING REMARKS

To emphasize the importance and relevance of economics to mineral processing, the subject was deliberately introduced in the first few pages of this book. However, it would be unrealistic to conclude without also emphasizing that there are often intangibles that override or appear to override economic aspects. These intangibles may pass under a variety of guises, such as political factors, social aspects, or environmental issues. To a considerable extent they are an extension of a concept we have repeatedly stated: no system can be considered in isolation. So-called political factors, for example, can be thought of as an effort to integrate a mineral operation into a larger social fabric. The difficulty is, of course, that this introduces a further layer of complexity, and the analysis simply becomes even less well defined.

REFERENCES

1. D. R. Woods, *Financial Decision Making in the Process Industry*, Prentice-Hall (1975).
2. F. C. Jelen (Ed.), *Cost and Optimization Engineering*, McGraw-Hill (1970).
3. M. S. Peters and K. D. Timmerhaus, *Plant Design and Economics for Chemical Engineers*, 2nd ed., McGraw-Hill (1968).
4. K. M. Guthrie, *Process Plant Estimating, Evaluation and Control*, Craftsman (1974).
5. A. L. Mular, *Mineral Processing Equipment Costs and Preliminary Capital Cost Estimates*, Special Vol. 18, CIM (1978).
6. A. L. Mular and R. B. Bhappu (Eds.), *Mineral Processing Plant Design*, AIME/SME (1978).
7. W. H. Yarroll and F. T. Davis, "The Economics of Small Milling Operations," *Min. Ind. Bull.*, **11**, No. 2 (March 1968); **18**, No. 6 (November 1975).
8. D. W. Gentry and M. J. Hrebar, "Procedure for Determining Economics of Small Underground Mines," *Min. Ind. Bull.*, **19**, No. 1 (January 1976).
9. "Reference Manual and Buyers Guide," *Can. Min. J.* (published annually).
10. A. L. Mular, "The Estimation of Preliminary Capital Costs," Ch. 3 in *Mineral Processing Plant Design*, A. L. Mular and R. B. Bhappu (Eds.), AIME/SME (1978).
11. R. J. Balfour and T. L. Papucciyan, "Capital Cost Estimating for Mineral Processing Plants." Presented at Canadian Mineral Processors Annual Meeting, 1972; also published in *Can. Min. J.*, **93**, 88-96 (June 1972).
12. K. M. Guthrie, "Data and Techniques for Preliminary Capital Cost Estimating," *Chem. Eng.*, **76**, 114-140 (March 24, 1969).
13. O. T. Zimmerman, "Use of Ratio Cost Factors in Estimating," *Cost Eng.*, **10**, 13-17 (October 1965).
14. H. J. Lang, "Engineering Approach to Preliminary Cost Estimates," *Chem. Eng.*, **54**, 130-133 (September 1947).
15. H. J. Lang, "Cost Relationships in Preliminary Cost Estimation," *Chem. Eng.*, **54**, 117-121 (October 1947).
16. H. J. Lang, "Simplified Approach to Preliminary Cost Estimates," *Chem. Eng.*, **55**, 112-113 (June 1948).
17. C. A. Miller, "New Cost Factors Give Quick Accurate Estimates," *Chem. Eng.*, **72**, 226-236 (Sept. 13, 1965).
18. R. H. Buchanan and C. G. Sinclair, *Costs and Economics of the Australian Process Industries*, West (1966).
19. H. E. Wessel, "New Graph Correlates Operating Labor Data for Chemical Processes," *Chem. Eng.*, **59**, 209-210 (July 1952).
20. F. P. O'Connell, "Chart Gives Operating Labor for Various Plant Capacities," *Chem. Eng.*, **69**, 150-152 (Feb. 19, 1962).
21. J. A. Barrat et al., "The Influence of Energy Conservation on Concentrator Design," *CIM Bull.*, **68**, 85-93 (December 1975).
22. *Minerals Yearbook*, volume entitled *Metal and Non-metal Industries*, U.S. Bureau of Mines (published annually).
23. F. M. Lewis and R. B. Bhappu, "Operating Cost," Ch. 4 in *Mineral Processing Plant Design*, A. L. Mular and R. B. Bhappu (Eds.), AIME/SME (1978).
24. F. J. Stermole, "Investment Decision-Making and Economic Evaluation of Mineral Projects," *Min. Congr. J.*, **59**, 59-64 (September 1973).
25. H. Popper (Ed.), *Modern Cost-Engineering Techniques*, McGraw-Hill (1970).

26. K. I. Mackenzie, "Sizing the Open Pit Mine and Concentrator to the Deposit for Best Profits." Presented at 108th AIME Annual Meeting, 1979.
27. H. R. Fraser, "New Trends in the Copper Industry," *CIM Bull.*, **69**, 25–27 (March 1976).
28. K. L. Clifford, E. J. Haug, and K. L. Purdy, "Galena-Sphalerite-Chalcopyrite Flotation at St. Joe Minerals Corp.," *Min. Eng. (Trans. AIME/SME)*, **31**, 180–182 (1979).
29. C. L. Lewis, "Application of a Computer to a Flotation Process," *CIM Bull.*, **64**, 47–50 (January 1971).
30. W. R. Bull and J. R. Roos, "Flotation Plant Research—The Determination of Priorities by Financial Analysis of Mill Performance," *Trans. AIME/SME*, **244**, 385–390 (December 1969).
31. H. A. Steane, "Coarser Grind May Mean Lower Metal Recovery But Higher Profits," *Can. Min. J.*, **97**, 44–47 (May 1976).
32. H. K. Taylor, "General Back Ground Theory of Cut-Off Grades," *Trans IMM (A)*, **81**, A160–A179 (1972).
33. K. F. Lane, "Choosing the Optimum Cut-Off Grade," *Colorado School Mines Q.*, **59**, 811–829 (1964).
34. W. Schaap, "Problems in Model Formulation for Optimal Mineral Extraction," *Proc. Australian Soc. for Operations Research, August 1979*, pp. 286–301, ASOR (1979).
35. W. Schaap, "Economic Rent Sharing with Optimal Extraction of a Disseminated Mineral," *N. Z. Operational Research*, **9**, 33–45 (January 1981).
36. W. Schaap, "Effect of Mineral Grainsize and Ore Hardness on Mill-Dump Cut-Off Grades," *Trans. IMM (A)*, **90**, A27–A33 (1981).
37. W. A. Vogely (Ed.), *Economics of the Mineral Industries*, AIME (1976).
38. G. J. S. Govett and M. H. Govett (Eds.), *World Mineral Supplies: Assessment and Perspective*, Elsevier (1976).

Appendix A

Minerals and Their Characteristics

Minerals and Their Characteristics

NAME	FORMULA	PERCENT METAL	COLOR	LUSTRE	STREAK	HARD-NESS	SPEC. GRAV.	CHARACTERISTICS—OCCURRENCE
ACTINOLITE	$Ca(MgFe)_3(SiO_3)_4$	No metal source	Green	Vitreous		5.0–6.0	3.0–3.2	Usually long crystals, columnar or fibrous
ALBITE	$NaAlSi_3O_8$	Al_2O_3 – 19.5%	White to blue	Vitreous	White	6.0–6.6	2.6–2.7	Occurs sometimes in platy masses. Otherwise like anorthite. See Anorthite
ALMANDITE	$Fe_3Al_2(SiO_4)_3$	No metal source	Red to black		White	6.5–7.5	3.1–4.3	Variety of garnet. See garnet
ALTAITE	$PbTe$	61.9% Pb	Tin white Yellow tinge	Metallic	Grayish Black	3.0	8.2	Associated with pyrite, galena, tetrahedite
ALUNITE	$K_2(Al_2OH)_6 \cdot (SO_4)_4$	K – 9.4% Al – 19.6%	Pink-red	Vitreous Pearly	White	3.8	2.7	Associated with kaolin and pyrite
AMOSITE	$(FeCaH_2Mn)OSiO_2$	No metal source	Gray to green			2.2–2.3		Long fibered asbestos
ANALCITE	$NaAlSi_2O_6 \cdot 2H_2O$	Al_2O_3 – 23.2%	White	Vitreous	White	5.0–5.5	2.2–2.3	Trapezohedral crystals in cavities in basic igneous rocks
ANDALUSITE	Al_2SiO_2	Al_2O_3 – 63.2%	White Red-green	Vitreous		7.5	3.2	Nearly square prisms; occurs with gneiss, mica, schists
ANDRADITE	$Ca_3Fe_2(SO_4)_3$	No metal source	Green Red-black	Adamantine	White	6.5–7.5	3.1–4.3	Variety of garnet. See garnet
ANGELSITE	$PbSO_4$	Pb – 68.3%	Yellow Green-gray	Adamantine, Vitreous	White	2.8–3.0	6.1–6.4	Occurs in oxidation zones of lead veins
ANORTHITE	$CaAl_2Si_2O_8$	Al_2O_3 – 36.7%	White, Gray-red	Vitreous	White	6.0–6.5	2.7–2.8	Tabular crystals in igneous rocks, with fine longitudinal lines on the better of two perfect cleavages at 90° to each other
ANTHOPHYLLITE	$(MgFe)SiO_3$	No metal source	Gray, Brown-green	Vitreous	Uncolored, Grayish	5.0	3.0–3.2	Found in crystalline schists
APATITE	$Ca_4(CaF)(PO_4)_3$	P_2O_5 – 42.3%	Green-blue	Vitreous	White	4.5–5.0	3.2	Usually granular or in 6-sided prisms
ARAGONITE	$CaCO_3$	CaO – 56%	White	Vitreous	White	3.5–4.0	2.9	Effervesces like calcite. Powder becomes lilac or purple when boiled in 10% solution of cobalt nitrate
ARGENTITE	Ag_2S	Ag – 87.1%	Black	Metallic	Shiny Black	2.0–2.5	7.2–7.4	Cuts like lead; with silver, cobalt and nickel
ARGYRODITE	$3Ag_2S \cdot GeS_2$	Ag – 73.5%	Steel gray red tinge	Metallic	Grayish Black	2.5	6.1	Occurs with sphalerite, siderite and marcasite
ARSENOPYRITE	$FeAsS$	Fe – 34.3% As – 46.0%	Steel Gray	Metallic	Gray, Black	5.5–6.0	5.9–6.3	Widely spread; yields sparks and garlic odor when struck slanting blows with steel
ATACAMITE	$Cu_2(OH)_3Cl$	Cu – 59.5%	Green	Adamantine, Vitreous	Apple Green	3.0–3.5	3.8	Always of secondary origin with copper ores
AZURITE	$2CuCO_3 \cdot Cu(OH)_2$	Cu – 55.0%	Blue	Vitreous, Dull	Blue	3.5–4.0	3.8–3.9	Oxidized mineral that effervesces vigorously in muriatic acid of any strength and temperature

Mineral	Formula	Composition	Color	Luster	Streak	Hardness	Sp. Gr.	Remarks
BARITE	BaSO₄	BaO – 65.7%	White, Blue-red	Vitreous	White	2.5–3.5	4.3–4.6	Found commonly as gangue of lead-zinc ores. Platy or granular masses or either diamond-shaped or rectangular crystals
BAUXITE	Al₂O₃·3H₂O	Al – 34.9%	White-red Brown-yellow	Dull	Like Color	1.0–3.0	2.6	Chief ore of aluminum; occurs massive. Completely soluble in salt of phosphorous bead
BENTONITE	(CaMg)O·SiO₂ (AlFe)₂O₃	No metal source	Blue	Dull	Light Gray	1.0	2.1	The clay of montmorillonite. Swells greatly when placed in water
BERYL	Be₃Al₂(SiO₃)₆	Be – 5% Al₂O₃ – 19%	White, Green-blue	Vitreous	White	7.5–8.0	2.6–2.8	Often imbedded in quartz; with mica, feldspar. Usually in 6-sided prisms with flat terminations in pegmatite. Gem varities
BERYLLONITE	NaBePO₄	Be – 7.1%	White-yellow	Vitreous, Brilliant	White	5.8	2.8	Found with beryl, feldspar, columbite
BIOTITE	(HK)₂(MgFe)₂ Al₂(SiO₄)₃	No metal source	Black-Brown	Pearly, Vitreous	White	2.5–3.0	2.7–3.1	Cleaves easily into very thin, flexible and elastic plates
BISMITE	Bi₂O₃	No metal source	Straw Yellow White	Pearly	Like Color		4.4	Of secondary origin resulting from oxidation
BISMUTH	Bi	Bi – 100%	Silver White	Metallic	Silver White	2.3	9.7	Native; with cobalt, nickel; brassy tarnish
BISMUTHINITE	Bi₂S₃	Bi – 81.2%	Lead gray	Metallic	Like Color	2.0	6.4–6.5	Occurs in form of thin coating
BISMUTITE	(BiO)₂·CO₃·H₂O	No metal source	Green-white	Vitreous, Dull	Greenish Gray-White	4.0	6.9–7.7	Incrusting fibrous, or earthy and pulverulent
BORAX	Na₂B₄O₇·10H₂O	B₂O₃ – 36.6% Na₂O – 16.2%	White	Vitreous, Dull	White	2.0–2.5	1.7	
BORNITE	Cu₅FeS₄	Cu – 63.3%	Reddish	Metallic	Blackish Gray	3.0–3.5	4.9–5.4	Associated with chalcocite. Usually massive. Quickly tarnishes iridescent blue
BOURNONITE	3(PbCu₂)S·Sb₂S₃	Pb – 24.7% Cu – 42.5%	Steel gray Iron black	Metallic	Like Color	2.5–3.0	5.7–5.9	Occurs fine-grained massive; brittle
BRAUNITE	3Mn₂O₃·MnSiO₃	Mn – 78.3%	Steel gray Brownish black	Submetallic	Like Color	6.0–6.5	4.8	Occurs in porphyry; brittle
BREITHAUPTITE	NiSb	Ni – 32.5% Sb – 67.5%	Copper red	Metallic	Reddish Brown	5.5	7.5	Occurs with other sulfides and other minerals
BROCHANTITE	CuSO₄·3Cu(OH)₂	Cu – 56.2%	Green	Vitreous	Green	3.5–4.0	3.9	Oxidized mineral. Dissolves quietly in nitric acid
BRUCITE	MgO·H₂O	MgO – 69%	White to gray, blue, green	Pearly, Vitreous	White	2.5	2.4	Associated with serpentine; secondary mineral
CALAMINE	H₂(Zn₂O)·SiO₄	Zn – 67.5%	White, blue, green, brown	Vitreous, Dull	White	4.5–5.0	3.4–3.5	Usually in crystal coatings; sometimes in cockscomb-like aggregates. Often with smithsonite
CALAVERITE	AuTe₂	Au – 43.6%	Bronze yellow Silver-yellow tinge	Metallic	Yellowish Gray	2.5	9.0	Similar to sylvanite, but never in crystals
CALCITE	CaCO₃	CaO – 56%	Many colors	Vitreous	White	3.0	2.7	Massive and 6-sided pointed or prismatic crystals. Effervesces vigorously in muriatic acid of any strength or temperature

Minerals and Their Characteristics

NAME	FORMULA	PERCENT METAL	COLOR	LUSTRE	STREAK	HARD-NESS	SPEC. GRAV.	CHARACTERISTICS—OCCURRENCE
CALOMEL	$HgCl$	Hg—85% Cl—15%	White, yellow	Adamantine	Pale Yellow, White	1.0–2.0	6.5	Associated with cinnabar
CARNALLITE	$KMgCl_3 \cdot 6H_2O$	K—14.1% Cl—38.3%	White	Shining	White	2.5	1.6	Strongly phosphorescent; taste—bitter
CARNOTITE	$K_2O \cdot 2U_2O_3 \cdot V_2O_5 \cdot 3H_2O$ Variable	Variable	Yellow	Vitreous, Dull	Yellow	1.5		Powder or earth in sandstone. Often concentrated around petrified wood
CASSITERITE	SnO_2	Sn—78.8%	Brown, black, red	Adamantine	White, Light Brown	6.0–7.0	6.8–7.1	Massive or squarish crystals
CELESTITE	$SrSO_4$	Sr—47.7%	Light blue, white, red	Vitreous	White	3.0–3.5	3.9–4.0	Same as barite
CERARGYRITE	$AgCl$	Ag—75.3%	Pearly gray	Waxy, greasy	White to Gray	1.0–1.5	5.6	Cuts like wax; exposure changes color to violet brown
CERUSSITE	$PbCO_3$	Pb—77.5%	White, gray	Adamantine	White	3.0–3.5	6.5–6.6	Oxidized mineral. Effervesces vigorously in warm concentrated or boiling dilute muriatic acid
CERVANTITE	$2Sb_2O_4 \cdot Sb_2O_3 \cdot Sb_2O_5$	Sb—79.4%	Yellow reddish white	Greasy, Pearly	White	4.0–5.0	4.1–5.3	Usually associated with stibnite
CHALCANTHITE	$CuSO_4 \cdot 5H_2O$	CuO—31.8%	Blue	Vitreous	White	2.5	2.1–2.3	Oxidized mineral. Tastes metallic and nauseating
CHALCEDONY	SiO_2	No metal source	Pale blue, gray, White to black	Waxy	White	7.0	2.6–2.7	Smoothly rounded fracture. Semi-precious gem varieties
CHALCOCITE	Cu_2S	Cu—79.8%	Black-Gray	Metallic	Like Color	2.5–3.0	5.5–5.8	Highly polished surface where cut
CHALCOMENITE	$CuSeO_3 \cdot 2H_2O$	Cu—28.1% Se—34.9%	Blue	Vitreous	Bluish White	2.5–3.0	3.8	With various selenides of silver, copper, and lead
CHALCOPYRITE	$CuFeS_2$	Cu—34.6%	Brassy yellow	Metallic	Greenish Black	3.5–4.0	4.1–4.3	Softer than pyrite; with pyrite, galena, sphalerite
CHERT	SiO_2	No metal source	White-gray	Dull	White	7.0	2.6	Impure, coarse-grained, opaque flint
CHLOANTHITE	$NiAs_2$ Variable	Ni—28.1% As—71.9%	Tin white, Steel gray	Metallic	Grayish black	5.8	6.5	Granular or in crystals like pyrite. Often associated with erythrite. See erythrite
CHROMITE	$FeO \cdot Cr_2O_3$	Cr—46.2%	Black	Vitreous	Dark Brown	5.5	4.3–4.6	Grains may look like black glass. Often with serpentine
CHRYSOBERYL	$BeAl_2O_3$	BeO—19.3%	Green	Vitreous	White	8.5	3.7–3.8	Usually in crystals or worn pebbles. Gem varieties
CHRYSOCOLLA	$CuOSiO_2 \cdot 2H_2O$	Cu—36.2%	Blue, green	Vitreous, Dull	White	2.0–4.0	2.0–2.2	Adheres to dry tongue; important ore of copper
CHRYSOLITE	$(MgFe)_2SiO_4$	No metal source	Green	Vitreous	White or Yellowish	6.5–7.0	3.3	In granular masses, glassy grains or crystals. Gem varieties

Mineral	Formula	Color	Luster	Streak	Hardness	Specific Gravity	Remarks
CHRYSOTILE	$H_4Mg_3Si_2O_9$	White, greenish		White	1.7	2.2	Best asbestos. Masses of tough, usually parallel, slender fibers
CINNABAR	HgS	Red	Metallic	Scarlet	2.0–2.5	8.0–8.2	Only important ore of mercury; tastes "chalky"
CLAUSTHALITE	$PbSe$	Lead gray	Adamantine, Submetallic	Lead Gray	2.8	8.0	Resembles granular galena
COBALTITE	$CoAsS$	Tin white, steel gray	Metallic	Grayish Black	5.5	6.0–6.3	Granular or in crystals like pyrite. Often with erythrite. See erythrite
COLEMANITE	$Ca_2B_6O_{11} \cdot 5H_2O$	White, yellowish	Brilliant, Vitreous	White	4.0–4.5	2.4	Usually occurs as geodes; brittle
COLUMBITE	$(FeMn)(NbTa)_2O_6$	Iron black	Submetallic	Dark Red, Black	6.0	6.3	Brittle, nearly pure niobate
COPPER	Cu	Copper red	Metallic	Copper-red	2.8	8.8	Tarnishes easily; malleable
CORUNDUM	Al_2O_3	All colors	Vitreous, Adamantine	White	9.0	3.9–4.1	In 6-sided crystals and masses that may break in three directions at nearly 90°. Gem varieties
COSALITE	$Pb_2Bi_2S_5$	Lead gray	Metallic	Black	2.8	6.5	In quartz veins; with pyrite, sphalerite
COVELLITE	CuS	Blue	Submetallic	Black	1.5–2.0	4.6	Platy or granular massive. Turns purple when moistened
CROCIDOLITE	$NaFe(SiO_3)_2 \cdot FeSiO_3$	Blue to green	Silky, Dull	Like Color	4.0–5.0	3.2–3.3	Fibrous masses, like asbestos, valuable
CROCOITE	$PbCrO_4$	Red	Adamantine	Orange Yellow	2.5	6.0	Found in quartz, galena, vanadirite
CRYOLITE	Na_3AlF_6	Snow white	Greasy to Vitreous	White	2.5	3.0	Appearance, hardness are distinctive
CUPRITE	Cu_2O	Red	Adamantine to dull	Red	3.5–4.0	5.9–6.2	Oxidized mineral. Often in crystals – usually cubical
CYANITE (or Kyanite)	Al_2SiO_5	White, to blue or green	Vitreous, Pearly		4.0–7.0	3.6	Bladed crystals with flat cleavage surfaces that are easily scratched longitudinally but not transversely
DESCLOIZITE	$4RO \cdot V_2O_5 \cdot H_2O$	Red, brown, black	Greasy	Orange	3.5	6.0	Associated with vanadinite
DIAMOND	C	White, gray	Adamantine, Greasy	Ash Gray	10.0	3.5	Occurs in crystals (usually rounded octahedrons) in a basic igneous rock, and in placers. Gem varieties
DIASPORE	$Al_2O_3 \cdot H_2O$	Many colors	Vitreous	White	6.5–7.0	3.4	Occurs in thin scales; very brittle
DIATOMACEOUS EARTH	$SiO_2 \cdot nH_2O$	Yellow to brown	Vitreous	White to Gray	2.0	2.2	Roughens glass. Uniformly very fine texture and light in weight
DOLOMITE	$CaMg(CO_3)_2$	White, gray, pink, yellow	Vitreous, Pearly	White	3.5–4.0	2.8–2.9	Effervesces vigorously in any condition of muriatic acid except cold dilute. Like calcite, but common in warped rhombohedrons
ENARGITE	$3Cu_2S \cdot As_2S_5$	Iron black	Metallic	Black	3.0	4.4	Color and streak both black, prismatic cleavage
EPIDOTE	$Ca_2(AlOH)(AlFe)_2(SiO_4)_3$	Green	Vitreous, Dull	White	6.0–7.0	3.2–3.5	Brittle, usually granular

Minerals and Their Characteristics

NAME	FORMULA	PERCENT METAL	COLOR	LUSTRE	STREAK	HARD-NESS	SPEC. GRAV.	CHARACTERISTICS—OCCURRENCE
EPSOM SALT	$MgSO_4 \cdot 7H_2O$	Mg—9.9%	White	Vitreous	White	2.3	1.7	Tastes bitter and saline, in mineral waters
ERYTHRITE	$Co_3As_2O_8 \cdot 8H_2O$	Co—29.5%	Usually pink, gray	Pearly	Paler than Color	1.5–2.5	3.0	Deposits of secondary origin; with cobalt ores
FERBERITE	$FeWO_4$	W—60.6%	Brown, black, gray	Metallic		5.0–5.5	7.2–7.5	Found with other tungsten ores
FLUORITE	CaF_2	F—48.9%	All colors	Vitreous	White	4.0	3.0–3.3	Octahedral cleavage, brittle
FRANKLINITE	$(ZnFeMn)O \cdot (FeMn)_2O_3$	Zn—14.2% Mn—35.7%	Iron black	Metallic	Brown to Black	5.5–6.5	5.2	Usually associated with zincite, sometimes magnetic
GALENA	PbS	Pb—86.6%	Lead gray	Metallic	Lead Gray	3.0	7.4–7.6	Very brittle, cubic cleavage
GARNET	Various	No metal source	Red, brown, yellow	Vitreous	White	6.5–7.5	3.2–4.3	Often in complete dodecahedral crystals, in schists or limestone. Gem varieties
GARNIERITE	$H_2(NiMg)SiO_4$	Ni—25% to 30%	Green	Dull, greasy	Greenish White	2.0–4.0	2.4	Amorphous; source of nickel; with serpentine, chromite
GENTHITE	$2NiO \cdot 2MgO \cdot 3SiO_2 \cdot 6H_2O$	Ni—22.6%	Green	Dull, greasy	Greenish White	2.0–4.0	2.4	Similar to garnierite
GIBBSITE	$Al(OH)_3$	Al—34.6%	White, green	Pearly		2.0–3.5	2.4	Occurs under same conditions as bauxite
GOLD	Au	Au—100%	Golden	Metallic	Golden Yellow	2.8	15.6–19.3	Malleable and sectile. Does not tarnish
GRAPHITE	C	C—100%	Black	Dull, Submetallic	Dark Gray, Iron Black	1.0–2.0	2.2	Soft; marks paper; feels greasy, often impure
GREENOCKITE	CdS	Cd—77.7%	Yellow	Adamantine	Yellow to red	3.0–3.5	5.0	Usually occurs as coating on zinc minerals
GROSSULARITE	$Ca_3Al_2(SiO_4)_3$	No metal source	White, green, yellow	Vitreous	White	6.5–7.5	3.4–3.7	Often imbedded in mica and schists, limestones. Variety of garnet. See Garnet
GYPSUM	$CaSO_4 \cdot 2H_2O$	CaO—32.6%	White, red	Vitreous	White to Gray	1.5–2.0	2.3	Earthy, fibrous, scaly, and crystals with perfect cleavage in one direction
HALITE	NaCl	Na—39.4%	White	Vitreous	White	2.5	2.1–2.6	Natural table salt. Perfect cubical cleavage
HALLOYSITE	$Al_2O_3 \cdot 2SiO_2 \cdot H_2O$	No metal source	White, green, blue, red	Pearly, Waxy, dull		1.0–2.0	2.0–2.2	Often occurs in veins of ore as secondary product
HAUSMANNITE	Mn_3O_4	Mn—72%	Black, brown	Metallic	Brown	5.3	4.7	Associated with other manganese minerals
HEMATITE	Fe_2O_3	Fe—70%	Brown, red, black	Metallic, Dull, Submetallic	Red, Brown	5.5–6.5	4.9–5.3	Becomes magnetic upon heating under reducing conditions

Name	Composition	Metal source	Color	Streak	Luster	Hardness	Sp. Gr.	Remarks
HESSITE	Ag_2Te	Ag—63%		Gray	Metallic	2.5–3.0	8.3–8.9	With chalcopyrite, pyrite, and sphalerite
HORNEBLENDE	Variable	Variable		White, green, black	Vitreous	5.0–6.0	3.2	Crystals have 6-sided or diamond-shaped cross sections. Two perfect cleavages at angle of about 124°
HUEBNERITE	$MnWO_4$	Mn–18.1% W–60.7%		Brown	Submetallic	5.0–5.5	7.2–7.5	Usually in bladed aggregates with rough, flat parting, in quartz
HYDROZINCITE	$ZnCO_3 \cdot 2Zn(OH)_2$	Zn–59.5%		White, gray, yellow	Dull	2.0–2.5	3.6–3.8	Usually associated with other zinc ores
HYPERSTHENE	$(FeMg)SiO_3$	No metal source		Black	Pearly	5.0–6.0	3.5	Occurs in foliated or platy masses
ILMENITE	$FeTiO_3$	Ti–31.6%		Iron black	Metallic, Submetallic	5.0–6.0	4.5–5.0	Magnetic; with pyrite, hornblende, feldspars
IODYRITE	AgI	Ag–46%		Yellow, green		3.0–4.0	5.6–5.7	Usually in thin plates; rare
IRIDIUM	Variable	Variable		White	Metallic	6.7	22.7	With platinum and allied metals
IRIDOSMENE	$IrOs(RhPtRu)$	Alloy–100%		Tin White	Metallic	6.0–7.0	19.3–21.1	Rare metals alloy
JAMESONITE	$2PbS \cdot Sb_2S_3$	Pb–50.8% Sb–29.5%		Gray	Metallic	2.0–3.0	5.5–6.0	Usually in parallel or divergent aggregates of narrow blades. Sometimes in hair- or needle-like crystals
JEFFERISITE	Variable	Variable		Yellowish brown	Pearly	1.5	2.3	A mica with flexible but not elastic cleavage plates that puffs out greatly when heated
KAINITE	$MgSO_4 \cdot KCl \cdot 3H_2O$	KCl–30.0%		White to red	Vitreous	2.8	2.1	See Cyanite
KAOLINITE	$H_4Al_2Si_2O_9$	Al_2O_3–39.5%		White, yellow	Pearly	2.0–2.5	2.6	Widespread; earthy odor; clay
KERMESITE	Sb_2S_2O	Sb–75.3%		Cherry	Adamantine Metallic	1.3	4.6	Occurs with stibnite
KIESERITE	$MgSO_4 \cdot H_2O$	Mg–17.6%		White, yellow	Vitreous	3.3	2.6	Often with gypsum and carnallite
LEPIDOLITE	$KLi[Al(OHF)_2] \cdot Al(SiO_3)_3$	Small amount of Li		Red, lilac, white	Pearly	3.0	2.8–3.3	A mica with flexible, elastic cleavage plates. Usually in pegmatites
LEUCITE	$KAl(SiO_3)_2$	K_2O–21.5% Al_2O_3–23.5%		Gray	Vitreous, Dull	5.5–6.0	2.5	Complete trapezohedral crystals in igneous rock
LIMONITE	$2Fe_2O_3 \cdot 3H_2O$	Fe–59.9%		Brown, yellow	Submetallic	5.0–5.5	3.6–4.0	Massive, fibrous or porous; magnetic after fusing
LINNAEITE	Co_3S_4	Co–58.0%		Steel gray	Metallic	5.5	4.8–5.0	Copper red tarnish; in gneiss with chalcopyrite
LIVINGSTONITE	$HgS \cdot 2Sb_2S_3$	Hg–22.0%		Lead gray	Metallic	2	4.81	Resembles stibnite; fuses easily
MAGNESITE	$MgCO_3$	Mg–28.9%		White to black	Vitreous	4.0–4.5	3.1	Effervesces vigorously in hot concentrated muriatic acid
MAGNETITE	$FeO \cdot Fe_2O_3$	Fe–72.4%		Iron black	Metallic, Submetallic	5.5–6.5	5.2	Strongly magnetic; many associations
MALACHITE	$CuCO_3 \cdot Cu(OH)_2$	Cu–57.5%		Green	Silky	3.5–4.0	4.0	Oxidized mineral. Effervesces vigorously in muriatic acid of any strength or temperature
MANGANITE	$Mn_2O_3 \cdot H_2O$	Mn–62.5%		Iron black, steel gray	Metallic, Submetallic	4.0	4.2–4.4	Hardens and streak are distinctive
MARBLE	Chiefly $CaCO_3$	Ca–40%		Variable	Vitreous, Earthy	3.0	2.7	Granular calcite. See calcite

Minerals and Their Characteristics

NAME	FORMULA	PERCENT METAL	COLOR	LUSTRE	STREAK	HARD-NESS	SPEC. GRAV.	CHARACTERISTICS—OCCURRENCE
MARCASITE	FeS_2	Fe—46.6%	Yellow	Metallic	Grayish, Brown, black	6.0–6.5	4.9	Deposited near earth's surface. Often in tabular crystals in coxcomb-like groups
MARMATITE	$(ZnFe)S$ Variable	Zn—46.5% to 56.9%	Yellow, brown, black	Adamantine	Brownish	5.0	3.9–4.2	Closely allied with galena; common zinc ore
MELACONITE	CuO	Cu—79.9%	Black	Earthy, Metallic		3.0–4.0	6.5	
MELILITE	$Ca_{12}Al_4Si_9O_{36}$		White, yellow, green, brown	Vitreous	White	5	2.9–3.1	Formed from magmas; common in Portland cement
MERCURY	Hg	Hg—100%	Tin white	Metallic			13.59	Liquid; rarely found in metallic state
METACINNABARITE	HgS	Hg—86.2%	Grayish black	Metallic	Black	3	7.7	Found in upper portions of mercury deposits
MILLERITE	NiS	Ni—64.8%	Yellow	Metallic	Grennish Black	3.0–3.5	5.3–5.7	Crusts with a radiating texture and hair- or needle-like crystals
MIMETITE	$(PbCl)Pb_4As_3O_{12}$	Pb—69.7%	Yellow to brown	Resinous	White	3.5	7.0–7.3	Often in crystals with 6-sided cross sections, which may taper
MOLYBDENITE	MoS_2	Mo—60%	Lead gray	Metallic	Greenish Gray	1.0–1.5	4.7–4.8	Feels greasy. Makes light greenish yellow mark on glazed paper
MOLYBITE	MoO_3	Mo—66.67%	Yellow	Adamantine, Pearly		1.5	4.5	Occurs with molybdenite
MONAZITE	$(CeLaDy)PO_4.ThSiO_4$	ThO_2—9%	Yellow, brown	Resinous	White	5.0–5.5	4.9–5.3	Rounded grains; with gold, chromite, iron
MOTTRAMITE	Variable	Variable	Black, yellow	Resinous	Yellow	3	5.8	A vanadate of lead and copper
MUSCOVITE	$H_2KAl_3(SiO_4)_3$	Variable	Yellowish white	Vitreous, Pearly	White	2.0–2.5	2.8–3.0	Cleaves easily into very thin, elastic, flexible leaves
NAUMANNITE	$(Ag_2Pb)Se$	Ag—43.0%	Iron black	Metallic	Iron Black	2.5	8	Malleable; in cubic crystals; selenide of silver and lead
NEPHELITE	$NaAlSiO_4$	No metal source	White, yellow	Vitreous, Greasy	White	5.5–6.0	2.5–2.7	Widely distributed in igneous rocks; usually massive
NICCOLITE	$NiAs$	Ni—44.1% As—55.9%	Copper red	Metallic	Brownish Black	5.0–5.5	7.3–7.7	Often found with a green coating; brittle, compact
NITRE	KNO_3	K—38.6% N—13.9%	White	Vitreous	White	2	2.1	Tastes saline and cooling; salt petre
OLIVINE	$(MgFe)_2SiO_4$	No metal source	Green	Vitreous	White or Yellowish	6.5–7.0	3.3	Same as Chrysolite
OPAL	$SiO_2.nH_2O$	No metal source	All odors	Greasy, Vitreous	White	5.5–6.5	1.9–2.3	Very smooth, curving fracture
ORPIMENT	As_2S_3	As—61%	Lemon yellow	Resinous	Lemon Yellow	1.5–2.0	3.5	Usually associated with realgar; seldom valuable

Mineral	Formula	Composition	Color	Luster	Streak	Hardness	Specific Gravity	Remarks
ORTHOCLASE	KAlSi$_3$O$_8$	Al$_2$O$_3$—18.4%	Red, gray, yellow, white	Vitreous, Dull	White	6.0–6.5	2.5–2.6	Common, often pinkish igneous rock mineral with two smooth right angle cleavages
PENTLANDITE	(FeNi)S	Fe—42.0% Ni—22.0%	Yellow-bronze	Metallic	Black	3.5–4.0	4.6–5.0	Associated with pyrrhotite, millerite, chalcopyrite, etc.
PETZITE	(AuAg)$_2$Te	Au—25.5% Ag—42%	Gray to black	Metallic	Gray	2.5	9.1	A rare but valuable ore of gold and silver; often tarnishes
PHOSPHATE ROCK	Ca$_3$(PO$_4$)$_2$	P$_2$O$_5$—32.1%	Gray	Dull	Gray	5	3.2	Occurs in massive deposits
PLATINUM	Pt	Pt—100%	Tin white, steel white	Metallic	Shiny Gray	4.5	17.0	Sometimes magnetic; with gold and chromite
POLIANITE	MnO$_2$	Mn—63.2%	Steel gray, iron gray	Metallic	Black	6.3	4.9	Looks like pyrolusite, but harded and dryer; rare
POLYBASITE	9Ag$_2$S·Sb$_2$S$_3$	Ag—75.6% Sb—9.4%	Iron black	Metallic	Black	2.0–3.0	6.0–6.2	With chalcopyrite, calcite, pyrargyrite, stephanite
POWELLITE	Ca(Mo,W)O$_4$	Variable	Greenish yellow	Resinous		3.5	4.5	Often associated with scheelite
PROUSTITE	3Ag$_2$S·As$_2$S$_3$	Ag—65.5%	Scarlet	Adamantine, Dull	Scarlet	2.0–2.5	5.6	Usually associated with other silver ores
PSILOMELANE	MnO$_2$·H$_2$O·K$_2$BaO$_2$		Black	Submetallic, Dull	Black, Brownish Black	5.0–6.0	3.7–4.7	Either powdery (Wad) or has smooth, curving fracture
PYRARGYRITE	3Ag$_2$S·Sb$_2$S$_3$	Ag—60% Sb—22.2%	Black, reddish	Adamantine, Metallic	Purplish Red	2.5	5.8–5.9	Often associated with argentite and proustite
PYRITE	FeS$_2$	Fe—46.7%	Brass yellow	Metallic	Greenish, Brown-Black	6.0–6.5	5.0	Often in crystals that are cubical or show prominently a form with 5-sided faces
PYROLUSITE	MnO$_2$	Mn—63.2%	Black, dark gray	Metallic, Dull	Black, Blue-Black	1.0–2.5	4.8	Soils fingers; hardness and streak are distinctive
PYROMORPHITE	Pb$_5$Cl(PO$_4$)$_3$	Pb—76.4%	Yellow	Greasy, Adamantine	White, Yel.-White	3.5–4.0	5.9–7.1	Alteration product of lead minerals. Occurs like mimetite
PYROPE	Mg$_3$Al$_2$(SiO$_4$)$_3$	No metal source	Red	Vitreous, Resinous	White	6.5–7.6	3.7	Variety of garnet. See garnet
PYROPHYLLITE	HAl(SiO$_3$)$_2$	Al$_2$O$_3$—28.3%	White, brown	Pearly, Dull	White	1.0–2.0	2.8–2.9	Feels greasy or soapy
PYROXENE	Ca(AlMgMnFe)(SiO$_3$)$_2$	No metal source	Green	Vitreous, Dull	White to Green	5.0–6.0	3.3	Commonly in igneous rocks in square or 8-sided crystals
PYRRHOTITE	Fe$_5$S$_6$ to Fe$_{16}$S$_{17}$	Fe—61.5% Variable	Brownish yellow	Metallic	Grayish Black	3.5–4.6	4.6	Only magnetic sulphide and therefore distinctive
QUARTZ	SiO$_2$	Si—46.9%	Colorless, all colors	Vitreous	White	7.0	2.65–2.66	Common in 6-sided prisms with pointed terminations. Gem varieties
REALGAR	AsS	As—70.1%	Orange	Resinous	Orange	1.5–2.0	2.6	Usually associated with Orpiment; flexible
RHODOCHROSITE	MnCO$_3$	MnO—61.7%	Usually red	Vitreous, Pearly	White	3.5–4.5	3.5–3.6	Blackens on exposure. Effervesces vigorously in hot, concentrated muriatic acid

Minerals and Their Characteristics

NAME	FORMULA	PERCENT METAL	COLOR	LUSTRE	STREAK	HARD-NESS	SPEC. GRAV.	CHARACTERISTICS—OCCURRENCE
RHODONITE	$MnSiO_3$	Mn—42.0%	Brownish red	Vitreous, Dull	White	5.5–6.5	3.4–3.7	With calcite, Zincite, tetrahedrite
ROSCOELITE	$H_2KAl_2V(SiO_4)_3$	Variable	Brown	Pearly		Soft	2.9	Vanadium mica in which vanadium replaced aluminum
RUBY	Al_2O_3	Al—52.9%	Many Colors	Adamantine, Vitreous		9.0	4.0	Brittle; when compact very tough; variety of corundum
RUTILE	TiO_2	Ti—60%	Brown, red, black	Adamantine, Submetallic	Light Brown	6.0–6.5	4.2	Commonly in crystals with longitudinally grooved faces, or needle or hair-like
SCHEELITE	$CaWO_4$	W—63.9%	White-Yellowish	Vitreous, Adamantine	White	4.5–5.0	5.9–6.1	Weight, hardness, and uneven fracture are distinctive
SENARMONTITE	Sb_2O_3	Sb—83.3%	Colorless, grayish	Vitreous, Dull	White	2	5.3	Formed by oxidation of stibnite
SERPENTINE	$H_4Mg_3Si_2O_9$	Mg—43%	Green, blackish, or yellow white	Wax-like, Silky	White	4.0	2.5–2.6	Feels smooth and sometimes slightly greasy
SIDERITE	$FeCO_3$	Fe—48.3%	Brown, gray	Vitreous, Pearly, Dull	White to Yellow	3.5–4.0	3.9	Magnetic after heating. Effervesces vigorously in hot concentrated muriatic acid
SILVER	Ag	Ag—100%	Silver white	Metallic	Silver-White	2.8	10.5	Malleable and sectile. Tarnishes quickly
SMALTITE	$CoAs_2$	Co—28.2%, As—71.8%	Tin white, steel gray	Metallic	Grayish Black	5.5–6.0	5.7–6.8	Granular or in crystals like pyrite. Often with erythrite. See erythrite
SMITHSONITE	$ZnCO_3$	Zn—52%	Green, gray, blue	Vitreous, Dull	White, grayish	5.0	4.3–4.5	Effervesces vigorously in any strength or temperature or muriatic acid except cold dilute
SODA NITRE	$NaNO_3$		White, reddish, brown; colorless	Vitreous	White	1.8	2.3	Taste-cooling; incrustations in beds; massive
SPERRYLITE	$PtAs_2$	Pt—56.6%, As—43.4%	Tin white	Metallic, Brilliant	Black	6.5	10.6	Found with gold-quartz, covellite, limonite
SPESSARTITE	$Mn_3Al_2(SiO_4)_3$	No meta source	Purplish, red	Vitreous		6.5–7.5	4.0–4.3	A form of garnet
SPHALERITE	ZnS	Zn—67.1%	Brown, yellow, reddish, black	Submetallic, Resinous	Light Brown, Yellow	3.5–4.0	3.9–4.1	Cleaves smoothly in six directions at angles of 60°, 90°, and 120°
SPINEL	$MgOAl_2O_3$	Al_2O_3—71.8%, MgO—28.2%	Black, gray, brown, red	Vitreous, Dull	White to Gray	8.0	3.5–4.1	Massive or in octahedral crystals. Gem varieties
SPODUMENE	$LiAl(SiO_3)_2$	Al_2O_3—27.4%, Li_2O—8.4%	White, grayish	Vitreous, Dull	White	6.5–7.0	3.1–3.2	Occurs usually in platy masses or chunky crystals, sometimes huge. Gem varieties
STANNITE	$Cu_2S \cdot FeS \cdot SnS_2$	Sn—27.5%, Cu—29.5%	Steel gray, iron black	Metallic	Blackish	4.0	4.5	Has appearance of bronze

Mineral	Formula	Metal Source	Color	Luster	Streak	Hardness	Specific Gravity	Remarks
STEPHANITE	$5Ag_2S \cdot Sb_2S_3$	Ag – 68.5%	Iron Black	Metallic	Iron Black	2.0–2.5	6.2–6.3	Associated with other silver ores
STIBNITE	Sb_2S_3	Sb – 71.8%	Lead gray	Metallic	Lead Gray, Black	2.0	4.5–4.6	Cleavage surfaces marked transversely with parallel lines
STRONTIANITE	$SrCO_3$	Sr – 59.3%	Yellow to brown, Green	Vitreous, Greasy	White to Gray	3.5–4.0	3.7	Effervesces vigorously in dilute cold, but not in concentrated cold, muriatic acid. Effervescing fragment colors alcohol flame red.
SULFUR	S	S – 100%	Yellow	Greasy, Adamantine	Pale Yellow	2.0	2.0	Burns with a characteristic odor
SYLVANITE	$(AuAg)Te_2$	Au – 24.5%, Ag – 13.4%	White to steel gray	Metallic	Same as Color	1.5–2.0	7.9–8.3	Occurs often in small, bladed or prismatic crystals
SYLVITE	KCl	K – 52.4%	White, yellowish red	Vitreous	White	2.0	1.98	Taste – saline, soluble; bitter
TALC	$H_2Mg_3(SiO_3)_4$	Mg – 19.2%, Si – 29.6%	Green to white	Pearly	White	1.0–1.5	2.7–2.8	Common; feels greasy; extensive beds
TANTALITE	$FeTa_2I_6$	Variable, Ta_2O_5 – 65.6%	Iron black	Submetallic, Greasy, Dull	Reddish Brown	6.3	5.3–7.3	Iron and manganese content variable; with columbite
TENNANTITE	$Cu_6As_2S_7$ Variable	Cu – 57.5% Variable	Steel gray, iron black	Metallic	Black, Reddish Brown	3.0–4.5	4.4–4.5	Occurs granular massive or in tetrahedral crystals
TENORITE	CuO	Cu – 79.9%	Black	Metallic		3.0	5.8–6.3	Sublimation product in volcanic regions
TEPHROITE	Mn_2SiO_4	No metal source	Red, ash gray	Vitreous		6.5–7.0	4.0–4.1	Rarely in small crystals; like chrysolite
TETRADYMITE	$Bi_2(TeS)_3$	Variable	Pale steel gray	Metallic		1.8	7.4	Soils paper; found in gold-quartz and igneous rocks
TETRAHEDRITE	$4Cu_2S \cdot Sb_2S_3$	Cu – 52.1%, Sb – 24.8%	Gray to black	Metallic	Black	3.0–4.5	4.4–5.1	Like tennantite but has a darker streak – not reddish
TITANITE	$CaTiSiO_5$	TiO_2 – 40.8%	Brown, gray, yellow, green	Adamantine	White	5.0–5.5	3.4–3.6	Occurs in platy massive or in wedge-shaped crystals
TOPAZ	$(AlF)_2SiO_4$	No metal source	Many	Vitreous		8.0	3.4–3.6	Often in prismatic crystals with diamond-shaped cross sections and a perfect transverse cleavage. Gem varieties
TOURMALINE	$HgAl_3(BOH)_{12}Si_4O_{19}$	No metal source	Black, brown, & many others	Vitreous to Resinous	White	7.0–7.5	3.0–3.2	Usually in prismatic crystals with spherical triangular cross sections. Gem varieties
TREMOLITE	$CaMg_3(SiO_3)_4$	No metal source	White to dark gray	Silky	White	5.0–6.0	2.9–3.4	Perfect cleavages in two directions at an angle of about 124°
TRIPHYLLITE	$LiFePO_4$	Li – 4.4%	Greenish gray, bluish gray	Vitreous, Resinous	White to Grayish White	4.8	3.5	A phosphate or iron, manganese and lithium
ULLMANNITE	NiSbS	Ni – 27.6%, Sb – 57.3%	Steel gray to white	Metallic	Grayish	5.3	6.4	With galena and chalcopyrite
URANINITE	$UO_3 \cdot UO_2$ Variable	Radium Source	Gray, green, brown	Submetallic to Greasy	Black, Gray, Green	5.5	9.0–9.7	Of primary and secondary origin, no definite formula
UVAROVITE	$Ca_3Cr_2(SiO_4)_3$	No metal source	Green	Vitreous	White	6.5–7.5	3.5	A form of garnet

Minerals and Their Characteristics

NAME	FORMULA	PERCENT METAL	COLOR	LUSTRE	STREAK	HARD-NESS	SPEC. GRAV.	CHARACTERISTICS—OCCURRENCE
VALENTINITE	Sb_2O_3	Sb—83.5%	White	Adamantine to Pearly	White	2.5–3.0	5.6	An oxidized mineral
VANADINITE	$(PbCl)Pb_4(VO_4)_3$	Variable	Red, brown, yellow	Resinous	White or Yellow	2.7–3.0	6.6–7.1	Like mimetite, but crystals usually very sharp and do not taper
VERMICULITE	$3MgO \cdot (FeAl)_2O_3 \cdot 3SiO_2$	Variable	Grayish	Talc-like	Uncolored	1.5	2.7	Becomes worm-like threads upon heating—exfoliates
WILLEMITE	Zn_2SiO_4	Zn—58.5%	Green, yellow, brown	Vitreous, Dull	White or Grayish	5.5	3.9–4.2	Massive to granular; valuable zinc ore
WITHERITE	$BaCO_3$	BaO—77.7%	Yellow, brown	Vitreous Pearly	White	3.4	4.4	Reacts like strontianite in muriatic acid, but effervescing fragment colors alcohol flame light yellowish green
WOLFRAMITE	$(FeMn)WO_4$	W—5˜3%	Gray, brown, black	Submetallic	Reddish-Brown	5.0–5.5	7.2–7.5	Differs from huebnerite in streak
WULFENITE	$PbMoO_4$	Pb—56.4%, Mo—26.2%	Yellow, grayish	Resinous, Adamantine	White	3.0	6.8	In square crystals, usually tabular with beveled edges
ZARATITE	$NiCO_3 \cdot 2Ni(OH)_2 \cdot 4H_2O$	Ni—46.8%	Green	Vitreous	Light Green	3	2.6	Emerald nickel; amorphous
ZINCITE	ZnO	Zn—80.3%	Red, yellow	Sub-Adamantine	Orange-Yellow	4.0–4.5	5.4–5.7	Associated with other zinc ores
ZIRCON	$ZrSiO_4$	ZrO_2—67.2%	Yellow, gray	Adamantine	Colorless	7.5	4.2–4.7	In sharp crystals with square cross sections and as pebbles. Gem varieties

Appendix B

Selected Plant Flowsheets

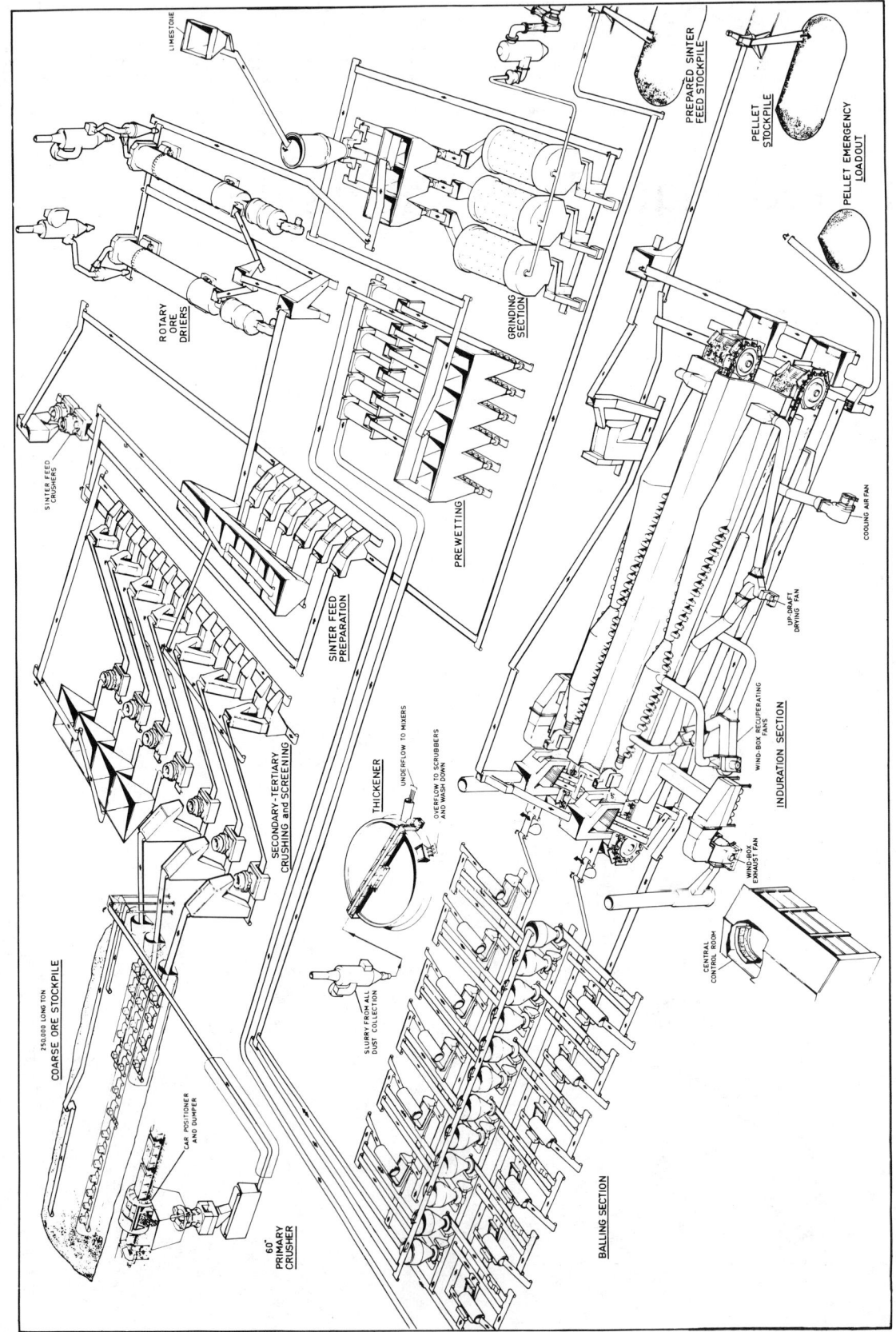

Figure B.1. An iron ore crushing and pelletizing plant. (Courtesy Cliffs Western Australia Ltd.)

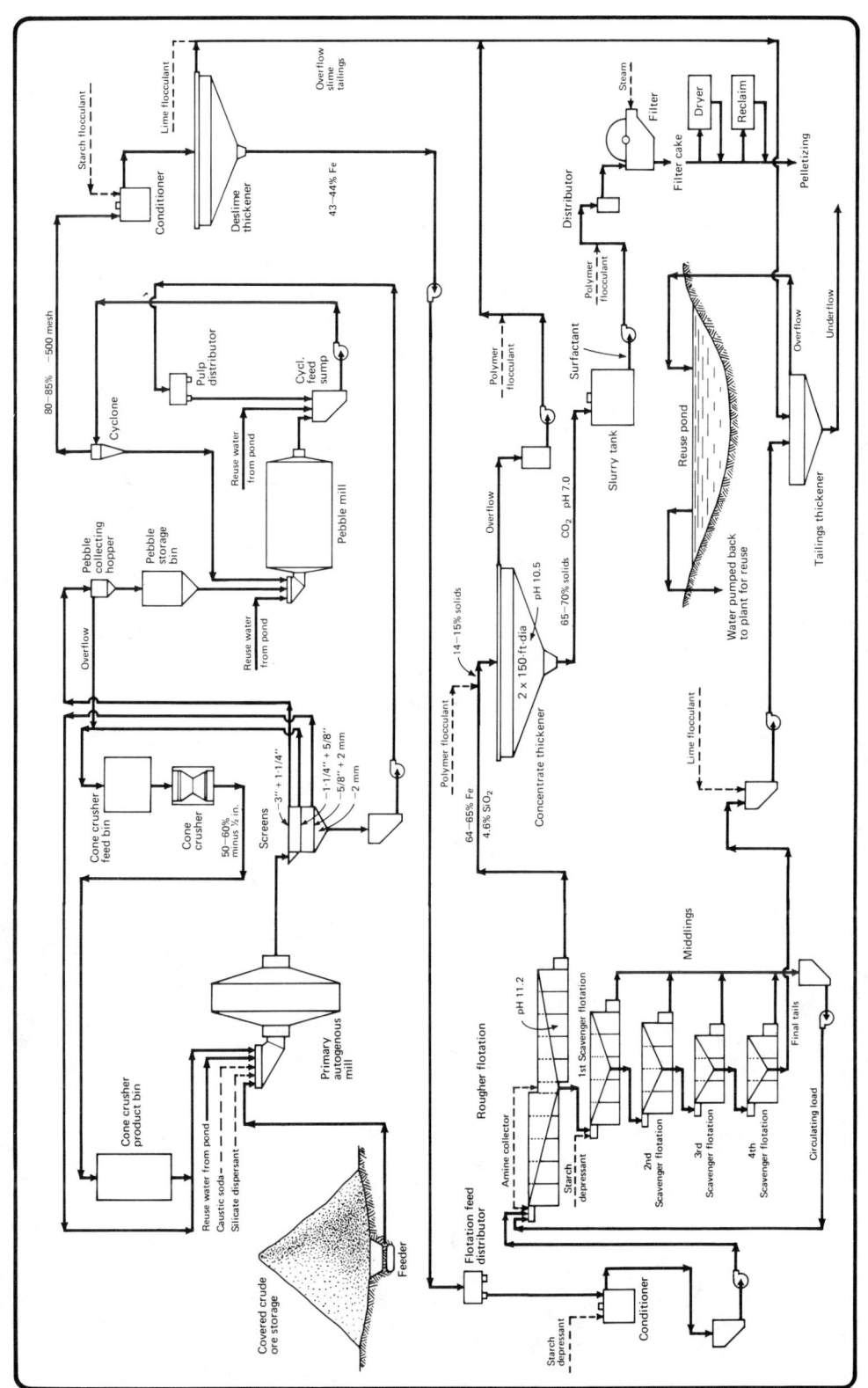

Figure B.2. The Tilden iron ore concentrator. (Courtesy Engineering and Mining Journal)

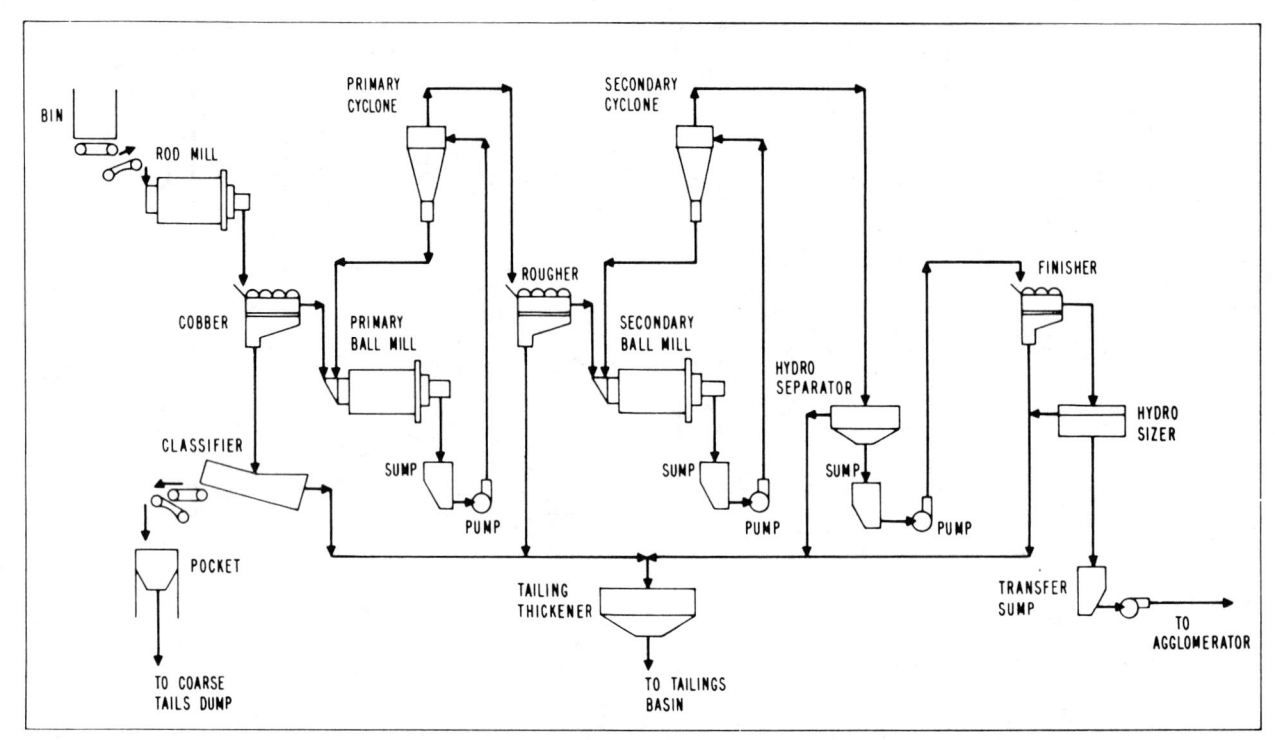

Figure B.3. The Minntac iron ore concentrator.

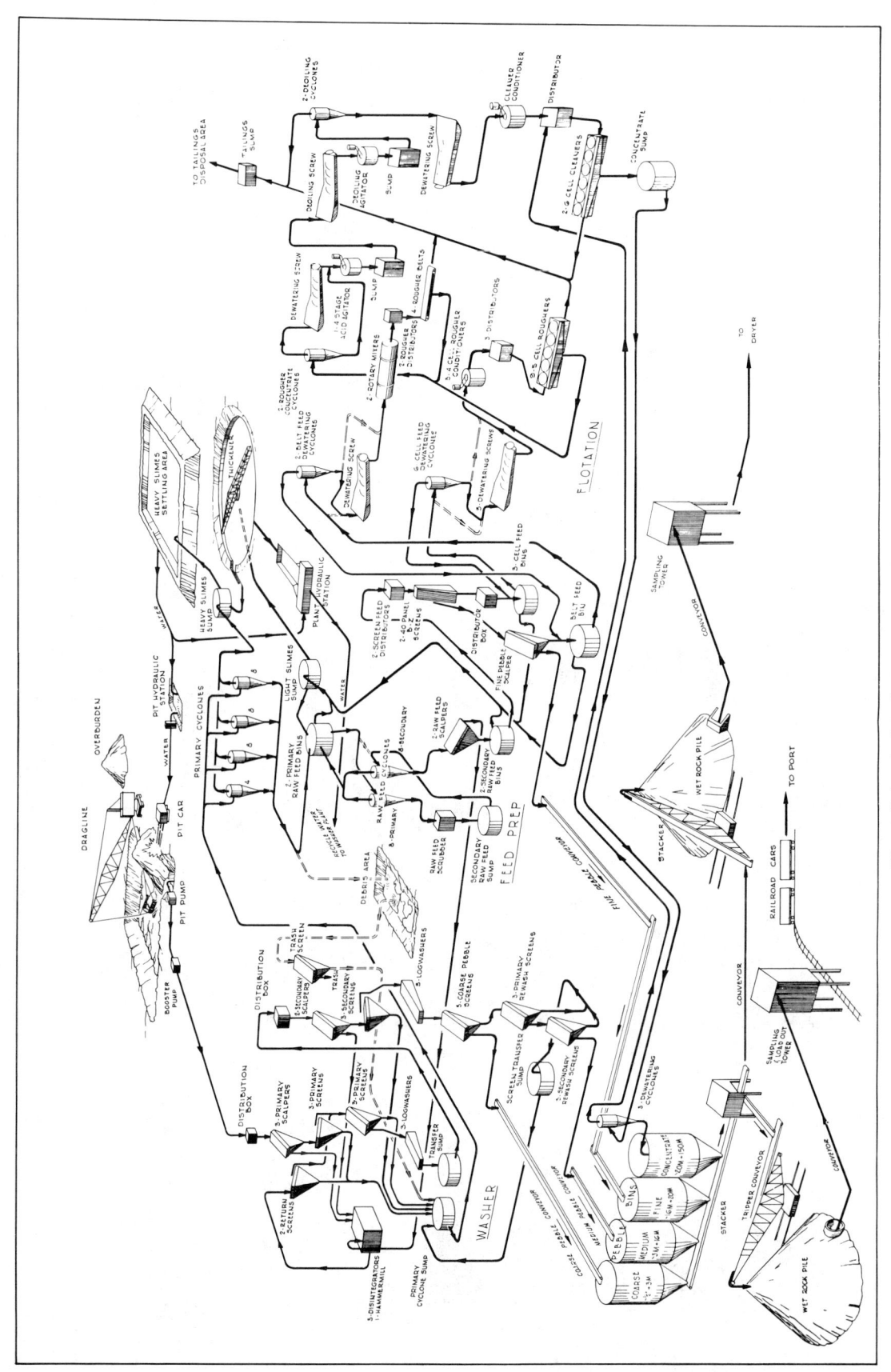

Figure B.4. A phosphate ore concentrator.

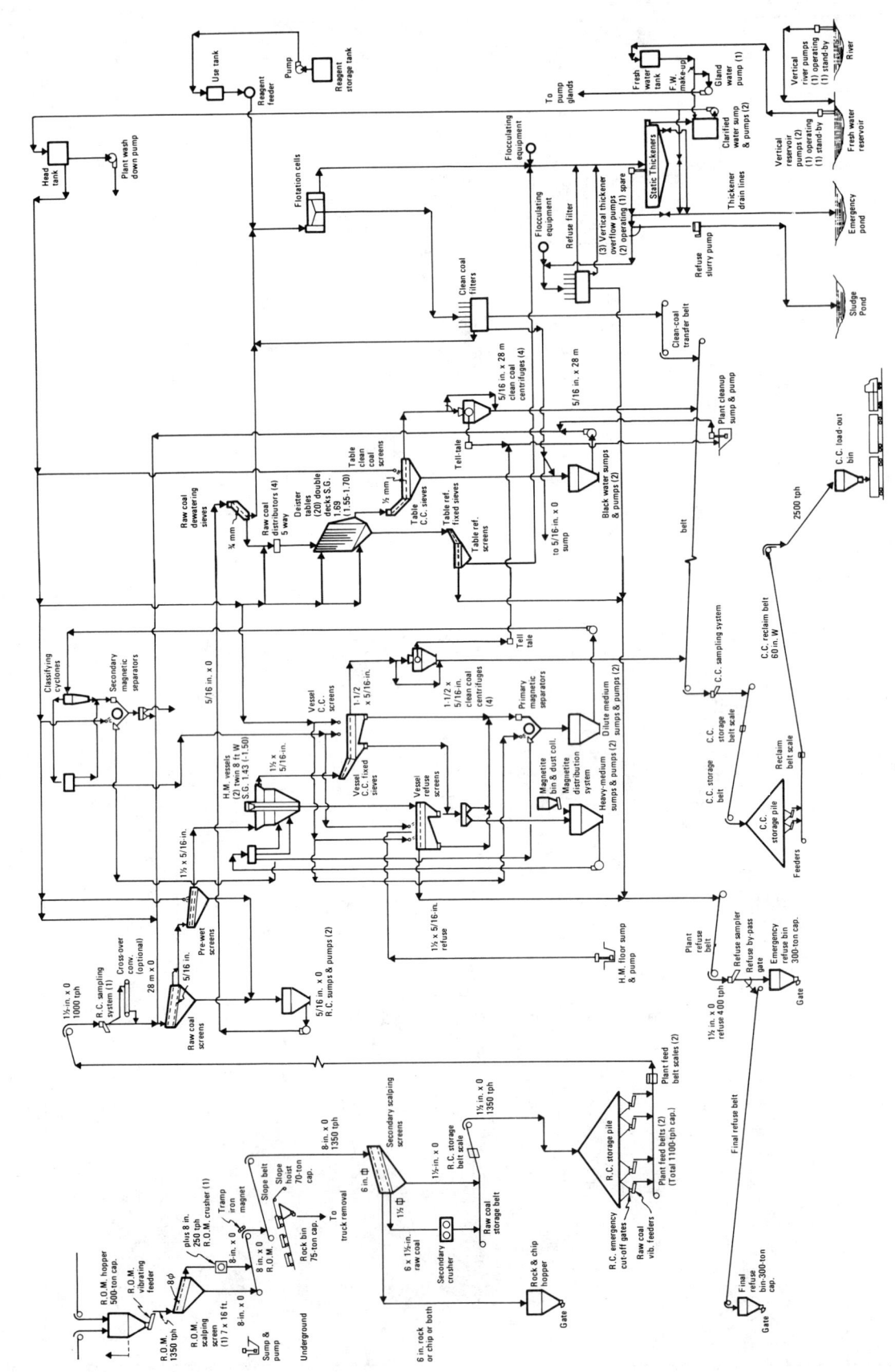

Figure B.5. Coal cleaning by dense medium separation, tabling, and flotation. (Courtesy Coal Age)

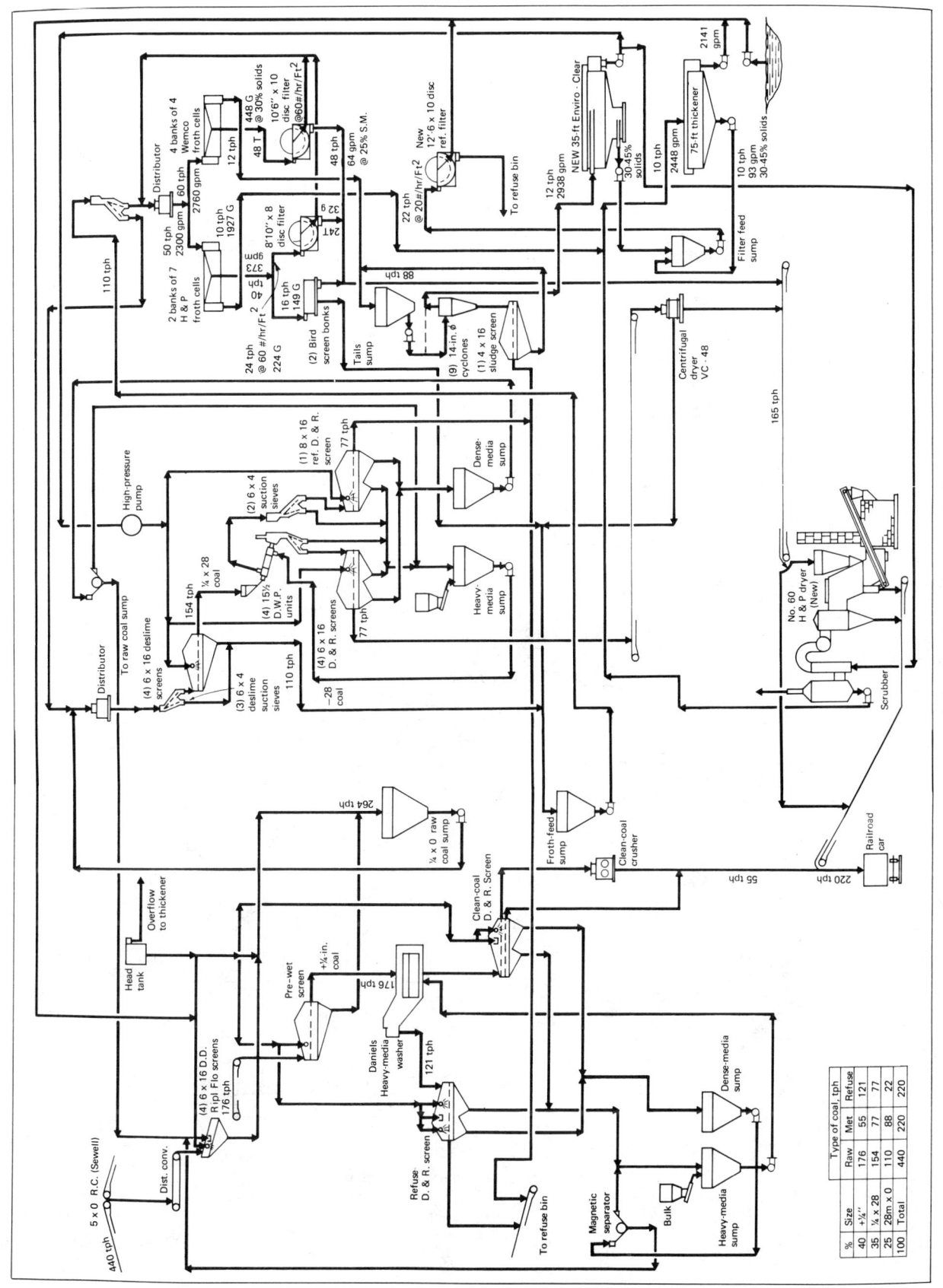

Figure B.6. A coal cleaning plant. (Courtesy Coal Age)

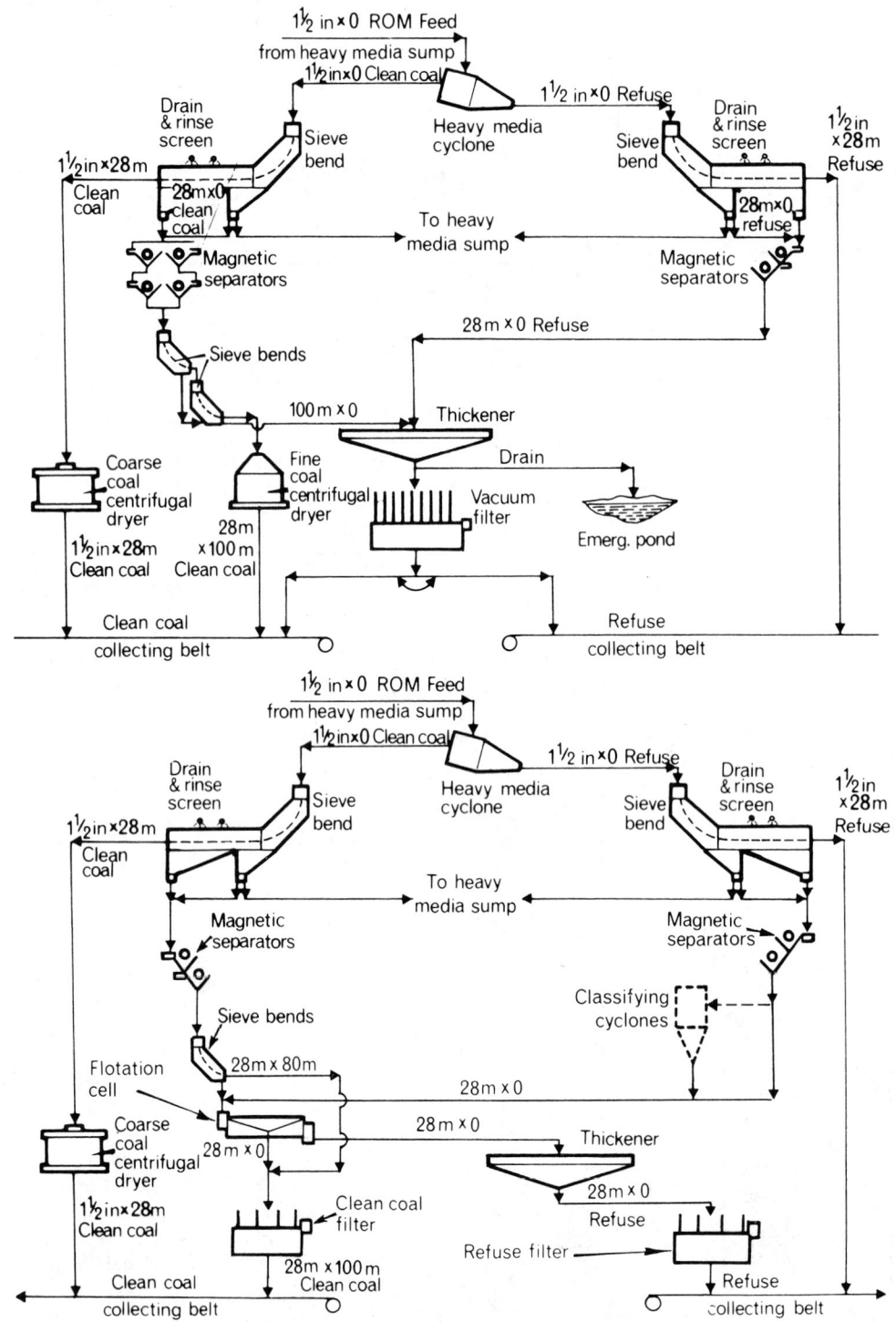

Figure B.7. Dense medium coal cleaning.

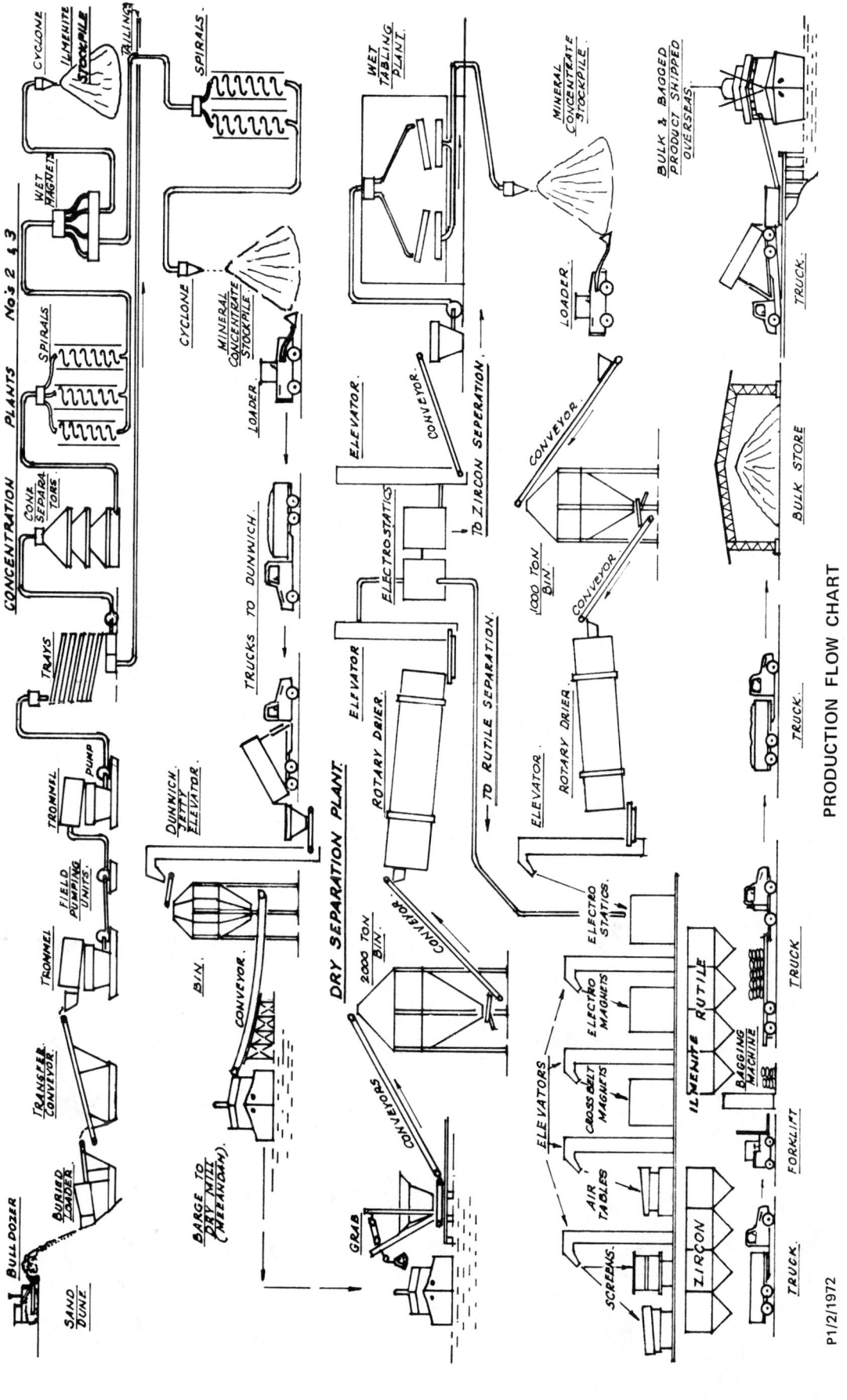

Figure B.8. The Consolidated Rutile beach sands concentrator.

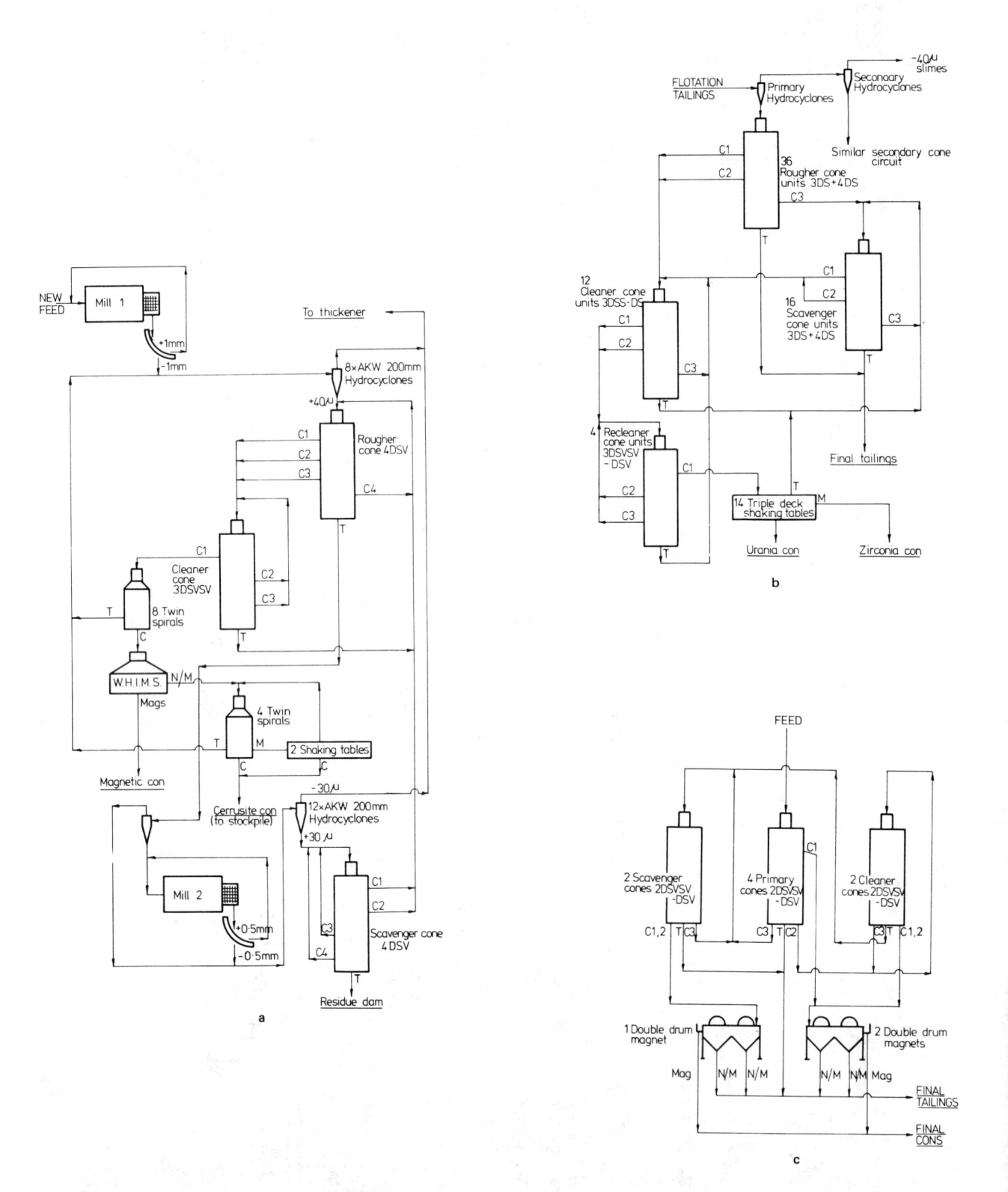

Figure B.9. (*a*) Concentrator for a complex cerrusite ore. (*b*) Concentrator for titaniferous magnetite. (*c*) Recovery of urania and zirconia from copper flotation plant tailings. (Courtesy Mineral Deposits Ltd.)

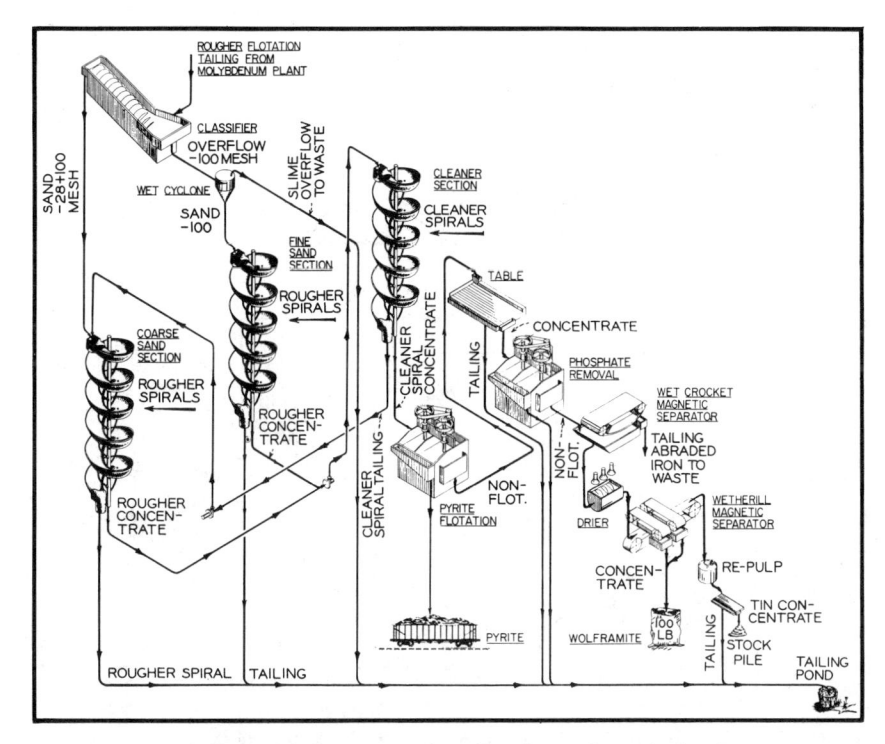

Figure B.10. Recovery of wolframite and cassiterite from flotation plant tailings.

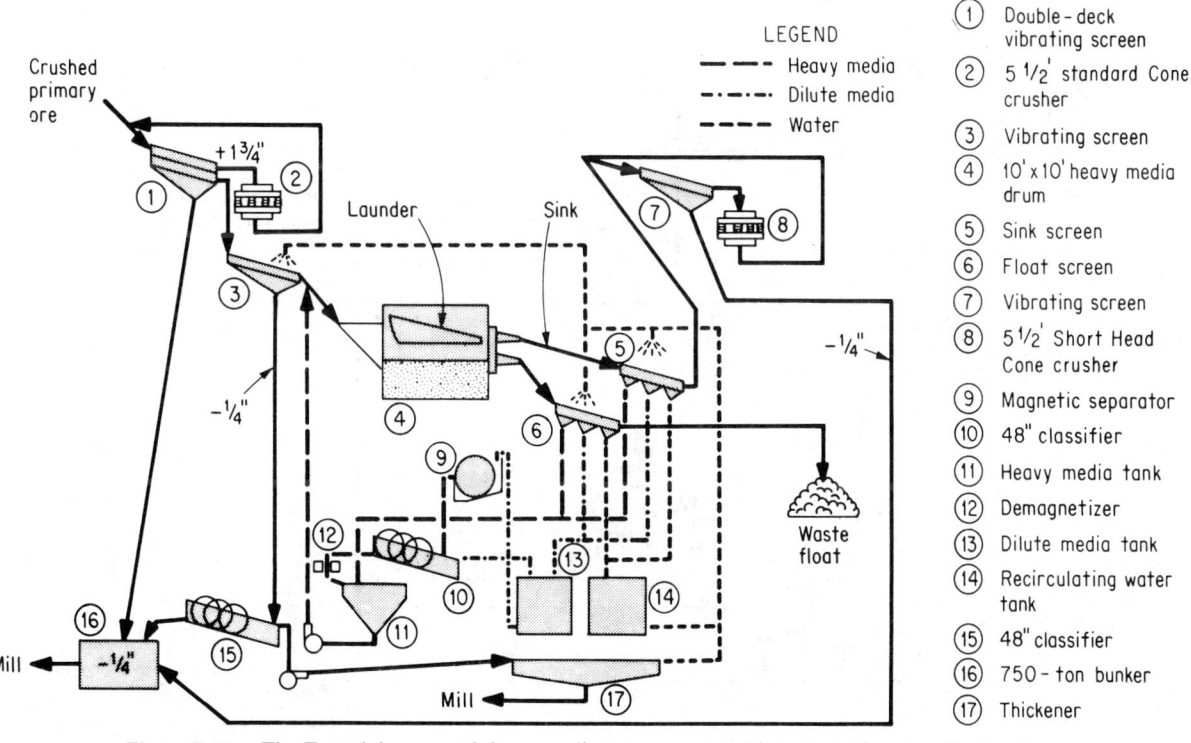

Figure B.11. The Tynagh base metal dense medium pre-concentration plant. (Courtesy Engineering and Mining Journal)

LEGEND
– – – Heavy media
–·–·– Dilute media
- - - - Water

① Double-deck vibrating screen
② 5 1/2' standard Cone crusher
③ Vibrating screen
④ 10' x 10' heavy media drum
⑤ Sink screen
⑥ Float screen
⑦ Vibrating screen
⑧ 5 1/2' Short Head Cone crusher
⑨ Magnetic separator
⑩ 48" classifier
⑪ Heavy media tank
⑫ Demagnetizer
⑬ Dilute media tank
⑭ Recirculating water tank
⑮ 48" classifier
⑯ 750-ton bunker
⑰ Thickener

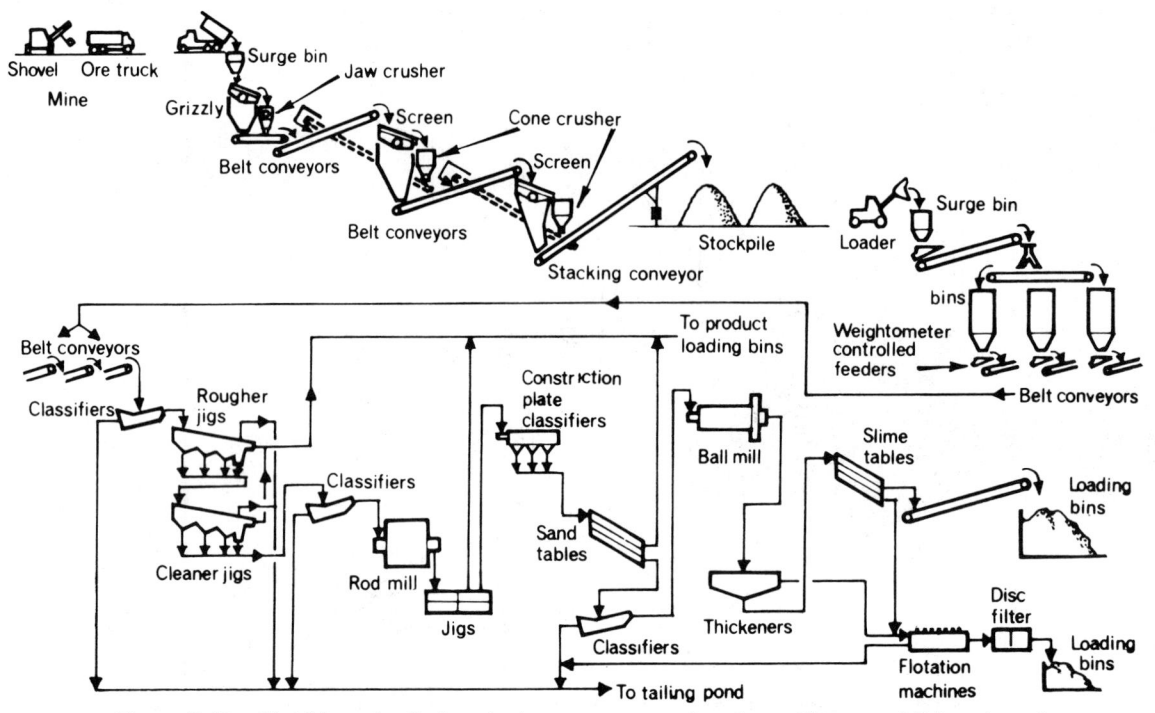

Figure B.12. The Mountain Springs barite concentrator flowsheet. (Courtesy Mining Annual Review)

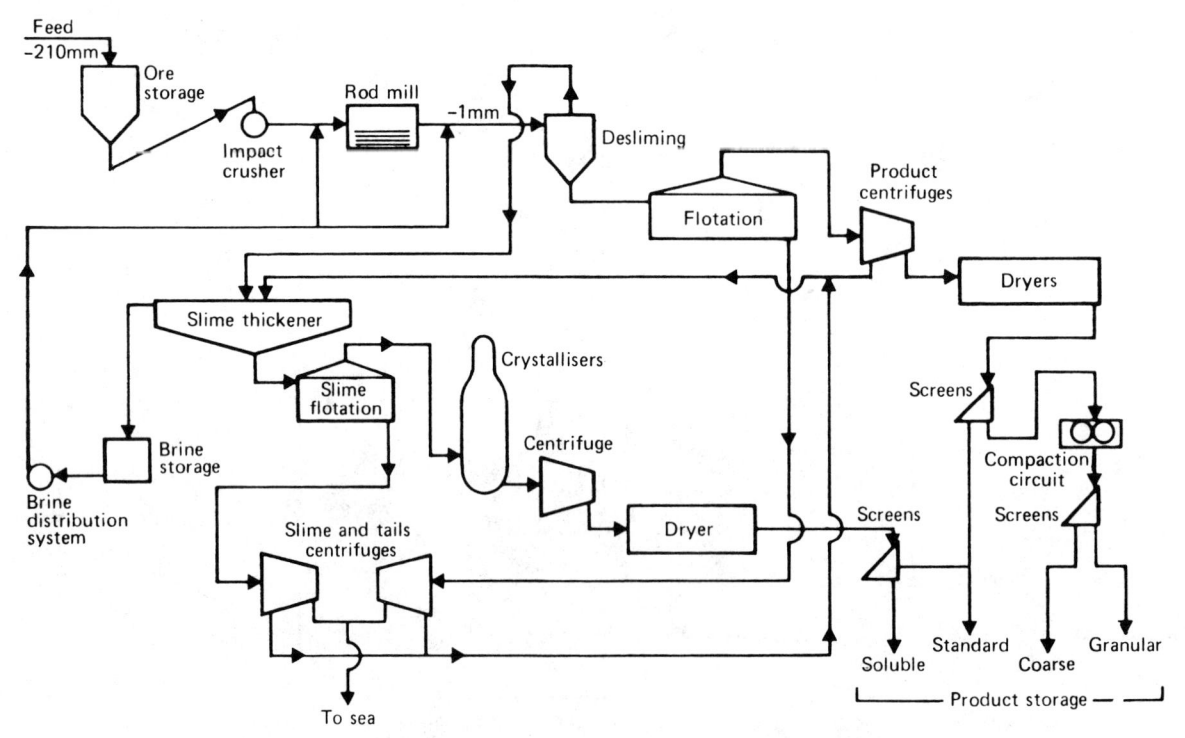

Figure B.13. The Boulby potash concentrator. (Courtesy Mining Annual Review)

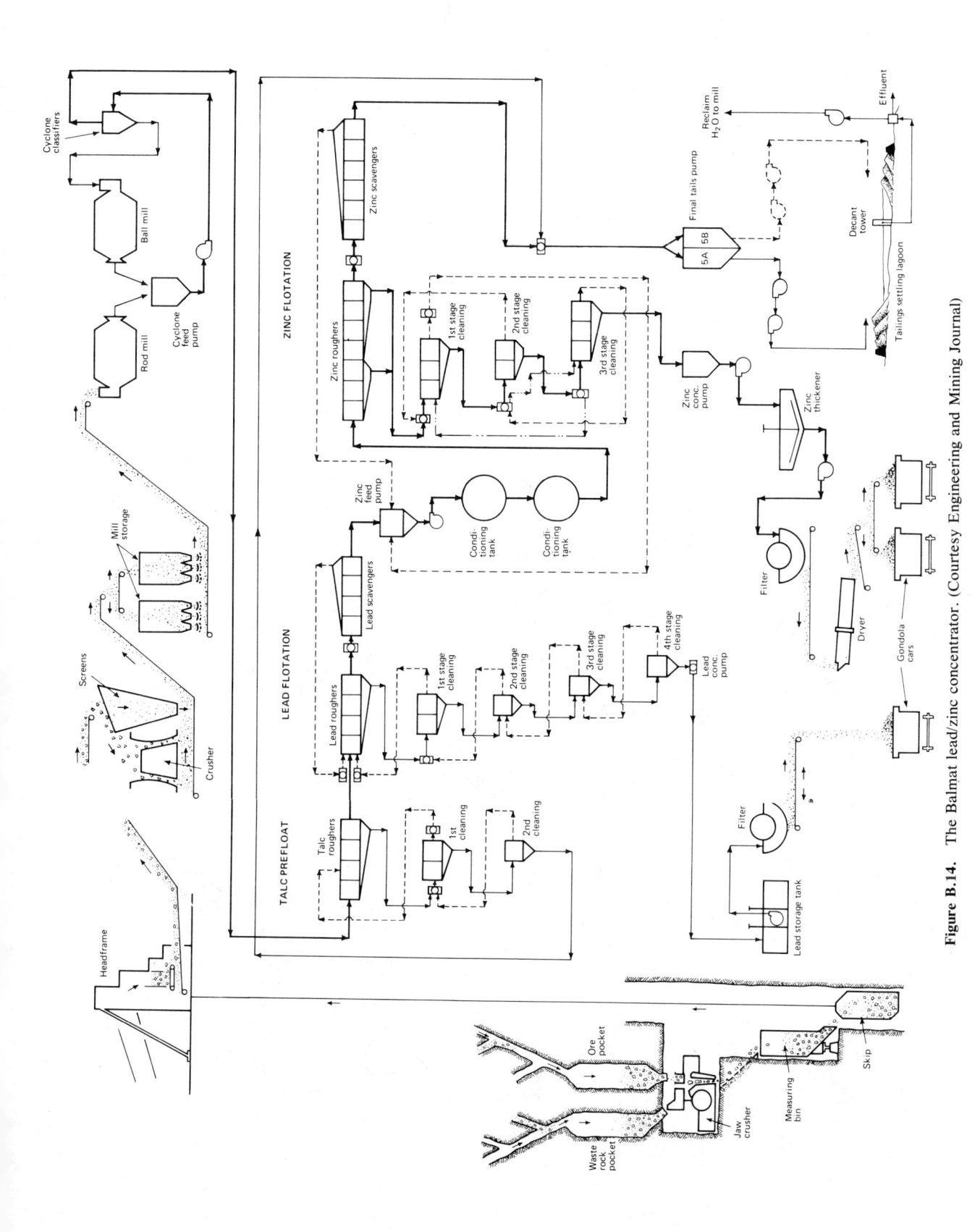

Figure B.14. The Balmat lead/zinc concentrator. (Courtesy Engineering and Mining Journal)

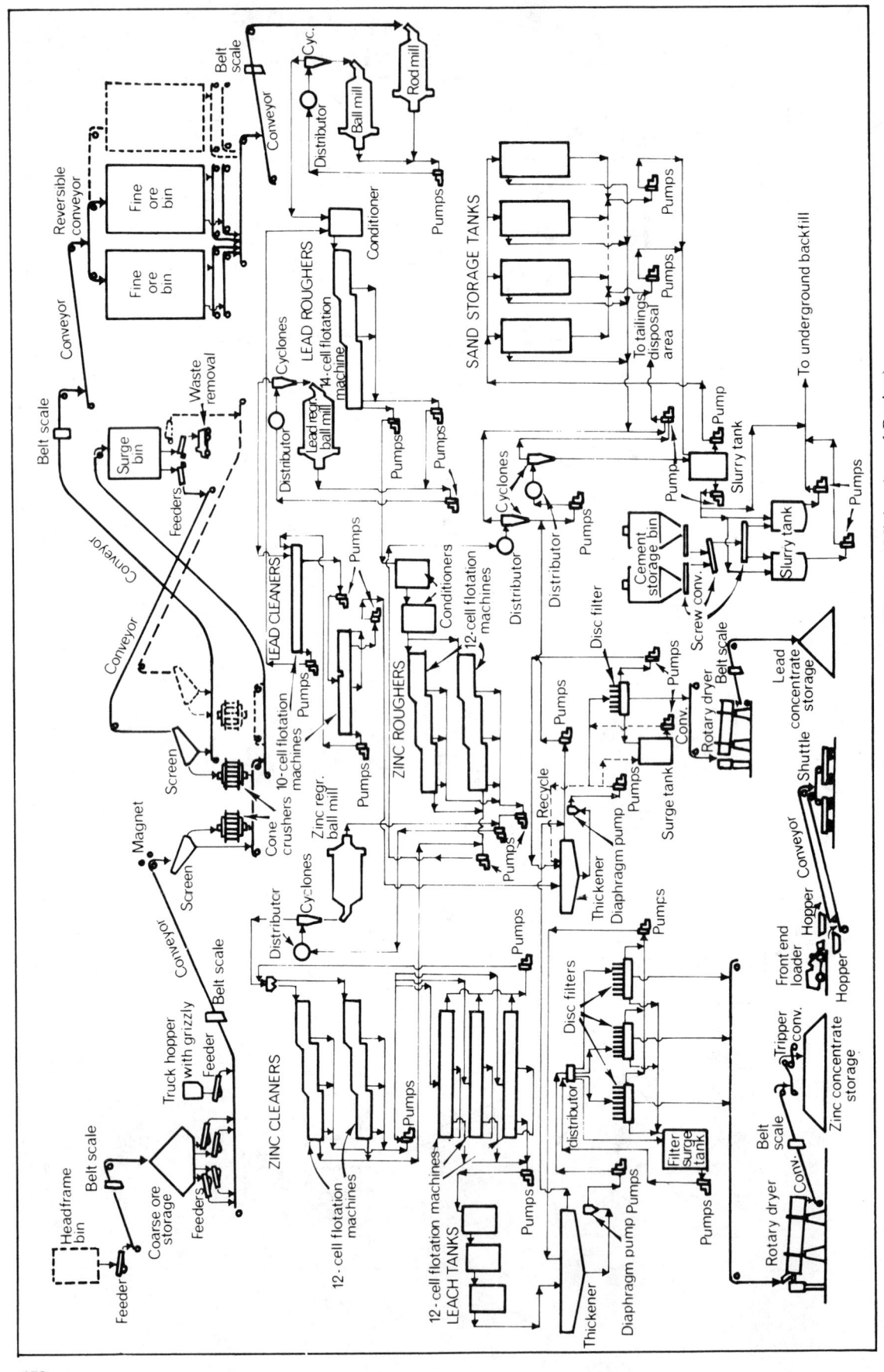

Figure B.15. The Navan lead/zinc concentrator. (Courtesy Mining Annual Review)

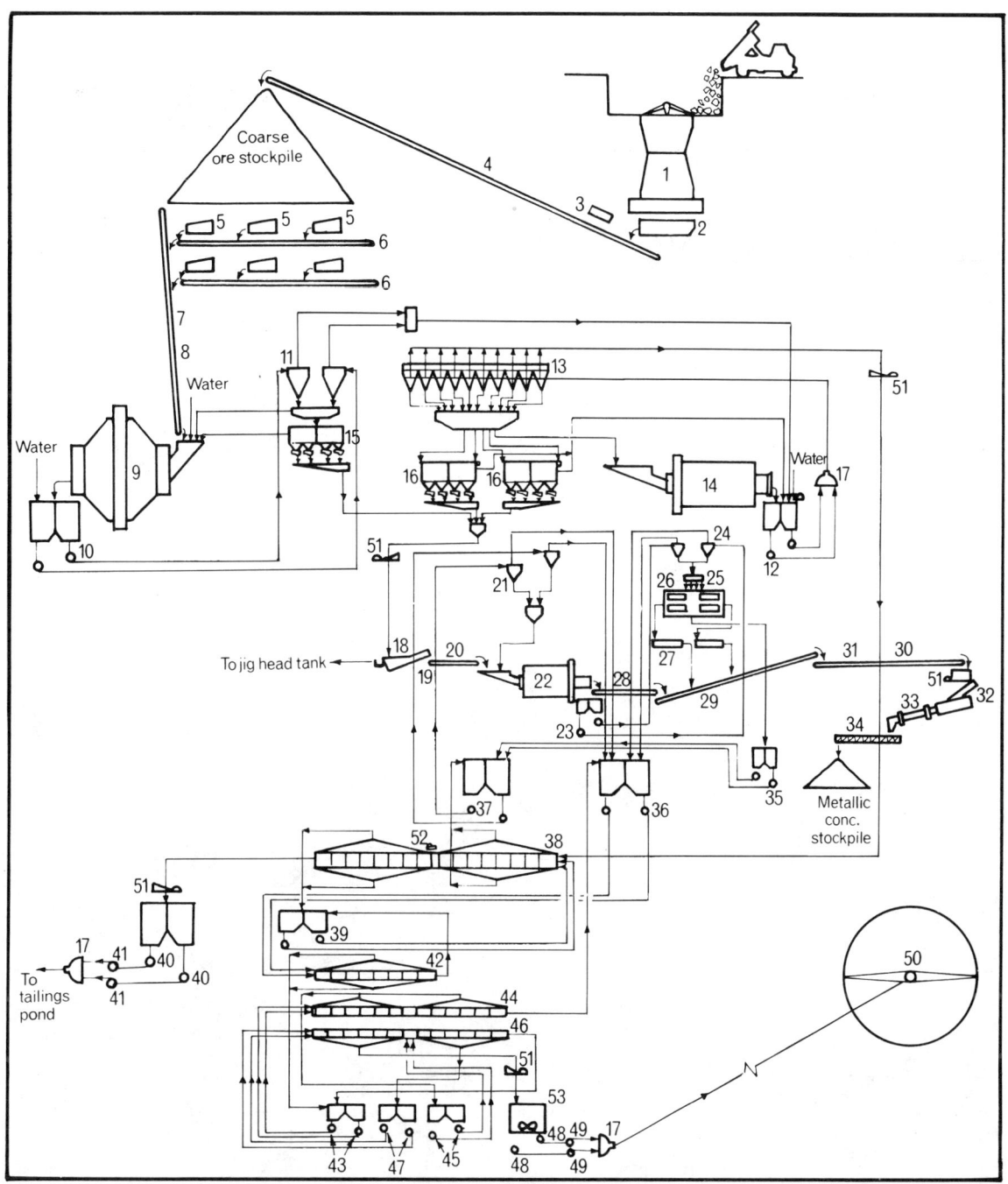

Afton concentrator flowsheet: 1. 42in × 65in Allis Chalmers gyratory crusher; 2. 72in × 16ft Nico TR 1100 hydrastroke feeder; 3. Asea metal detector; 4. No. 1 conveyor system (48in); 5. 6 – 48in × 10ft Nico 550 hydrastroke feeders; 6. No. 2 and No. 3 conveyor systems (42in); 7. No. 4 conveyor system (42in); 8. Merrick 440 DS-1 weightometer; 9. 28ft × 12ft Koppers 1y mill; 10. 2 – 10 × 12 Georgia ironworks pumps, 250hp; 11. 2 – D50 Krebs cyclones; 12. 2 – 16 × 14 × 34 SRL-C Allis Chalmers 350hp pumps; 13. 10 – D20 LB Krebs cyclones; 14. 16ft 6in × 29ft Koppers 2y ball mill; 15. 2 – 24in × 36in Denver 1y mineral jigs; 16. 4 – 24in × 36in Denver 2y mineral jigs; 17. 3 – Tech-Taylor valves; 18. 1 – 30in × 20ft 3in triple pitch Wemco spiral classifier; 19. No. 5 conveyor system (18in); 20. Merrick 440 Ds-1 weightometer; 21. 2 – Krebs D-15 cyclones; 22. 1 – 9ft 6in × 12ft Koppers regrind mill; 23. 2 – 5 × 4 × 14 SRL-C Allis Chalmers 30hp pumps; 24. 2 – Krebs D-10 L cyclones; 25. Deister 12-way distributor; 26. 4 – Concenco 999 triple deck concentrating tables; 27. 2 – straight line concentrate filters; 28. No. 6 conveyor system (18in); 29. No. 7 conveyor system (18in); 30. No. 8 2-speed conveyor system (18in); 31. Merrick 440 DS-1 weightometer; 32. 1 – 8in × 14ft 6in dryer feed screw conveyor; 33. 1 – 3ft × 24ft Lochhead Haggerty rotary dryer; 34. 1 – 9in × 25ft 9in dryer discharge screw conveyor; 35. 2 – 5 × 4 × 14 SRL-C Allis Chalmers 5hp table tailings pumps; 36. 2 – 12 × 10 × 25 SRL-C Allis Chalmers 30hp regrind cyclone overflow pumps; 37. 2 – 10 × 8 × 21 SRL-C Allis Chalmers 60hp regrind mill cyclone feed pumps; 38. 8 + 8 Denver 600H rougher and scavenger flotation cells; 39. 2 – 14 × 12 × 29 SRL-C Allis Chalmers 50hp scavenger conc. pumps; 40. 2 – 14 × 12 × 29 SRL-C Allis Chalmers 250hp flotation tailings pumps; 41. 2 – 14 × 12 × 29 SRL-C Allis Chalmers 250hp variable speed flotation tailings pumps; 42. 8 – Denver 300V cleaner flotation cells; 43. 2 – 6 × 6 × 15 SRL Allis Chalmers 15hp 1st cleaner conc. pumps; 44. 12 – Denver 30 DR cleaner flot. cells; 45. 2 – 6 × 6 × 15 SRL Allis Chalmers 15hp 2nd cleaner conc. pumps; 46. 12 – Denver 30 DR cleaner flot. cells; 47. 2 – 6 × 6 × 15 SRL Allis Chalmers 15hp 3rd cleaner conc. pumps; 48. 2 – 5 × 5 × 14 SRL Allis Chalmers 30hp final conc. pumps; 49. 2 – 5 × 5 × 14 SRL-C Allis Chalmers 30hp final conc. pumps; 50. 50ft flot. conc. thickener; 51. 5 – Denver automatic samplers; 52. 4 – Clarkson reagent feeders; 53. 1 – final conc. conditioner tank.

Figure B.16. The Afton copper concentrator flowsheet. (Courtesy Mining Annual Review)

Appendix C

Standard Sieve Sizes

Comparison Table of U. S. A., Tyler, Canadian, British, French, and German Standard Sieve Series

U.S.A. (1)		TYLER (2)	CANADIAN (3)		BRITISH (4)		FRENCH (5)		GERMAN (6)
*Standard	Alternate	Mesh Designation	Standard	Alternate	Nominal Aperture	Nominal Mesh No.	Opg. M.M.	No.	Opg.
125 mm	5"		125 mm	5"					
106 mm	4.24"		106 mm	4.24"					
100 mm	4"		100 mm	4"					
90 mm	3½"		90 mm	3½"					
75 mm	3"		75 mm	3"					
63 mm	2½"		63 mm	2½"					
53 mm	2.12"		53 mm	2.12"					
50 mm	2"		50 mm	2"					
45 mm	1¾"		45 mm	1¾"					
37.5 mm	1½"		37.5 mm	1½"					
31.5 mm	1¼"		31.5 mm	1¼"					
26.5 mm	1.06"	1.05"	26.5 mm	1.06"					
25.0 mm	1"		25.0 mm	1"					25.0 mm
22.4 mm	⅞"	.883"	22.4 mm	⅞"					
19.0 mm	¾"	.742"	19.0 mm	¾"					20.0 mm
16.0 mm	⅝"	.624"	16.0 mm	⅝"					18.0 mm
13.2 mm	.530"	.525"	13.2 mm	.530"					16.0 mm
12.5 mm	½"		12.5 mm	½"					12.5 mm
11.2 mm	⁷⁄₁₆"	.441"	11.2 mm	⁷⁄₁₆"					
9.5 mm	⅜"	.371"	9.5 mm	⅜"					10.0 mm
8.0 mm	⁵⁄₁₆"	2½	8.0 mm	⁵⁄₁₆"					8.0 mm
6.7 mm	.265"	3	6.7 mm	.265"					
6.3 mm	¼"		6.3 mm	¼"					6.3 mm
5.6 mm	No. 3½	3½	5.6 mm	No. 3½					
4.75 mm	4	4	4.75 mm	4			5.000	38	5.0 mm
4.00 mm	5	5	4.00 mm	5			4.000	37	4.0 mm
3.35 mm	6	6	3.35 mm	6	3.35 mm	5			
2.80 mm	7	7	2.80 mm	7	2.80 mm	6	3.150	36	3.15 mm
2.36 mm	8	8	2.36 mm	8	2.40 mm	7	2.500	35	2.5 mm
2.00 mm	10	9	2.00 mm	10	2.00 mm	8	2.000	34	2.0 mm
1.70 mm	12	10	1.70 mm	12	1.68 mm	10	1.600	33	1.6 mm
1.40 mm	14	12	1.40 mm	14	1.40 mm	12			
1.18 mm	16	14	1.18 mm	16	1.20 mm	14	1.250	32	1.25 mm
1.00 mm	18	16	1.00 mm	18	1.00 mm	16	1.000	31	1.0 mm
850 µm	20	20	850 µm	20	850 µm	18			
710 µm	25	24	710 µm	25	710 µm	22	.800	30	800 µm
							.630	29	630 µm
600 µm	30	28	600 µm	30	600 µm	25			
500 µm	35	32	500 µm	35	500 µm	30	.500	28	500 µm
425 µm	40	35	425 µm	40	420 µm	36			
355 µm	45	42	355 µm	45	355 µm	44	.400	27	400 µm
							.315	26	315 µm
300 µm	50	48	300 µm	50	300 µm	52			
250 µm	60	60	250 µm	60	250 µm	60	.250	25	250 µm
212 µm	70	65	212 µm	70	210 µm	72			
							.200	24	200 µm
180 µm	80	80	180 µm	80	180 µm	85			
							.160	23	160 µm
150 µm	100	100	150 µm	100	150 µm	100			
125 µm	120	115	125 µm	120	125 µm	120	.125	22	125 µm
106 µm	140	150	106 µm	140	105 µm	150			
							.100	21	100 µm
90 µm	170	170	90 µm	170	90 µm	170			90 µm
75 µm	200	200	75 µm	200	75 µm	200	.080	20	80 µm
									71 µm
63 µm	230	250	63 µm	230	63 µm	240	.063	19	63 µm
									56 µm
53 µm	270	270	53 µm	270	53 µm	300			
							.050	18	50 µm
45 µm	325	325	45 µm	325	45 µm	350			45 µm
							.040	17	40 µm
38 µm	400	400	38 µm	400					

(1) U.S.A. Sieve Series - ASTM Specification E-11-70
(2) Tyler Standard Screen Scale Sieve Series.
(3) Canadian Standard Sieve Series 8-GP-1d.
(4) British Standards Institution, London BS-410-62.
(5) French Standard Specifications, AFNOR X-11-501.
(6) German Standard Specification DIN 4188.

*These sieves correspond to those recommended by ISO (International Standards Organization) as an International Standard and this designation should be used when reporting sieve analysis intended for international publication.

Appendix D

Rosin-Rammler (Weibull) Graph Paper

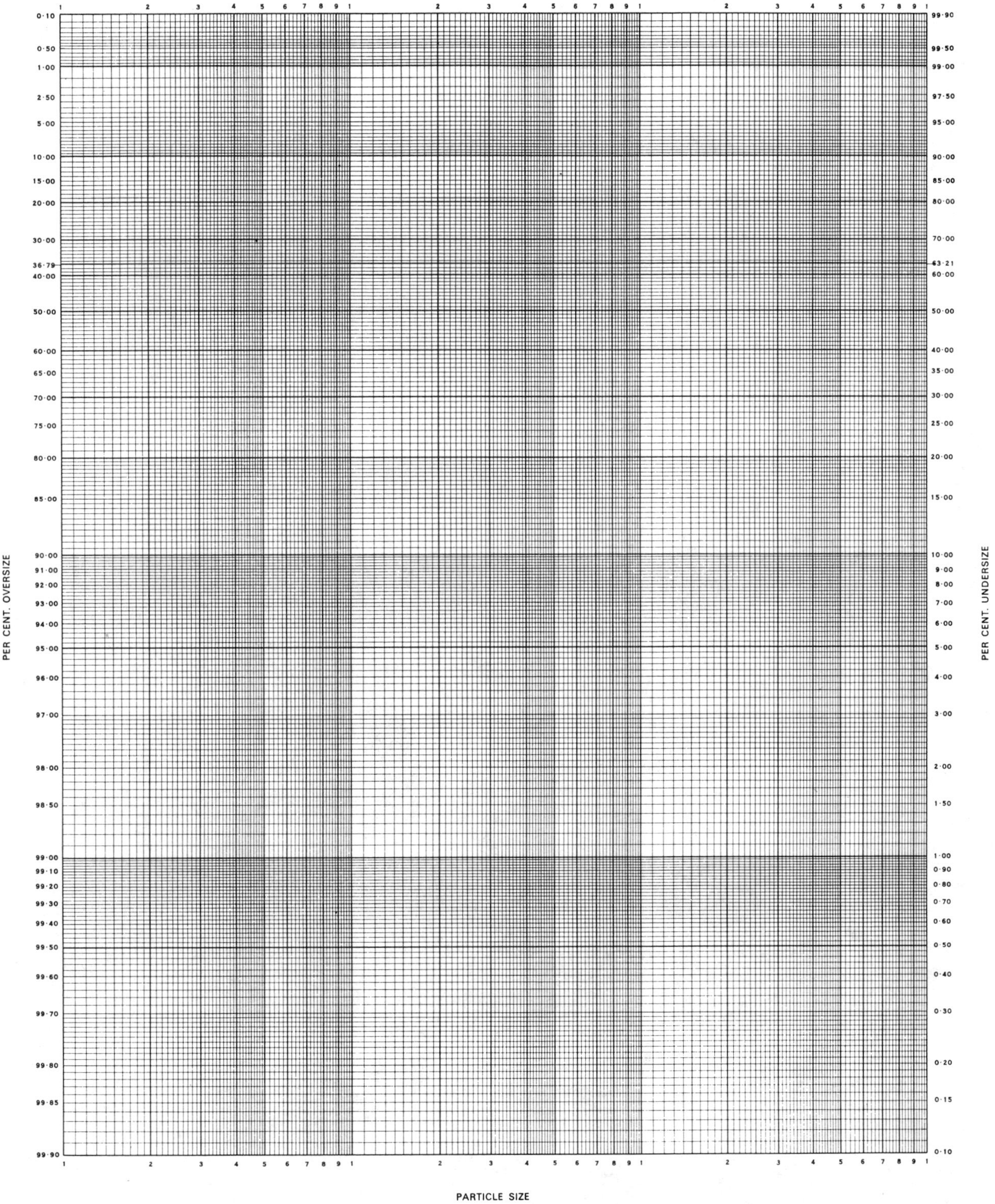
PER CENT. OVERSIZE
PER CENT. UNDERSIZE
PARTICLE SIZE

Appendix E

Magnetic Attractibility of Minerals

TABLE OF MAGNETIC ATTRACTIBILITY OF MINERALS

MINERAL	SOURCE OF SAMPLE	RELATIVE ATTRACTIBILITY
GROUP 1 — FERRO-MAGNETIC		
Iron	Unknown	100.00
Magnetite	Port Henry, N.Y.	48.000
Magnetite	Franklin Furnace, N.J.	14.862
Franklinite	Franklin Furnace, N.J.	13.089
GROUP 2 — MODERATELY MAGNETIC		
Ilmenite	Edge Hill, Pa.	9.139
Mica, spotted, ruby	Bengal, India	5.880
Pyrrhotite	Sudbury, Ontario	2.490
Franklinite	Franklin Furnace, N.J.	1.480
GROUP 3 — WEAKLY MAGNETIC		
Hematite	Lake Superior district	0.769
Siderite	Roxbury, Conn.	0.743
Rhodonite	Franklin Furnace, N.J.	0.560
Hematite	Iron Mountain, Minn.	0.531
Limonite	Nova Scotia	0.314
Pyrolusite	Thuringia	0.280
Corundum	Gaston County, N.C.	0.264
Hematite	Cumberland, England	0.257
Pyrolusite	Bartow County, Ga.	0.248
Pyrite	French Creek, Pa.	0.203
Manganite	Bridgeville, Nova Scotia	0.194
Calamine	Friedensville, Pa.	0.187
Sphalerite	Frieburg, Germany	0.182
Dolomite	Cumberland, England	0.178
Rutile	Maine	0.175
Siderite	Magnet Cove, Ark.	0.168
Garnet	Unknown	0.160
Serpentine, green	Unknown	0.149
Zircon	Hendersonville, N.C.	0.140
Molybdenite	Frankford, Pa.	0.134
Mica, spotted	Bengal, India	0.118
Huebnerite	Henderson, N.C.	0.115
Cararapyrite	Silver City, N.M.	0.105
Wolframite	Chochiwon, Korea	0.105
Argentite	Guanajuato, Mexico	0.102
Ferberite	Malaya	0.101
Wolframite	Climax, Colo.	0.100

MINERAL	SOURCE OF SAMPLE	RELATIVE ATTRACTIBILITY
Rutile	Graves Mountain, Ga.	0.095
Orpiment	Felsobanya, Hungary	0.089
Bornite	New South Wales, Australia	0.086
Apatite	Eganville, Ontario	0.083
Tetrahedrite	Peru	0.080
Willemite	Franklin Furnace, N.J.	0.076
Bornite	Union Bridge, Maryland	0.067
Sphalerite	Iowa	0.057
Cerrusite	New South Wales, Australia	0.057
Dolomite	Sing Sing, N.Y.	0.054
Fluorite	Jefferson County, N.Y.	0.054
Arsenopyrite	Acton, York Co., Maine	0.054
Chalcopyrite	South Australia	0.051
Cuprite	Bisbee, Arizona	0.051
Molybdenite	New South Wales, Australia	0.048
Celestite	Strontium Island, Ohio	0.038
Chalcocite	Butte, Montana	0.038
Cinnabar	New Almaden, California	0.038
Gypsum	Derbyshire, England	0.038
Zincite	Franklin Furnace, N.J.	0.038
Orthoclase	Elam, Pa.	0.035
Epidote	Unknown	0.033
Fluorite	Rosiclare, Ill.	0.032
Smithsonite	Mineral Point, Wis.	0.029
Augite	Unknown	0.027
Talc	Marietta, Ga.	0.026
Hornblende	Unknown	0.025
GROUP 4 — FEEBLY MAGNETIC		
Pyrite	Rio Tinto, Spain	0.022
Smithsonite	Kelly, N.M.	0.022
Sphalerite	Joplin, Mo.	0.022
Stibnite	Germany	0.022
Cryolite	Greenland	0.019
Enargite	Butte, Montana	0.019
Galena	Joplin, Mo.	0.019
Magnesite	Lancaster County, Texas	0.019
Senarmontite	Unknown	0.019
Gypsum	Grand Rapids, Mich.	0.016
Niccolite	Bebra Hesse, Germany	0.016

MINERAL	SOURCE OF SAMPLE	RELATIVE ATTRACTIBILITY
Serpentine, red	Unknown	0.016
Calcite	Joplin, Mo.	0.013
Stibnite	Juab County, Utah	0.013
Dioptase	Unknown	0.012
Cuprite	Cornwall, England	0.0096
Galena	Galena, Ill.	0.0096
Pyrite	Unknown	0.008
Witherite	Cumberland, England	0.0064
Rutile	Unknown	0.0034
Mica, ruby, clear	Bengal, India	0.0032
Orthoclase	Alexandria, N.Y.	0.0032
Cobaltite	Unknown	0.0023
Sapphire	Unknown	0.0023
Pyrite	Unknown	0.002
Tourmaline	Unknown	0.0012
Dolomite	Unknown	0.0011
Spinel	Unknown	0.001
Beryl	Unknown	0.0008
Feldspar	Unknown	0.0006
Sphalerite	Jefferson City, Tenn.	0.0005
Zircon	Unknown	0.0002
GROUP 5 — NON-MAGNETIC AND DIAMAGNETIC		
Barite	Bartow County, Ga.	0.0
Adularia	Unknown	— 0.0004
Calcite	Unknown	— 0.0004
Fluorite	Unknown	— 0.0004
Halite	Unknown	— 0.0004
Sphalerite	Unknown	— 0.0004
Celestite	Unknown	— 0.0005
Quartz	Unknown	— 0.0005
Corundum	Unknown	— 0.0006
Topaz	Unknown	— 0.0006
Galena	Unknown	— 0.0011
Antimony, native	Unknown	— 0.0023
Bismuth	Unknown	— 0.0032
Apatite	Unknown	— 0.0034
Argonite	Unknown	— 0.0048
Graphite	Ceylon	— 0.0056
Graphite	Ceylon	— 0.032

Appendix F

Separation Characteristics of Minerals

SEPARATION CHARACTERISTICS OF MINERALS

SP. G.	NON-CONDUCTORS (High Tension Pinned)			CONDUCTORS (High Tension Thrown)				SPECIFIC GRAVITY
	MAGNETIC	WEAKLY MAGNETIC	NON MAGNETIC	HIGHLY MAGNETIC	MAGNETIC	WEAKLY MAGNETIC	NON MAGNETIC	
Over 8.0							Gold Copper	Over 8.0
8.0								8.0
7.5					Ferberite	Wolframite	Galena Cassiterite	7.5
7.0								7.0
6.5								6.5
6.0			Scheelite			Columbite — Tantalite		6.0
5.5								5.5
5.0	Monazite	Bastnasite		Magnetite Ilmenite — (High Iron)	Ilmenite Davidite	Samarskite Euxenite	Pyrite	5.0
4.5	Xenotime		Zircon Barite			Hematite Chromite	Molybenite	4.5
4.0	Garnet Siderite Staurolite		Corundum Celestite Perovskite				Rutile Chalcopyrite Brookite Limonite	4.0
3.5		Epidote Olivine Apatite	Kyanite Topaz Sphene				Diamond	3.5
3.0		Hornblende Tourmaline Mica (Biotite)	Sillimanite Fluorite Anhydrite Mica (Muscovite)					3.0
2.5			Beryl Feldspars Calcite Quartz					2.5
2.0			Gypsum Chrysotile Sulphur				Graphite	2.0
Under 20								Under 2.0

EXPLANATION OF TABLE:

Starting with a mixture of any of the above minerals it may be determined whether or not they can be separated by high tension, magnetic, or gravity methods and whether any one, or a combination of methods is required. If the minerals appear in different columns they may be separated by high tension and/or magnetic methods alone. Two or more minerals appearing in the same column can be separated by gravity concentration if they have sufficient difference in gravity (usually a difference of approximately 1.0).

It should be noted that grain shape and/or size may alter separation characteristics. This is sometimes a detriment and other times useful. As an example, mica and quartz may in many cases be separated by high tension due to their grain shape.

Minerals behavior characteristics shown are from tests made in our laboratories rather than theoretical. Mineral characteristics and behaviors sometimes vary from different deposits. The behavior of minerals not shown can usually be predicted by the behavior of similar minerals in the above table.

Appendix G

Conveyor Belt Design Data

TABLE G-1: BULK MATERIAL CHARACTERISTICS

MATERIAL	CLASS	AVG. WT. LBS. CU. FT.	MAX. CONVEYING ANGLE	MATERIAL	CLASS	AVG. WT. LBS. CU. FT.	MAX. CONVEYING ANGLE
Ashes, Coal, Dry, Minus 3"	D46T 40°	35-40	20°-25°	Iron Ore	D36 35°	100-200	18°-20°
Ashes, Coal, Wet, Minus 3"	C46T 50°	45-50	23°-27°	Kaolin Clay, Minus 3"	D36 35°	63	19°
Barite	D36	180	18°	Lignite, Air Dried	D25	45-55	20°
Barite, Crushed Minus 3"	D36	75-85	20°	Lime, Ground Minus 1/8"	B45X 43°	60-65	23°
Bentonite, Minus 100 Mesh	A26XY	50-60	20°	Lime, Pebble	D35 30°	53-56	17°
Borax, Fine	B26T	45-55	20°-22°	Lime, Over 1/2"	D35	55	18°
Cast Iron Chips	C46	130-200	20°	Limestone, Agricultural	B26	68	20°
Cement, Portland	A26M 39°	94	20°-23°	Limestone, Crushed	C26X 38°	85-90	18°
Cement, Clinker	D37 30°-40°	75-95	18°-20°				
Charcoal	D36Q 35°	18-25	20°-25°				
Cinders, Coal	D37T 35°	40	20°	Manganese Ore	D37 39°	125-140	20°
Coal, Anthracite, Sized, 3/8" to 6"	C26 27°	55-60	16°	Marble, Crushed, Over 1/2"	D27	80-95	20°
Coal, Bituminous, Slack	C45T 40°	43-50	22°	Mica, Ground, Minus 1/8"	B36 34°	13-15	23°
Coal, Bituminous, Run of Mine	D35T 38°	45-55	18°	Phosphate Rock	D26 25°-30°	75-85	12°-15°
Coffee, Bean	C25Q 25°	32	10°-15°				
Coke, Loose	D47QVT	23-35	18°	Salt, Coarse Dry	C25TU	40-45	18°-22°
Coke, Petroleum	D36V	35-45	20°	Salt, Fine, Dry	D26TUW 25°	70-80	11°
Coke, Breeze, Minus 1/4"	C37Y 30°-45°	25-35	20°-22°	Sand, Bank, Damp	B47 45°	110-130	20°-22°
Concrete, Wet:				Sand, Bank, Dry	B37 35°	90-110	16°-18°
6" Slump	D26	110-150	12°	Sand, Foundry, Prepared	B47	80-90	24°
4" Slump	D26	110-150	20°-22°	Sand, Foundry, Shakeout	D37 39°	90-100	22°
2" Slump	D26	110-150	24°-26°	Sand, Silica, Dry	B27	90-100	10°-15°
Copper Ore	D27	120-150	20°	Sand, Saturated	B27	110-130	15°
Coral, Crushed	D26	40-45	20°	Shale, Crushed	C36 39°	85-90	22°
Corn, Shelled	C25NW 21°	45	10°	Slag, Furnace, Crushed	A27 25°	80-90	10°
Cullet, Crushed	D37Z	80-120	20°	Slag, Furnace, Granulated	C27 25°	60-65	13°-16°
Culm, Minus 3/64", Damp	B25TVY	45-60	20°	Slate, Crushed, Minus 1/2"	C26 28°	80-90	15°
Dolomite, Lumpy	D26	90-100	22°	Slate, Ground, Minus 1/8"	A36Y 35°	70-80	20°
Earth, Common, Loam, Dry	B36 35°	70-80	20°	Soda Ash, Light	A36Y 37°	20-35	22°
Earth, Clay, Dry	B36 35°	65	20°	Soda Ash, Heavy	B36 32°	55-65	19°
Earth, Moist	B46 45°	100-110	23°	Stone, Crushed	D36V	85-90	20°
Feldspar, Ground, Minus 1/8"	B36 38°	70-85	18°	Stone, Screenings	C36	85-90	18°
Fluorspar	D46	110-120	20°	Stone, Dust	B36Y	75-85	20°
Fuller's Earth, Burnt	B26 35°	40	20°	Sulphate, Crushed, Minus 1/2"	C25NS	50-60	20°
Fuller's Earth, Raw	B26 35°	35-40	20°	Sulphate, Lumpy, Minus 3"	D25NS	80-85	18°
Glass, Batch	D27Z	80-100	20°-22°	Sulphate, Powdered	B25NW	50-60	21°
Granite, Broken	D27	95-100	20°	Traprock, Crushed	D37	100-110	20°
Gravel, Average, Blended	D27 38°-40°	90-100	20°				
Gravel, Sharp	D27 40°	90-100	15°-17°	Vermiculite Ore	D36Y	70-80	20°
Gravel, Pebble	D36 30°	90-100	12°				
Gypsum, Calcined	C36 40°	70-80	21°				
Gypsum, Crushed	D26 30°	70-80	15°	Wheat	C25N 28°	45-48	12°
Gypsum, Powdered	A36Y 42°	60-70	23°	Wood Chips	E45WY	10-30	27°

KEY TO CLASSIFICATION OF MATERIAL

Size Characteristics
- A - Very fine, under 100 mesh
- B - Fine, under 1/8"
- C - Granular, 1/8" to 1/2"
- D - Lumpy, over 1/2"
- E - Irregular, stringy, interlocking, mats together

Flow Characteristics
- 2 - Free flowing, angle of repose 20° to 30°
- 3 - Average flowing, angle of repose 30° to 45°
- 4 - Sluggish, angle of repose over 45°

Abrasive Characteristics
- 5 - Non-abrasive
- 6 - Abrasive
- 7 - Very abrasive

Miscellaneous Characteristics
- N - Contains explosive dust
- Q - Degradable, affecting use or saleability
- S - Highly corrosive
- T - Mildly corrosive
- U - Hygroscopic
- V - Interlocks or mats
- W - Oils or chemical present, may affect rubber products
- X - Packs under pressure
- Y - Very light and fluffy, may be wind swept
- Z - Elevated temperature

Example: Limestone, Crushed - C26X 38°
- C - Granular, 1/8" to 1/2"
- 2 - Free flowing, angle of repose 20° to 30°
- 6 - Abrasive
- X - Packs under pressure
- 38° - Angle of repose

TABLE G-2

CAPACITY IN TPH OF EQUAL ROLL TROUGH BELT CONVEYORS
(1) FOR 100 LB./CU. FT. MATERIAL
AT 100 FPM BELT SPEED

(4) MAXIMUM SIZE LUMPS FOR 20° SURCHARGE

BELT WIDTH	(3) ANGLE OF SURCHARGE												UNIFORM SIZE	(2) MIXED WITH 50% FINES
	20° TROUGH IDLER			35° TROUGH IDLER				45° TROUGH IDLER						
	20°	25°	30°	10°	20°	25°	30°	5°	10°	20°	25°	30°		
18"	50	56	63	53	65	70	75	55	60	70	75	80	4"	4"
24"	96	108	120	102	122	132	142	106	115	132	140	150	5"	7"
30"	157	175	195	167	200	215	232	175	187	215	230	244	6"	10"
36"	230	260	290	248	295	318	343	258	278	318	340	360	7"	12"
42"	320	360	400	344	408	442	475	358	386	440	470	500	8"	14"
48"	430	480	530	457	540	585	630	475	510	584	623	660	10"	16"
54"	547	612	678	585	693	750	806	608	655	748	797	845	11"	18"
60"	680	762	844	730	863	933	1000	758	815	930	992	1050	12"	20"

1. FOR MATERIAL OTHER THAN 100#/CU. FT. — CAPACITY = TABLE CAPACITY $\times \dfrac{WT/CU. FT.}{100}$
2. 50% FINES IS DEFINED AS 50% SHALL BE LESS THAN ½ THE MAXIMUM.
3. ANGLE OF SURCHARGE IS ABOUT 15° LESS THAN ANGLE OF REPOSE.
4. FOR 30° SURCHARGE, USE ½ LUMP SIZE OF 20° SURCHARGE.

TABLE G-3

MAXIMUM RECOMMENDED BELT SPEEDS IN FPM

MATERIAL			BELT WIDTH							
	CHARACTERISTICS	EXAMPLE	18"	24"	30"	36"	42"	48"	54"	60"
Maximum Size Lumps	Non-Abrasive	Coal, Earth	350	400	450	500	550	600	600	600
	Smooth - Abrasive	Gravel	300	350	400	450	500	550	550	550
	Sharp and Jagged - Abrasive	Stone, Ore	250	300	350	400	450	500	500	500
Half-Max. Size Lumps	Non-Abrasive	Coal, Earth	400	450	500	550	600	650	700	750
	Smooth - Abrasive	Gravel	350	400	450	500	550	600	650	700
	Sharp and Jagged - Abrasive	Stone, Ore	300	350	400	450	500	550	600	650
Granular ⅛" to ½"		Sand, Grain, Wood Chips	400	500	600	700	800	900	900	900
Aerating Powders		Cement, Flue Dust	200 to 300							
Conveyors With Plow Discharge			200							

TABLE G-4 FACTOR (x)

HORSEPOWER AT HEADSHAFT FOR EMPTY CONVEYOR

AT 100 FPM BELT SPEED

(For other speeds use direct proportion)

CONVEYOR CENTERS FEET	BELT WIDTH					
	18"	24"	30"	36"	42"	48"
25	.44	.53	.62	.72	.82	.98
50	.47	.57	.67	.77	.89	1.06
100	.52	.63	.76	.87	1.02	1.21
150	.57	.69	.85	.97	1.15	1.36
200	.62	.76	.93	1.08	1.28	1.50
250	.67	.82	1.02	1.18	1.41	1.65
300	.72	.89	1.11	1.29	1.54	1.80
350	.77	.95	1.20	1.39	1.67	1.95
400	.82	1.02	1.28	1.50	1.80	2.10
450	.87	1.08	1.37	1.60	1.93	2.25
500	.92	1.15	1.46	1.71	2.06	2.40

TABLE G-5 FACTOR (y)

HORSEPOWER AT HEADSHAFT TO MOVE LOAD HORIZONTALLY

ANY BELT SPEED — ANY MATERIAL

CONVEYOR CENTERS FEET	CAPACITY - TONS PER HOUR									
	50	100	150	200	250	300	350	400	500	600
25	.25	.50	.76	1.01	1.26	1.51	1.77	2.02	2.52	3.03
50	.28	.57	.85	1.14	1.42	1.70	1.99	2.27	2.84	3.41
100	.35	.69	1.04	1.39	1.74	2.08	2.43	2.78	3.47	4.17
150	.41	.82	1.23	1.64	2.05	2.46	2.87	3.28	4.10	4.92
200	.47	.95	1.42	1.89	2.37	2.84	3.31	3.79	4.73	5.68
250	.54	1.07	1.61	2.15	2.68	3.22	3.75	4.29	5.36	6.44
300	.60	1.20	1.80	2.40	3.00	3.60	4.20	4.80	6.00	7.20
350	.66	1.32	1.98	2.65	3.31	3.97	4.64	5.30	6.63	7.95
400	.72	1.45	2.17	2.90	3.63	4.35	5.08	5.81	7.26	8.71
450	.79	1.58	2.36	3.16	3.94	4.73	5.52	6.31	7.89	9.47
500	.85	1.70	2.55	3.41	4.26	5.11	5.96	6.82	8.52	10.23

NOTE: The above HP values are for idlers with anti-friction bearings.

TABLE G-6 FACTOR (z)

HORSEPOWER AT HEADSHAFT TO LIFT LOAD VERTICALLY

ANY BELT SPEED — ANY MATERIAL

VERTICAL LIFT FEET	CAPACITY - TONS PER HOUR									
	50	100	150	200	250	300	350	400	500	600
5	.25	.51	.76	1.01	1.26	1.51	1.76	2.02	2.52	3.03
10	.51	1.01	1.52	2.02	2.52	3.03	3.53	4.04	5.05	6.06
20	1.01	2.02	3.03	4.04	5.05	6.06	7.07	8.08	10.10	12.12
30	1.52	3.03	4.55	6.06	7.57	9.09	10.60	12.12	15.15	18.18
40	2.02	4.04	6.06	8.08	10.10	12.12	14.14	16.16	20.20	24.24
50	2.53	5.05	7.58	10.10	12.62	15.15	17.67	20.20	25.25	30.30
60	3.03	6.06	9.09	12.12	15.15	18.18	21.21	24.24	30.30	36.36
70	3.54	7.07	10.60	14.14	17.67	21.21	24.74	28.28	35.35	42.42
80	4.04	8.08	12.12	16.16	20.20	24.24	28.28	32.32	40.40	48.48

THE TOTAL HP AT THE HEADSHAFT IS THE TOTAL OF FACTORS (x) + (y) + (z)

NOTE: If factor (z) exceeds ½ the sum of (x + y), backstop is necessary.

Appendix H

Pump Selection Chart

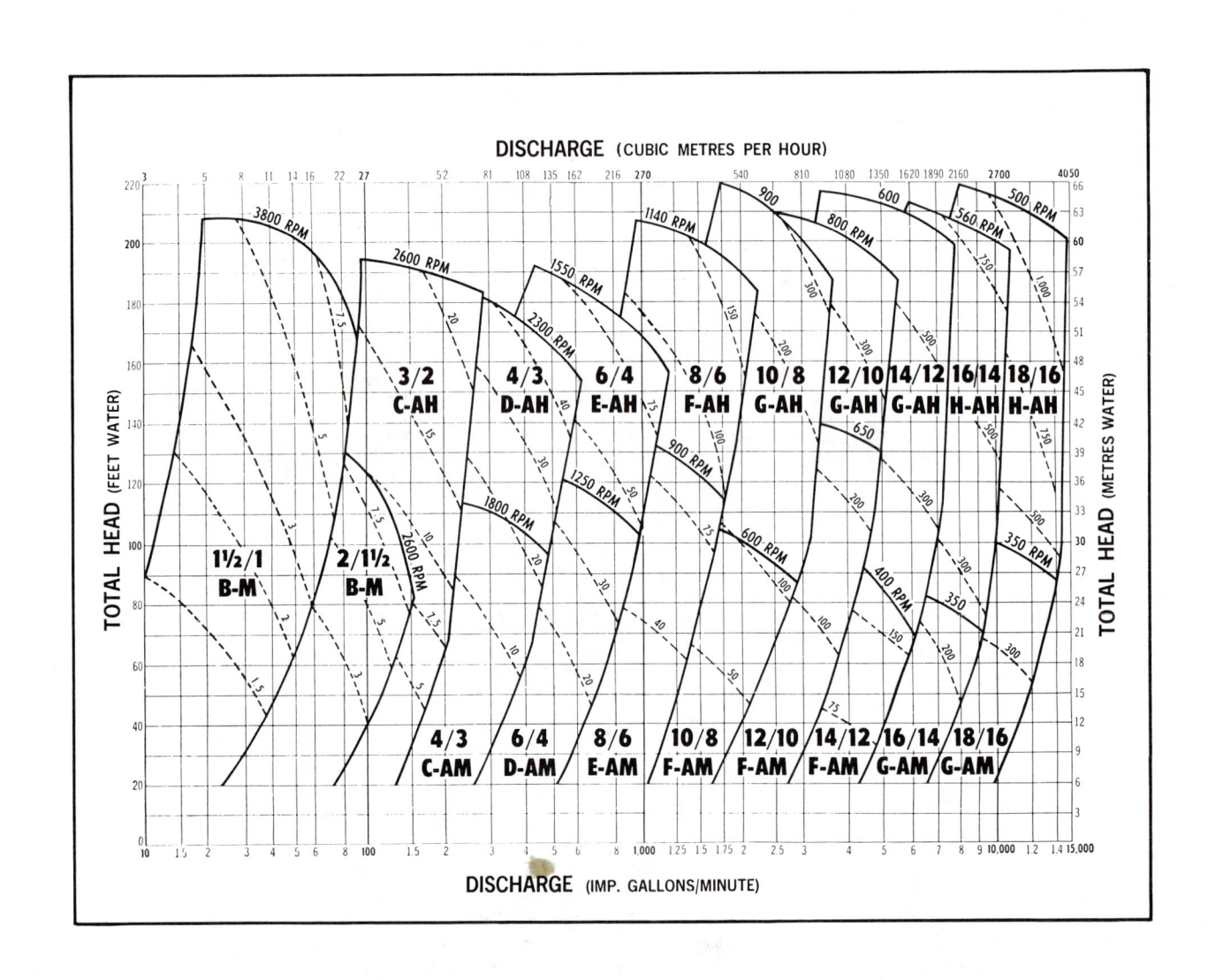

Appendix I

SI Units and Conversion Factors

TABLE I-1: CONVERSION FACTORS

Mass
1 kilogram (kg)	=	2.2046 pounds
1 tonne (t)	=	1000 kilograms
	=	0.9842 tons (long)
	=	1.1023 tons (short)

Length
1 meter (m)	=	39.3701 inches
	=	3.2808 feet
	=	1.0936 yards

Area
1 square meter (m^2)	=	10.7639 square feet

Volume
1 cubic meter (m^3)	=	35.3147 cubic feet
1 liter (l)	=	0.2642 gallons (US)
	=	0.2200 gallons (Imp.)

Force
1 newton (N)	=	1 kg.m/s^2
	=	0.2248 pounds force
	=	100,000 dynes

Pressure
1 pascal (Pa)	=	1 N/m^2
	=	1.4504×10^{-4} pounds force/square inch
	=	7.5006×10^{-3} mm Hg (0°C)
	=	0.9869×10^{-5} atmospheres

Density
1 kilogram/cubic meter (kg/m^3)	=	0.001 gram/cubic centimeter
	=	0.06243 pound/cubic foot

Energy (Work and Heat)
1 joule (J)	=	1 N.m
	=	0.7376 foot pound force
	=	9.478×10^{-4} British thermal units (Btu)
	=	2.388×10^{-4} calories

Power
1 watt (W)	=	1 J/s
	=	1.341×10^{-3} horsepower

Viscosity
1 pascal second	=	1 N.s/m^2
	=	1 kg/m.s
	=	1000 centipoise

Magnetic Flux
1 weber (Wb)	=	1 V.s
	=	10^8 maxwell

Magnetic Flux Density
1 tesla (T)	=	1 Wb/m^2
	=	10^4 gauss
1 ampere/meter (A/m)	=	0.01256 oersted

Gas Constant
8.314 (N/m^2) (m^3)/mol K	=	1.987 cal/mol K
	=	1.987 Btu/lb mole °R
	=	0.7302 atm ft^3/lb mole °F

Gravitational Acceleration
9.8066 m/s^2	=	32.174 ft/s^2

TABLE I-2: SI UNITS

Physical Quality	Unit	Symbol	Derivation
Base Units			
length	meter	m	
mass	kilogram	kg	
time	second	s	
electric current	ampere	A	
thermodynamic temperature	kelvin	K	
luminous intensity	candela	cd	
amount of substance	mole	mol	
Derived Units			
area	square meter	m^2	
volume	cubic meter	m^3	
density	kilogram per cubic meter	kg/m^3	
velocity	meter per second	m/s	
angular velocity	radian per second	rad/s	
acceleration	meter per second squared	m/s^2	
angular acceleration	radian per second squared	rad/s^2	
volume rate of flow	cubic meter per second	m^3/s	
moment of inertia	kilogram meter squared	$kg.m^2$	
moment of force	newton meter	N.m	
intensity of heat flow	watt per square meter	W/m^2	
thermal conductivity	watt per meter kelvin	W/m.K	
luminance	candela per square meter	cd/m^2	
Derived Units with Special Names			
frequency	hertz	hz	s^{-1}
force	newton	N	$kg.m/s^2$
pressure, stress	pascal	Pa	N/m^2
work, energy, quantity of heat	joule	J	N.m
power	watt	W	J/s
electric charge	coulomb	C	A.s
electric potential	volt	V	W/A
electric capacitance	farad	F	C/V
electric resistance	ohm	Ω	V/A
electric conductance	siemens	S	Ω^{-1}
magnetic flux	weber	Wb	V.s
magnetic flux density	tesla	T	Wb/m^2
inductance	henry	H	Wb/A

Index

abrasion fracture, 117
acceleration, differential, 88, 263-264
acceleration of particle, centrifugal, 69
  in fluid, 69, 72, 264
  gravitational, 69
activator, 311, 313
adhesion, tenacity of, 104
  work of, 104
adsorption, 96-107
  chemisorption, 98, 101, 107, 307, 310, 312
  electrostatic, 312
  free energy of, 102
  gas, for area determination, 42
  monolayer, 103
  multilayer, 103
  physical, 101, 307, 312
  specifically adsorbed ions, 100
adsorption density, 97
adsorption isotherm, 98, 102
agglomeration, 107
  selective, 319
aggregate, 252
aging, of samples, 334, 405
agitated tank, *see* tank, agitated
agitation, 82
  scale of agitation, 386
air flow number, 82, 314
air separators, 201
alcohol frothers, 309
algorithms, control, 411
amine collectors, 309
analysis, on-stream chemical, 42, 416
  visual, on table, 257
analytical complexity, 203, 405
angle of repose, 367, 376, Appendix G
apatite, 313
aperture shape, 182
arching, conditions for in bin, 371
arching in bins, 368
area principle, 81

arsenopyrite, 310
asbestos, 238
attachment, bubble-particle, 105, 108
attraction, coulombic, 100
  electrostatic, 100
autogenous grinding, *see* grinding, autogenous
average size (diameter) of particles, 23, 27-29

backfill, 359, 390
baffles, 82, 387
Bagnold shear mechanisms, 84, 90-91, 265-269
Bahco microparticle classifier, 40
balance, energy, 13
  force, *see* force balance
  materials, 9
  water, 392
ball mill, 132, 133, 411
  breakage function, 145-146
  breakage rate, 146-150
  media, *see* grinding media
  representation of flow through, 54, 151, 155-156
  see also grinding; mill
bare module, 425
barite, 237, 313, Appendix B12
Bartles cross belt concentrator, 252, 258
Bartles-Mozley concentrator, 257, 258
basis, for expressing quantity in size fraction, 23
  of materials balance, 10
batch grinding, 152
batch sedimentation test, 213, 330, 332
beach sands, 252, 259, 291, Appendix B8
bed dilation (expansion), 250, 261, 263, 264, 276
beryllium ore, 238
bin, 368
  arching in, 268
  arching condition in, 371
  design, 369-374
  flow in, 368
  friction in, 370

function of, 368
funnel-flow, 369
mass-flow, 369
mass-flow condition in, 371
multiple outlet, 373
piping condition in, 371
segregation in, 91, 369, 373
see also hopper
Bingham plastic, 65, 80, 248
Blaine permeameter, 41
blending, 367
blinding, 171, 180, 185, 190
Bond equation, 116
Bond's method, 159-162
boron ore, 238
boundary layer, 62
breakage, 150
abnormal (nonideal), 148
see also fracture
breakage function, 119, 145-146, 150
autogenous grinding, 120
crushing, 120
cumulative, 119
breakage in tumbling mills, 139-142
breakage parameters, 119-120
breakage processes, summary of, 150
see also fracture
breakage rate, 120, 146-150
brittle materials, 113
Broadbent and Callcott equation, 118
bubble size, 108
bubble-particle attachment, 105, 108
bubble-particle contact, 105
bubble-particle detachment, 108
Buckman tilting concentrator, 260
Burke-Plummer equation, 76

calcite, 313
capacity, classifier, 225
crushers, 157
fluidized bed classifier, 223-226
live, 367
mills, 159
optimum operating, 430
pump, 381
screne, 180, 193-196
sedimentation classifier, 225
size reduction equipment, 157-162
surge, 338, 406
thickener, 335-337
capital, fixed, 422
working, 421, 425
capital costs, 421-425
carboxylate collectors, 308
Carman-Kozeny equation, 41, 76, 349
carrier flotation, 106

cash flow, 428
cassiterite, 238, 252, 254, 257, 275, 291
center of gravity lowering, 92-93, 263, 265, 267
centrifuge, cut size, 358
equipment, 356
perforated basket, 357
solid bowl, 357
stress in bowl, 358
chain length, of hydrocarbon, 308
chalcocite, 310
chalcopyrite, 310, 311, 397
charge volume, 137
Charles equation, 117
chemical potential, 97
chemisorption, 98, 101, 307, 310, 312
chemistry, solution, 307
chutes, 179, 376
circuit analysis, 404
circuits, 397, 400, 405, Appendix B
classifier, 226
control of, 417-419
flotation, 317
size reduction, 143-144
see also flow sheet
circulating load, 11, 228, 292, 402-404
clarification, 327, 328, 329
theory, 207-213, 330-331
clarifier, rectangular, 330
classification, 399
equipment, 199-205
heterogeneous solids, 228
theory, 207-213, 223-225
two-stage, 227-228
classification/discharge function, 145, 151
classifier, 193, 195-205, 250
air, 199, 224
air separator, 201
bowl, 200
capacity, 225
changing of cut size, 212, 229
cone, 200
counter current, 201
cylindrical tank, 200
drag, 200
d_{50} correlations (fluidized bed), 223
d_{50} correlations (sedimentation), 211-213
elutriator, 201, 224-225
flows in, 210
fluidized bed, 199, 223-225
hydraulic, 199
hydraulic bowl, 200
hydraulic cone, 200
hydraulic cylindrical tank, 200
hydrocyclone, see hydrocyclone
improving performance of, 226-228
log washer, 200, 202
operation, 229

Index

partition concept, 208
performance of, 202-207
pocket, 201
rake, 200
Reid representation, 209
representation of flow through, 54
residence times in, 54, 209-210
sands discharge mechanism, 199-201, 212, 226
scrubber, 201
sedimentation, 199, 207-223
sloping tank, 200
solid bowl centrifuge, 201
spiral, 200, 212, 213
tapping concept, 209
theory (fluidized bed), 223-225
theory (sedimentation), 207-213
clay, china, 275
cleaners, 267
cleaning, 9, 292, 400-404
cleavage fracture, 117
closed circuit grinding, 193, 202, 207, 226
coal, 243, 250, 251, 252, 256, 259, 278, 301, Appendix B5-B7
cobbing, 274
collector, 98, 307-309
  adsorption mechanisms, 310
  adsorption on oxide and silicate minerals, 312
  adsorption on sulfide minerals, 310
  anionic, 308
  cationic, 309
  hydrocarbon chain length, 308
  non-ionic, 309
  selectivity, 308
  see also specific types
colloid stability, 107
columbite, 275, 291
comminution, *see* size reduction
complexity, analytical, 230, 405
complex ores, 301
concentrate, 6
concentration, 8
  particle size effect, 235
  preconcentration, 236, 259, Appendix B11
concentration separation, 235-323
conditioning, in flotation, 316
  separate, in flotation, 316
conductivity, 238, Appendix F
cone (Reichert), 252, 259, 261
  theory of, 267
cone crusher, 127, 128, 131
consolidating pressure, 371
contact, bubble-particle, 105
  interparticle, 106
  three phase, 103
contact angle, 103
contact angle data, 311
contamination, groundwater, 390

contingencies, costing of, 425
control, 408-420, 430
  cascade, 413
  crushing circuits, 417
  feed-back, 409
  flotation, 419
  grinding circuits, 418
  multivariable, 413
  principles of, 409
  system stability, 412
control algorithms, 411
control element, final, 409
controller, 409, 411
control loop, 409
control objectives, 408
conveyors, 367, 376-377
  mechanical, 376
  pneumatic, 377
copper, native, 238
corona, 297
correlation, d_{50} of hydrocyclone, 217-219
  d_{50} of sedimentation classifier, 211
  flow split of hydrocyclone, 219
  pressure drop of hydrocyclone, 219
cost estimate methods, 425-426
cost estimates, 421
cost factors, 424
cost indexes, 422
cost module, 425
costs, 6, 406
  capital, 421
  direct, 422
  of equipment, 423
  indirect, 422
  labor, 426
  lowest, 406
  operating, 426
  production, 7, 426
Coulter Counter, 40
counterions, 98
  diffuse layer of, 100
covalent bonding, 100
covellite, 311
crack propogation, 114
crack tip, 113
cresylic acid frother, 309
critical speed, 141
crusher, 127-132
  Blake, 128
  breakage function of, 146
  cage disintegrator, 129
  cone, 127, 128, 131
  control of, 417
  Dodge, 128
  double toggle, 128
  Gyradisc, 129
  gyratory, 128

hammer mill, 129
impact, 129, 131
impactor, 129, 132
openings, definitions, 131
overhead eccentric, 128, 130
overhead pivot, 128
roll, 129, 131
rotary breaker, 129, 132
Short Head, 128, 131
single toggle, 128
vertical spindle, 129
crushing, 8, 116, 120, 127, 397
theory of, 144-157
cut-off grade, 431
cut point, 57. *See also* separation point
cut size, 205
apparent, 206
changing of in classifiers, 229
see also separation size
cyanide depression, 311
cyclone, *see* hydrocyclone
Cyclosizer, 25, 40

dam, *see* tailings dam
Davis tube, 279
deactivation, 312
dead time, 410
demagnetizing, 279
dense liquids, 55
dense medium, 243-249
applications, 243-246, Appendix B5-B7, B11
autogenous separator, 246
Belknap calcium chloride washer, 243
centrifugal separators, 245
Chance Cone, 243
cone separator, 244
drum separator, 244
dry separator, 246
Dynawhirlpool, 245
equipment, 243-246
gravity separators, 244
hydrocyclone, 245
media, 243
media consistency, 246
media control, 246
media density, 246
media recovery, 246
performance curve, 248
static bath separators, 243
trough separator, 245
Vorsyl separator, 245
water-only separator, 246
density, apparent, 79, 87, 218
bulk, 92, Appendix G
of minerals, Appendix A, F
density gauges, 414

depreciation, 427
depressant, 311, 313
design, flexibility in, 405
plant, 405-407
detachment, bubble-particle, 108
detection, 239
dewatering, 171, 192, 325-363, 399
filter cake, 354
hydrocyclones, 358
screens, 358
systems, 358
thermal, *see* drying
diamagnetic, 274
diameter, 21-22, 27-29, 37-38
arithmetic mean, 23, 27, 169
average, 27-29, 37
Feret's, 37
geometric mean, 27
harmonic mean, 27
Martin's, 37
median, 27
mill, 145-149
nominal, 21-22, 73-74, 169
list of, 22
particle, 21
scale-up of mill diameter, 145-149
statistical, 27, 37
statistical average, 27-29
diamond, 237
dielectric separation, 295
differential acceleration, 88, 263
differential hindered settling, 87, 263
differential velocity in a flowing film, 88, 265
difficult to settle solids, 340, 391
diffuse layer, 100
diffusion, 57, 190, 208, 209, 212, 218, 230, 270
dimensional analysis, 117
discharge/classification function, *see* classification/discharge function
discharge rate from hopper, 373
discharge from size reduction equipment, 133, 134, 151, 160
discharges, electrode, 297
dispersant, 307
disposal of tailings, 390-393
distribution, slurry, 389
distribution curve 57. *See also* performance curve
dithiophosphate collectors, 308
dixanthogen collectors, 308
DLVO theory, 107
double layer, electrical, 98
drag, 64
drag chart, 68, 68-74
modified, 70, 71
drag coefficient, 68
dressability, 270
driers:
flash, 360

Index

fluidized bed, 360, 361
  rotary, 359, 360
dry grinding, 135
drying, 359-363
  equipment, 359-362
  theory, 360-363
d_{50}, 206
  apparent, 206
d_{50} correlation, fluidized bed classifier, 223
  hydrocyclone, 217-219
  sedimentation classifier, 211

earth's crust, 5, 13
economics, 5, 390, 421-431
EDAX, 19
efficiency, 56, 59, 162, 180, 181, 194, 402, 404
  areal, 81
  gravity concentrators, 269
  limits in size reduction equipment, 143
  screening, 180, 191
  see also performance curve
ejection, from ore sorting equipment, 240
electro-breaker, 128, 135
electrochemical oxidation mechanism of collector adsorption, 310
electroflotation, 306
electrokinetic, 100
electrolyte, indifferent, 100
electron microprobe, 19
electrophoretic mobility, 107
electrophoretic separation, 295
electrostatic separation, 291-300
  circulating load in, 292
  cleaning stages, 292
  corona, 297
  discharges, 297
  effect of environment, 296
  effect of particle coatings, 295
  effect of particle size, 292
  effect of surface moisture, 296
  effect of temperature, 296
  equipment, 291-295
  multistage operation, 299
  other media, 295
  particle charging mechanisms, 296-297
  particle trajectory, 298
  theory, 296-300
electrostatic separator, applications, 291-295, Appendix B8, F
  electro-dynamic types, 292, 293
  electro-static types, 294
  high tension, 294
  operating variables, 293
  plate-type, 294
  rotor-type, 294
  screen-type, 294

  vibrated bed, 295
elevators, 377
elutriation, 40
elutriator, 201, 224-225. See also classifier
energies, interfacial, 103
energy balances, 13
energy density, 114
energy-size reduction relationships, 116
entrainment, 315
entrapment ratio, 285
equipment, choice of, 400
  costs, 423-424
  large, 405
  selection philosophy, 397-404, Appendix F
  versatility of, 400
  see also specific and generic names
equipment applications, Appendix B
  classifiers, 200-201
  concentration, 235
  dewatering, 325
  electrostatic, 291-295
  gravity, 250-261
  heavy medium, 243-246
  magnetic, 274-279
  screening, 171, 173
  size reduction, 125, 128, 129, 131
  sorting, 237-238
equisettling particles, 87, 228
Ergun equation, 75
extraction, liquid-liquid (two-liquid), 314
extractive metallurgy, 13

factor, see specific factor
fatty acid collectors, 308
feasibility study, 405
feedback control, 409
feeders, 374-375
feed rate, 368, 430
feed well, 327
feldspar, 275
ferromagnetic, 274
ferrosilicon, 243
field strength, specification of, 277
film concentrators, 259-261
  theory, 267
filter, 343-348, 358
  Artisan, 347
  belt, 347
  continuous pressure, 347
  disc, 345
  drum, 344
  horizontal, 346
  media, 347
  Moore, 347
  pan, 346
  pressure, 346

vacuum, 343-346
filtration, batch, 353
  cake dewatering, 354
  cake washing, 353
  centrifugal, 353
  compressibility coefficient, 350
  continuous, 351
  equipment, 343-348
  incompressible cake, 349
  interpretation of data, 350
  media resistance, 349
  modes of, 343
  pressure drop through cake, 349
  slurry concentrations, 349
  specific cake resistance, 349
  theory, 348-354
first-order process, 410. *See also* Rate law, first order
Fisher subsieve sizer, 41
fixed capital, 421, 422
flaws, in initiating fracture, 113
flexibility, 408
floatability, 55
  natural, 307
float product, 243
float-sink, 243-249. *See also* dense medium
flocculation, 107-109, 327, 329, 339, 354
  flotoflocculation, 314
  magnetic, 279
  mechanical syneresis, 109
  orthokinetic, 108
  perikinetic, 108
  rate of, 109
  selective, 319
  theory, 108-109
flotation, 97, 257, 259, 290, 301-322, 399, 400-404
  adsorption, *see* collector
    agglomerate, 314
    carrier, 106, 314
    cell banks, 53-54, 301, 316-317, 402-404
    chemistry, 307
    circuits, 317, Appendix B
    coarse particles, 313
    collector adsorption, 310
    collectorless, 307
    conditioning, 316
    control of, 419
    determination of separability curves, 55
    dissolved air, 303, 306
    effect of pH with sulfide mineral, 311
    electroflotation, 306
    emulsion, 314
    entrainment, 315
    equipment, 301-306
    fine particles, 313, 315
    froth in, 315
    ion, 314
    kinetics, 53, 108, 316-317, 402-404
    mechanics, 314
    modifiers, 307, 310-313
    oil, 314
    oxides, 312
    precipitate, 314
    probabilities, 108, 317-319
    probability of adhesion, 108
    probability of attachment, 108
    probability of collision, 108
    reagents, 307-314. *See also specific and generic types*
    salt-type, 313
    selectivity, 311-313
    silicates, 312
    skin, 314
    soluble salt, 313
    sulfides, 310
    ultraflotation, 314
    vacuum, 306
    variables, 301
    water flow rate, 315
flotation cell, 82
flotation machine, Agitair, 302, 304, 305
  characteristics of, 305
  column, 304, 305, 306
  components of, 304, 306
  Cyclo-cell, 303, 305
  Davcra, 304, 305
  Denver, 302, 305
  design, 314
  Flotaire, 305
  functions of, 301
  Hallimond tube, 307
  Krupp, 305
  laboratory, 306
  Maxwell, 304, 305, 306
  mechanical, 301, 305
  microflotation cell, 307
  Nagahm, 305, 306
  Outokumpu, 303, 305
  pneumatic, 301, 305
  Sala, 305
  scale-up, 314
  Wemco-Fagregren, 303, 305
flow, cash, 428
  laminar, 63, 69
  turbulent, 63, 69
flow-aid devices, 368
flow of bodies through fluid, 68-75
flow factor, 371
flowing film, fluid mechanics of, 84
  particle behavior in, 88-90
flow meters, 414
flow number, 82
flow through packed bed, 75
flow patterns, 52-55

Index

flow in pipes, 63, 66-68
flow properties of solids, 369
flow sheet, 8
  case study, 397-399
  examples, Appendix B
  size reduction, 143, 226
  see also circuits
flow of slurries in pipes, 83
flow of solids through size reduction equipment, 134
flow split correlations for hydrocyclone, 219
fluidization, 76-81
  minimum velocity, 77
fluid rheology, 64-65
fluorite, 313
flux, magnetic, 280
flux curve, of solids, 81, 229, 331
flux envelope, 340
force, centrifugal, 298
  competing, 93, 404
    in magnetic separators, 284-286
  detachment, 285
  drag, 64, 68
  electrical, 298
  entrapment, 285
  external accelerating, 69
  gravitational, 284, 298
  image, 297
  interparticle, 280
  magnetic, 280, 282-286
  secondary, 280, 400
  separating, 47, 400
force balance, electrostatic separation, 298
  magnetic separation, 282-284
  particle in fluid, 69
forces on particle in fluid, 68
fracture, abrasion, 101
  cleavage, 101
  effect of additives, 122, 147, 149
  energy density, 114
  environment, 122
  intergranular, 32
  mechanisms, 117
  product size distributions, 117
  shatter, 101
  single particle, 113-122
  stress distributions, 115
  transgranular, 32
  types, 32, 117
  see also breakage
fracture mechanics, 113
fracture stress, 114
Frantz isodynamic separator, 279
free energy, surface, 95
free energy of adsorption, 102
friction, angle of internal, 370
  effective angle of, 370
  wall angle of, 370

friction factor, 65-66, 82, 83
  agitator, 82
  Fanning, 66
  fluidization, 77
  packed bed, 75
  particle, 68
  pipe, 66
    with slurry, 83, 388
  sedimentation, 77
friction factor charts, 66
froth, 97, 315
  breakdown, 315, 316
  drainage, 315
  height, 315
frother, 97, 309
Froude number, 82, 83, 388
funnel-flow bin, 369

gain, 410
galena, 310, 311, 404
gangue, 8, 17, 47, 315
Gaudin-Schuhmann equation, 117
Gaudin-Schuhmann plot, 23, 24
Gibbs adsorption equation, 96-98
glycol ethers, 309
gold, 238, 257, 259
Gouy layer, 100
grade, 8, 33, 264, 430
  cut-off, 430
grade/recovery curve, 59, 402-403, 430
grain size, 32, 430
  effective, 33
graphic representations of size distribution data, 23-26
gravity concentration, 250-273, 399
  equipment, 250-261. *See also* cone; jig; pinched sluice; spiral; table
  theory, 261-268
gravity concentrators, classification of, 250
gravity separation, determination of separability curves, 55
Griffith theory of fracture, 114-115
grindability, 160
  limit of, 114
grinding, 8, 116, 127, 399
  autogenous, 132, 150, 162
  breakage function, 120, 146
  batch, 152-154
  media segregation, 138
  media shape, 145, 147-149
  perfect mixing, 154-155
  plug flow, 152-154
  residence times, 151
  see also mill
grinding media:
  action, 139-142
  segregation, 138

shape, 149
size, 137, 148
wear, 139
grinding mill, *see* mill
gypsum, 237, 313
gyratory crusher, 127, 128

halide, 313
Hamaker constant, 106
handling, materials, 365-393
hand sorting, 237
hardness, ore, 431. *See also* mass-size balance; work index
head, liquid, 381
net positive suction, 382
heavy liquids, 55
heavy media, 243-249. *See also* dense medium
Helmholtz plane, 100
hematite, 275, 278, 292
heteropolar, 307, 309
hindered settling, 76-80, 107, 109, 209, 211, 225, 248, 263, 331
differential, 87
holdup in mills, 149
hopper, design, 369-374
discharge rate, 373
slot size, 373
see also bin
Humphreys spiral, 252. *See also* spiral
hydrocarbon chain length, 308
hydrocyclone, 213-223, 411
concentrator, 267
correlations, 217-220
significance, 219
dense medium, 254
dewatering with, 358
flows in, 213-216
geometry, 221
pressure drop, 219
relative merits of, 222
Reynolds number, 213
underflow discharge, 222
hydrophilic, 97
hydrophobic, 97
hysteresis, 281

ideal settling pool, 81, 207
ilmenite, 259, 291
impact crusher, 129, 131
impellers, 387
impulse tracer, 52, 411
incremental investment, 429
induction time, 106, 108
industrial minerals, 274, 278
instrumentation, 414-416

instruments, *see generic types*
interest, 428
interface, 95-112
adsorption at, 96
air-water, 97, 309
chemistry of, 95
determination of properties, 107
mechanisms of attachment at, 107
mineral-water, 101
thermodynamics of, 95-102
interfacial chemistry, 95
interfacial energies, 103
interfacial tension, 96
interparticle contact, 106
investment worth, 427-430
iron ores, 238, 252, 259, 274, 292, Appendix B1-B3
isoelectric point (IEP), 100
isotherm, adsorption, 102

jaw crusher, 127, 128
Jernqvist construction, 333
jig, Batac, 251-253
Baum, 251, 252, 254
circular, 252, 255
Denver, 251, 252
diaphragm, 252
equipment, 251-255
feldspar, 251, 252
Harz, 252
I.H.C. Cleaveland, 252, 255
plunger mineral, 252
pneumatic, 252, 256
Wemco-Remer, 252
jiggability, 92, 270
jigging, theory of, 261-265

Kick equation, 116
kinetics, 50-52, 93, 402-404
electrostatic separation, 299
flotation, 53, 316-317
screening, 186-189
size reduction, 146-156
Klimpel and Austin equation, 118

laminar flow, 63, 69
launders, 260
"laws" of size reduction, 116
liberation, 32-36, 43, 46, 235, 397, 401
intergranular, 32
transgranular, 32
lifting factor, 298
limiting reduced performance curve, 57, 58
limiting size, 204, 207
liners, 138

Index

profiles of, 138, 140, 142
rubber, 138-139
wear of, 139
liquid, dense, 55
liquid-liquid extraction, 314
live capacity, 367
locked particles, *see* middlings
log-probability plot, 24, 25
log washer, 201, 202
long tube test, 213

magnesite, 237, 313
magnetic attractability of minerals, Appendix E
magnetic field, applied, 280
magnetic flux, 280
magnetic induction, 280
magnetic remanence, 281
magnetic separation, 274-290, 399
 circuits, Appendix B
 cobbing, 274
 determination of separability curves, 55
 equipment, 274-279
 theory, 280-286
 tramp removal, 274
magnetic separator, Ball-Norton type, 275
 canister type, 275, 278
 carousel type, 275, 277
 cobbing drum, 275, 276
 concurrent, 275, 276
 counter current, 275, 276
 counter rotation, 275, 276
 cross belt, 275
 Davis tube, 279
 drum, 286-287
 dry, 274, 278
 Frantz isodynamic, 279
 grate magnet, 275
 high gradient, 278
 high intensity, 277, 288
 high speed drum, 275
 low intensity, 276, 286
 magnetic pulley, 275
 matched, 284-286
 matrix, 282-286
 particle size limits, 285-287
 performance, 286
 plate, 275
 suspended magnet, 275, 276
 tank design, 286
 wet, 274
 wet drum, 275, 276, 277
 WHIMS, 278
magnetic susceptibility, 274, 280
magnetite, 243, 275
magnetization, 281
 saturation, 281

magnetohydrodynamic separation, 279
magnetohydrostatic separation, 279
maintenance, 427
marble, 237
marcasite, 311
mass balance, *see* materials balances
mass flow, conditions for in bin, 370
mass-flow bin, 369
mass-size balance equations, 144-157
mass-size balance parameters, determination of, 156
materials balances, 9-13
 excess data, 13
materials handling, 8, 365-393
 slurries, 380-389
 solids, 367-379
matrix, ferromagnetic, *see* magnetic separator, matrix
media, filter, 347
 grinding, 349. *See* grinding media
 see also dense medium
mercaptan collectors, 308
mesh number, 169, Appendix C
metal prices, 13-14, 59-60
methyl isobutylcarbinol (MIBC), 309
micelle, critical concentration, 103
 hemimicelle, 103
micellization, 309
microscopy, 18, 36, 43
middlings (middling particles), 9, 32, 292
 behavior of, 35
 treatment of, 401-404
 types, 35
mill, 132-143
 autogenous, 132, 135
 ball, 132, 133, 411
 charge volume, 137, 147
 compartment, 131, 133, 138
 control of, 418
 Hardinge, 138
 holdup, 149
 liners, 138-142
 media action, 139
 power draft, 142
 residence times, 151
 ring-roller, 135, 136
 rod, 132
 semi-autogenous, 134
 tumbling, discharge methods, 143
  feed methods, 143
 vibrating, 135, 137
 see also grinding
mill diameter, 145-149
 scale-up of, 159
mineralogical examination, 17
mineralogy, 17, 397, 405
mineral processing, need for, 5
mineral properties, Appendix A, E, F, G. *See also* property

minerals, 5, 17, Appendix A
  trace, 259
mineral surfaces, 95
miners pan, 258, 259
mining, 14-16, 127, 241
mining costs, 430
misplaced particles, 9, 181, 400
mixer combinations for flow representation, 54, 155
mixing, 52, 82, 93
mixing patterns, 52-55
modern control theory, 413
modifier, 307, 310-313
module, bare, 425
module method of capital costing, 425
modulus, distribution, 26
  size, 26
moisture, in electrostatic separations, 296
  in screening separations, 185
monazite, 275, 291

near-mesh, see particle, near-mesh
negative response, 46
neutron activation, 416
Newtonian behavior, 64
node, 9
nominal diameter, 21, 73, 169
non-Newtonian behavior, 64-65, 80, 83, 84
normalized performance curve, see performance curve

on-line analysis, 42, 416
on-stream analysis, 42, 416
operating plants, 404
operating point, 59
optimum design, 429
optimum investment, 429
optimum operating capacity, 430-431
ore, 7, 13-17
ore body, 5, 17, 241
ore hardness, 431. See also mass-size balance, equations; work index
organic sulfate collectors, 308
overburden, 390
overflow, 199
overloading, of circuit, 404
  of crusher, 127
  of separator, 58
  of thickener, 338
oversize, 199
oxyhydryl anionic collectors, 308

packed bed, flow through, 75-76
paramagnetic, 274
particle, acceleration in fluid, 21, 69, 72
  average size, 27-29, 37. See also diameter

characterisation, 21-45
  complete, 43
  determination of, 36-43
detachment from bubble, 108
diameter, 21. See also diameter
difficult to settle, 340, 391
effect of coating in electrostatic separations, 295
equisettling, 87, 228
fine, 106, 108
  flotation of, 313
  flotation of by entrainment, 315
irregularly shaped, 65, 73
limiting size, 81
measurement of size, 36-40
middling, see middlings
misplaced, 9, 400
motion in fluid, 68-75
near-mesh, 181, 189, 190, 196
Reynolds number, 69
segregation, 369. See also segregation
segregation of dry particles, 91
shape, 29-32, 73
size, 21, 37
  determination of, 36-40
  on-line measurement of, 42, 414
  real-time analysis of, 42, 414
  see also diameter
specific surface, 30
stress distributions for fracture, 115
suspension of, 82, 383-387
terminal velocity, 68-75
partition curve, 57. See also performance curve
payback time, 428
payments, penalty, 6
pebble mill, 132
penetration theory, 105
perfect mixing, 52-53, 151, 410
perfect mixing grinding, 154
performance curve, 51, 57-58, 205, 207, 217-221,
    223-224, 226, 227, 230, 241, 248, 404
  bulk, 404
  corrected, 206
  dense medium, 248
  fluidized bed classifier, 223-224
  gravity concentrators, 268-271
  limiting reduced, 57, 58
  magnetic separation, 288
  screening, 195
  reduced (normalized), 57, 195, 206
  screening, 187-188
  typical of sedimentation classifier, 206
permeability, 41
pH, 98, 311, 393
phosphate ore, 252, Appendix B4
phosphate slimes, Florida, 340, 391
physical adsorption, 101
pilot plants, 405

Index

pinched sluice, 259-260
  theory of, 267
pine oil frother, 309
pinning factor, 298
pipe, non-circular, 68
pipe flow, 66-68
  heterogeneous slurry, 387
  homogeneous slurry, 387
piping, 381
piping, conditions for in bin, 371
planilla, 259
plant design, 405-407
plant layout, 406, 407
plant practice, 395-432
plot, Gaudin-Schuhmann, 23, 24
  log-log, 23, 24
  log-probability, 24, 25
  Rosin-Rammler, 23, 24, Appendix D
  Weibull, 23, 24, Appendix D
plug flow, 52, 53, 151, 207
plug flow grinding, 152
pneumatic jack hammers, 135
point of zero charge (PZC), 99
pond, 390. *See also* tailings pond
pool, ideal settling, 81, 207
porosity, 75, 77-80
positive response, 46, 187, 199, 207
potash, 313, Appendix B8
potential-determining ion, 98-100
power, for agitation, 82
  for agitation in tanks, 384
  for size reduction, 159-162
power draft of tumbling mills, 142, 143
power number, 82, 314
preconcentration, 236, 259, Appendix B11
pressure, consolidating, in bins, 371
  disjoining, 105
pressure drop, 77
pressure drop correlations for hydrocyclone, 219
pressure drop in filtration, 349
pressure drop in pipe, 66, 389
pretreatment, 259
prices, metal, 13-14, 59-60
probability, 51, 108, 186, 299, 317, 318
  electrostatic separations, 299
  flotation, 108, 317-319
probability of breakage, 120
probability of passage, through aperture, 181-182
probability screening, theory, 190-192
probable error, 58
process dynamics, 409
profitability, 427-430
properties, 235, 239
  bulk, 240
  internal, 240
  of materials, for conveying, 376, Appendix G
  of minerals, Appendix A, E, F, G

property, 23, 46, 47. *See also* basis
pseudo-plastic behavior, 65
pump, 380-381
  centrifugal, 380
  equipment, 380
  positive-displacement, 380
  reciprocating, 380
  rotary, 380
  selection, 383, Appendix H
  sizing of, 381-383
  see also head, liquid
pumping, 68, 380-383
pumping number, 82
pump operating characteristics, 381
pyrite, 278, 311
pyrrhotite, 310, 312, 397

quartz, 238
quebracho, 313

radioactivity, 240
ragging, 254, 265
rate, production, 426
rate of breakage, *see* breakage rate
rate constant, 50
  flotation, 317
  screening, 186-190
  size reduction, 146
rat-holing, in bins, 264
  in thickeners, 341
rate law, first order, 50, 93, 120, 146, 263, 317
  second order, 52, 263
rate of return, 427-430
reagents, 307-314
  influence of, 104
reclaimers, 367
recovery, 8, 33, 264
recycle, 11, 392, 406
reduced performance curve, 57, 195
reduction ratio, 127, 143
  optimum, 143
regrinding, 401, 403
Reichert cone, 252. *See also* cone
release analysis, 55, 400
reserves, mineral, 17
residence time, 52-55, 151, 410
  distribution of, in classifier, 210
  in mills, 151
  factors effecting, 151
  in tailings ponds, 392
response, negative, 46
  positive, 46, 187, 199, 207
return, incremental, 429
return on investment, 427-430
Reynolds number, 63

agitator, 82
hydrocyclone, 213
particle, 69
pipe, 66
rheolaveur, 261
rheology, 64-65
Richardson and Zaki equation, 79-80, 229
ring-roller mill, 135, 136
Rittinger equation, 116
river bend action, 85, 267
rod mill, 132. *See also* mill
roll crusher, 129, 131. *See also* crusher
Rosin-Rammler, *see* Weibull
roughing, 9, 400-404
rutile, 259, 291

safety, 390
salt, rock, 237
sample size, for sieving, 171
sampling, 36, 378, 405
sands, 199
saturation magnetization, 281
scales, belt, 414
scalping, 171
scanning electron microscope (SEM), 19, 37, 43
scavengers, 267
scavenging, 9, 400-404
scheelite, 238, 252, 254, 257, 313, 397-399, 405
screen, ball deck, 180
  blinding, *see* blinding
  casting, 173
  centrifugal, 172
  conveying, 173
  dewatering, 358
  dynamic, 173
  fixed, 173
  Flipflow, 178
  grizzly, 172
  gyrating, 173
  horizontal, 172
  inclination, 183
  inclined, 172
  movement, 171, 173, 176
  oscillating, 173
  performance, 180
  probability, 172
  reciprocating, 172
  revolving, 172
  rotary sifter, 172
  shaking, 172
  sieve bend, 172
  static (stationary), 173
  trommel, 172
  vibrating, 172
screenability characteristics, 195
screening, 169, 171-198, 399
  area, 183, 193-196
  bed depth, 183, 184-185, 195
  calculation of area, 193-196
  crowded, 187
  equipment, 171-180
  equipment classification, 171-173
  factors affecting, 181
  feed chutes, 179
  heated surfaces, 180
  moisture, 185
  probability, 175, 179
  probability of passage, 181-182, 186, 190-192
  separated, 187
  stratification during screening, 184
  theory, 185-193
  wet, 179
screen surface, 173-179, 182-183
  percentage open area, 182
  perforated plate, 173
  probability, 179
  profile bar, 178
  wear, 173-178, 193
  woven wire cloth, 178
secondary forces, 400
sedimentation, 76
  classification of, 327, 328
  equipment, 327-330
  theory, 330-340
sedimentation behavior, 330-341
sedimentation classifier, 199. *See also* classifier
sedimentation particle sizing, 38-39
segregation (stratification), 91, 184, 190, 367, 369, 373
  in gravity concentrators, 265, 266
segregation of dry particles, 91, 369, 373
selection function, 120. *See also* breakage rate
selective agglomeration, 319
selective flocculation, 319
selectivity, oxide and silicate flotation, 312
  salt-type flotation, 313
  sulfide flotation, 311
selling price, 6, 14
semi-autogenous mill, 134
sensing systems for sorting, 329
separability, 93
separability curve, 47, 50, 51, 57, 181, 241, 288, 400
  determination of, 55-56, 288
  flotation, 47
  gravity, 49
  magnetic, 49, 288
  sizing, 47
  sorting, 241, 242
separating force, 400
separation, 8, 43, 167-322
  factors involved, 46-61
  review of mechanisms, 93
separation characteristics of minerals, Appendix F

Index

separation density, 269
separation efficiency, 56. *See also* efficiency
separation limits, 93, 266, 400-404
separation point, 57
separation sharpness, 58. *See also* performance curve
separation size, 204, 207
separator, 8, 46, 87. *See also* dense medium; electrostatic separation; electrostatic separator; flotation; gravity concentration; magnetic separation; magnetic separator; sorting
separator limits, 400
separator setting, 47-52, 430
serpentine, 237
set, of particles, 46
set point, 50, 409
setting, separator, 47-52, 430
settling, hindered, *see* hindered settling
shaking surface concentrators, equipment, 252, 254-258
　theory of, 265-266
shape factor, 29
sharpness of separation, *see* performance curve
shatter fracture, 117
shear mechanisms, Bagnold, 90-91, 265-267
shear plane, 100
shear stress, 62
sieve bend, theory, 192
sieve series, 179, Appendix C
sieve shakers, 170
sieving, 36, 169-171
　by hand, 171
　end point, 171
　theory, 185-190
　wet, 171
　see also screening
single particle fracture, 113-122
singulation, 238
sink-float, 55, 243-249. *See also* dense medium
sink product, 243
S.I. units, Appendix I
size, average, 27-29
size distribution, 22
size distribution data, presentation of, 22-27
size distribution formulae, 26, 118
size distributions from fracture, 117
size fractions, 22, 169
size reduction, 8, 113, 125-165
　Bond's method, 159-162
　circuits, 143
　equipment, 127-135
　flow of solids through equipment, 134
　mass-size balance analysis of, 144-157
　matrix analysis of, 154
　purpose of, 125
　scale-up, 145-149, 156, 159-162
　size distribution from, 149
　theory, 144-157
　see also crusher; grinding; mill
size reduction "laws", 116
size separation, 8, 167-232, 405
sizing, of particles, industrial, 167-231
　laboratory, 36-40, 169-171
slags, 275
slimes, 199
sludge blanket, 327
sluice, 259, 260, 262
slurry, flow behaviour, 65, 82-84, 380-389
slurry distribution, 389
slurry transport, 387-389
smelter schedule, 6-7, 59, 430
smelting, 6
solids, flow properties of, 367
　transport of, 375-378
solids flux, 81
solution chemistry, 307
solvation effects, 100
sortability, 241
sorting, 237-242
　detection, 239
　ejection, 240
　equipment, 237-238
　equipment capacity, 237
　fluorescence, 238
　hand, 237
　mechanics, 238
　photometric, 237
　radiometric, 238
　singulation, 238
specific cake resistance, 349
specific surface, 30, 349
sphalerite, 310, 311, 312
sphalerite activation, 311
sphericity, 20, 68, 71, 73, 76
spiral, 252, 259, 262, 314
　theory of, 267
spiral flow, 84
spreading coefficient, 104
stackers, 367
starch, 313
state space, 413
statistical screening, *see* probability screening
step change, 411
Stern layer, 100
Stern model, 98
Stern plane, 100
stockpiles, 367
Stokes' law, 69
stratification, 265. *See also* segregation
streaming potential, 107
strength, unconfined yield, 370
stress distributions in particle, 115
stress for fracture, 114
sub-sieve sizing, 36
sulfide ores, 301

sulfides, collector adsorption on, 310
   flotation selectivity, 311
sulfonate collectors, 308
sump, 411, 413
surface, fracture, 114
surface area, determination of, 41-42
surface charge, 98
surface energy, 114
surface excess, 97
surface free energy, 95
surface potential, 99
surfaces, 95-117
surface tension, 96, 97
surge, 358, 368
surge capacity, 338, 367, 406
sylvite, 313

table, 257, 314, 399
   bumping, 256
   Holman slimes, 252, 258
   operating variables, 257
   shaking, 252, 254
   theory of, 265-266
   Wilfley, 256
tailings, 7, 8
   treatment of, for valuables, 259
tailings dam, 390
   design, 391
tailings disposal, 390-393
tailings pond, 390
tailings water, 391-393
talc, 237
tall oil, 308
tank, agitated, 82, 327, 383-385
   baffled, 82
   baffles, 387
   conditioning, 385
   design of, 384-387
   scrubbing, 385
tannin, 313
teetering, 76
terminal velocity, 68-75, 207-209
test, batch sedimentation, 330, 332
   batch settling, 213
   filtration, 344, 350, 351
   laboratory, 405
   long tube, 213
thermal dewatering, see drying
thermodynamic aspects, 93, 400-401
thermodynamics, of interfaces, 95
thickener, 327-329, 358, 393
   cylindrical, 327
   deep cone, 329
   design, 333, 336
   flocculated pulps, 339
   flow patterns in, 335, 341
   lamella, 329, 340
      theory, 340
   maximum capacity, 335-337
   operation of, 341
   overdilute feed, 339
   overload, 338
   raking in batch tests, 340
   sludge blanket fed, 327
   surge in, 338
   underload, 337
   see also clarifier
thickening, 327
   channels in sludge, 340
   compression in sludge, 340
   ideal, 336
   non-ideal, 336, 339-340
thin layer flow, 84
thionocarbamate collectors, 308
time constant, 410. See also residence time
tin, 259, 291
trace elements, 95, 101
trace minerals, 259
tramp removal, 274
transfer function, 410
transportation lag, 410
transport of bulk solids, 375-378
transport of slurries, 387-389
trickling, 88, 263, 266, 267
Tromp area, 58
Tromp curve, see performance curve
trough washer, 260
tumbling mill, see mill, tumbling
turbine, see impellers
turbulent flow, 63
two-product formulas, 13
two-stage classification, 227-228

ultraflotation, 314
ultramicroscopy, 38
unconfined yield strength, 370
underflow, 199
unit operations, 395
utilities, costing of, 425, 427

valuable, 17, 47
valuation of concentrates, see smelter schedule
valves, 381
vanner, 258
variable, controlled, 409
   disturbance, 409
   manipulated, 409
velocity profile, 63
vibrating mill, 135, 137
vibrator, 368
viscosity, 62

Index

apparent, 65, 79, 218, 246
viscous, *see* laminar flow
viscous friction, coeficient of, 57

wall effect, 74, 79
washability, 270
washability curve, *see* separability curve
water, flow rate of in flotation, 315
water balance, 392
water quality, 392
water recycle, 392
Weibull, 23, 58, 115, 118, 209, Appendix D
wet grinding, 135
wetting, 103
wolframite, 238, 275, Appendix B10

work index, 159
working capital, 421, 425

xanthate collectors, 308
X-ray diffraction, 19
X-ray flourescence, 42, 416

Young equation, 103

zeta potential, 79, 100
zircon, 259, 291
zone settling, 333